T0189196

Lecture Notes on Data Engineering and Communications Technologies

Volume 59

Series Editor

Fatos Xhafa, Technical University of Catalonia, Barcelona, Spain

The aim of the book series is to present cutting edge engineering approaches to data technologies and communications. It will publish latest advances on the engineering task of building and deploying distributed, scalable and reliable data infrastructures and communication systems.

The series will have a prominent applied focus on data technologies and communications with aim to promote the bridging from fundamental research on data science and networking to data engineering and communications that lead to industry products, business knowledge and standardisation.

Indexed by SCOPUS, INSPEC, EI Compendex.

All books published in the series are submitted for consideration in Web of Science.

More information about this series at http://www.springer.com/series/15362

Jennifer S. Raj · Abdullah M. Iliyasu ·
Robert Bestak · Zubair A. Baig
Editors

Innovative Data Communication Technologies and Application

Proceedings of ICIDCA 2020

 Springer

Editors
Jennifer S. Raj
Department of Electronics and
Communication Engineering
Gnanamani College of Technology
Namakkal, Tamil Nadu, India

Robert Bestak
Telecommunications
Czech Technical University in Prague
Prague, Czech Republic

Abdullah M. Iliyasu
College of Engineering
Prince Sattam Bin Abdulaziz University
Al-Kharj, Saudi Arabia

Zubair A. Baig
School of Information Technology
Deakin University
Geelong, Australia

ISSN 2367-4512   ISSN 2367-4520 (electronic)
Lecture Notes on Data Engineering and Communications Technologies
ISBN 978-981-15-9653-7   ISBN 978-981-15-9651-3 (eBook)
https://doi.org/10.1007/978-981-15-9651-3

This Springer imprint is published by the registered company Springer Nature Singapore Pte Ltd.
The registered company address is: 152 Beach Road, #21-01/04 Gateway East, Singapore 189721,
Singapore

We are honored to dedicate this conference proceedings to all the participants and editors of ICIDCA 2020.

Foreword

It is with deep satisfaction that I write this Foreword to the Proceedings of the ICIDCA 2020 held in RVS College of Engineering and Technology, Coimbatore, Tamil Nadu, India, on September 3 and 4, 2020.

This conference brought together researchers, academics, and professionals from all over the world, experts in Innovative Data Communication Technologies.

This conference particularly encouraged the interaction of research students and developing academics with the more established academic community in an informal setting to present and discuss new and current work. The papers contributed the most recent scientific knowledge known in the field of distributed operating systems, middleware, databases, sensor, mesh, and ad hoc networks, quantum and optics-based distributed algorithms, Internet applications, social networks, and recommendation systems. Their contributions helped to make the conference as outstanding as it has been. The local organizing committee members have given their full effort into ensuring the success of the day-to-day operation of the meeting.

We hope that this program will further stimulate research in theory, design, analysis, implementation, and application of distributed systems and networks. We feel honored and privileged to serve the best recent developments to you through this exciting program.

We thank all authors and participants for their contributions.

Dr. V. Gunaraj
Conference Chair, ICIDCA 2020

Principal, RVS College of
Engineering and Technology
Coimbatore, India

Preface

This conference proceedings volume contains the written versions of most of the contributions presented during the conference of Innovative Data Communication Technologies and Application. The conference provided a setting for discussing recent developments in a wide variety of topics including dynamic, adaptive, and machine learning distributed algorithms, game-theoretic approaches to distributed computing, security in distributed computing, cryptographic protocols fault-tolerance, reliability, self-organization, self-stabilization, and so on. The conference has been a good opportunity for participants coming from various destinations to present and discuss topics in their respective research areas.

ICIDCA 2020 tends to collect the latest research results and applications on Innovative Data Communication Technologies. It includes a selection of 71 papers from 274 papers submitted to the conference from universities and industries all over the world. All of the accepted papers were subjected to strict peer-reviewing by 2–4 expert referees. The papers have been selected for this volume because of quality and the relevance to the conference.

We would like to thank the keynote speakers and technical program chairs for taking part in our conference. We would like to express our sincere appreciation to all authors for their contributions to this book. We would like to extend our thanks to all the referees for their constructive comments on all papers especially, we would like to thank organizing committee for their hard work. Finally, we would like to thank Springer publications for producing this volume.

Technical Program Chairs
Dr. Jennifer S. Raj
Professor
Gnanmani College of Engineering and Technology
Namakkal, India

Dr. Abdullah M. Iliyasu
Professor
School of Computing
Tokyo Institute of Technology
Yokohama, Japan

Dr. Robert Bestak
Czech Technical University in Prague
Prague, Czech Republic

Dr. Zubair A. Baig
Division of Cyber Security
Faculty of Science
Engineering and Built Environment
School of Information Technology
Deakin University
Geelong, Australia

Acknowledgments

ICIDCA 2020 would like to acknowledge the excellent work of our conference organizing committee, keynote speakers for their presentation on September 3 and 4, 2020. The organizers also wish to acknowledge publicly the valuable services provided by the reviewers.

On behalf of the editors, organizers, authors, and readers of this conference, we wish to thank the keynote speakers and the reviewers for their time, hard work, and dedication to this conference. The organizers wish to acknowledge Dr. V. Gunaraj for the discussion, suggestion, and cooperation to organize the keynote speakers of this conference. The organizers also wish to acknowledge speakers and participants who attended this conference. Many thanks are given to all persons who helped and supported this conference. ICIDCA 2020 would like to acknowledge the contribution made to the organization by its many volunteers. Members have contributed their time, energy, and knowledge at a local, regional, and international levels.

We also thank all the chairpersons and conference committee members for their support.

Contents

About the Editors

Dr. Jennifer S. Raj received the Ph.D. degree from Anna University and Master's Degree in Communication System from SRM University, India. Currently, she is working in the Department of ECE, Gnanamani College of Technology, Namakkal, India. She is a life member of ISTE, India. She has been serving as Organizing Chair and Program Chair of several international conferences and in the Program Committees of several international conferences. She is book reviewer for Tata Mc Graw hill publication and publishes more than fifty research articles in the journals and IEEE conferences. Her interests are in wireless health care informatics and body area sensor networks.

Abdullah M. Iliyasu (aka Abdul M. Elias) received the M.E., Ph.D. and Dr. Eng. degrees in computational intelligence and intelligent systems engineering from the Tokyo Institute of Technology, Japan, where he is presently a Research Professor with the School of Computing. Concurrently, Prof. Elias is the Principal Investigator and Team Leader of the Advanced Computational Intelligence and Intelligent Systems Engineering (ACIISE) Research Group at the College of Engineering, Prince Sattam Bin Abdulaziz University in the Kingdom of Saudi Arabia. He is also a Professor with the School of Computer Science and Engineering, Changchun University of Science and Technology, China. He has to his credit more than 100 publications traversing the areas of computational intelligence, quantum cybernetics, quantum image processing, quantum machine learning, cyber and information security, hybrid intelligent systems, health informatics and electronics systems reliability.

Robert Bestak obtained PhD degree in Computer Science from ENST Paris, France (2003) and MSc degree in Telecommunications from Czech Technical University in Prague, CTU, Czech Republic (1999). Since 2004, he has been an Assistant Professor at Department of Telecommunication Engineering, Faculty of Electrical Engineering, CTU. He participated in several national, EU, and third party research projects. He is the Czech representative in the IFIP TC6 organization, and chair of working group TC6 WG6.8. He annually serves as Steering and

Technical Program Committee member of numerous IEEE/IFIP conferences (Networking, WMNC, NGMAST, etc.) and he is member of editorial board of several international journals (Computers & Electrical Engineering, Electronic Commerce Research Journal, etc.). His research interests include 5G networks, spectrum management and big data in mobile networks.

Dr. Zubair A. Baig is a cyber security academic and expert who has over 85 research publications and 2 US patents to his credit. He has worked on national and international research projects such as the application of machine learning for design of intrusion detection systems, design of Blockchain solutions for securing national food supply chains for the United Nations/CSIRO, and the design and deployment of honeypot-based solutions for national-scale network security. He has delivered several keynote talks on cyber security both nationally and internationally and has appeared on Australian TV and Radio to provide expert opinion on numerous occasions.

DIBSM: Dominance on Incomplete Big Data Using Scoring Method

Anu V. Kottath and Prince V. Jose

Abstract Big data is a collection of data which increases exponentially; generally, big data is complex in nature due to its dimensional characteristics. Present data managing tools do not efficiently process and store huge data. In an incomplete data set, there will be missing nodes, which will be randomly distributed in its dimensions. When the data set is large, it is very difficult to get the information. So the dominance value in the data set is considered as most significant value. An in-depth study is essential to obtain these k-dominant values from a data set. Algorithms such as skyline-based, bitmap index guided, upper-bound-based and pairwise comparisons are some of the familiar models available to identify the dominance values. Due to its slow data processing characteristics, those models are suitable to process small data sets and it requires numerous comparisons between values, complexity increases with increasing data, respectively. Considering these issues, this research work proposed a novel algorithm dominance on incomplete big data using scoring method (DIBSM) to obtain the dominance values quickly using a bit string representation of each user and the scores. Based on the score, we will get the top-k dominance values. MapReduce method is used to process the data in the proposed algorithm which helps to enable parallel data processing and obtains the results more quickly. As compared to the existing method, using this algorithm dominance values got more quickly.

Keywords High-dimensional data · Big data · Dominance value · Incomplete data

1 Introduction

World running behind the data and big data is an unavoidable portion in data storage and processing, though processing complex data is a difficult process. In big data,

A. V. Kottath (✉) · P. V. Jose
Computer Science and Engineering, St. Josephs College of Engineering and Technology, Palai, Kerala, India
e-mail: anukottath@gmail.com

P. V. Jose
e-mail: Prince.v.jose@sjcetpalai.ac.in

© The Author(s), under exclusive license to Springer Nature Singapore Pte Ltd. 2021
J. S. Raj et al. (eds.), *Innovative Data Communication Technologies and Application*,
Lecture Notes on Data Engineering and Communications Technologies 59,
https://doi.org/10.1007/978-981-15-9651-3_1

1

data are unstructured and structured forms. Institutions are generating data in an alarming amount, and the studies are showing that 250 TB (Terabytes) data collected by Facebook in a day. Large data sets hold incomplete data. Incomplete data represent the missing dimensions in a data and it occurs due to privacy preservation policies, data loss, etc. Suppose an object A in the data set, size of A (1, 7, -, 4). The given object has four dimensions and some missing values in the dimension. Figure 1 shows incomplete data sets.

In a data set, dominance values are the most important values. Finding the dominance value is a difficult task for incomplete data sets. Identifying dominant values on data sets which are incomplete is useful in big data studies [1]. Definition of dominance: Two terms $N1$ and $N2$ are given, then say $N2$ dominates $N1$ if the values in the $N2$ is greater than the value on $N1$, apart from the lost data from the data set.

Dominance on the data set can be stated as shown below:

$$\forall N1[\forall di] \text{ and } N2[\forall dj] : N1[\forall di] > N2[\forall dj]$$

To learn more, consider the example, if $R1 = (1; -; 2; 0)$ and $R2 = (-; -; 4; 2)$. Can be said that $R2$ is dominating $R1$, because the value in $R2[d3] = 4$ is larger compared to $R1[d3] = 2$. Compared with other types of data sets, incomplete data sets require different processing methods. One of the main problems when dealing with incomplete data sets is how to deals with lost values. To effectively deal with incomplete data sets, innovative methods are needed. From the incomplete data set, in order to identify the dominance value, mainly three methods such as skyline-based algorithms, upper-bound-based algorithms and bitmap index guided algorithms are generally used.

Skyline queries are used in the skyline-based algorithm and it retrieves the essential dominant objects and rejects the unnecessary objects [2]. The major idea in the upper limit-based algorithm (UBB) algorithm is to determine the number of times the data item dominated by the currently selected value.

The rest of this article is divided into four parts. Section 2 details the existing methods for finding the advantages of incomplete data sets. Section 3 explains the proposed algorithm (DIBSM algorithm). In Sect. 4 experiment and analysis of the proposed algorithm. Finally, Sect. 5 provides the conclusion of this paper.

	d_1	d_2	d_3	d_4		d_1	d_2	d_3	d_4
m_1	–	1	2	–	m_6	–	–	–	3
m_2	1	–	3	2	m_7	1	1	–	–
m_3	3	1	–	–	m_8	–	3	2	–
m_4	–	–	–	1	m_9	2	–	2	2
m_5	–	2	1	–	m_{10}	3	2	–	–

Fig. 1 Incomplete data set sample

2 Different Techniques Used to Identify Dominance on Incomplete Data

2.1 Skyline-Based Algorithm

Khalefa et al. introduced the first work on processing incomplete data sets in 2008. From the data, the dominance values are obtained using the skyline queries. Replacement, bucket and ISkyline are used in this method. The main concept of the replacement algorithm is to use the α symbol to replace missing values n incomplete data so that the data set forms a complete table. This method is impractical in different situations. To solve the difficulties in the replacement algorithm, a new algorithm called bucket algorithm is introduced. In this method, every data item is separated into multiple bucket sets, and every single bucket consists of data sets with a similar missing value. In the next step, for every bucket, the skyline algorithm is applied in order to obtain local skyline points. The obtained skyline points are compared with local skyline points in the final step. The main disadvantage of the bucket algorithm is it contains different useless pairwise comparisons; as a result, time to complete execution is high. ISkylin algorithm was introduced to identify and decrease local skyline points. In skyline algorithm, the primary step uses bucket algorithm, bucket process to identify the local skyline points. Then using two optimizations such as virtual and shadow skylines, the number of local skyline points is reduced. Further the local skylines are reduced using ISkylin algorithm. A major problem in skyline algorithm is pairwise comparison which increases the complexity when the data set size increases. Due to this reason, the ISkylin algorithm is not widely adopted for cloud database and distributed database [3].

Arefin and Morimoto proposed an algorithm to process the skyline queries as Replacement-Based Skyline Sets Queries (RBSSQ) in 2012. Data processing is the primary step in this algorithm, and then the skyline set calculation is the next step. The data processing is done by changing incomplete data set to complete data set by taking the place of all lost data with data outside of the field values. In this, if the larger values are selected for processing, then it automatically changes all the other values into smaller values. Using the function named oracle, the resulting skyline can be identified. This method is safe and secure than other existing methods. The main disadvantages of RBSSQ are that it is not suitable for incomplete data sets which are centralized [4].

Bharuka and Kumar introduced a new algorithm named Incomplete Data Skyline algorithm (SIDS) in 2013 to process the skyline queries. In SIDS algorithm, primary step is to identify input data and process it in a decreasing manner. The data values are processed cyclically to find dominant value faster and reduce the comparison number to find the sky point. In this method, data set is considered as candidate skylines. In the last, skyline points are identified based on the remaining available data. This algorithm works well on small data sets, but this method does not work well on huge data sets due to the lack of using parallel execution, which leads an increase in time of execution as the data set increases [5].

In 2015, Jiang et al. proposed a method which includes four steps. The primary step is to place data with the similar dimensions in the same bucket, and next step is to process data in decreasing manner, which keep data at the forefront, that may become the skyline. In the second step, create an array similar to the number of missing values. Next step is to data cyclically read in each array. During the scanning phase, at least once every data set should be scanned. In the third stage, delete data with a number less than 2. Finally, identify the skyline query to the leftover data sets. In the last stage, the data sets are compared to identify the skyline [6].

Lee proposed two algorithms such as sorting-based bucket skyline (SOBA) and BUCKET algorithm in 2016. BUCKET algorithm and Iskyline algorithm are the same. Unwanted pairwise comparison is considered as the major drawback of BUCKET algorithm. To overcome this issue, a new method named SOBA is introduced. These techniques are used to reduce the pairwise comparisons. In this method, the first step is, each bucket is rearranged in ascending order. In second step, item is compared to each data set with a common value. Descending order arrangement is followed in point-level optimization data in each bucket. In SOBA, the main benefit of this technique is it can avoid dominant data sets quickly and reduce unnecessary proportional comparisons between data. Major disadvantages of this technique are that this is suitable for databases which are centralized [1].

To deal with the skyline needs on incomplete data sets, Gulzar and Alwan proposed another method in 2017. This technical data is stored on the cloud databases, and the data distributed vertically to store in various places. In this method, use filtering to delete unnecessary data items. This helps reduce the time of execution. In order to obtain the skyline points, parallel processing is used in this technique. Initially, all the dimensions are divided to find the skyline points, then sorting the data in a decreasing manner, and the next step is reading and circularly counting the data values. The appearance of every data in the array continues this method every data item is processed almost once. The next step is to identifying data values whose count is less than 2 and remove it. This method helps delete unwanted values. Next step is to obtain the necessary process points along with local skyline points. Those identified skyline points are connected for every dimension in the fourth step. From the local skyline points and stationary skyline points, the difference value is observed in the fifth step. But, the issue in this model is, it requires a sorting function in order to sort the observed points before joining operation. Another issues is, this model requires more comparisons, which results in an increased time of execution [11].

The final method proposed by Babanejad et al. in 2017, introduces an algorithm named DInSkyline. The method works well in dynamic databases. Dynamic databases are the database whose data often changes due to updates of data sets. Main advantages of this method are that you can find the advantages without having to rescan the entire database when updating it. This algorithm primary determines local skylines from every data sets, the next step is data set updated, then find skylines from the data sets, the last step is to combine results and find the final skyline point. This method contains three lists to identify skyline points. That are Dominating Bucket (BDG), Dominating Bucket (BDD), Dominating History (DH). The dominated local skyline points are included in BDD. BDG has a list that dominates the local skyline.

DH contains data items that are dominating the resulting skyline. For the latest added data sets, the local skyline values are identified to obtain the final values by comparing with BDG model. Even if the data is removed, the main data sets are obtained in the DH list, then the items on the BDG are compared and get resulting skyline [7].

2.2 Upper-Bound-Based Algorithms

Upper-bound-based algorithms are introduced to solve the issues in skyline algorithms.

An upper-bound-based algorithm was proposed by Buji Babu and Swapna, in which dominant values are obtained based on a TKD query. Identifying the upper bound from the data sets to obtain the necessary dominant values is the key idea in upperbound-based algorithms. In each dominance, the major goal is to identify the various values. The score represents a number that helps quickly identify an advantage value by reducing pairwise comparisons and improve the speed of execution. In this algorithm, the data set identifies the recognition frequency is dominated by the current value. Since full data sets are used, this method finds the dominance on the incomplete data sets well. If the data is incomplete, compare the data with similar dimension. UBB changing data set for faster processing. In the definition of advantage, the result is returned based on the score. Major disadvantages of UBB is it needs a large number of pairwise comparisons. UBB algorithm is complex for huge data sets [8].

2.3 Bitmap Index Guided Algorithm

In the previous section, saw that the existing method has various problems with finding dominant values on incomplete data sets. Use of the skyline algorithm mainly depends on the amount of data in the data set. If the data set is large, it will create a large number of data buckets. One of the main problems of the upper-bound-based method is it needs large number pairwise comparison. Therefore, to overcome this, a bitmap index guidance algorithm is introduced. A bitmap index table is created in the initial step which improves the processing speed for the given data set. Identifying dominance values using a function named scoring is the next step in this process.

2.3.1 Single Machine Algorithm

Single machine algorithm was proposed in 2018 by Yang et al., which identifies the dominance valyes in the data set. In this algorithm, creating bitmap index table is the primary step. When creating, the bitmap index table is set to zero. Details about bitmap index creation given below:

- Every missing data: left the field in the corresponding row without making any changes. Therefore, each vi (variable represents a range of values), its missing values are all 0.
- For all non-missing data, place the values 1 to the corresponding position.

Creating bitmap index table for huge incomplete data is complex which is considered as a main disadvantages of this method. Another disadvantage is that it takes more time to use stand-alone processing [9].

2.3.2 MapReduced Modified Algorithm

To solve problems on the single machine algorithm, Payam Ezatpoor introduced a method named MapReduce Enhanced Bitmap Index Guidance Algorithm (MRBIG) in 2018. Proposed model used MapReduce technique to obtain the dominance values. Parallel processing is possible in MRBIG by dividing the task into subtasks. Compared with the previously proposed algorithm, this method provides twice the processing time when finding TKD query results. The MRBIG algorithm uses the concept of MapReduce. Using of MapReduce method reduces the time need for execution. MRBIG algorithm uses three values such as P, Q and $nonD$ to obtain the dominance values in which the value of Q is not greater than m in the data set. $nonD$ represent values that are not dominated to set of value m. The job of the mapper is to find P, Q and $nonD$ values. In the reducer stage, the score function is calculated, then sorted it in a decreasing manner, then the dominant value is obtained. If the data size increases, the length of bitmap index table is also increases which is considered as a drawback of this model [10].

3　Problem Statement

Identifying dominance values is a complex task and different methods are available to obtain the different values.

On skyline-based algorithms, the main problems are:

- Comparing data sets will increase the time needed for execution.
- When the data set is large, this method is not suitable.

Main problems in the upper-bound algorithm (UBB) is that it needs to be more compared with a large number of data sets, and this method is not suitable for large amounts of data sets.

The bitmap index is a convenient way to deal with incomplete data. In case of single machine algorithm, the processing time is high. Bitmap index table is large for huge data sets is the main disadvantage of the MRBIG algorithm.

To overcome those all problems, an algorithm named dominance on incomplete big data using scoring method (DIBSM) is being developed which identify dominance on data sets which are incomplete more quickly as compared to existing methods.

4 DIBSM (Dominance on Incomplete Big Data Using Scoring Method)

Our proposed algorithm mainly focused on finding dominance on incomplete big data by using a scoring method. In this using MapReduce method to find dominance. In the MapReduce method, the process can be executed parallel; this leads to an increase in the time of execution. To execute large data set, using a single machine leads into increased execution time. But in MapReduce having two main functions that are called mapper and reducer, which are used to divide each of the work between different nodes and process them quickly. In mapper, different tasks are executed parallel using different mappers. The data sets are divided based on their internal patterns. The function of the reducer is to combine the results from the different mappers and get the final result.

4.1 DIBSM Structure

The detailed steps in DIBSM algorithm is given below.

Input:

Movie Lens data set

Mi : users; d: Movies

Output:

Top-k Dominating values

Mapper:

1   For all users mi ,where i=0 to n users

2   For each movie di ,rating r1 from mi

3   Create pi and qi

4   Binstring=create[positional presence of

                each users]

5   if(current rate<= ri)

6   qi=put 1's to corresponding   in Binstring positions otherwise 0.

7   Binstr=create[positional presence of each users]

8   if(current rate< ri)

9   pi=put 1's to corresponding

                    positions   in Binstr otherwise 0

10   End for

11   q*=$\sum$_di qi&qi+1

12   q sum=count(number of 1s in q*)

13   p*=Pn di pi&pi+1

14   p sum=count(number of 1s in p*)

15   End for

Reducer:

16   For each user mi to

17   Score =qsum-psum

18   End for

Finding top-k

19   K=Z

20   Sort(descending(Score)

21   Return top-k

In DIBSM algorithm, scoring function is used to identify the dominance values in the data sets. These scoring function is the main feature that influences the algorithm performance. The scoring method allows us to compare the values they get from the data set. The higher scores become the resulting dominances. The score shows how powerful an object to be top-k-dominant value. Based on the characteristics of MapReduce, the most logical method is to calculate the score for each particular users. In order to find the top-k dominance values from the incomplete data sets, there are two internal sets used in the DIBSM algorithm. First internal set is Q represent the values which is greater than or equal to Mi. The second set is p which is greater than Mi. For better understand DIBSM algorithm, have a data set for movie rating site. In this, data set contains a set of users and their movie ratings. From theses users which users are most dominance are needed to identify. For this, by using DIBSM algorithm first need to identify the P and Q values of each user. To identify Q values of each user, an array called Bitsting for each user is needed to be created. The length of Bitstring is depending on the number of users. For example, if the number of users is 20, the length of Bitsting is also 20. The result of this array is a set of bits. Then next step to create Q value for each user by comparing the movie rating of the current user is greater than or equal to the rating given by other users for the same movie. If it is greater, then put 1 to the corresponding positions of the users in Bitsting, otherwise, put 0. Then to identify P-value, create a Binstr which length is equal to the number of users and result in the form of bits. Then the next step is to compare the movie rating of the current user is greater than the rating given by other users for the same movie. If it is greater, then put 1 to the corresponding positions of the users in Bitsting, otherwise put 0. This process continues till each users having P and Q values. Then next step on DIBSM algorithm is to identify Q^* and P^* by doing AND operation of each Q values and P values respectively. Then count the number of 1's in the P^* and Q^* then store the result into Psum and Qsum respectively. To understand this concept more clearly, consider the data set shown in Fig. 2, which consist of movies, users and ratings. From this data set, the P and Q values of user $C2$ can be found as follows:

$P1 = 1111111111001111001;$
$P2 = 11,1111111111111111111;$
$P3 = 1111111111111111111111;$
$P4 = 10111101111011111011.$
$Q1 = 1111111111111111111111;$
$Q2 = 1111111111111111;$
$Q3 = 1111111111111111111111;$
$Q4 = 11111111111111111.$

Doing AND operation $P1$ to $P4$ and doing AND operation $Q1$ to $Q4$ to get the P and Q values of user $C2$.

So the P and Q values of user $c2$ are:

$\cap_i^4 = Qi = 1111111111111111111111;$
$\cap_i^4 = Pi = 10111011100111110011.$

Those all process are done by using mappers. Then next step is to identify the score of each user by doing Psum-Qsum. This is done by using a reducer. Then to find most dominance values by sorting the score in descending order. In top-k the

User Id	Movie id			
	V1	V2	V3	V4
A1	-	3	1	3
A2	-	1	2	1
A3	-	1	3	4
A4	-	7	4	5
A5	-	4	8	3
B1	-	-	1	2
B2	-	-	3	1
B3	-	-	4	9
B4	-	-	3	7
B5	-	-	7	4
C1	2	-	-	3
C2	2	-	-	1
C3	3	-	-	2
C4	3	-	-	3
C5	3	-	-	4
D1	3	5	-	2
D2	2	1	-	4
D3	2	4	-	1
D4	4	4	-	5
D5	5	5	-	4

Fig. 2 Sample data set

K represent the number of dominance values. Using K values the top-k-dominant values could be identified quickly.

5   Experiment and Analysis

In the experimentation to obtain the dominance values, Movie Lens data set is used, which contains a set of movies with a rating from a set of users. This data set is used to build the movie recommender systems. The size of the data set is 27,000,000 ratings for 58,000 movies by 280,000 users. The data set lastly updated on 09/2018. The experiment conducted in Intel core i3-4005U CPU@1.70 GHz and 8 GB RAM running Linux Ubuntu 16.04 LTS. By using the clustering method, the execution speed can be increased. DIBSM algorithm implemented using Apache spark 2.3.2 using scala language. The programming interface used is Intelli J IDEA 2018.2.

5.1   Analysis Graph

The below graph Fig. 3 shows the run time comparison between existing MRBIG algorithm and proposed DIBSM algorithm. To plot this graph, 6000 users are being taken from the Movie Lens data set and run using two algorithms and identify the time taken to find the dominance values. In this graph, it can be seen that our proposed algorithm finding dominance values quickly as compared to the existing method. This graph shows that DIBSM algorithm is two times faster and efficient than the existing system. From this graph, it can be proved that our proposed algorithm is faster.

Fig. 3 Run time comparison graph between MRBIG and DIBSM

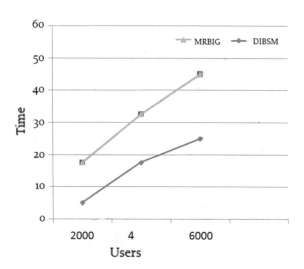

6 Conclusion

This paper proposed an algorithm named DIBSM, which is capable to find top-k dominance values on incomplete big data sets more quickly as compared to existing systems. In this algorithm, a scoring function is used to identify dominance values. By using MapReduce technique, parallel data processing is performed and this increases the execution speed. To find the score of each user, two variables called P and Q are being used. Based on the P and Q values, find the score of each user. To get the dominance values to sort the score of each user in descending order. The sample data set used to find the dominant value is Movie Lens data set. It contains a set of movies rated by a set of users.

As compared to the existing MRBIG algorithm our proposed DIBSM algorithm finding the dominance values more quickly. Because in MRBIG algorithm a bitmap index table is created along with MapReduce technique to obtain the dominance values. Major drawback with this technique is table creation as it takes more time and more space as a result it needs more processing time. But, in the proposed DIBSM algorithm, the dominance values are identified quickly without bitmap index table which reduces the processing time.

References

1. Lee J, Im H, You GW (2016) Optimizing skyline queries over incomplete data. Inform. Sci. 361:14–28
2. Koh J-L, Chen C-C, Chan C-Y, Chen ALP (2017) MapReduce skyline query processing with partitioning and distributed dominance tests. Inf Sci 375:114–137
3. Khalefa ME, Mokbel MF, Levandoski JJ (2008) Skyline query processing for incomplete data. In: Proceedings of the IEEE 24th international conference on data engineering, 7–12 Apr 2008. IEEE Xplore Press, Cancun, Mexico
4. Arefin MS, Morimoto Y (2012) Skyline sets queries for incomplete data. Int J Comput Sci Inform Technol
5. Bharuka R, Kumar PS (2013) Finding skylines for incomplete data. In: Proceedings of the twenty-fourth australasian database conference, 29 Jan–01 Feb
6. Jiang T, Zhang B, Lin D, Gao Y, Li Q (2015) Incremental evaluation of top-k combinatorial metric skyline query. Knowl-Based Syst 74:89–105
7. Babanejad G, Ibrahim H, Udzir NZ, Sidi F, Alwan AA (2017) Deriving skyline points over dynamic and incomplete databases. In: Proceedings of the 6th international conference of computing
8. Bujji Babu D, Swapna G (2017) Enhancing query efficiency using pruning techniques on incomplete data. Int J Comput Sci Inf Technol 8(3):413–416
9. Yang Z, Li K, Zhou X, Mei J, Gao Y (2018) Top k probabilistic skyline queries on uncertain data. Neurocomputing 317:1–14
10. Ezatpoor P, Zhan J, Wu, JM-T, Chiu C (2018) Finding top-k dominance on incomplete big data using MapReduce framework. https://doi.org/10.1109/ACCESS.2018.2797048. IEEE Access
11. Informatics Gulzar Y, Alwan AA, Salleh N, Al Shaikhli IF (2017) Skyline query processing for incomplete data in cloud environment. In: Proceedings of the 6th international conference on computing and informatics (ICOCI'17), Kuala Lumpur, Malaysia

Stability Region of Trapezoidal Rule for Dynamic Software Dependability Assessment Model for Open-Source Cloud Computing

K. Selvakumar and S. V. M. G. Bavithiraja

Abstract This paper presents a simple way to find influential papers to research the future development on the topic stability region of trapezoidal rule for dynamic software dependability assessment model of open-source cloud computing using citation network analysis. Using co-citation network analysis, influential papers are also identified for three subdomains of applications; finding influential papers will save the time of active researchers and will avoid repetition of research works. Also, the stability region of a trapezoidal rule on applying to a dynamic software dependability assessment model is obtained and noticed that there is a gap in the neighborhood of (2, 0). So the trapezoidal rule is not A-stable. To design a trapezoidal rule, a new approach is suggested. The influential papers direct, to get higher-order methods, not only the order of convergence but also one must improve the stability region of the numerical solution of the numerical method. Experimental results are provided to show the applicability and performance of citation network analysis and trapezoidal rule.

Keywords Citation network analysis · Co-citation network analysis · Stability region · Trapezoidal rule · Dynamic software dependability · Open-source cloud computing

1 Introduction

A network service consists of sharing IT resources, such as storage, servers, networks, services, and applications [1, 2]. This is cloud computing. The OSS cloud is the

K. Selvakumar (✉)
Department of Mathematics, Anna University, University College of Engineering, Nagercoil, Tamil Nadu, India
e-mail: selvakumaruce@gmail.com

S. V. M. G. Bavithiraja
Department of Computer Science and Engineering, Shri Eshwar College of Engineering, Coimbatore, Tamil Nadu, India
e-mail: bavithira2000@yahoo.com

© The Author(s), under exclusive license to Springer Nature Singapore Pte Ltd. 2021
J. S. Raj et al. (eds.), *Innovative Data Communication Technologies and Application*, Lecture Notes on Data Engineering and Communications Technologies 59, https://doi.org/10.1007/978-981-15-9651-3_2

paradigm for next-generation software services due to cost reduction, fast delivery, and labor savings. Much open-source software (OSS) is currently being developed worldwide, the Apache HTTP server, the Linux operating system, etc. However, poor quality and poor customer support management prevent the OSS process. Software quality issues are focused on that prevents free software from advancing due to poor quality management.

A mathematical model for the cloud service is based on the number of faults identified over time. Let $N(t)$ be the number of faults that were recognized in the OSS cloud by the operating time t ($t \geq 0$). Assume that $N(t)$ takes continuous real values. Latent faults in the OSS cloud are detected and eliminated during the operational phase, and, therefore, $N(t)$ gradually increases as the operational processes progress. With general assumptions for the dynamic modeling of software reliability growth, the following linear rigid differential equation is considered [1]:

$$\frac{dN(t)}{dt} = b(t)[D(t) - N(t)], \quad N(0) = 0. \tag{1}$$

Here, $b(t)$ is the rate of software fault detection at operating time t and a non-negative function and $D(t)$ means the number of specification changes that are subject to OSS faults. Also, defined $D(t)$ as $D(t) = \alpha \exp(-\beta t)$, where α is the number of latent defects in the OSS and β is the rate of variation in the specifications that are subject to defects.

The fault-prone specifications of the OSS grow exponentially concerning t, as shown in Fig. 1 is assumed. Thus, the OSS shows a trend of reliability regression when β is negative. On the other hand, the OSS shows a trend toward increasing reliability when β is positive.

Fig. 1 Fault-prone specification domain

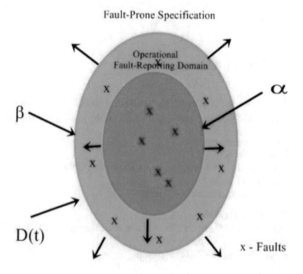

A general form of dynamic software dependability assessment model (SDAM) is considered based on the stiff differential equation (SDE)

$$\frac{dN(t)}{dt} + a(t)N(t) = f(t), \quad t \in [p, q], \quad N(p) = \eta. \tag{2}$$

where $a(t)$ and $f(t)$ are continuous and $a(t) \geq \alpha > 0$ for all t in the domain $[p, q]$. Equation (2) has a unique and bounded solution [3–12]. For simplicity take $f(t, N(t)) = -a(t)N(t) + f(t)$ in Eq. (2) and so Eq. (2) gets the form

$$\frac{dN(t)}{dt} = f(t, N(t)), \quad t \in [p, q], \quad N(p) = \eta. \tag{3}$$

The problem (3) is a linear initial value problem for the first-order stiff differential equation. The problem (3) can be solved for the theoretical or analytical solution $N(t)$ if the problem is of simple functions. If the problem (3) is not integrable, then this problem can be solved numerically using the implicit Euler method [3–11] to give order one approximation to the exact solution. To get order two approximation, trapezoidal rule has to be used [3–12],

$$N_{n+1} = N_n + \frac{h}{2}\left[f(t_n, N_n) + f(t_{n+1}, N_{n+1}) \right], \quad n \geq 0, \quad N_0 = \eta_=. \tag{4}$$

Similarly, to get order two, the implicit method of order two have to be applied

$$N_{n+1} = N_n + k_1, n \geq 0, \quad N_0 = \eta. \tag{5}$$

where h is the step size and $k_1 = hf\left(t_n + \frac{h}{2}, N_n + \frac{k_1}{2}\right)$.

In the literature, there are many papers with methods of orders one, two, and higher orders. For future development, a citation network analysis is used to identify influential articles. The citation network analysis (CNA) is applied to find the influential papers motivated by the works of Zhang et al. [13, 14].

In the literature, the stability region of the numerical solution due to a method (4) and (5) is the entire right half of the complex plane. It is not so. In this paper, the stability region is identified as different from the entire right half of the complex plane and it is noted that to get higher-order methods, one must improve the stability region of the numerical solution.

For ta stiff equation $N' = f(t, N(t)), t \in [p, q], N(p) = \eta$, subject to the condition $-f_N \geq \alpha > 0$, the methods (4) and (5) are not A-stable. The numerical solutions of the methods (4) and (5) are not A-stable.

In Sect. 2 of this paper, influential papers on the topic stability region of trapezoidal rule for dynamic software dependability assessment model for open-source cloud computing are identified using CNA with the help of the tools Science square and Pajeck. In Sect. 3, three subdomains are identified by CNA. Using the Project tool, the CNA is done to identify influential papers. In Sect. 4, the stability region of the trapezoidal rule is obtained and it is shown that there is a gap in the neighborhood of $z = 2$, z is a complex number. The stability range is not the entire right half of the complex plane and so the trapezoidal rule is not A-stable. Finally, in Sect. 5, the experimental results are presented to show the applicability of CAN and co-CAN and stability regions are also plotted to show trapezoidal rule is not A-stable.

2 Influential Papers

In the Web of science database, using the keyword stability region of trapezoidal rule for dynamic software dependability assessment model for open-source cloud computing, collect the papers, and get the extracted paper citation network. Then, use the Pajeck tool to draw the network (Fig. 2).

There are two-star graphs that indicate there are two influential papers. The central node refers to the influential paper and the end nodes connecting the central node are reference papers in it. Now, with the help of these two papers, researchers can think about future development in this topic. Influential papers are given in Table 1.

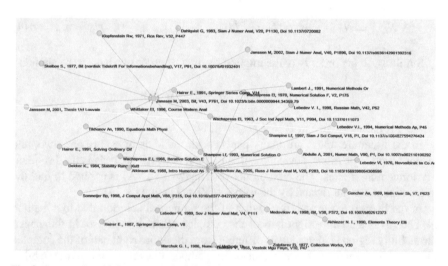

Fig. 2 Star graphs which represent the influential papers

Table 1 Influential papers

List	Influential papers
Paper 1	Janssen and Van Hentenryck [4]
Paper 2	Medovikov and Lebedev [5]

3 Influential Papers to Subdomains

On using co-citation network analysis, there are three subdomains and it is identified in noticed and in Fig. 3.

3.1 Mathematics Applied

As in the previous section, collect the papers from the Web of science database then extract paper citation networks to the subdomain and then draw network using the Pajeck tool. The output is given in Fig. 4. There are two-star graphs in Fig. 4 representing two influential papers. The center nodes of the two-star graphs refer to influential papers and their respective nodes are their references. Influential papers are Paper 1 and Paper 2 listed in Table 1.

3.2 Computer Science Software Engineering

First, collect the papers from the Web of science database then extract paper citation networks for this subdomain and then draw network using the Pajeck tool. The output is given in Fig. 5. There is a one-star graph in Fig. 5 which represents one influential paper. The center node of the star graph refers to an influential paper. The influential paper is Paper 1 listed in Table 1.

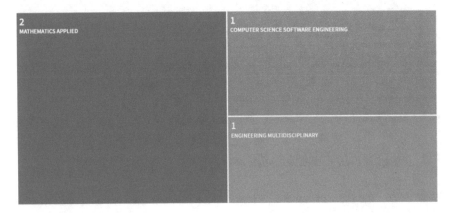

Fig. 3 Subdomains of applications

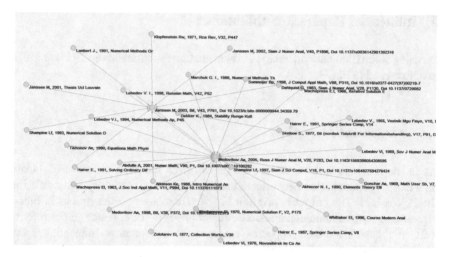

Fig. 4 Star graphs for mathematics applied

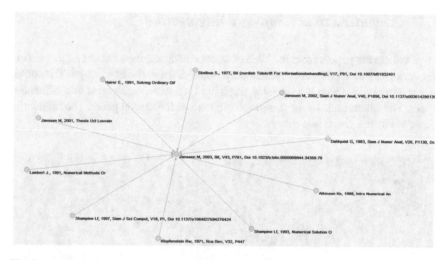

Fig. 5 Star graphs for computer science software engineering

3.3 Engineering Multidisciplinary

First, collect the papers from the Web of science database then extract paper citation networks for the subdomain and then draw network using the Pajeck tool. The output is given in Fig. 6. There is only one one-star graph in Fig. 6 which represents one influential paper. The central node of the star graph refers to influential papers. The influential paper is Paper 2 listed in Table 1.

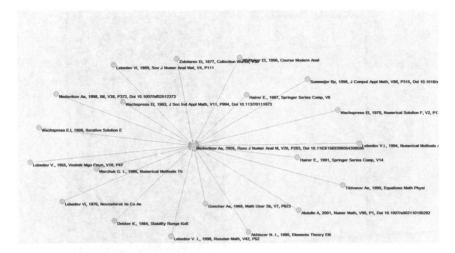

Fig. 6 Star graphs for engineering multidisciplinary

4   Stability Region

In this section to study the stability, the Cauchy problem is considered [12],

$$N' = -\lambda N, \quad t \in [0, 1], \quad N(0) = 1, \quad \lambda > 0, \tag{6}$$

whose exact solution is $N(t) = \exp(-\lambda t)$.

4.1   Stability Region of Exact Solution

The stability region (range) of the exact solution is defined from the stability function and the stability function is defined as

$$\text{Stability function} = Q(h\lambda) = \frac{N(t_{n+1})}{N(t_n)} = \exp(-\lambda h).$$

And, hence,

$$\text{Stability region} = \text{abs}(Q(h\lambda)) < 1.$$

Here, the stability region (range) of the exact solution is the entire right half of the complex plane. It is plotted in Fig. 7.

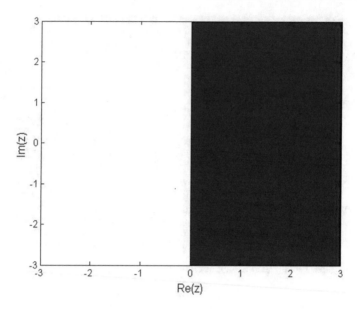

Fig. 7 Stability region of exact solution

4.2 Stability Region of Numerical Solution

The stability region of the numerical solution is defined as

$$\text{Stability function} = Q(h\lambda) = \frac{N_{n+1}}{N_n}$$

And

$$\text{Stability region} = \text{abs}(Q(h\lambda)) < 1.$$

4.3 A-Stable

If the stability region of the Cauchy problem is the same as that of the numerical method, then the method is said to be A-stable. The stability region of the Cauchy problem is the entire right half complex plane.

4.4   Trapezoidal Rule

Trapezoidal rule is an average of an explicit non-A-stable method and an implicit A-stable method. The explicit non-A-stable method is

$$N_{n+1} = N_n + hf(t_n, N_n), n \geq 0, \quad N_0 = \eta. \tag{7}$$

and the implicit A-stable method is

$$N_{n+1} = N_n + hf(t_{n+1}, N_{n+1}), \quad n \geq 0, \quad N_0 = \eta. \tag{8}$$

It is a linear combination of an A-stable and a non-A-stable method. The trapezoidal rule as A-stable cannot be exposed.

The stability function of the method (7) is $Q(h\lambda) = 1 - \lambda h$. And,

$$\text{Stability region} = \text{abs}(Q(h\lambda)) < 1$$

and it is plotted in Fig. 8. The stability region(range) is a unit circular region(range) in the right half complex plane with center $(1, 0)$ and radius 1.

The stability function of the method (8) is $Q(h\lambda) = 1/(1 + \lambda h)$. And,

$$\text{Stability region} = \text{abs}(Q(h\lambda)) < 1$$

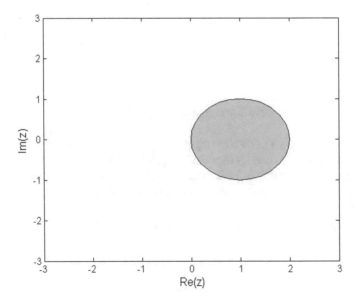

Fig. 8 Stability region of numerical t solution of (7)

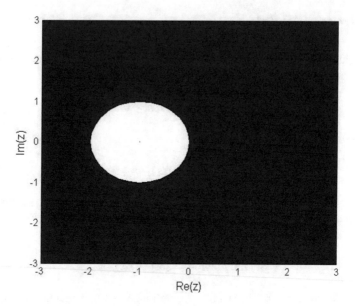

Fig. 9 Stability region of numerical t solution of (8)

and it is plotted in Fig. 9. The stability region is a region outside the unit circular region in the left half complex plan 9 with center $(-1, 0)$ and radius 1.

On applying the trapezoidal rule (4), had,

$$\text{Stability function} = Q(h\lambda) = \frac{N_{n+1}}{N_n} = \frac{1 - \lambda h/2}{1 + \lambda h/2}$$

And

$$\text{Stability region} = \text{abs}(Q(h\lambda)) < 1,$$

it is plotted in Fig. 10. The stability region (range) is in the right half complex plane with a small gap in the neighborhood of the point $(2, 0)$. So the testability region of the trapezoidal rule is not the entire right half of the complex plane and so it is not A-stable. The stability region of the trapezoidal method is not the same as that of the exact solution of the Cauchy problem.

Similarly, on applying the implicit method (5), got,

$$\text{Stability function} = Q(h\lambda) = \frac{N_{n+1}}{N_n} = \frac{1 - \lambda h/2}{1 + \lambda h/2}$$

And

$$\text{Stability region} = \text{abs}(Q(h\lambda) < 1,$$

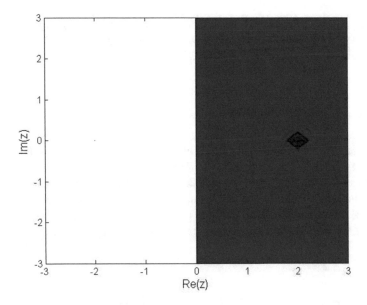

Fig. 10 Stability region of numerical t solution of (4) and (5)

it is plotted in Fig. 10. The stability region of the numerical solution of the Cauchy problem is the same as the trapezoidal rule (4). And so, both the methods are not A-stable. Using the stability region, have shown both the methods (4) and (5) are not A-stable.

To get an A-stable trapezoidal rule, take the average of two A-stable methods. It is observed that the trapezoidal rule is a linear combination of an A-stable and a non-A-stable method. The linear combination of the non-A-stable methods is again a non-A-stable method. And, the linear combination of A-stable methods is again an A-stable method. To get A-stability in the trapezoidal rule, one must take the average of the two A-stable methods only.

5   Experimental Results

In this section, the experimental results for the Sects. 2 and 3 are presented using Science square and Pajeck tools. The experimental results for Sect. 4 using MATLAB.

To the topic of this paper, the influential papers are identified by collecting papers from the Web of science databases. The extract paper citation network is got using a science square tool. Using the Project tool, from the extracted paper citation file, a network is drawn in the form of star graphs. It consists of a node at the center and other end nodes are connected with the center node. Each node is marked by the information about the paper such as the author name, journal name, volume number, issue number, year of publication, and DOI number. The output is given in Fig. 2.

Using co-citation network analysis, Fig. 3 is obtained to identify the subdomains. Again, each subdomain is treated separately, and to find the influential papers in each subdomain citation network analysis has been done.

5.1 Mathematics Applied

Taking the first subdomain again the citation network analysis is performed and the output in Fig. 4. It is noted that to get higher-order methods, not only the order of convergence but also one must improve the stability region of the numerical solution of the numerical method. The first influential paper [3] and the second influential paper [4] direct the numerical analysts to design numerical methods to models in ordinary and partial differential equations, respectively, by improving the stability region.

5.2 Computer Science Software Engineering

Taking the second subdomain again, the citation network analysis is performed and the output in Fig. 5. It is noted that to get higher-order methods, not only the order of convergence but also one must improve the stability region of the numerical solution of the numerical method. The first influential paper [3] directs the numerical analysts to design numerical methods to models in ordinary differential equations, by improving the stability region.

5.3 Engineering Multidisciplinary

Taking the third subdomain again, the citation network analysis is performed and the output in Fig. 6. It is noted that to get higher-order methods, not only the order of convergence but also one must improve the stability region of the numerical solution of the numerical method. The second influential paper [4] directs the numerical analysts to design numerical methods to models in partial differential equations, by improving the stability region.

5.4 Stability Region (Range) of Exact Solution

The stability region of the solution of the Cauchy problem (6) is plotted in Fig. 7. It is the entire region on the right of the complex plane from the imaginary axis.

5.5   Stability Region of the Solution of the Method (7)

The stability region of the numerical solution of the Cauchy problem (6) using the numerical method (7) is plotted in Fig. 8. It is the unit circular region on the right of the complex plane from the imaginary axis. A circle with center (1, 0) and radius 1.

5.6   Stability Region of the Solution of the Method (8)

The stability region of the numerical solution of the Cauchy problem (6) using the numerical method (8) is plotted in Fig. 9. It is the entire complex plane except for a unit circular region with center $(-1, 0)$. It is the union of the entire right half of the complex plane and on the left, a unit circular region is not included.

5.7   Stability Region of the Trapezoidal Rule (4)

The stability region (range) is in the right half complex plane with a small gap in the neighborhood of the point (2, 0). The stability region of the trapezoidal method (4) is not the same as that of the exact solution of the Cauchy problem. It is plotted in Fig. 10. So the testability region of the trapezoidal rule is not the entire right half of the complex plane and so it is not A-stable.

5.8   Stability Region of the Method (5)

The stability region (range) is in the right half complex plane with a small gap in the neighborhood of the point (2,0). The stability region of the method (5) is the same as the trapezoidal method (4). It is plotted in Fig. 10. So the stability region of the method (5) is not the entire right half of the complex plane and so it is not A-stable.

6   Conclusions

Information about the research papers is stored in many databases all over the world. With the help of citation network analysis from the database Web of science, researchers can identify influential papers in the field of their research. This will save the time and effort of researchers and will avoid the repetitions of the past works. The need of the hour is only present and future developments in the field of research for the welfare of society and nations. One can identify influential papers and then do innovative research.

In this paper also, on the topic of the paper using citation network analysis, influential papers are identified. And, for the subdomains also influential papers are identified. Then, toward the present and future development of trapezoidal rules, an important observation is done by stability regions and concluded the stability region due to the numerical solution of Cauchy problem on applying the trapezoidal rule, will not cover the entire right half of the complex plane. There is a gap in the neighborhood of (2, 0). So the trapezoidal rule is not A-stable. The trapezoidal method is a linear combination of an A-stable and a non-A-stable method. To get an A-stable trapezoidal rule, one must take the linear combination of two A-stable methods.

It is observed that to get higher-order methods, not only the order of convergence but also one must improve the stability region of the numerical solution of the numerical method. The first influential paper [3] directs the numerical analysts to design numerical methods to models in ordinary differential equations, by improving the stability region. The second influential paper [4] directs the numerical analysts to design numerical methods to models in partial differential equations, by improving the stability region.

Acknowledgements This research work was carried out in HRDC, University of Kerala, Thiruvananthapuram, India and Cochin University of Science and Technology, Cochin, India.

References

1. Tamura Y, Yamada S (2012) Dependability analysis and optimal maintenance problem for open source cloud computing. IEEE international conference on systems, man, and cybernetics, COEX, Seoul, Korea, pp 1592–1597, 14–17 Oct 2012
2. Tamura Y, Yamada S (2012) Dependability analysis and optimal maintenance problem for open source cloud computing. In: IEEE international conference on systems, man, and cybernetics, COEX, Seoul, Korea, pp 1592–1597, 14–17 Oct 2012
3. Adesanya AO, Sunday J, Momoh AA (2014) A new numerical integrator for the solutions of stiff first order ordinary differential equations. Int J Pure Appl Math 97(4):431–445
4. Janssen M, Van Hentenryck R (2003) Precisely A (alpha)-stable one-leg multistep methods. BIT Numer Math 43(4):761–774 (2003). 10.1023/B:BITN.0000009944.34359.79
5. Medovikov AA, Lebedev VI (2005) Variable time steps optimization of L-omega-stable Crank-Nicolson method, Russian J Numer Anal Math Model **20**(3):283–393. https://doi.org/10.1163/1569398054308595
6. Selvakumar K (1992) Uniformly convergent difference for differential equations with a parameter. Bharathidasan University, Ph.D. Thesis
7. Sivakumar K (1994) A computational method for solving singularly perturbation problems using exponentially fitted finite difference schemes. Appl Math Comput 66:277–292. https://doi.org/10.1016/0096-3003(94)90123-6
8. Selvakumar K (1994) Optimal uniform finite difference schemes of order two for stiff initial value problems. Commun Numer Methods Eng 10:611–622. https://doi.org/10.1002/cnm.1640100805

9. Selvakumar K (1995) A computational procedure for solving a chemical flow-reactor problem using shooting method. Appl Math Comput 68:27–40. https://doi.org/10.1016/0096-300 3(94)00082-F
10. Selvakumar K (1997) Optimal uniform finite difference schemes of order one for singularly perturbed Riccati equation. Commun Numer Methods Eng 13:1–12. https://doi.org/10.1002/(sici)10990887(199701)13:1
11. Selvakumar K, Ramanujam N (1996) Uniform finite difference schemes for singular perturbation problems arising in gas porous electrodes theory. Indian J Pure Appl Math 293–305
12. Selvakumar K (2018) Numerical method for non-stiff differential equations. DJ J Eng Appl Math 4(2):21–33. https://doi.org/10.18831/djmaths.org/2018021003
13. Zhang S, Zhao D, Cheng R, Cheng J, Wang H (2016) Finding the influential paper in citation networks. In: IEEE first international conference on data science in cyber space. IEEE Comput Netw 658–662. https://doi.org/10.1109/dsc.2016.5
14. Zhang S, Zhao D, Cheng R, Cheng J, Wang H (2016) Finding the influential paper in citation networks. In: IEEE first international conference on data science in cyber space. IEEE Computer Networks 658–662. https://doi.org/10.1109/dsc(2016)

A Study of Blending Ensembles for Detecting Bots on Twitter

Sanjay Kumar, Ryan Bansal, and Raghav Mehta

Abstract In recent years, AI is taking over the world and such rapid advancement has bought many concerns with it, one of them being—deployment of bots on social media platforms to spread propaganda and gain a political advantage. Moreover, nowadays on Twitter, politics has taken the front seat, and hence, it has become more important than ever to tackle this problem by taking a more comprehensive approach. In this work, two blending ensembles are being studied, one of them based on LSTM and the other one based on CNN and will be comparing their performance for a text classification problem on users tweets—which is our primary sub-problem when detecting bots. Alongside, also be shown exactly how insightful the tweet text and metadata may turn out to be for getting a new set of state-of-the-art results as opposed to not considering these at all.

Keywords Ensemble blending · Deep learning · Machine learning · Bot detection

1 Introduction

The leading technology in mobile phones and computers along with blazing fast internet speeds with the uninterrupted connection has aided social networking. However, the recent boom of social media has also seen a rise in misuse of these platforms by various organizations, from political parties to wealthy corporations to sway the mindset of millions of people in one direction. One such example could be seen through the 2016 US elections where bots and software were developed

S. Kumar (✉) · R. Bansal · R. Mehta
Department of Computer Science and Engineering, Delhi Technological University, Main Bawana Road, Shahbad Daulatpur, Delhi 110042, India
e-mail: Sanjay.kumar@dtu.ac.in

R. Bansal
e-mail: ryanbansal_bt2k16@dtu.ac.in

R. Mehta
e-mail: raghavmehta_bt2k16@dtu.ac.in

29

to automatically inflate the internet traffic to make voters believe in the likelihood of a politician coming into power. 'Bots' are essentially software program scripts that naturally produce or repost content and communicate with people in informal organizations. There have been reports of bots spreading deceiving data about the coronavirus, unconfirmed fear inspired notions about the Australian bushfires and endeavouring to impact a year ago Gulf emergency. Different examinations have featured and dissected the enormous scope of interfering by bots to impact elections around the globe [1] in countries like Japan [2], Finland [3] and France [4]. At the point when utilized on a huge enough scale, such bots have had a huge effect on this present reality—from spreading counterfeit news during decisions, impacting stock costs, influencing conclusion about an organization or an item, hacking and cybercrimes, stealing data and advancing and contorting the prevalence of people and gatherings. There is a rise of social bots in the cyber world [5]. The approaches like convolution networks and long short-term memory networks are aimed to bring to compare and propose the best out of them. The technique of blending is worked on by taking the predictions from tweet data and combining them as a feature with user data. The organization of this paper is as follows: In Sect. 2, the related works are presented. Datasets and pre-processing activities are discussed in Sect. 3. In Sect. 4, the proposed methodology is discussed. Section 5 presents the results and analysis. Finally, the work is concluded in Sect. 6.

2 Related Work

Before getting the inspiration for our work, many previous pieces talking about this problem is being read and most of them also showcased innovative ideas coupled with outstanding results. For instance, in [6], the best accuracy was **98.39%** without synthetic oversampling at the user level and **96.33%** at the tweet level. The two more previous research approaches such as [11] gave the best accuracy of **97.6%** whereas [12] measured the best accuracy of **79.64%**. Both of these use the latest deep learning techniques like Bi-LSTM-based approach for textual analyses for bot detection.

But to our astonishment, all of those works either worked only with the user activity, i.e. tweet level, or they only worked with the profile information, i.e. user level—but never both. And even though the results were impressive, it is clear that the approach could be improved further as quickly recognized that either of two levels fails to tell the whole story of the said user account.

So naturally, the first thought that popped into our heads was '*Why not combine the two levels?*' Using something like blending.

3 Data

The dataset was procured from [7]. It comprises the account authenticity information utilized in our research. It has eight manually annotated CSV files. Six files constitute fake user tweets and profile data. While two files consist of genuine tweets and their profile data. There are a total of more than 6 million tweets accumulated from 8386 users, including both fake and genuine ones. The details of the dataset are shown in Table 1.

3.1 Data Pre-processing

The imbalanced class distribution had to be removed. There was an acute difference in the number of fake and genuine tweets. The number of bot tweets observed in our data was equal to 3,457,143 and the number of human tweets was 2,839,362. Thus, the model tends to bias towards classifying an instance as bot rather than a genuine account. Figure 1 shows this distribution through a pi-chart.

To solve this problem, synthetic sampling is used to make instances of both classes equal. Rows are randomly selected from genuine class and replicated their occurrences to make equal distribution of classes.

Table 1 Details of the dataset

Class	Users	Tweets
Fake	4912	3,457,143
Genuine	3474	2,839,362

Fig. 1 Bar graph showing class imbalance

No. of tweets

● BOT ● HUMAN

Fig. 2 Snapshot of tweets
before cleaning

RT @sayingsforgirls: i can send a text at 12:0...	0
Dogs come when they're called; cats take a mes...	1
RT @JulianBeatsCom: We have Free Beats http://...	1
...	...
"Maestra, come mai ti sei fatta il piercing e ...	1
RT @htTweets: PM @narendramodi launches key #L...	0
"Mamma perché non credi nei sogni? Io sono sta...	1
RT @i_D: It's time to talk about the future of...	0
@pam_qt kinantahan kita neto http://t.co/TfhHqGBL	0

After normalizing the classes, the data is preprocessed. As you know, to get any sort of results, every NLP task calls for comprehensive data cleaning. Since our data comprises more than 6 million tweets, had to keep in mind various aspects of cleaning of data to maximize the learning of our models. Start with replacing the occurrences of special Twitter vocables which could be then tokenized and fed into GloVe embedding. This step is very crucial as a word like 'RT', '@user', '#word', etc. do not mean anything in English vocabulary. So if do not consider these, our model would not learn valuable information from these Twitter vocables.

To summarize, the following changes were made to our text data:

- Hyperlinks were tagged as 'url'.
- Numeric values were tagged as 'number'.
- '#' was changed to 'hashtag'.
- '@user' was changed to 'user'.
- 'RT' was changed to 'reply'.
- All the special characters and anything other than an English word was removed.

Figures 2 and 3 help to distinguish the before and after semantics of tweets.

After cleaning the data, proceeded towards making a word cloud of the genuine tweets and find prominent usage of words like 'user', 'reply', 'hashtag' and a large vocabulary of words and can be inferred that genuine tweets are usually based on people communicating with each other through tagging, commenting and replying on social platforms, which was not the case with the fake users.

The fake ones for the most part just showed a repeated usage of some particular words too much, a lot of curse words and trolling. Others showed a lot of repeated usage of hashtags as though trying to fulfil a defined agenda (Fig. 4).

Fig. 3 Snapshot of tweets
after cleaning

reply user i can send a text at number:number ...	0
Dogs come when they're called; cats take a mes...	1
reply user We have Free Beats url url	1
...	...
"Maestra, come mai ti sei fatta il piercing e ...	1
reply user PM user launches key hashtag promis...	0
"Mamma perché non credi nei sogni? lo sono sta...	1
reply user It's time to talk about the future ...	0
user kinantahan kita neto url	0

Fig. 4 Word cloud showing
prominently used words by
people on Twitter

4 Methodology

The main aim of this piece is to do a comparative analysis of various blending
ensemble configurations for improving the accuracy of detecting bots on the Twitter
platform. Having two models, a base classifier and a meta classifier, train the base
classifier on the tweet input data and get the predictions for our test data. These
predictions are joined or 'blended' as a feature during the training of the meta classifier. Although the blending techniques are really powerful in and of itself, and for
the most part give great results irrespective of the kind of problem they are applied

to, they are particularly useful here. The premise of using blending for this comes from the fact that a model which works solely on tweet level or solely on user profile level is prone to error. With online social media platforms becoming increasingly mainstream, there is a whole new set of problems which comes into the picture—one of the main ones being that huge amounts of traffic to one's profile can nowadays be bought for just over the price of a lunch. With keeping that in mind, any method which works solely on 'the numbers generated by a user' can fail catastrophically. Using ensembling, can 'blend' the users tweeting behavioural patterns with the amount of interactions the user is generating. This approach can help us make a better judgement on whether the user is of a fake kind. As it is pretty intuitive that for any fake, user the hardest thing to fake is the text, it is tweeting daily, consequently, the purpose of this work is to study various neural networks—particularly LSTMs and CNN's—and compare how they perform when given a text classification problem for detecting a bot user. Simply put a better answer to the question—'*What the user is tweeting about?*'—will help us better differentiate a real user from a malicious one—or at least it will provide a more 'thorough' analysis of the Twitter feed in question. The performance of these blended predictive structures is measured through a series of metrics like F-measure, precision, recall and accuracy. For an imperative study, our research work is divided on the basis of base classifiers used. Figure 5 represents the methodological approach used in this study. The tweet data is started with. On the text (tweets) base, classifier 1 (CNN or LSTM) is applied. On the tweet metadata, the base classifier 2 (Ada boost) is applied. The predictions from these two classifiers are taken and merged them as features with user data and applied classic machine learning models to obtain the final predictions.

4.1 Data Splitting

As the blending ensembles are being used, the splitting of data gets a little tricky as to have separate datasets for training models on different levels. In our study, first, the data is split into a 70/30 distribution, i.e. the 70% is used for training our models at level 0 and the remaining 30% for training level 1 model.

Moreover, at level 0, 20% of its training set is used for validating level 0 models. And at level 1, 30% of its training set is used for final testing and obtaining results.

4.2 Level 0 or Tweet Level

For our level 0, two models have been used separately for tweets textual and numeric tweet metadata features separately using two different base classifiers as explained previously in Fig. 5.

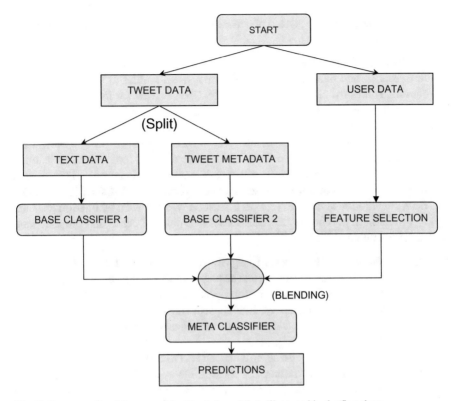

Fig. 5 Our general architecture of the blended models is illustrated in the flowchart

4.2.1 Handling Tweet Metadata

Tweet metadata consists of numeric features like number of hashtags, number of hyperlinks, number of mentions to the other users, like count and retweet count.

AdaBoost: The purpose of integrating an AdaBoost classifier was to analyze the numeric tweet metrics including various tweet interaction metrics such as number of likes, number of comments and number of retweets, coupled along with numerous other features like number of hashtags, number of URLs and number of mentions to other users to help us capture information about user's interactions. Although, as already mentioned, this information can be bought but still it can prove to be insightful in many cases.

4.2.2 Handling Tweets Text

As mentioned before, recognizing textual analysis was the most important factor for us to get any kind of decent result. Both CNN's and LSTMs seem to have their positives and negatives, so instead of choosing both, one by one is used and compared their performance.

Convolutional Neural Network: To tackle the classic text classification problem with tweets textual data, a CNN model is first used, similar to [8]. Also, stacked it with the AdaBoost classifier at tweet level. CNN is used for its ability to extract semantic features when dealing with text data, and unlike LSTM, it has a lower dependence on the order of tokens in any given sequence.

Our architecture for CNN was of relatively lower complexity to prevent overfitting. The configuration particulars for our CNN include first and foremost an embedding layer with GloVe, followed by four layers of one-dimensional convolutions each having 100 filters but of different sizes, increasing from 2, then 4, then 6, and finally 8. Each convolution layer had 'relu' as activation was followed by one-dimensional global maximum pooling layer. After these, two pairs of dense and dropout were attached to complete the model.

Recurrent Neural Network with Long Short-Term Memory: One of the limitations of CNN is that it fails on longer text sequences and although tweets can officially go up to 280 char, which can mean a total of more than 50 words in a single tweet. To handle longer tweet sequences, RNNs particularly LSTMs are used to capture relationships between words which may lie at a distance within the given tweet but might be related.

So, this time, everything else being equal, an LSTM instead of a CNN was stacked with the same AdaBoost classifier. The most suitable LSTM configurations for this study consisted of an embedding layer of varying dimensions followed by a 1D spatial dropout of 0.2, then the LSTM layer with 32 hidden units and 0.2 dropouts to avoid overfitting and finally the dense layer with sigmoid activation, as for our task, the number of classes wanted to classify an instance was only 0 and 1.

4.3 Embedding

GloVe or Global Vectors for Word Representation is an unsupervised learning algorithm for obtaining vector representation for words. For our study, the publicly available pre-trained embeddings provided by Stanford University is used which is trained on 6 billion tokens and has a vocabulary size of 400 thousand words. It is available in four different dimensional categories namely 50d, 100d, 200d, 300d. For further information, refer to [9].

4.4 Level 1 or User Profile Level

At this point, done with the tweet level or level 0 classification and now comes the user account level or level 1 classification. For studying the blending methods, the predictions from the level 0 classifier obtained on the test data will be used as an additional feature apart from the already existing profile features such as number tweets, follower count, following count, etc.

However, as there many tweets for a single user consequently, have a set of predictions for a single user. For simplicity, the mean of these predicted probabilities is taken. This new feature indicates what the level 0 models 'think' about the possibility of the user being a bot by just analysing its tweeting behaviour. All of these combined features are put out into popular machine learning classifiers such as decision tree, random forest, linear discriminant analysis and compared their performance in a tabular form.

5 Results

Firstly, our base-level AdaBoost classifier is tested solely without the textual analysis; it only gave an accuracy of **76.63%** on the level 1 training set, which to some extent proves our premise that the user interactions such as likes and retweet counts are easy to fake. Secondly, our base level neural networks are tested solely without the metadata analysis; the LSTM produced an accuracy of whooping **95.91%**, whereas CNN was also able to perform pretty well, giving an accuracy of **94.97%**. These results also go along the lines of our initial premise that the textual part of Twitter is the hardest one to fake. Third, the effect of embeddings on the performance is studied. While increasing the dimensions from starting from 50d and going all the way up to 300d, saw a steady growth of ~**1%** on every step for both our base deep learning models.

Finally, the graphical Fig. 6 illustrates a clear difference of how managed to observe higher accuracy by blending the predictions from textual features with tweet metadata and user data. Compared to using only a user's profile data, which yields low accuracy, a model that incorporates both textual and profile features observes much better results.

The highest accuracy of **99.84** and lowest accuracy of **99.05** are observed with LSTM as a base classifier and a reasonably similar highest accuracy of **99.38** and lowest accuracy of **98.94**. The blended models performed fairly consistently across all machine learning models with a mean accuracy of **99.26** with LSTM as base and **99.13** with CNN as the base classifier. On the other hand, for a predictive model consisting only of tweet metadata and user's profile information like follower count, following count, friends count, listed count etc., could only manage to yield a maximum accuracy of **98.11**. While predictive models with no tweet data as features could only give the highest accuracy of **97.87**. Thus, could safely be concluded that blending tweet level and user features have shown a promising increase in the accuracy of classifying bots. Moreover, coming to a comparison between LSTM and CNN as base classifier [10], slightly higher accuracy achieved by the LSTM blended model is observed across all meta classifiers.

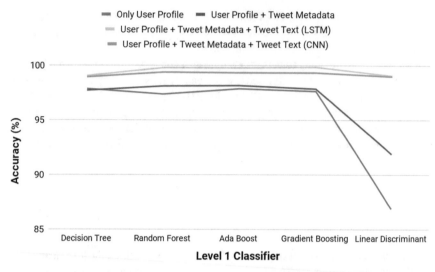

Fig. 6 Comparison of accuracy between various blending approaches for bot detection

This result is attributed to the ability of LSTM to make meanings out of sentences which are passed as input. Plus, as mentioned before, our tweets represent sentences which basically make conversational sense in case of genuine tweets and propagation in the case of fake tweets. Figure 6 helps us distinguish the accuracies of various approaches used in this study (Table 2).

Table 2 Tabulation of results obtained when CNN was used as the base text classifier and when LSTM was used for the same task

Level 0 classifiers	Level 1 classifiers	Accr (%)	Precision	Recall	*F*-score
Convolution neural network with 300d embeddings for text data combined with AdaBoost for metadata	Linear discriminant	98.99	0.990	0.989	0.990
	Decision tree	98.94	0.989	0.989	0.989
	Random forest	**99.38**	**0.993**	**0.993**	**0.993**
	AdaBoost	99.33	0.993	0.993	0.993
	Gradient boosting	99.33	0.993	0.993	0.993
Recurrent neural network with 300d embeddings for text data combined with AdaBoost for metadata	Linear discriminant	99.05	0.991	0.990	0.990
	Decision tree	99.05	0.992	0.992	0.992
	Random forest	99.78	0.997	0.997	0.998
	AdaBoost	99.78	0.997	0.996	0.997
	Gradient boosting	**99.84**	**0.998**	**0.998**	**0.998**

6 Conclusion

Through our study on the problem of Twitter bots by the means of blending ensembles were able to conclude the following: whenever an account is tried to classify as a human or a bot, a good look is needed to take at the tweeting text of the user. Solely classifying the user by using only the numbers can fail catastrophically, as seen in the case of linear discriminant classifiers.

For the problem of textual analysis of tweets in the context of bot detection, LSTM seems to outperform CNN, although the difference is not that significant.

Speaking about the future scope, the strategies for this exploration can be extrapolated to other online life stages, for example, Instagram, Facebook and electronic mail administrations like Gmail for distinguishing social spambots, political bots, bogus purposeful publicity bots, commercial bots, clinical bot. This examination endeavours to recognize social spambots found on Twitter. A progressively conventional bot location model can be worked upon that would effectively distinguish various classes of bots.

References

1. Hanouna S, Neu O, Pardo S, Tsur O, Zahavi H (2019) Sharp power in social media: patterns from datasets across electoral campaigns. Simo Hanouna Ben-Gurion University of the Negev Omer Neu Sharon Pardo Oren Tsur Hila Zahavi Ben-Gurion University of the Negev, vol 11, pp 95–111 (2019)
2. Mintal JM, Vancel R (2020) (Un)trendy Japan: Twitter bots and the 2017 Japanese general election. Polit Cent Eur 15:497–514. https://doi.org/10.2478/pce-2019-0027
3. Rossi S, Rossi M, Upreti B, Liu Y (2020) Detecting political bots on twitter during the 2019 Finnish Parliamentary election. In: Proceedings of 53rd Hawaii international conference on system sciences, vol 3, pp 2430–2439 (2020). https://doi.org/10.24251/hicss.2020.298
4. Ferrara E (2017) Disinformation and social bot operations in the run up to the 2017 French presidential election. First Monday 22(8)
5. Ferrara E, Varol O, Davis C, Menczer F, Flammini A (2016) The rise of social bots. Commun ACM 59(7):96–104
6. Kudugunta S, Ferrara E (2018) Deep neural networks for bot detection. Inf. Sci. 467:312–322. ISSN 0020-0255. arXiv
7. Cresci S, Di Pietro R, Petrocchi M, Spognardi A, Tesconi M (2017) The paradigm-shift of social spambots: evidence, theories, and tools for the arms race. In: Proceedings of the 26th international conference on world wide web companion
8. Kim Y (2014) convolutional neural networks for sentence classification. In: Proceedings of the 2014 conference on empirical methods in natural language processing (EMNLP 2014), pp. 1746–1751
9. Pennington J, Socher R, Manning CD (2014) Glove: global vectors for word representation, pp 1532–1543. https://doi.org/10.3115/V1/D14-1162
10. Yin W, Kann K, Yu M, Schütze H (2017) Comparative study of CNN and RNN for natural language processing. Cite arxiv:1702.01923

11. Wei F, Nguyen UT (2019) Twitter bot detection using bidirectional long short-term memory neural networks and word embeddings. In: 2019 First IEEE international conference on trust privacy and security in intelligent systems and applications, pp 101–109
12. Luo L, Zhang X, Yang X, Yang W.: Deepbot: A deep neural network based approach for detecting Twitter Bots. IOP conference series material science in engineering, p. 719 (2020)

Design and Development of the Robot for Aquatic Debris Using a Smartphone

Subash Chandra Bose, Azath Mubarakali, Ninoslav Marina, and Reshmi

Abstract Project aims to decrease the contamination in the water which is a very serious concern both for humans and water organisms. It uses a smart fish-type robotic sensor network which continuously monitors the contamination and sends the data to the control unit. The data include the contamination level and the GPS location of the contamination area. The benefits of this project are to protect the lives of water species, save lives of humans, low cost, highly reliable, easy to install and from anywhere can get the information about the contamination level. The main applications of the project are fresh water conserving and monitoring system, sewage disposal system.

Keywords Smartphone · Aquatic debris · Robot · Sensors · GPS

1 Introduction

Intelligent robots can perform themselves in uncertain environments, need new design desires for recent engineers. The idea of making an aquatic debris robot is a useful technology to help the environment in any place that is not clean causes disease and the spread of epidemics, especially in the aquatic environments, has imposed

S. C. Bose (✉)
Dean of Faculty of AITMIR, University of Information Science and Technology (UIST) "St. Paul the Apostle", Ohrid, Republic of Macedonia
e-mail: subash.jaganathan@uist.edu.mk

A. Mubarakali
Department of CNE, College of Computer Science, King Khalid University, Abha, Saudi Arabia
e-mail: mailmeazath@gmail.com

N. Marina
Rector, University of Information Science and Technology (UIST) "St. Paul the Apostle", Ohrid, Republic of Macedonia

Reshmi
School of Computing Science and Engineering, VIT, Chennai, India

© The Author(s), under exclusive license to Springer Nature Singapore Pte Ltd. 2021
J. S. Raj et al. (eds.), *Innovative Data Communication Technologies and Application*,
Lecture Notes on Data Engineering and Communications Technologies 59,
https://doi.org/10.1007/978-981-15-9651-3_4

new challenges on interaction modeling in robotics. Over the years, robotics with artificial intelligence research has taken many shapes and forms.

Robot in current use performs most tasks that require little interaction with casual users, and another side with heavy load continuous work in fire extinguish industries [], factories, power plants, etc. In recent years, however, robotic assistants are more common in environments such as restaurants, offices and in houses, where robots often need to communicate with users much less exposed to technology than their previous operators. In the situation of this type, the design and development of the robot for aquatic debris using a smartphone will be a promising approach.

Therefore, the causes of pollution have been researched, which found many and multiple for the existence of factories and because of human waste and to reduce diseases and anything that harms living organisms, whether in the sea or on land, reduce it and clean it wherever it is and clean it must know the identification of places of waste and so came the idea of our project, which will be the work to search for and send its location by latitude and longitude.

This project aims to decrease the contamination in the freshwater which is a very serious concern both for humans and water organisms. It uses a smart fish-type robotic sensor network which continuously monitors the contamination and sends the data to the control unit. The data include the contamination level and the GPS location of the contamination area.

2 Literature Survey

A. *Related works*

Many have tried and research on this subject, here listed a few shown in Table 1.

3 System Analysis and Requirements

A. *Software Requirement*

- MPLAB IDE
- Proteus ISIS
- HONESTECH TVR 2.5

B. *Hardware Requirement*

- Pic16f877A MICROCONTROLLER
- PH SENSOR
- ULTRASONIC SENSOR
- L298 DUAL H-BRIDGE MOTOR DRIVER
- BUZZER

Table 1 Related works

S. No.	Method	Description
1	Wang et al. [1]	Developed smartphone-based aquatic robot using rotational scheduling algorithms
2	Yang et al. [2]	Developed a robot fish using rotational scheduling algorithms and image-processing techniques
3	Chen and Yu [3]	Used meanshift algorithm with an embedded system
4	Chao and Yaping [4]	C-SpiralHill algorithm for environment monitoring robot fish
5	Jadaliha and Choi [5]	Developed optimal sampling strategies
6	Toda and Kubota [6]	Used multiresolution maps
7	Yang and Ryuh [7]	Developed Tchthus v5.6. for self-plotting
8	Floyd and Sitti [8]	Developed basilisk lizard's robot
9	Yuo et al. [9]	Developed robot using a smartphone
10	Ratsame [10]	Used maximum power point tracking (MPPT) method
11	Shi and Jia [11]	Used mode-selectable synchronous buck DC-DC converter

- BATTERY
- ZIGBEE MODULE
- GPS
- LCD
- CAMERA

C. *MicroController*

See Fig. 1.

D. DC Motor

L298 Dual H-Bridge motor driver had been used to drive the motor. This driver will drive two motors at the same time. The motor used for our project has been shown in Fig. 2

Fig. 1 PIC16F877A

Fig. 2 DC motor

Fig. 3 Power supply

E. *Power Supply*

For power supply to the circuit, lead acid storage battery is used. The lead acid storage battery is shown in Fig. 3.

F. *PH Sensors*

The PH sensor shown in Fig. 4 is used to sense the PH values in the water. Then the PH value through the microcontroller is displayed in the ZigBee software and the hardware LCD.

G. *Ultra-Sonic Sensor*

It transmits and detects the waves in the air and sends a signal to the microcontroller. The ultrasonic sensor is shown in Fig. 5.

H. *GPS device*

The GPS device is used to identify the wastes in the water. The GPS device is shown in Fig. 6

I. *ZigBee*

This circuit is used as a remote control and wireless security. The operating range is 30 m without any antenna. The ZigBee circuit is shown in Fig. 7.

Fig. 4 PH sensor and circuit
diagram

Fig. 5 Ultra-sonic sensor

Fig. 6 GPS device

Fig. 7 ZigBee

J. *LCD and Buzzer*

2X16 basic LCD is used in our project. The output in the hardware like the PH value of water, Gas pollution value and obstacle values and the alarm sound for the respective will be displayed in LCD. The LCD and alarm device is shown in Figs. 8 and 9.

Fig. 8 2 × 16 LCD

Fig. 9 Buzzer

4   System Design and implementation

A. *Block diagram*

Figure 10 shows the block diagram and it was designed using smartphone-based robot and it helps to monitor aquatic debris system. The major blocks included as microcontroller, PH sensor, buzzer system, DC driver, ultra-sonic sensor, GPS and Zigbee wireless technology.

B. *System testing and results*

1. Simulation

The simulation is shown in Fig. 11. The simulation had been done using PROTEUS (ISIS).

The external structure front view, back view and the internal structure of the robot are shown in Figs. 12, 13, and 14.

C. *Output and Results*

Here, the output result is being discussed as shown in Figs. 15, 16, 17, 18, 19, 20, 21, 22, 23, 24 and 25.

(1) Inputs and Outputs.

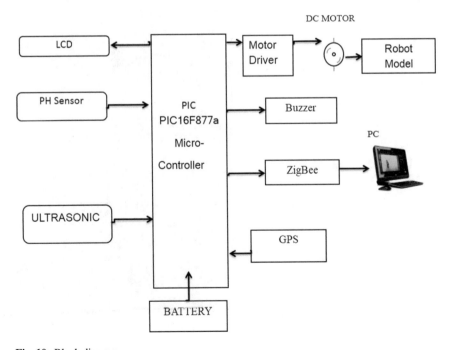

Fig. 10 Block diagram

Fig. 11 Simulation

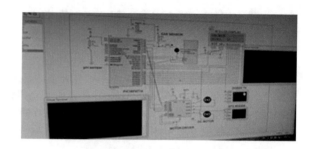

Fig. 12 The external structure of the robotic fish front view

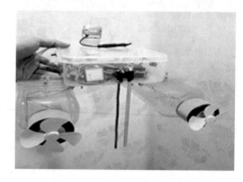

Fig. 13 External structure of robotic fish back view

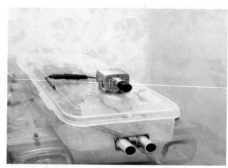

Fig. 14 Internal structure of smart robotic fish

Fig. 15 Initialization of software

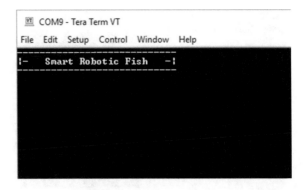

Fig. 16 Initialization of hardware

Fig. 17 Testing gas sensor

1. Operating voltage range: 2.5 to 5.5 V
2. High sensitivity (type −104 dBm)
3. Programmable output power −20 to 1 dBm
4. Operation temperature range: −40 to + 85 °C
5. Operation voltage: 1.8–3.6 V.
6. The available frequency at 2.4–2.483 GHz
7. Health and safety

Fig. 18 Output for gas sensor displayed in software

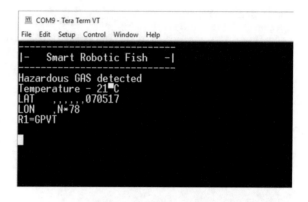

Fig. 19 Output for gas sensor displayed in robot hardware

Fig. 20 Water PH value tested

The comparison of performance results for the existing system and the proposed system has been shown in the Figs. 26 and 27.

5 Conclusion

The UIST, "St. Paul the Apostle," Ohrid, North Macedonia have done studies on the aquatic environment, and its problems and found that the most important problem is

Fig. 21 Output for water PH value tested displayed in software

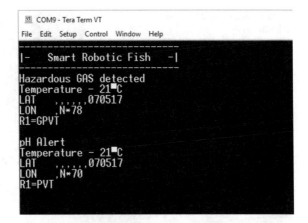

Fig. 22 Output for water PH value tested displayed in the hardware

Fig. 23 Output for the ultra-sonic sensor during obstacles

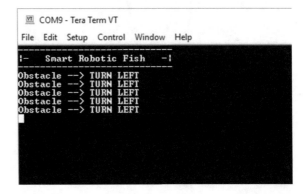

pollution and its causes, whether from factories, or human residues, so have done this robot to reduce this pollution, which helps us to detect the source of pollution and thus concluded that have achieved success in this robot special sensors helps to know whether there is pollution or not?!, And also the way to move over the water as well as a camera to see the waste in kind that accumulates on the water such as plastic and

Fig. 24 Output for obstacles
in hardware

Fig. 25 Output for obstacles
in hardware

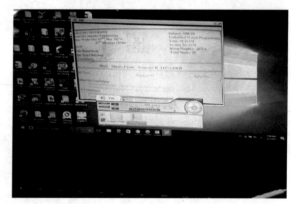

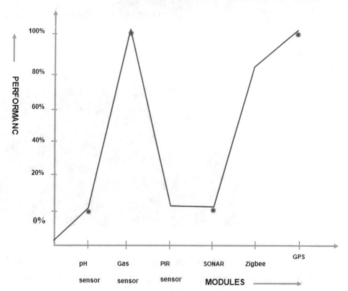

Fig. 26 Comparison graph for the existing system

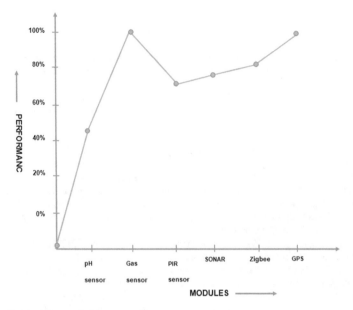

Fig. 27 Comparison graph for proposed system

other, and shortly added sensors, remote control, and a long-term communication to reach this idea to perfection.

6 Further Work

Have achieved success in this project, which is to find a good idea to help us to reduce pollution and live in a clean environment free of pollution, and this idea is excellent, and its practical experience is wonderful and interesting, and propose shortly to develop the robot by putting additional sensors, image-processing techniques for corrupted image restoration, classification of data, losses image compression techniques and control device full and intensifying the work of robot and this must be done after consulting experienced engineers from all departments of engineering for high quality to cost and must always hope the best. In future, there exist a chance to use this aquatic debris robot with lots of image-processing techniques, blockchain technology, cybersecurity [12–22]. Nanomonitoring aquatic debris robots can be created too.

References

1. Wang Y, Tan R, Xing G, Wang J, Tan X, Liu X, Chang X (2014) Aquatic Debris monitoring using smartphone-based robotic sensors. In: 14 Proceedings of the 13th international symposium on Information processing in sensor networks, Apr 2014
2. Yang G-H, Kim K-S, Lee S-H, Cho C, Ryuh Y (2013) Design and control of 3-DOF robotic fish 'ICHTHUS V5.5', In: Proceedings of the 6th international conference on intelligent robotics and applications, Sept 2013
3. Chen S-F, Yu J-Z (2014) Underwater cave search and entry using a robotic fish with embedded vision. In: Proceedings of the 33rd Chinese control conference, July 2014
4. Chao W, Yaping D (2014) A cruise route design of robot-fish for the pollution source location. In: Proceedings of the 33rd Chinese control conference, July 2014
5. Jadaliha M, Choi J (2013) Environmental monitoring using autonomous aquatic robots: sampling algorithms and experiments. IEEE Trans. Control Systems Technol
6. Toda Y, Kubota N (2013) Self-localization based on multiresolution map for remote control of multiple mobile robots. IEEE Trans Ind Inf
7. Yang G-H, Ryuh Y (2013) Design of high speed robotic fish 'ICHTHUS VS.6'. In: 13th international conference on control, automation and systems, Oct 2013
8. Floyd S, Sitti M (2008) Design and development of the lifting and propulsion mechanism for a biologically inspired water runner robot. IEEE Trans Robot
9. You C-W, Lane ND, Chen F, Wang R, Chen Z, Bao TJ, Montes-de-Oca M, Cheng Y, Lin M, Torresani L, Campbell AT (2013) CarSafe app: alerting drowsy and distracted drivers using dual cameras on smartphones. In: 11th international conference on mobile systems, applications, and services, June 2013
10. Ratsame C (2012) A new switching charger for photovoltaic power system by soft-switching. In: 12th international conference on control, automation and systems, Oct 2012
11. Shi L-F, Jia W-G (2014) Mode-selectable high-efficiency low-quiescent-current synchronous buck DC–DC converter. IEEE Trans Ind Electron
12. Sebastian L, Subash Chandra Bose J (2015) Efficient restoration of corrupted images and data hiding in encrypted images. J Ind Eng Res
13. Subash J, Bose C, Changalasetty SB et al (2016) A hybrid GA/kNN/SVM algorithm for classification of data. Biohouse J Comput Sci (BJCS)
14. Subash Chandra Bose J, Babu Changalasetty S et al (2016) The ensemble technique uses variable hidden neurons hierarchical fusion and tenfold cross validation for detection of breast masses. Biohouse J Comput Sci (BJCS)
15. Subash Chandra Bose J, Kavitha S, Mohan Raj S (2015) Lossless medical image compression using set partitioning in hierarchical trees (SPIHT) algorithm. J Ind Eng Res
16. Subash Chandra Bose J, Saranya B, Monisha S (2015) Optimization and generalization of Lloyd's algorithm for medical image compression. Middle-East J Sci Res
17. Mubarakali A, Chandra Bose S, Srinivasan K, Elsir A, Elsier O (2019) Design a secure and efficient health record transaction utilizing blockchain (SEHRTB) algorithm for health record transaction in blockchain. J Ambient Intell Hum Comput. ISSN 1868-5137. https://doi.org/10.1007/s12652-019-01420-0,2019
18. Subash Chandra Bose J, Gopinath G (2017) A survey based on image encryption then compression techniques for efficient image transmission. J Ind Eng Res 1(1):15–18. ISSN: 2077-4559
19. Subash Chandra Bose J, Monisha S, Pradeepa S, Saranya B (2017) Detection of similar mammogram image data using various file attributes. J Ind Eng Res 1(2):35–37. ISSN: 2077-4559
20. Subash Chandra Bose J, Gopinath G (2015) An ETC system using advanced encryption standard and arithmetic coding. Middle-East J Sci Res 23(5):932–935. ISSN 1990-9233, © IDOSI Publication. https://doi.org/10.5829/idosi.mejsr.2015.23.05.22233

21. Chandra Bose S, Veerasamy M, Mubarakali A, Marina N, Hadzieva E (2014) Analysis of feature extraction algorithm using two dimensional discrete wavelet transforms in mammograms to detect microcalcifications. Comput Vis Bio Insp Comput. https://doi.org/10.1007/978-3-030-37218-7_4

22. Subash Chandra Bose J, Yamini J (2014) Optimization of bit rate in medical image compression. Int J Innov Res Comput Commun Eng. ISSN (Print): 2320–9798 ISSN (Online): 2320-9801-06-07

Multi-crop Selection Model Using Binary Particle Swarm Optimization

V. Kaleeswaran, S. Dhamodharavadhani, and R. Rathipriya

Abstract The multi-crop selection model is the method by which crops are selected to cultivate according to certain requirements such as accessibility of water, soil accessibility, climate variables, market price, net profit, etc. In general, a different sort of soil, different kinds and quantities of nutrients, and water are needed for each crop. Furthermore, the amount of water required by the crop depends on the crop and the climate in which the crop is cultivated. It is, therefore, classified as an optimization problem. The purpose of this research article is to identify the optimal crop set based on water availability propose multi-crop selection model using a binary particle swarm optimization (MCS-BPSO). The primary aim is to maximize the farmer's net profit. In the MATLAB toolbox, the proposed algorithm is developed. The performance of this work is studied on the Indian agriculture dataset. MCS-BPSO revealed that it is an effective multi-crop selection algorithm at specified conditions like total cropping area and water availability. These findings assist farmers in selecting the most appropriate crop, thus enhancing net profit and productivity in the agriculture region with available water.

Keywords Crop selection · Water availability · Binary PSO · Agricultural data · Crop selection optimization

1 Introduction

India is the second-largest producer of agricultural products worldwide. Agriculture is the widest economic sector in India, and it plays a significant part in India's

V. Kaleeswaran (✉) · S. Dhamodharavadhani · R. Rathipriya
Department of Computer Science, Periyar University, Salem 636011, Tamil Nadu, India
e-mail: k04656@gmail.com

S. Dhamodharavadhani
e-mail: vadhanimca2011@gmail.com

R. Rathipriya
e-mail: rathipriyar@gmail.com

© The Author(s), under exclusive license to Springer Nature Singapore Pte Ltd. 2021
J. S. Raj et al. (eds.), *Innovative Data Communication Technologies and Application*,
Lecture Notes on Data Engineering and Communications Technologies 59,
https://doi.org/10.1007/978-981-15-9651-3_5

socioeconomic fabric. Agricultural production depends on many things relating to weather and economy. The factors such as soil, climate, cultivation, irrigation, manure, temperature, precipitation, harvesting, pesticide weed, and others have influence agriculture more. However, the main consideration when choosing the crops that are best suited to smallholder production is market demand—something is not worth producing unless somebody wants to buy it. For this, the crop selection problem plays a vital role in choosing the most demanded crops based on the availability of water. An accurate selection of crops for cultivation helps farmers and companies in planning their agricultural-related schedule, which is based on the availability of cultivatable land and water. Therefore, a multi-crop selection system is proposed using BPSO in this work. It helps farmers in providing the optimal set of crops for available land and water with a forecast of increasing the net profit and reducing risk management.

2 Literature Review

The various data mining techniques were discussed forecasting crop production. In the big data, models with numerical methods were applied to the significance of farmers in reforming the cropping pattern [1]. The implementation of mathematical models [2] such as fuzzy logic designs in crop production optimization, artificial neural networks in validation studies, genetic algorithm designs in accessing the fitness of the implemented system, decision trees, and supporting vector machines was discussed in detail to analyze soil, weather environments, and water regimes related to crop production and pest control in agriculture. In this paper, authors have discussed the status of differentiating past data to the present to identify optimum conditions in favor of increased crop production and considered the impact of the right selection of crops depending on the crop seasons and weather factors that support increased crop production. The regression analysis was used in [3, 4] to determine the relationship between the variables [5], and authors have presented a case study (Nebraska (USA) and nationally for Argentina and Kenya) on the implementation of a clear expressed baseline design approach to the identification of data sources that approximate maize crop production and assists to measure maize yield gaps. Authors have recommended comprehensive guidance to examine crop production differences, access environment, and international land-use adjustments to resolve crop production issues. The current system, i.e., modeling of nearest neighbors, has developed in [6] to calculate and predict crop production calculate on the big data. Authors have discussed the different methods used to quantify local to global production differences. Some models reported were standard modes of operation used to quantify crop production perspective on results collected from West Kenya, Nebraska in the USA, and Victoria in Australian farmers. It was observed that most of the works for crop selection are based on its yield. A very few research works were proposed based on the optimization model (such as genetic algorithm, linear programming) for crop pattern selection. Mostly, these works have used crop production as the priority for the crop selection. In the crop production forecast, the study proposed using reliable and present yield data with optimized and justified plant models and upscaling methods [7, 8]. Therefore, in this work, optimal crop selection by using binary PSO is proposed and developed in the MATLAB toolbox. It is the main objective to achieve a net profit of the farmer by considering the limitations such as water accessibility and accessibility of the cropping area [9–15].

3 Materials and Methods

3.1 Particle Swarm Optimization

In 1995, particle swarm optimization was developed by Dr. Eberhart and Dr. Kennedy. It is nature-inspired evolutionary and pattern that may be analyzed statistically but may not be predicted precisely. And this optimization technique is going to solve computationally and optimize the problem. It is a robust optimization technique on the movement and intelligence of nature swarms. This technique is widely applied successfully to a combination of search and optimization. It has been built by separating the jobs of natural phenomena, and it is much inspired by nature swarms. In this algorithm, a swarm of an individual communicates either directly or indirectly with others using search gradients, and the algorithm adopted uses a set of individuals flying over a search to locate a global optimum. The interaction of PSO has each individual its position according to its previous experience and experience of its neighbors [9, 10].

An individual is composed of three vectors are as follows:

1. The X-vector records the current location (i.e., crop set) of the individual in the search space,
2. The P-vector (personal best) records the location (i.e., crop set) of the best solution which has found so far by the individual, and
3. The V-vector contains a direction for which the individual will travel if is undisturbed.

The global best records the location (i.e., crop set) of the best solution found so far by all individuals in the population. In BPSO, position and velocity of the particles are defined as the change of probabilities, i.e., the particle is initialized as the binary string. Therefore, the search space is restricted to zero and one. Similarly, the velocity of the particle is also restricted to zero and one. The velocity and position updation is responsible for the optimization ability of the BPSO algorithm [11, 12]. The velocity of each position is updated using the following equation:

$$\text{Vel}_i(t+1) = wt * \text{Vel}_i(t) + c1r1(\text{personal best}(\text{Pos}(t))$$
$$(- \text{Pos}_i(t)) + c_2r_2(\text{global best}(t) - \text{Pos}_i(t)) \tag{1}$$

where $Vel_i(t)$—velocity of the ith position at a time 't'.
$Pos_i(t)$—position of the ith position at time 't'.
personal best—personal best of the particle.
global best—global best among all the position.
c_1, c_2—social and cognitive constant ($c_1 = c_2 = 2$).
r_1, r_2—random numbers.
w—inertia weight.
Positions are given in Eqs. (1)–(3), respectively.

$$S_{sig}(Vel_i(t+1)) = 1/(1 + e^{((Vel_i(t+1)))}) \tag{2}$$

The particle's new position obtained using Eq. (3) is explained below:

$$Pos_i = \begin{cases} 1, & \text{if } r_3 < S_{sig}(Vel_i(t+1)) \\ 0 & \text{otherwise} \end{cases} \tag{3}$$

where r_3 is a standard random number inside the range [0, 1].

3.2 Model Formulation

The model is designed to discover the best crop set to maximize the total profit-based surface water irrigation system. Therefore, the model has the goal of maximizing net profits by selecting the optimal crop set while limiting the total cultivable area available and water availability.

3.3 Fitness Function

The fitness of the particle is to maximize the profit of the agriculturists. It defined as (Fig. 1):

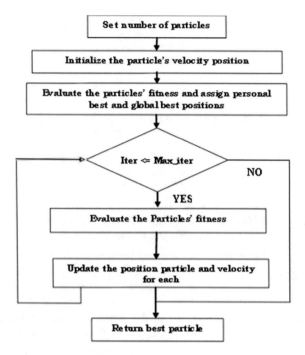

Fig. 1 Flowchart for the process of BPSO

$$f(z) = \text{Max} \begin{cases} \sum_{i=1}^{n} (R_i - C_1)x_i & \text{if} & \sum_{i=1}^{nX_1*X_2} X_i \leq \text{TA} \text{ and } \sum_{i=1}^{n} Wr_1 \leq W_a \\ 0, & \text{otherwise} \end{cases}$$

where
$i = i$th crop.
$X_i = $ cropping area of ith crop.
$W_{ri} = $ water requirement of ith crop/hector.
$C_i = $ cultivation cost of the crop/hector.
$R_i = $ market price of ith crop/hector.

4 Results and Discussions

4.1 Data Availability

The data taken for the study is available on the following Web sites:

1. https://lib.uconn.edu/ [16]
2. https://agropedia.iitk.ac.in/content/water-requirement-different-crops [17]

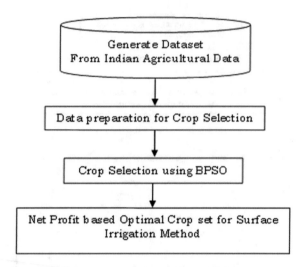

Fig. 2 Methodology framework

Figure 2 represents the framework of the proposed research work. The study area of the proposed work is Tamil Nadu, India. Table 1 [10] describes dataset 1 yields the crop production (kilogram/hector), crop market price (rupees/hector), and cultivation cost (rupees/hector) for the crop in column 1. But, the crops taken for the study are rice, maize, sugarcane, cotton, and groundnut. Table 2 [11] describes the crop minimum water requirement dataset 2. Table 3 describes that to select the crop based on water availability. Table 4 describes showing the result based on the optimized result. But, the crops taken for the study are rice, maize, sugarcane, cotton, and groundnut.

Pseudocode: BPSO-based Crop Selection

(continued)

(continued)

```
function [gbest, OptCropCost, OptArea,        if rndval < (1/(1 + (1/exp(ivel(i,j)))))
CropSelid] = PSOCrop(pop, Y, R, C, Ta, nC,    ipos(i,j) = 1;
Wr, Wa, Crop)                                 else
[Ar,Ac] = size(pop);                          ipos(i,j) = 0;
popsize = Ar;                                 end;
Threshold = Ta;                               end;
vmin = -20;vmax = 20;                         end;
xmin = -10;xmax = 10;                          [cval] = fitcrop1(ipos,Y,C,R,Ta,nC,Wa,Wr);
fval = fitcrop1(pop,Y,C,R,Ta,nC,Wa,Wr);       idx = find(cval > 0);
for i = 1:popsize                             pbest(idx,:) = ipos(idx,:);
for j = 1:Ac                                  fval(idx) = cval(idx);
ivel(i,j) = vmin + (vmax-vmin) *rand;         [gbestval,id] = max(fval);
% ipos(i) = xmin + (xmax-xmin)*rand;          end;
end;                                          [fval,B] =
end;                                          fitcrop1(pbest,Y,C,R,Ta,nC,Wa,Wr);
ipos = pop;                                   [gbestval,id] = max(fval);
ifval = fval;                                 gbest = pbest(id,:);
pbest = ipos;                                 selid = id;
ipbest = pbest;                               cropOptimalArea = B(selid,:)
[gbestval,id] = max(fval);                    cropOptCost = cropOptimalArea.*(R–C)
gbest = pbest(id,:);                          OptCropCost = sum(cropOptCost')
c1 = 2;c2 = 2;w = 0.7;                        [val,maxid] = max(OptCropCost);
for kk = 1:100                                OptArea = cropOptimalArea(maxid,:)
for i = 1:popsize                             OptCost = OptCropCost(maxid);
for j = 1: Ac                                 CropSelid = find(OptArea > 0);
ivel(i,j) = w*ivel(i,j) +                     Crop(CropSelid)
(c1*rand*(pbest(i,j)-ipos(i,j))) +
(c2*rand*(gbest(:,j)-ipos(i,j)));
rndval = rand;
```

The experimental study is conducted with different water availability (mm) capacity for total area 5500 ha. Table 3 summarizes the findings of the proposed work for total available cultivable land with different water availability in mm. The selected crop and their corresponding optimal cropping area are tabulated. It is observed that, as water capacity increases, the cropping area of the rice also increases, which is represented in Table 3.

After the close observation of Table 3, it came to know that the rice is the major crop for Tamil Nadu, India, based on the cropping area and water availability. But the proportion of the cultivatable area of the rice crop is varying at various total available cropping areas. The second major crop for this area is groundnut. Figure 3 shows the cropping area of the selected crop against the different levels of water availability. The rice crop occupies most of the total available cropping area to increase the net profit [18].

The second experimental study is conducted to identify the net profit using the proposed multi-crop selection model for the study area Tamil Nadu, India. The proposed MCS-BPSO model is run with different total cultivatable area thresholds, which ranges from 4500 to 6500 ha. Table 4 lists the optimal crop set with their cropping area and net profit of that crop set. From Table 4, it is exciting to know that

Table 1 Parameter of the crop selection model [16]

Crop	Crop production (kg/hector)	Crop market price (rupees/kg)	Cost of cultivation for crop (rupees/ha)
Green gram	562	60	7760
Potato	4000	7	10,160
Tomato	8000	10	7520
Chili	511	25	7400
Field beans	120	25	7040
Cabbage	14,000	15	9760
Cauliflower	14,000	15	9760
Carrot	1600	15	5960
Brinjal	320	24	7120
Ladyfinger	320	25	7400
Bottle gourd	36	26	5760
Bitter guard	320	30	6360
Sponge guard	320	25	5760
Rice	3687	35	8200
Jawar-K	1301	16	7560
Bajra	2616	14	7360
Maize	7132	19	7560
Groundnut	2574	50	7960
Soya bean	280	30	7720
Cotton	375	38	7720
Maize-R	1400	19	7560
Horse gram	588	50	7560
Onion-R	3200	10	10,160
Sugarcane	101	4.2	10,160

the non-rice crop set/crop selection gives a very high net profit. Therefore, sugarcane is the highest profit crop irrespective of water availability and total cropping area.

5 Conclusions

A model for the optimal crop selection by using binary PSO has been developed for Tamil Nadu to achieve net profit by considering the limitations such as water accessibility and accessibility of the cropping area and water. This crop selection model is based on the water requirements of crops in surface irrigation methods. The experimental decision is maximum net profit from the BPSO-based crop selection

Crop name	Minimum water requirement (in mm)
Table 2 Minimum amount of water required by a crop for the whole cultivation period [17]	
Green gram	600
Potato	700
Tomato	900
Chili	850
Field beans	1400
Cabbage	1400
Cauliflower	1400
Carrot	1400
Brinjal	800
Ladyfinger	600
Bottle gourd	600
Bitter guard	600
Sponge guard	600
Rice	1200
Jawar-K	600
Bajra	600
Maize	600
Groundnut	600
Soya bean	500
Cotton	600
Maize-R	600
Horse gram	600
Onion-R	400
Sugarcane	900

is 419.45 million rupees for the cropping area of 6000 ha with water availability of 7 million (mm).

Table 3 Crop selection based on different water availability

Water availability (Wa)	Selected crop	Optimal area at Ta = 5500 Ha
9,000,000	Rice	1485
	Sugarcane	1760
	Groundnut	1870
	Maize	385
5,000,000	Rice	4070
	Sugarcane	275
	Groundnut	1155
10,000,000	Rice	3740
	Maize	1760
8,000,000	Maize	2255
	Sugarcane	2035
	Groundnut	1210
7,000,000	Rice	3685
	Maize	825
	Groundnut	990

Table 4 Net profit of the optimal crop set for different total cultivatable area

Total area (Ta)	Selected crop	Optimal area at Wa = 7,000,000	Net profit (million rupees)
6000	Sugarcane	4320	419.45
	Cotton	1680	
6500	Sugarcane	1365	212.33
	Maize	2080	
	Groundnut	3055	
5500	Rice	1400	105.92
	Maize	2660	
	Groundnut	2490	
5000	Rice	1350	70.30
	Groundnut	1650	
	Maize	1150	
	Cotton	8500	
4500	Groundnut	2970	74.03
	Maize	405	
	Cotton	1125	

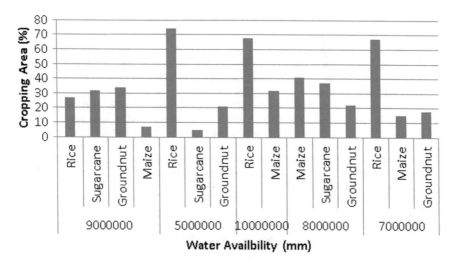

Fig. 3 Water availability-based crop selection with cropping area

Acknowledgements The second author thankfully acknowledges the UGC-Special Assistance Programme for the financial support of her research under the UGC-SAP at the level of DRS-II (Ref.no.F.5-6/2018/DRS-II (SAP-II)), 26 July 2018 in the department of computer science, Periyar University, Salem, Tamil Nadu, India.

References

1. Dakshayini Patil M (2017) Rice crop production prediction using data mining techniques: an overview. Int J Adv Res Comput Sci Softw Eng 7(5):427–431
2. Daniel J (2008) A survey of artificial neural network-based modeling in agroecology. Soft Comput Appl Ind 247–269 (2008)
3. Dhivya BH, Manjula R (2017) A survey on crop production prediction based on agricultural data. Int J Innov Res Sci Eng Technol 6(3):4177–4182
4. Evan J (2014) Coopersmith: machine learning assessments of soil drying for agricultural planning. Comput Electron Agric 10:93–104
5. Van Evert FK, Fountas S (2017) Big data for weed control and crop protection. Wiley on behalf of European Weed Research Society, pp 218–233
6. Grinblat G (2016) Deep learning for plant identification using vein morphological patterns. Comput Electron Agric 127:418–424
7. Huang X (2017) A multiple crop model ensemble for improving broad-scale yield prediction using Bayesian model averaging. Field Crops Res 211:114–124
8. Kung H-Y (2016) Accuracy analysis mechanism for agriculture data using the ensemble neural network method. Sustainability 8:735
9. Wu W-C, Tsai M-S (2008) Feeder reconfiguration using binary coding particle swarm optimization. Int J Control Autom Syst 6(4):488–494
10. Kour VP, Arora S (2019) Particle swarm optimization based support vector machine (P-SVM) for the segmentation and classification of plants. IEEE Access 7: 29374–29385
11. Rathipriya R, Thangavel K, Bagyamani J (2011) Binary particle swarm optimization based biclustering of web usage data. Int J Comput Appl 0975–8887 25(2)

12. Dhamodharavadhani S, Rathipriya R, Chatterjee JM (2020) COVID-19 mortality rate prediction for India using statistical neural network models. Front Pub Health 8:441. https://doi.org/10.3389/fpubh.2020.00441
13. Dhamodharavadhani S, Rathipriya R (2016) A pilot study on climate data analysis tools and software. In: 2016 Online international conference on green engineering and technologies (IC-GET). https://doi.org/10.1109/get.2016.7916863
14. Dhamodharavadhani S, Rathipriya R (2020). Enhanced logistic regression (ELR) model for big data. In: Garcia Marquez FP (ed) Handbook of research on big data clustering and machine learning, IGI Global, pp 152–176. http://doi.org/10.4018/978-1-7998-0106-1.ch008
15. Dhamodharavadhani S, Rathipriya R (2020) Variable selection method for regression models using computational intelligence techniques. In: Ganapathi P, Shanmugapriya D (ed), Handbook of research on machine and deep learning applications for cyber security, IGI Global, pp. 416–436. http://doi.org/10.4018/978-1-5225-9611-0.ch019
16. https://lib.uconn.edu/
17. https://agropedia.iitk.ac.in/content/water-requirement-different-crop
18. Dhamodharavadhani S, Rathipriya R (2020) Forecasting dengue incidence rate in Tamil Nadu using ARIMA time series model. Mach Learn Healthc 187–202. https://doi.org/10.1201/978 0429330131-13

A GUI for Multimodal Sentiment Analysis on Gurmukhi Script

Ramandeep Kaur and Vijay Bhardwaj

Abstract To provide smart access to the results of a sentiment analysis system, a graphical user interface has been developed. A text comment along with an image from Twitter is taken in the existing work by assembling thousands of posts spread on different classes. Twitter found more effective where one can express their feelings in a few words and post it as a text or along with pictures also. It founds more interesting platform where people post their views and others comment on it which makes them related and interactive to each other. It has been carried out by collecting Gurmukhi data from Twitter using Twitter API. Collected content has been filtered, and four categories of data have been generated named as happy, love, sad and angry moods of people. For text feature extraction, TF-IDF and N-gram features are used which are reduced further using mutual information (MI). For image feature extraction, hybrid approach using two different LBP variants is used. The resultant values were recorded to obtain the recall, precision, f-measure, and accuracy to check the efficiency of the proposed system in which three types of classifiers, i.e., Naïve Bayes, k-nearest neighbor and decision tree, have been used. The proposed modal shows decision tree classifier well performed for classifying the emotions of image and text data.

Keywords Term frequency · N-gram · Gurmukhi script · Mutual information · Particle swarm optimization · Local binary pattern

1 Introduction

In the era of social media, multimodal sentiment analysis has recently captured more interest in the research field which became a challenging task to deal with, because

R. Kaur (✉) · V. Bhardwaj
Department of Computer Science and Engineering, Guru Kashi University, Talwandi Sabo 151302, India
e-mail: ramandhillon14@gmail.com

V. Bhardwaj
e-mail: Prof.dr.bhardwaj@gmail.com

there are both types of data which are available on the social platforms that are in the form of image and text. As time is an important resource in everyone's life, utilizing it at its best is much needed [1]. A research shows that 73% of adults use social network sites such as WhatsApp, Snapchat, and Google Plus. From various shopping and communication forums, the Web site contains numerous opinions, emotions, views, and sentiments. Twitter users generate nearly 340 million of tweets daily in different languages and have around 140 million users. The MSA is the analysis of emotions, views, and thoughts from audio-visual-text format [2]. Firstly, the availability of huge image dataset became a hunt for mining the knowledge pattern of visual sentiment [3]. Secondly, the intuition of human intelligence forces us to unfold the difficulties of human perception of visual sentiment. Thirdly, the procedure of understanding the human sentiment from low-level features to high-level features is being enhanced by various advanced computer vision techniques. Hence, exploration of multi-modal sentiments based on text and images becomes our motivation of research in sentiment analysis. MSA on Gurmukhi script is the beginning of new era.

The Gurmukhi script is consistent by Guru Angad Dev Ji in the sixteenth century. The Gurmukhi script comprises 35 letters, and the leading three letters are vowels. Six additional consonants are formed by retaining a dot at the base of the consonant and are utilized typically for loan words. The entire Guru Granth Sahib is transcribed in Gurmukhi writing. In this, first, the feature extraction is done on both image and text separately. After that, those features are combined for the further processing of the system. With the long time interval, sentiment exploration research has achieved various milestones in the improvement of techniques, tools, and algorithms, but still a long way to cover in the development for multimodal sentiment examination. For major problem in multimodal sentiment, survey task is the fusion of all the modalities of samples, including video, text, speech, or image. The main techniques used for detecting feelings from data mined from text, video, and audio need to be elaborated. The study on these techniques will provide an optimal outcome for emotion identification and also discusses the pros and cons of these methods.

Figure 1 illustrates an example of the proposed multi-model system. In this as per the text, the keywords are fetched, and from the image, the objects are found. Here, from the image, three keywords can be easily found that are ਸੋਠੇ, ਵਾਧਾ, and ਸੁਸਤ that are trained using the classifiers models. Lack of words and trend of adding images to the post, majority of users want to convey their thoughts and views by posting images. All the information about the above post is concealed in the photo which leads to the problem of traditional text-based sentiment analysis. If only text is taken into consideration, half of the information will get missing, but following the hybrid feature extraction approach by concatenating text-image both the actual meanings of the post will get revealed. Therefore, MSA becomes more valuable for marketing analysis, product analysis, and government sentiment analysis in microblog.

Fig. 1 An illustration of the image-text pair

ਸੋਨੇ ਚ ਹਲਕਾ ਵਾਧਾ ਕਰੂਡ ਦੀ ਚਾਲ ਸੁਸਤ

2 Related Work

A framework for acknowledgment of machine-printed Gurmukhi content is introduced. The acknowledgment framework introduced works at a sub-character level [4]. The division procedure breaks a word into sub-characters and the acknowledgment stage comprises of grouping these sub-characters and joining them to frame Gurmukhi characters. A lot of extremely basic and simple to PC highlights is utilized and a crossover order plot comprising of paired choice trees and closest neighbors is utilized [5]. Hybrid technique for the opinion/emotion mining of the Punjabi text is proposed. N-grams and Naïve Bayes hybrid approach is adapted. Results acquired are also compared with already existing methods. This paper provides a deep discussion of various techniques used to accomplished sentiment analysis for Indian languages [6]. Hindi Subjective Lexicon is used to propose an algorithm. The proposed method provides better performance on the testing data [7]. Scoring method is offered to discover the sentiment orientation, i.e., polarity of the Punjabi reviews. Separate positive and negative results will be useful for the clients in decision making. The results are contrasted with the already existing methodologies [8]. Support vector machine is used to analyze sentiment analysis on Punjabi News Articles. It helps in categorizing the data mainly focuses on revealing positive or negative sentiment from Punjabi content. The results of the proposed modal show exceptional accuracy.

Kaur et al. [9] use random forest classifier to present the recognition results for newspaper text printed in Gurmukhi script. Random forest classifier is fed with the feature of characters extracted by using various extraction techniques. For exploratory work standard k-fold cross-validation and dataset, partitioning policy has been used. Using fivefold cross-validation and dataset partitioning method maximum recognition accuracy of 96.9 and 96.4% has been obtained. Goyal et al. [10] modified A-star algorithm, and combining it with strip-based projection technique is used for line segmentation. Character segmentation technique is also proposed based on horizontal and vertical projections combining with the aspect ratio of characters. Accuracy of 94.28 and 99.78% with pixel count for line segmentation, and 95.70% for character segmentation is achieved. Above all the proposed method, it is widely recognized much work need to be done on multimodal sentiment analysis in Gurmukhi script. Unlike their text-only counterparts, image posts have only been studied in a few works. These previous works merely studied the characteristics of text-image tweets, and many aspects of this new class of social multimedia have not been explored. Not enough sentiment corpora for the Gurmukhi script is available, and this problem is being resolved by collecting our dataset. NLP rulemaking in understanding the meaning of phrases and sentences is the main stage on which has been worked upon. Sentiment analysis will be carried out using text features and image features individually as well as by concatenation of both feature sets.

3   Our Model

3.1   Methodology

Complete working of the proposed model is shown in Fig. 2.

- Extraction of Gurmukhi text and image tweets from social media sites using Twitter API using 'UTF-8' character encoding settings for fonts.
- Pre-processing of the extracted text by word tokenization [11]. A token is a small unit of characters, and a sequence grouped characters that makes a sensible sentence and it useful for semantic processing. Remove punctuation marks, special symbols, numerical values (':', '*', ',', ':', '|', '/', '[', ']', '{', '}', '(', ')', '^', '&', ' + ', '?', ' = ', ' < ', ' < ', '|', '.', '?', '_', '-', '\', '%', '""', '!', '@', '#', '$', '%', '1', '2', '3', '4', '5', '6', '7', '8', '9', '0', etc.) from the Gurmukhi script document.
- For feature extraction of text, TF-IDF and N-gram is used. To evaluate the significance of a word in a document TF-IDF is used which is a well-defined method widely used in sentiment analysis and text categorization. Term frequency (TF) of a particular term (t) finds out as a no. of time that particular term occurs in a review or a document to total terms in the document. Inverse document frequency (IDF) is used to evaluate the significance of the word. Inverse document frequency is evaluated as IDF (t) = log (N/DF) where N is no. of documents and DF is no. of documents which have that term. IF_IDF is effective in converting the textual information in a vector-space model. N-gram is also used as a feature extraction method when machine learning classifiers are used for text classification process [11].To maximize the feature-relevancy and to minimize redundant features, a filter based approach has been introduced in [12] which use mutual information (MI) in between discrete or continuous feature set. To generate a candidate list of features that cover a wider spectrum of characteristic features an optimal first-order incremental selection is used.

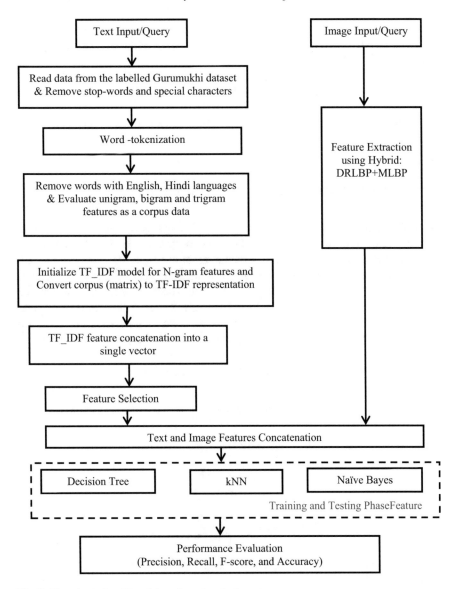

Fig. 2 Flowchart of multimodal sentiment analysis system

- From images, the features were extracted using the hybrid of dominant rotated
  local binary pattern (DRLBP) and median robust extended local binary pattern
  (MLBP) as this showed the highest performance in image sentiment analysis. The
  result of image rotation can be analyzed by rotating the image which leads to the
  rotation of image pixels at a particular angle and the angle of rotation is unknown.
  To deal with this problem, an adaptive method known as dominant rotated local

binary pattern (DRLBP) and MLBP has been applied which computes a reference direction that is used to rotate the image pixels by the same angle with which the image is rotated.

- Sentiment analysis will be carried out by text and image feature concatenation.
- Classifiers, i.e., k-nearest neighbor, decision tree, and Naïve Bayes, etc., will be used.
- Performance evaluation will be carried out in terms of f-measure, precision, recall, and accuracy metrics.

3.2 Classifiers and Performance Metrics

Three classifiers are used, namely decision tree, Naïve Bayes, and kNN. The performance metrics were evaluated for all these classifiers [13, 14]. The decision tree is a difficult algorithm in training and computation of results as for training the data to match the features to the existing tree nodes, it consumes time. But the outcomes of the decision tree algorithms are high in accuracy. There are many standard performance measurements and needed to select them based on the problem on hand [15]. Following are common performance metric for classification problem. The most commonly used metrics for information retrieval effectiveness are recall and precision [16].

Precision (P) is the fraction of retrieved documents that are relevant:

$$Precision = \frac{TP}{(TP + FP)} \tag{1}$$

Recall (R) is the fraction of relevant documents that are retrieved

$$Recall = \frac{TP}{(TP + FN)} \tag{2}$$

$$F\_Measurement = \frac{2 * Precision * Recall}{(Precision + Recall)} \tag{3}$$

Accuracy can be defined as equation f true positive, true negative, false positive, and false negative.

$$Accuracy = \frac{TP + TN}{(TP + TN + FP + FN)} \tag{4}$$

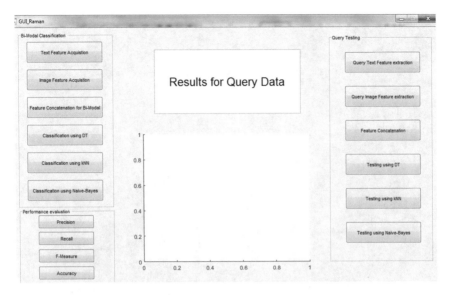

Fig. 3 Graphical user interface designed for a Bi-modal emotion classification system

3.3 GUI (Graphical User Interface)

A GUI has been developed in MATLAB for proposed Bi-modal emotion classification system which gives results for evaluation metrics, i.e., F-measure, precision, recall, and accuracy along with query testing by three different classifiers as shown in Fig. 3, and the result of the query data from anger category is shown in Fig. 4.

4 Experiments and Results

4.1 Dataset Details

There are no enough sentiment corpora available for the Gurmukhi script. Our own dataset has been created obtaining a text comment along with an image from Twitter that is taken in the existing work by assembling thousands of posts spread on different classes. The whole fetching process is carried out by using **IDLE** (Python GUI). In case of text, a total of 4237 tweets have been categorized into six classes. 70% of documents are used for training in both evaluating the objective function for PSO and final classification by different classifiers. In the case of an image, a total of 3446 tweets have been composed of four classes.

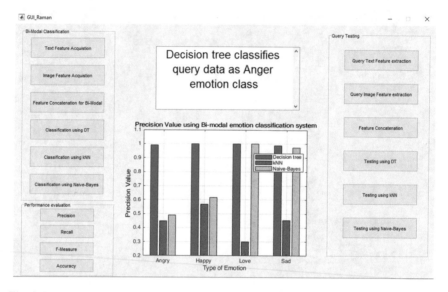

Fig. 4 Results for query data from anger category tested by decision tree classifier

4.2 Results and Analysis

This segment describes the results of multimodal sentiment analysis system obtained after combined the methodology of both text and image-based sentiment analysis systems. A total of 3446 tweets have been categorized into four classes as shown in Table 1. Training and testing have been carried out using a hybrid of DRLBP and MLBP method in which three types of classifiers, i.e., Naïve Bayes, decision tree, and k-nearest neighbor, have been used. The results show the efficiency of decision tree classifier for classifying the text and image correctly as per the sentiment category. Table 2 shows the results obtained for different classifiers and using the text and image input for the proposed system.

Figures 5, 6, 7, and 8 show the results of performance metrics for a bi-model system of emotion classification. The results have shown that the decision tree classifier shows highest results as compares to the other two classifiers. For the categories happy and love, it obtains maximum value for precision, recall, f-measure, and accuracy, i.e., 1.00. For the other two categories (angry and sad) as well, it obtained

Table 1 Table showing percentage of samples used for testing

Emotion class	Number of documents
Happy	1687
Love	377
Sad	712
Angry	670

Table 2 Performance evaluation of proposed Bi-modal emotion classification system

Parameters	TP	TN	FP	FN	Precision	Recall	F-measure	Accuracy
Category	Decision tree results							
Angry	658	2770	6	10	0.9910	0.9850	0.9880	0.9954
Happy	1687	1757	0	0	**1.0000**	**1.0000**	**1.0000**	**1.0000**
Love	377	3067	0	0	**1.0000**	**1.0000**	**1.0000**	**1.0000**
Sad	706	2722	10	6	0.9860	0.9916	0.9888	0.9954
	kNN Results							
Angry	195	2535	241	473	0.4472	0.2919	0.3533	0.7927
Happy	1374	704	1053	313	0.5661	0.8145	0.6680	0.6034
Love	66	2913	154	311	0.3000	0.1751	0.2211	0.8650
Sad	164	2535	197	548	0.4543	0.2303	0.3057	0.7837
	Naive Bayes results							
Angry	293	2472	304	375	0.4908	0.4386	0.4632	0.8028
Happy	1429	867	890	258	0.6162	0.8471	0.7134	0.6667
Love	179	3067	0	198	**1.0000**	0.4748	0.6439	0.9425
Sad	339	2722	10	373	0.9713	0.4761	0.6390	0.8888

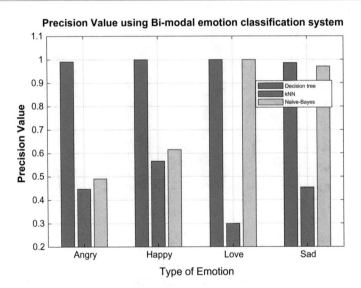

Fig. 5 Precision using Bi-model emotion classification system

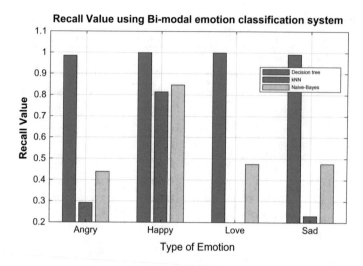

Fig. 6 Recall using Bi-model emotion classification system

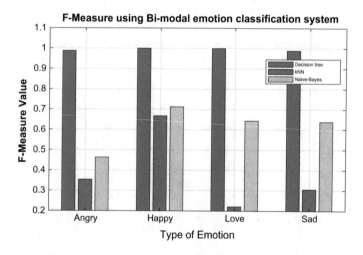

Fig. 7 *F*-measure using Bi-model emotion classification system

more than 0.98 values for each performance metric. This shows the significance of the decision tree classifier as compared to the other classifiers. The Naïve Bayes classifier got the maximum precision as 1.00 for love category and lowest in the angry category, i.e., 0.49. *k*NN classifier did not show high results for any category. The highest precision it got is 0.56 that was in the happy category and lowest precision in love category, i.e., 0.30. For accuracy, decision tree got the highest results for all the categories as compared to the other two classifiers. It scored 0.9954, 1.00, 1.00, and 0.9954 for emotions angry, happy, love, and sad, respectively. The highest

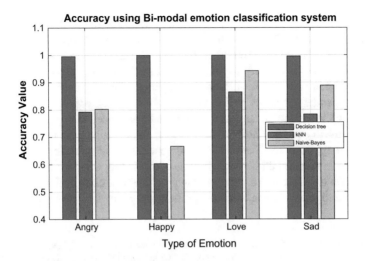

Fig. 8 Accuracy using Bi-model emotion classification system

accuracy scored by Naïve Bayes classifier is for category love, i.e., 0.9425, and lowest is for happy category, i.e., 0.66. In the case of *k*NN, the maximum accuracy was obtained in the love category (0.86), and the lowest accuracy was obtained in the happy category (0.60). Overall, it can be said that decision tree classifier is well performed for classifying the emotions of image and text data.

5 Conclusion

In this proposed work, two features N-grams (value of $n = 2$ and $n = 3$) and TF-IDF (word level) are considered to carry out emotion classification on a Twitter dataset collected for the Punjabi language with Gurmukhi script that has been posted for both text and image. Dataset has been collected using Twitter API in which Gurmukhi script filtering has been applied which eliminates Hindi, English, or other language data and provides only Gurmukhi content from the tweets. Further, this data has been filtered, and only content having four emotion categories named as happy, love, sad, and angry has been considered. Classification results are shown for four presentation parameters, i.e., accuracy, precision, f-score, and recall in which four classification methods Naive Bayes, decision tree, and *k*NN are used for classification. In the proposed method, first, the feature extraction was done by using the hybrid of MLBP and DRLBP. The hybrid of DRLBP and MLBP was done due to their higher performance on the mentioned classifiers as compared to the other variants of the LBP. After the feature extraction, the classification was done by using the supervised classifiers named as decision tree, Naïve Bayes, and *k*NN that classified the data in the respective category of emotions. The resultant values were recorded to obtain the precision, recall, f-measure, and accuracy to check the efficiency of the proposed system. Thus, overall, it can be said that decision tree classifier is well performed for classifying the emotions of image and text data. Present work uses text and image content to classify the emotion and sentiment of a person in which data corpus has been collected for Gurmukhi language. This approach can be extended to other Indian languages as well in which multi-language corpus can be designed. Current work is a research approach which can be advanced to use it as a desktop or mobile platform such that android or java languages can be used to make it usable in real-time applications. The third amendment can be done by involving artificial intelligence in it such that if it classifies data into wrong category or category does not exist then it can use that data in training providing true class by the user. Involving artificial intelligence can make it more perfect, and more data can be included in the training.

Acknowledgements I would like to express here the very thanks to my supervisor Dr. Vijay Bhardwaj. This paper would have been not possible without his immense guidance and inspiration. He has provided me extensive personal and professional knowledge and taught me a great deal about scientific research. Nobody has been more important to me in the pursuit of this paper than the members of my family. I would like to thank my parents. They are the ultimate role models. Most importantly, I wish to thank my supportive husband Gurniwaz Singh Sandhu, who provided an unending inspiration.

References

1. Subarna S (2020) Analysis of artificial intelligence based image classification. In: JIIP **2**(1):44–54. https://doi.org/10.36548/jiip.2020.1.005
2. Ullah MA, Islam MM, Azman NB, Zaki ZM (2017) An overview of multimodal sentiment analysis research: opportunities and difficulties. In: icIVPR, pp 1–6, Dhaka Bangladesh. https://doi.org/10.1109/ICIVPR.2017.7890858
3. Ji R, Cao D, Zhou Y, Chen F (2016) Survey of visual sentiment prediction for social media analysis. Front Comput Sci 10(4):602–611. https://doi.org/10.1007/s11704-016-5453-2
4. Lehal GS, Singh C (200) A Gurumukhi script recognition system. In: 2. https://doi.org/10.1109/ICPR.2000.906135
5. Kaur A, Gupta V (2014) N-gram based approach for opinion mining of Punjabi text. In: MIWAI 8875, pp. 81–88 (2014). https://doi.org/10.1007/978-3-319-13365-2_8
6. Kaur A, Gupta V (2014) Proposed algorithm of sentiment analysis for Punjabi text. In: JETWI 6(2):180–183. https://doi.org/10.4304/jetwi.6.2.180-183
7. Arora P, Kaur B (2015) Sentiment analysis of political reviews in Punjabi language. IJCA 126(14):20–23. https://doi.org/10.5120/ijca2015906297
8. Kaur G, Kaur K (2017) Sentiment detection from Punjabi text using support vector machine. In: IJSRCSE 5(2):39–46 (2017). https://doi.org/10.26438/ijsrcse/v5i6.3946
9. Kaur RP, Kumar M, Jindal MK (2019) Newspaper text recognition of Gurumukhi script using random forest classifier. Multimed Tools Appl 79(5):7435–7448. https://doi.org/10.1007/s11042-019-08365-8
10. Goyal R, Narula DR, Kumar DM (2019) A novel approach for segmentation of typewritten Gurumukhi script. In: IJAST 27(1), 77–96. https://sersc.org/journals/index.php/IJAST/article/view/99

11. Kaur R, Bhardwaj DV (2019) Gurumukhi text emotion classification system using TF-IDF and N-gram feature. Int J Emerg Technol (IJET) 10 (3):352–362
12. Peng H, Long F, Ding C (2005) Feature selection based on mutual information criteria of max-dependency, max-relevance, and min-redundancy. TPAMI 27(8):1226–1238. https://doi.org/10.1109/TPAMI.2005.159
13. Mitchell TM (1997) Machine learning, vol 45(37), pp 870–877. McGraw Hill, Burr Ridge
14. Friedl MA, Brodley CE (1997) Decision tree classification of land cover from remotely sensed data. Remote Sens Environ 61(3):399–409.://doi.org/https://doi.org/10.1016/S0034-4257(97)00049-7
15. Vijayakumar T, Vinothkanna R (2002) Capsule network on font style. In: JAICN 2(2):64–76 (2020). https://doi.org/10.36548/jaicn.2020.2.001
16. Hailong Z, Wenyan G, Bo J (2014) Machine learning and lexicon based methods for sentiment classification: a survey. In: WISA, pp 262–265. https://doi.org/10.1109/WISA.2014.55

A Framework for Data Analytics-Based Healthcare Systems

V. Muneeswaran, P. Nagaraj, U. Dhannushree, S. Ishwarya Lakshmi,
R. Aishwarya, and Boganatham Sunethra

Abstract In this modern techno-world, the term data is unavoidable and certainly, nothing is possible without its usage. The trends about how to analyse the data are the need of the hour. Data analytics is becoming a future escalating tool of all industries including medicine, robotics, etc. This article briefly explains how data analytics is used in healthcare systems. Health care is the process of maintaining and improving the health of an individual by preventing, diagnosing and treating the diseases, illness and other physical and mental imbalances in people. Data analytics is classified into four types and they are descriptive, diagnostic, predictive and prescriptive analysis. Health care makes use of prescriptive analysis to arrive at the best results and make better decisions. Big data plays a major role in data analytics. It helps the data analysts to collect data from the patients and store them efficiently. After the completion of this whole article, the reader will be able to get the collective idea about health care analytics.

V. Muneeswaran
Department of Electronics and Communication Engineering, Kalasalingam Academy of Research and Education, Krishnankoil 626126, Tamil Nadu, India
e-mail: munees.klu@gmail.com

P. Nagaraj (✉) · U. Dhannushree · S. Ishwarya Lakshmi · R. Aishwarya · B. Sunethra
Department of Computer Science and Engineering, Kalasalingam Academy of Research and Education, Krishnankoil 626126, Tamil Nadu, India
e-mail: nagaraj.p@klu.ac.in

U. Dhannushree
e-mail: dhannushreeudhayakumar06@gmail.com

S. Ishwarya Lakshmi
e-mail: ironygirl7@gmail.com

R. Aishwarya
e-mail: rajaaishu3@gmail.com

B. Sunethra
e-mail: sunethraboganatham9@gmail.com

© The Author(s), under exclusive license to Springer Nature Singapore Pte Ltd. 2021
J. S. Raj et al. (eds.), *Innovative Data Communication Technologies and Application*,
Lecture Notes on Data Engineering and Communications Technologies 59,
https://doi.org/10.1007/978-981-15-9651-3_7

Keywords Data analytics · Health care · Big data · Prediction · Diagnosis ·
Decision support

1 Introduction

Data analytics stands for all the qualitative and quantitative procedures through which
data can be extorted, categorized and analysed to make better business decisions. It
uses statistical and/or coherent techniques to depict, illustrate and weigh up data.
The five major processes in data analytics are collect, measure, analyse, improve and
control. It is shown in Fig. 1. Data analytics is a science of analysing raw data for
making terminations about that information. It is the process of grouping data sets
to portray conclusions regarding the information they enclose, escalating with the
abet of specialized systems and software. Data analytics techniques are mostly used
in companies and industries to make good business decisions and by scientists to
verify theories and hypocrisies. There is a step by step procedure in data analytics
which aids the business to know their vacant strategies, i.e. loopholes and lapses.
Data analysis plays a major role in the medical field such as medical data analysis.
[1–3], Data analytics is broken down into four crucial types. They are:

- Descriptive analytics
- Diagnostic analytics
- Predictive analytics
- Prescriptive analytics

1.1 Descriptive Analytics

This analytics illustrates what has happened over a given period of time. It demonstrates that the number of views gone up or not. And are the sales stronger this month
or previous month.

Fig. 1 Five major processes
in data analytics

1.2 Diagnostic Analytics

Diagnostic analytics focuses on why something happened. The palpable successor descriptive analytics is diagnostic analytics. This entails more diverse data inputs and a bit of hypothesizing. In a configured business environment to tools for both descriptive and diagnostic analytics go hand-in-hand [4].

1.3 Predictive Analytics

Predictive analytics abets business to forecast trends based on contemporary events. It enlightens what is likely going to happen shortly.

1.4 Prescriptive Analytics

This type of analytics elucidates the step by step process in a circumstance. Prescriptive analytics prescribes us what action should be taken to prevent an upcoming disease or illness. Prescriptive analytics uses advanced tools and technologies like AI/ML, etc.

2 Previous Works

The research method was built upon a review protocol to review the literature survey systematically. In the study, Kitchenham's systematic review procedure was employed by the steps followed:

- Determining the topic of research
- Extraction of the studies from the literature
- Evaluation of the quality of the studies
- Analysis of the data
- Report on the results

Literature survey started with conducting researches on the most popular database [5].

This review revealed the challenges and opportunities that big data is offering to the health care industry. This survey also mentioned the possibilities of increased quality, better management of people's health, finding the diseases much earlier. These findings help us to take the correct decisions for future research. The flow is illustrated in Fig. 2.

Fig. 2 A process adopted in
the literature survey for data
analytics in health care

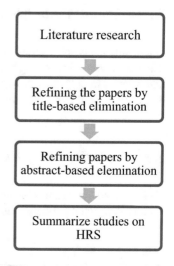

Healthcare data management is the process of collecting, organizing and analysing all the data collected from several patients. This method will help the healthcare organizations to take care of their patients in an effective way, provide good care and the required treatments for every individual and increase the health outcome. For a long time, hospitals and healthcare organizations have spent a lot of money on developing data warehouses for and creating expert analysts' for the betterment of the people. The problem with the analytics is that what is going to happen can just be predicted but cannot come to a decision why that had happened and no one can say what the right thing to do is. But now this condition has changed and could predict why that particular thing had happened and what should do to prevent it. This is made possible by focusing on particular data by using predictive analysis. Lately, the need for accountability and transparency has increased in health care. These forces are becoming very strong and powerful. Henceforth, data analytics is important. Hospitals face the same cost that airlines have faced for years. Due to this, healthcare centres are expected to minimize their continuous spending of money and optimally use the existing resources. To minimize the cost, providers should make correct decisions. In the upcoming years, big data and the other analytics concepts may grow up to a larger extent so that hospitals can make better decisions. The large amounts of information generated yearly within hospitals must be assembled, organized and arranged to make it possible to be accessed by the particular patient from anywhere in the world. By enabling this many things such as cost reduction, better decision-making systems, time-efficient systems, etc., can be made possible.

2.1 A. Benefits of Data Analytics

Business with foresight that uses data analytics has the following benefits:

1. Design enhanced business initiatives through depth. Sympathizing the needs of the customer and the organization.
2. Enlarge a competitive frame over competitors from the insights originated from data.
3. Improve customer contentment by paving approach for personalization.
4. Optimize business manoeuvre and operational efficiency prevents racket data breach and other trends.
5. It assets businesses optimize their recital.
6. A company can also employ data analytics to make good decisions and help the analytics customer tendency and satisfaction which can escort to new and better products and service.

2.2 B. Big Data

Big data is the large, diverse locations where information is stored. It encompasses the volume of information, velocity or speed at which it is formed and composed and the variety or scope of the data points being enclosed.

3 Existing Approaches

3.1 Electronic Health Records

EHRs are the most popular application of big data used in medicine. The goal of healthcare analytics is to guide doctors to make decisions within seconds and improve the health of the patients who have a complex unsolvable medical history or multiple disorders. Certain systems are used to predict the symptoms of a particular disease way earlier so that the patient can take precautions. Every patient has his or her digital record that includes medical history, test results, etc. Records are shared through secure information systems. Every record comprises of one changeable file. The EHRs first collect and identify the data and then repairs the system flaws, then follows the information and then finally audit or take the survey and it is shown in Fig. 3.

3.2 Convolutional Neural Networks

A convolutional neural network (CNN or ConvNet) is used for analysing visual imagery. They have another name that is shift-invariant or spaces-invariant artificial neural networks (SIANN). Can be even said that they are modified versions of multilayer perceptrons. Multilayer perceptrons are fully connected networks (in

Fig. 3 Connection between
EHRs and the nursing
process

neurons). This compatibility of these networks makes them okay for overfitting data. CNNs have a different approach: they gather more complex patterns using smaller and easier patterns. They can recognize a video, classify the images and analyse a medical image [6].

An example of the usage of data analytics in health care:

IBM had reported that some un-named healthcare provider is organizing the data in EMR to reduce the cost and develop the medical care given to an individual patient. Healthcare analytics can be used to improve medical care guide caregivers to better caretaking [7].

3.3 Challenges

3.3.1 Big Data Analytics

Health care needs real-time big data analysis. A healthcare analytics platform must support the basic functions needed for accessing and processing the data and they need to be menu-driven and user friendly. They may include availability, reliability, ease of use, privacy and security and most of the platforms that are currently available are open source. The delay between the data collection and the data processing must be noted. The important problems of ownership and control have to be solved. Healthcare data is infrequently organized and produced in systems in compatible formats.

- Big data sources
- Big data transformations
- Big data platforms
- Big data analytics applications

3.3.2 In CNN

As the big data applications in healthcare advances, more attention has been paid to predicting the diseases with more accuracy. The works at present mostly consist of structured data. For unstructured data, CNN can be used. Healthcare analytics changes from place to place. The following challenges remain:

1. How can the missing data be accessed?
2. How should the main lethal diseases in a certain region be identified?
3. How the disease can be analysed completely using big data analytics?

To solve these problems, structured and unstructured are combined to find the risks of the diseases. In the beginning, latent factor models were used to reconstruct the missing data from the medical records in central China. Then, by using statistical knowledge, the major diseases in the region were discovered and thirdly, to consultations are made with the healthcare analytics experts to extract the useful information. For unstructured features will already be selected by using CNN algorithm. In recent years, a novel CNN-based multimodal disease risk prediction algorithm for structured and unstructured data has been proposed for structured and unstructured data.

3.4 Databases and Cloud Storage

Enterprise data warehouses are used for the data collection. They collect data from various sources and the collected data are stored in a common, specific data repository. The most used database in healthcare analytics is OLTP known as Online Transaction Processing. It can be used to create many applications like EHRs, laboratory systems, billing systems, etc. The OLTP database is constrained to a single application.

Traditionally, healthcare analytics has avoided cloud storage. But currently, cloud computing has been implemented by healthcare organizations. It has become more popular in healthcare analytics in the past years. Here, in healthcare analytics, the EHRs act as clouds for data storage. The data are collected from the patients and are stored in the cloud so that both the patient and the doctor can access it. Cloud computing has benefitted for both healthcare organizations and patients [8].

4 Implementation

Data analysis is the process of analysing the data for better growth of industries, a profit of companies, the health of people in hospitals, etc. Nowadays, the implementation of new methods in the healthcare system is a complex issue because people are afraid of following new methods as they are adapted to old methods in the health system.

4.1   A. Data Analytics Techniques

Data analytics needs extraordinary techniques to process a large volume of data within a limited time. Data analytics is driven by specified applications/techniques. Two of the techniques are:

1. Text analysis technique
2. Statistical analysis technique

4.2   B. Text Analysis Techniques

Text analysis also called DATAMINING. Data mining is the extraction of useful data from a dustbin of data. In this process, the data is first preprocessed, transformed into text, the important features are selected, data mining is undertaken, data is evaluated and the applications are made and is reflected in Fig. 4. Data mining provides:
  To manage health care at different levels.
  It evaluates the effectiveness of the treatment.
  It saves the lives of patients using predictive analysis.
  It detects waste, fraud and abuse done in the health care system.
  Data mining is used in the healthcare industry because it offers benefits to all doctors, patients, healthcare organizations and health insurance. Doctors scam using data mining technique for predicting the reason why the patient is suffering and what treatment one has to do for her/his better care. Patients who are suffering from severe diseases can get better treatment, more affordable healthcare services, and the use of appropriate medicines. Healthcare organizations use data mining to improve their profit, satisfaction of the patient, and to take more care on patients. By data mining

Fig. 4 Processes involved in text mining

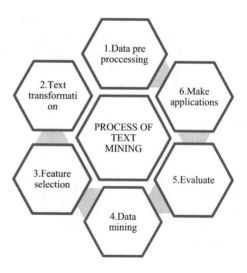

technique, one can detect fraud and abuse and can reduce losses in the healthcare system [9, 10].

4.3 Statistical Analysis Technique

Statistics is defined as the set of mathematical equations that are used to analyse data and predict the things. Statistics plays an important role in decision making in the healthcare system. It tells us about what is happening in the world around us. By using this technique, how many people were in hospital last year, how many got cured of the disease and how many died because of diseases, which medicine is sold more can be determined. This technique may help to figure out who is at the risk of disease, to find ways to control diseases and deciding which disease should be studied.

5 Methodology

The main theme of this article is to study the data analytics and its relation to the healthcare system. Health is known to be a very important aspect of our life. So to provide information well about hospitals and its institutes to the society and to improve healthcare system features and facilities a system is need of which satisfies all these things. Data analytics plays a major role here. A data analyst collects whole information about the healthcare institutes in different parts. A data analyst collects information from the past and based on that future predictions are done which helps health care management [11].

5.1 Descriptive Analytics Related to Healthcare System

The first step is that analyst collects whole information about the health issues which are faced by the people, by small surveys and from the records and they also collect information from the hospitals, laboratories and all health checkup organizations that from which disease people are suffering more, how many people have cured their disease by using medicines and how many are suffering from the same disease and have not cured their disease on the usage of medicines, etc. Like this, they collect the information and to understand this is an easy way this information is represented in the form of pie charts, graphs, pictograms, etc. This type of information helps the organizations to improve its facilities.

Fig. 5 Effects of data
analytics in healthcare

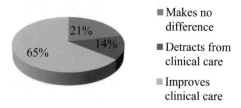

- Makes no
  difference

- Detracts from
  clinical care

- Improves
  clinical care

Fig. 6 Integrated clinical
analysis

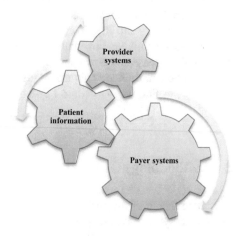

5.2 Diagnostic Analytics Related to Healthcare System

From the data collected, analysts starts to analyse the reasons for the issues that
happened in the past. This helps the hospitals to improve their services to the society
by changing some methods based on the report. It helps both people and hospitals to
have a better healthcare system [12, 13]. Figure 5 shows the effects of data analytics
in the healthcare system.

The information is collected from people and hospitals and it is stored through
some storage network technologies into the cloud. So the people can know the information that which hospital is near to them, location of all the specialized hospitals
and they can choose easily where they can get good treatment for the disease they
are suffering from [14, 15]. The process is shown in Fig. 6

5.3 Prescriptive Analytics Related to Healthcare System

From all the information collected from the previous phases, future predictions are
made. For example which medicines have to be produced, which diseases will occur
more in future, how to reduce the patients in the hospitals, how to provide medicines
to rural areas, in which areas hospitals are not available, where to construct the

Fig. 7 Prescriptive analysis
in healthcare

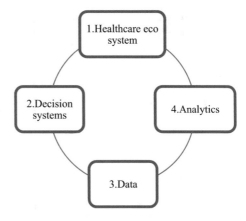

hospitals for the social welfare and also to get the profit for the hospitals, use of digital instruments, which instruments must be produced more, etc [16, 17].

5.4 Prescriptive Analytics Related to Healthcare System

It is the final stage of data analytics. Prescriptive analytics combines the application of both mathematical and computational strategies and suggests decisions. By the information collected in all these phases not only helps the society and hospitals but also helps the government to provide better health schemes to the poor, and to take much better steps ahead. It also helps the government in issuing health cards, etc. The data in the healthcare system is identified first, then decisions are taken, then the data set is formed and then the data is analysed. The process is shown in Fig. 7.

The data analysts follow these different methods and provide us with the information ensuring better livelihood. Not only in the healthcare system, has data analytics played a major role in all the aspects.

6 Future Enhancement

Data analytics will change the way to live and do work in future and expect data analytics to make the impossible possible. Analytics programs in the past delivered results in days or weeks, but in future, the days or weeks may reduce to seconds or even fractions of a second. Many of these algorithms not only help in data preparation but also now enable the customers to visualize and figure out the data and find predictions. They also have enhancements such as automatically exposing correlations, and end up with predictions within the data without having built up models or writing algorithms. This may lead to an increase in data scientists. In future, healthcare

Fig. 8 Showing the certainty in prediction

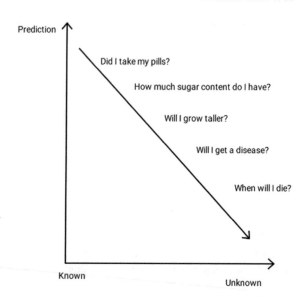

analytics may have a larger amount of data for healthcare organizations to arrange, manage and manipulate. New technologies will emerge as customer requests over self and personal health increases over time. If the features of healthcare analytics increase, the increasing demand of people in self-care can be reached. Predictive analysis is used and will be used to make the best decisions for personalized healthcare for every individual.

By implementing predictive analysis, can come to a trustable solution in the end in any field. In the present, some medical questions cannot be answered. For example, asking, "Have I had my breakfast today?" can be answered. But asking an advanced question like, "How much height will I grow?" is harder to predict with a high level of confidence. [Fig. 8] this is where the predictive analysis comes and reduces the waste in the healthcare system. Many more predictions might be possible using quantum computing. If quantum computing power makes the current supercomputers into instant machines, it might change the healthcare systems. For instance, it may sequence our entire DNA results under minutes like doing the urine tests or blood tests and after that, the quantum computers may do the reliable predictions and find out whatever is wrong with the person. It may also be used in creating the perfect medicine for every disease [18].

7 Conclusion

"Where there is data smoke, there is business fire". According to a survey, 90% of the people admitted that analytics is emerging to be a must to healthcare centres in the upcoming years. Analytics is emerging to be a trend in the healthcare industry.

Advanced analytics is used in every healthcare system to perform operations such as arranging data and predicting the results in every sector of the healthcare industry. Big data and predictive analytics help in making better healthcare decisions leading to healthier and more motivational relationships between healthcare staff and patients. This has an important role in generating better results to reduce the loss of money, data and patients and prevents dangerous and chronic diseases. By studying and analysing structured and unstructured data organization can predict illness, prevent epidemics. These are some of the applications of data analytics used in health care. By using these techniques, the number of medicines required can be predicted and the pharmacy losses can be decreased and can also predict where there is the recursion of diseases and take measures to minimize them. The healthcare organizations use prescriptive analytics the most so that better decision can be made by accessing the data from the EHRs. Thus, the use of data analytics in health care is known.

References

1. Nagaraj P, Muneeswaran V, Veera Reddy L, Upendra P, Vishnu Vardhan Reddy M (2020) Programmed multi-classification of brain tumor images using deep neural network. In: 2020 4th international conference on intelligent computing and control systems (ICICCS), pp 865–870. IEEE
2. Nagaraj P, Muneeswaran V, Sudheer Kumar A (2020) Competent ultra data compression by enhanced features excerption using deep learning techniques. In: 2020 4th international conference on intelligent computing and control systems (ICICCS). IEEE, pp 1061–1066
3. Pramanik PKD, Pal S, Mukhopadhyay M (2019) Healthcare big data: a comprehensive overview. In: Intelligent systems for healthcare management and delivery, pp 72–100. IGI Global
4. Cortada JW, Gordon D, Lenihan B (2012) The value of analytics in healthcare: from insights to outcomes. IBM Global Business Services Executive Report. Annual Institute for Business Value
5. Coyle D (1996) Statistical analysis in pharmacoeconomic studies. Pharmacoeconomics 9(6):506–516
6. Mehta N, Pandit A, Kulkarni M (2020) Elements of healthcare big data analytics. In: Big data analytics in healthcare, pp 23–43. Springer, Cham
7. Weinstein MC, Russell LB, Gold MR, Siegel JE (1996) Cost-effectiveness in health and medicine. Oxford University Press
8. Wang L, Alexander CA (2015) Big data in medical applications and health care. Am Med J 6(1):1
9. Eybers S, Mayet R (2020) From data to insight: a case study on data analytics in the furniture manufacturing industry. In: International conference on integrated science, pp 392–405. Springer, Cham
10. Shafqat S, Kishwer S, Rasool RU, Qadir J, Amjad T, Ahmad HF (2020) Big data analytics enhanced healthcare systems: a review. J Supercomput 76(3):1754–1799
11. Hailey DM (2007) Health technology assessment in Canada: diversity and evolution. Med J Aust 187(5):286–288
12. Ambigavathi M, Sridharan D (2020) A survey on big data in healthcare applications. In: Intelligent communication, control and devices, pp 755–763. Springer, Singapore
13. Banerjee A, Chakraborty C, Kumar A, Biswas D (2020) Emerging trends in IoT and big data analytics for biomedical and health care technologies. In: Handbook of data science approaches for biomedical engineering. Academic Press, pp 121–152

14. Islam M, Usman M, Mahmood A, Abbasi AA, Song OY (2020) Predictive analytics framework for accurate estimation of child mortality rates for Internet of Things enabled smart healthcare systems. Int J Distrib Sens Netw 16(5):1550147720928897
15. Naqishbandi TA, Ayyanathan N (2020) Clinical big data predictive analytics transforming healthcare: an integrated framework for promise towards value based healthcare. In: Advances in decision sciences, image processing, security and computer vision, pp 545–561. Springer, Cham
16. Dremel C, Herterich MM, Wulf J, Vom Brocke J (2020) Actualizing big data analytics affordances: a revelatory case study. Inf Manag 57(1):103121
17. Nagaraj P, Saiteja K, Abhishek D, Ganesh M, Manikanta K (2020) Strategies of analysis for the improvement of business analytics using collaborative data mechanism. In: Test engineering and management. The Mattingly Publishing Co., Inc., pp 9567–9571
18. Nagaraj P, Deepalakshmi P (2020) A framework for e-healthcare management service using recommender system. Electron Gov Int J 16(1–2):84–100

IoT Application for Real-Time Condition Monitoring of Voltage Source Inverter Driven Induction Motor

R. Jyothi, Sumitgupta, K. Uma Rao, and R. Jayapal

Abstract Induction motors are highly reliable, robust and efficient with low maintenance. Hence, they are preferred in high power industrial drives. The revenue loss due to failure of electric drives is huge and a major concern in the mining and manufacturing sectors. In drives, failure need not lead to a complete shutdown of motor. It is feasible that the motor is running, but at reduced efficiency due to incipient faults occurring in the motor and drives. These faults lead to abrupt changes in the voltage and current values at the motor terminals, causing severe vibrations and insulation failure of the motor thus reducing the lifespan of the motor. Eventually if unattended these faults can cause the breakdown of the motor.

Various electrical faults on variable voltage variable frequency converters such as open diode fault in the rectifier, open IGBT fault in the inverter, also open phase fault at the motor terminal lead to drastic changes in voltage and current THD resulting in power quality disturbance affecting the performance of the induction motor load. So, there is a need to identify the severity and type of the fault in the VSI fed induction motor drive by feature extraction in real time. This paper presents a real time monitoring system for VSI driven induction motor. A prognostic diagnostic tool is presented with a minimal number of sensors deployment for effective feature extraction of different electrical faults.

Keywords Voltage source inverter (VSI) · Condition monitoring · Stator currents · Induction motor

R. Jyothi (✉) · Sumitgupta · K. U. Rao
Department of Electrical and Electronics Engineering, RV College of Engineering, Bengaluru, India
e-mail: jyothir@rvce.edu.in

Sumitgupta
e-mail: sumitg650@gmail.com

K. U. Rao
e-mail: umaraok@rvce.edu.in

R. Jayapal
RV Institute of Technology and Management, Bengaluru, India
e-mail: jayapalr@rvce.edu.in

© The Author(s), under exclusive license to Springer Nature Singapore Pte Ltd. 2021
J. S. Raj et al. (eds.), *Innovative Data Communication Technologies and Application*,
Lecture Notes on Data Engineering and Communications Technologies 59,
https://doi.org/10.1007/978-981-15-9651-3_8

1 Introduction

Three-phase AC machines are considered as crucial components for electrical utilities. Motor failure can result in the blackout of a production line. There is always a constant pressure on the operators to reduce maintenance prevents spontaneous downtimes, which result in less production. Timely detection of an incipient fault within an induction motor drive prior to complete failure provides an opportunity for maintenance to be performed without interruption in production.

Condition monitoring is a process wherein sensors are deployed in a system to assess the health of the system. Two major issues in condition monitoring are the choice of the parameter to be measured and fault identification algorithm. For real-time monitoring, it is necessary to minimize the parameters that can capture the features of the faults for developing an algorithm suitable for real-time application.

Though several techniques and various commercial tools are accessible to monitor induction motors to guarantee a high degree of reliability uptime, many industries still have unforeseen system failures and reduced motor life. Environmental factors and operating conditions such as on no load, load and also installation issues may jointly increase motor failure much earlier than the intended motor lifetimes.

It gives an overview of VSI fed induction motor drive and explains various electrical faults such as open-circuit and short-circuit faults on rectifier and inverter side and faults at machine terminals [1]. It also analyses the severity and impact of these faults by performing harmonic analysis on the stator current using MATLAB/Simulink. Design of DC link filter and inverter output filter for induction motor drive system to reduce the harmonic content of the stator current is explained [2].

The intrinsic failures in induction motor due to unavoidable electrical stresses result in stator faults, rotor faults and unbalanced voltage faults. If these faults are not recognized at the primary stage, they may turn out to be catastrophic to the operation of the motor. Current signals data collection of induction motors under different fault conditions was carried out by using laboratory experiments and was analysed.

Condition monitoring of induction motor and hybrid intelligent models is developed for condition monitoring from the real-time data samples [3]. For condition monitoring, MCSA techniques are applied in which the current signals are processed to produce a set of harmonic-based features for classification of faults.

The Internet of things for initial detection and monitoring of single-phase induction motor system failures remotely was discussed [4]. By combining various parameter measurements in real time, the system was developed to improve the delectability of different faults. A factory induction motor (IM) was monitored with wireless TCP/IP protocol to detect and predict deviations from normal operating parameters before the occurrence of motor failure to carry out maintenance of the machine with least possible disruption [5].

For monitoring and controlling of three-phase induction motor parameters, ZigBee protocol (highly reliable protocol for communication) and MATLAB GUI were used [6]. Average absolute values as principal quantities were used to develop real-time diagnostics of multiple open-circuit faults in VSI-driven AC machine [7].

During the literature survey, it was found that quality work is carried out on the fault analysis of induction motor and design of LC filter at the inverter terminals to reduce harmonics in VSI fed induction motor drive. But there is a lacuna in the fault diagnosis of VSI fed three-phase induction motor drive for prognostic diagnosis using online monitoring. Hence, the current research work focuses on the effect of various faults in VSI fed induction motor drive for condition monitoring to increase the reliability of the induction motor in industrial drives.

2 VSI Fed Induction Motor Drive

Figure 1 represents the block diagram of VSI fed induction motor drive [8, 9]. To obtain a DC voltage, the three-phase uncontrolled diode bridge rectifier is used. The DC link is connected to regulate the DC voltage supplied to the three-phase voltage source inverter. Six IGBT power semiconductor switches with conductive features like high input impedance, low saturation voltage, simple control, high switching frequency are used in the three-phase inverter to supply AC voltage to the motor. The PWM technique is used to trigger the six-pulse inverter for easy implementation and control.

2.1 Faults in VSI-Driven Induction Motor

An event that causes abnormality in parameters like currents, voltages, speed, vibration and disturbs the performance of the system is termed as a fault. The common faults in the induction motor drive are mainly categorized into three groups: electrical, mechanical and environment-related faults.

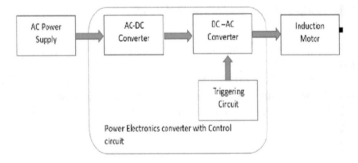

Fig. 1 Block diagram of VSI-driven induction motor

Electrical Faults

The malfunctioning of semiconductor switches such as diodes, IGBTs, gate drive circuits leads to open-circuit or short-circuit electrical faults mainly in the voltage source inverter. Also, the electrical faults can occur at the motor terminal. Open-circuit electrical faults affect the power quality by introducing the harmonics into the system affecting the performance of the system over the long run. Primary protection circuits will be included in the system to overcome the problems due to short-circuit faults. Faults at the motor terminals result in increased vibration of the motor leading to motor bearing damage.

Mechanical Faults

The mechanical faults mainly cause damage in the bearings of the motor, broken rotor bar, air gap eccentricity, stator winding failure due to stress, manufacturing defect and insulation failure. Out of all mechanical failures, broken rotor bar failure is considered the most severe type of fault.

Environmental-Related Faults

Environmental factors such as ambient temperature, external moisture also affect the performance of the induction motor. Vibrations of the machine, due to any reason such as installation defect, foundation defect, also will affect the performance.

In an induction motor, multiple faults may occur instantaneously [6]. Effects of such faults in induction motor result in unbalanced stator currents and voltages, reduction in efficiency, oscillations in torque, overheating, excessive vibration degrading the performance of the machine over the long run. Hence, it is essential to monitor the parameters to avoid the unpredicted failure and damage to the whole process using IoT and inform the operator about the type, severity and location of the fault to take corrective measures. The objectives of the proposed work involve real-time analysis of various electrical incipient faults with a minimum number of parameters mainly stator currents occurring in an induction motor drive using an IoT platform.

3 Block Diagram of the Proposed System

Figure 2 represents the block diagram of an online monitoring system for voltage source inverter fed induction motor drive. The proposed system consists of Hall effect current sensors to monitor the stator currents of three-phase squirrel cage induction motor, temperature sensor to monitor change in temperature upon the occurrence of incipient faults in the system and intimate the operator about the change in parameters to take corrective measures without complete shutdown of the system.

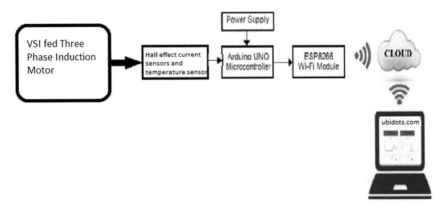

Fig. 2 Block diagram of an online monitoring system for VSI fed induction motor drive

4 Hardware and Software Specifications

The hardware specifications of the components used in the voltage source inverter are as follows: The diodes used in uncontrolled rectifier circuit are rated as 1200 V and 30 A at 25 °C, and IGBTs used in inverter circuit is rated as 1200 V and 50 A at 25 °C. Passive LC filter ratings are as 120 and 100 μF. Hardware implementation of VSI fed induction motor drive involves the following steps. The first stage is an uncontrolled three-phase diode bridge rectifier followed by LC filter [10]. The second stage is a three-phase inverter with six power IGBT switches [11]. The gating pulses to six IGBT switches are generated using gate driver circuit [2, 2]. The gate driver circuit consists of six-step down transformers, bridge rectifier and an optocoupler. The gate pulses for 180° conduction mode are generated by Arduino Mega 2560 microcontroller. The signals are fed to optocoupler to operate each IGBT. The optocoupler IC MCT6 is used to isolate the power circuit and control circuit. Also, to prevent the short circuit of the IGBTs of the same leg, a dead band of 10% is maintained. The three-phase 415 V, 50 Hz, 0.5 HP squirrel cage induction motor is used as a load for the voltage source inverter. Further, ACS716w Hall effect current sensors are used for measurement of stator currents at the motor terminal and a DTH11 temperature sensor is placed on the motor yoke. Node MCU is used to acquire data from the Arduino UNO which is connected to the sensors and for LCD for a visual representation to communicate to the cloud server and alert the user by emailing the reports of raw data collected when the system is running and also sending a message to an authorized operator for preventive maintenance.

4.1 Cloud Storage

Current and temperature data obtained from sensors are transferred wirelessly to the local and cloud server for analysis when the system is under operation. On receiving the raw data, a system devised will process and analyse the raw data and further stores data to the cloud with open-source Ubidots computing platform. This saved data is accessible via the Internet to the user anywhere.

5 Hardware Implementation:

Figure 3 shows the complete hardware setup involving the three-phase 415 V to the rectifier supplied from 415 V, 50 Hz autotransformer, passive LC filter, IGBT power inverter and the triggering circuit, switches for creating faults at the rectifier end and motor terminals, voltmeters at the supply and the three-phase induction motor load, analog DC ammeter to measure the rectifier current, an LCD display is used for the visual representation of current sensors, temperature sensor placed on the motor, PC for online monitoring the parameters.

Four main parameters namely stator currents (IR, IY, IB) in Amps and temperature in degree Celsius are displayed on the Ubidots platform. The gauges are used at the design end with current rating ranging from 0 to 2 A and the thermometer is used to display the temperature of the motor during the working condition. The data is logged every 4 s and the system is programmed to send the data to the user. Current values indicated by the gauges online are validated with the LCD in the current measuring circuit. The raw data is acquired and the report is generated in the form of PDF and mailed to the operator. Feature extraction of the parameters for various faults can be carried out from the report to develop algorithms. An alert message can also be sent

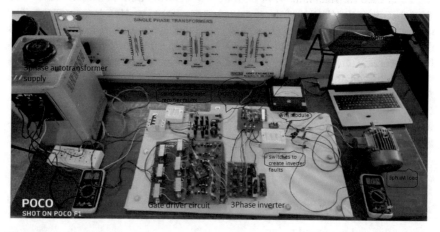

Fig. 3 Hardware implementation setup

Table 1 Fault analysis of stator current and temperature data

Condition of the system	I_R (A)	I_Y (A)	I_B (A)	Temperature (°C)
Healthy system	1.33	1.33	1.33	25.1
One diode open fault (D1)	1.39	1.25	1.36	25.1
Two cross diodes open fault (D1, D6)	1.33	1.24	1.30	25.2
One IGBT open fault (I1)	0.89	1.69	1.41	25.4
Two cross IGBTs open fault (I1, I6)	0.83	1. 49	1.33	25.7

to the authorized mobile number to inform the operator about the condition of the system.

5.1 Case Studies

(a) *Healthy condition*: stator currents I_R, I_Y, I_B are found to be balanced and equal.

(b) *One diode open fault at the rectifier*: results in the reduction of DC voltage at inverter input and inverter output voltage. Hence, there is a reduction in the I_R, I_Y, I_B stator currents of the induction motor.

(c) *Two cross diodes open at the rectifier*: results in a reduction of DC voltage at inverter input and inverter output voltage. Hence, there is a reduction in the magnitude of I_R, I_Y, I_B stator currents of the induction motor and are found to be unbalanced.

(d) *One IGBT open fault at the inverter*: results in a reduction of respective line value of the stator current and reduction in the speed of the motor, and operating temperature of the motor.

(e) *Two cross IGBTs open fault at the inverter*: results in significant reduction of respective line value of the stator currents in which IGBTs are open and also reduction in the speed of the motor, increasing vibration and operating temperature of the motor.

(f) *One diode and one IGBT open fault*: results in a reduction of DC output voltage and reduction of respective line value of the stator current and reduction in the speed of the motor, and increased operating temperature of the motor.

The severity and effect of faults on the stator currents are tabulated in Table 1.

6 Results and Discussions

The system is made to operate at the rated voltage of 415 V under normal operating condition. The LCD and the gauges on Ubidots display the three stator currents magnitudes and are found to be balanced. Upon creation of single and cross diode

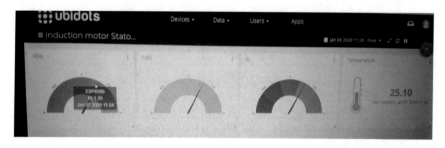

Fig. 4 Real-time monitoring of stator currents of VSI-driven induction motor under healthy condition

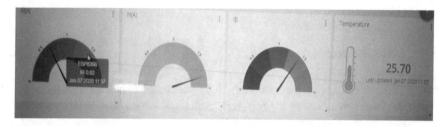

Fig. 5 Real-time monitoring of stator currents of VSI fed induction motor during cross IGBTs open fault

faults at the rectifier end by opening the switches provided in the circuit, there is a reduction in the rectifier output voltage and the speed of the motor in comparison with the healthy condition. On creation of open-circuit faults at the inverter by removing the gating pulses to the single IGBTs and cross IGBTs, the stator currents are found to be unbalanced and there is an abrupt change in the vibration of the motor. For the creation of faults at the motor terminals, switches provided at the motor terminals are opened. The current in the line which is open falls to zero and the load is shared by the other two lines hence leading to a reduction in speed of the motor by around 10%; also, there is an increase in temperature of the motor which may result in insulation damage over the long run.

Figure 4 represents the online data monitoring of stator currents using ubidots platform under healthy condition. Similarly, Fig. 5 represents stator current data under cross leg IGBTs (1, 6) open faults. It can be noted that the faults result in an imbalance of stator current.

7 Conclusion

In the proposed work, application of IoT using Ubidots platform for real-time monitoring of stator currents for identification, classification of the electrical faults are discussed. Mechanical faults such as eccentricity and rotor broken faults can be created and the data acquired by real-time monitoring can be used to identify the occurrence of multiple faults simultaneously. The number of broken rotor bars can also be predicted. With online monitoring, the changes in stator current can be used as a reference, and after fault analysis, the user or operator should be informed about the condition of machines. In case of faults, it will aid the user to take necessary action in prior without affecting the performance of the machine.

References

1. Jyothi R, Jayapal R, Uma Rao K (2016) Severity and impact of faults on current harmonics in inverter-fed AC drive. In: IEEE international conference on innovative smart grid technologies—Asia (ISGT-Asia) Melbourne, Australia, pp 401–405, 28 Nov–1 Dec 2016
2. Swetha SG, Patel N (2013) Design of DC link filter for induction motor drive system. Int J Eng Sci Invent 2(1)
3. Lim CP, Nahavandi S, Seera M, Loo CK (2014) Condition monitoring of induction motors. A review and an application of an ensemble of hybrid intelligent models, 41(10):4891–4903
4. Rekha VSD, Srinivasa Ravi K (2017) Induction motor condition monitoring and controlling based on IoT. Int J Electron Electr Comput Syst 6(9):75–81. ISSN 2348–117X
5. Şen M, Kul B (2017) IoT-based wireless induction motor monitoring. In: Proceedings of XXVI international scientific conference electronics—ET2017, Sept 2017, Sozopol, Bulgaria, 978-1-5090-2883-2
6. Khairnar VC, Kamal Snadeep K (2018) Induction motor parameter monitoring system using MATLAB GUI. In: 4th international conference on advance in electrical, electronics, information, communication and bio-informatics
7. Estima JO, Marques Cardoso AJ (2011) A new approach for real-time multiple open-circuit fault diagnosis in voltage source inverter. IEEE Trans Ind Appl Nov 2011
8. Fernandes BG (2009) Power electronics. Available: https://nptel.ac.in
9. Shanthanu R, Karthikeyan G (2015) International based GUI monitoring and control system. Int J Adv Res Educ Technol (IJARET) 2(2)
10. Ahmed KH, Finney SJ, WilliamS BW (2007) Passive filter design for three-phase inverter interfacing in distributed generation. J Electr Power Qual Util 2:919–929
11. Bhardwaj M (2017) Voltage source inverter design. Texas instrument, application note TIDUAY6C, Nov 2016, Revised 2017
12. Balogh L (2018) Fundamentals of MOSFET and IGBT gate driver circuit. Texas Instrument, Application Report SLUA618A, Mar 2017, Revised October 2018

Application Developed on Data Hiding Using Cryptography and Steganography

Yash Gupta and Komal Saxena

Abstract In today's world, technology is fast-moving. Humans are more dependent on computer networks to communicate with each other. A computer network accessible to all has lead to many unauthorized activities around the globe. These activities are basically controlled by the attackers who are either the envy employees or the people misusing their rights for personal gains. Vulnerabilities and security breaches occurred due to poor layout, misuse by the user or the defects of the system. In recent years, many authors proposed various approaches to enhance the security of secret data over a transmission but every one of them have their own limitations. In this paper, a model is researched and proposed by implementing enhanced tiny encryption algorithm embedding (ETEA), combining both the features of cryptography and steganography together to further improve the level of security to data.

Keywords Cryptography · Steganography · Vulnerabilities · ETEA · TEA

1 Introduction

The users of the Internet, use various online services to collect, send or receive private information. Transforming the data in different forms or sizes is the most familiar technique to protect the data. The resultant data is only accessible to who knows how to retrieve the original data [1]. Cryptographic techniques and algorithms are used to scramble the data so, that if it is intercepted by any unauthorized user, it cannot be studied. As the data is scrambled, it is not easy to discover and decode the original data. To hide the existence of the original data, the user can use steganography so that it cannot attract any suspicion at all [1]. Distinctive styles of steganographic

Y. Gupta (✉) · K. Saxena
Amity Institute of Information Technology, Amity University, Noida, Uttar Pradesh, India
e-mail: yashgupta1327@gmail.com

K. Saxena
e-mail: ksaxena1@amity.edu

strategies had been used that employ invisible inks, microdots, man or woman association, virtual signatures, covert channel, and unfold spectrum communications [2]. Watermarking intention is to guard the quilt medium against any change without a real prominence on secrecy. It could be found as steganography that concentrates on excessive robustness too really little or nearly no security [3]. Many straightforward to practice steganography device are available to cowl mystery messages on one side of the exchange and find unseen data at the other Way [4]. Encryption, when combined with steganography, provides a high level of reliability on the system to the user [5]. In this paper, the author has implemented a model which will allow the user to encrypt the data and embed that into another file and can perform the reversal of actions to retrieve the original data at the other end. The whole application is developed to achieve ease to get the idea of GUI, which should be in an autodidactic model for the end client. The structure shall give all the utilitarian rules of the appropriate route inside the system, which makes it quite practical for the clients to have obstruction free flow while using the system. The general structure of the system should give genuine menu-based directions for the proper flow of data and its operations.

2 Related Work

As of late, data security has turned into a critical issue. Encryption and steganography have come up as an answer and play a vital job in data security. Many strategies are required to ensure and conceal information. The current work centers around cryptography to confirm the information while transferring in the system and hiding the information in another document to make it unbreakable. In symmetric keys encryption or secret key encryption, just one key is used to convert plaintext into ciphertext and vice-versa. In asymmetric keys, two keys are taken into consideration: private and public keys [6]. Asymmetric encryption methods are more or less 1000 times slower than symmetric methods, as they require progressively computing and operating powers [7]. Steganography is not new, it has been since the times of ancient Rome and Greece [8]. The exploration talked about in [9] proposes a piece of information concealing procedure in the DWT space. DWT with the principal level is utilized to deteriorate both confidential and concealed pictures, where each one of them is divided into disjoint (4 × 4). In [10, 11], a gathering of authors talk about a steganography strategy, in light of wavelet pressure strategies that joins assigned data to pictures to diminish the measure of data put away in a database of pictures. Bailey and Curran [12] proposed a stego shade cycle (SSC) approach for shade images that hides information in exceptional channels of an image in a cyclic manner. Karim [13] offered an ultra-modern technique to enhance the security of the current LSB substitution approach via way of inclusive of one spare obstacle of thriller key. Inside the stated method, secret key and pink channel are used as an indicator on the equal time as inexperienced and blue channels are facts channels. Gutub proposed an excessive (PIT) payload pixel indicator technique [14] in which individual channel is used as a hallmark and the other channels are used as information channels. The suggested process embeds the name of the concealed data in a single or each of the data channels in a predefined cyclic way.

3 Proposed Work

In this paper, a model has been implemented using enhanced tiny encryption algorithm (TEA) with concealing (ETEA) utilizing bundles of Java [15]. Enhanced TEA gives reliability during the transferral of information over the system. In TEA, just conversion of plaintext into ciphertext can be carried out. Yet, in our actualized model uses ETEA, encoding, decoding, and concealing of data is joined to give an abnormal state of security to the information with the mission that it cannot be hacked.

The encoded information utilizing TEA is concealed in a video document utilizing steganography and integrated Java bundles. This document can be exchanged by the system with a high level of reliability to another client. The beneficiary can de-mask the video record and unscramble the first information applying the same key which is used at the very beginning of encryption.

4 Proposed Methodology

This is very useful to the user; a feasibility study has been conducted to know its technical, economical, and operational feasibleness. The objective of the feasibility study is acquiring a sense of the scope of the system in the future implementation.

4.1 Technical Feasibility

By minimizing the need for human effort in the system, the work can be done easily since the only authorized person can connect through the system.

4.2 Economical Feasibility

The main purpose of this system is minimizing the manipulation of the data which occurs by the personnel in the organization.

4.3   Operational Feasibility

GUI of this system is prepared as a user-friendly platform. User's easy access to all
the features is the main consideration of this system.

4.4   Data Can Be Used by Different Users

As per the final results, can be said it is very useful for defence services, multiple
users at a given period of time, which need private information or data to be shared in
a most secured manner without getting intercepted by an anonymous user or hacker
(Fig. 1).

This will be the first screen the user will see when accessed the application. In
Fig. 2, it has two options for the user which is either to exit the application or continue
using the application (Fig. 3).

After clicking the continue button on the welcome screen, the user is directed
toward the login screen of the application. To improve the security of the application,
a login screen is implemented, so that only authorized user can use the application.
The user, however, can exit the application, login into the application by using the
credentials provided at the very beginning, and clear the fields if wants to do so.

After logging in successfully into the application, the user is directed toward
the home page of the application supporting multiple options for the users such as
options, security, steganography, and help.

Fig. 1 Login phase of the
application

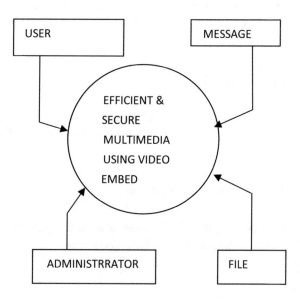

Fig. 2 Screen appearance of application

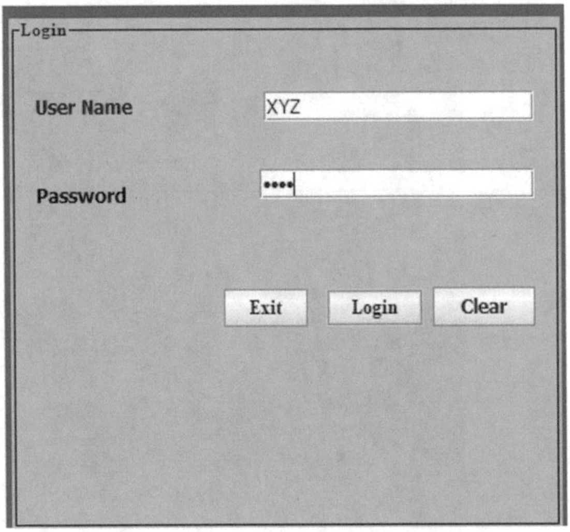

Fig. 3 Login screen of the application

To encrypt a file, the user can click the security option from the upper ribbon of options as shown in Fig. 4.

The user here can browse for the desired file needed for encryption. For successful encryption of the file, the user has to enter a key which is used at the decryption time also. After that, the user can exit from the screen with the help of the close button placed on the same screen itself (Figs. 5, 6 and 7).

After encrypting the file, the user has to browse the encrypted file and video file to conceal the data in the video file browsed by the user. The user can embed the first file into the second file to obtain concealed file by the clicking on the button

Fig. 4 Home screen of the application

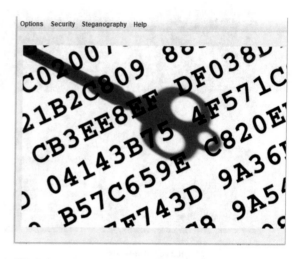

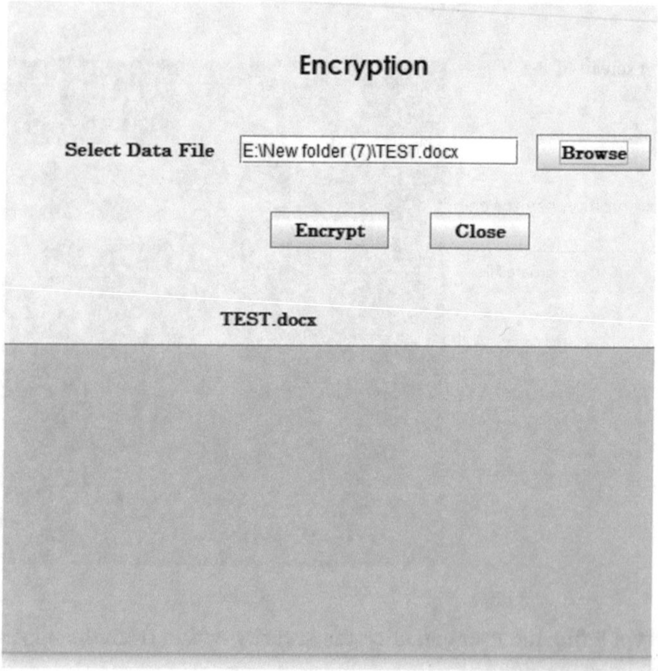

Fig. 5 Encryption screen of the application

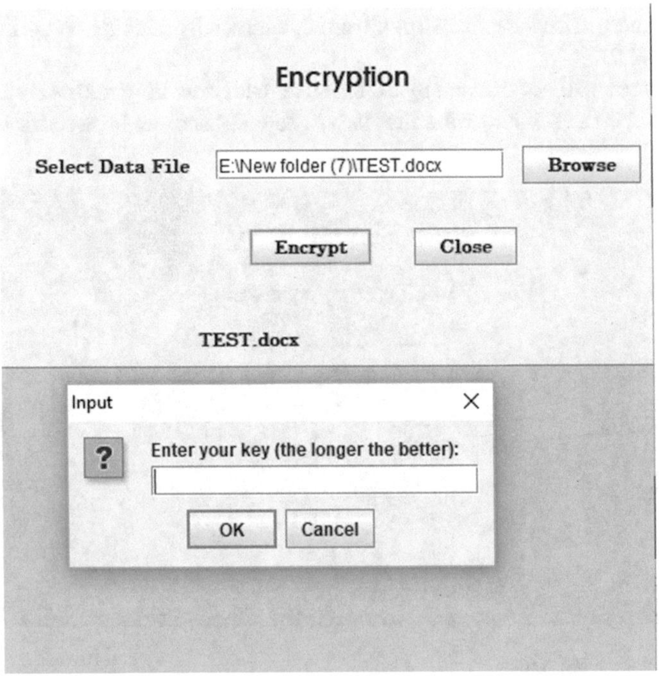

Fig .6 Enter common key for encryption and decryption

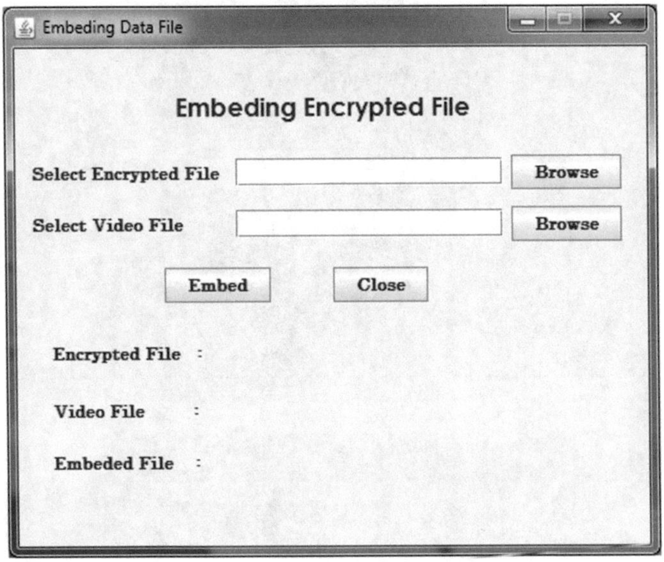

Fig. 7 Embedding screen

named Embed. The user can also exit from this screen by clicking on the button close (Figs. 8 and 9).

This screen will be accessed by the end-user who is waiting for the concealed data transferred by the sender or the first user. The end-user can browse for the desired

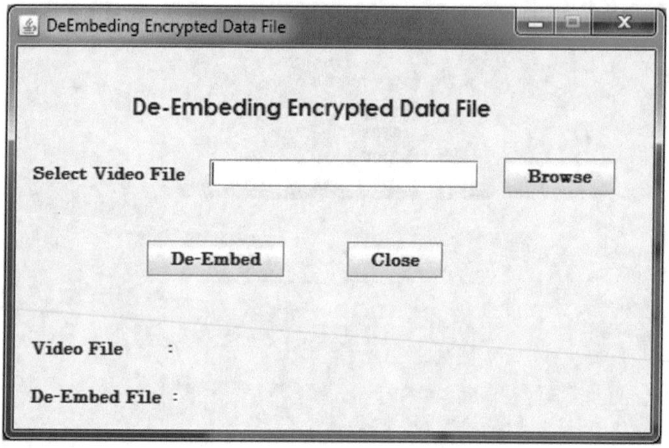

Fig. 8 De-embedding screen

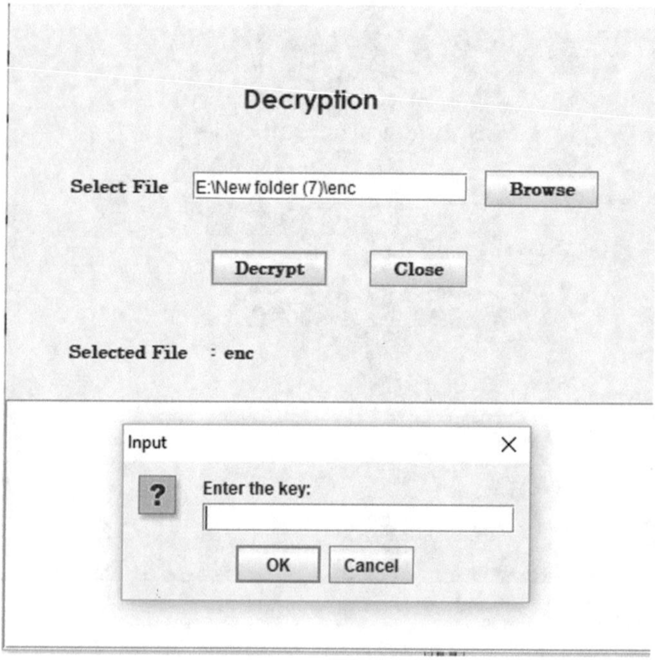

Fig. 9 Decryption screen

file in the system and de-embed the data from the video by the clicking on the button named as De-Embed and after that can close the screen by clicking on the button Close.

After de-embedding the data from the video file, the user can decrypt the data by using the decrypt option from the Security Tab. For successful decryption, the key which is used for encryption is again entered to retrieve the original data from the encrypted file.

A. There are four ways to implement steganography: Text, Images, Audio, and Video. In the case of a Text file, the user can conceal the data in text files having extensions such as (.doc) and (.docx). In the case of Image file, the user can conceal the data in image file having extension such as (.jpg) and (.jpeg). In the case of Audio File, the user can conceal the data in a file having extension such as (.mp3). In the case of Video File, the user can conceal the data in a file having extension such as (.mp4).

B. Some terms are used on context of steganography: Image which is used as a hauler for concealed information is called as Cover Image. The process of hiding information in text, image, audio, and video is called Embedding. The reverse process of embedding is called Extraction. After embedding a message into image, audio and video, the resultant file is called as Stego-File. Study of detecting hidden message is called as steganalysis (Figs. 10 and 11).

In Fig. 12. It is shown that Alpha, Delta, Beta, and Gamma have been assigned a value and that value is stored in the mentioned variables. The values of these variables are further shifted to the right or left by using Right Shift and Left Shift operators.

5 Conclusion

The TEA, DES, RES, and AES are considered and our applied system have joined up the highlights of cryptography and steganography [16]. So, presumed that even though the key size for ETEA is on a lesser side yet it has increasingly quantity of cycles which improves it giving the most noteworthy security. The real attribute of ETEA is that it is both reliable and progressively impervious to attacks.

As ETEA experiences proportional keys and can be cracked in bits making use of a linked-key attack the answer is incorporated for this issue in our future work. The entire system has been created and conveyed according to the requirements expressed by using the client, it is far found to be free of any malicious bug in step with the trying out measures that are completed. Any element or untraced errors will be accumulated in the coming framework that is meant to be created in a not so distant future. As consistent with the present frame the system created could be very a good deal prepared to cope with the focal record association of an organization on a server and provide access to the clients with distinct advantages as advocated through the higher specialists in the report.

Fig. 10 Flow diagram of
data hiding application

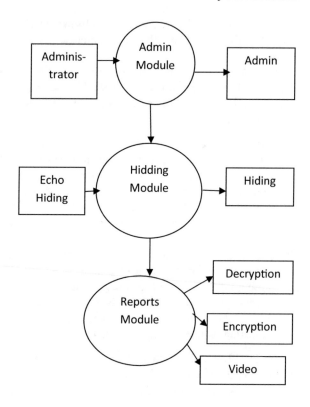

6 Future Enhancement

The present application was implemented as a stand-alone application; in the coming years, a similar Web-based system is going to put in force to develop also. The present system supports only *.mpeg, *.dat formatted files for embedding. In the future, a system is going to be implemented that will support all format files for embedding.

Fig. 11 Working process of data hiding application using steganography and cryptography

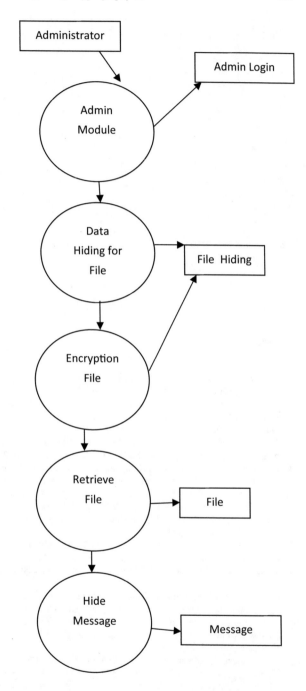

```
delta = 0x9e3779b9;
alpha = 0x7f2637c6;
beta  = 0x5d656dc8;
gamma = 0x653654d9;

// Shift the bits slightly (>> and << bitwise operators)
// as determined by characters in the key
    // ----------------------------------------------------------

sumA = (long)(alpha >> key.charAt(0));
sumB = (long)(beta << key.charAt(1));
sumC = (long)(gamma >> key.charAt(2));
sumD = (long)(delta >> key.charAt(3));

// Sorts out the problem of having an odd or even number of bits
```

Fig. 12 Class encryption

References

1. Bailey K, Curran K (2006) An evaluation of image-based steganography methods. Multimedia Tools Appl 30(1):55–88. https://doi.org/10.1007/s11042-006-0008-4
2. "Anderson RJ (1996) Stretching the limits of steganography. In: 1st information hiding workshop, Springer Lecture notes in computer science, vol 1174, pp 39–48
3. Gutub A, Ghouti L, Amin A, Alkharobi T, Ibrahim MK (2007) Utilizing extension character 'Kashida' with pointed letters for arabic text digital watermarking. In: International conference on security and cryptography—SECRYPT, Barcelona, Spain, 28–31 July 2007
4. Neeta D, Snehal K, Jacobs D (2006) Implementation of LSB steganography and its evaluation for various bits. In: 1st international conference on digital information management, pp 173–178
5. Halim SA, Sani MFA (2010) Embedding using spread spectrum image steganography with GF (2m). In: Proceedings of IMT-GT-ICMSA, pp 659–666
6. Menezes A, van Oorschot PC, Vanstone SA (2017) Handbook of applied cryptography. CRC Press. ISBN 0-8493-8523-7
7. Hamid N, Yahya A, Badlishah Ahmad R, Al-Qureshi M (2012) Image steganography techniques: an overview. Int J Comput Sci Secur (IJCSS) 6(3)
8. Terungwa YS, Moses A (2017) Development of a data security model using steganography, p 6
9. Abdelwahab AA, Hasan LA (2008) A discrete wavelet transform based technique for image data hiding. In: Proceedings of 25th National Radio Science Conference, pp 1–9
10. . Areepongsa S, Kaewkammerd N, Syed YF, Rao KR (2000) Exploring on steganography for low bit rate Wavelet-based coder in the image retrieval system. In: Proceedings of IEEE TENCON, pp 250–255
11. . Areepongsa S, Kaewkammerd N, Syed YF, Rao KR (2000) Steganography for low bitrate Wavelet-based image coder. In: Proceedings of IEEE ICIP, pp 597–600
12. "K. Bailey and K. Curran, "An evaluation of image based steganography methods,"Multimedia Tools and Applications, vol. 30, pp. 55–88, 2006. Article (CrossRef Link)"
13. Karim M (2011) A new approach for LSB based image steganography using secret key. In: 14th international conference on computer and information technology (ICCIT 2011), pp 286–291

14. Gutub AAA (2010) Pixel indicator technique for RGB image steganography. J Emerg Technol Web Intell 2:56–64
15. www.stackoverflow.com
16. Kumar A, Pooja K (2010) Steganography—a data hiding technique. Int J Comput Appl 9(7):23–34 (0975–8887), Nov 2010

Automated ERP System with Internet of Things

Vivek Sethia and Komal Saxena

Abstract This Paper focuses on Internet of things in ERP systems. In worldwide, Internet of things (IOT) has been basically transformed how the items are designed, produced, delivered, and sold to the customer. In this paper we have researched absolutely new procedures and methodology for designing the products as well as automated manufacturing systems that supports to achieve high sales point, productivity, and inventory. With IOT, IP networks, and examination, producers can turn out to be increasingly productive in task, improves security form and offer new strategies of action. IOT associates all devices and takes abrupt arrangements. Producers embrace this new unique change will have an collection of new Wide open opportunities for revenue expansion and cost reserve funds.

Keywords ERP · IOT · Manufacturing · IP networks · Inventory

1 Introduction

Nowadays, there is a buzz among organizations and undertakings to jump on to the IoT fleeting trend. More on every exchange will come down to concentrate on innovation and maximize the output by accepting the IOT. But now many of them doesn't know about IOT. IOT refers to the internet of things. In this there are some gadgets which collects the data through embedded sensors in the system [1]. Similarly as IoT, is one of the ERP's key target to serve the clients properly. The thought of 'servitization' ties these two unique tech contributions to converge into a maximum output for associations over verticals. (M2M) Machine-to-machine advances over a tremendously extended in Internet are the subsequent stage in gathering the data and information. This is an idea that conceives a reality where machines, individuals,

V. Sethia (✉) · K. Saxena
Amity Institute of Information Technology, Amity University, Noida, Uttar Pradesh, India
e-mail: Viveksethia007@gmail.com

K. Saxena
e-mail: ksaxena1@amity.edu

© The Author(s), under exclusive license to Springer Nature Singapore Pte Ltd. 2021
J. S. Raj et al. (eds.), *Innovative Data Communication Technologies and Application*,
Lecture Notes on Data Engineering and Communications Technologies 59,
https://doi.org/10.1007/978-981-15-9651-3_10

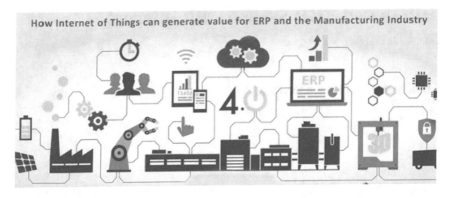

Fig. 1 IOT in ERP system working [3]

ERP, and customer relationship management (CRM) are all joined with sources like social sites and media and the capacities to investigate and utilize that information where it is required [2].

Here Internet of things will play an important role for ERP system and it will help in forming major offering by the user access devices like mobile, computers, laptops, wristband, etc. IoT fortifies ERP due to its capacity to help with information gathering (Fig. 1).

2 Literature Review

According to Lianos and Douglas [4] that the improvement and selection of IoT is a simple element of manufacturing. According to Gigli and Koo [5] that for the situation of manufacturing, high regard creation gear has been strongly instrumented for quite a while in a sealed, hard-wired system condition. According to Aggarwal and Lal Das [6] that there are a few purposes behind these inheritance structures, including noteworthy security issues. According to Jayavardhana et al. [7], in his research that the IoT will associate and share information from inanimate articles. According to Yehia et al. [8] that the Internet of things (IoT) depicts the future where steadily physical objects will be associated with Web and they have capacity to differentiate themselves to different devices. According to Payero et al. [9] that many sensor types and information logging frameworks have been created for this reason throughout the years, yet their far-reaching reception in handy water system booking is still constrained because of an assortment of components.

3 Proposed Work

To help manufacturing change and give a stage to advancement, ground breaking enterprise will received a progression of innovation like cloud, analytics, and Internet of things (IOT). The outcome is a capacity to combine intelligent automation with investigation, enabling more sale and high productive environment.

Enterprise can start their work with something new like conveying sensors on their system and observe the knowledge they get, like their productivity, sales point, and inventory.

Information assembled from IoT—associated gadgets empowers organization to develop from responsive to proactive selling. Organization can observe the progressing execution of the product they sell to the customer for patterns that may demonstrate other item needs. A system equipment manufacturer may see customer utilization themes that are stressing the clients current infrastructure and utilize that information to sell the organization more suitable equipment to anticipate future failures.

Retailer can use sensors or check points in which when the customer will leave the shop through that check point, the check point will automatically checks each items purchased by the customer and generate a bill for the customer and then automatically it will take that cost from the mobile wallet app. This will reduce the time for the customer and increase the sale of retailer.

A standout among the most tedious parts of any business that bargains in physical merchandise originates from stock related assignments. Retailers rely upon knowing the status of their stock for every thing. In addition to the fact that they need to know which things to restock, yet they likewise need to know which things are taking off the racks, and which ones are gathering dust.

An IoT framework can convey constant data to retailers. From the minute, an item touches base from a distribution center to the second a client exits with their buy, an IoT structure can guarantee that the stock following procedure is smooth, precise, and forward-thinking. Retailers would then be able to settle on the right choice with respect to which items to restock and which ones to keep away from. Making an application in which there will be record of every item which comes from the distribution center, till the item sale. It would be told by the application or software. There will be the sensors on each and every item which is connected through the application or the software which will give the information about the inventory and manage the inventory.

4 Proposed model

A. Sales

Retailer can use sensors or check points in which when the customer will leave the shop through that check point, the check point will automatically checks each items

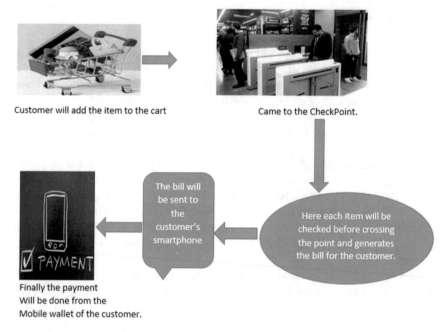

Customer will add the item to the cart   Came to the CheckPoint.

The bill will be sent to the customer's smartphone

Here each item will be checked before crossing the point and generates the bill for the customer.

Finally the payment
Will be done from the
Mobile wallet of the customer.

Fig. 2 Model for automated sale system

purchased by the customer and generate a bill for the customer and then automatically it will take that cost from the mobile wallet app. This will reduce the time for the customer and increase the sale of retailer (Fig. 2).

B. Inventory

An IoT framework can convey constant data to retailers. From the minute, an item touches base from a distribution center, to the second a client exits with their buy, an IoT structure can guarantee that the stock following procedure is smooth, precise, and forward-thinking. Retailers would then be able to settle on the right choice with respect to which items to restock and which ones to keep away from. Making an application in which there will be record of every item which comes from the distribution center, till the item sale. It would be told by the application or software. There will be the sensors on each and every item which is connected through the application or the software which will give the information about the inventory and manage the inventory (Fig. 3).

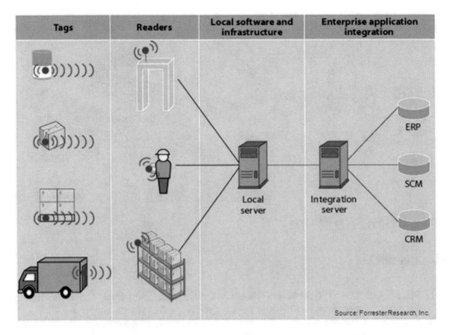

Fig. 3 Model for automated inventory management [12]

5 Conclusion

The conclusion the paper is that the existing writing on Future of Factory (FoF) center mainly around special issues, for example, machines self -rule, small-scale controller, and other digital and physical difficulties. On the other hand, ERP systems are viewed as spine for the fourth-industrial unrest, be that as it may, there will be additionally a hole with regards to the ERP readiness for Future of Factory (FOF). In light of the outcome displayed in this examination, ERP system will be considered as mechanically and operationally prepared for this unrest [10]. In any case, it has seen difficulties with regards to the M2M, machine to the ERP correspondence, as there will no bound together standard and the convention. Furthermore, many of the machine merchants depends on their restrictive correspondence conventions, which could be prompt merchants lock-in and could hinder the entire essential [11] (Fig. 4).

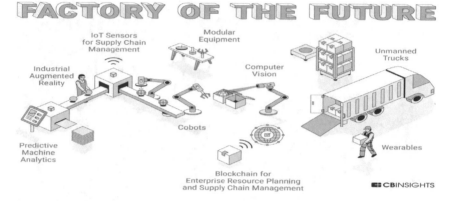

Fig. 4 Future of IOT in ERP [13]

References

1. Rupasinghee TD (2017) Internet of Things (IoT) embedded future supply chains for industry 4.0: an assessment from an ERP-base fashion apparel and footwear industries, Jan 2017
2. Dhayia PK, Choaudhary P, Saini JS, Shakti K (2007) Intelligent systems: feature, challenges, techniques, applications and future scopes
3. Image referece (2017) How IOT can generate value of ERP and the manufacturing system. https://sailotech.com/blog/how-internet-of-things-can-generate-value-for-erp-and-the-manufacturing-industry/
4. Madakam S, Ramaswamy R, Tripathi S (2015, May 25) Internet of Things (IoT): a literature review. J Comput Commun 3(5)
5. Gigli M, Koo S (2011) Internet of things, services & application categorizations. Adv. Internet Things 1:27–31. https://doi.org/10.4236/ait.2011.12004
6. Aggarwal R, Lal Dass M (2012) RFID security in the contexts of "internet of things". In: First International Conferences on Security of Internet of Things, in Kerala, on 17–19 Aug 2012, pp 51–56. https://doi.org/10.1145/2490428.2490435
7. "Jayavardhana G, Rajkumar B, Marusic S, Palaniswami M (2013) Internet of things: vision, architectural elements, & future direction. Future Generation
8. Yehia L, Khedr A, Darwish A (2015) Hybrid security techniques for the internet of things healthcare application. Adv Internet Things 5:21–25. https://doi.org/10.4236/ait.2015.53004
9. Payero J, Mirzakhani-Nafchi A, Khalilian A, Qiao X, Davis R (2017) Development of low-cost internet-of things (IoT) system for monitoring soil water potential using watermark 200SS sensors. https://www.scirp.org/(S(czeh2tfqyw2orz553k1w0r45))/journal/PaperInforma tion.aspx?PaperID=77413
10. Edward D (2016) Smart ERP moves businesses into the fourth revolution. Published: 29 Nov 2016. https://www.syspro.com/blog/owning-or-running-erp/smart-erp-moves-busine sses-into-the-fourth-revolution/
11. Jawahar, Dharminder K. Overview of systems analysis & design. https://www.ddegjust.ac.in/studymaterial/pgdca/ms-04.pdf
12. Image reference, Model of automated inventory management system. https://www.peerbits.com/blog/warehouse-smart-inventory-management-solution.html
13. Image reference, Future of IOT in ERP. https://www.cbinsights.com/research/future-factory-manufacturing-tech-trends

Predicting Students Performance Using Educational Data Mining and Learning Analytics: A Systematic Literature Review

Amita Dhankhar⑩, Kamna Solanki⑩, Sandeep Dalal⑩, and Omdev⑩

Abstract With technological advancement, data is proliferating. Therefore, a need has been felt for advanced analytic techniques, which can help in extracting useful information, patterns, and relationships for decision making. The same is true for the educational domain. For analyzing the data relating to the educational field, educational data mining, and learning analytics techniques have been used. One of the crucial objectives of educational research is to predict the performance of students to bring about changes in educational outcomes. The research in this field aims to find the techniques/methods that can improve prediction, to identify the attributes that can be used to make a prediction or to assess performance. This paper aims to conduct a systematic literature review of predicting student's performance using educational data mining and learning analytics to identify techniques, attributes, metrics used, and to define the current state of the art.

Keywords Systematic literature review · Educational data mining · Predicting performance · Learning analytics

1 Introduction

In today's era, due to technological advancement, big data becomes an integral part of every field [1]. The educational domain is no exception. Advanced analytical techniques can be applied to this enormous amount of educational data to extract

A. Dhankhar (✉) · K. Solanki · Omdev
University Institute of Engineering and Technology, Maharshi Dayanand University, Rohtak, Haryana, India
e-mail: Amita.infotech@gmail.com

K. Solanki
e-mail: Kamna.mdurohtak@gmail.com

S. Dalal
Department of Computer Science and Applications, Maharshi Dayanand University, Rohtak, Haryana, India
e-mail: Sandeepdalal.80@gmail.com

© The Author(s), under exclusive license to Springer Nature Singapore Pte Ltd. 2021
J. S. Raj et al. (eds.), *Innovative Data Communication Technologies and Application*,
Lecture Notes on Data Engineering and Communications Technologies 59,
https://doi.org/10.1007/978-981-15-9651-3_11

useful information, unknown patterns, and relationships, which in turn help teachers in making decisions to provide instructions and allocate resources more accurately. Educational data mining (EDM) and learning analytics (LA) are the techniques that can be used to extract useful information from these enormous educational data [2]. One important task to use this extracted information is to predict students' performance. Educational institutes have been implementing EDM and LA techniques to predict students' performance that would help these institutes to facilitates students to improve academic performance, enables instructors and decision-makers to identify student at-risk, keep track of individual students, and to take timely actions [3]. In the last few years, various researchers used EDM and LA to predict students' performance, engagements, or at-risk of dropout/retention. Considering the importance that prediction of student's performance plays in today's educational domain, detailed insight into this would be valuable. This paper presents a systematic literature review (SLR) of students' performance prediction using educational data mining and learning analytics. It is engaged at "identifying, evaluating, and interpreting all available research" [4] applicable to this research area. The following sections are arranged listed:

Section 2 describes the methodology; Sect. 3 summarizes the pieces of evidence, i.e., is results, Sect. 5 summarizes the conclusion.

2 Methodology

The purpose of this SLR is to address problems by "identifying, critically evaluating and integrating the findings of all relevant, high- quality individual studies addressing one or more research questions" and it follows the guidelines published by Kitchenhams [4] which are (i) frame the research questions (ii) search strategy (iii) search process (iv) inclusion/exclusion criteria (v) data extraction (vi) summarize the evidence (vii) conclusions/interpret the findings. The following sections discuss these guidelines in detail.

2.1 Research Questions

The first step in the review process is to formulate the research questions that SLR will address. The research questions identify which relevant papers are to be selected, what data is to be retrieved from these selected papers, and how to analyze this data to answer the research questions. This SLR follows the criteria specified by Kitchenhams [5] to formulate the research questions, which consist of PIOC (Population, Intervention, Outcome, Context). PIOC criteria for research questions are shown in Table 1.

Table 1 Research questions criteria

Criteria	Details
Population	University/Engineering Institutions (Traditional and Online, Blended, MOOC, Moodle, LMS, VLE)
Intervention	Models/methods/techniques used for prediction of student's performance
Outcome	Effectiveness of the student's prediction Models/Techniques/Methods, i.e., its accuracy
Context	University/Engineering Institutions (Traditional and Online) with a focus on empirical studies

Therefore, with the purpose to determine and assess all the research done on predicting students' performance, the research questions purposed in this SLR are as follows:

RQ1: What are the predictive Models/Techniques/Methods used for predicting student's performance?

RQ2: What are the attributes used in the Predictive Models for predicting student's performance?

RQ3: What types of metrics used by the Predictive Models for describing student's performance?

2.2 Search Strategy

A search strategy plays an essential role in SLR so that the relevant literature can be identified in the search result. Therefore, it is vital to identify the keywords/search items from the research question to conduct an extensive search for research papers. This review used the search string, as shown below. For each database, the syntax of the search string was tailored.

(Predictive analytics **OR** Prediction **OR** Predict)
AND
(modeling **OR** techniques **OR** framework **OR** method **OR** model)
AND
(student's performance **OR** at-risk **OR** persistence **OR** attrition **OR** retention)
AND
(higher Education **OR** engineering **OR** Computer Science **OR** programming)
AND
(educational data mining OR learning analytics)

Search String used in SLR

Table 2 Outline of search results

Database names	Search results	Selected articles
Science direct	11	10
IEEE	35	18
Scopus	170	56
Total	216	84

Search process

The search process is conducted after the search string is compiled. In this review, three databases are searched, which are Science Direct, IEEE, and Scopus, and considered articles published from 2016 to June 2020. Table 2 shows the search results along with selected articles.

Inclusion and Exclusion Criteria along with Data Extraction Process

The articles identified in the search process are assessed using inclusion/exclusion criteria so that those articles can be selected, which can address the research questions. Articles were included if:

- Must discuss predictive model/technique/ method for predicting students' performance
- Predicted values must be relevant to progress through a program/assessment/course
- Only journal articles are included
- Studies are in the domain of university, college, engineering, computer science, programming.

Papers were excluded if:

- No abstract or full paper available
- The predictive model does not relate to the student's performance
- Not focus on the prediction of performance
- Articles not written in English.

The number of selected articles after applying inclusion and exclusion criteria is 84, as shown in Table 2. Based on the articles selected in the search process and research questions, the data extraction form was constructed. For this purpose, spreadsheets were used to input relevant information. The data that was extracted were:

- Paper title
- Publication year
- Model/technique/method used
- Features/attributes used
- Metrics used
- Domain/mode of learning
- Dataset used.

3 Results and Analysis

After data is extracted, it is compiled to synthesized for each research question. Figure 1 shows the number of publications per year. It can be observed from Fig. 1 that there is an increase in work published each year in the current research area.

3.1 Research Question 1

Table 3 shows the various models/techniques/methods that have been used for predicting student's performance using educational data mining and learning analytics. These include decision tree(CART, C4.5, CHAID, J48, ID3, etc.), Neural networks (Multilayer perceptron, deep ANN, long short-term memory (LSTM), deep belief network, etc.), Regression (Linear regression, logistic regression, etc.), Naïve Bayes (Gaussian NB), random forest, support vector machine (RTV-SVM, RBF-SVM), K-nearest neighbor, clustering, probabilistic graph model (HL-MRF, Bayesian network), ensemble methods (Stacking, boosting, bagging, dagging, grading), statistical, and rule-based classification. Some articles use more than one technique for comparison. The decision tree is the most commonly used method for prediction. Forty-two out of 84 articles have used decision tree for predicting students' performance.

3.2 Research Question 2

Figure 1 shows the attributes/features that have been used for predicting student's performance using EDM and LA. Most often attributes/features being used are demographics (which include nationality, place of birth, gender, age, region, ethnicity,

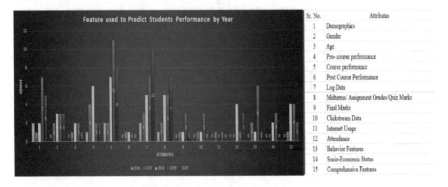

Fig. 1 Features used to predict students performance by year

Table 3 Models/techniques/methods used for prediction of student's performance

S. No.	Technique/method/model	References	Advantages	Disadvantages
1	Decision tree	[6–47]	Easy to implement, easily interpreted, and very intuitive	Can results in different decision tree due to a small change in the data, the time required for searching is large
2	Neural network	[6, 16, 17, 19, 20, 23–25, 27, 28, 32, 37, 38, 43–45, 48–64]	No reprogramming needed as it learns events, applicable to real world problems, as few parameters to adjust so easy to use	For large network high processing time required, difficult to identify how many neurons and layers required
3	Regression	[6, 12, 16, 17, 20, 28, 29, 31, 36, 37, 42, 44, 45, 55–58, 61, 65–78]	Perform better for continuous attributes, perform better for linearly separable data	Effected by outliners, not suited for nonlinearly separable data. overfitting
4	Naïve Bayes	[9, 14, 16, 17, 21, 25, 27–31, 33, 34, 37, 43, 46, 49, 56, 61, 62]	Generate better results for categorical input variable, perform well in multi-class prediction	For small data, precision of algorithm decreases

(continued)

Table 3 (continued)

S. No.	Technique/method/model	References	Advantages	Disadvantages
5	Random forest	[9, 12, 15–17, 20, 28, 30, 36, 43, 47, 49, 53, 54, 58, 59, 68, 73, 79]	Perform efficiently for large data set	Random forest are difficult to interpret as compared to decision tree, overfitting
6	Support vector machine	[6, 12, 13, 16, 23–25, 29, 30, 36–38, 45, 49, 53, 57, 61, 62, 68, 73]	Well suited for nonlinearly separable data in base feature space	Requirements of extensive memory for classification, high complexity
7	K-nearest neighbor	[9, 13, 15, 17, 24, 31, 33, 37, 43, 49, 61]	Perform well in multimodal classes, work well with nonlinearly separable classes	In large training data set, the excessive time required for finding the nearest neighbor
8	Clustering	[25, 80]	Automatic recovery from failure	High complexity
9	Probabilistic graph model	[8, 44, 46, 81–83]	Provide probabilistic prediction	Either underfit or overfit the data, not well suited for problems with many features

(continued)

Table 3 (continued)

S. No.	Technique/method/model	References	Advantages	Disadvantages
10	Ensemble (stacking, boosting, bagging, dagging, grading)	[6, 16, 17, 28, 38, 39, 54, 70, 84–87]	As compared to a single model, the ensemble method gives better prediction	Lack of interpretability
11	Statistical	[76, 77]	Reliable and easy to implement	Required large data set for better results
12	Rule-based classification	[61]	Perform better for categorical features	In terms of prediction quality, they are not the best performers

Table 4 Metrics used by the predictive models for describing student's performance

Metrics used for describing students performance	Count
Course grade range (A, B, D, E/pass-fail/division)	25
Course grade/marks/scores	19
Assignments (performance/time to complete)	3
GPA/CGPA	9
At-risk of dropout/attrition	18
Graduate/retention	3
Vague performance	8

family background, disability, etc.), gender, age, pre-course performance (which include pre-entry features, pre-course grades, etc.), course performance (which include grades/marks/assessment in the current course), post-course performance, log data, midterms/assignment grades/quiz marks, final marks, clickstream data, Internet usage, attendance, behavior features (which include self-regulatory features, the behavior of student during learning and education, discussion, announcement view, visited resources, raised hands), socio-economic Status (which include parents income, family SES, financial background), Comprehensive features (which include learning state of an individual, students learning activity, student study habits, cognitive, self-efficacy, etc.). Thirty out of eighty-four have used course performance as their primary attribute. Next, most often attribute being used is log data, and then midterm/assignment marks/quiz marks.

3.3 Research Question 3

Table 4 shows the metrics used by the predictive models for describing student's performance. These include Course grade range (A, B, D, E/pass-fail/division), course grade/marks/scores, assignments (Performance/time to complete), GPA/CGPA, At-risk of dropout/attrition, graduate/retention, vague performance. Some papers predicted more than one metric. Twenty-five has used course grade range as the primary metrics for predicting student's performance. Next, most often, metrics being used are course grade/marks/score.

4 Threats to Validity

As with systematic literature review, several limitations can limit the validity of the drawn conclusion which are discussed below:

- The absence of an exhaustive search is the major limitation of this SLR. It might be possible that the authors have not reviewed enough articles or included irrelevant studies.
- It is possible that the search string used has not covered other search terms like "modeling" or "forecast" that would otherwise uncover more articles.
- Data extracted from the selected articles are based on research questions. This data would extract other data with different research questions.

5 Conclusion

This paper presents the SLR of student's performance prediction using educational data mining and learning analytics. As discussed before, the search process returned 216 results, out of which 84 were selected. Relevant data were extracted from the selected studies and analyzed to answer our research questions. This review used three databases, namely Science Direct, IEEE, Scopus, and covered the period of the year 2016–2020 (till June). After the analysis of the extracted data, it is found that there has been an increase in the number of papers published in this area.

Furthermore, most frequently used prediction techniques were decision tree, regression, and neural network. It is also observed that most studies tend to predict course performance (pass, fail), course scores/marks, student at-risk of dropout/attrition/retention in traditional/online/blended context. Moreover, course performance, log data, midterm marks/assignments/quiz marks, demographics were the most commonly used attributes used to predict the performance. The majority of selected papers used academic data for prediction, and self-reported data were rarely used. As far as quality is concerned, the majority of selected papers have clear research questions and discussed validity issues. However, the majority of selected papers used a single institute or single course and not verified in other populations. Some work is published implicitly, like the attributes ruled out during the feature selection process were not reported. Based on the findings of this SLR, it is suggested that the researcher should report both the attributes that work and the attributes that did not work; it will provide a holistic understanding of the topic. A more comprehensive approach toward the prediction of student's performance can be taken into research.

References

1. Cook DJ, Das SK (2012) Pervasive computing at scale: transforming the state of the art. Pervasive Mob Comput 8(1):22–35
2. Ferguson R (2012) The state of learning analytics in 2012: a review and future challenges (Technical Report KMI-2012)
3. Dhankhar A, Solanki K (2020) State of the art of learning analytics in higher education. Int J Emerg Trends Eng Res 8(3):868–877

4. Kitchenham BA (2007) Guidelines for performing systematic literature reviews in software engineering (version 2.3). Software Engineering Group, School of Computer Science and Mathematics, Keele University, and Department of Computer Science, The University of Durham
5. Kitchenham B, Pretorius R, Budgen D, Pearl Brereton O, Turner M, Niazi M, Linkman S (2010) Systematic literature reviews in software engineering—a tertiary study. Inf Softw Technol 52(8):792–805 (2010)
6. Coussement K, Phan M, De Caigny A, Benoit DF, Raes A (2020) Predicting student dropout in subscription-based online learning environments: the beneficial impact of the logit leaf model. Dec Support Syst 113325
7. Rizvi S, Rienties B, Khoja SA (2019) The role of demographics in online learning; a decision tree based approach. Comput Educ 137:32–47
8. Xing W, Chen X, Stein J, Marcinkowski M (2016) Temporal predication of dropouts in MOOCs: reaching the low hanging fruit through stacking generalization. Comput Hum Behav 58:119–129
9. Ashraf M, Zaman M, Ahmed M (2020) An Intelligent prediction system for educational data mining based on ensemble and filtering approaches. Procedia Comput Sci 167:1471–1483
10. Ortigosa A, Carro RM, Bravo-Agapito J, Lizcano D, Alcolea JJ, Blanco O (2019) From lab to production: lessons learnt and real-life challenges of an early student-dropout prevention system. IEEE Trans Learn Technol 12(2):264–277
11. Figueroa-Cañas J, Sancho-Vinuesa T (2020) Early prediction of dropout and final exam performance in an online statistics course. IEEE Revista Iberoamericana De Tecnologias Del Aprendizaje 15(2):86–94
12. Moreno-Marcos PM, Pong TC, Muñoz-Merino PJ, Kloos CD (2020) Analysis of the factors influencing learners' performance prediction with learning analytics. IEEE Access 8:5264–5282
13. Rubiano SMM, Garcia JAD (2016) Analysis of data mining techniques for constructing a predictive model for academic performance. IEEE Latin Am Trans 14(6):2783–2788
14. Baneres D, Rodríguez-Gonzalez ME, Serra M (2019) An early feedback prediction system for learners at-risk within a first-year higher education course. IEEE Trans Learn Technol 12(2):249–263
15. Wakelam E, Jefferies A, Davey N, Sun Y (2020) The potential for student performance prediction in small cohorts with minimal available attributes. Br J Edu Technol 51(2):347–370
16. Huang AY, Lu OH, Huang JC, Yin CJ, Yang SJ (2020) Predicting students' academic performance by using educational big data and learning analytics: evaluation of classification methods and learning logs. Interactive Learn Environ 28(2):206–230
17. Guerrero-Higueras ÁM, Fernández Llamas C, Sánchez González L, Gutierrez Fernández A, Esteban Costales G, González MÁC (2020) Academic success assessment through version control systems. Appl Sci 10(4):1492
18. López-Zambrano J, Lara JA, Romero C (2020) Towards portability of models for predicting students' final performance in university courses starting from moodle logs. Appl Sci 10(1):354
19. Bedregal-Alpaca N, Cornejo-Aparicio V, Zárate-Valderrama J, Yanque-Churo P (2020) Classification models for determining types of academic risk and predicting dropout in university students. (IJACSA) Int J Adv Comput Sci Appl 11(1)
20. Alhassan A, Zafar B, Mueen A (2020) Predict students academic performance based on their assessment grades and online activity data. Int J Adv Comput Sci Appl 11(4)
21. Alhakami H, Alsubait T, Aliarallah A (2020) Data mining for student advising. Int J Adv Comput Sci Appl 11(3)
22. Hoque MI, Kalam Azad A, Tuhin MAH, Salehin ZU (2020) University students result analysis and prediction system by decision tree algorithm. Adv Sci Technol Eng Syst J 5(3):115–122
23. Xu X, Wang J, Peng H, Wu R (2019) Prediction of academic performance associated with internet usage behaviors using machine learning algorithms. Comput Hum Behav 98:166–173
24. Al-Sudani S, Palaniappan R (2019) Predicting students' final degree classification using an extended profile. Educ Inf Technol 24(4):2357–2369

25. Francis BK, Babu SS (2019) Predicting academic performance of students using a hybrid data mining approach. J Med Syst 43(6):162
26. Buenaño-Fernández D, Gil D, Luján-Mora S (2019) Application of machine learning in predicting performance for computer engineering students: a case study. Sustainability 11(10):2833
27. Mimis M, El Hajji M, Es-Saady Y, Guejdi AO, Douzi H, Mammass D (2019) A framework for smart academic guidance using educational data mining. Educ Inf Technol 24(2):1379–1393
28. Adekitan AI, Noma-Osaghae E (2019) Data mining approach to predicting the performance of first year student in a university using the admission requirements. Educ Inf Technol 24(2):1527–1543
29. Rodrigues RL, Ramos JLC, Silva JCS, Dourado RA, Gomes AS (2019) Forecasting Students' performance through self-regulated learning behavioral analysis. Int J Dist Educ Technol (IJDET) 17(3):52–74
30. Sadiq HM, Ahmed SN (2019) Classifying and predicting students' performance using improved decision tree C4.5 in Higher Education Institutes. J Comput Sci 15(9):1291–1306
31. Yaacob WFW, Nasir SAM, Yaacob WFW, Sobri NM (2019) Supervised data mining approach for predicting student performance. Indones J Electr Eng Comput Sci 16:1584–1592
32. Jorda ER, Raqueno AR (2019) Predictive model for the academic performance of the engineering students using CHAID and C 5.0 algorithm. Int J Eng Res Technol 12(6): 917–928. ISSN 0974–3154
33. Livieris IE, Tampakas V, Karacapilidis N, Pintelas P (2019) A semi-supervised self-trained two-level algorithm for forecasting students' graduation time. Intell Dec Technol 13(3):367–378
34. Helal S, Li J, Liu L, Ebrahimie E, Dawson S, Murray DJ, Long Q (2018) Predicting academic performance by considering student heterogeneity. Knowl-Based Syst 161:134–146
35. Nguyen HQ, Pham TT, Vo V, Vo B, Quan TT (2018) The predictive modeling for learning student results based on sequential rules. Int J Innov Comput Inf Control 14(6):2129–2140
36. Bucos M, Drăgulescu B (2018) Predicting student success using data generated in traditional educational environments. TEM J 7(3):617
37. Zhou Q, Quan W, Zhong Y, Xiao W, Mou C, Wang Y Predicting high-risk students using Internet access logs. Knowl Inf Syst 55(2):393–413
38. Adejo OW, Connolly T (2018) Predicting student academic performance using multi-model heterogeneous ensemble approach. J Appl Res High Educ
39. Shanthini A, Vinodhini G, Chandrasekaran RM (2018) Predicting students' academic performance in the university using meta decision tree classifiers. J Comput Sci 14(5):654–662
40. Jamil JM, Mohd Pauzi NF, Shahara Nee INM (2018) An analysis on student academic performance by using decision tree models. J Soc Sci Res (6):615–620. ISSN(e): 2411–9458, ISSN(p): 2413–6670
41. Mishra S, Panda AR (2018) Predictive evaluation of student's performance using decision tree approach. J Adv Res Dyn Control Syst 14(special issue):511–516
42. Kostopoulos G, Kotsiantis S, Pierrakeas C, Koutsonikos G, Gravvanis GA (2018) Forecasting students' success in an open university. Int J Learn Technol 13(1):26–43
43. Asif R, Merceron A, Ali SA, Haider NG (2017) Analyzing undergraduate students' performance using educational data mining. Comput Educ 113:177–194
44. Seidel E, Kutieleh S (2017) Using predictive analytics to target and improve first year student attrition. Austr J Educ 61(2):200–218
45. Tran TO, Dang HT, Dinh VT, Phan XH (2017) Performance prediction for students: a multi-strategy approach. Cybern Inf Technol 17(2):164–182
46. Evale D (2016) Learning management system with prediction model and course-content recommendation module. J Inf Technol Educ Res 16(1):437–457
47. Dhankhar A, Solanki K, Rathee A, Ashish (2019) Predicting student's performance by using classification methods. Int J Adv Trends Comput Sci Eng 8(4):1532–1536
48. Waheed H, Hassan SU, Aljohani NR, Hardman J, Alelyani S, Nawaz R (2020) Predicting academic performance of students from VLE big data using deep learning models. Comput Hum Behav 104:106189

49. Injadat M, Moubayed A, Nassif AB, Shami A (2020) Systematic ensemble model selection approach for educational data mining. Knowledge Based Syst 105992
50. Qu S, Li K, Zhang S, Wang Y (2018) Predicting achievement of students in smart campus. IEEE Access 6:60264–60273
51. Olive DM, Huynh DQ, Reynolds M, Dougiamas M, Wiese D (2019) A quest for a one-size-fits-all neural network: early prediction of students at risk in online courses. IEEE Trans Learn Technol 12(2):171–183
52. Yang TY, Brinton CG, Joe-Wong C, Chiang M (2017) Behavior-based grade prediction for MOOCs via time series neural networks. IEEE J Select Topics Signal Process 11(5):716–728
53. Hung JL, Shelton BE, Yang J, Du X (2019) Improving predictive modeling for at-risk student identification: a multistage approach. IEEE Trans Learn Technol 12(2):148–157
54. Al-Shabandar R, Hussain AJ, Liatsis P, Keight R (2019) Detecting at-risk students with early interventions using machine learning techniques. IEEE Access 7:149464–149478
55. Tsai SC, Chen CH, Shiao YT, Ciou JS, Wu TN (2020) Precision education with statistical learning and deep learning: a case study in Taiwan. Int J Educ Technol High Educ 17:1–13
56. Qu S, Li K, Wu B, Zhang X, Zhu K (2019) Predicting student performance and deficiency in mastering knowledge points in MOOCs using multi-task learning. Entropy 21(12):1216
57. Aljohani NR, Fayoumi A, Hassan SU (2019) Predicting at-risk students using clickstream data in the virtual learning environment. Sustainability 11(24):7238
58. Pal VK, Bhatt VKK (2019) Performance prediction for post graduate students using artificial neural network. Int J Innov Technol Explor Eng (IJITEE). ISSN 2278-3075
59. Crivei LM, Ionescu VS, Czibula G (2019) An analysis of supervised learning methods for predicting students' performance in academic environments. ICIC Exp Lett 13:181–190
60. Ramanathan L, Parthasarathy G, Vijayakumar K, Lakshmanan L, Ramani S (2019) Cluster-based distributed architecture for prediction of student's performance in higher education. Cluster Comput 22(1):1329–1344
61. Kokoç M, Altun A (2019) Effects of learner interaction with learning dashboards on academic performance in an e-learning environment. Behav Inf Technol 1–15
62. Vora DR, Rajamani K (2019) A hybrid classification model for prediction of academic performance of students: a big data application. Evol Intell 1–14
63. Raj JS, Ananthi JV (2019) Recurrent neural networks and nonlinear prediction in support vector machines. J Soft Comput Paradigm (JSCP) 1(01):33–40
64. Bashar A (2019) Survey on evolving deep learning neural network architectures. J Artif Intell 1(02):73–82
65. Gašević D, Dawson S, Rogers T, Gasevic D (2016) Learning analytics should not promote one size fits all: the effects of instructional conditions in predicting academic success. Internet High Educ 28:68–84
66. Burgos C, Campanario ML, de la Peña D, Lara JA, Lizcano D, Martínez MA (2018) Data mining for modeling students' performance: a tutoring action plan to prevent academic dropout. Comput Electr Eng 66:541–556
67. Qiu L, Liu Y, Liu Y (2018) An integrated framework with feature selection for dropout prediction in massive open online courses. IEEE Access 6:71474–71484
68. Gitinabard N, Xu Y, Heckman S, Barnes T, Lynch CF (2019) How widely can prediction models be generalized? Performance prediction in blended courses. IEEE Trans Learn Technol 12(2):184–197
69. Sothan S (2019) The determinants of academic performance: evidence from a Cambodian University. Stud High Educ 44(11):2096–2111
70. Raveendran Pillai B, Gautham J (2019) Deep regressor: cross subject academic performance prediction system for university level students. Int J Innov Technol Explor Eng (IJITEE) 8(11S). ISSN: 2278-3075
71. Rajalaxmi RR, Natesan P, Krishnamoorthy N, Ponni S (2019) Regression model for predicting engineering students academic performance. Int J Recent Technol Eng 71–75
72. Singh K, Maloney T (2019) Using validated measures of high school academic achievement to predict university success. New Zealand Econ Pap 53(1):89–106

73. Zhang X, Sun G, Pan Y, Sun H, He Y, Tan J (2018) Students performance modeling based on behavior pattern. J Ambient Intell Human Comput 9(5):1659–1670
74. Yang SJ, Lu OH, Huang AY, Huang JC, Ogata H, Lin AJ (2018) Predicting students' academic performance using multiple linear regression and principal component analysis. J Inf Process 26:170–176
75. Christensen BC, Bemman B, Knoche H, Gade R (2018) Pass or fail? Prediction of students? Exam outcomes from self-reported measures and study activities. ixD&A 39:44–60
76. Ellis RA, Han F, Pardo A (2017) Improving learning analytics–combining observational and self-report data on student learning. J Educ Technol Soc 20(3):158–169
77. Strang KD (2017) Predicting student satisfaction and outcomes in online courses using learning activity indicators. Int J Web-Based Learn Teach Technol (IJWLTT) 12(1):32–50
78. Gershenfeld S, Ward Hood D, Zhan M (2016) The role of first-semester GPA in predicting graduation rates of underrepresented students. J Col Stud Reten Res Theory Pract 17(4):469–488
79. Gutiérrez L, Flores V, Keith B, Quelopana A (2019) Using the Belbin method and models for predicting the academic performance of engineering students. Comput Appl Eng Educ 27(2):500–509
80. Gutiérrez F, Seipp K, Ochoa X, Chiluiza K, De Laet T, Verbert K (2020) LADA: a learning analytics dashboard for academic advising. Comput Hum Behav 107:105826
81. Delen D, Topuz K, Eryarsoy E (2020) Development of a Bayesian belief network-based DSS for predicting and understanding freshmen student attrition. Eur J Oper Res 281(3):575–587
82. Ramesh A, Goldwasser D, Huang B, Daume H, Getoor L (2018) Interpretable engagement models for MOOCs using Hinge-loss markov random fields. IEEE Trans Learn Technol
83. Lan AS, Waters AE, Studer C, Baraniuk RG (2017) BLAh: boolean logic analysis for graded student response data. IEEE J Select Topics Signal Process 11(5):754–764
84. Wan H, Liu K, Yu Q, Gao X (2019) Pedagogical intervention practices: improving learning engagement based on early prediction. IEEE Trans Learn Technol 12(2):278–289
85. Xu J, Moon KH, Van Der Schaar M (2017) A machine learning approach for tracking and predicting student performance in degree programs. IEEE J Select Topics Sig Process 11(5):742–753
86. Bhagavan KS, Thangakumar J, Subramanian DV (2020) Predictive analysis of student academic performance and employability chances using HLVQ algorithm. J Ambient Intell Human Comput 1–9
87. Kamal P, Ahuja S (2019) An ensemble-based model for prediction of academic performance of students in undergrad professional course. J Eng Des Technol
88. Chui KT, Fung DCL, Lytras MD, Lam TM (2020) Predicting at-risk university students in a virtual learning environment via a machine learning algorithm. Comput Hum Behav 107:105584
89. Zollanvari A, Kizilirmak RC, Kho YH, Hernández-Torrano D (2017) Predicting students' GPA and developing intervention strategies based on self-regulatory learning behaviors. IEEE Access 5:23792–23802
90. Almutairi FM, Sidiropoulos ND, Karypis G (2017) Context-aware recommendation-based learning analytics using tensor and coupled matrix factorization. IEEE J Select Top Sig Process 11(5):729–741
91. Popescu E, Leon F (2018) Predicting academic performance based on learner traces in a social learning environment. IEEE Access 6:72774–72785
92. Ramanathan L, Geetha A, Khalid L, Swarnalatha P (2016) A novel genetic nand paft model for enhancing the student grade performance system in higher educational institutions. IIOABJ 7(5)
93. Siemens G, Baker RSJD (2012) Learning analytics and educational data mining: towards communication and collaboration. In: Buckingham Shum S, Gasevic D, Ferguson R (eds) Proceedings of the 2nd international conference on learning analytics and knowledge. ACM, New York, NY, pp 252–254
94. Chatti MA, Dyckhoff AL, Schroeder U, Thüs H (2012) A reference model for learning analytics. Int J Technol Enhanced Learn 4(5):318–331

Probability Density Analysis of Non-orthogonal Multiple Access Over Rayleigh Channel Using AF Relay

K. Ashwini and V. K. Jagadeesh

Abstract In this paper, non-orthogonal multiple access (NOMA) over Rayleigh channel using cooperative relay technique is presented. Using non-orthogonal multiple access concepts, the probability density function (PDF) analysis is studied over Rayleigh channel. The relaying strategy used is amplify and forward (AF) technique which is simpler than other relaying technique. A scenario with two users with one user acting as a relay and one base station are studied. The combined signal from the relay and base station obtained at the receiver using a maximal combining ratio (MRC) combiner is derived to obtain the PDF. A response of PDF versus the signal-to-noise ratio of the source is plotted that can be used to characterize different properties of the wireless communication channel.

Keywords NOMA · Maximal combining ratio · Probability density function · Rayleigh channel · Amplify and forward

1 Introduction

In the coming generations, one of the rising technologies for enhancing the performance of mobile communication is non-orthogonal multiple access (NOMA) [1, 2]. NOMA, in comparison with orthogonal frequency division multiple access (OFDMA) technique [3] presents many desirable benefits, like greater spectrum efficiency, enormous connectivity, and better user fairness. In NOMA, there are power-domain NOMA and code-domain NOMA techniques. In power-domain NOMA [4] superposition coding (SC) is used at the sender side and the signal is transmitted.

K. Ashwini (✉)
Department of Electronics and Communication Engineering, NMAM Institute of Technology, Nitte, Udupi, India
e-mail: ashwinik@nitte.edu.in

V. K. Jagadeesh
Department of Electronics and Communication Engineering, National Institute of Technology, Patna, India
e-mail: vkjagadeesh@gmail.com

© The Author(s), under exclusive license to Springer Nature Singapore Pte Ltd. 2021
J. S. Raj et al. (eds.), *Innovative Data Communication Technologies and Application*,
Lecture Notes on Data Engineering and Communications Technologies 59,
https://doi.org/10.1007/978-981-15-9651-3_12

At the receiver side, successive interference cancellation (SIC) [5, 6] is used for retrieving the desired signal. The signal transmitted from the transmitter travels in multiple paths which reduces the strength of the signal as it approaches the receiver end, and such a phenomenon is named as multipath propagation. Multipath propagation [7] leads to multipath fading which exhibit randomness in the received signal strength, phase of the signal, and the angle of arrival (AOA). Different types of channel models illustrate different fading envelope. They are the Rayleigh model, Rician model, Nakagami-m model, Normal shadowing model, Weibull fading model, Nakagami-q model, and many others. The signal propagating in a wireless communication environment experiences quick changes in its level which is mostly in Rayleigh distribution form. This is mainly due to the interference by scattering in many separate paths between the base station and the end receiver. The signal propagation of refracted paths can be modeled using Rayleigh distribution. Hence, Rayleigh fading channel is selected for the analysis. Besides increasing the coverage area in systems that use wireless communication, cooperative relaying techniques [8] promise gains in energy efficiency and capacity of the wireless system by taking benefit of higher data rates in intermediate cooperative relaying nodes [9]. The relaying strategies used can be classified into three types. They are amplify and forward (AF), decode and forward (DF), and compress and forward strategies (CF) [10]. In AF strategy, the intermediate node amplifies the received signal and forwards it to the next node. In DF relay strategy, the intermediate node overhears transmissions from the base station, decodes them, and then forwards them to the next node. In CF strategy, the intermediate node will compress the signal it receives and forwards it to the next node.

Probability density function (PDF) is a familiar method for distinguishing a fading channel. The probability density of the strength of a signal received is explained by its PDF. The power received by a receiver is proportional to the square of the received envelope [7]. This envelope can be used to calculate the capacity of a fading wireless link. This paper discusses the use of a power-domain NOMA technique over Rayleigh channel using AF relaying strategies. A system with one base station and two users, one which is nearer to the base station the other user has been analyzed. The PDF of the signal at the destination side is obtained and its curve is plotted by considering the maximal ratio combining (MRC) technique [11–13]. The paper is organized as follows. The second section provides the details of the system considered for the analysis. The third section contains the analysis of PDF over Rayleigh model. The fourth section shows the simulation of the PDF. The fifth section gives the conclusion.

2 System Model

A system model which implements a cooperative relay-based NOMA [14–16] is considered. It comprises of a base station (BS) and two users (UE1) and (UE2) as shown in Fig. 1. UE2, which is nearer to the base station than UE1 functions as relay and using AF relay strategy forwards the signal to UE1. Moreover, since

Fig. 1 a Traditional
cooperative relay network,
b NOMA-based cooperative
relay network

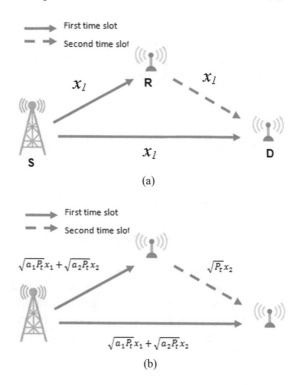

(a)

(b)

UE1 is far from the base station, it needs help from UE2 to receive a signal with adequate strength. In the system, each node uses a single antenna and the relay functions using an AF protocol with a half-duplex mode of propagation. The cooperative NOMA system comprises of two-time slots. Figure 1a shows the cooperative relaying network with a half-duplex relay and Fig. 1b shows NOMA-based cooperative relaying systems.

2.1 Amplify and Forward

In a cooperative system, a user not only transmits its information but also helps other users to transmit their information by relaying it to the destination. One of the common cooperative signaling techniques is the AF method. The data sent by the user is combined by the base station and relate and settles on a final choice on the transmitted bit. Although, in this method, the noise is also amplified by cooperation method, the base station gets two independently faded variants of the signal and the station can settle on better options on the data received [17]. The BS transmits the superimposed signal to UE2 and UE1. Consider two signals x_1 and x_2, the signals destined for UE1 and UE2, respectively. According to the NOMA protocol, let variables a_1 and a_2 be the power allocation coefficients. Assume $a_1 < a_2$ and $a_1 + a_2 =$

1. The total power transmitted from the BS is indicated as $P_t = a_1 P_t + a_2 P_t$ where P_t is the total transmit [18, 19]. Let h_{SR}, h_{SD}, and h_{RD} indicate the channel coefficients between source and relay, source, and destination and between relay and destination respectively. The superimposed signal [1] is given as

$$t = \left(\sqrt{a_1 P_t} x_1 + \sqrt{a_2 P_t} x_2 \right) \tag{1}$$

The signals that are received by the relay station and destination stations in the initial time slot are given as follows

$$R_{SR} = h_{SR} \left(\sqrt{a_1 P_t} x_1 + \sqrt{a_2 P_t} x_2 \right) + n_{SR} \tag{2}$$

$$R_{SD} = h_{SD} \left(\sqrt{a_1 P_t} x_1 + \sqrt{a_2 P_t} x_2 \right) + n_{SD} \tag{3}$$

where n_{SR} and n_{SD} are the additive white Gaussian noise (AWGN) with zero mean μ and variance σ^2 [8]. The signal x_1 is decoded by the destination by treating the signal x_2 as noise, while the user that is acting as relay decodes signal x_2 from Eq. (1). SIC technique is used for decoding the signal. Hence, signal-to-noise ratios (SNRs) obtained for signals x_1 and x_2 that is received at the relay station can be achieved, respectively, as

$$\gamma_{SR} \text{ for signal } x_1 = \frac{|h_{SR}|^2 a_1 P_t}{|h_{SR}|^2 a_2 P_t + \sigma^2} \tag{4}$$

$$\gamma_{SR} \text{ for signal } x_2 = \frac{|h_{SR}|^2 a_2 P_t}{\sigma^2} \tag{5}$$

The SNR for signal x_1 received at the destination station is achieved as

$$\gamma_{SD} \text{ for signal } x_1 = \frac{|h_{SD}|^2 a_1 P_t}{|h_{SD}|^2 a_2 P_t + \sigma^2} \tag{6}$$

The relay strategy amplifies the signal obtained at the relay to the destination during the next consecutive time slot is conveyed as

$$R_{RD} = \eta |h_{RD}| \sqrt{P_t} x_2 R_{SR} + n_{RD} \tag{7}$$

where n_{RD} is the AWGN, $\eta = \sqrt{\frac{P_t}{P_t |h_{SR}|^2 + \sigma^2}}$ is the amplifying factor for AF relaying strategy and the received SNR [16] for signal x_1 can be expressed as

$$\gamma_{RD} = \frac{\eta^2 |h_{RD}|^2 |h_{SR}|^2 a_1 P_t}{\eta^2 |h_{RD}|^2 |h_{SR}|^2 a_2 P_t + \eta^2 |h_{RD}|^2 \sigma^2 + \sigma^2} \tag{8}$$

In the proposed system, the destination station is capable of receiving two signals in two different time slots. Therefore, it performs better than the regular cooperative relay system concerning throughput.

3 Probability Density Function Analysis

The nature of Rayleigh distribution is characterized by a continuous probability distribution for non-negative random variables. The Rayleigh distribution emerges naturally as the signal travels in multiple paths system from BS. The PDF [20] of SNR is very essential to obtain the outage probability, probability of symbol error, and many other performance metrics of a wireless system [6]. For this reason, the approximate expression of the PDF formula for SNR is derived and confirmed by simulation. Let us consider the MRC at the receiver end. Hence, the total SNR at the combiner output is given by

$$\gamma_{\text{MRC}} = \sum_{l=1}^{L} \gamma_l \tag{9}$$

where L is the number of branches. The average combined SNR at the MRC combiner output with two branches is described as

$$\gamma_{\text{MRC}} = \gamma_1 + \gamma_2 \tag{10}$$

The PDF of the sum of the first and second branch can be expressed as [11]

$$P_{\gamma_{\text{MRC}}}(\gamma_{\text{MRC}}) = \int_{0}^{\gamma_{\text{MRC}}} P_{\gamma_2}(\gamma_{\text{MRC}} - \gamma_1) P_{\gamma_1}(\gamma_1) d\gamma_1 \tag{11}$$

The PDF of the signal envelope at the receiver, $f(r)$ [21], where r is a random variable, exhibit Rayleigh distribution and is given by

$$f(r) = \frac{r}{\sigma^2} e^{\left(-\frac{r^2}{2\sigma^2}\right)} \tag{12}$$

Therefore the probability density function over the Rayleigh channel at the receiver end of the proposed system is given in Eq. (13).

$$P_{\gamma_{\text{MRC}}}(\gamma_{\text{MRC}}) = \int_{0}^{\gamma_{\text{MRC}}} \frac{(\gamma_{\text{MRC}} - \gamma_{\text{SD}})}{\sigma^2} e^{\frac{(\gamma_{\text{MRC}} - \gamma_{\text{SD}})^2}{2\sigma^2}} \frac{(\gamma_{\text{SD}})}{\sigma^2} e^{\frac{\gamma_{\text{SD}}^2}{2\sigma^2}} d\gamma_{\text{SD}} \tag{13}$$

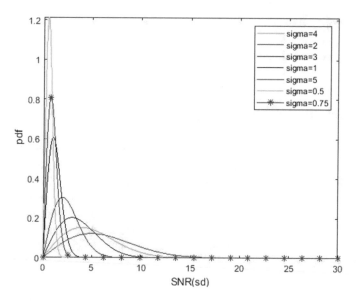

Fig. 2 PDF of Rayleigh channel in cooperative NOMA

4 Results and Discussions

Figure 2 shows the PDF of the Rayleigh distribution at the receiver end plotted against the average combined SNR at the MRC output at the destination. PDF for different values of variance σ^2 by changing the value of standard deviation, σ is plotted against the SNR. The PDF indicates the probability of appearance of SNR over a particular range. From the graph, it is observed that the PDF attains its maximum peak which is 1.2 at $\sigma = 0.5$. It has been observed that as σ value increases, the distribution starts to flatten, that is the curve spreads over high SNR. It can also be observed that the probability is high at lower SNR and tends to zero for high SNR depending on the value of σ. From the graph, it is analyzed that the PDF curve of the MRC output is similar to the Rayleigh distribution. Probability density analysis of the received signal is used for computing the outage performance, which indicates the probability that the signal strength falls below a particular threshold value.

5 Conclusion

In this paper, the cooperative NOMA is investigated for its performance over the Rayleigh channel. The probability density function over Rayleigh channel is derived for the proposed system. These random variables contribute to land mobile radio as they precisely express the instantaneous amplitude and power, respectively, of a multipath fading signal. These can now be used to characterize, derive various properties of the wireless channel. As future work, the system can be realized over different channels and other relay strategies for its PDF curve which helps to analyze the outage probability. One of the limitations of the proposed system is that the SIC technique introduces complexity in computation. Hence, techniques which overcome this would be of great value.

References

1. Jiao R, Dai L, Zhang J, MacKenzie R, Hao M (2017) On the performance of NOMA-based cooperative relaying systems over rician fading channels. IEEE Trans Veh Technol 66(12)
2. Liu Y, Xing H, Pan C, Nallanathan A, Elkashlan M, Hanzo L (2018) Multiple-antenna-assisted non-orthogonal multiple access. IEEE Wirel Commun 25(2):17–23
3. Gong X, Yue X, Liu F (220) Performance analysis of cooperative NOMA networks with imperfect CSI over Nakagami-m fading channels. Sensors 20(2):424
4. Islam SMR, Avazov N, Dobre OA, Kwak K (2016) Power-domain non-orthogonal multiple access (NOMA) in 5G systems: Potentials and challenges. IEEE Commun Surv Tutorials 19(2):721–742
5. Haci H (2017) A novel precoding technique for non-orthogonal multiple access. Procedia Comput Sci 120:676–683
6. Rai R, Jain A (2018) Result analysis of efficient MIMO framework for NOMA downlink and uplink transmission based on signal alignment. Int J Comput Appl 180(37): 0975–8887
7. Sadeque G, Mohonta SC, Ali F (2015) Modeling and characterization of different types of fading channel. Int J Sci Eng Technol Res 4(5):1410–1415
8. Kim J, Lee I (2015) Capacity analysis of cooperative relaying systems using non-orthogonal multiple access. IEEE Commun Lett 19(11)
9. Ding Z, Dai H, Poor HV (2016) Relay selection for cooperative NOMA. IEEE Commun Lett 5(4):416–419
10. Chen J, Yang L, Alouini M (2018) Performance analysis of cooperative NOMA schemes in spatially random relaying networks. IEEE Access 6
11. Nikolic P, Krstic D, Stamenovic G, Stefanovic D, Stefanovic M (2009) The performance analysis of MRC combiner output signal in the presence of Weibull and log-normal fading. In: WSEAS international conference. Proceedings. Mathematics and computers in science and engineering, no 13. World Scientific and Engineering Academy and Society
12. Men J, Ge J (2015) Non-orthogonal multiple access for multiple-antenna relaying networks. IEEE Commun Lett 19(10):1686–1689
13. Al-Juboori S, Fernando XN (2018) Multiantenna spectrum sensing over correlated Nakagami-m channels with MRC and EGC diversity receptions. IEEE Trans Veh Technol 67(3):2155–2164. https://doi.org/10.1109/TVT.2017.2764387
14. Ding Z, Liu Y, Choi J, Sun Q, Elkashlan M, Chih-Lin I, Poor HV (2017) Application of non-orthogonal multiple access in LTE and 5G networks. IEEE Commun Mag 55(2):185–191
15. Liang X, Gong X, Wu Y, Ng DWK, Hong T (2018) Analysis of outage probabilities for cooperative NOMA users with imperfect CSI. arXiv preprint arXiv:1809.09411
16. Yang L, Jiang H, Ye Q, Ding Z, Lv L, Chen J (2018) On the impact of user scheduling on diversity and fairness in cooperative NOMA. IEEE Trans Veh Technol 67(11):11296–11301
17. Mohammed AA, Li Y, Al-Kali M, Wang D (2016) Performance analysis of amplify and forward cooperative networks over Rayleigh and Nakagami-m channels based relaying selection. Int J 15(5)
18. Abbasi O, Ebrahimi A, Mokari N (2018) NOMA inspired cooperative relaying system using an AF relay. IEEE Wireless Commun Lett

19. Liang X, Wu Y, Kwan Ng DW, Zuo Y, Jin S, Zhu H (2017) Outage performance for cooperative NOMA transmission with an AF relay. IEEE Commun Lett 21(11)
20. Ding Z, Schober R, Poor HV (2016) On the design of MIMO-NOMA downlink and uplink transmission. In: IEEE international conference on communications (ICC). IEEE, pp 1–6
21. Kumar S, Gupta PK, Singh G, Chauhan DS (2013) Performance analysis of Rayleigh and Rician fading channel models using Matlab simulation. Int J Intell Syst Appl 5(9):94

Web Scraping: From Tools to Related Legislation and Implementation Using Python

Harshit Nigam and Prantik Biswas

Abstract The Internet is the largest database of information ever built by mankind. It contains a wide variety of self-explanatory substances obtainable in varied designs such as audio/video, text, and others. However, the poorly designed data that largely fills up the Internet is difficult to extract and hard to use in an automated process. Web scraping cuts this manual job of extracting information and organizing information and provides an easy-to-use way to collect data from the webpages, convert it into some desired format, and store it in some local repository. Owing to the vast scope of applications of Web scraping ranging from lead generation to reputation and brand monitoring, from sentiment analysis to data augmentation in machine learning, many organizations use various tools to extract useful data. This study deals with different Web scraping tools and libraries, categorized into (i) Partial tools, (ii) Libraries and frameworks, and (iii) complete tools that have been developed over the last few years and is extensively used to collect data and convert into structured data to be used for text-processing applications. This paper explores the terms Web scraping and Web crawling, categorizes the tools available in the current market, and enables the reader to make their Web scraper using one such tool. The paper also comments on the legality associated with Web scraping at the end.

Keywords Web scraping · Legislation · Web data extraction · Scraping tool · DOM tree

1 Introduction

The Internet is the biggest source of information in the twenty-first century, wherein almost every day, several billion gigabytes of data is printed in dissimilar forms, such

H. Nigam (✉) · P. Biswas
Jaypee Institute of Information Technology, Noida, India
e-mail: nigamharshit1712@gmail.com

P. Biswas
e-mail: prantik.biswas@jiit.ac.in

© The Author(s), under exclusive license to Springer Nature Singapore Pte Ltd. 2021
J. S. Raj et al. (eds.), *Innovative Data Communication Technologies and Application*,
Lecture Notes on Data Engineering and Communications Technologies 59,
https://doi.org/10.1007/978-981-15-9651-3_13

Fig. 1 Amount of data created each day. *Source* blog.microfocus.com

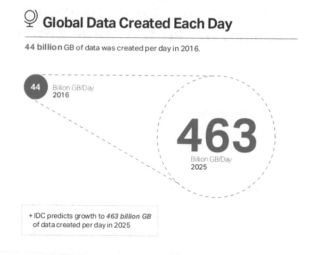

Global Data Created Each Day

44 billion GB of data was created per day in 2016.

44 Billion GB/Day 2016

463 Billion GB/Day 2025

+ IDC predicts growth to *463 billion GB* of data created per day in 2025

as CSV, PDF, and HTML among others (Fig. 1). Data in multiple disciplines is posted on the Internet every day on different Web sites. Most of the professionals typically use copy-and-paste technique to get the data and export it to spreadsheets, a local database or some other data format like CSV, XML, or any other format available. Web scraping and Web crawling are two major concepts in the world of Web data extraction that makes the retrieval of data from millions of webpages easy, less time consuming, automated, and useful for many different applications.

A wide variety of uses of Web scraping and crawling includes business intelligence [1, 2], competitive intelligence [3, 4], Bioinformatics [5, 6], market sentiment analysis [7], understanding how a customer feels about a product or a service [8], information retrieval from e-commerce Web sites [9] and agriculture and food economics [10]. Social media is the best platform to gauge consumer trends and individual behavioral patterns and the way people react to campaigns, thus analyzing human behavior and their relationships with other humans in this virtual civilization of social media. Digital touches produced by the users on social media can be recognized using Web scraping [6, 11, 12].

The Internet is the richest source of 'Big Data' and data being an all-encompassing feature of human lifestyle in the twenty-first century, several studies [13] focus on analyzing this data using time-efficient deep learning models. However, 'Data Collection' forms the very first stage in this process of data analysis. This is where Web scraping and Web crawling come in handy but specifically, both the techniques have some slight differences. Web crawling [14–16] is an extension of Web scraping that not only performs Web scraping on a certain webpage but also follows up the links present on that webpage and scrape the data simultaneously. In simpler words, Web crawling is something that a search engine like Google does. It hops from one link to others and index the webpages to include them in its search database. On the other hand, Web scraping is a term used when scraping a certain Webpage, thus being specific to a Web site.

The below sections goes on describing Web scraping in detail, mentioning about the tools used to perform this task, and then, allowing readers to witness financial data scraping from Yahoo finance page using some Python libraries for better understanding on practical aspects of Web scraping as well.

2 About Web Scraping

There are many definitions for Web scraping; however, the most appropriate one that the author of this paper finds is as defined by Osmar Castrillo-Fernández [17, p. 6],

> A Web scraping tool is a technology solution to extract data from Web sites, in a quick, efficient and automated manner, offering data in a more structured and easier to use format, either for B2B or for B2C processes.

2.1 The Need to Scrape Web Sites and PDF

Approximately, around 70% of the data searched using Google Web browser is in PDF format. It is an unstructured and hard-to-handle format. On the other hand, webpages are composed of HTML and CSS which are structured and contain valuable and desired content placed between various tags.

Some tools mentioned in the later sections of the report may be used for PDF scraping which returns a structured form of data, however in a very basic and rudimentary way. Also, most tools mainly are capable of extracting the data tables.

Contrary to PDFs, the structured design of the HTML document behind a webpage increases the possibilities explored via Web scraping. Web Scraping became easy when users got tools to parse a more structured design of the HTML of a Webpage. This structured design opened up ways to parse and match specific elements in the HTML and extract data placed between them, thus relieving professionals of tiresome, tedious, and error-prone duty of data extraction [17, 18].

2.2 How Do Web Scrapers Work?

A conventional Web scraper is provided with one or more URLs to the webpages that are targeted for extracting data. Once the Web scraper loads the entire HTML of the webpages, the extraction process begins. Ideally, the user decides to extract a relevant part of the data rather than the entire data published on the site. For instance, you might want to scrape Amazon's Webpage for the products and its prices but not the product review. Then, the data is outputted in some friendly and usable format for further analysis: CSV, PDF, XML, JSON, XLS or RSS.

Numerous studies [19, 20] focuses on different methodologies and tools available to perform Web content extraction. Four major approaches used in this task have been identified—(i) The DOM tree-based approach [21–24], (ii) The string-based approach [23–26], (iii) Semantic framework approach [27], and (iv) Machine learning-based Web scraping [24, 26, 28, 29, 30, 31, 32, 33]. These methodologies are prominently used in the task of Web data extraction and form the theoretical foundation for the practical tools available in the market for use.

This study categorizes these various tools available for Web scraping and showcases the implementation of one such tool, the 'libraries and frameworks' specific to the Python programming language, to let the readers build their Web scraper based on DOM tree-based approach.

Web scraping processes can be written in different programming languages and the respective libraries provide access to the site, maintaining the main characteristics of the HTTP protocol, along with SSL certificates. Web scrapers can also come in many different forms with very different features on a case-by-case basis. For example, Web scraper can come as plugins which categorize in 'Partial tools for Web scraping,' or are designed as a Web site extension or more powerful desktop application that is downloaded to your computer. Web scrapers also can scrape sites locally from your computer using your computer resources and Internet connection or can be deployed on the cloud. This type of scraping tools has been categorized as 'Complete tools.'

3   Tools for Web Scraping

This section broadly classifies the tools which can be used to accomplish the process of Web scraping. Basic information and their functionality are provided so that one may choose a scraping tool for his concerned task.

- **Partial tools**: These are basically plug-in or add-on to third-party software. No APIs are provided by them and they focus on specific scraping techniques like extracting a data table.
- **The third-party libraries and frameworks**: These come with general-purpose programming languages which allow access to sites, sustaining the characteristics of the HTTP protocol and helps to extract Web content.
- **Complete tools**: They provide a general scraping service in the form of an extension to the Web browser or a full-fledged native desktop application. They provide features such as strong GUI, computer vision-enabled visual scraping utility, query catching and recording, APIs or audit dashboards.

3.1 Partial Tools

Partial tools for PDF extraction. There were times when data was available in PDF format such as on Government Web sites, which was least usable or kind of as good as dead. Luckily, many tools were developed to extract the data tables from PDF and export it in some structured format (see Table 1).

Partial tools to extract from HTML

'IMPORTHTML' function in Google spreadsheet. It is a simple yet sufficiently effective solution to obtain the tables of data from an HTML of a webpage. It imports tables or lists into the Google Sheets from blogs or external Web sites. When the data on the Webpage changes at any point of time in the future, the same changes will also reflect on the imported data in the Google Sheets. The actual formula of the function is as follows:

```
IMPORTHTML(url, query, index)
```

Here, 'url' represents the URL of the Webpage that encloses the table or list, 'query' means whether you want to import a table or a list and 'index' represents the index of the table or the list, that is 1 denotes the first table in the Webpage, 2 denotes the second table or list and so on.

Table Capture. Table Capture is a Google Chrome extension. Table Capture [34] allows copying the content of a table easily. It allows two options for every table it detects on a Webpage—Copy to Clipboard or paste the content of the table in a Google Doc.

Table to Clipboard. Similar to the previous case, Firefox Web browser also provides add-on (similar to Google Chrome's extensions). An example is Table2Clipboard 2 (https://addons.mozilla.org/en-US/firefox/addon/table2clipbo ard2/). It allows copying to clipboard an HTML table rows/columns selection correctly formatted.

Table 1 List of popular PDF extraction tools

Extraction tool	Web site	Usage
PDFTables	pdftables.com	Provides 75 free page credits to use it for free to extract tables from PDF
CometDocs	cometdocs.com	Works online to convert PDFs into Excel, HTML, Word, and other formats
Tabula	tabula.technology	A desktop tool useful in extracting tables trapped in PDF documents
Nitro cloud	nitrocloud.com/pricing	Useful in merging, editing, and converting PDFs

3.2   The Third-Party Libraries and Frameworks

The third-party libraries. The common strategy followed by many professionals to scrape data is to construct a Web scraper for their application using programming languages they are comfortable with. This case witnesses Web scraper as a conventional software program that uses the functionalities and data structures provided by the language used. Some third-party libraries are dedicated to implement client-side of the HTTP protocol to provide access to the site, while other focus on extracting Web data using tokenization, trimming, and string-based searching algorithms such as Skip Search, Horspool, Optimal Mismatch, and indexOf method for searching specific patterns [25] or more advanced DOM tree parsing and extracting data using CSS selectors and/or Xpath matching.

libcurl (https://curl.haxx.se/) is one of the most favored site access library. HTTP POST, FTP uploading, cookies, HTTP form-based upload, HTTP PUT, proxies, HTTP authentication and other major features of the HTTP protocol are provided by this library. It also provides useful connections to many other programming languages.

Perl is one of the programming languages that is widely used in bioinformatics, which includes the WWW: Mechanize module for programmatic Web browsing and automated interaction with Web sites. Also, it allows Xpath expression matching for data extraction around supplementary modules.

Java includes jsoup (https://jsoup.org/), a HTML parsing library and the Apache httpClient package (https://hc.apache.org/httpcomponents-client-ga/index.html) for HTTP key features implementation. XPath matching is supported by Java and html-cleaner (https://htmlcleaner.sourceforge.net) like cleaning libraries are also provided by this programming language.

Similarly, Libraries namely, BeautifulSoup (https://www.crummy.com/software/BeautifulSoup/) and Requests (https://pypi.org/project/requests/) can be used for HTML parsing and site access respectively.

In environments like UNIX, piping OS command-line programs inside shell scripts enables the users to build Web scrapers easily. HTTP client layer is implemented via programs like libcurl and wget (https://www.gnu.org/software/wget/), while functionalities like sed (https://www.gnu.org/software/sed/), awk (https://www.gnu.org/software/gawk/) and grep may be used to parse and output contents suitably [5, 35].

Frameworks. The libraries are separate pieces with their individual functionalities which need to be integrated to serve fully for Web scraping and robots based on DOM tree parsing, offered by such libraries tends to break when HTML code behind the Webpage changes and thus, needs to be maintained continuously. In Java-like compiled languages, any change in the bot forces recompilation and oftentimes, redeployment of the entire application.

Frameworks for Web scraping tends to be a more integrated solution. For instance, Python's fast and efficient Web scraping application framework, namely Scrapy (https://scrapy.org) provides for all-round features for scraping and crawling, by allowing the user to create its custom class 'Spider' and parse and transform content conveniently along with the bypassing of restrictions imposed on scraping by the Web sites.

Web Harvest (https://web-harvest.sourceforge.net/) and jARVEST (https://sing. ei.uvigo.es/jarvest) are Web data scraping frameworks for Java. They present Domain Specific Language (DSL) used for a specific purpose. Several functionalities like loops and defining variables and many primitives such as 'html-to-xml' (to clean HTML), 'http' (to retrieve Web content), 'Xpath matching' (for extraction purposes) among others are also supported by DSL [5, 35].

3.3 Complete Tools

With more and more information being published every day on the Internet, the market needed robust Web scraping tools to extract the data and use it further in lead generation, reputation and brand monitoring, data processing, data augmentation in ML, opportunities extraction, content comparators, and other applications.

These tools were robust in the sense that it provided,

- A powerful and responsive GUI.
- API that supports linking and easily integrating of data.
- Allows selecting desired content from the Webpage in an interactive manner, avoiding any specifications of XPath queries, regular expression, parsing the HTML DOM tree, or any other technicalities.
- Pipelines to output data in multiple formats.
- Data caching and storage.

The following briefly describes two of the many widely used Complete tools available in the market.

ScrapeBox. ScrapeBox is a native desktop software that provides many functionalities from Search Engine Harvester to Keyword Harvester, from Proxy Harvester to numerous tools such as downloading videos, creating RSS feeds, extracting Emails, and dozens of more time-saving features (Fig. 2). A custom search engine comes with ScrapeBox which can be trained to collect URLs from virtually any Web site that contains a search feature. You may want to harvest all the URLs on a certain keyword or number of keywords on a major search engine like Google. The custom scraper already contains about 30 trained search engines, so that you can start simply by plugging in your keywords or using the included Keyword Scraper (Fig. 3).

Powerful features like Keyword Scraper, help gain valuable intuition into related products services and search phrases which users often search when using services like YouTube, Wikipedia, and others. It contains many interesting features to be explored and exploited by the organizations depending on the data and is a good option for SEO professionals and agencies [20, 36].

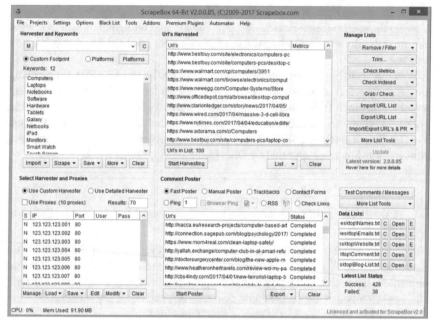

Fig. 2 ScrapeBox. *Source* www.scrapebox.com

Pros:

- Run-on your local machine.
- Low cost (one-time payment).
- Feature-rich including Keyword Research and Blogger Outreach Research.
- Can be used for legit and whitehat tactics in a very easy manner.

Cons:

- Slow for large-scale scraping.
- A feature of Mass comment on blogs is something that should not be facilitated by ScrapeBox.
- Often abused as people may use it for spamming and blackhat tactics.

Diffbot. Diffbot has developed machine learning and computer vision algorithms to extract the content of a Webpage on a visual basis (see Fig. 4). It works differently from most of the conventional HTML DOM tree parsing-based Web scrapers and uses computer vision to extract relevant information from the Webpage and returns it in a structured format. It has built a large database of structured Web data (kind of Semantic Web), which they call as their 'Knowledge Graph' which has grown to include over 2 billion entities and 10 trillion facts [37]. It is pretty clear that Diffbot makes use of machine learning-based Web scraping methodology.

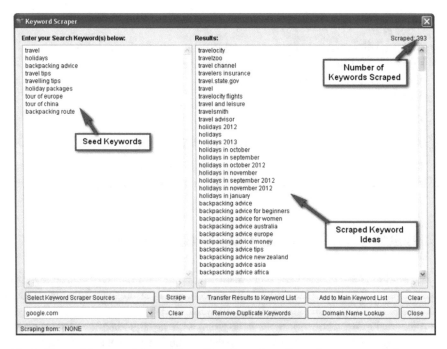

Fig. 3 Keyword Scraper: a feature in ScrapeBox. *Source* https://www.jasonmun.com/using-scrape box-for-good-not-evil

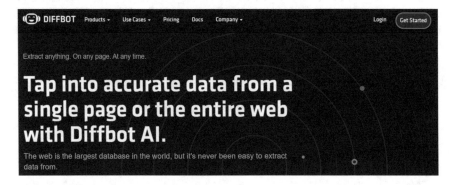

Fig. 4 DIFFBOT. *Source* www.diffbot.com

Diffbot has also constructed its Page Classifier API, which classifies webpages into specific 'page types' in an automated manner. These products allow to analyze webpages with different layouts and extract relevant information while ignoring the irrelevant part such as advertisements. Given that the webpages look the same visually, the algorithms behind the working of Web data extraction does not break even if the HTML structures behind the Webpage changes.

Pros:

- The incredible feature of screen scraping using computer vision.
- Highly useful for particularly those enterprises who scrape Web sites that have their HTML structure changing often.
- Unaffected by a change in HTML structure of the webpage.

Cons:

- A bit expensive (the most economical plan is $299/month).

Other complete tools that provide scalable Web scraping for all types of data extraction include Mozenda, ScrapingHub, Import.io, ScrapingBee, and Scraper API.

4 Building a Web Scraper Using Python Libraries[1]

Many studies [38–41] have been focused on predicting future stock prices, taking into consideration the stock's historical data and technical measures and not considering the financial information existing in the financial market.

Financial information sentiment value along with the augmentation of stock price data through Web scraping can help in predicting the future price trend of the stocks more reliably [42]. Thus, in this section, the construction of a Web scraper using Python libraries Requests and BeautifulSoup has been shown and we have extracted data from Yahoo Finance webpage (https://finance.yahoo.com/) and have created a dataset of all Indian stocks with relevant information such as Market Cap, PE Ratio, Profit Margin, and others presented on the first page of the website. The below steps briefly describes the workflow of the entire process of scraping using Python libraries. The subsequent sections describe these steps in detail.

```
Step1.Import Requests and BeautifulSoup Libraries.
Step2.Use requests.get() method to get access to the site and
assign the value returned, to a variable(say 'page').
Step3.Use page.content to get the raw HTML document of the
webpage accessed and assign it to a variable (say 'content').
Step4.Define BeautifulSoup object (say 'soup') with parameters
including the raw HTML that is stored in 'content' and the Html
parser.
Step5.Use soup.find_all() a method with CSS tags and attributes
as parameters and store the returned value in a list.
Step6.Use text()or get_text() method to get the text part of
the HTML code returned by find_all() method.
Step7.Collect all individual list of data fields extracted into
a Dictionary.
Step8.Use Python library Pandas to convert the dictionary into
a data frame and export it into a.csv file.
```

---

[1] Source code for this project is available on GitHub under the MIT License at https://github.com/NightmareNight-em/Financial-Data-Scraper.

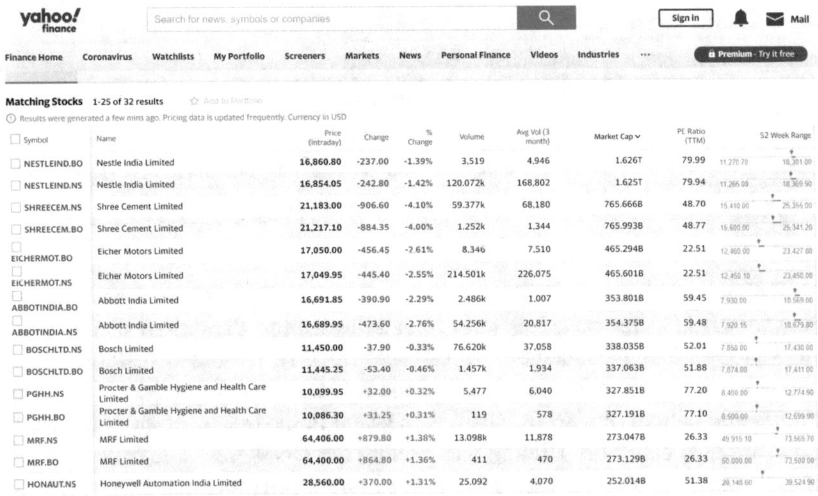

Symbol	Name	Price (intraday)	Change	% Change	Volume	Avg Vol (3 month)	Market Cap	PE Ratio (TTM)	52 Week Range	
NESTLEIND.BO	Nestle India Limited	16,860.80	-237.00	-1.39%	3,519	4,946	1.626T	79.99	11,270.70	18,301.00
NESTLEIND.NS	Nestle India Limited	16,854.05	-242.80	-1.42%	120.072k	168,802	1.625T	79.94	11,266.00	18,369.90
SHREECEM.NS	Shree Cement Limited	21,183.00	-906.60	-4.10%	59.377k	68,180	765.666B	48.70	15,410.00	25,356.00
SHREECEM.BO	Shree Cement Limited	21,217.10	-884.35	-4.00%	1.252k	1,344	765.993B	48.77	15,600.00	25,341.20
EICHERMOT.BO	Eicher Motors Limited	17,050.00	-456.45	-2.61%	8,346	7,510	465.294B	22.51	12,460.00	23,427.80
EICHERMOT.NS	Eicher Motors Limited	17,049.95	-445.40	-2.55%	214.501k	226,075	465.601B	22.51	12,450.10	23,450.00
ABBOTINDIA.BO	Abbott India Limited	16,691.85	-390.90	-2.29%	2.486k	1,007	353.801B	59.45	7,930.00	18,569.00
ABBOTINDIA.NS	Abbott India Limited	16,689.95	-473.60	-2.76%	54.256k	20,817	354.375B	59.48	7,920.15	18,679.80
BOSCHLTD.NS	Bosch Limited	11,450.00	-37.90	-0.33%	76.620k	37,058	338.035B	52.01	7,850.00	17,430.00
BOSCHLTD.BO	Bosch Limited	11,445.25	-53.40	-0.46%	1.457k	1,934	337.063B	51.88	7,874.00	17,411.00
PGHH.NS	Procter & Gamble Hygiene and Health Care Limited	10,099.95	+32.00	+0.32%	5,477	6,040	327.851B	77.20	8,400.00	12,774.90
PGHH.BO	Procter & Gamble Hygiene and Health Care Limited	10,086.30	+31.25	+0.31%	119	578	327.191B	77.10	8,500.00	12,699.90
MRF.NS	MRF Limited	64,406.00	+879.80	+1.38%	13.098k	11,878	273.047B	26.33	49,915.10	73,565.70
MRF.BO	MRF Limited	64,400.00	+864.80	+1.36%	411	470	273.129B	26.33	50,000.00	73,500.00
HONAUT.NS	Honeywell Automation India Limited	28,560.00	+370.00	+1.31%	25,092	4,070	252.014B	51.38	20,148.60	39,524.90

Fig. 5 Yahoo finance page screener. *Source* https://finance.yahoo.com/screener/new

4.1 Site Access

The Webpage used for scraping is the Yahoo finance page. The Yahoo Finance API has been deprecated since 2017 but the site is still active, however, the data cannot be relied upon. Thus, it is safe to scrape it without breaking any law or consent.

Firstly, a new screener on Yahoo Finance with the following parameters had been created: Region: India; Volume > 10,000 and had found the stocks as per the parameters (Fig. 5).

The HTTP protocol is implemented by the Web scraper to gain access to the Web site to be scrapped. In HTTP, GET is the most common request method used for resource request and POST is used in file uploading and form submission. Requests (https://pypi.org/project/requests/) is an easy-to-use HTTP library for Python. It lets you send HTTP/1.1 requests very easily. There is no need to manually add query strings to your URLs or to form-encode your POST data.

Steps 1, 2, and 3 allows us to access the site of the screener on Yahoo finance page and get the raw HTML document of the webpage. The link to the webpage is given as a parameter to requests,get() method.

4.2 The DOM Tree-Based Approach of Scraping

After the HTML document retrieval, the Web data scraper may extract desired content.

Table 2 HTML parsers

Programming language	Corresponding HTML parser
Java	Jsoup, Jerry, Lagarto, htmlcleaner
.NET	HAP, AngleSharp, Microsoft HTML document
Python	HTML Parser of the Standard Library, html5lib, Html5-parser, lxml
Node.js	Cheerio, Htmlparser2, Jsdom, Parse5

HTML parsing involves DOM tree-based approach. DOM tree (https://dom.spec. whatwg.org/) is a tree-based hierarchy for representing the design of the HTML of a Webpage. The Document Object Model (DOM) tree-based method involves building a tree structure involving tags of the HTML of the Webpage and searching for a specific pattern of CSS or Xpath expression in the tree. Moreover, the HTML parsers are also based on building and parsing the DOM tree. The following table (Table 2) presents several parsers for four different programming languages, as is mentioned by Uzun [25].

The HTML parser of the Standard Library in Python has been used along with the BeautifulSoup library (https://www.crummy.com/software/BeautifulSoup/bs4/doc/) to extract the desired content from the DOM tree structure of the HTML document of the page. Step 4 shows the creation of the BeautifulSoup object and defines the use of the Html parser of the Standard Library in Python.

Steps 5 and 6 involve the extraction of desired data using CSS tags and attributes matching while parsing the DOM tree of the raw HTML document stored in 'content' by the Html parser. The data fields scrapped from Yahoo Finance Page include Company URL, its Symbol, name, current Price, Volume, Average Volume, Market capital, and PE ratio. All of these content were available on the HTML document of the site accessed. Tags like 'td', 'a' and attributes like 'class', 'aria-label' are given as parameters inside the find_all() method.

Specifically, find_all() method returns a list of all content and find() returns the first matched content, matching with the tag and corresponding attributes. The attributes and tags for particular data have to be seen from the HTML document of the webpage (right-click on the Webpage and select 'inspect'), therefore these parameters ought to be different for different data and webpages.

In step 7, the individual data in the form of 'List' is collected and all lists are accumulated in a 'Dictionary'. 'Lists' and 'Dictionaries' are data structures provided in Python [43].

Name	Profit Margin	Market Cap(in rupees)	Return on Equity (ttm)	Diluted EPS (ttm)
Nestle India Limited	15.92%	1.642T	70.27%	204.28
Tasty Bite Eatables Limited	9.60%	28.243B	27.69%	159.24
Eicher Motors Limited	22.00%	475.048B	10.38%	757.49
Shree Cement Limited	11.94%	764.332B	13.44%	435.35

Fig. 6 Extracted content from Yahoo Finance page (as on 2nd June, 2020)

4.3 Output Building[2]

Our main objective is to transform the extracted contents into a format that is suitable for further analysis and storage. Step 8 involves the use of Pandas Library (https://pandas.pydata.org/docs/) for Python, that lets us convert dictionary of data into a data frame and then export that dataframe into.csv file and save it in our local system (Fig. 6).

5 Scraping and Legislation

Web scraping has been making data extraction easier, faster, and reliable than ever. Web scrapers or bots come in two forms—Good bots and bad bots. Good bots are those that allow Web data indexing, price comparison to save customer's money, acknowledge sentiment on social media, and many other highly useful applications whereas, bad bots can be used for a variety of harmful activities such as competitive data mining, retrieval of consumers' private data such as phone numbers and emails, account hijacking, spam and digital data fraud. Thus, the question to be made is clear: **Is Web scraping legal?**

The answer is complicated. However, scraping or crawling one's Web site is not to question in legal terms. Big companies exercise Web Scraping for their benefit but at the same time, do not want their data to get scrapped. Also, startups exercise Web scraping to extract desired content from the web, which is an easy and economical way to get the data without having partnerships.

One such example of the use of a Web robot that went against a company was a well-known case between **hiQ labs** and **LinkedIn.** In Summers 2017, LinkedIn sued hiQ Labs, a San Francisco-based startup. hiQ undertook Web scraping to scrape publicly available LinkedIn profiles to render to its clients the data revealing the probability of turnover risks in the future and also the skill-gaps in the employees. In simpler words, the employer can use one's public LinkedIn profile against him. However, the US district court decided that this was okay.

---

[2]The complete dataset in .csv format is available at https://github.com/NightmareNight-em/Financial-Data-Scraper.

However, there is a need to take care of the legislation specific to each country regarding Web scraping. Europe is less permissive than the USA. Spain has its laws fairly restrictive where Web scraping is considered legal under specific conditions. Author rights and legal conditions on certain Web sites are other legal characteristics that should be given special attention before scraping the information from the Web sites.

Thus, it is a good practice of going through the legal implications related to Web scraping. With regards to the company whose data is still lifted, misused, and the business logic of their Web sites abused, they need to start solving this problem using anti-bot and anti-scraping technologies until the court tries to further decide the legality of scraping and come up with just laws [10, 17, 44].

6 Conclusion

A huge amount of data freely available on the Web enhances the possibilities opened up by the Web scraping. This term involves the use of many different methodologies forming a solid theoretical foundation and multiple tools to get the job done in a practical scenario. For years, it has been used by numerous professionals, researchers, and organizations that require the data for further analysis. Throughout this paper, revolved around Web scraping from explaining its necessity to the variety of its tools and then, to the making of our Web scraper, keeping in mind the legalities associated with it. However, the Web scraper is not powerful enough to bypass the restrictions imposed on Web scraping by the Web sites, such as IP Blocking, Honeypot trapping. This would be our future work to improve the Web scraper and make it a versatile tool for extracting not only the financial data but also any relevant data for any application.

References

1. Ferrara E, De Meo P, Fiumara G, Baumgartner R (2014) Web data extraction, applications and techniques. Knowl Based Syst 70:301–323. https://doi.org/10.1016/j.knosys.2014.07.007
2. Baumgartner R, Frölich O, Gottlob G, Harz P, Herzog M, Lehmann P, Wien T (2005) Web data extraction for business intelligence: the lixto approach. In: Proceedings of 12th conference on Datenbanksysteme in Büro, Technik und Wissenschaft, pp 48–65
3. Menczer F, Pant G, Srinivasan P (2004) Topical web crawlers: evaluating adaptive algorithms. ACM Trans Internet Tech 4:378–419
4. Anica-Popa I, Cucui G (2009) A framework for enhancing competitive intelligence capabilities using decision support system based on web mining techniques. Int J Comput Commun Control 4:326–334
5. Glez-Pena et al (2013) Web scraping technologies in an API world. Briefings in Bioinformatics Advance Access. https://doi.org/10.1093/bib/bbt026, published 30 Apr 2013
6. Calabrese B, Cannataro M, Ielpo N (2015) Using social networks data for behavior and sentiment analysis. In: Lecture notes in computer science, pp 285–293. https://doi.org/10.1007/978-3-319-23237-9_25

7. Lin L, Liotta A, Hippisley A (2005) A method for automating the extraction of specialized information from the web. In: Hao Y et al (eds) Computational intelligence and security. CIS 2005. Lecture notes in computer science, vol 3801. Springer, Berlin, Heidelberg

8. Suganya E, Vijayarani S (2020) Sentiment analysis for scraping of product reviews from multiple web pages using machine learning algorithms. In: Abraham A, Cherukuri A, Melin P, Gandhi N (eds) Intelligent systems design and applications. ISDA 2018 2018. Advances in intelligent systems and computing, vol 941. Springer, Cham

9. Nguyen-Hoang B-D, Bao-Tran P-H, Yiping J, Phu TV (2018) Genre-oriented web content extraction with deep convolutional neural networks and statistical methods. PACLIC

10. Hillen J (2019) Web scraping for food price research. Br Food J Ahead-Of-Print. https://doi.org/10.1108/BFJ-02-2019-0081

11. Catanese SA, De Meo P, Ferrara E, Fiumara G, Provetti A (2011) Crawling Facebook for social network analysis purposes. In: Proceedings of the international conference on web intelligence, mining and semantics (WIMS '11). Association for computing machinery, New York, NY, USA, Article 52, pp 1–8. https://doi.org/10.1145/1988688.1988749

12. Traud AL, Kelsic ED, Mucha PJ, Porter MA (2008) Comparing community structure to characteristics in online collegiate social networks. SIAM Rev 53(3):17

13. van den Broucke S, Baesens B (2018) Managerial and legal concerns. In: Practical web scraping for data science. Apress, Berkeley, CA. https://doi.org/10.1007/978-1-4842-3582-9_7

14. Stein L (2002) Creating a bioinformatics nation. Nature 417(6885):119–120. https://doi.org/10.1038/417119a

15. Heydon A, Najork M (1999) Mercator: a scalable, extensible Web crawler. World Wide Web 2(4):219–229. https://doi.org/10.1023/A:1019213109274

16. Chakrabarti S, Berg M, Dom B (2000) Focused crawling: a new approach to topic-specific Web resource discovery. Comput Netw 31:1623–1640. https://doi.org/10.1016/S1389-1286(99)00052-3

17. Castrillo-Fernández O (2015) Web scraping: applications and tools. European Public Sector Information Platform Topic Report No. 2015/10, Dec 2015

18. Dhaya R, Kanthavel R (2020) Comprehensively meld code clone identifier for replicated source code identification in diverse web browsers. J Trends Comput Sci Smart Technol (TCSST) 2(02):109–119

19. Najafabadi MM, Villanustre F, Khoshgoftaar TM et al (2015) Deep learning applications and challenges in big data analytics. J Big Data 2:1. https://doi.org/10.1186/s40537-014-0007-7

20. Laender AHF, Ribeiro-Neto BA, da Silva AS, Teixeira JS (2002) A brief survey of web data extraction tools. SIGMOD Rec 31(2):84–93. https://doi.org/10.1145/565117.565137

21. Ferreira T, Vasconcelos GC, Adeodato P (2005) A new evolutionary method for time series forecasting. In: ACM proceedings of genetic evolutionary computation conference-GECCO. ACM, Washington, DC, pp 2221–2222

22. Feng Y, Hong Y, Tang W, Yao J, Zhu Q (2011) Using HTML Tags to improve parallel resources extraction. In: 2011 international conference on Asian language processing, Penang, pp 255–259. https://doi.org/10.1109/IALP.2011.23

23. Uzun E (2020) A novel web scraping approach using the additional information obtained from web pages. IEEE Access 8:61726–61740. https://doi.org/10.1109/ACCESS.2020.2984503

24. Uzun E, Agun HV, Yerlikaya T (2013) A hybrid approach for extracting information content from Webpages. Inf Process Manage 49(4):928–944. https://doi.org/10.1016/j.ipm.2013.02.005

25. Kumar A, Paprzycki M, Gunjan VK (eds) (2020) ICDSMLA 2019. In: Lecture notes in electrical engineering. https://doi.org/10.1007/978-981-15-1420-3

26. Uzun E, Güner ES, Kılıçaslan Y, Yerlikaya T, Agun HV (2014) An effective and efficient Web content extractor for optimizing the crawling process. Softw Pract Exper 44(10):1181–1199. https://doi.org/10.1002/spe.2195

27. Zhou Z, Mashuq M (2014) Web content extraction through machine learning, 2014. 1.1, 3.8.1

28. Liu B (2011) Structured data extraction: wrapper generation. In: Web data mining. Data-centric systems and applications. Springer, Berlin. https://doi.org/10.1007/978-3-642-19460-3_9

29. Dzitac I, Moisil I (2008) Advanced AI techniques for web mining
30. Fernández-Villamor J, Blasco-García J, Iglesias C, Garijo M (2011) A semantic scraping model for web resources—applying linked data to web page screen scraping. In: ICAART 2011—proceedings of the 3rd international conference on agents and artificial intelligence, vol 2, pp 451–456
31. Broucke SV, Baesens B (2018) Practical web scraping for data science, 1st edn. Apress, Berkeley, CA. https://doi.org/10.1007/978-1-4842-3582-9
32. Table Capture—Chrome Web Store—Google Chrome. 19 July 2020. https://chrome.goo gle.com/webstore/detail/table-capture/iebpjdmgckacbodjpijphcplhebcmeop. Accessed 22 July 2020
33. Joby PP (2020) Expedient information retrieval system for web pages using the natural language modeling. J Artif Intell 2(02):100–110
34. Data Structures—Python 3.8.3 documentation. https://docs.python.org/3/tutorial/datastruc tures.html, last accessed 2020/6/23
35. Singrodia V, Mitra A, Paul S (2019) A review on web scraping and its applications. In: 2019 international conference on computer communication and informatics (ICCCI), Coimbatore, Tamil Nadu, India, 2019, pp 1–6. https://doi.org/10.1109/ICCCI.2019.8821809
36. Mun J (2020) Personal website. https://www.jasonmun.com/using-scrapebox-for-good-not-evil/, last accessed 2020/5/22
37. ScrapeBox homepage. https://www.scrapebox.com/, last accessed 2020/6/10
38. Chen H, Chau M, Zeng D (2002) CI Spider: a tool for competitive intelligence on the Web. Dec Support Syst 34(1):1–17. ISSN 0167-9236. https://doi.org/10.1016/S0167-9236(02)000 02-7
39. Chanduka B, Bhat SS, Rajput N, Mohan BR (2020) A TFD approach to stock price prediction. In: Bhateja V, Satapathy S, Zhang YD, Aradhya V (eds) Intelligent computing and communication. ICICC 2019. Advances in intelligent systems and computing, vol 1034. Springer, Singapore
40. Skabar A, Cloete I (2002) Neural networks, financial trading and the efficient markets hypothesis. In: ACSC 02: proceedings of the twenty-fifth Australasian conference on computer science. Australian Computer Society, Inc., Darlinghurst, Australia, pp 241–249
41. Nayak R, Braak P (2007) Temporal pattern matching for the prediction of stock prices. In: Proceedings of the second international workshop on integrating artificial intelligence and data mining. Australian Computer Society, Inc, Darlinghurst, Australia, pp 95–103
42. Diffbot homepage. https://en.wikipedia.org/wiki/Diffbot, last accessed 2020/6/10
43. Cao R, Liang X, Ni Z (2012) Stock price forecasting with support vector machines based on web financial information sentiment analysis. In: Zhou S, Zhang S, Karypis G (eds) Advanced data mining and applications. ADMA 2012. Lecture notes in computer science, vol 7713. Springer, Berlin. https://doi.org/10.1007/978-3-642-35527-1_44
44. Imperva Blog. https://www.imperva.com/blog/is-web-scraping-illegal, last accessed 2020/6/10

Improved Synonym Queries Bolstered by Encrypted Cloud Multi-watchword Ranked Search System

Madhavi Katamaneni, SriHarsha Sanda, and Praveena Nutakki

Abstract Cloud storage becomes more prominent in recent years due to its benefits such as scalability, availability, and low price carrier over standard storage solutions. Organizations are stimulated to migrate their information from neighborhood sites to the central business public cloud server. By outsourcing information on the cloud, customers receive alleviation from storage maintenance. Although there are many benefits to migrate facts on cloud storage, it brings many security problems. Therefore, the statistics owners hesitate to migrate the sensitive data. In this case, the manager of information is going toward a cloud carrier provider. This safety trouble induces fact owners to encrypt records at the client-side and outsource the data. By encrypting data improves the fact's safety, however, the fact effectivity is diminished because searching on encrypted information is difficult. The search strategies which are used on undeniable textual content cannot be used over encrypted data. The existing options help only equal keyword search, fuzzy keyword search, but the semantic search is now not supported. In the project, the Semantic Multi-keyword Ranked Search System (SMKRSS) is proposed. It can be summarized in two factors: multi-keyword ranked search to acquire additional accurate search results and synonym-based search to support synonym queries. The experimental result shows that our proposed system is better than the unique MRSE scheme.

Keywords Multi-keyword search · Ranking · Synonym-based search

M. Katamaneni (✉) · S. Sanda · P. Nutakki
Department of IT, VR Siddhartha Engineering College, Vijayawada, India
e-mail: itsmadhavi12@gmail.com

S. Sanda
e-mail: sriharsha.sanda@gmail.com

P. Nutakki
e-mail: praveena.4u@gmail.com

© The Author(s), under exclusive license to Springer Nature Singapore Pte Ltd. 2021
J. S. Raj et al. (eds.), *Innovative Data Communication Technologies and Application*,
Lecture Notes on Data Engineering and Communications Technologies 59,
https://doi.org/10.1007/978-981-15-9651-3_14

165

1 Introduction

Cloud computing is being seen as an emerging field in the IT change due to the fact it presents scalable "on-demand service" to the customers over the net. Cloud computing objective is to supply the sources to the consumers as per their order. Cloud allows the purchaser to pay as per their requirement and needs no longer pay for the unused required resources. It is an efficient way to use or share sources due to the fact it gives commercial enterprise to various people remotely above the Internet. Thus, patron and vendor do not come face to face to make the deal and each can be served for their purposes remotely over the net with the aid of cloud computing. There had been many groups that have their personal clouds such as Google, Amazon, Microsoft, IBM, Oracle, and so on, to assist the human beings to get property of cloud services. The cloud provides three extraordinary layers, namely Software as a Service (SaaS), Platform as a Service (PaaS), and Infrastructure as a Service (IaaS). Also, the cloud is of a variety of types inclusive of Private Cloud, Public Cloud, and Hybrid Cloud [1].

2 System Feasibility

2.1 Existing System

Security is most widely utilized in many cloud applications [2–4]. In the cloud, the required data is a search based on the keyword given by the user [5–7]. The proposed system mainly focuses on providing security for the data in the cloud [8]. After searching the data by the user, it is important to provide the security for the document which is stored in the cloud server [9]. The multi-keyword search is also integrated into this system [10].

2.2 Disadvantages of Existing System

The existing system has many drawbacks like the system supports only a single keyword search, the data owner can be upload in only one file, and limitation to the size of the data file.

2.3   Proposed System

The proposed system is mainly focused on security for the data in the cloud with the integration of privacy-preserving ranked multi-keyword search protocol in a multi-owner cloud model. The multi-owner model is developed to provide data privacy and data search with keyword based on encrypted cloud data. In this paper, a user-friendly data access authentication is developed to prevent the data from attackers with the provided secret keys. The details of the users are collected based on data usage, searching, data with the integration of user authentication and revocation.

The proposed system gathers a unique secure inquiry convention, which will enable the cloud server to firmly shut positioned catchphrase look aside from understanding the correct information of every watchword and trapdoors. It enables information owners to access the unique key expressions which are self-collected keys and supports the validated insights, the clients can further raise question still knowing these keys.

2.4   Advantages

The PRISM awards multi-catchphrase look for over mixed records which would be encoded with unique keys for undeniable estimation owners.

The PRISM contrive underwrites new information owners to enter this device other than impacting assorted data owners or information customers, i.e., the arrangement helps bits of knowledge owner flexibility in a fitting and play illustrate. The PRISM guarantees that lone verified records clients can work right hunts. Besides, when a records client is renounced, the owner can no longer capture right hunts over the scrambled cloud information.

To empower a cloud servers to play out a protected hunt without knowing the genuine expense of every watchword and trapdoors. A novel strong pursuit convention is developed. Subsequently, particular information proprietors utilize explicit keys to scramble their documents and watchwords. Verified measurement clients can disconcert a question other than knowing the mystery keys of this one of a sort data proprietor.

To rank the inquiry impacts and keep the protection of pertinence rankings among catchphrases and documents, yet another added substance request and security safeguarding capacity are introduced. This enables the cloud server to restore the most material query items to measure clients despite uncovering any harmful data.

To keep the attackers from spying mystery keys and claiming to be criminal realities, clients encompass a novel powerful mystery key time convention and other realities individual validation convention.

3   Framework Analysis

3.1   The Study of the System

The reliable game plan of a gadget identifies with a kind of depicting the bits of knowledge streams, data sources, and yields of the structure. This routine is driven through illustrating, the use of an over-hypothetical (and on occasion graphical) model of the authentic structure. Concerning structures, the reasonable graph combines the ER diagrams, for instance, component relationship diagrams.

a. **Java Knowledge**
b. **Java is the most popular programming language and runs on an independent platform. The interpreting process**

The Java encoding language is a high-level language that can be characterized by all of the following buzzwords like simple, object oriented, portable, robust, secure, and so on.

The programming languages gather the instructional application to run the applications in the centralized computer. A.class file does not contain code that is native to your processor; it instead contains *bytecodes. T*he Java launcher tool then runs the application with an instance of the Java Virtual Machine. A variety occurs just before; remark arrives each one time the machine is expert.

4   Results and Observations

Welcome page of the cloud environment Fig. 1 shows the welcome page of the Cloud environment. By using this cloud environment, the data can be created, uploaded, and executed in the project (Fig. 2).

Figure 3 shows the data owner home page; the data owner can upload the data, view the uploaded file, review the user request, and download the data (Figs. 4, 5, 6, 7, and 8).

The above figures show that the data owner can upload the file with different keywords in the cloud. The data user downloads the file by the same process. In this process, a file can be searched any number of times and the downloads are numbered and ranked (Fig. 9).

5   Conclusion

The Semantic Multi-keyword Ranked Search System (SMKRSS) is proposed in the current project. It can be summarized in two factors: multi-keyword ranked search to acquire additional accurate search results and synonym-based search to support

Fig. 1 Homepage of the cloud environment

Fig. 2 User is a successful login for searching the data

synonym queries. This system provides security for the data present in the cloud and this can be achieved by providing certain access privileges.

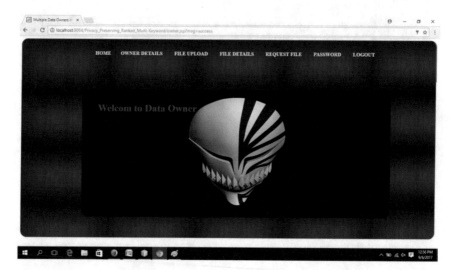

Fig. 3 Data owner home page

Fig. 4 Data owner has uploaded the data

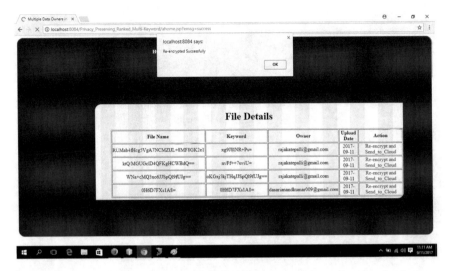

Fig. 5 Admin is re-login and data owner upload data are re-encryption and send to the cloud

Fig. 6 The data user is before login the password is sent to his registered mail id

Fig. 7 Data user sending request sees the data owner and gives the response for user downloading the file

Fig. 8 Data user receiving the decryption key by registered mail

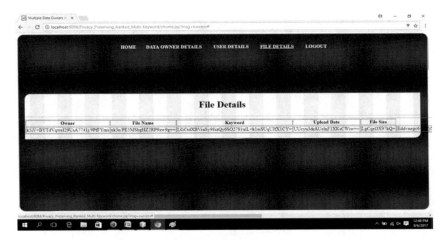

Fig. 9 Cloud server displays the data owners/users/files details

References

1. https://searchcloudcomputing.techtarget.com
2. https://iitd.vlab.co.in/?sub=66&brch=184&sim=1147&cnt=1
3. Armbrust M, Fox A, Griffith R, Joseph AD, Katz R, Konwinski A, Lee G, Patterson D, Rabkin A, Stoica I, Zaharia M (2010) A view of cloud computing. Commun ACM 53(4):50–58
4. Wang C, Chow SS, Wang Q, Ren K, Lou W (2013) Privacy preserving public auditing for secure cloud storage. Comput IEEE Trans 62(2):362–375
5. Song D, Wagner D, Perrig A (2000) Practical techniques for searches on encrypted data. In: Proceedings of IEEE international symposium on security and privacy (S&P'00), Nagoya, Japan, pp 44–55
6. Goh E (2003) Secure indexes. [Online]. Available: https://eprint.iacr.org
7. Curtmola R, Garay J, Kamara S, Ostrovsky R (2006) Searchable symmetric encryption: improved definitions and efficient constructions. In: Proceedings of ACM CCS'06, VA, USA, pp 79–88
8. Boneh D, et al (2004) Public key encryption with keyword search secure against keyword guessing attacks without random oracle. EUROCRYPT 43:506–522
9. Golle P, Staddon J, Waters B (2004) Secure conjunctive keyword search over encrypted data. In: Proceedings of applied cryptography and network security (ACNS'04), Yellow Mountain, China, pp 31–45
10. Mounika Lakshmi V, Mallikarjunamallu K (2019) A novel approach for dynamic query processing in web search engines. Int J Dev Comput Sci Technol V-6, I-7

Public Distribution System Based on Blockchain Using Solidity

C. Devi Parameswari and Venkatesulu Mandadi

Abstract Blockchain is an emerging technology which is applied in all real-world transactions to proved security and transparency. Public distribution system is one of the key areas in which blockchain plays a major role. In this paper, an automated public food distribution system is presented using solidity language. Here, a blockchain-based system is developed that can be used to record all transactions/and log details of these events. The system uses a contract-oriented language framework that has most of the features and functionalities required to carry out any kind of transactions between the central government, state government, the district level offices, retail ration shops, and the beneficiaries/customers.

Keywords Blockchain · Public distribution system · Smart contracts · Solidity · Security

1 Introduction

Bitcoin was developed by Nakamoto [1], and since then, the first example of using bitcoin's digital exchange (digital currency) financial transactions has been measured. At first, bitcoin's high volatility and its complexity slowed its growth to some extent, but the benefits of bitcoin, the underlying technology of bitcoin, began to gain more attention. While some advances have been made since the first bitcoin, today many challenging crypto currencies are based on a similar technology core of blockchain [2]. As for printed money, the use of printed currency is controlled and verified by a central authority, usually a bank or government—but no such thing in bitcoin. Instead,

C. Devi Parameswari (✉)
Department of Computer Applications, Kalasalingam Academy of Research and Education, Anandnagar, Krishnankoil, Srivilliputtur, Virudhunagar 626126, Tamilnadu, India
e-mail: deviparameswari@klu.ac.in

V. Mandadi
Department of Information Technology, Kalasalingam Academy of Research and Education, Anandnagar, Krishnankoil, Srivilliputtur, Virudhunagar 626126, Tamilnadu, India
e-mail: m.venkatesulu@klu.ac.in

transactions made in bitcoin are verified by a computer network. Blockchain has some advantages such as distributed ledger, decentralized system, transparency, and tamper-proof construction. The growth of blockchain has been great in recent times. There are different versions of blockchain include its blockchain 1.0, 2.0, and 3.0 in terms of applications. Blockchain can be used in decentralized immutable ledger [3]. Blockchain can effectively share a record between any two parties. A ledger keeps a verifiable document of every transaction. Security is essential to eliminate unauthorized harm to valuable assets by preventing and responding to harm, misuse, threat, and safety risks [4].

The Public Distribution System (PDS) was launched in India in June 1947. Blockchain technology can be used to implement these schemes to completely eliminate unauthorized activities during the distribution process. The system for PDS consists of a network having the central government, Food Corporation of India (FCI), state government, district, fair price shops and customers, all acting as nodes of a blockchain network and contains records of all transactions that have taken place between any two parties during the distribution chain. The central government adds food grains to its central stock and is supplied to the FCI. FCI can add the commodity details (rice, wheat, pulses, oil, sugar, etc.) and include state, district information also. The FCI supplies food grains to various states. The state government distributes the food grains to every one of its districts. The state government can add a new beneficiary and include authorized new ration shops. Finally, the food grains are distributed through ration shops to the beneficiaries.

2 Literature Review

Blockchain is finding applications across various industries [5]. Blockchain has more features as distributed, verified, and immutable [6]. Presently, blockchain is mostly used in cryptocurrency and the crypto economy [7]. Historically, bitcoin was an early use of blockchain, and since then, there have been many applications beyond bitcoin [8]. Blockchain can be used in many applications related to various non-financial sectors, such as agriculture [9], pharmaceutical industry [10, 11], health care [12], and education [13]. Hybrid (melt) web-based code clone identification tool is engaged in the task of identifying the cloning of codes in various web browsers. The visualized tool is the most powerful tool for accurately and efficiently detecting clones [14]. This paper emphasizes the use of natural language model-based information retrieval to retrieve semantic insights from vast data [15].

Blockchain applications and usability in supply chains are growing with the help of the Internet (IoT) and devices that provide functionality information. It is progressively successful in tracking and controlling the objects from their starting point to the destination by using IoT [16]. Blockchain is now applied in many utility areas and is not fully explored in agri-food chain management. There is a wide gap between blockchain applications and supply chain management. Also, many businesses have little information about blockchain and are not ready to use.

Blockchain technology can strengthen supply chain management to a large extent [16]. One of the key potential functions is to use blockchain to follow all the functions in the supply chain, such as who initiates operations, the time of operation, and the location of every activity [17]. All parties in the distribution chain can follow food grains stock, deliveries, and requirement. They can also monitor all activities in the distribution chain and quality of food grains during shipment [18]. For this reason, a blockchain reduces the established supply chain faults and strengthens transparency [17].

3 Existing System

In the current systems in use, security is available to some extent in public distribution system yet there is a large number of nodes which could not be associated simultaneously, and several unauthorized activities can occur along the distribution chain. It is mostly due to the reason that the exact transactions done at lower levels (till the ration shops) cannot be tracked at the upper level of the distribution chain, especially by the government, and this drawback gives raise a chance for the unauthorized/fraud activities to happen in the public distribution system.

4 Proposed System

The proposed system is developed by using solidity language in Ethereum platform and ensures security with the use of blockchain technology. Data can be viewed by anyone in the distribution chain and enable the authorities at all levels, especially the central government to track the transactions//money and commodities at any time along the distribution chain. The smart contract system can be used by all nodes simultaneously at all times and all levels.

Merits of proposed system

- All nodes along the distribution chain can track the activities simultaneously.
- It completely eliminates fraud and thefts.
- Cost and time-efficient.
- The central government can anytime verify transfers of money and commodity.
- The transaction details cannot be altered.

5 Blockchain Technology

Blockchain is basically a chain of blocks. Every block holds information on a transaction or contract. Transaction details are disseminated among nodes of the network

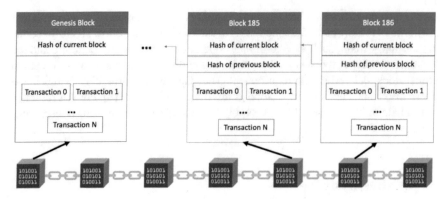

Fig. 1 Blockchain architecture

without the need of a middle man. Each block holds transaction information in the system for a period of time. Every block contains timestamps, hash information, and transaction details [18, 19] (Fig. 1).

Benefits of blockchain technology

Blockchain innovation can be used in several real-life applications and mostly in the financial tech industry.

- As a public, open ledger framework, the blockchain records and validates each and every transaction, which makes it secure and reliable.
- All the transactions are validated by the authorized miners only, which make them immutable and cannot be altered by others, thus keeping it from the risk of hacking.
- Peer-to-peer exchanges can be done without a middle man or a third party.
- And finally, decentralization of the activities.

5.1 Ethereum

It is an open software platform and so a public blockchain. It is based on a decentralized computing platform which permits the designers to construct Decentralized Applications (DApps). Ethereum runs on a virtual machine known as Ethereum Virtual Machine (EVM), implementing scripts and logic of the application. Ethereum has its value token called "ether," a cryptocurrency exchange value similar to bitcoin. Ethereum has a live network of millions of nodes throughout the world which runs and deals with all the contracts deployed by the people using the platform. Ethereum was introduced by Vitalik Buterinin in 2013; Later, in July 2015, this platform was enhanced and released to the public as open-source.

5.2 Smart Contract

A smart contract is an actual logic used to build blockchain applications. Smart contracts are like real contracts which enclose the contract between two participants instantaneously for a specific value. Smart contracts are computer programs which can be used to make similar contracts to carry out transactions between any two parties using the logic programmed in it. Smart contracts in Ethereum are used to store the blockchain state. The contracts can be used to prove, authenticate, and modify the state of the blockchain. A state change occurs in the blockchain when a transaction is executed with the support of functions of the smart contract. The smart contracts can also be used to send/receive the cryptocurrency ether between any two parties instantaneously; of course after the transaction is mined in the blockchain.

Smart contracts can be written by using solidity, serpent, viper, and LLL languages. The smart contracts are compiled using Ethereum Virtual Machine (EVM) and converted into byte code and then deployed onto the public Ethereum blockchain.

6 PDS Design Model

The Public Distribution System (PDS) consists of central government, Food Corporation of India (FCI), and state government, district office, ration shops, and customers/beneficiaries. The central government is the highest authority of the network. All the components of PDS can be authenticated through the login. After successful authentication, every component level transaction can be carried out. At the central government level, the commodities are added/deleted into/from blockchain network and state; district information is added. The price details of each commodity and authorized ration shops details can be added at the higher levels. The food grains are stocked with FCI. FCI distributes commodities to various states. The state government supplies food grains to various districts. The district office distributes food grains to various ration shops located at different places within its jurisdiction. Finally, the ration shop sells/distributes the food grains to the beneficiaries. At each level, the commodity transaction details are recorded on the blockchain network. The central government can track all the food grain distribution details until it reaches the last beneficiary (Fig. 2).

7 Results

Our system design has been implemented on the Intel® Pentium® 3558U @ 1.70 GHz processor with 4 GB RAM. The system is implemented using solidity language in Remix IDE. The remix is a browser-based compiler for building smart contracts in Ethereum platform. It also offers to test, debugging, deploying code.

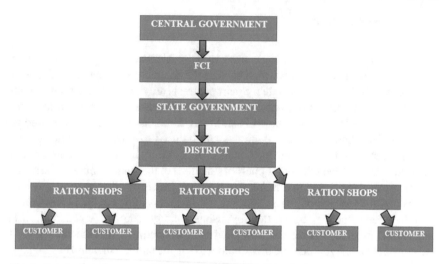

Fig. 2 Supply chain from central government to customer

7.1 REMIX Layout-Solidity Compiler

This panel offers compiler, solidity language, and a version of EVM. Compiling starts when the compile button is clicked on. After successful compilation, deploy and run the smart contracts.

Below presents the screenshots of solidity compiler, deploy and run smart contracts (Fig. 3).

In this JavaScript VMmodule, all transactions are executed in the browser. When the deploy button is clicked on, it sends the transaction that deploys selected contract.

Fig. 3 Solidity compiler—solidity deploy and run

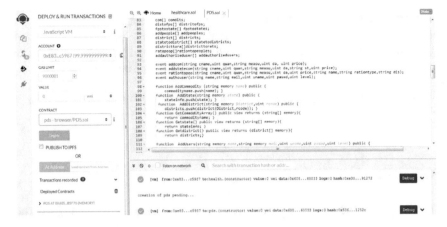

Fig. 4 Deploy the smart contract

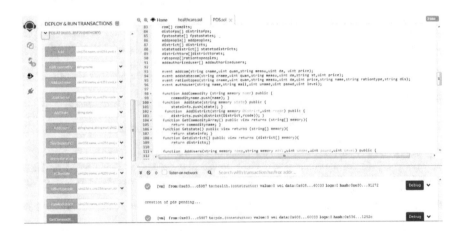

Fig. 5 Run the smart contract

The JVM has five accounts associated with the current environment. Gas limit sets the limit (money/commodities) for all transactions (Figs. 4 and 5).

8 Conclusion

In this work, a blockchain-based system for public distribution system has been proposed. The system ensures security, transparency, and immutability, thereby eliminating fraud and corruption along the distribution chain. The system supports the public distribution framework having the central government, the FCI, the state government, the district office, the ration shops, and beneficiaries as nodes in the blockchain network in a decentralized fashion to cooperate and coordinate with one another and carry out the required transactions of food grains distribution. The system (PDS framework) is presently deployed on the Java Virtual Machine (JVM) in the browser. Future enhancement is to make an interactive web application.

Acknowledgements The first author thanks the management of Kalasalingam Academy of Research and Education for providing a scholarship to carry out the research.

References

1. Nakamoto S (2008) Bitcoin: A peer-to-peer electronic cash system. https://bitcoin.org/bitcoin.pdf
2. Sato T, Himura Y (2018) Smart-contract based system operations for permissioned blockchain. In: 9th IFIP international conference on new technologies, mobility and security (NTMS 2018), pp 1–6
3. Agbo CC, Mahmoud QH, Eklund JM (2019) Blockchain technology in healthcare: a systematic review. Healthcare (Basel) 7(2):56
4. Luther WJ (2016) Cryptocurrencies, network effects, and switching costs. Contemp Econ Policy 34(3):553–571
5. Wright A, De Filippi P (2015) Decentralized blockchain technology and the rise of lex cryptographia
6. Hackius N, Petersen M (2017) Blockchain in logistics and supply chain: trick or treat?. In: Hamburg international conference of logistics (HICL), pp 3–18
7. Efanov D, Roschin P (2018) The all-pervasiveness of the blockchain technology. Procedia Comput Sci 123:116–121
8. Underwood S (2016) Blockchain beyond bitcoin. Commun ACM 59(11):15–17
9. Tian F (2016) An agri-food supply chain traceability system for China based on RFID and blockchain technology. In: International conference on service systems and service management (ICSSSM), pp 1–6
10. Apte S, Petrovsky N (2016) Will blockchain technology revolutionize excipient supply chain management? J Excipients Food Chem 7(3):76–78
11. Bocek T, Rodrigues BB, Strasser T, Stiller B (2017) Blockchains everywhere—a use-case of blockchains in the pharma supply-chain. In: IFIP/IEEE symposium on integrated network and service management (IM), pp 772–777
12. Esposito C, Santis AD, Tortora G, Chang H, Choo KR (2018) Blockchain: a panacea for healthcare cloud-based data security and privacy? IEEE Cloud Comput 31–37
13. Chen G, Xu B, Lu M, Chen NS (2018) Exploring blockchain technology and its potential applications for education. Smart Learn Environ 5(1):2–10
14. Dhaya R, Kanthavel R (2020) Comprehensively meld code clone identifier for replicated source code identification in diverse web browsers. J Trends Comput Sci Smart Technol (TCSST) 2(02):109–119
15. Joby PP (2020) Expedient information retrieval system for web pages using the natural language modeling. J Artif Intell 2(02):100–110
16. Francisco K, Swanson D (2018) The supply chain has no clothes: technology adoption of blockchain for supply chain transparency. Logistics 2(1):2–13
17. Kshetri N (2018) Blockchain's roles in meeting key supply chain management objectives. Int J Inf Manage 39:80–89
18. Chen S, Shi R, Ren Z, Yan J, Shi Y, Zhang J (2017) A blockchain-based supply chain quality management framework. In: International conference e-business engineering (ICEBE), pp 172–176

19. Esposito C, Santis AD, Tortora G, Chang H, Choo KR (2018) Blockchain: a panacea for healthcare cloud-based data security and privacy? IEEE Cloud Comput 31–37

Minimization of Sidelobe Level of Uniformly Excited Concentric Circular Array Antenna Using Evolutionary Technique

Kaialsh Pati Dutta, Sashikant Sundi, Sumit Sourav Khalkho, Jimmy Tirkey, and Karan Kumar Hembrom

Abstract This work proposes the minimization of sidelobe levels of circular antenna array with concentric rings using the evolutionary technique. The well-known evolutionary algorithm, namely quantum particle swarm optimization has been implemented with a new approach to solve the problem under consideration. An attempt has been made to advance the performance of the projected array antenna using constraints like ring radii along with/number of elements in each ring. The array elements have been considered to be uniformly excited with unit excitation amplitude. The results revealed that the application of QPSO overtakes the prior arts in term of SLL minimization. The statistical results shown in the paper validates the effectiveness of this technique.

Keywords Circular array antenna · Evolutionary algorithm · Quantum particle swarm optimization · Sidelobe levels (SLL) · Synthesis

1 Introduction

Antennas play a key role in wireless communication and have wide applications in not only local, civilian and defence communications but they are also equally indispensable in use of radar, satellite as well as underwater communications [1]. Though conventional applications involve linear arrays, however, in the case where 360° azimuthal and omnidirectional scanning with symmetry are required, circular array antennas are used [2–6]. Such circular array antennas consist of various array elements on its circular ring. Moreover, for better array performance, multiple

K. P. Dutta · S. Sundi (✉) · S. S. Khalkho · J. Tirkey · K. K. Hembrom
Department of Electronics and Communication Engineering, Cambridge Institute of Technology, Ranchi, Jharkhand 835103, India
e-mail: sasszx@gmail.com

K. P. Dutta
Department of Electronics and Communication Engineering, National Institute of Technology Durgapur, Durgapur, West Bengal 713209, India

© The Author(s), under exclusive license to Springer Nature Singapore Pte Ltd. 2021
J. S. Raj et al. (eds.), *Innovative Data Communication Technologies and Application*,
Lecture Notes on Data Engineering and Communications Technologies 59,
https://doi.org/10.1007/978-981-15-9651-3_16

numbers of rings with a common centre called concentric circular array antenna are applied.

Concentric circular array antenna (CCAA) has good directivity [6, 7] but they undergo with the problem of high sidelobes levels. Hence, the synthesis of CCAA for better performance marked by lower SLL has always been a challenging task for antenna designers. For such tasks, applications of various evolutionary algorithms [8] are preferred because of their robust nature towards dynamic changes, conceptual simplicity and broad applicability [9].

In past, many CCAA synthesis techniques have been proposed. For examples, an investigation of concentric ring arrays using Chebyshev radiation pattern function was proposed by Stearns and Stewart [2]. In [3], suppressing undesired echoes entering the sidelobes of a radar antenna were demonstrated. A differential evolution algorithm (DEA) [4] was described by Kurup et al. in 2003 for synthesizing the antenna arrays with uniform features of two classes, i.e. inequitably spaced arrays with phases of equal and unequal value. In the same paper, the existence of a trade-off between unequally spaced array size and phase ranges for the same radiation characteristics was successfully established. In 2004, Fondevila et al. proposed optimization of linear arrays having uniformly exciting elements with the help of time modulation method [5]. In 2008, Haupt proposed binary genetic algorithm (GA) [6] in the optimal synthesis of multiple CCAA and established that by varying the ring gap and the no. of elements in each ring of the CCAA, better performance can be obtained in terms of low sidelobe level and high directivity. A hybrid algorithm is known as artificial bee colony–differential evolution (ABC-DE) [10] which was proposed by Yang et al. in 2013 where promising performance was claimed in terms of the convergence rate and exploration ability of the said technique in synthesizing the concentric circular array antenna. In 2014, Maldonado and Panduro used harmony search algorithm (HAS) and particle swarm optimization [11] for solving concentric ring array problems for wide coverage pattern. Paper [12] presented an algorithm called invasive weed optimization (IWO) for minimizing sidelobe levels (SLL) with explicit half-power beamwidth (HPBW) of multiple CCAA having uniformly excited elements isotropic in nature. In 2017, Guo et al. [7] proposed a hybrid approach (HA) towards synthesizing uniformly excited multiple ring arrays for minimizing the peak sidelobe levels taking various constraints including minimum spacing between elements, total element's number and aperture size into consideration. Authors in [9] demonstrated the application of improved discrete cuckoo search algorithm (IDCSA) and cuckoo search invasive weeds algorithm (CSIWA) [9] for solving similar kind of problems. Massive multiple-input-multiple-output (MIMO) systems [13] and maximal ratio combining (MRC) along with equal gain combining (EGC) diversity techniques [14] for multi-antenna sensing system are discussed in past to add further feathers in the advancements of antenna performances.

Though the above techniques are useful in the synthesis of concentric circular array antenna, however performance improvement by use of different approaches always attracts.

In this work, the use of a well-known evolutionary algorithm called quantum particle swarm optimization (QPSO) [15] towards a new approach of application for optimizing the CCAA performance in terms of minimized SLL is proposed. Since its commencement, QPSO has been successfully applied to numerous engineering problems of optimization [16–18]. For example, authors in [16] applied the quasi-Newton (QN) method based on the QPSO algorithm for electromagnetics with the help of HFSS simulation software. Recently, QPSO with the teamwork evolutionary strategy (TEQPSO) was applied in [17] for the multi-objective optimization towards the optoelectric platform. Jamming resources allocation of radar was discussed in [18] using hybrid QPSO-genetic algorithm (HQPSOGA). Moreover, QPSO has only one tuning parameter which makes it easy to use with lesser computational complexities. Due to these advantages and popularity, QPSO has been used in this research. Here, three constraints have been kept in consideration—1. variable ring radii; 2. variable number of array elements; and 3. variable number of both ring radii and array elements. A unique cost function has been formulated to achieve the objective. The results obtained in this work have been contrasted with other states of the art result prevailing in the literature to prove the effectiveness of the work.

This communication is structured as per the following details: Sect. 2 reflects the array geometry and mathematical outlines of the CCAA. Section 3 represents the details of the evolutionary algorithm applied in this particular problem. The results achieved after simulation and their discussions are presented in Sect. 4. The conclusion is displayed in Sect. 5.

2 Array Geometry and Mathematical Outline

Uniformly excited CCAA has its elements that are arranged in different circular rings that have a common centre, different radius and different no. of antenna array elements in each ring. Taking the central element feed into account, CCAA's far-field pattern laid down on the x–y plane is given by (1). The CCAA array geometry may be seen in Fig. 1.

$$E(\theta, \varphi) = 1 + \sum_{m=1}^{M} \sum_{n=1}^{M_n} I_{mn} e^{j[kr_m \text{Sin}\theta \text{Cos}(\emptyset - \emptyset_{mn}) + \varphi_{mn}]} \tag{1}$$

where M, N_m and d_m are the number of concentric circles, no. of elements isotropic in nature in mth circle and inter-element spacings of mth circle, respectively. $r_m = N_m d_m / 2\pi$ represents the radius of the mth circle; $\emptyset_{mn}$ is the angular position of nth element of mth circle amid $1 < n \leq N_m \theta$ and $\emptyset$ signify polar and azimuthal angle, respectively. Also, k stands for a wave no. which equals $2\pi/\lambda$ wheres as λ signifies wavelength in operation, j symbolizes complex number and φ_{mn} gives the element's phase; I_{mn} represents current excitation amplitude of mnth elements of the CCAA which is set to the unit value. All elements are considered to have uniform phase excitation of $0°$.

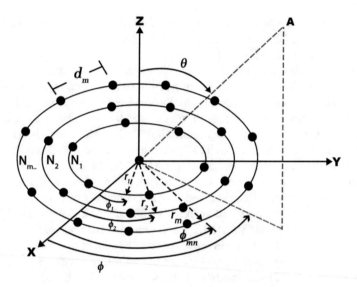

Fig. 1 Array geometry of concentric circular array antenna

There are six rings in the proposed problem. The number of elements in any particular ring is given by (2).

$$N_m = \frac{2\pi r_m}{d_m}. \tag{2}$$

The normalized absolute power pattern presented by $P(\theta, \varphi)$ can be expressed in (3). It gives the power pattern in dB.

$$P(\theta, \phi) = 10 \log_{10}\left[\frac{|E(\emptyset, \varphi)|}{|E(\theta, \varphi)|_{\max}}\right]^2 = 20 \log_{10}\left[\frac{|E(\emptyset, \varphi)|}{|E(\theta, \varphi)|_{\max}}\right] \tag{3}$$

The objective (cost) function is designed in such a way that the sidelobe level (SLL) get optimized using uniform inter-section arrangement in rings/or inter-ring radii, many arrays present in every ring. In other words, the SLL is maximally reduced while searching the optimally arranged radius of each ring as well as the optimal number of array elements.

After deciding the optimization parameters, the subsequent task is to write the objective function to be minimized which is expressed as $F^q_{\text{Objective}}$ and given in (4).

$$F^q_{\text{Objective}} = \text{SLL}^q_{\max}\big|_{\text{freq}'} \tag{4}$$

where q denotes the qth cycle of the iteration, freq$'$ is the operating frequency, and $\text{SLL}^q_{\max}$ is the maximum reduced SLL.

3 Quantum Particle Swarm Optimization (QPSO) Algorithm

The QPSO [15] is the state of the art algorithm for various types of optimization problems. QPSO functions depending upon the features of quantum mechanics and behaviour of the swarm particles available in nature. In QPSO, the velocity vector is absent, and it has position vector only. Unlike various other evolutionary algorithms [19] having a large number of tuning parameters, QPSO has only one tuning parameter. Due to this, QPSO has a much lesser computation cost. This also makes the proposed algorithm simpler and more attractive as far as computation complexity of the problem under consideration, and its convergence rate is concerned.

The process of implementation of the QPSO through flow chart is illustrated in Fig. 2.

Following are the steps of implementation of QPSO directly taken from [15].

Step 1. Initialization of particle positions in defined search space.
Step 2. Evaluation of fitness (objective) function.
Step 3. Comparison of the present fitness value with personal best (*p*best) for each particle. For better present fitness value, it is assigned to (*p*best).
Step 4. Evaluating the present value of best fitness among the complete population.

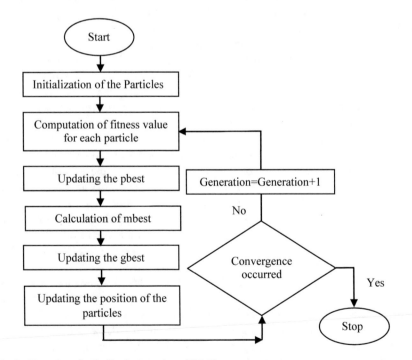

Fig. 2 Flow chart for the implementation of QPSO

If the present best fitness is better than global best (*g*best), then this best fitness assigned to best.

Step 5.   Finding mean of best position (*m*best) for every particle.

Step 6.   Updating the position 'X_{id}' for the *i*th particle by use of Eq. (5):

$$X_{id}^t = P_{id}^t + (-1)^{\text{ceil}(0.5+r3_{id}^t)} \times \alpha \times \left| m\text{best} - X_{id}^{t-1} \right| \times \log_e\left(1/r4_{id}^t\right) \quad (5)$$

where P parameter acts as 'local attractor' for every particle. α represents the coefficient of contraction–expansion. This is the only tuning control parameter in QPSO. It controls the speed of convergence.

Step 7.   Selection of Optimum value by repeating the process until the completion of the total number of iterations.

4   Simulation Results

Here, the antenna elements taken into consideration are isotropic. Two optimizing parameters have been considered which include inter-ring radii and the number of array elements in each ring. The excitation amplitude for each element in every ring is made unity. By considering the array factor for CCAA with a single element at the centre, minimization of sidelobe level at boresight is achieved by optimizing the element placement in the proposed concentric ring array.

Basically, ring radii (r_m), total no. of array elements (N) are initially optimized individually to get minimized SLL. Again, both are optimized together which gives the peak SLL. Six rings have been randomly considered as an example only. For sake of convenience towards execution time, the algorithm is run for 5 runs and 50 iterations with each run. It has been found that the algorithm converges in 5 runs only. So, increasing the number of runs beyond this figure will be simply wastage of time. The size of the population is considered to be 20. The SLL and the optimal number of array element along optimized ring radii were recorded depending upon the value of the best cost achieved at the end of all runs. The values of all the control parameters were applied directly based on the earlier experience with standard functions and the similar type of synthesis problems available in the literature.

All the simulation works have been done in our personal computer using 2015b version of MATLAB software. The general configuration of our PC may be mentioned here which has Intel core (i5 duo 7th G) processor having 3.1 GHz and 8 GB of clock frequency and DDR4 RAM, respectively.

The simulation results are displayed in Table 1. When the ring radii of the individual ring were varied, the no. of the variable used was 5. In this situation, the array that was obtained consisted of 202 elements whereas the SLL was −22.76 dB. However, when the number of array elements was optimized, the array performance was improved in terms of SLL up to −26.76 dB. In this case, the array geometry so obtained comprises of 183 elements. The result so obtained here was about 4

Table 1 Comparison of key performance parameters of the proposed algorithm using various constraints to prior arts [6] and [9]

Algorithm	Constraints	No. of array elements	No. of rings	No. of variables	SLL (dB)
IDCSA [9]	Opt.d_m	221	10	–	−25.62
GA [6]	Opt. N_m and $r_m(\lambda)$	142	9	18	−27.82
QPSO (proposed)	Opt. $r_m(\lambda)$	202	6	5	−22.76
	Opt.N_m	183	6	8	−26.76
	Opt. $r_m(\lambda)$ and N_m	140	6	15	−28.61

and 1.14 dB better than that obtained with optimized ring radii using QPSO and optimized inter-element spacing using IDCSA [9], respectively.

However, finally, QPSO with optimized ring radii and the number of array elements yielded the best peak SLL of −28.61 dB with only 140 number of array elements. This result was far better than the case of optimized ring radii and optimized no. of elements using the same algorithm. Specifically, the results with optimized ring radii and no. of array elements generated 6 and 2 dB better SLL as compared to the same algorithm using optimized ring radii and optimized no. of array elements, respectively. Also, the peak SLL in our case is about 3 dB and about 0.8 dB better than that obtained in case of [9] and [6], respectively. Apart from this, IDCSA displays 221 elements distributed over ten rings, and QPSO gives 140 array elements distributed over six rings to reach up to their excellent levels. Hence, QPSO saves a total of 81 antenna array element with three lesser rings to achieve the optimization of SLL. This will save the designing cost and complexity as well which is an added advantage with the proposed technique. The optimal results and analysis produced in this work are based on a fitness function that have been designed in such a way that the SLL is minimized to the maximum possible extent while searching optimal ring radii and/or the optimal number of array elements but it does not consider other characteristics like return loss, directivity, coupling effect, etc. as that is not the purpose of this synthesis problem in the present work.

Figure 3 displays the outlay of the geometry of an array. Figure 4 shows the peak sidelobe levels (SLL) versus iterations using QPSO obtained during the best run by optimized ring radii (r_m) and an optimal number of antenna array elements (N). Also, Table 2 presents the ring radii and the number of array elements in each ring. Figure 5 displays the fitness value versus the individual number of the population size obtained during the simulation process.

Fig. 3 Array geometry of concentric circular array antenna with optimized r_m and N_m

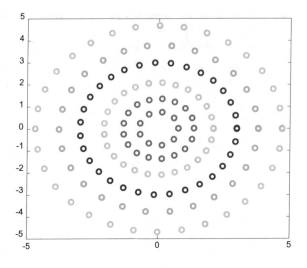

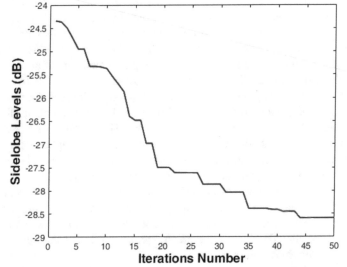

Fig. 4 Sidelobe levels (SLL) versus iterations using QPSO obtained by optimized ring radii (r_m) and the optimal number of antenna array elements N_m obtained during best run

Table 2 Ring radii and array elements in each ring with optimized r_m and N_m

M	Ring no. 1	Ring no. 2	Ring no. 3	Ring no. 4	Ring no. 5	Ring no. 6
$r_m(\lambda)$	0.76	1.36	2.09	2.99	3.78	4.7
N_m	9	17	25	31	26	32

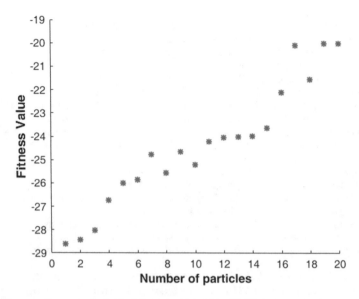

Fig. 5 Fitness value versus individual number obtained by QPSO during best run

5 Conclusion

A new approach in implementing the popular evolutionary technique called quantum particle swarm optimization is demonstrated here to obtain minimized sidelobe with constraints. The results prove the worth of the proposed algorithm. The optimized ring radii along with the optimal number of array elements yield the best result giving sidelobe levels of −28.61 dB with 140 array elements considering the central element. This overtakes the results obtained in [6] and [9]. Further efforts may be made in future to realize better SLL taking first null beamwidth, directivity, return loss and non-uniform time modulated array excitation into considerations.

Disclosure Statement
The authors declare the non-existence of conflict of interest.

References

1. Ballanis A (1997) Antenna theory analysis and design, 2nd edn. Willey Inc., New York
2. Stearns C, Stewart A (1965) An investigation of concentric ring antennas with low sidelobes. IEEE Trans Antennas Prop 3(6):856–863
3. Lewis BL, Evins JB (1983) A new technique for reducing radar response to signals entering antenna sidelobes. IEEE Trans Antennas Prop 31(6):993–996
4. Kurup DG, Himdi M, Rydberg A (2003) A synthesis of uniform amplitude unequally spaced antenna arrays using the differential evolution algorithm. IEEE Trans Antennas Prop 51(9):2210–2217

5. Fondevila J, Bregains JC, Ares F, Moreno E (2004) Optimizing uniformly excited arrays through time modulation. IEEE Antennas Wirel Prop Lett 3:298–301
6. Haupt RL (2008) Optimized element spacing for low sidelobe concentric ring arrays. IEEE Trans Antennas Prop 56(1):266–268
7. Guo Q, Chen C, Jiang Y (2017) An effective approach for the synthesis of uniform amplitude concentric ring arrays. IEEE Antennas Wirel Prop Lett 16:2558–2561
8. Sotirios K, Christos K, Mittra R (2016) Evolutionary algorithms applied to antennas and propagation: emerging trends and applications. Int J Antennas Prop 2016:1–12
9. Sun G, Liu Y, Chen Z, Liang S, Wang A, Zhang Y (2018) Radiation beam pattern synthesis of concentric circular antenna arrays using hybrid approach based on cuckoo search. IEEE Trans Antennas Prop 66:4563–4576
10. Yang J, Li W, Shi X, Li X, Yu J (2013) A Hybrid ABC-DE algorithm and its application for time-modulated array pattern synthesis. IEEE Trans Antennas Prop 61:5485–5495
11. Maldonado AR, Panduro MA (2015) Synthesis of concentric ring antenna array for a wide isoflux pattern. Int J Numer Model 28:433–441
12. Wu H, Liu C, Xie X (2015) Thinning of concentric circular antenna arrays using improved binary invasive weed optimization algorithm. Math Probl Eng 2015:1–8
13. Noha H, Xavier F (2017) Massive MIMO wireless networks: an overview. MDPI J Electron 6:63
14. Al-Juboori S, Fernando XN (2018) Multiantenna spectrum sensing over correlated Nakagami-m channels with MRC and EGC diversity receptions. IEEE Trans Veh Tech 67(3):2155–2164
15. Mikki SM, Kishk AA (2006) Quantum particle swarm optimization for electromagnetic. IEEE Trans Antennas Prop 54:2764–2775
16. Liu BJ, Shen ZX, Lu YL (2014) Optimal antenna design with QPSO–QN optimization strategy. IEEE Trans Magn 50(2):645–648
17. Liu G, Chen W, Chen H (2019) Quantum particle swarm with teamwork evolutionary strategy for multi-objective optimization on electro-optical platform. IEEE Access 7:41205–41219
18. Jiang H, Zhang Y, Xu H (2017) Optimal allocation of cooperative jamming resource based on hybrid quantum-behaved particle swarm optimisation and genetic algorithm. IET Radar, Sonar Navig 11(1):185–192
19. Mahto SK, Choubey A (2015) A novel hybrid IWO/WDO algorithm for interference minimization of uniformly excited linear sparse array by position-only control. IEEE Antennas Wirel Prop Lett 15:250–254

Early Screening Framework to Detect Keratoconus Using Artificial Intelligence Techniques: An Extensive Review

D. Priya and G. S. Mamatha

Abstract Keratoconus detection and diagnosis has become a crucial step of primary importance in the preoperative evaluation for the refractive surgery. With the ophthalmology knowledge improvement and technological advancement in detection and diagnosis, artificial intelligence (AI) technologies like machine learning (ML) and deep learning (DL) play an important role. Keratoconus being a progressive disease leads to visual acuity and visual quality. The real challenge lies in acquiring unbiased dataset to predict and train the deep learning models. Deep learning plays a very crucial role in upturning ophthalmology division. Detecting early stage keratoconus is a real challenge. Hence, our work aims to primarily focus on detecting an early stage and multiple classes of keratoconus disease using deep learning models. This review paper highlights the comprehensive elucidation of machine learning and deep learning models used in keratoconus detection. The research gaps are also identified from which to obtain the need of the hour for detecting keratoconus in humans even before the symptoms are visible.

Keywords Keratoconus · Artificial intelligence · Machine learning · Deep learning

1 Introduction

India is a highly populated country wherein as many as 550 million people are estimated to be suffering from vision problems and diseases. Ophthalmology is a breakneck improving field with very much improved new technologies for eye diseases diagnosis and treatment is getting implemented every year. It is very high time now for doctors to adopt the latest diagnostic methods to ease the complications in detection and identification of vision-related problems. Hence, the involvement of advanced

D. Priya (✉) · G. S. Mamatha
Department of Information Science & Engineering, RV College of Engineering®, Bengaluru, India
e-mail: priyad@rvce.edu.in

G. S. Mamatha
e-mail: mamathags@rvce.edu.in

© The Author(s), under exclusive license to Springer Nature Singapore Pte Ltd. 2021   195
J. S. Raj et al. (eds.), *Innovative Data Communication Technologies and Application*,
Lecture Notes on Data Engineering and Communications Technologies 59,
https://doi.org/10.1007/978-981-15-9651-3_17

corneal imaging instruments and techniques continues to grow. Artificial intelligence (AI) and machine learning (ML) have entered several avenues of modern life, and health care is just one of them. Ophthalmology is a field with a lot of imaging and measurable data, thus impeccable for AI and ML applications. Many of these are still in the research stage but show promising results.

The most prominent ophthalmic diseases where AI is being used are diabetic retinopathy, glaucoma, age-related macular degeneration, retinopathy of prematurity, retinal vascular occlusions, keratoconus, cataract, refractive errors, retinal detachment, squint, and ocular cancers. It is also useful for intraocular lens power calculation, planning squint surgeries, and planning intravitreal antivascular endothelial growth factor injections. In addition, AI can detect cognitive impairment, dementia, Alzheimer's disease, stroke risk, and so on from fundus photographs and optical coherence tomography. Surely, many more innovations will be seen in this rapidly growing field. The most common eye diseases that a person suffers from are keratoconus, cataract, glaucoma, retinopathy, red eyes, cross eyes, colour blindness, vision changes, pink eyes, blurred vision, teary eyes, night blindness, etc. To identify most of these diseases at the early stage may prove to be inherently challenging. Especially keratoconus is a deadly eye disease that percolates through stages, and only at the last stage, it will be identified. The inference is that to overcome this lacuna, designing and developing a doctor (ophthalmologist)-friendly machine intelligence framework for early diagnosis of keratoconus in humans is need of the hour.

1.1 Market Survey

According to the Times of India (ToI) survey in 2019, a Bengaluru hospital sees over 20% rise in keratoconus (KC) patients [1]. In 2017–18, approximately 400 keratoconus patients were found in Agarwals Eye Hospital, Bengaluru. In 2018–19, approximately 480 patients were found in Bengaluru Agarwals Eye Hospital. One in every 1500 children in India suffers from the disease according to the ToI survey conducted in May 2019. All the above studies are depicted in Fig. 1.

According to the National Programme for Control of Blindness (NPCB) estimates, there are currently 120,000 corneal blind persons in the country [2]. According to this estimate, there is the addition of 25,000–30,000 corneal blindness cases every year in the country. The burden of corneal disease in our country is reflected by the fact that 90% of the global cases of ocular trauma and corneal ulceration leading to corneal blindness occur in developing countries. Eye images of patients are captured by various instruments and one of the most popular is Pentacam which works on the Scheimpflug principle. The instrument results in a three-dimensional mapping of the cornea from 25,000 actual lifting points, anterior and posterior surfaces of the cornea, corneal thickness, and angle anterior chamber.

Figure 2 shows the Pentacam 4 Map Report [3]. A pachymetry map illustrates a colour map that shows the axial curvature, anterior and posterior float, and corneal

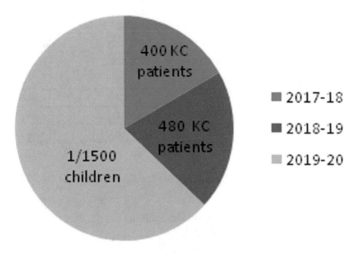

Fig. 1 Market survey of keratoconus patients in India

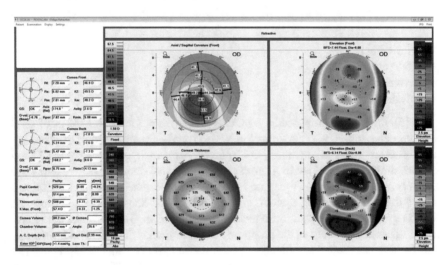

Fig. 2 Pentacam 4 maps report of a patient

thickness. Dataset for keratoconus is not available on the Internet like other diseases. Hence, the challenge lies in acquiring and preprocessing the dataset.

2 Promising Features of Keratoconus

In this section, the features are discussed which are considered to declare whether a patient has keratoconus or not. The vast majority of patients can be quickly and safely

Table 1 Grading system for ABCD keratoconus classification

Stages	A (in mm)	B (in mm)	C (in μm)	D
0	>7.25	>5.90	>490	>= 20/20
1	>7.05	>5.70	>450	<20/20
2	>6.35	>5.15	>400	<20/40
3	>6.15	>4.95	>300	<20/100
4	<6.15	<4.95	<=300	<20/400

screened with the following four feature maps for the early diagnosis of keratoconus disease:

- Corneal thickness map(Pachymetric distribution)
- Anterior and posterior corneal elevation
- Anterior sagittal curvature (power dioptric).

The corneal thickness of <450 μm is abnormal based on the severity of corneal curvature and the kerataconus disease is classified into various stages as shown below [4].

- Mild < 45D
- Moderate 45–52D
- Severe > 52D
- Advanced > 62D.

In [5], the authors proposed an ABCD grading system for various stages of keratoconus detection. ABCD classification is based on A—Anterior radius of curvature, B—Posterior radius of curvature, C—Thinnest pachymetry, and D—BDVA (Best-corrected disease visual acuity). From early stage keratoconus to very severe stage keratoconus, i.e. Stage 0–IV can be identified based on the classification. Table 1 shows the ABCD grading system with the ranges for Stage 0–IV. The proposed ABCD grading system should be merely available in the future Pentacam instruments. The major benefit of the proposed ABCD system is that it provides information on both anatomical and functional data. It also covers the anterior and posterior elevation of the cornea, thinning, corneal scars, etc. and this allows for the timely and accurate decision for treatments.

3 Related Work

This section provides the literature work done in recent years on various machine learning and deep learning models used for detection of keratoconus eye disease.

3.1 Machine Learning Models Used for Keratoconus Detection

In [6], authors have used unsupervised learning algorithm on corneal images and the developed prototype was able to detect multiple stages of keratoconus and how severe the disease. Optical coherence tomography (OCT) images were collected from multiple hospitals in Japan. Between the institute and hospitals in Japan, the data agreement was signed, i.e. the details of the patients will not be disclosed by the authors. Authors used the R software for implementation and principal component analysis (PCA) for reducing the dimension of the dataset and also to generate the prominent principal components. A density-based clustering algorithm was used to split the dataset into different clusters. The obtained results are measured with the specificity and sensitivity of 94.1 and 97.4%.

In [7], support vector machine algorithm is used to develop the framework to detect keratoconus in Weka software. Authors considered images taken from Pentacam instrument and the classes of five keratoconus division are used in their work. The classes of keratoconus that were used are KC, forme fruste (FF), astigmatic, post-refractive surgery, and normal eyes. The prominent twenty-two parameters were identified for detection of all the five classes of keratoconus. The developed framework was also cross-validated with other classification algorithms. The accuracy of binary classification of whether keratoconus or normal is 98.9%; the specificity and sensitivity are measured as 98.5 and 99.1%. For multiple class classification, i.e. five groups classification are measured and the results show a little decrease in accuracy, specificity, and sensitivity. The results that are achieved by the developed model are, 88.8% of accuracy, 89% of sensitivity, and 95.2% of specificity.

In [8], a framework using machine learning algorithms was implemented to identify the candidates suitable for corneal refractive surgery. An ensemble classifier was also developed to improve the performance of the model. There are five difference algorithms were evaluated to predict the patients who need to undergo corneal refractive surgery. The authors used the data of patients who visited the hospitals in Korea during the year 2016 and 2017. The models were evaluated with the help of area under curve (AUC) and the results were proven that the combination of machine learning algorithms performed better than the single classifier. The classifiers used were SVM, random forest, neural networks, AdaBoost, etc. and the AUC of the algorithms are 96.3, 97.2, 98.1, 96.2, and 93.8%. The accuracy of the ensemble classifier is 98.3% which is better than the other models.

3.2 Deep Learning Models Used for Keratoconus Detection

In [9], Valdés-Mas et al. proposed a framework to predict the patients who gained the vision after ring implantation using various machine learning models and the artificial neural networks outperformed all the other models. The model made use of corneal curvature and astigmatism for evaluation. The neural network achieved the best results of 97% for corneal curvature and 93% for astigmatism.

In [10], the convolution neural network is used to detect keratoconus from corneal topography. The paper also stressed upon the image quality, i.e. the resolution of an image that needs to be considered during image preprocessing stage. The dataset was initially divided into three sets, one for training the neural network, one for validation, and one for testing the trained model. The trained model detects the two categories of images which are keratoconus or normal eyes. The accuracy of keratodetect on the test dataset is 99.33%.

In [11], customized neural network architecture was implemented to segment healthy and keratoconus affected eyes. The proposed algorithm was proven to be 50 times faster than the previous algorithms used by authors. Corneal thickness was generated both in healthy and keratoconus patients, and as a result, corneal thickness maps are generated. The accuracy of validation dataset is 99.56%. The authors used high-performance computing GPUs, keras, and TensorFlow for implementation of the deep convolutional neural network model.

In [12], authors used deep learning model to detect keratoconus and its stages using colour-coded maps of anterior segment optical coherence tomography (OCT). For different stage classification, Amsler Krumeich classification method is used. The research focussed on the standard six colour code maps which are anterior elevation, anterior curvature, posterior elevation, posterior curvature, total refractive power, and pachymetry map. Images were captured and authors excluded the colour scale bars in the image and fed into the deep learning model. Pytorch was used for implementation and Resnet-18 model is used as a deep learning model. Four stages of keratoconus according to Amsler Krumeich classification and normal were classified by the developed model. This work used the six colour-coded maps instead of using the numerical indices of topographic or tomographic images. According to authors, arithmetic mean without weighted averaging resulted in the finest accuracy and fact that is a limitation where there no weighted averaging is been used in the work. The six colour-coded maps were proven to be efficient not only in the detection of keratoconus but also for the detection of the grade of the disease. The arithmetic mean of six colour maps achieved an accuracy of 99.1%, and for a single map, posterior elevation achieved the highest accuracy of 99.3%. Similar research work can be found in papers [13–23].

4 Research Challenges and Gaps Identified

Machine learning techniques are successful in identifying keratoconus through supervised and semi-supervised techniques though utilizing inputs from the clinician risks biasing the framework. These models are limited by the training set it sees and are often at the risk of overfitting the dataset. The notion of applying machine learning or deep learning models to ophthalmology imaging data is a very challenging and improving research area; however, several challenges degrade its progress. The identified challenges are lack of dataset, privacy and legal issues, dedicated medical experts, etc. hence the detection of an early stage and multiple class keratoconus identification is an unexplored area of research in our country.

5 Conclusion

The prototypes developed in literature were tested for the standard set of images and their sensitivity, specificity, and accuracy. With the help of an automated framework to detect keratoconus, the laboratory costs incurred can be drastically reduced, patients travelling costs are narrowed down, and timeline of identification of the infection is also reduced by using a deep learning model. Despite many benefits, some important points have to be considered are the key indices used in keratoconus detection. Also, the understanding of necessary key features by the researchers plays an important role in developing an automated framework for keratoconus detection. To conclude, this paper provides an extensive review of deep learning and machine learning models that detect the keratoconus eye disease and the possible research gaps are also identified in this domain.

References

1. https://timesofindia.indiatimes.com/city/bengaluru/bengaluru-hospital-sees-20-rise-a-year-in-eye-disease/articleshow/69349830.cms
2. Burden of Corneal Blindness in India. https://www.ncbi.nlm.nih.gov/pmc/articles/PMC383 1688/
3. https://webeye.ophth.uiowa.edu/eyeforum/tutorials/Corneal-Imaging/Index.htm
4. Bariah MA, et al (1994) Ethnicity related to keratoconus: a study with clinical implications. Int Med J 18(3):237–240
5. Belin MW, Duncan JK (2016) Keratoconus, the ABCD grading system. Klin Monatsbl Augenheilkd
6. Yousefi S, et al (2018) Keratoconus severity identification using unsupervised machine learning. www.ncbi.nlm.nih.gov/pmc/articles/PMC6219768/
7. Hidalgo R (2016) Evaluation of a machine-learning classifier for keratoconus detection based on scheimpflug tomography. Cornea. https://www.ncbi.nlm.nih.gov/pubmed/27055215
8. Yoo TK, Ryu IH, Lee G et al (2019) Adopting machine learning to automatically identify candidate patients for corneal refractive surgery. Digit Med 2:59. https://doi.org/10.1038/s41 746-019-0135-8
9. Valdés-Mas MA (2014) A new approach based on Machine Learning for predicting corneal curvature (K1) and astigmatism in patients with keratoconus after intracorneal ring implantation. Comput Methods Programs Biomed 116(1):39–47
10. Lavric A et al (2019) KeratoDetect: keratoconus detection algorithm using convolutional neural networks. Comput Intell Neurosci
11. dos Santos VA (2019) CorneaNet: fast segmentation of cornea OCT scans of healthy and keratoconic eyes using deep learning. Biomed Opt Exp. https://www.ncbi.nlm.nih.gov/pmc/articles/PMC6377876/

12. Kamiya K et al (2019) Keratoconus detection using deep learning of colour-coded maps with anterior segment optical coherence tomography: a diagnostic accuracy study. BMJ Open 9:e031313. https://doi.org/10.1136/bmjopen-2019-031313
13. Maeda N, Klyce SD, Smolek MK, Thompson HW (1994a) Automated keratoconus screening with corneal topography analysis. Invest Ophthalmol Vis Sci 35(6):2749–2757
14. Pinero DP, Alio JL, Aleson A, Escaf Vergara M, Miranda M (2010) Corneal volume, pachymetry, and correlation of anterior and posterior corneal shape in subclinical and different stages of clinical keratoconus. J Cataract Refract Surg. 36(5):814–825. https://doi.org/10.1016/j.jcrs.2009.11.012
15. Maeda N, Klyce SD, Smolek MK, Thompson HW (1994b) Automated keratoconus screening with corneal topography analysis. Invest Ophthalmol Vis Sci 35:2749–2757
16. Twa MD, Parthasarathy S, Roberts C et al (2005) Automated decision tree classification of corneal shape. Optom Vis Sci 82(12):1038–1046
17. Chastang PJ, Borderie VM, Carvajal-Gonzalez S et al (2000) Automated keratoconus detection using the EyeSys videokeratoscope. J Cataract Refract Surg 26(5):675–683
18. Goebels S, Eppig T, Wagenpfeil S, Cayless A, Seitz B, Langenbucher A (2017) Complementary keratoconus indices based on topographical interpretation of biomechanical waveform parameters: a supplement to established keratoconus indices. Comput Math Methods Med Article ID 5293573:7
19. Platt J (1998) Fast training of support vector machines using sequential minimal optimization. In: Scholkopf C, Burges C, Smola A (eds) Advances in kernel methods: support vector learning. MIT Press, Cambridge, MA, USA
20. Lee A, Taylor P, Kalpathy-Cramer J, Tufail A (2017) Machine learning has arrived! Ophthalmology 124(12):1726–1728
21. Toutounchian F, Shanbehzadeh J, Khanlari M (2012) Detection of keratoconus and suspect keratoconus by machine vision. In: Proceedings of the international multiconference of engineers and computer scientists, Hong Kong, China
22. Belin MW, Khachikian SS (2009) An introduction to understanding elevation-based topography: how elevation data are displayed–a review. Clin Exp Ophthalmol 37(1):14–29
23. Rozema JJ, Rodriguez P, Ruiz Hidalgo I, Navarro R, Tassignon MJ, Koppen C (2017) SyntEyes KTC: higher order statistical eye model for developing keratoconus. Ophthalmic Physiol Opt 37(3):358–365

Analysis of Farm Data Using Artificial Intelligence

K. Sujatha, Manjula Sanjay Koti, and R. Supriya

Abstract The population of the world is escalating rapidly, and hence, the necessity of food production is a very serious concern these days. Agricultural productivity can be enhanced by using artificial intelligence. In recent days, many jobs have been replaced by robots, drones and expert systems, but robotics require much more artificial intelligence and advanced programming to replace all the work of farmers. Since the system will not be able to handle the situation effectively as humans, there is less possibility for the system to replace human work completely. Instead of a complete replacement, they can complement each other. The objective of this paper is to discuss some of the problems faced by the farmers, how artificial intelligence is used in the agricultural sector, need for weather predictions and proposed algorithm for irrigation.

Keywords Artificial intelligence · Agriculture · Robotics · Expert systems · Irrigation

1 Introduction

With continuous growth in world population and scarcity of land, people need to be creative enough and efficient in farming. More crop production using less land enhances the yield and productivity of the farmed acres. Across the globe, agriculture is a $5 trillion industry and these days, the agricultural sector is moving toward

K. Sujatha (✉)
Department of Computer Science, Wenzhou Kean University, Wenzhou, China
e-mail: sujatha@wku.edu.cn

M. S. Koti
Department of MCA, Sir M Visvesvaraya Institute of Technology, Bengaluru, India
e-mail: manjula.dsce@gmail.com

R. Supriya
Department of Computer Applications, Shri Shankarlal Sundarbai Shasun Jain College for Women, Chennai, India
e-mail: r.supriya@shasuncollege.edu.in

© The Author(s), under exclusive license to Springer Nature Singapore Pte Ltd. 2021
J. S. Raj et al. (eds.), *Innovative Data Communication Technologies and Application*,
Lecture Notes on Data Engineering and Communications Technologies 59,
https://doi.org/10.1007/978-981-15-9651-3_18

artificial intelligence (AI) technologies to support yield healthier crops, control pests, soil monitoring with emergent constraints and progress in a huge range of agriculture-related tasks in the entire food chain. Hence, there is huge progress made in the agricultural sector [1]. According to the UN Food and Agriculture Organization, the population will increase by 2 billion and according to FAO, the production of food should be increased up to 70% by 2050 [2]. AI and the latest technological solutions can be made use to make efficient farming and improved cultivation. AI is an algorithmic computer model which tries to mimic human behavior that can be considered to make use in farming for seeding, weeding, fertilization, irrigation and harvesting. Microsoft in collaboration with United Phosphorous (UPL) India's largest producer of agrochemicals has created Pest Risk Prediction API which will take the AI in the agricultural sector to a higher level. The pest prediction app has been initiated which helps farmers to predict the pest attack in advance and take preventive measures before the pest spoils the crop completely [3]. Some of the problems faced by farmers during cultivation and due to climatic changes are discussed. Review of various papers was done which includes discussion about using robots and drones for harvesting and picking fruits, Weeding and hoeing using slug robots. Proposed an irrigation algorithm which is used to water the field after checking the soil moisture level.

2 Literature Survey

A thorough literature survey has been done pertaining to the topic of study and nowhere in the literature has this kind of work existed. Many authors have done a study on various aspects of Agriculture.

In [4], the authors have reviewed about precision autonomous farming, where the authors have presented four current abilities of automatic agricultural service vehicles which include guidance (navigation in the field), detection (biological feature extraction), action (task execution) and mapping (agricultural field) for the design and implementation of the vehicle for precision agricultural tasks. The propagation of positioning errors was minimized by using SLAM algorithms. The robotic service unit must interact with field workers in various situations. To make this possible several robotic tools could be used like human–robot interaction, cooperative and collaborative work and control systems, etc.

In [5], those authors have discussed defining and implementing human–robot collaboration levels for target recognition in agricultural environments. Robots perform well in industrial environments if the working conditions are constant, structured and predictable. But if the environment is unstructured and unpredictable developing a robot for the agricultural environment is a difficult task. Hence, introducing Human Operator (HO) into the system can help to improve the performance. When HO-Rr (Human Operator-Robot) used in melon harvesting the HO acknowledge the robots detections and also marks the target missed by the robot, hence the performance was increased in detecting the melons [6].

In [7], the authors highlight the importance and availability of drones in agriculture for monitoring and observing the quality of crop yielding and preventing fields from damage. Currently, 85% of drones are used by the military and the remaining 15% for civilians for other applications. Drones are categorized into two types: Fixed Wing Airplanes and Rotary Motor Helicopters. Advantages of using drones include Agricultural farm analysis, Time Saving, Higher Yield, GIS mapping integration, Imaging of crop health status. Different types of drones used for various agricultural methods are Honeycomb AgDrone system, DJI Matrice 100, EBEE SQ-Sensefly, Lancaster 5 precision hank and SOLO AGCO edition.

In [8], the author has discussed agricultural robots in hoeing the fields in rows. In agriculture using chemical herbicides and insecticides has taken cultivation to a higher level. But due to the increase in the cost of herbicides many minority crops like sugar beet or cauliflower is not worthwhile to develop suitable herbicides. Moreover, certain crops need to be hoed between the rows routinely. Hence, some form of automation which includes accurate, fast, robust and able to work for long duration required for hoeing. Silsoe have developed vision-guided hoe for cereals and sugar beet.

In [9], the authors highlight the wireless path following system in the irrigation process of land. Sensors were used to detect the soil moisture level of the land. Here sensors were used and connected to the Arduino board to control all the self-calibrating motion of robot using an algorithm. In this irrigation algorithm robot is placed in the starting point of the field which checks the soil moisture at every third step if the water is required, using call function the water dispenses through motor connected to pump pours water.

In [10], the authors have discussed that using big data in smart agriculture will improve many dimensions of agriculture, viz. crop, livestock, fish and agribusiness. The collection of big data in agriculture can be done by drones or aerial vehicles using cameras. AgriPrice Book is an application initiated by North America Strategic Institute in 2014 for helping farmers in enhancing transparent supply chain in agriculture and CGIAR developed Citizen Science for promoting climatic changes and food security management.

3   Problems Faced by Farmers

Some of the problems faced by farmers are (i) Degradation of soil quality (ii) Manures, Fertilizers and Biocides (iii) Irrigation (iv) Lack of mechanization (v) Soil Erosion (vi) Agricultural marketing (vii) Inadequate storage facilities (viii) Inadequate transport and (ix) Scarcity of capital (x) Transportation.

All these problems could be rectified by using different methods and applications of artificial intelligence. For example; Degradation of soil quality could be caused by using a large amount of fertilizers and this happens if the cause of the problem is not analyzed effectively. To overcome this particular problem, AI assists various apps, which is developed to check the actual cause of the problem and for checking the quality of soil and treating the soil by using natural fertilizers instead of chemical fertilizers.

4 Artificial Intelligence in the Agricultural Industry

AI plays a major role in the agricultural industry which is applied in almost all the methods of farming such as harvesting, weeding, hoeing, and watering.AI in agriculture can be divided into 3 major categories.

4.1 Robots Used for Agricultural Task

Agricultural Robots need to travel in a challenging and dynamic environment. Current agricultural—robots combine the existing platforms (e.g., autonomous tractors) with other technologies (e.g., drones). The weight and locomotion of robots is an important consideration when it comes to ground and crops. Based on the user's requirements and task such as heavy or light crop harvesting different platforms have been used (e.g., multiple wheeled robots). Many companies are using robots for the agricultural task which is done in high volume. Some of the robots are mentioned below.

Slugbot. This robot is used in agriculture for identifying and disposing of the slugs which cause damage to the leaves of crops like lettuce and tomatoes. Slugbot uses image sensors to detect the slugs. Ultrasound sonar and touch sensors are used to navigate and avoid the obstacles [11].

Legged Robot. It can move sideways which is essential in uneven fieldwork, it works by minimizing its footprint and maximizing the flexibility [12]. Figure 1 represents Legged Robot.

Fig. 1 Legged robot

Fig. 2 Crop weeding robot

Crop Weeding Robots. This robot reduces the need for herbicides by using camera guided hoes, precision sprayers or lasers to manage weeds [13]. Figure 2 represents Crop Weeding Robot.

4.2 Monitoring Crop and Soil

Soil is the basic requirement for agriculture. But Deforestation and degradation of soil quality is the threat to food security and a negative impact on the economy. To overcome this problem Berlin-based agricultural tech startup PEAT, have developed a deep learning app called Plantix.

Crop Monitoring App (Plantix). This app is used to identify potential defects and nutrient deficiency in the soil through pattern detection method and checks the quality of the crop. The image recognition app identifies various possible defects through the image captured by the user's smartphone camera [14]. This app helps the farmers to identify the quality of crops and improve productivity by reducing the defects seen in crops.

Crop Planting and Harvesting Drones. Drones also are known as UAV (Unmanned Aerial Vehicles) are used in agriculture to help farmers from planting the crops till harvesting. Drones handle the crops carefully which helps to reduce the damage caused to crops during planting and harvesting, this will help to increase the crop yield compared to manual farming. Some of the drones are EBEE SQ-SenseFly, DJI Matrice 100, SOLO AGCO Edition and Agras MG-1- DJI, etc.[15]. Figure 3 represents Planting and Harvesting Drone.

Crop Health and Harvest Size. If the drones are equipped with a camera and other required equipment they can help the farmers by monitoring the health of plants like checking the temperature, chlorophyll, thickness of leaves and various pest

Fig. 3 Planting and Harvesting Drone

vulnerabilities. All these preventive measures will not only grow the crops healthy it will also increase the size of harvest [16].

Supplying Crop to Market. Applying various methods to improve the crop quality will help the farmers to yield productivity. As the production increases, it is necessary to supply the crop to market on time. If the supply is not quick enough to send to the market there will be a delay in distribution which may spoil the crop and leads to financial loss. Hence, equal importance should be given for the production of quality crops and fast supply to the market. Crowed Sourced is the name given for big data in agriculture [10] which can be used to store the information required for the supply of crops to market based on production date.

Soil Monitoring app. There are different apps available to monitor the soil-related task. Soil quality monitors an IoT application is used to check the soil quality with the help of few materials like moisture sensor, 12v pump, TIP120 Transistor, Breadboard and an iOS device.

Soil Moisture Sensor. These types of apps have a database of different types of soils and plants. It works with the help of smart soil sensors. Some of the apps are Planlink, Mist, ODO, Koubachi Wifi plant sensor, etc., used to analyze the soil and will water the plants when necessary [17].

Soil Quality Analyzer. The quality of soil could be analyzed using various mobile apps which include SQAPP a project of iSQAPER. The soil ph value will be evaluated by comparing the soil quality indicators provided in the app [18].

4.3 Weather Predictions

Climatic changes are the major unpredictable and changing problem faced by farmers. During heavy rain the crops may be spoiled due to stagnant water and crops may be dried due to heavy summer.

aWhere. A company which uses a machine-learning algorithm along with satellite which predicts weather changes, the sustainability of crops and identifies the presence of pest and other diseases in plants [19].

5 Proposed System

Most of the farmers have huge hectares of agricultural land, and hence, they face a lot of problems to perform the cultivation. Problems such as (i) Requires a large number of labors to do cultivation (ii) Takes time to reach from one part of the land to another (iii) Water wastage if not used sparingly (iv) They may not be able to look after all the crops properly and (v) Climatic changes may damage all the crops and cause financial loss to the farmers.AI plays an important role to overcome all these kind of problems by using existing methods like using robots and drones. In this paper, methods are proposed to automate the irrigation process.

5.1 Irrigation Problem

If the land is geographically large, farmers may cultivate different varieties of crops in each field, which requires a varying amount of water for irrigation. So it takes a lot of time for the farmers to check each and every field and supply the water.

5.2 Solutions for Irrigation Problem

Step 1. Divide the fields based on the crop variety.
Step 2. Using image recognition software gets the field to co-ordinate values (along x and y-axis) for each division. By doing this the system can differentiate the fields based on their coordinate values.
Step 3. Soil moisture sensor will identify the level of water requirement of various fields. If water required control will move to the sprinkler, If not require then moves to the next field
Step 4. Water requirement report will be generated by the sensor and it will be intimated to the plant irrigation water sprinkler robot.
Step 5. The data collected in steps 3 and 4 will be stored in a database for farmer's reference (Fig. 4).

The farmers are aware of the quantity of water supply to various kinds of crops. But to investigate all the fields' individually they require quite a lot of time. Another possibility is to use labors; again it may be a waste of money to appoint a large

Fig. 4 Depicts the solution
that could be provided for
irrigation

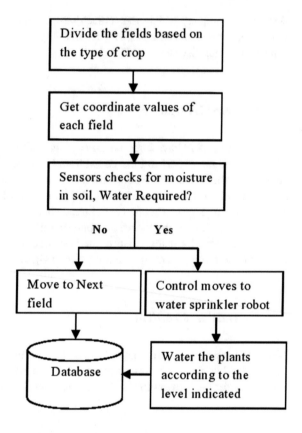

number of labors for irrigation. By making irrigation automatic the farmers will not
only avoid water wastage, but they will also be benefited economically.

6   Conclusion

Artificial intelligence could be used much effectively in the field of agriculture by
innovative ideas and also by combining the existing methods. After a few decades,
the growth in the population may be very high and the demand for food reaches the
highest peak, and hence, it is appropriate to use the latest technologies. By using
artificial intelligence, farmers will utilize the tools so that they get most from every
acre. An agricultural revolution could be driven by artificial intelligence, where the
world produces sufficient food by using fewer resources. The problems faced by
farmers could be rectified by using different methods and applications of artificial
intelligence. Most of the farmers have huge hectares of agricultural land, and hence,
they face a lot of problems to perform the cultivation. Hence, a method is proposed

to solve the irrigation problem. Artificial intelligence solutions enable farmers to improve the quality of crops and ensure faster supply of these crops to markets.

References

1. Wolfert S, Ge L, Verdouw C, Bogaardt M-J (2017) Big data in smart farming–a review. Agric Syst 153:69–80
2. FAO (2017) E-agriculture in action. Rome, Italy
3. https://krishijagran.com/news/artificial-intelligence-for-success-in-agriculture-ram-nath-kov ind/
4. Cheein, FAA, Carelli R (2013) Agricultural robotics: U manned robotic service units in agricultural tasks. IEEE Ind Electron Mag 7(3):48–58
5. Bechar A, Edan Y (2003) Human-robot collaboration for improved target re ognition of agricultural robots. Ind Rob Int J 30(5):432–436
6. Bechar A, Edan Y, Meyer J, Rotman M, Friedman L (2000) Human-machine coll boration for melons detection. In: Human-machine collaboration for melons dete tion 1–7
7. Puri V, Nayyar A, Raja L (2017) Agriculture drones: a modern brea through in precision agriculture. J Stat Manag Syst 20(4):507–518
8. Hollingum J (1999) Robots in agriculture. Ind Rob Int J 26(6):438–446
9. Alam I, Antony James R, Padmanabhan V, Sunny A (2018) A low cost automated irrigation system with soil moisture sensor. Int Res J 5:4146–4148
10. Sarker, MNI, Wu M, Chanthamith B, Yusufzada S, Li D, Zhang J (2019) Big data driven smart agriculture: pathway for sustainable development. In: 2019 2nd international conference on artificial intelligence and big data (ICAIBD), pp 60–65. IEEE
11. https://www.wired.com/2001/10/slugbot-enemy-of-slugs/
12. Kajita S, Espiau B (2008) Legged robots. In: Springer handbook of robo ics, pp 361–389
13. Fennimore SA, Cutulle M (2019) Robotic weeders can improve weed control options for specialty crops. Pest Manag Sci 75(7):1767–1774
14. https://plantix.net/en/
15. Doknić V (2014) Internet of things greenhouse monitoring and automation system. In: Internet of things: smart devices, processes, services
16. https://emerj.com/ai-sector-overviews/ai-agriculture-present-applications-impact/
17. Chandra A (2019) Diagnosing the health of a plant in a click. In: Research into design for a connected world. Springer, Singapore, pp 593–601
18. Reinecke M, Prinsloo T (2017) The influence of drone monitoring on crop health and harvest size. In: 2017 1st international conference on next gener tion computing applications (NextComp). IEEE, pp 5–10
19. https://www.awhere.com/

Enhanced Compression Model for Brain MRI Images Using Genetic Algorithm and Clustering Techniques

M. Thilagam and K. Arunesh

Abstract In general, analyzing the information present in biomedical images is a very challenging task in the wireless communication domain. Image retrieval is one of the most emerging technologies for telemedicine to handle medical data present in the images (MRI and CT). The core intention of this process is to utilize a genetic algorithm (GA) and clustering algorithm in order to perform the transformed computing of vector quantizer (VQ) design in biomedical image compression such as brain MRI images. In this proposed work, the genetic algorithm and clustering is applied to the optimal design of codebook generation in VQ, where a momentous improvement is observed in the decompression performance. The proposed method uses PSNR for evaluation method as fitness function in the system. The proposed method has used a combined scheme that makes complete use of the stochastic searching capability in GA and the computational complexity reduction in clustering to compute the codebook.

Keywords Vector quantization · Genetic algorithm · Clustering · Image retrieval · Optimization

1 Introduction

The process of sinking the data present in the digital image without data loss is called image compression. This technique can be broadly bifurcated, namely lossless compression and lossy compression [1]. Lossless compression aims at removing redundancies from the source images and makes sure that the relevant information will be maintained in the compressed images. In the case of lossy compression, much of the relevant information can be lost while compressing the images. Lossless compression provides a very low compression ratio in the range of 2–10, while lossy compression provides a better compression ratio up to 30 [2].

M. Thilagam (✉) · K. Arunesh
Department of Computer Science, Sri S Ramasamy Naidu Memorial College, Sattur, Virudhunagar, India
e-mail: thilagamresearch@gmail.com

© The Author(s), under exclusive license to Springer Nature Singapore Pte Ltd. 2021
J. S. Raj et al. (eds.), *Innovative Data Communication Technologies and Application*,
Lecture Notes on Data Engineering and Communications Technologies 59,
https://doi.org/10.1007/978-981-15-9651-3_19

Medical image compression is the technique used to compress the biomedical images [3]. Medical image compression aims to safeguard important information in the image and remove the less important features in order to reduce the volume of the image. The ultimate aim of compression methods is to reduce the number of binary space required to save and transmit the image data without any considerable loss of useful information [4]. Medical image compression can be carried out using both lossless and lossy compression, but lossless compression is preferred as it preserves the relevant information. While using lossy image compression techniques along with some algorithms or techniques may overcome the drawbacks of lossy compression to a great extent [5].

2 Literature Review

Jau-Ji et al. [6] present the image compression based on vector quantization technique. It can be seen from the experimental results that, although additional computation and measurements are needed to be done, the quality of the compressed images can be improvised in a better way.

Yerva et al. [7] proposed a simple algorithm for compression known as lossless image folding technique by providing improved quality of the compressed image. For prediction, the property of contiguous neighbor redundancy is being used, where the image size reduction is carried out to a smaller value, which is pre-defined using column and row folding. On comparison with the traditional lossless image compression mechanisms, a better performance is obtained.

A technique called neural networks is deployed for compression process in Sahami et al. [8] for comparing the commonly used lossy image compression methods, where it can be shown that this method has several advantages over the later. The compression ratio is high, and the noise factors affecting the image are also reduced. For further modifications, guided scales and some complicated calculations can be used to increase its flexibility.

A new international image encoding technique is used for encrypting the peak sized images in a useful and safe manner thereby simplifying the process of conversion to other image formats. The obtained sectored image files are further integrated using the Huffman algorithm, and the entire sectored image is integrated to compress into a single image. Finally, an advantageous way is used for transferring the encoded images to multipath routing techniques to build a secure transmission.

3 Proposed Methodology

In the proposed methodology, it combines and makes the merits of GA and clustering for the well-organized preparation of codebooks with ideal image compression

Fig. 1 Proposed
implementation steps

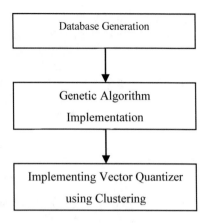

routine. The overall design of the proposed system is divided into three sections as shown in Fig. 1.

3.1 Database Generation

Database generation is simply performed by using the biomedical images like brain MRI images that tend to be used in this proposed research work. The ultimate goal of compressing and analyzing a medical image is to abstract the useful information for achieving a better interpretation of the image to make an important description to the end user [9]. The image is analyzed based on the following parameters.

(1) Mean Square Error (MSE)—Peak-to-noise signal ratio (PSNR) is defined as the ratio of the maximum possible power of the signal, and the power of noise signal alters the integrity of its representation. It is used to measure the reconstruction quality of the compressed image. PSNR is measured with the help of MSE [10]. Further, MSE is defined as,

$$\mathrm{MSE} = \frac{1}{MN} \sum_{j=1}^{M} \sum_{K=1}^{N} \left(x_{j,k} - x'_{j,k} \right)^2$$

where $m * n =$ Size if the grayscale image;
$K =$ Image approximation.
The PSNR (in DB) is measured as,

$$\mathrm{PSNR} = 10\log \frac{(2^n - 1)}{\mathrm{MSE}} = 10\log \frac{255^2}{\mathrm{MSE}}$$

(2) Image Fidelity—Image fidelity measures how much the compressed image approximates to the original image in terms of real source distribution [8]. This mainly depends on the thermal noise, but other factors such as imaging artifacts amplitude and phase will also affect the measure of image fidelity.

(3) Average Difference (AD)—Average difference is simply the difference between the original (reference) image [11], i.e., $x(i, j)$, and the compressed (test) image, i.e., $y(i, j)$, and is defined as,

$$AD = \sum_{J=1}^{M} \sum_{K=1}^{N} (Xj, k - X'j, k)/MN$$

(4) Structural Content (SC)—It measures the similarity between the original (reference) image [7], i.e., $x(i, j)$, and the compressed (test) image, i.e., $y(i, j)$, and is derived as,

$$SC = \sum_{i=1}^{M} \sum_{j=1}^{N} ((y(i, j))^2) / \sum_{i=1}^{M} \sum_{j=1}^{N} ((x(i, j))^2)$$

(5) Maximum Difference (MD)—As the name suggests, the MD is the maximum dissimilarity between the original (reference) image [12], i.e., $x(i, j)$, and the compressed (test) image, i.e., $y(i, j)$, and is defined as,

$$MD = MAX|x(i, j) - y(i, j)|$$

For the evaluation of medical image compression techniques, several standards are used. Table 1 shows some of these main image compression standards. Figure 2 shows a few examples of the brain MRI medical images [13].

Brain MRI Representation

See Fig. 2.

Table 1 Image compression standards

Modality	Image dimension	Gray level (bits)	Avg. no. of images per evaluation	Avg. MBs per evaluation
Brain MRI	512 × 512	12	30	2.0
	236 × 212	12	30	6.5
	128 × 128	8	50	0.5

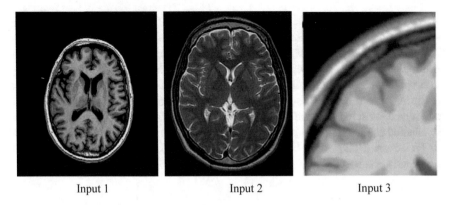

| Input 1 | Input 2 | Input 3 |

Fig. 2 Input 1, Input 2, Input 3

3.2 Genetic Algorithm Implementation

The advantage of genetic algorithm (GA) is its ability to acquire the universal ace elucidation; hence, it is an effective technique that could be used in codebook generation. It is a stochastic search algorithm having many advantages over the traditional single point search algorithms. These advantages include gradient-free and parallel optimization to achieve the global optimum solution.

Figure 3 explains the GA cycle. Initially, GA starts with the initialization of the population which is the randomly generated blocks where every block contains a

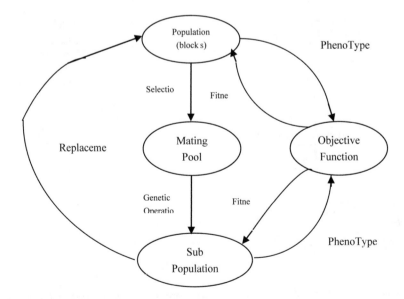

Fig. 3 Genetic algorithm cycle

solution to the raised problem. Then the evaluation of the blocks is carried out by some fitness function (we are using PSNR in the proposed system), and a new population is formed from a collection process based on the evaluated fitness values. In each generation, a loop from one generation to another, the crossover and mutation operations keep upgrading the individuals to form a new population known as offspring. The selection process again selects the blocks to form a new population and repeat the same until the required criterion is satisfied. The best solution (block) is selected as the optimal solution to be added in the codebook.

Flowchart 1 shows the initial phase and GA phase to be implemented in the proposed system.

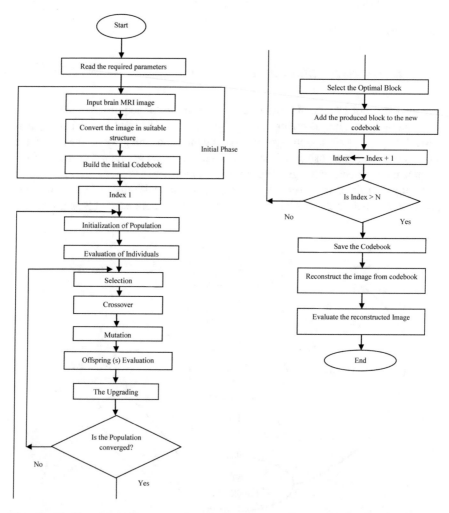

Flowchart 1 Flowchart of the proposed system for image compression using genetic algorithm

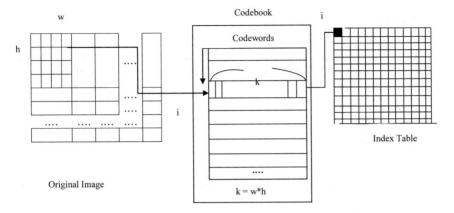

Fig. 5 Vector quantization process

3.3 *Implementing Vector Quantization Using Clustering*

As the genetic algorithm is based on lossy image compression technique [7], the quality of the decompressed image is degraded to some extent. Also, vector quantization cannot be applied alone as a lossy image compression technique. Therefore, we implement vector quantization using clustering and apply it to the image compression using genetic algorithm. In vector quantization, some prototype images are selected as training vectors for generating the codebook. Each training vector has a standard dimension k. Vector quantization is a drafting Q of a k-dimension Euclidean space $<k$ into a finite subset C of $<k$, where

(1) Codebook $C = \{c_1, \ldots, c_N\}$
(2) N = size of codebook
(3) Codeword in C, $c_i = (c_{i1}, \ldots, c_{ik})$
(4) K = dimension of C (Fig. 5).

The quantizer design is a function of codebook C and subspaces $s = \{s_1, s_2, \ldots, s_n\}$. The mapping function is given by,

$$Q(X) = c_i \,(\text{if } X \in s_i)$$

For encoding an image, the brain MRI image is spliced into sets of blocks (vectors) and appropriate codebook ci is assigned to the vectors $X = \{x_1, x_2, \ldots, x_n\}$ such that the distortion between c_i and X is minimized. Now, the sets of vectors are formed using the principle of clustering. The similar vectors are placed in similar clusters while the dissimilar clusters are placed in other clusters. An important concept of clustering is the distance measure between data points (vectors). Euclidean distance can be useful to group the vectors in similar groups if the integral of the data instance vectors are all in the same physical units.

4 Results and Discussions

In this research work, we compressed two images using the proposed system in MATLAB 2013. Time required for compression and decompression of images was very low in both cases. The recital of the system is measured in terms of PSNR. Figures 6 and 7 show the results,

(a) Brain MRI image Input 1
   Size of the image $= 512 * 512$.

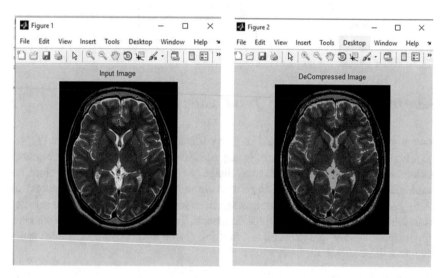

Fig. 6 Results for brain MRI image Input 1

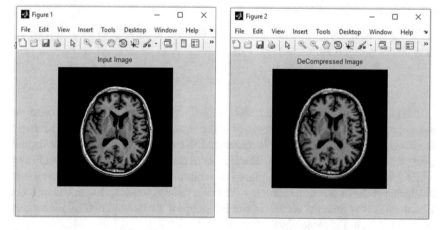

Fig. 7 Results for brain MRI image Input 2

Table 2 Comparison of MSE and compression ratio for different codebook sizes for brain MRI image using proposed vector quantizer design

Codebook size	32	64	128	256
MSE	8.3433e+006	8.1923e+006	7.9255e+006	6.8495e+006
Compression ratio	7.4862	5.2000	4.120	3.5620

    PSNR $= 29.4279$ dB.
    Time required for compression and decompression $= 1.45$ s.

(b) Brain MRI image Input 2
    Size of the image $= 236 * 212$.
    PSNR $= 22.5484$ dB.
    Time required for compression and decompression $= 0.26$ s.

The compression ratio depends upon the codebook size we are using to reconstruct the image. The greater codebook size is used; greater is the number of bits used to represent the block and lesser is the MSE. Therefore, the compression ratio decreases as the codebook size is increased. Initially, the codebook size for VQ is 256. By using genetic algorithm, the codebook size is reduced to 128, and by applying the genetic algorithm and clustering on VQ, the codebook size is reduced up to 64, 32 and so on. The comparison of MSE and compression ratio for different codebook sizes for MRI image using proposed vector quantizer design are showed in the following Table 2.

Thus, by using proposed vector quantizer design, the compression ratio can be obtained between 3.5 and 8.7, where only a negligible amount of data will be lost. Though a higher compression ratio can be achieved by using differential matrix for computing the codebook. As a result, the MSE will be reduced to some extent. But the compression ratio is also compromised by a negligible margin. The following table shows the comparison of vector quantization, genetic algorithm and the proposed algorithm (Table 3).

5 Conclusion

The VQ-based image compression is time-consuming, and the computational complexity is also increasing at an unprecedented rate. The proposed algorithm reduces the computational complexity, as well as the image compression time is comparatively less. Also, the compression is higher when compared only to VQ-based compression. The genetic algorithm helps in the better reconstruction of the image, while clustering reduces the computational complexity of the entire system. This paper proposes that the algorithm can be combined with some lossless image compression technique to enhance the quality of the reconstructed image. The proposed algorithm can also be applied to color images for compression. Various other mechanisms can be clubbed with the proposed algorithm to form a versatile image compression system.

Table 3 Comparison analysis of vector quantization, genetic algorithm and the proposed algorithm

Parameters	Vector quantization [14]	Genetic algorithm applied on Vector quantization [6]	Proposed vector quantizer design using genetic algorithm and clustering
Codebook size	It designs very large super codebook	It designs a small codebook as compared to VQ	This algorithm designs codebook with the smaller size as compared to the genetic algorithm applied on VQ
Storage space	It needs more storage space for the codebook	It needs less storage space for the codebook	It requires less storage space as compared to genetic algorithm applied on VQ
Computational complexity	A complete design requires a large number of computation	This design requires less number of computation	This algorithm reduces the computations dramatically
Convergence time	Convergence time is very large	Convergence time is less than the VQ algorithm	Convergence time is less than that of the genetic algorithm applied on VQ

References

1. Rani MLP, Rao GS, Rao BP (2020) An efficient codebook generation using firefly algorithm for optimum medical image compression. J Ambient Intell Hum Comput. https://doi.org/10.1007/s12652-020-01782-w
2. Kala R, Deepa P (2019) Adaptive fuzzy hexagonal bilateral filter for brain MRI denoising. Multimed Tools Appl. https://doi.org/10.1007/s11042-019-7459-x
3. Zhao M, Yin X, Yue H (2017) Genetic simulated annealing-based kernel vector quantization algorithm. Int J Pattern Recogn Artif Intell
4. Sheejakumari V, Sankara Gomathi B (2015) MRI brain images healthy and pathological tissues classification with the aid of improved particle swarm optimization and neural network. Comput Math Methods Med 1–12
5. Omari (2015) Image compression based on genetic algorithm optimization. Web applications and networking (WSWAN). In: 2nd world symposium, pp1–5
6. Shen J-J, Huang H-C (2010) An adaptive image compression method based on vector quantization. IEEE, pp 377–381
7. Yerva S, Nair S, Kutty K (2011) Lossless image compression based on data folding. IEEE, pp 999–1004
8. Sahami S, Shayesteh MG (2012) Bi-level image compression technique using neural networks. IET Image Process 6(5):496–506
9. Mittal M, Lamba R (2013) Image compression using vector quantization algorithms: a review. Int J Adv Res Comput Sci Softw Eng 3(6):354–358
10. Sanyal N, Chatterjee A, Munshi S (2013) Modified bacterial foraging optimization technique for vector quantization-based image compression. In: Computational intelligence in image processing. Springer, Berlin, pp 131–152
11. Miller C, Moore F, Babb B, Peterson MR (2011) Improved reconstruction of quantized CT scans via genetic algorithms. In: 2011 IEEE congress evolutionary computation (CEC), pp 2293–2299

12. Liu S-H, Hou H-F (2009) A combination of mixture genetic algorithm and fuzzy C-means clustering algorithm. In: IEEE international symposium IT in medicine and education. ITIME '09, pp 254–258. https://doi.org/10.1109/ITIME.2009.5236422
13. https://openfmri.org/. Last accessed on Jan 2020
14. Teja SP, Modi R (2013) Image compression using DWT and vector quantization international. J Innov Res Comput Commun Eng 1(3):651–659

RETRACTED CHAPTER:
A Cluster-Based Distributed Cooperative Spectrum Sensing Techniques in Cognitive Radio

N. Shwetha, N. Gangadhar, L. Niranjan, and Shivaputra

Abstract The optimization techniques are used in various fields to save transmit power with the shortest distances and achieve energy-efficient clusters while restricting interference to primary users. Spectrum sensing among the multi-users grouping in cognitive radio (CR) systems using Distributed Swarm Intelligent-Based Clustering (DSIBC) shows superiority in power saving, convergence time, and sensing error. The proposed DSIBC algorithm is used to develop energy-efficient distributed cluster-based sensing with an optimal number of clusters on their connectivity. In this work, the primary users (PUs) and secondary users (SUs) with random

The original version of this chapter was retracted: The Volume Editors and authors retract this conference paper [1] because it has substantial overlap with a PhD thesis by a different author [2]. Shwetha, N. Gangadhar, and Shivaputra agree to this retraction; L. Niranjan has not responded to any correspondence from the publisher about this retraction.

[1] Shwetha N., Gangadhar N., Niranjan L., Shivaputra (2021) A Cluster-Based Distributed Cooperative Spectrum Sensing Techniques in Cognitive Radio. In: Raj J.S., Iliyasu A.M., Bestak R., Baig Z.A. (eds) Innovative Data Communication Technologies and Application. Lecture Notes on Data Engineering and Communications Technologies, vol 59. Springer, Singapore. https://doi.org/10.1007/978-981-15-9651-3_20

[2] Babu G., (2019) Analysis and Evaluation of Distributed Cooperative Spectrum Sensing Techniques in Cognitive Radio Sensor Networks, St. Peter s Institute of Higher Education and Research; 2019 http://hdl.handle.net/10603/280064

The retraction note to this chapter is available at https://doi.org/10.1007/978-981-15-9651-3_72

N. Shwetha (✉) · Shivaputra
Department of ECE, Dr. Ambedkar Institute of Technology, Bangalore, Karnataka 560056, India
e-mail: shwethaec48@gmail.com

Shivaputra
e-mail: putrauvce@gmail.com

N. Gangadhar
Department of ME, Dr. Ambedkar Institute of Technology, Bangalore, Karnataka 560056, India
e-mail: ngangadhar4@gmail.com

L. Niranjan
Department of ECE, R R Institute of Technology, Bangalore, Karnataka 560090, India
e-mail: niranjanl1983@gmail.com

© The Author(s), under exclusive license to Springer Nature Singapore Pte Ltd. 2021, corrected publication 2021
J. S. Raj et al. (eds.), *Innovative Data Communication Technologies and Application*, Lecture Notes on Data Engineering and Communications Technologies 59, https://doi.org/10.1007/978-981-15-9651-3_20

waypoint mobility are considered for implementation. DSIBC has increased the speed of convergence by grouping among multi-users clustered communication than other optimization techniques. The results proved that the reduction in average node power is superior by 9.646% compared to existing in primary nodes. Similarly, DSIBC is superior by 24.231% in SUs average node power. In DSIBC, the detection performance is superior to the existing method. For small signal-to-noise ratio (SNR) <2 dB, the probability of detection is high. In primary detection, the proposed DSIBC is yielding a low false alarm rate compared to other optimized techniques. It is used to solve the problem of multimodal optimization and compare performance with various optimization methods to maximize network bandwidth.

Keywords Cognitive radio (CR) · Distributed Swarm Intelligent-Based Clustering (DSIBC) · Energy-efficient distributed cluster-based sensing · Multimodal optimization

1 Introduction

Accurate spectrum sensing in cognitive radio (CR) is an essential task in modern communication scenarios [1]. Different approaches are already implemented for the best identification of spectrum holes. Cooperative sensing among all secondary users helps to acquire accurate sensing information with shortening sensing time and maximizes the system reliability [2]. Each secondary user shares the locally sensed information to other secondary user without any centralized control unit [3]. In distributed cooperative sensing, all cognitive users attempt to find out the unused primary channel dependently. Hence, this will reduce the burden on data transmission and the time delay to make the final decision on the availability of the primary signal [4]. There are clustering techniques already studied in other works such as Groupwise Constrained Agglomerative Clustering, k-neighborhood clustering, k-means clustering, and Distributed Spectrum-Aware Clustering (DSAC) [5, 6]. Fine-tuning of clustering is made by particle swarm optimization (PSO), firefly algorithm (FA), and jumper firefly algorithm (JFA). In [7], the authors discussed the best energy-efficient protocol for the hierarchy of low-energy adaptive cluster (LEACH) to reduce energy consumption, and it can extend the life of wireless sensor networks (WSNs). Clustering procedures can be used to interconnect with cluster head and base station. If the sink station is away from the cluster head (CH), energy consumption will be raised, and it can diminish the lifetime of WSN. To overcome these, the PSO strategy is actualized with this protocol to achieve the most astounding lifetime of WSN. PSO is used to augment the adaptable and energy efficiency. It is not difficult to complete, and the change estimation rate is to a great degree rapidly. PSO technique is used to improve the lifetime performance of the network. Using optimization techniques, the clusters and cluster head selection based on energy is generated. After this whole process, data transfer begins for this node on the shortest path. Wang et al. [8] discussed that clustering using the firefly technique can be divided into two types: One is hierarchical, and the other is partitioned. In the hierarchical approach, a large number of hierarchical clusters are divided into a small number of clusters with a nearby centroid. It has two methods: The first is the agglomeration method, consisting of two or more small clusters combined into a large cluster, and the second is the division method, which divides a longer cluster into two or more smaller clusters. Sectional clustering attempts to separate a set of disjoint clusters from a data set without creating a hierarchical structure. The prototype-based partitioned clustering is creating cluster centers, and further, it is used to classify the data set. In [9], the authors discussed JFA at the base station instead of FA. Among the population in every living creature, there is diversity in quality and fitness. In general, low-level members with low quality are not able to reach high-quality achievements.

Each population quality is estimated with respective members and quality probability situation to obtain the eligibility. To avoid that problem, the JFA is developed to improve appropriate solutions by making the changes to eligible situations and find the optimal solution by a status table. From the status table, it is observed that all the current situation records help to change the new suitable situation by the jumping process. This process executes search agents (fireflies) to jump the option to make the decision process. In the status table, every firefly location is situated in a particular search space at the ith stage, and fitness maintains solution quality at the ith stage by the fireflies. Every firefly's worst solution is attained by each firefly at searching phases. After the search process, the cost of each firefly qualification is investigated from the status table. The above various optimized clustering problem is solved in the present work. To propose a new optimization algorithm that will increase the lifetime of cognitive radio sensor networks by forming energy-efficient clusters. This work mainly focuses on cooperative sensing among all subordinate users. The main objective of the paper to acquire accurate sensing information with shortening sensing time maximizes the system reliability, reductions in the number of wrong panics, and increases the detection amount. In this paper, different heuristic optimization algorithms such as Distributed Swarm Optimized Clustering (DSOC), Distributed Firefly Optimized Clustering (DFOC), Distributed Jumper Firefly Optimized Clustering (DJFOC), and Distributed Swarm Intelligent-Based Clustering (DSIBC) techniques are simulated and performances are compared.

In the first method of DSOC proposed, every group of clustering node moves randomly toward their best swarm particle having the least neighborhood distance [10]. Each particle with its best position and speed is evaluated by the objective function until an optimal global best position is achieved. The DSOC convergence rate is similar to the genetic algorithm (GA). Since DSOC is having the drawback of low convergence in refined search space and weak local searchability. The second proposed method of Distributed Firefly Optimized Clustering (DFOC) is adopted for the best grouping of nodes [11]. All cognitive nodes move to a brighter firefly with random speed, forming an organized cluster with the least computation time. In DFOC, fireflies at critical positions disappear while doing the clustering without using the status table, and it cannot memorize any history of the past positions. To overcome this problem, the third proposed method of the Distributed Jumper Firefly Optimized Clustering (DJFOC) technique is presented [11]. DFOC and DJFOC are nonlinear optimization tools based on the random attractiveness of firefly intensity behavior.

DJFOC is used to collect the whole situation in the current records and support to change the new appropriate situation by the status table. This work shows how to employ a firefly algorithm to implement dynamic spectrum access (DSA) using energy-efficient cooperative distributive algorithms. DJFOC effectively improves dynamic spectrum access for PU and SU than DFOC. The convergence rate of DJFOC is better than the DSOC and DFOC. The DJFOC is having an optimal number of cluster communication and a high probability of detection.

Since DJFOC cannot be similar to all the portions that determine the next movement of a firefly. The final proposed method of the Distributed Swarm Intelligent-Based Clustering (DSIBC) technique is mainly aimed to reduce the average sensing time of the primary users by cooperative secondary users. It does not require light intensity using the objective function, and the whole population is not automatically subdivided into sub-swarms. The results proved that with the inter user channel conditions, the reduction in average node power of the primary and secondary user nodes compared to the existing DSAC method. The performance comparison is analyzed for various proposed optimized clustering techniques such as DSOC, DFOC, DJFOC, and DSIBC. The following performance metrics are considered for comparison such as (i) average converge time for various CRSN sizes, (ii) average node power for different cluster numbers, (iii) average node power for PUs and SUs, and (iv) detection performance for spectrum sensing scheme. The rest of this document is organized as follows: Sect. 2 discusses Distributed Spectrum-Aware Clustering (DSAC) in cognitive radio network and issues. Section 3 provides a description of proposed dynamic channel allocation in the clustering model based Distributed Swarm Intelligence-Based Clustering (DSIBC) algorithm and reduction of cluster communication power. The simulation results and discussion are presented in Sect. 4, and finally, the document ends in Sect. 5.

2   Existing Distributed Spectrum Aware Clustering (DSAC) in Cognitive Radio Networks and Issues

DSAC performs the spectrum measurement process at three different stages: channel determination, beacon, and coordination. Each secondary user checks the free available channels in the environment and compares them with previously detected results in the case of channel detection. In beaconing, according to the channel sensing results, SU nodes beacons each node information to all the nodes in the available vacant channel. If any dynamic activity in the PU is detected in the environment, either the node assigns a new cluster through the beacon of the new cluster identifier, or the node remains with the same cluster. Once the node beaconing is complete, it revises cluster observed information such as the size of the cluster, cluster head (CH), and commonly available channels. In the intra-coordinate clustering, each node in the cluster first evaluates neighboring beacon signal strength and formerly coordinates the pairwise distances. From that, CH determines the inter-cluster distance of complete link rule. In the inter-cluster coordination, each CH transmits complete merge invitations to the shortest neighboring cluster in their maximum transmission range. If any of the two different clusters receive merge requisition, either it will form a cluster by cluster new ID, selects commonly available channels and new CH with maximum energy, or the cluster selects a new CH selected by the cluster with the same topology. If secondary users or PU changes its position, the re-clustering process occurs in DSAC. Consequently, the network offers less sustainable stability and requires more management overhead. During re-clustering, the position of the remaining nodes and the clustering structure remain unchanged. When any PU's status is changed, only 3 out of 50 SU nodes are affected [6]. In this case, the network once again converges the stability by two merges.

3 Proposed Clustering Model for Dynamic Channel Allocation in Cognitive Radio Networks

In Fig. 1, a novel Distributed Swarm Intelligence-Based Clustering (DSIBC) is proposed to form energy-efficient clusters while restricting interference to PU and save transmit power with shortest distances. Cooperative SU are also called cognitive radio sensor nodes (CRSNs). The number of CRSNs is divided into 'K' number of clustered structures based on connectivity in a dynamic spectrum environment.

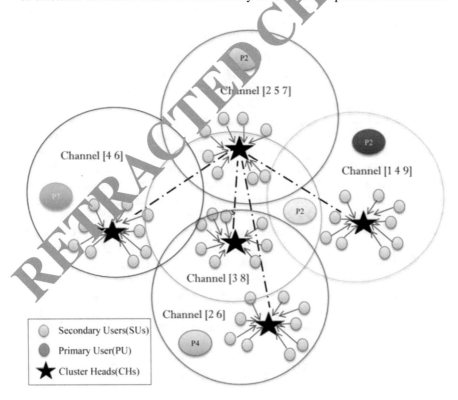

Fig. 1 An example of cooperative DSIBC clustering structure

Every CRSN can discover the available channels. While PUs and SUs are operating in certain frequency bands, the frequencies are assigned continuously and consecutively. To overcome the re-clustering process problems in the existing method, all the CRSN nodes are grouped using a swarm intelligence-based clustering technique. For a sensor network with the number of nodes N, it is possible to form the optimal number of clusters 'K'. The optimal number of clusters can be found using the following equation [6]

$$K_{\text{opt}} = \left[\frac{N}{d_{\max}\sqrt{3\rho}} + 0.5 \right] \tag{1}$$

where N is the total number of nodes, ρ is the number of CRSN nodes per unit area, and $d_{\max}$ is the maximal transmission range of CRSN nodes. The base station (BS) selects one among the 'K' number of cluster head (CH) in its communication range randomly. The base station sends the information about the list of selected CHs along with the position of CHs to the remaining CRSN nodes. Each CRSN node finds the distance between itself and all the CH nodes. DSABC ranked CHs with the least cost fitness function and obtained best among the CRSN nodes, with a less converging time for large CRSN size. To increase the lifetime of cognitive radio sensor networks, energy-efficient clusters and dynamic channel allocation are formed. Channels are not accessible when SUs are placed within the PUs transmission range. The dynamic spectrum access is used in next-generation networks to access the spectrum opportunistically without any spectrum space. Each secondary user shares the locally sensed information to other secondary users without any centralized control unit. In distributed cooperative sensing, all cognitive users attempt to find out the primary unused channel dependently.

CR uses the best available bandwidth for communication. Spectrum management determines the characteristics of the spectrum, the choice of user requirements, and CR reconfiguration functions. The xG network uses the most suitable spectral range to improve the overall quality of spectral service (QoS) requirements. The availability of channels is detected either by the secondary users or by the primary users to avoid interference. The primary user resides in different channels is represented by corresponding colors. One common channel that is not occupied by other PUs should be available for CRSNs that belong to the same cluster. So, each CH senses the commonly available channels and selects one channel from that and assign that to all its members. During the channel selection period, the cluster leader must check the condition that the selected channel should not be used by any of the nearest PUs. The different clustering of best CHs is communicating to allocate the channel to its members with the least clustering node power and solved multimodal optimization problems. The following hypothesis model is used to detect PUs.

$$r(t) = \begin{cases} n(t) & H_0, \\ hs(t) + n(t) & H_1, \end{cases} \tag{2}$$

where $r(t)$ is the received signal of the secondary user, $s(t)$ is the transmitted signal PU, $n(t)$ is the additive white Gaussian noise with zero mean (AWGN), and h is the amplification of the channel amplitude. On the one hand, 'H_0' is the absence of PU by channel in a certain band of the spectrum (null hypothesis). On the other hand, '$H1$' is the presence of PU over the channel (the presence of a hypothesis).

3.1 Clustering Power Calculations in DSIBC Algorithm

All the CRSN nodes are linked to one of the CH with minimum distance, and then the base station re-elects the CH for each cluster with the following condition:
The CH should be inside the communication range of the base station.

- The CH should have high residual energy.

Sum of the transmitting total power for the communication is given by

$$P_{tx} = \sum_{k=1}^{K} \sum_{i=1}^{N} \text{Dist}_{\min}\left(n_i^k, \text{Center}^k(CH)\right) \tag{3}$$

To find the two points, minimum distance between coordinates is (x_0, y_0) and (x_1, y_1).

$$\text{Dist}_{\min} = \sqrt{(x_0 - x_1)^2 + (y_0 - y_1)^2} \tag{4}$$

In the intra-cluster communication, cluster members (CMs) send the source node information to the center position of CH through the common available channel with the shortest communication distance. Minimum distance for an ith node of the kth cluster (c_k) and the center position of CH in the kth cluster (c_k) is given by

$$P_{\text{intra}}(\text{Tot}) = \sum_{k=1}^{K} \text{Dist}_{\min}\left(n_i^k, \text{Center}^k(\text{CH})\right)c_k$$

$$= \sum_{k=1}^{K} \text{Dist}_{\min}\left(n_i^k, n_j^k\right)c_k \tag{5}$$

In the inter-cluster communication, the center position of CH gathered source node information compressively to communicate the nearby CH and prove that the shortest communication power across the cluster centers. Minimum distance for the ith center position of CH in the kth cluster (c_k) and the jth center position of another CH in the kth cluster (c_k) is given by

$$P_{\text{inter}}(\text{Tot}) = \sum_{k=1}^{K}\left(\sum_{i=1}^{N}\sum_{j=1}^{M}\text{Dist}_{\min}\left(\text{Center}^k(\text{CH}_i), \text{Center}^k(\text{CH}_j)\right)\right)c_k$$

$$= \sum_{k=1}^{K} \left(\sum_{i \neq j} \mathrm{Dist_{min}} \left(\mathrm{Center}^k(\mathrm{CH}_i), \mathrm{Center}^k(\mathrm{CH}_j) \right) \right) c_k \qquad (6)$$

where $\mathrm{Center}^k(\mathrm{CH}_i)$ is the ith center position of CH in kth cluster (c_k), and $\mathrm{Center}^k(\mathrm{CH}_j)$ is the jth center position of CH in kth cluster (c_k).

The minimum distance between coordinates is found. Sum of the total power for intra- and inter-clustering communication is:

$$P(\mathrm{Tot}) = \sum_{i=1}^{N} (P_{\mathrm{intra}} + P_{\mathrm{inter}}) n_i \qquad (7)$$

3.2 Proposed DSIBC Algorithm

In Fig. 2, DSIBC sensor network consists of 'N' number of nodes and 'K' predetermined number of clusters form as follows:

(i) Set 'S' elements to comprise 'K' arbitrarily chosen CHs among all the suitable cluster head candidates.

    a. For each node point n_i, $i = 1, 2, ..., N$. estimate the distance $d(n_i, \mathrm{CH}_{\mathrm{p,k}})$ between each node and all CHs point position.

    b. Allocate each node point n to CH where: $d(n_i, \mathrm{CH}_{\mathrm{p,k}}) = \min\{d(n_i, \mathrm{CH}_{\mathrm{p,k}})\}$ for $k = 1, 2, ..., K$

(ii) Estimate the cost function of each arbitrarily chosen CH:
Find the best CH for transmission using FA. All the clustering sets of rules will be ensured at the base station by the centralized algorithm [12].

The BS runs FA to finalize the best CHs and minimize the cost function.

$$\cos t = f_1 \times \beta + f_2 \times (1 - \beta) \qquad (8)$$

$$f_1 = \max_{k=1,2,...,K} \left\{ \sum_{\forall n_i \in C_{p,k}} d(n_i, \mathrm{CH}_{p,k}) / |C_{p,k}| \right\}$$

$$f_2 = \frac{\sum_{i=1}^{N} E(n_i)}{\sum_{k=1}^{K} E(\mathrm{CH}_{p,k})} \qquad (9)$$

Here, the β is the user-defined constant. Let $\beta = 0.5$, f_1 is the maximum average distance between the nodes with related cluster heads (CHs), and $|C_{p,k}|$ is the cluster particle p (i.e., the node). Function f_2 is the ratio of the average nodes energy to average CHs energy. $E(n_i)$ is the energy of ith node and $E(\mathrm{CH}_{p,k})$ is the energy of kth CH of particle p.

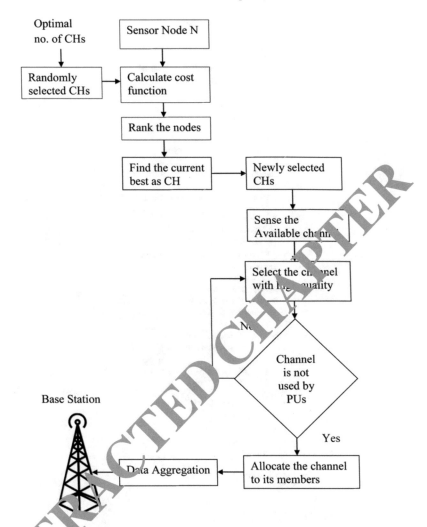

Fig. 2 Flow chart of the DSIBC

(iii) The chosen CHs to sense the available channel in its range.
(iv) Select the channel with high channel quality with the condition that the selected channel should not be used by the nearby PUs.
(V) CHs aggregate the data from the cluster members through the local common available channel.
(vi) The CHs transmit the collected information to the base station.

The pseudo code for DSIBC algorithm is described:

Step 1: Random node creation
  Foreach {n_i} {node(i)}
  {
  d= d(n_i,$CH_{p,k}$)
  Assign node n_i to cluster head $CH_{p,k}$ where:
  d(n_i,$CH_{p,k}$)=$^{min}_{\forall k=1,2,...k}${d($n_i$,$CH_{p,k}$)}
  }
Step 2: Channel selections
  foreach {ch} {CHL}
  {
  cluster(cl_no) [list]
  foreach {cm} {Clustermember(ch)}
  {
  cluster(cl_no) $cm
  }
Step 3: To find the best cluster head selection cost function
  $f_1 = {}^{max}k=1,2,..,K$ {$\sum \forall n_i \in C_{p,k} d(n_i,CH_{p,k})$ / $C_{p,k}$ }

  $f_2 = sum_{(i=1,2,...,N)} E(n_i)$ / $sum_{(i=1,2,...,K)} E(CH_{p,k})$

  fitness=f1*β+f2*(1- β)
Step 4: Assign random channels to primary and
    secondary users
  foreach pu $PU {
  cha($pu) [lrandom $chList]
    $node($pu) color $c($cn($cha($pu)))"
    $node($pu) label $cn($cha($pu))"
  }
Step 5: Neighbour node detection
  for {i 0} {$i<[expr $nonode-1]} {incr
  {
  v [expr x_pos*x_pos+y_pos*y_pos]
  d [expr sqrt($v)]
  nd($i,$j) $d ;
  }
Step 6: Find Nearest
  foreach $r
  {
  npu($i) [list]
  v [expr x_pos*x_pos+y_pos*y_pos]
  d [expr sqrt($v)]
  and npu($i) $j
  }
Step 7: Reclustering
  foreach cm $cluster($i)
  {
  if {$chead($i)!=$NCHv($i)} {
  NCHv($i) $cm ;#NCHv-new cluster head
  }
  }
Step 8: Data Aggregation
  foreach cl $cluster($i)
  {
  if {($cl!=$chead($i)) && ($energy($cl)>0)}
  {
  Reclustering()
  }
  }

4 Simulation Results and Discussion

Let us implement the performance evaluation of proposed DJFOC with NS2 simulation for CR networks [13]. The parameter specification is briefly described in Table 1. Figure 3 shows the number of PUs and SUs dynamically accesses the channels with 'K' optimal number of clustered structures connectivity. The simulation was investigated by 10 primary users and 90 secondary user nodes to validate the dynamic spectrum access performance of the proposed model. Each user node randomly placed on a 1000×1000 m field and 10 common available channels marked by maroon, hot pink, cyan, yellow, yellow-green, deep pink, sky blue, violet red, green, and blue colored selection. The notations for channel occupied by SUs are $C0, C1, …, C9$ and PUs are $0.1, …, 0.9$ in the NS2 (Ver. 2.34) simulation. Each PU chooses any one of the common 10 channels and the fortification range is 200 m which the remaining CRSN neighbors cannot access the occupied channel. The analysis is carried over from the time of simulation 131.0 s, and the constant packet size is 512 bytes.

The performance of the proposed scheme is analyzed by using the metrics of average converge time for various CRSN sizes, average node power for different cluster numbers, average node power for PUs and SUs, and detection performance techniques. The CRSN size plays a major role in communicating with the base station. Hence, the main objective of this work is to select the best CHs which are optimum among multi-users in the grouped clusters which helps to reduce further in converging time, node power, and interferences. From Table 2, the average converge

Table 1 Parameters specification of optimized clustering techniques

Parameter type	Specification	Parameter type	Specification
Channel type	Wireless channel	Data packet interval	0.0625 s
MAC layer	80.11	Routing protocol	AODV
Network interface type	Phy/wireless phy	Simulation software	NS-2, version 2.34
Interface queue type	Queue/drop tail/pri-queue	Simulation coverage area	1000 m × 1000 m
Radio propagation model	Two-ray ground	Simulation time	131 s
Antenna model	Omni antenna	Number of SUs	90
Mobility model	Random way point	Transmission range radius of SUs	150 m
Mobility speed	5 m/s	Number of PUs	10
Number of channels	10	Transmission range radius of PUs	200 m
Data traffic model	CBR over UDP	PU activity checking interval	0.2 s
Initial energy	50 J	Receive power	0.375 W
Transmit power	0.75 W	Sense power	0.25 W

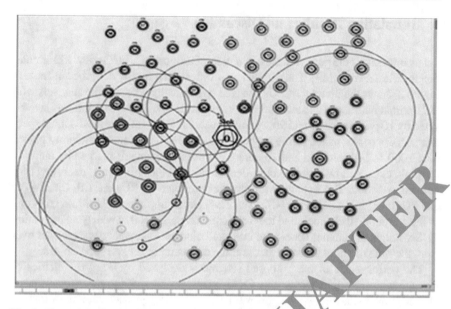

Fig. 3 Channel distribution for PUs and SUs in DSIBC

Table 2 Comparison of average converge time for various CRSN sizes

CRSN size	Average convergence time (s)				
	DSAC	DSOC	DFOC	DJFOC	DSIBC
20	1.440	4.533	2.720	1.943	1.658
40	11.520	9.733	5.840	4.171	3.560
80	31.519	20.133	12.080	8.629	7.365
120	51.519	35.733	21.440	15.314	13.073
160	71.599	46.133	27.680	19.771	16.878
200	91.19	61.733	37.040	26.457	22.585
240	1.4.	77.333	46.400	33.143	28.292
280	131.520	87.733	52.640	37.600	32.097

time for DSIBC is comparatively less with respect to the other optimized clustering techniques. The proposed DSIBC average converge time is 47.2 s lesser than existing DSAC for various CRSN sizes with a percentage of performance '74.98%'. Figure 4 shows the overall comparisons of average converge time for various CRSN sizes in different optimized clustering techniques. As the CRSN size increase, the average converge time increases linearly. Converge time is the indication to measure the fast factor of the algorithm. In this work maximum, CRSN size of 280 is considered.

From Table 3, it is observed that the average node power for proposed DSIBC is less compared to other optimized clustering techniques. The proposed DSIBC average node power is 2585.774 μW lesser than the existing DSAC for various

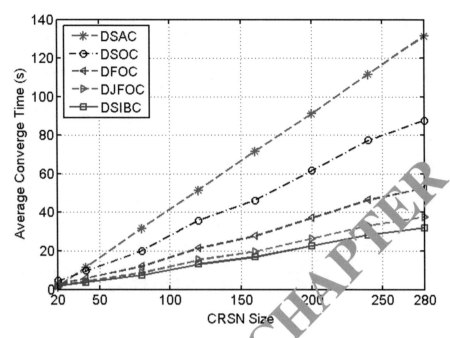

Fig. 4 Comparison of average converge time for various CRSN sizes in different optimized clustering techniques

Table 3 Comparison of average converge time for various CRSN sizes

Cluster number	Average node power (μW)				
	DSAC	DSOC	DFOC	DJFOC	DSIBC
2	10,106.955	6506.954	5142.954	2421.954	1098.955
4	5045.989	3245.989	2563.989	1203.490	541.992
8	2511.826	1615.826	1274.826	594.576	263.830
12	1672.646	1072.646	845.313	391.813	171.317
16	1251.078	801.078	630.578	290.453	125.083
20	997.857	637.858	501.458	229.357	97.062
24	829.381	529.381	415.714	188.964	78.719
28	708.832	451.689	354.261	159.903	65.408

cluster numbers, with a performance percentage of '89.44%'. Figure 5 shows the overall comparisons of average node power for various cluster numbers in different optimized clustering techniques. The graph shows the power observed for different cluster number from 2 to 28. Average node power for CRSN is defined as the ratio of the sum of the total energy of PUs and SUs to the total number of nodes in clusters, often expressed in watt (W).

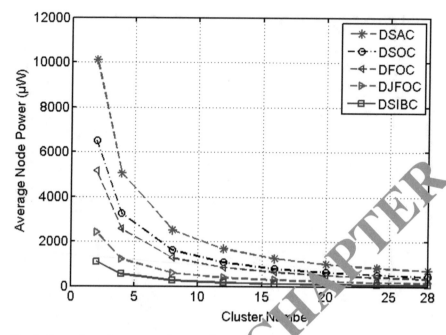

Fig. 5 Comparison of average node power for various cluster numbers in different optimized clustering techniques

$$\text{Average Node Power for CRSN} = \frac{\sum_{i=1}^{N} PU_i(\text{Energy}) + \sum_{i=1}^{N} SU_i(\text{Energy})}{\text{Total number of nodes in clusters}} \quad (10)$$

From Fig. 5, it is observed that the power remains constant for cluster number which is above 28, so the simulation was stopped at cluster number 28. The graph shows cluster number versus average node power values of DSAC, DSOC, DFOC, DJFOC, and DSIBC during the simulation analysis for a combination of PUs and SUs in the network. Cluster number is the indication to reduce the high node power of the DSIBC algorithm.

Average node power for PUs is defined as the ratio of the sum of the total energy in PUs to the total number of PUs, often expressed in watt (W).

$$\text{Average Node Power for PUs} = \frac{\sum_{i=1}^{N} PU_i(\text{Energy})}{\text{Total number of PUs}} \quad (11)$$

As the number of PUs increases, average node power increases linearly. Table 4 shows the PU number versus average node power values of different optimized clustering techniques during the simulation analysis for 10 PU nodes. The DSIBC average node power of the primary user is 75.3 μW lesser than DSAC, with a performance percentage of '9.646%'.

Figure 6 shows the overall comparisons of average node power for various PUs in

Table 4 Comparison of average node power for various PU numbers

PU number	Average node power (μW)				
	DSAC	DSOC	DFOC	DJFOC	DSIBC
1	806	759	753	740	732
2	806	759	753	740	732
3	806	759	753	740	732
4	806	759	753	740	732
5	806	759	753	740	732
6	814	766	759	750	740
7	823	774	764	758	747
8	832	780	769	768	755
9	838	788	777	774	
10	846	794	786	782	767

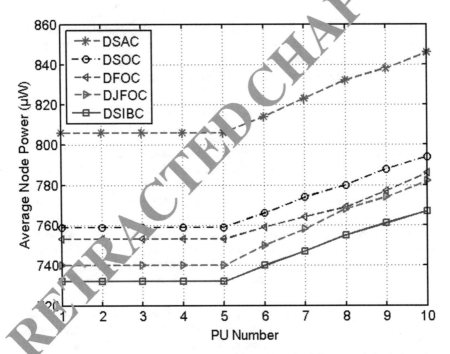

Fig. 6 Comparison of average node power for PU numbers in different optimized clustering techniques

different optimized clustering techniques. If more PUs are considered in the transmission range, the clustering process involved more spectrum resources opportunistically. So, clustering results affects in terms of energy consumption [6]. Hence, in the simulation, 10 primary user nodes only are considered within the fortification

Table 5 Comparison of average node power for various SU numbers

SU number	Average node power (μW)				
	DSAC	DSOC	DFOC	DJFOC	DSIBC
11	806	759	753	740	732
20	866	825	803	793.5	768
30	983	943	905	895.5	840
40	1111	1075	1022	998.5	907
50	1235	1201	1127	1097.5	978
60	1362	1332	1234	1198.5	1048
70	1478	1461	1338	1293.5	1129
80	1603	1585	1438	1388.5	1207
90	1732	1706	1539	1485.5	1270
100	1861	1829	1642	1584.5	1343

range of 200 m. The DSIBC technique is the best among the five different optimized clustering techniques by improving the spectrum utilization among the primary users with minimum average node power.

Similarly, average node power for SUs is calculated as follows

$$\text{Average Node Power for SUs} = \frac{\sum_{i=1}^{n} SU_i(\text{Energy})}{\text{Total number of SUs}} \quad (12)$$

As the number of SUs increases, average node power increases linearly. Table 5 summarizes the various optimized clustering results obtained for 90 SU nodes.

The DSIBC average node power of the secondary user is 281.978 μW lesser than DSAC, with a performance percentage of '24.231%' compared to DSAC in SUs average node power. Figure 7 shows the overall comparisons of average node power for various SUs in different optimized clustering techniques. In the simulation, 90 secondary user nodes are considered within the fortification range of 150 m. The DSIBC technique is the best among the five different optimized clustering techniques by improving the spectrum utilization among the secondary users with minimum average node power.

The single threshold detector performs well in cooperative spectrum sensing network by high detection probability with a less false rate. At the detection stage, the sensing error (noise) in cooperative nodes over the channel is removed with reliable decisions. The detection performance of a spectrum sensing technique can be evaluated using the probability of false alarm, detection, and missed detection [14]. Estimate the SNR for the detection of a received signal and decide output from the detection performance of spectrum sensing techniques. The threshold $\lambda = 4$ dB is based on the experimental results and observations [15].

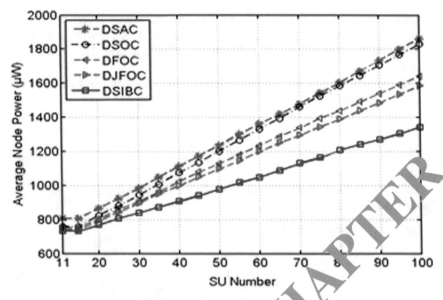

Fig. 7 Comparison of average node power for SU numbers in different optimized clustering techniques

a. If threshold value λ is greater than SNR (the primary user over a channel is falsely detected 'H_1'), then the hypothesis model is performed by the probability of false alarm technique, i.e., if $\lambda > $ SNR, Accept $H = H_1|H_0$
b. If threshold value λ is greater than SNR (the primary user over a channel is correctly detected 'H_1'), then the hypothesis model is performed by the probability of detection technique, i.e., if $\lambda > $ SNR, Accept $H = H_1|H_1$
c. If threshold value λ is lesser than SNR (the primary user over a channel is not detected 'H_0'), then the hypothesis model is performed by the probability of missed detection technique, i.e., if $\lambda \leq $ SNR, Accept $H = H_0|H_1$.

Implementation is ported using MATLAB R2013a. To determine if the channel is being used by the primary user, the detection test statistics for output Y are compared with a predefined threshold λ. The probability of false alarm (PFA) is the likelihood that a hypothesis test will choose the H_1, although in fact H0.

$$P_{FA} = P(Y > \lambda|H_0) = \frac{\Gamma(m, \lambda/2)}{\Gamma(m)} \tag{13}$$

Probability of detection (P_D) is the probability that correctly decides H_1 when it is H_1;

$$P_D = P(Y > \lambda|H_1) = Q_m\left(\sqrt{2\gamma}_{avg}, \sqrt{\lambda}\right) \tag{14}$$

where λ is the detection threshold, $\Gamma(.)$ is the complete gamma functions, $\Gamma(.,.)$ is the incomplete gamma functions, γ_{avg} is the average SNR, $Q_m()$ is the generalized Marcum Q-function, and $m = TW$ is the time-bandwidth product. Here, consider as $m = 5$.

The equation for the probability of detection is calculated using 'marcumq()' function in MATLAB.

$$Q(x) = \frac{1}{\sqrt{2\pi}} \int_x^\infty e^{-\frac{u^2}{2u}} du \tag{15}$$

The equation for the probability of false alarm is calculated using 'gamma()' and 'gammaincinv()' function in MATLAB.

gamma(x) represents the gamma complete function

$$\Gamma(x) = \int_0^\infty e^{-t} t^{x-1} dt \tag{16}$$

gamma(a, x) represents the gamma incomplete function

$$\Gamma(a, x) = \int_x^\infty e^{-t} t^{a-1} dt \tag{17}$$

i.e., igamma(a, x) = gamma($a, 1 - $ gammainc(x, a)).

The objective of the probability of missed detection (P_{MD}) is to reduce the P_{FA} and to increase P_D. In general, the performance of P_{MD} is the probability that a PU is present over the channel but not able to detect the primary transmission signal. In terms of hypothesis, it is written as

$$P_{MD} = 1 - P_D = \Pr(\text{Signal is not detected} | H_1) \tag{18}$$

From Fig. it is observed that the performance of detection under various values of probability of false alarm for SNR $= 4$ dB. However, the level of SNR $= 4$ dB is a little high for a proper range in the spectrum sensing [15]. The probability of a SU that decides that a PU access over the channel in the spectrum band. Hence, the SUs missed the opportunity for efficient channel utilization. In the existing method DSAC, the probability of detection is less for low values of false alarm (<0.1). In the proposed method of DSIBC, the probability of detection is high when the P_{FA} value is >0.1 compared to other optimized clustering techniques.

The performance of detection is assumed that SNR varied from 0 to 30 dB values, and the probability of false alarm is 0.1. As the SNR value increases, the probability of detection increased linearly and reached constant at '1'. From Fig. 9, it is noted

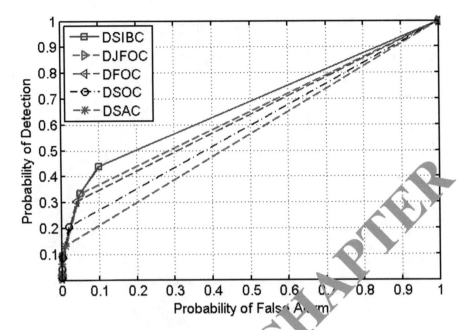

Fig. 8 Comparison of probability of false alarm versus probability of detection in different optimized clustering techniques

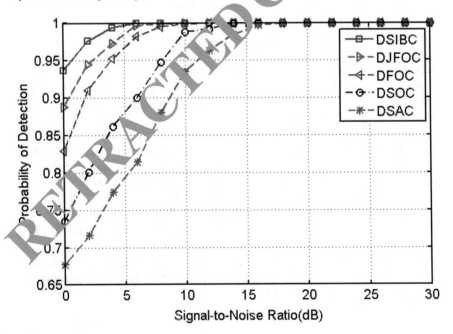

Fig. 9 Comparison of signal-to-noise ratio versus probability of detection in different optimized clustering techniques

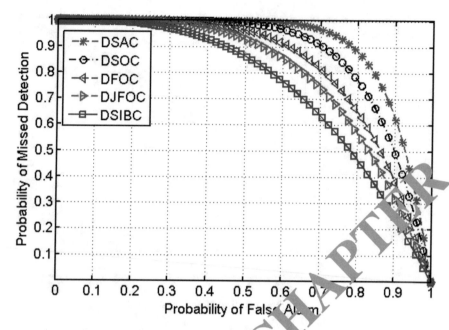

Fig. 10 Comparison of probability of false alarm versus probability of missed detection in different optimized clustering techniques

that the performance of detection under various values of SNR. In a CR network, a higher probability of detection corresponds to less interference with PUs. In the proposed method of DSIBC, the detection probability is about '0.937' compared to the existing method DSAC with '0.677' when SNR is at '0 dB'. The DSIBC curve is converted to '1' faster when the value of SNR is above 4 dB compared to other optimized clustering techniques.

From Fig. 10, it is observed that the performance of missed detection under various values of probability of false alarm for the complementary curve of ROC. As the probability of false alarm rate increases, the probability of missed detection rate decreases gradually for the complementary curve of ROC. The channel is active in the PUs but not able to detect the primary transmission. This causes harmful interference to both PUs and SUs. In the existing method DSAC, P_{MD} value remained at '1' up to '0.3' value of P_{FA} and reached '1' linearly. In the proposed method of DSIBC, P_{MD} value is '1' when the P_{FA} value is '0.1' and P_{MD} falls to '0' as P_{FA} is increased to '1'. Proposed DSIBC is good in detecting the primary transmission with its availability when the false alarm rate is high compared to other optimized clustering techniques (e.g., at $P_{FA} = 0.7$, $P_D = 0.372$, and $P_{MD} = 0.628$).

5 Conclusion

The DSIBC technique is the best among the five different optimized clustering techniques which is used to save transmit power with the shortest distances and achieve energy-efficient clusters while restricting interference to primary users. The simulation performance shows superior scalability and constancy for DSIBC. The performance analysis shows that there is an 89.440% reduction in power compared to the existing method. It also obtained the best CHs among the CRSN nodes, with a converging time of 32.09 s for 280 CRSN Size. DSIBC is superior by 9.646% compared to DSAC in PUs average node power and also by 24.231% compared to DSAC in SUs average node power. The re-clustering process offers steadier stable and requires less control overhead with minimum average node power and achievable limited cluster numbers. From the results obtained in the previous sections, the DSIBC algorithm is optimized for power savings and best for accurately detecting the spectrum white space. From the simulation of detection performance, it is observed that the proposed DSIBC is yielding a low false alarm rate with a high probability of detection and reduces sensing error compared to other optimized clustering techniques. In the extension work, other optimization algorithms such as a convolutional neural network may be considered for better clustering.

References

1. Krishnan R, Ganesh Babu R, Kaviya S, Pragadeesh Kumar N, Rahul C, Santhana Raman S (2018) Software defined radio (SDR) foundations, technology tradeoffs: a survey. In: IEEE international conference on power, control, signals and instrumentation engineering Chennai, India, pp 2677–2682
2. Ganesh Babu R, Amudha V (2018) A survey on artificial intelligence techniques in cognitive radio networks. In: 1st international conference on emerging technologies in data mining and information security in association with Springer advances in intelligent systems and computing series, Kolkata, West Bengal, India, pp 99–110
3. Rehmani M, Dhaou R (2018) Cognitive Radio, mobile communications and wireless networks. EAI Springer innovations in communication and computing
4. Zikria YB, Ishmanov F, Afzal MK, Kim SW, Nam SY, Yu H (2018) Opportunistic channel selection MAC protocol for cognitive radio ad hoc sensor networks in the internet of things. Future Gener Comput Inf Syst 18:112–120
5. Wierzchon ST, Kłopotek M (2015) Algorithms of cluster analysis. Institute of Computer Science Polish Academy of Sciences
6. Zhang H, Zhang Z, Dai H, Yin R, Chen X (2011) Distributed spectrum aware clustering in cognitive radio sensor networks. In: IEEE international conference on global telecommunications, Houston, Texas, USA, pp 5266–5271
7. Yadav A, Kumar S, Vijendra S (2018) Network life time analysis of WSNs using particle swarm optimization. In: 4th international conference on computational intelligence and data science, vol 132, pp 805–815
8. Wang H, Wang W, Cui L, Sun H, Zhao J, Wang Y (2018) A hybrid multi-objective firefly algorithm for big data optimization. Appl Soft Comput 69:806–815
9. Bidar M, Kanan HR (2013) Jumper firefly algorithm. In: IEEE 3rd international conference on computer and knowledge engineering, Mashhad, Iran, pp 1–5

10. Ganesh Babu R, Amudha V (2018) Comparative analysis of distributive optimized clustering techniques in cognitive radio networks. Int J Eng Tech 7(3.27):504–507
11. Ganesh Babu R, Amudha V (2018) Comparative analysis of distributive firefly optimized spectrum sensing clustering techniques in cognitive radio networks. J Adv Res Dynam Cont Syst 10(9):1364–1373
12. Wierzchon ST, Kłopotek M (2018) Modern algorithms of cluster analysis. Springer, Cham
13. Issariyakul T, Hossain E (2012) Introduction to network simulator NS2. Springer, US
14. Shanmugavel S, Bhagyaveni MA, Kalidoss R (2017) Cognitive radio: an enabler for internet of things. River Publishers, Netherlands
15. Lavanya S, Sindhuja B, Bhagyaveni MA (2015) Implementation of an adaptive spectrum sensing techniques in cognitive radio networks. In: IEEE international conference on computing and communications technologies, pp 344–349

Inconspicuous Perception: The Solution for the Neglected!

Yash Malhotra, Tilsy Thankachan, Prajyoti D'Silva, Robin Anthony, and Shruti Pednekar

Abstract Building a web application is the main aim and focus that connects all the orphanages and the NGOs to one website. There are numerous shelters contemporaneously accessible that are as of now working for the prospering of road kids or Orphaned, Abandoned, Surrendered (OAS) youngsters yet it is highly unlikely that crosses over any barrier between the road kids, OAS, and the orphanages. The web application works in the way of the utilization of the logistic regression algorithm alongside image analysis and recognition to characterize the kid experienced living in the city or an OAS classification into either a Juvenile or Juvenile Delinquent. The prototype also takes care of the lost and found category. Despite the arrangement of the child and the lost and found category, there is one more part of the medical diagnosis. By taking the child's data as the input, the web application will produce the result, wherein the necessary steps would be taken. The result would be either a Juvenile child or a Juvenile Delinquent. After taking a looking at the contemporaneous situation, thus advances a proposed solution "Inconspicuous Perception"—by the name itself "Inconspicuous" signifies something imperceptible or misty and "Perception" means the ability to become aware of something.

Keywords Juvenile Justice Act · Logistic regression · Image analysis and recognition

1 Introduction

Imminent guardians, in the interim, are obstructed by confounded tenets, long postponements, overcautious administration, and unlawful dealing [1]. According to one estimation, from in more than 30 million vagrants in the nation, just about quarter roughly of them were embraced a year ago, down from 5700 four years prior.

Y. Malhotra (✉) · T. Thankachan · P. D'Silva · R. Anthony · S. Pednekar
Department of Information Technology, St. Francis Institute of Technology, Mumbai 400103, India
e-mail: yashmalhotra124@gmail.com

© The Author(s), under exclusive license to Springer Nature Singapore Pte Ltd. 2021   247
J. S. Raj et al. (eds.), *Innovative Data Communication Technologies and Application*,
Lecture Notes on Data Engineering and Communications Technologies 59,
https://doi.org/10.1007/978-981-15-9651-3_21

Presently, a few authorities from the Ministry of Women and Child Development need to be changed [2].

Up to the present time, a lot of vagrants are as yet not conveyed to the halfway house or have not been received. The appropriation rate is conversely corresponding to the number of vagrants on the planet. Be that as it may, the road youngsters are never considered in the image [3]. The present world emphasizes on the vagrants. No one in this day and age centers around the surrendered and relinquished youngsters or the kids living in the city [4]. Even though some road youngsters remain with their families, they are revealed into a ton of criminal exercises and are named as Juvenile Delinquents. Every one of those kids and youngsters less than 16 years, perpetrating wrongdoing or an offense, is named as Juvenile Delinquents [5]. These Juvenile Delinquents should be accounted for an NGO. Moreover, underscoring at the territory of lost and found, the kids who are lost are never found and the number of policemen is not exactly the cases that are pending for lost youngsters, subsequently expanding the remaining task at hand of the policemen. The present situation principally centers around just a portion of the vagrants, the lost individuals who might be found because of a lot of pending cases or negligible instances of road youngsters who are disregarded or treated in all respects gravely [6]. As indicated by an investigation, the vast majority of the vagrant kids have been deserted by their folks. Indeed, estimation depicts that just 0.3% of these vagrants are kids whose guardians have passed on [7]. The road youngsters who might have families are disregarded by both their families and individuals. These road youngsters are either constrained into child work, child dealing, kid misuse or are revealed into wrong exercises because of the encompassing they invest the greater part of the energy with [8].

The point of this venture is to plan a web application that will interface every one of the shelters with the NGOs to embrace a road kid or Orphaned, relinquished, surrendered (OAS) kids along these lines utilizing machine learning calculations, where every one of the youngsters will be ordered either as a Juvenile or Juvenile reprobate, and further, consideration will be taken to form them better.

2 Motivation

In one part of India, an old young lady was discovered dead, in the wake of being explicitly abused by some men, one of whom evidently was an adolescent reprobate coordinating in kidnapping and at last the old lady was murdered; this was the case recorded according to the chargesheet by the crime branch. Besides, some years prior, December 2012, after a 23-year-elderly person's torment and gangrape on board a transport in the capital of India, an adolescent was blamed for dispensing the most extreme harm on the lady, who kicked the bucket of her wounds. Besides, an investigation was directed that lead an investigation of male overcomers of youngster sexual maltreatment. It was presumed that there is a culture of lack of concern in the nation about "the end result for men as young men." Moreover, a study of 200 Indian men approximately depicted that 75% of them were explicitly manhandled as kids. All the young men who are explicitly manhandled as kids go through their time on earth peacefully because of the shame appended to the male survivors standing up. This is extremely a significant issue that should be tended to.

There are a lot of plans and welfares that are kept running by the Government where they are making a decent attempt to connect with the road kids or to those youngsters who are explicitly manhandled. In any case, the quantity of Juvenile Delinquents is simply expanding step by step. This makes up one of the standard motivations to take up the activity to fabricate a model of the present situations referenced previously.

3 Related Survey

Authors Dr. G. Singh and Er. A. Singh in "Using Data Mining Clustering Technique to Predict Criminal Behavior" centers around the criminal conduct saw among the general population, particularly the young people. It elucidates the reasons for wrong-doing and its unfriendly impact on the general population. The paper refers to how unique kinds of data mining strategies like classification and clustering are utilized to anticipate criminal behavior. And also the likewise features that the procedures referenced do not function admirably with the groups of various sizes and diverse density. This constraint is considered as a future degree and the authors are taking a shot at its usage [1].

Authors Rebecca P. Ang and Dion H. Goh in "Predicting Juvenile Offending: A Comparison of Data Mining Methods" indicate about calculated relapse and prescient information mining procedures, for example, decision trees (DTs), artificial neural systems (ANNs), and support vector machines (SVMs). The examination of these strategies helps us to choose an appropriable technique as there are many machine learning algorithms which are utilized request to arrange or anticipate information. To choose any of them is a troublesome errand. In this manner, the authors have inspected these techniques on whether they could segregate between teenagers who were charged or not charged for starting adolescent culpable in an extensive Asian example [2].

Authors B. Gray, M. D. Birks, T. Allard, M. J. Ogilvie, and A. Lewis from Griffith University in "Exploring the Benefits of Data Mining on Juvenile Justice Data" stress on the best way to apply strategies from the field of neural networks and decision trees to criminal equity information to decide if these procedures could be utilized to improve the prescient exactness of models created to anticipate danger of re-insulating over base cases and commonly applied statistical methods. The real impediment experienced here is that it does not work [3]. Table 1 given does a comparative analysis on review of literature among the three authors.

Table 1 Comparative analysis on review of literature survey among the three research papers

S. No.	Author	Algorithms	Limitations
1.	B. Gray, M. D. Birks, T. Allard, M. J. Ogilvie, A. Lewis	(i) Data mining techniques-knowledge discovery in data (ii) Neural networks and decision trees	It does not work well with clusters (in the original data) of different size and different density
2.	Rebecca P. Aug and Dion H. Goh	(i) Logistic regression (ii) Decision tree (iii) Artificial neural networks (iv) Support vector machine	Different types of predictive data mining methods may be applicable
3.	Dr. G. Singh and Er. A Singh	(i) Knowledge data discovery (ii) Neural networks (iii) Decision tree	Decision tree showed less accuracy as compared to neural networks

4 Proposed Methodology

The main phase of the task is to connect every one of the shelters (orphanages) with the NGO. Any halfway house can associate by utilizing the connection in the site. The working begins with the client who encounters a kid. Presently, if the client is utilizing the web application for the absolute first time, at that point, he/she needs to enroll. The client enlistment comprises of client confirmation. Confirmation is finished utilizing SMS One Time Password validation authentication procedure and Gmail One Time Password check process to guarantee that the client is confirmed client. If the client has officially enrolled before, he/she simply needs to sign in utilizing their qualifications. At that point, he/she will enroll the youngster that they have experienced and the kid will be enlisted. The child is then checked by the NGO dependent on their skin pigmentation and the photograph of the kid which the client needs to transfer it. After the photograph of the kid has been transferred with the assistance of image analysis and recognition, it will check whether the subtleties and the photograph of a similar youngster exist in the police database or not. If the information of the kid is not in the police database, the client will enroll this child or else it will go to the lost and found. The admin of the NGO database will check the record and confirm it against the police database. The kid experienced by the client will be either a Juvenile or a Juvenile Delinquent by utilizing logistic regression as shown in Fig. 1.

If the kid is a Juvenile, that youngster will be sent to the orphanage, closest from the kid's area. On the off chance that the child is observed to be a Juvenile Delinquent, at that point the child will be accounted for to the NGO with the goal that the child can be molded in a better way. Presently, on the off chance that the kid is a lost kid that the client has experienced, at that point the kid will specifically be submitted to the closest police station. The police will begin looking for his folks/watchmen.

Fig. 1 Logistic regression

In the event that the police neglect to report the child and convey that child to the classification of found, at that point an alert will be sent, which will go about as an update that the lost child case is as yet pending. This following of the kid will be observed by the admin, which will know the present circumstance of the lost child. If the lost child is submitted to their folks/gatekeepers, at that point that kid will fall under the class of found.

This lost and found category of a youngster can again be classified by utilizing the unsupervised learning method. Toward the finish of half a year perhaps, a chart can be acquired of the number of kids that were lost, that has now fallen under the classification of found or the number of kids are as yet lost. This can be utilized by D3.js. Besides the classification of kid and the lost and found category, likewise have one more part of medical diagnosis. The kid submitted to the halfway house will be followed by the admin concerning whether the therapeutic analysis of the child is done in the stipulated time or not. If not, an alert will be sent to the shelter. The figure underneath demonstrates a definite outline of the usage plan as a stream graph as shown in Fig. 2.

5 Observation/Results

When the verified client enlists the youngster experienced in the city, following are the output:

The output as shown in Fig. 3a is delivered once the client fills in the child's subtleties. This is the output produced when the kid is neither in lost and found database nor police database. This youngster is enlisted possibly out of the blue or does not have any criminal records and is put away in the Child Registration Information database.

The output as shown in Fig. 3b is created when the child's information is discovered as of now in the lost and found database or the police database. This implies if a client enrolls a child and on the off chance that the child's information was at that point present in these databases, at that point it gives the desired output (Fig. 4).

Taking care of the databases is less demanding for the designer with the Django system utilizing Python and every database utilized is SQLite. If a kid's information was at that point there in the police database or lost and discovered, then the output would be diverse as found in the outcomes.

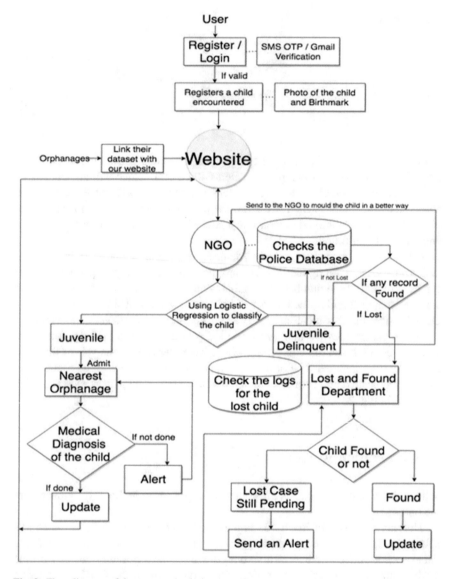

Fig. 2 Flow diagram of the proposed solution

For using logistic regression to train the data, the system requires a dataset. So a survey was conducted, covering parts of Mumbai which were fed to the algorithm and this is how the results were produced based on independent variables. The figure above that is Fig. 5a–c is displayed which shows the graph of the children encountered as either male or female, or either as a Juvenile or Juvenile Delinquent, or either as an orphan, lost or poor child.

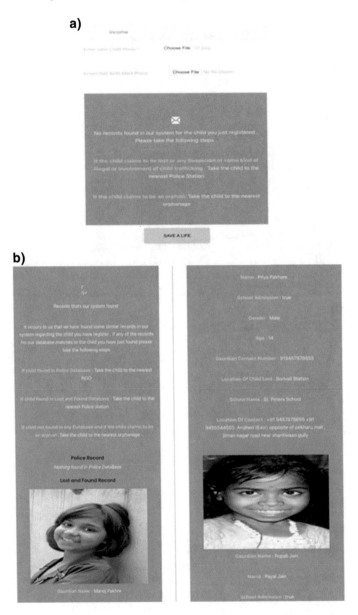

Fig. 3 **a** Output when the child's data is entered into the database for the first time. **b** Output when the child's data is found either in lost and found database or police database

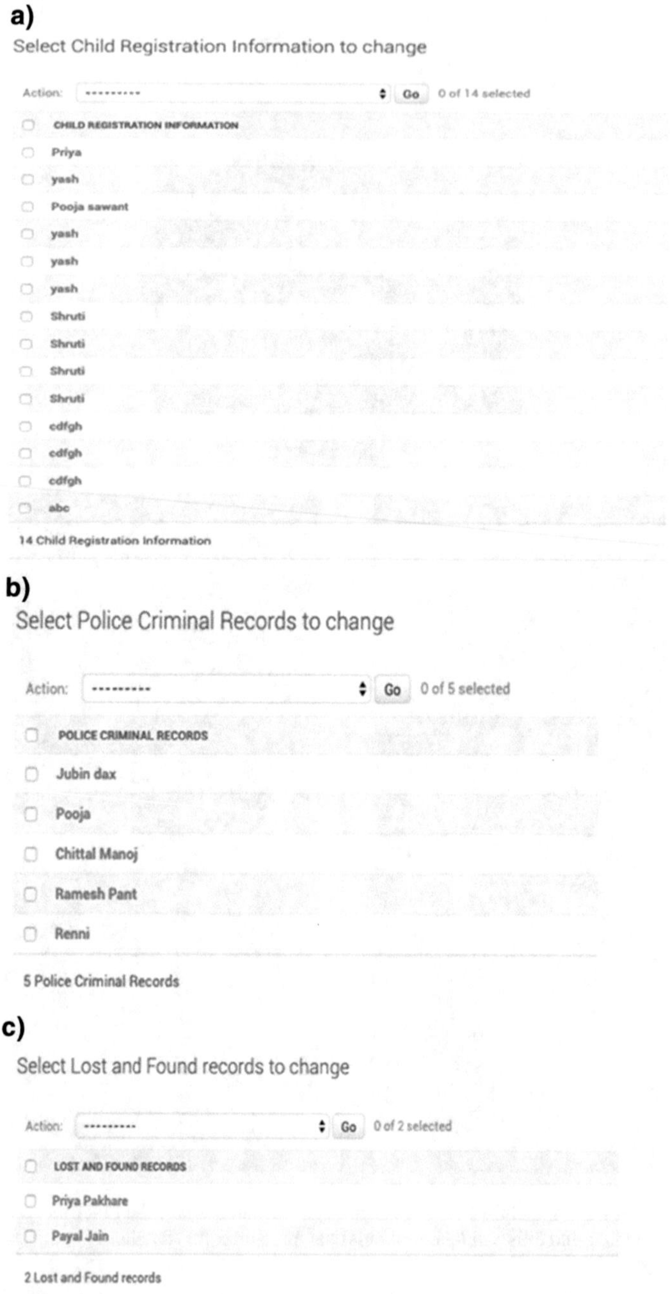

a)

Select Child Registration Information to change

| Action: | .... | ‡ Go | 0 of 14 selected |

- [ ] CHILD REGISTRATION INFORMATION
- [ ] Priya
- [ ] yash
- [ ] Pooja sawant
- [ ] yash
- [ ] yash
- [ ] yash
- [ ] Shruti
- [ ] Shruti
- [ ] Shruti
- [ ] Shruti
- [ ] cdfgh
- [ ] cdfgh
- [ ] cdfgh
- [ ] abc

14 Child Registration Information

b)

Select Police Criminal Records to change

| Action: | .... | ‡ Go | 0 of 5 selected |

- [ ] POLICE CRIMINAL RECORDS
- [ ] Jubin dax
- [ ] Pooja
- [ ] Chittal Manoj
- [ ] Ramesh Pant
- [ ] Renni

5 Police Criminal Records

c)

Select Lost and Found records to change

| Action: | .... | ‡ Go | 0 of 2 selected |

- [ ] LOST AND FOUND RECORDS
- [ ] Priya Pakhare
- [ ] Payal Jain

2 Lost and Found records

Fig. 4 **a** Child registration information database. **b** Police database. **c** Lost and found database

Fig. 5 a Real-time data for the child encountered in the street either as a male or female. b Real-time data for the child encountered in the street either as a Juvenile or Juvenile Delinquent. c Real-time data for the child encountered in the street either as a poor/street child, orphan child or lost child

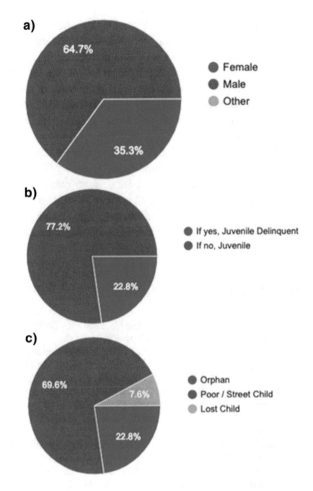

6 Conclusion and Future Scope

There is a lot of orphanages contemporaneously accessible working for the prospering of the road kid or OAS youngsters. Indeed, even today, plenty of vagrants and road kids are feeble. Kids live and deal with the road because their communities are poor, that are vagrants, or they have fled from home, frequently to escape misuse. They are constantly malnourished, get deficiency training and treatment, and are complex in kid work since the beginning. Along these lines, there is a need to build up this proposed framework. To begin executing the framework, it is extremely important to have some ongoing information. So before executing the framework, an overview was embraced in different pieces of Mumbai beginning from Borivali. The study was taken for the road youngsters to comprehend and to know the proportion of road kids living with families to the kids living without families. Additionally, the overview included concerning whether the child has carried out any wrongdoing or not, for example, a Juvenile Delinquent or a Juvenile.

Aside from studies, a ton of shelters was considered, yet sharing of information was restricted because of security issues. Consequently, for the beginning, the review was and is imperative to test the cases in the wake of actualizing the undertaking. The normal outcome is to get the information from the shelters, to consider the overview nearly in all pieces of Mumbai, and after that to test and sift through cases after the execution of the venture.

One of the greatest future scopes could be the zone of the orphanages. Since this framework is only a model, just pieces of Mumbai are considered for both surveys and orphanages. The future scope could contain different parts of India. Additionally, there exist many verification strategies to confirm the client. Finally, just an unsupervised machine learning algorithm is considered. Thus, another real future extension that can be considered is the reinforcement learning.

References

1. Gray B, Birks MD, Allard T, Ogilvie MJ, Lewis A (2008) Exploring the benefits of data mining on Juvenile justice data. Griffith University
2. Ang RP, Goh DH (2013) Predicting Juvenile offending: a comparison of data mining methods. Int J Off Ther Comp Criminol 57(2):191–207. https://doi.org/10.1177/0306624X11431132
3. Singh G, Singh A (2017) Using data mining clustering technique to predict criminal behaviour
4. Manoharan S (2019) Image detection classification and recognition for leak detection in automobiles. J Innov Image Process (JIIP) 1(02):61–70
5. Bindhu V (2019) Biomedical image analysis using semantic segmentation. J Innov Image Process (JIIP) 1(02):91–101
6. Manoharan S (2019) A smart image processing algorithm for text recognition information extraction and vocalization for the visually challenged. J Innov Image Process (JIIP) 1(01):31–38
7. Pandian AP (2019) Identification and classification of cancer cells using capsule network with pathological images. J Artif Intell 1(01):37–44
8. Vijayakumar T, Vinothkanna R (2020) Retrieval of complex images using visual saliency guided cognitive classification. J Innov Image Process (JIIP) 2(02):102–109

Optimizing Decision Tree Classification Algorithm with Kernel Density Estimation

A. Rajeshkanna and K. Arunesh

Abstract Data mining is defined as a knowledge discovery process, which uses several algorithms for various applications. Classification of data objects based on an efficient classification algorithm is a very important task to be carried out on a huge dataset. It can also be defined as a supervised learning algorithm in which classification is performed on a predefined knowledge of the objects and assigns class labels based on the relationship. Decision trees are the mainly used classification algorithm. In decision trees, entropy is calculated and maximum likelihood has been used which assumes probability distributions for the input labels. The influence of prior probabilities in non-parametric classification methods has proven to be useful only for particular classes of interest. To accomplish the objective such as better classification accuracy and to reduce the elimination of sparse data from classification, a non-parametric kernel function with modified entropy is implemented in the classification algorithm. This proposed methodology based on estimation proves that it provides good classification outcome for the frequency distribution of dataset irrespective of the decision boundaries and complexity of the datasets. The relative comparisons of the results and the consistency were examined with different datasets, and the results performed well in all the experiments conducted.

Keywords Non-parametric · Kernel density estimation · Classification tree · Entropy

1 Introduction

As data is almost enormous everywhere in the digital field, new terms for data are growing at a faster pace. These new terms include data science, data mining and

A. Rajeshkanna (✉) · K. Arunesh
Sri S. Ramasamy Naidu Memorial College, Sattur, Virudhunagar, India
e-mail: srnmcrajesh@gmail.com

K. Arunesh
e-mail: arunesh_naga@yahoo.com

machine learning. Data science becomes the umbrella for several techniques such as data mining and machine learning, a kind of artificial intelligence. Data science is a field which includes all the techniques and deals with both structured and unstructured data. A data scientist is responsible for creating new data products with the techniques in data science. Data mining is one of the techniques, which relates to searching for data to find patterns or knowledge from the whole data. It turns out that with the data mining algorithms, a huge dataset can generate data whose results can provide new knowledge information. Data mining is thus a confluence of various fields like artificial intelligence, data room virtual base management, machine learning and statistical studies. Classification is an important task in data mining which is also distinguished as machine learning classifiers. Classification is a data analysis task and is used to find a model that distinguishes both the data classes and its concepts. The major task of classification is the problem of identifying to which set of categories (subpopulations) a new observation belongs to. This is decided based on the existing training set of input data containing observations and on the categories of membership. Hence, a classification problem has an input dataset referred to as the training set which contains the existing information with the number of attributes. The attribute values may be ordered or unordered. Application domains where classification is successfully performed are retail target marketing, design of telecommunication service plans, fraud detections and so on. Several classification models like decision trees, Naive Bayes, neural networks and genetic algorithms have been recently used in many applications. Decision trees are probably the most important foundation of data mining classification task for complex models. Decision tree construction algorithms are performed by data decomposition at the root node with the training data. The performance of the algorithm can also be influenced by the selection of the splitting attributes. Decision trees are usually non-parametric in terms of statistics inside the algorithm, and there are no specific weight parameters [1, 8]. Shannon proposed statistical entropy as the uncertainty of a (discrete) random variable. ID3 the most popular algorithm used this Shannon entropy. The Renyi entropy was proposed with the aid of Alfred Renyi which also generalized the Shannon entropy. The fundamental principle of our proposed work is a non-parametric plug-in opportunity to Renyi entropy proposed by J. C. Principle. Encouraged by means of the above aspect, our work changed into layout kernels which are non-parametric plug-in entropy estimators, and it satisfied the situations. Further, the embedding of kernel distribution functions is used for gaining more knowledge in a decision tree framework. Hence, in the proposed work, estimating the kernel density estimation with non-parametric strategies along with entropy performs a better classification when compared with traditional decision tree classification. The organization of this paper is as: Sect. 2 describes the preliminaries related to the work, Sect. 3 describes the proposed work with its methodology and analysis, and Sect. 4 concludes the results.

2 Preliminaries

The decision tree is one of the models in supervised learning that does not require any parameter checking for knowledge discovery. A decision tree is a classifier that is expressed as a recursive partitioning of the instance space. It generalizes a directed rooted tree with no incoming edges and all other nodes with exactly one incoming edge. A node which has outgoing edges is also called as internal nodes or test nodes. The terminal nodes are also called as leaves or decision nodes. The instance space is divided into two or more subspaces. The internal nodes split the instance space based on the input attribute values on a certain discrete function. The partition is based on the value of the attribute which may be a certain specific range if it is a numeric value.

Decision tree inducers in existing literature represent the algorithms that automatically construct a decision tree from a given dataset. There are many algorithms such as ID3, Cart and C4.5 decision tree algorithms. The objective of the algorithm is to minimize the error rate and to construct the optimal decision tree. Finding a minimal decision tree with the training set and classifying the unseen instance is an NP-hard problem. The splitting measures, selection of the most appropriate function and stopping criteria are the important measures to be considered in the top-down recursive divide and conquer algorithm. There are two conceptual phases such as growing phase and the pruning phase. The splitting functions may be based on univariate or multivariate criteria. The node splitting may be on the number of possible splits such as binary split, categorical split and numerical split which depends on the number of possible probability ways. The optimal decision tree depends on the strength of the splitting rule. The non-parametric theory is useful in decreasing the bias and is an attempt to minimize the mean square error with good model complexity. The density estimation solutions such as density functions (univariate and multivariate), conditional distribution functions and density derivatives, regression functions and other statistical functions are used in non-parametric approach. The kernels, splines and series are some other main methods. Itani et al. in their work proposed a one-class model which relies on density estimation as part of a tree-based learning algorithm, called one-class decision tree (OC-Tree) [2, 7]. Van Loi Cao et al. in their work proposed a one-class classification with kernel density estimation and genetic programming for anomaly detection and proved their accuracy with support vector machines [3, 8, 9].

3 Proposed Work

A kernel is a non-negative real-valued integral function K satisfying the following two requirements:

$$1. \int_{+\infty}^{+\infty} K(u)du = 1 \qquad (1)$$

$$2.\ k(-u) = K(u) \qquad (2)$$

The kernel definition results in a probability density function, and the average of the distribution is the sample used. Kernel density estimation is used to find the unobservable probability density function in the input dataset. The kernel function is symbolized as K, the bandwidth is represented as h, and the estimated density function at any point x is [4, 10]

$$f(x) = \frac{1}{n} \sum_{i=1}^{n} k\left(\frac{x - x(i)}{h}\right) \qquad (3)$$

The general *probability distribution function* is

$$f(x) = \frac{1}{\sigma\sqrt{2\pi}} e^{-\frac{1}{2}\left(\frac{x - \mu}{\sigma}\right)^2} \qquad (4)$$

In probability theory, the normal or Gaussian distribution is a continuous probability distribution for a random variable. Under certain conditions, the average of observations with definite mean and variance converges to the normal distribution when the number of samples increases. These types of probability density functions are used with kernels in the classification of data.

Entropy

Entropy is related to information as an increase in uncertainty. Information entropy refers to the average rate at which information is produced. Shannon discovered the idea of entropy to measure the disorder in the data. Shannon entropy is given by

$$s = -\sum_{i} p_i \log p_i \qquad (5)$$

Based on the probability, if an unlikely outcome of an event is observed, a high amount of information is associated with it. In contrast, if a more likely expected outcome is observed, the output is associated with a smaller amount of information. In the classification task of decision tree learning, entropy is a very important factor to build the tree. Constructing a decision tree procedure starts at the initial root node, thereby splitting the dataset S into number of subsets based on the possible values of the "best" attribute. The best value here refers to the one that minimizes the (combined) entropy in the resulting subsets. This recursive procedure is repeated until if there are no more other attributes left out to split. The working of the classification algorithm (ID3) is that.

1. It begins with the dataset as the root node.
2. Every unused attribute is chosen, and the entropy is calculated using the following Eq. (6) as

$$\text{Entropy} = \text{classcounts[instances]} * \log_2(\text{classCounts[instances]}) \quad (6)$$

3. The attribute with the smallest entropy of the largest information gain is chosen for a further split.
4. Recursion of the subset takes place only on the never selected attributes.
5. The algorithm terminates in any of the cases:

    1. Every element belongs to the same class, i.e., leaf node.
    2. There are no more unused attributes to be selected.
    3. There is no matching attribute such that the particular data is created with a leaf node.

Iterative Dichotomiser (ID3) algorithm uses the above procedure for the construction of the tree. Hence, entropy characterizes the dataset, and if the value of entropy is equal to zero, it is perfectly classified. Upon measuring, a random variable entropy determines how much information is expected to be gained. It is used to enumerate the amount to which the distribution of the quantity's value is unknown. ID3[6] is thus a greedy heuristic algorithm performing the best first search with optimal entropy values. In the proposed *non-parametric entropy modified Iterative Dichotomiser algorithm (NEMID)*, the existing entropy equation in ID3 is further optimized by substituting the kernel density estimation function. Kernel density estimators center a kernel function at each data point to reduce the dependency on the endpoints of the bins. For smoother kernel function, there is a smooth density estimate over a local neighborhood of that data point [4]. In the proposed *NEMID algorithm*, the following kernel function is substituted with the regular Shannon entropy value in the ID3 classification algorithm, which estimates based on the non-parametric approach.

$$E = \left(\text{Math.}\frac{\log\left(\text{Math.abs}\left(\frac{x-a}{x+a}\right)\right)}{2a}\right) + \text{Math.}\frac{\log_{10}(\text{Math.E})}{c} \quad (7)$$

where X represents the class count and the value of a is chosen to be 2 for the splitting of pair of attributes. For fine grain of values, the binary logarithm in the existing ID3 algorithm has been replaced by log base 10. The proposed non-parametric kernel-based entropy better classifies than the existing entropy-based ID3 classification. Regular decision tree classifiers are based only on the default entropy function. The proposed kernel substituted function calculates the density of the distribution which will be very constructive to enumerate the unused selection of attribute. It supports the information gain by not eliminating data in the form of sparse data. The classification rate shows better performance which proves that the entropy value is accurate with the kernel function and the misclassification rate reduces as all the data has undergone exact test irrespective of the dense region of consideration.

3.1 Methodology and Data Analysis

The entire coding of the proposed decision tree classification with non-parametric kernel-based entropy (*NEMID algorithm*) experimented with Weka 3.7, the Waikato University open source data mining tool based on JAVA. Weka has the class for constructing an unpruned decision tree based on the ID3 algorithm which can only deal with nominal attributes. The seven training datasets are chosen from UCI repository with different types of attributes for data analysis. The results are tested and compared with the existing ID3 algorithm. The following table represents the performance analysis and comparison of the algorithm. *NI* represents the number of instances, *CCI* represents the correctly classified instances, and *UI* represents the unclassified instance. Classification accuracy is represented in %. Accuracy is defined as the ratio of the sum of true positive and true negatives to the sum of the total number of positive and negative cases.

$$A = \frac{t_p + t_n}{P + N} \tag{8}$$

MAE represents the mean absolute error, and RMSE represents the root mean square error. Mean absolute error is the quantity used to measure how close predictions are to the outcomes and is given by

$$\text{MAE} = \frac{1}{n} \sum_{i=1}^{n} |f_i - y_i| \tag{9}$$

where f_i represents the prediction and y_i represents the true value. RMSE represents the quadratic scoring rule which measures the average magnitude of the error. The difference between the predicted and observed values is each squared, and then, average over the sample is taken. Finally, the square root of the average is taken. RMSE usually gives a relatively high weight in large errors. Monk 1, Monk 2 and Monk 3 test data and train data are tested with the given instances and attributes. The accuracy % for Monk 1 test data is 18.98 for *NEMID* algorithm and it varied for ID3 algorithm as 5.09. For the train data, the accuracy again increased as 43.54 while ID3 has 29.83. The accuracy % increased gradually while the diabetes dataset outperformed well with accuracy % 63.02 when compared with the accuracy % of ID3 as 26.43. The number of instances of diabetes was high as 768 when compared with other datasets, and our proposed *NEMID* algorithm outperformed well in all the test cases. The error rate is also considerably low for the Monk 1 and Monk 2 datasets when compared with the existing algorithm. The other datasets showed minimal variation, and RMSE is found to be a good score in all the test cases when compared with the ID3 algorithm. In all the statistical measures, the *NEMID algorithm* outperformed well with better classifier accuracy with minimal error rate. The following figure depicts the classification accuracy of *NEMID algorithm* with the ID3 algorithm (Table 1; Fig. 1).

Table 1 Statistical measures of NEMID algorithm versus ID3 algorithm

Dataset	NI		NI	ID3 algorithm %	MAE	RMSE	NI	NEMID algorithm %	MAE	RMSE
Monk 1 test data	432	CCI	22	5.09	0.94	0.95	82	18.98	0.57	0.59
		ICI	400	92.52			350	81.01		
		UI	12	02.31			00	00.00		
Monk 1 train data	124	CCI	37	29.83	0.67	0.80	54	43.54	0.56	0.63
		ICI	77	62.09			69	55.64		
		UI	10	08.06			01	0.80		
Monks-problems-2 test	32	CCI	47	10.87	0.84	0.91	119	27.54	0.55	0.58
		ICI	293	67.82			312	72.22		
		UI	92	21.29			01	0.23		
Monks-problems-2_train	169	CCI	60	35.03	0.58	0.75	71	42.01	0.54	0.61
		ICI	87	51.47			91	53.84		
		UI	22	13.01			7	4.14		
Monks-problems-3_test	432	CCI	21	04.86	0.94	0.96	86	19.90	0.56	0.57
		ICI	399	92.36			346	80.09		
		UI	12	02.77			00	00.00		
Monks-problems-3_train	122	CCI	39	31.96	0.64	0.79	46	37.70	0.54	0.61
		ICI	71	58.19			70	57.37		
		UI	12	09.83			6	4.19		
Diabetics	68	CCI	203	26.43	0.39	0.62	484	63.02	0.44	0.47
		ICI	130	16.92			248	32.29		

(continued)

Table 1 (continued)

Dataset	NI		ID3 algorithm				NEMID algorithm				
			NI	%	MAE	RMSE	NI	%	MAE	RMSE
Glass	241	UI	435	56.65	0.12	0.35	36	04.68	0.21	0.32
		CCI	38	17.75			76	35.51		
		ICI	29	13.55			138	64.48		
		UI	147	68.69			00	00.00		

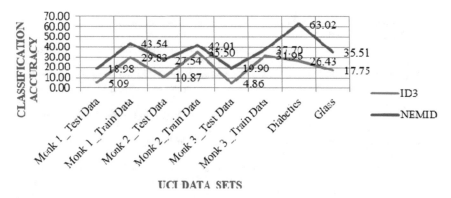

Fig. 1 Comparison of classification accuracy

4 Conclusion

The usage of prior probabilities in non-parametric classification methods has thus proven to be useful when compared with other approaches. The proposed non-parametric entropy modified ID3 (*NEMID algorithm*) thus better classifies than the existing entropy-based ID3 classification. Regular decision tree classifiers are mainly based only on the default entropy function. The proposed kernel substituted function calculates the density of the distribution which will be very positive to specify the unused selection of attribute and supports the information gain. The classification task is performed with better results with the proposed non-parametric decision tree classification approach based on kernel density estimation with entropy. Hence, it is concluded that the classifier works better than regular ID3 algorithms. As future work, it will be tested for all the datasets and the other parameters associated with the function with different entropy measures will be analyzed.

References

1. Rajeshkanna A, Arunesh K (2018) Role of decision tree classification in data mining. Int J Pure Appl Math 119(15):2533–2543
2. Itani S, Lecron F, Fortemps P (2018) A one-class decision tree based on kernel density estimation. arXiv preprint arXiv:1805.05021
3. Nicolau M, McDermott J (2016) One-class classification for anomaly detection with kernel density estimation and genetic programming. In: European conference on genetic programming. Springer, Cham, pp 3–18
4. Rajeshkanna A et al (2019) Estimation of probability density function using kernel density estimators for classification tree. J Adv Res Dynam Control Syst 11(11):344–348
5. Rajeshkanna A, Preetha V, Arunesh K (2019) Experimental analysis of machine learning algorithms in classification task of mobile network providers in Virudhunagar District. https://doi.org/10.1007/978-3-030-24322-7_43
6. Quinlan JR (1986) Induction of decisiontrees. Mach Learn 1:81–106

7. An introduction to information theory and entropy—Tom arter astarte.csustan.edu/~tom/SFI-CSSS/info-theory/info-lec.pdf
8. Leonenko N, Prozanto L, Savani V (2008) A class of Renyi information estimators for multidimensional densities. Ann Stat 36:2153–2182
9. Majdara A, Nooshabadi S (2019) Non-parametric density estimation using copula transform, Bayesian sequential partitioning and diffusion-based kernel estimator. IEEE Trans Knowl Data Eng
10. Maronna RA, Martin RD, Yohai VJ, Salibián-Barrera M (2019) Robust statistics: theory and methods (with R). Wiley
11. Izenman AJ (1991) Review papers: recent developments in nonparametric density estimation. J Am Stat Assoc 86(413):205–224
12. Exploratory data analysis: kernel density estimation. https://chemicalstatistician.wordpress.com/2013/06/09/exploratory-data-analysis-kernel-density-estimation-in-r-on-ozone-pollution-data-in-new-york-and-ozonopolis/

A Review on Word Embedding Techniques for Text Classification

S. Selva Birunda and R. Kanniga Devi

Abstract Word embeddings are fundamentally a form of word representation that links the human understanding of knowledge meaningfully to the understanding of a machine. The representations can be a set of real numbers (a vector). Word embeddings are scattered depiction of a text in an n-dimensional space, which tries to capture the word meanings. This paper aims to provide an overview of the different types of word embedding techniques. It is found from the review that there exist three dominant word embeddings namely, Traditional word embedding, Static word embedding, and Contextualized word embedding. BERT is a bidirectional transformer-based Contextualized word embedding which is more efficient as it can be pre-trained and fine-tuned. As a future scope, this word embedding along with the neural network models can be used to increase the model accuracy and it excels in sentiment classification, text classification, next sentence prediction, and other Natural Language Processing tasks. Some of the open issues are also discussed and future research scope for the improvement of word representation.

Keywords Word embeddings · Natural language processing · Bag of words · Contextualized word embeddings · Text classification · Transformer

1 Introduction

Word embedding is a technique used to map words from vocabulary to vector of real numbers. This mapping causes the words that emerge from a similar context that can be correlated with each other.

S. Selva Birunda (✉) · R. Kanniga Devi
Department of Computer Science and Engineering, School of Computing, Kalasalingam Academy of Research and Education, Anandnagar, Krishnankoil, Srivilliputtur, Virudhunagar, Tamil Nadu 626126, India
e-mail: sbirunda@gmail.com

R. Kanniga Devi
e-mail: rkannigadevi@gmail.com

© The Author(s), under exclusive license to Springer Nature Singapore Pte Ltd. 2021
J. S. Raj et al. (eds.), *Innovative Data Communication Technologies and Application*,
Lecture Notes on Data Engineering and Communications Technologies 59,
https://doi.org/10.1007/978-981-15-9651-3_23

The free text words of the vocabulary are converted into numeric values (vectors). This transformation is necessary as many Machine Learning models can understand only the vector representation. One of the elementary transformation approaches is one-hot encoding, where each word determines one dimension and the binary value (1 or 0) indicates the presence or absence of the word. One-hot encoding is computationally impossible when dealing with the entire dictionary, as the representation demands thousands of dimensions. Besides, this encoding has no hidden relationships among the words.

Word embedding is different from the conventional embedding in a way that it represents phrases and words in vectors of numeric values (which is non-binary). It refers to a dense representation of words in low dimensional vector space. For example, the words "Boy," "grapes," "man" is distributed in *n-dimensional* vector space. The distance between the words boy and man is smaller than between boy and grapes; thus, there is a similarity between the words "boy" and "man." This embedding provides a hidden semantic relationship among the words.

Currently, word embeddings are one of the successful applications of unsupervised learning as they do not need pricey annotation. Word embeddings are represented in a way that similar words have similar encoding. Word embeddings are useful in feature generation and Natural Language Processing tasks like text classification, document clustering, and sentiment classification. Word embedding technique can be categorized into Traditional or Frequency word embedding, Static word embedding, and Contextualized word embedding.

Frequency-based embeddings are categorized into count vector, TF-IDF, and co-occurrence. Static word embeddings are categorized into Word2Vec, glove, and fast text. Contextualized word embeddings are categorized into Elmo, GPT-2, and BERT.

This paper focuses on the review of different word embedding techniques, how they are used to represent the text, capture meanings, similarity among the text, and importance of the text.

2 Literature Survey

It is found from the literature survey that there exist three major word embedding techniques namely, Traditional word embedding, Static word embedding, and Contextualized word embedding. The following subsections present survey on these techniques.

2.1 Traditional Word Embedding

Traditional word embedding is based on the frequency that considers the whole document and discovers the significance of rare words in the document, count occurrence of each word, and co-occurrence of words.

El-Din [1] proposed a new enhancement bag of words method to solve the issue of manual evaluation of words. This embedding automatically evaluated the sentiment polarity and score it by using words weight. The proposed technique obtained a sentiment classification accuracy of 83.5% and could be extended to concentrate on phrases for the detection of the polarity.

Soumya et al. [2] proposed an approach to identify co-occurrence features from Wikipedia pages and incorporated this feature with a bag of words model. Co-occurrence was added to solve the issue of neglecting the word order. These features along with the naive bayes classifier brought about an F1 score of around 95% and could be enhanced further to use different classifiers.

Enríquez et al. [3] showed a word representation that is based on vectors, obtained through Word2Vec based on a bag of words. Bag of words (BoW) model is relevant to word embedding techniques as both techniques transform the word/text to numerical form. In the BoW approach, each document is depicted by vectors, where each dimension agrees to a word or group of words, producing vectors of high dimensionality. The BoW is a lexical representation relying on frequency-based metrics that describes the presence of a particular word in the sentence resulting in huge, sparse vectors. Word embedding is an alternative approach that converts text to numbers. This technique represents the words of vocabulary in a vector space, which has dimensions lower than the vocabulary size. Unlike BoW, word embedding technique traverses the lexical boundary as this word representation capture syntactic and semantic aspects. The combined representation of both BoW and Word2Vec provides higher accuracy of around 75%.

Tripathy et al. [4] compared Naive Bayes and Support Vector Machine (SVM) classifiers. To depict textual data as a numerical form, count vectorizer and term frequency-inverted document frequency (TF-IDF) embeddings were used. These embeddings, along with the SVM classifier achieved a maximum classification accuracy of 94% and could be boosted further to compare the maximum entropy classifier and stochastic gradient classifier.

Qu et al. [5] proposed a new TF-IDF method which added an improvement weight formula, as TF-IDF embedding studied only the relationship between the documents and not between the characters. The new TF-IDF along with the distance vector classifier obtained classification effect with an F1 score of 80–90% and might be enhanced to obtain more classification accuracy.

Dadgar et al. [6] proposed a novel text mining approach to classify news based upon TF-IDF embedding. This embedding calculated the weight of the word in the document. TF-IDF along with the SVM classifier achieved better performance using BBC and 20 newsgroup datasets, with precision around 97 and 94%. This could be augmented using different classifiers.

Jing et al. [7] proposed a new TF-IDF approach based on the improved TF-IDF. TF-IDF embedding used the IDF function to adjust the feature, whereas, new TF-IDF used the mutual information evaluation function. This new TF-IDF along with the vector space model achieved a classification precision of 88% and could be enhanced further to improve feature selection.

Kuang et al. [8] proposed a new feature weighting method, TF-IDF. Ci, an improvement of TF-IDF embedding. TD-IDF could not reflect the importance of degree and categorical differences. This issue got solved using the feature weighting method, which achieved a large macro F1 value of around 92.806. This could be extended further to take the fingerprint of the text for a copy test study.

Matsuo et al. [9] proposed an approach to extract the keyword from a single document using word co-occurrence embedding. Co-occurrence between frequent terms and each term were generated. This representation showed the importance of a term in the document. The proposed method achieved precision around 0.5 and further applied to domain-independent keyword extraction.

Albathan et al. [10] presented a co-occurrence embedding that extracted the high-quality patterns from the text by using weighted patterns. A pattern co-occurrence matrix was used to identify the relationship among extracted patterns and to reduce the noisy and closed sequential patterns. As a result, the number of extracted patterns was 89.04 and the noise could be reduced further.

Kadhim et al. [11] proposed a TF-IDF and co-occurrence embedding which adopted the cosine similarity to extract the features. The cosine similarity score function is calculated among the pair of vectors that depend on the number of common words and the frequency of the word in the document. TF-IDF and TF-IDF Global were compared and the results showed that features were reduced by 10–20% and could be improved further to reduce the high dimensionality of feature space.

Lott [12] surveyed the TF-IDF embedding which assigns weight to a given term to decide how the term characterizes an individual document inside a corpus. TF-IDF embedding could be used along with the naive bayes classifier to discover key phrases present in the document. This might be extended to include semantic analysis, lexical analysis, or co-occurrence measures.

Wang [13] suggested feature extraction by calculating the weight based on word co-occurrence. The degree of co-occurrence was calculated by the degree of relationship among each text. When the classification was high, the improved method seemed to be higher with an accuracy of 74% and could be enhanced further to consider synonyms, and the changed words.

Wartena et al. [14] proposed a co-occurrence embedding that measured the semantic relations among the words by calculating the co-occurrence distribution of words. This embedding excels in terms of precision, recall, and F1 score. Improvement in percents for top 5 and top 10 keywords in ACM dataset afford 9.2 and 6.9% precision, 10 and 8% recall, 9.7, and 7.3% F1 score and could be extended to improve precision and recall furthermore.

From the detailed survey of Traditional word embedding techniques, it is found that the "Co-Occurrence matrix" provides better results as this technique maintains the semantic relationship between the words.

2.2 Static Word Embedding

Static word embedding is a prediction-based that provides probabilities to the words and maps each word into a vector. Static embeddings learn by training the lookup tables which converts words into dense vectors. This embedding is static in the way that they do not alter the context once been learned and also the embedding tables do not change among different sentences.

Ge et al. [15] proposed a word embedding, Word2Vec that analyzed the semantic similarity between the words at a high precision rate. Feature reduction was achieved by loosely clustering similar features using a graph search technique. Multinomial Naïve Bayes classifier, SVM classifier, K nearest neighbor and random forest classifiers were compared. Word2Vec embedding with the SVM classifier achieved the highest accuracy of 93.93% and extended to clarify the loose clustering technique and a refined text corpus.

Pennington et al. [16] proposed a word embedding model called "Glove," Global vectors for word representation. The glove was a global log bilinear regression model based on the count data, it captured the linear substructures. Glove model performed well in the word similarity, word analogy, and named entity recognition tasks. The glove was based on the unsupervised model to generate vectors, and thus, provided accuracy and F1 score around 90% and could be improved to perform additional tasks.

Alrashdi et al. [17] proposed a model for processing the twitter tweets, it got preprocessed by removing hashtags, punctuation, emojis, and stop words. Glove embedding and crisis embedding and two architectures CNN and Bi–LSTM. Glove word embedding along with the Bi-LSTM reported the highest performance with the F1 score of 62.04%. This could be instructed further to use N-gram, CNN for crisis response in classifying the tweets.

Zhou et al. [18] proposed a Fast Text word embedding along with the KNN classifier provided high performance and preserved with pruning. This approach was suitable for online text classification applications. TC-Apte (Reuters version 3 corpus) and TC-PARC (Reuters version 4) generated a pruning ratio of about 82.8 and 72.1%. This might be extended for Chinese text corpus to knob multi-label and multi-class text classification.

Joulin et al. [19] proposed a word embedding approach "Fast Text," which was evaluated on two different tasks such as sentiment analysis and tag prediction. Fast text was trained on more than one billion words within a limited period. This method was able to classify half a million sentences within a minute. The maximum test accuracy of Fast Text evaluated with and without bigrams was 98.1 and 98.6%, and it could be improved furthermore.

Kuyumcu et al. [20] proposed a new approach Fast Text word embedding developed by Facebook. Fast Text embedding took into account the internal structure of words in the Turkish language. Fast text embedding assumed a word to be n-gram of characters. Fast Text embedding classified the text achieving an accuracy of 94% and might be enhanced further to different language texts.

Rezaeinia et al. [21] proposed a word embedding method improved word vector (IWV) which enhanced the accuracy of word embedding vectors by combining the POS tagging, lexicon-based approaches, and Word2Vec/Glove methods. Pre-trained word embeddings were trained on a large text corpus. The experimental result showed that the IWV method achieved accuracy around 80% and it could be considered as the base for sentiment analysis techniques.

Lilleberg et al. [22] suggested an approach based on the combination of word embeddings Word2Vec and TF-IDF. This combination without stop words outperformed the individual performance of themselves, and along with the SVM classifier, it provided 90% accuracy on text classification. This could be improved further to combine TF-IDF with Word2Vec along with the meaning of the words.

Vora et al. [23] proposed a word embedding approach to classify tweets based on emotions using the random forest classifier. The twitter tweets were classified into emotion categories. Word2Vec, Glove, and Fast Text had been used to generate word representations for the text. These word embeddings along with the random forest classifier achieved 91% precision and embellished further by classifying the text as more fine-grained classes.

Stein et al. [24] analyzed word embeddings along with the classifiers showed improvement in automatic document classification tasks. Word embeddings such as Glove, Word2Vec, and Fast Text were compared. The experimental results showed that the Fast Text with a classifier achieved an LCA F1 score of 0.893. Furthermore, applying Fast Text to the PubMed data analyzed whether this embedding extended to the medical text context.

Tezgider et al. [25] improved the word representations by tuning the Word2Vec parameters for Turkish text. For measuring vector size, word count, and window size, parameters were used. Word embedding quality was improved by properly selecting the parameters. TF-IDF, Fast Text, Glove, and Word2Vec embeddings along with the deep learning classifier model improved the accuracy of classification around 89% and could be enhanced to other language text.

Elsaadawy et al. [26] proposed Fast Text and Word2Vec embedding that create a vector using a weighted average of words representation. Naïve Bayes log count ratio was used to find the weight of each word. These embeddings were compared with the traditional representation such as TF-IDF and Naïve Bayes (SVM). Both the models provided a maximum classification accuracy of 90.26 and 89.93% and might be extended to use the ensemble classifiers.

Mikolov et al. [27] focused on the importance of pre-trained word representations predicted from large text corpus like Wikipedia, Web Crawl, and news collections. Fast Text and Glove embeddings were compared and pre-trained on text classification tasks along with the above corpus. Fast Text embedding resulted in achieving the maximum classification performance on an average accuracy of 82.7% and enhanced further to use this model in various tasks.

Raunak et al. [28] presented a novel technique that combined principal component analysis (PCA) based dimensionality reduction along with the post-processing algorithm. This algorithm could improve the pre-trained word vectors. Word embeddings used in memory-constrained devices could be improved by reducing their size. The reduced embeddings achieved maximum accuracy around 90% and explored further to use the word embeddings influenced by the contextualized embeddings.

From the detailed survey of Static word embedding techniques, it is found that the "Fast Text" provides better results as this technique can efficiently handle a lot of rare words that are present in the text corpus.

2.3 Contextualized Word Embedding

Contextualized word embedding is based on the context of a particular word, in which, similar words will have contrast context representations. These representations dynamically vary based upon the context in which a word appears.

Peters et al. [29] proposed deep Contextualized word representations, embeddings from language models (ELMo). ELMo embedding was pre-trained on a large text corpus. It depicted the syntactic and semantics information and word that had different contexts. ELMo embedding along with the neural network achieved classification accuracy around 97%. This might be improved further by applied to different NLP tasks.

Manoharan [30] proposed a smart algorithm for image processing by recognizing the text and vocalization for the visually challenged. LattePanda Alpha system on board was used to process the scanned images. Following the pre-processing, segmentation, extraction of features, images were classified into its similar alphanumeric characters. The textual data was converted into audio output by making use of a speech synthesizer. This model provided accuracy of about 97% and enhanced further to use Machine Learning algorithm to process the textual information and to boost the accuracy of the system and minimize the time delay for processing the information.

Maslennikova [31] compared the performance of ELMo and Word2Vec embedding. ELMo embedding increased the quality of prediction and to solve a binary classification problem. By using ELMo embedding, the F1 score was increased by 10% than the Word2Vec embedding. ELMo embedding could be enhanced by combining it with contrasting pre-trained models adopting different datasets.

Alsentzer et al. [32] explored Bidirectional Encoder Representations for Transformers (BERT) models for clinical text. Clinical researchers got benefitted from these embeddings. Some models trained on the clinical text and some others finetuned as BioBERT. BioBERT trained the BERT model over PubMed which was the source of biomedical research article corpus. This clinically trained BERT model provided high accuracy and an F1 score above 90% and could be further use for other clinical NLP research.

Wu et al. [33] proposed a novel data augmentation method called BERT for labeled sentences. BERT was a deep bidirectional model that is more powerful than the unidirectional and shallow bidirectional model. The conditional masked language model, an improvement of the masked language model to which conditional constraint was applied. This model outperformed other embeddings with an accuracy of around

79.60%. This might be enhanced further by applying BERT contextual augmentation to paragraph or document level.

Chang et al. [34] proposed X-BERT, a solution to fine-tune BERT embedding. This solved the difficulty in capturing dependencies that occurred in BERT. X-BERT denoted eXtreme multi-label text classification. X-BERT was used to abduct the contextual relations between the induced label cluster and the input text. This embedding achieved a precision rate of 68% and further extended by many fine-tuned BERT models to achieve high performance.

Schwartz et al. [35] explored the combination of two-word level embeddings for performing binary classification tasks. ELMo and BERT combination did not improve the classification ability. ELMo or BERT combined with the Glove embedding could improve the classification performance with maximum x^2 value as 12 and 16.69 and improved further by evaluating their performance in multiple datasets.

Jwa et al. [36] adopted BERT embedding to design an automatic fake news detection model "BAKE" (BERT applied to FAKE). Extra unlabeled news information added to the BAKE is "exBAKE." Both models were an automated process to detect fake news. These models worked on the FNC-1 dataset by examining the relationship between headlines and the body text of news articles and exceeded the performance by 0.125 and 0.137 F1 scores. This could be enhanced further to add additional news in the pre-training phase.

Han et al. [37] proposed domain adaptive fine-tuning, a simple approach for the unsupervised labeling applied to new domains. The contextualized embeddings were adapted through masked language modeling in the text. Fine-tuning of ELMo and BERT embeddings gave ample improvements with impressive results. The proposed model achieved an accuracy of 83% and extended further to focus on the source and the target domains.

Chakraborty et al. [38] analyzed the performance of sparse vectorizer that outperformed the neural word embeddings like Word2Vec, Glove, and Fast Text and character embeddings like ELMo. This embedding improved their performance and increased their F1 score by 3–5% and could be extended further by using neural classifiers like CNN and RNN instead of static embeddings.

Felipe et al. [39] compared the different models. Prediction-based embeddings predicted the probability of the next word. Count-based models leveraged global co-occurrence statistics in word. High-Level word embeddings encoded multiple relationships between words. Word2Vec, Glove, and Fast Text resulted in more accurate and faster embeddings. This might be improved further for high-level entities like sentences and documents.

Ethayarajh [40] examined the Contextualized word embeddings such as ELMo, BERT, and GPT-2 and how contextual they were. The upper layers of these embeddings produced more context-specific representations than the lower layers. Less than 5% of the variance on average, in contextualized representations, might be defined by static embedding. This could be extended to accomplish static embeddings as even more isotropic.

From the detailed survey of Contextualized word embedding techniques, it is found that the "BERT" provides better results as this technique can perform various

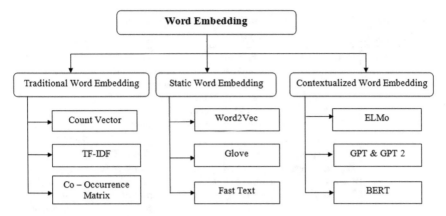

Fig. 1 Types of word embedding techniques

NLP tasks such as finding similarity between the texts, prediction of next sentence sequence, text classification.

Figure 1 shows the types of word embedding techniques which is found from the literature survey.

Table 1 Depicts the description, merits and demerits of each type of word embedding technique.

3 Future Scope

Contextualized word embeddings have attained incredible achievement in major NLP tasks. Even though, there are still a variety of problems that remain unexplored. In this section, some challenging problems are presented and future research directions.

Multi-Task Learning (MTL) is an active approach in Machine Learning and Natural Language Processing which desires to improve the performance of learning representations by making use of the information efficiently. In MTL, multiple learning tasks are figured out at the same time; i.e., multiple related tasks are studied together, and advantageous information is shared among the tasks. Liu et al. [41] conferred a multi-task deep neural network, to combine BERT language model pre-training and MTL for learning language representations obtaining state of the art performance on general language understanding evaluation (GLUE) benchmark at 82.2%. From the above perspective, it is better to manipulate further research by combining MTL and language model pre-training as both are integral to each other.

Few Shot Learning refers to the application of providing a learning model with a limited amount of training data. Conventional methods rely on hand-crafted features. Currently, it focuses on metric learning; an approach that depends upon the distance metric which finds similarity or dissimilarity between the objects. This metric method

Table 1 Merits and demerits of the types of word embedding techniques

Word embedding		Description	Merits	Demerits
Traditional	Count vector	Count vector is a method of counting the number of times each word occurs in the document	Count vectors can be used on real text data to provide accurate counts of the word content	Count vector did not afford semantic information and similarity between two documents
	TF-IDF (term frequency-inverse document frequency)	TF-IDF is calculated by the product of the frequency of the word in a particular sample and frequency of the word in the whole document	(i) Insists on the importance of a particular term by the rare occurrence of the word in all the documents (ii) It can easily compute the similarity between two documents	(i) Could not capture semantics, co-occurrences in different documents (ii) In the task of retrieving, it could not differentiate between singular and plural words
	Co-occurrence vector	The co-occurrence matrix is based on the concept of similar words in the same context that tend to occur together	It can maintain the semantic relationship between the words and is foster	Needed a large memory size to store the co-occurrence matrix
Static	Word2Vec	Word2Vec method can generate word embedding by using dense representation. It is a predictive model and it provides probabilities to the words which excel in word similarity tasks	(i) It can alter unlabeled corpus into labeled data by mapping the target word to its context word (ii) Sub-linear relationships are not defined implicitly	(i) It does not use global information (ii) Sub-linear relationships are not defined explicitly

(continued)

Table 1 (continued)

Word embedding		Description	Merits	Demerits
	Glove	The Glove is a count-based model. It is an unsupervised model for generating word vectors. Glove combines two features namely, local context window method and global matrix factorization	Glow relies on both local context information of words and global statistics (word co-occurrence). It can derive semantic relationships. It can predict the neighboring words by maximizing probability through the log bilinear model with a weighted least square objective. Word vectors can apprehend sub linear relationships	The Glove is framed on the co-occurrence man ix. It needs more memory for storage
	Fast text	Fast text is an extension of Word2Vec. It identifies the words as n-gram of characters. It provides an efficient vector representation of rare words	Fast Text can supply the vector representations for the words that are not present in the OOV words (dictionary), It can handle unseen words and it is word Fragment-Based	It does not add contextual information
Contextualized	ELMo	ELMo embedding is context-dependent and is character-based. A word may have dissimilar meanings depending upon the context where it is being used	ELMo provides multiple word embed dings for a single word	ELMo is shallow bidirectional as it cannot take advantage of both left and right contexts simultaneously

(continued)

Table 1 (continued)

Word embedding		Description	Merits	Demerits
	GPT-2	GPT-2 denotes Generative Pre-trained Transformer 2. It is a decoder only transformer	GPT-2 can predict the next word by looking at parts of a sentence. It can give more than ten possible predictions that what will be the next word using the probability score	GPT-2 requires heavy computation, chance to create false information as it is trained over millions of websites
	BERT	BERT is the first unsupervised deep bidirectional system with a multi-layer bidirectional Transformer encoder	BERT is used to learn the contextual relations between the words or sub words. It holds both syntactic and semantic meanings of a text. It can make assumptions for the blank word in between the sentence	BERT can only handle sentences of minimum length

studies the sample representation, that acts as a word embedding for token level Natural Language Processing tasks. This leads to the demand for designing embedding committed to few-shot learning samples. Hou et al. [42] reduced representation ambiguity using a special query support arrangement in a few-shot learning and pairwise embedding.

Pre-training models like ELMo, BERT, and GPT are trained by **self-supervision**. Here, unlabeled corpus is reassembled to get the labeled data and these tasks depend upon the co-occurrence of sentences and words. Syntactic, semantic, and lexical information have been ignored which is very important. Sun et al. [43] suggested learning the tasks through MTL using a continual pre-training framework. Various pre-training tasks concerning semantic, structure, and word-level are examined which yields auspicious results. Thus, imposing this model to learn strong representations by designing challenging tasks.

4 Conclusion

From the detailed literature survey, it is found that there exist three major word embedding techniques namely, Traditional word embedding, Static word embedding, and Contextualized word embedding. Each word embeddings constitutes three techniques. Traditional word embeddings are frequency-based; Static word embeddings are Prediction-Based, and Contextualized word embeddings are contextually-based. Further, transformer-based Contextualized word embedding along with the neural network can be applied to tasks such as text classification, document clustering, and sentiment analysis to improve the accuracy of classification and clustering. Finally, open issues and promising future scope in word representations have been discussed.

References

1. El-Din DM (2016) Enhancement bag-of-words model for solving the challenges of sentiment analysis. Int J Adv Comput Sci Appl (IJACSA)7(1):244-252
2. Soumya George K, Joseph S (2014) Text classification by augmenting bag of words (BOW) representation with co-occurrence feature. IOSR J Comput Eng (IOSR-JCE) 16(1):34–38
3. Enríquez F, Troyano JA, López-Solaz T (2016) An approach to the use of word embeddings in an opinion classification task. Exp Syst Appl 66:1–6
4. Tripathy A, Agrawal A, Rath SK (2015) Classification of sentimental reviews using machine learning techniques. Proc Comput Sci 57:821–829
5. Qu S, Wang S, Zou Y (2008) Improvement of text feature selection method based on TFIDF. In: 2008 international seminar on future information technology and management engineering. Leicestershire, United Kingdom, pp 79–81
6. Dadgar SMH, Araghi MS, Farahani MM (2016) A novel text mining approach based on TF-IDF and support vector machine for news classification. In: IEEE international conference on engineering and technology (ICETECH), Coimbatore, pp 112–116
7. Jing L-P, Huang H-K, Shi H-B (2002) Improved feature selection approach TFIDF in text mining. In: Proceedings of the first international conference on machine learning and cybernetics. IEEE, vol 2, pp 944–946
8. Kuang Q, Xu X (2010) Improvement and application of TF·IDF method based on text classification. In: 2010 international conference on internet technology and applications. Wuhan, pp 1–4
9. Matsuo Y, Ishizuka M (2004) Keyword extraction from a single document using word co-occurrence statistical information. Int J Artif Intell Tools 13(01):157–169
10. Albathan M, Li Y, Algarni A (2012) Using patterns co-occurrence matrix for cleaning closed sequential patterns for text mining. In: Li Y, Zhang Y, Zhong N (eds) Proceedings of the IEEE/WIC/ACM international conference on web intelligence and intelligent agent technology, vol 1, pp 201–205
11. Kadhim AI, Cheah Y-N, Ahamed NH, Salman LA (2014) Feature extraction for co-occurrence-based cosine similarity score of text documents. In: 2014 IEEE student conference on research and development, pp 1–4
12. Lott B (2012) Survey of keyword extraction techniques. UNM Educ 50:1–11
13. Wang LH (2014) An improved method of short text feature extraction based on words co-occurrence. Appl Mech Mater 519:842–845 (Trans Tech Publications Ltd.)
14. Wartena C, Brussee R, Slakhorst W (2010) Keyword extraction using word co-occurrence. In: 2010 workshops on database and expert systems applications. IEEE, pp 54–58

15. Ge L, Moh T-S (2017) Improving text classification with word embedding. In: 2017 IEEE international conference on big data (big data), Boston, MA, pp 1796–1805
16. Pennington J, Socher R, Manning CD (2014) Glove: global vectors for word representation. In: Proceedings of the 2014 conference on empirical methods in natural language processing (EMNLP), pp 1532–1543
17. Reem A, O'Keefe S (2019) Deep learning and word embeddings for tweet classification for crisis response. In: The 3rd national computing colleges conference. arXiv preprint arXiv:1903.11024.
18. Zhou S, Ling TW, Guan J, Hu J, Zhou A (2003) Fast text classification: a training-corpus pruning based approach. In: Eighth international conference on database systems for advanced applications, 2003 (DASFAA 2003). Proceedings, Kyoto, Japan, pp 127–136
19. Joulin A, Grave E, Bojanowski P, Mikolov T (2017) Bag of tricks for efficient text classification. In: Proceedings of the 15th conference of the European chapter of the association for computational linguistics, vol 2, Short Papers, pp 427–431
20. Kuyumcu B, Aksakalli C, Delil S (2019) An automated new approach in fast text classification (fastText). A case study for Turkish text classification without pre-processing. In: Proceedings of the 2019 3rd international conference on natural language processing and information retrieval, pp 1-4
21. Rezaeinia SM, Ghodsi A, Rahmani R (2017) Improving the accuracy of pre-trained word embeddings for sentiment analysis. arXiv preprint arXiv:1711.08609
22. Lilleberg J, Zhu Y, Zhang Y (2015) Support vector machines and Word2Vec for text classification with semantic features. In: IEEE 14th international conference on cognitive informatics & cognitive computing (ICCI*CC), Beijing, pp 136–140
23. Vora P, Khara M, Kelkar K (2017) Classification of tweets based on emotions using word embedding and random forest classifiers. Int J Comput Appl 178(3):1–7
24. Stein RA, Jaques PA, Valiati JF (2019) An analysis of hierarchical text classification using word embedding. Inf Sci 471:216–232 (Elsevier)
25. Tezgider M, Yıldız B, Aydın G (2018) Improving word representation by tuning Word2Vec parameters with deep learning model. In: 2018 international conference on artificial intelligence and data processing (IDAP), Malatya, Turkey, pp 1–7
26. Elsaadawy A, Torki M, Ei-Makky N (2018) A text classifier using weighted average word embedding. In: 2018 international Japan-Africa conference on electronics, communications and computations (JAC-ECC). Egypt, pp 151–154
27. Mikolov T, Grave E, Bojanowski P, Puhrsch C, Joulin A (2017) Advances in pre-training distributed word representations. arXiv preprint arXiv:1712.09405
28. Raunak V, Gupta V, Metze F (2019) Effective dimensionality reduction for word embeddings. In: Proceedings of the 4th workshop on representation learning for NLP (RepL4NLP-2019), pp 235–243
29. Peters ME, Neumann M, Iyyer M, Gardner M, Clark C, Lee K, Zettlemoyer L (2018) Deep contextualized word representations. arXiv preprint arXiv:1802.05365 [cs.CL]
30. Manoharan S (2019) A smart image processing algorithm for text recognition, information extraction and vocalization for the visually challenged. J Innov Image Process (JIIP) 1(01):31–38
31. Maslennikova E (2019) ELMo word representations for news protection. In CLEF (Working Notes).
32. Alsentzer E, Murphy J, Boag W, Weng W-H, Jindi D, Naumann T, McDermott M (2019) Publicly available clinical BERT embeddings. In: Proceedings of the 2nd clinical natural language processing workshop, pp 72–78
33. Wu X, Lv S, Zang L, Han J, Hu S (2019) Conditional BERT contextual augmentation. In: Rodrigues J et al (eds) Computational science—ICCS 2019. Lecture notes in computer science, vol 11539. Springer, Cham
34. Chang W-C, Yu H-F, Zhong K, Yang Y, Dhillon I (2019) X-BERT: eXtreme multi-label text classification using bidirectional encoder representations from transformers. arXiv preprint arXiv:1905.02331 [cs.LG]

35. Schwartz A (2020) Combining word embeddings for binary classification tasks
36. Jwa H, Oh D, Park K, Kang JM, Lim H (2019) exBAKE: automatic fake news detection model based on bidirectional encoder representations from transformers (BERT). Appl Sci 9:4062
37. Han X, Eisenstein J (2019) Unsupervised domain adaptation of contextualized embeddings for sequence labeling. In: Proceedings of the 2019 conference on empirical methods in natural language processing and the 9th international joint conference on natural language processing, pp 4238–4248. arXiv:1904.02817 [cs.CL]
38. Chakraborty R, Elhence A, Arora K (2019) Sparse victory—a large scale systematic comparison of count-based and prediction-based vectorizers for text classification. In: Proceedings of the international conference on recent advances in natural language processing (RANLP 2019), pp 188–197
39. Felipe A, Xexéo G (2019) Word embeddings: a survey. arXiv preprint arXiv:1901.09069
40. Ethayarajh K (2019) How contextual are contextualized word representations? Comparing the geometry of BERT, ELMo, and GPT-2 embeddings. arXiv preprint arXiv:1909.00512
41. Liu X, He P, Chen W, Gao J (2019) Multi-task deep neural networks for natural language understanding. arXiv preprint arXiv:1901.11504
42. Hou Y, Zhou Z, Liu Y, Wang N, Che W, Liu H, Liu T (2019) Few-shot sequence labeling with label dependency transfer. arXiv preprint arXiv:1906.08711
43. Sun Y, Wang S, Li Y, Feng S, Tian H, Wu H, Wang H (2019) Ernie 2.0: a continual pre-training/framework for language understanding. arXiv preprint arXiv:1907.12412

Perceiving Social Computing via Social Networks on Cloud

Jana Shafi and P. Venkata Krishna

Abstract Online social networks growing rapidly in the industry as well as in the academia. Nowadays, it is easy for people to share their facilities, data, and resources through social networks. Social network made a drastic change in the method of interaction and communication. The digital communities can represent, document, and explore the social relationship by establishing and contributing. With this development of social networking, computational applications also further develop social interaction and collaboration beyond personal computing. The compute networks facilitate the intersection of social studies and human associated which is the part of information technology further administered by social computing. The key findings of this work converse for the service of social networks control by social computing also demonstrate the model of social computing of the social networks which clarifies the allocation of resources in the cloud.

Keywords Social cloud · Cloud computing · Resource allocation · OSN

1 Introduction

Public science and computer science intersection has emerged into a computing prototype of social computing [1]. This leads to multi-disciplinary tactics in studying the model and analyzing social behavior to construct intelligent communicating apps. The results of these analyses play a vital role in diverse media and platforms [2]. Systems of trust/reputation, recommender systems, peer networks, Wikipedia's, social networks, and online auctions are some of the group of services provided by social computing and further handled by a group of people. The advancement of Web–Internet technologies as well as computing applications and technology has

J. Shafi · P. V. Krishna (✉)
Sri Padmavati Mahila Visvavidyalayam, Tirupati, India
e-mail: parimalavk@gmail.com

J. Shafi
e-mail: janashafi09@gmail.com

© The Author(s), under exclusive license to Springer Nature Singapore Pte Ltd. 2021   283
J. S. Raj et al. (eds.), *Innovative Data Communication Technologies and Application*,
Lecture Notes on Data Engineering and Communications Technologies 59,
https://doi.org/10.1007/978-981-15-9651-3_24

progressed widely in past few years. With an increase in the usage of mobile devices and computing resources, convenience of accessing media content has become easy with an outcome of social and cultural changes. Computing has changed from centralized to a decentralized network allowing a single user with an authorization to use the Web in social interaction and sharing content, expertise, and information. As a result, linking people, sharing knowledge and social networks has become a part of computer networks [3]. Connectivity establishment, social collaboration, and social community are the main attributes of computational social science [4]. Decision making and information overloading are some of the problems occurring in social computing services because of its invasive dominant usage. Users are confronted with selecting products and reliable services in any type of dealings, and finally, users want to take the opinion of trusted parties or friends in social networks.

The full paper is organized as Sect. 2 explains service of social computing in social networks. Section 3 clarifies the allocation of resouces, and Sect. 4 demonstrates the cloud architecture of social computing. Section 5 concludes and Sect. 6 depicts the future work with their improvement.

2   Social Networks: Service of Social Computing

The communication between the OSN and users is established within individuals or objects to facilitate information exchange among a group or individual user based on the relationship among them. Graphs and matrices are used to represent data. Social networks are widely analyzed by using graph theory because of its presentational simplicity and capacity. Study analysis about properties of social network graphs is extensively carried out [5]. The progressing research regarding possessions of OSNs graph and the features related to the community effects outlining the network structure as well as influence information systems. In social networks, the edges are called relationships, and nodes are termed as actors. The graphical presentation is also called a "sociogram," and relationships are of two types non-directional or directional. The foremost goal of the social network is analyzing a user's behavior and characterizing the relationship exit among user's social group. Social networks are locally intense and globally spare [6]. Social network snapshot, deducing innovative communications between its participants which are probably to happen in upcoming time, is dignified as a link forecast issue. The link prediction issue requests toward the growth of a modeled social network using structures inherent to the networks. It is associated with the inferring problem of lost links after a perceived network. In numerous domains, a Web network of communications grounded on viewable information is fabricated, and after that, existing links are concluded [7]. Everyway can be observed as computing an amount of immediacy or resemblance among vertices comparative to the topology of the network. In common, the approaches are revised from graph theory methods and analysis of the social network which are the vibrant graph theory power is not only in its vocabulary also in its theorems. Following are the two groups of link prediction approaches:

- Neighborhood node ways: These methods are grounded on the notion of the two nodes which are expected to create a link in the upcoming future if their neighboring circles consume a big overlay.
- Shortest path means: It grades distance of their shortest route between two nodes. Such a degree practices association networks to which people are connected via small chains. Few approaches improve the idea of shortest route by indirectly seeing the collaboration of altogether routes among binary nodes. An extraordinary social network type is termed as a recognized network, in which nodes are as actors who belong to events. Associated networks are also recognized as groups of subgroups of objects. Every incident labels the subgroup of performers who are united with it, and respectively every actor terms the subclass of events. In the hypergraph approach, inspecting an associated network is inspected in this way. A hypergraph is a generalized graph, in which an edge can connect sum of nodes. These so-called nodes are edges and actors which are measured as the set of events. Moreover, simple graphs or directed graphs are used to denote the compound networks which do not give a whole explanation of the practical systems under examination. For instance, the linked network signifies the structure of the graph.

2.1 Model

Social relation exists among dual users concluded grounded on a transitive or a mutual belief relation among them. In this method, the presence of the trusted network is represented by solid thick edges which aid to drag social network relations (hyperlinks) and outcomes a smaller amount scarcity. The same stuff score forms among two clients which encourages social relationships among them [7]. Consequently, product–product edges, which are the same among them in a recommender-based system are represented by dotted, dashed edges, can be cast-off to build a socialize relationship represented by hyper ends among the clients who score alike for those items. In the naivest method, two users are linked in case of who has scored a mutual product.

3 Resource Allocation: Social Cloud Compute via Social Networks

3.1 Social Cloud

A social cloud is defined as a service and resource which is used to share context exploiting relations recognized among members of social networks [8]. It is active environments by which (novel) cloud-like provisioning situations can be recognized created upon the implied trust levels that excel the inter-personal relations which is digitalized in social networks. Controlling social networking platforms such as intermediaries or the attainment of a cloud structure can be inspired by their common acceptance, scope, and the degree of their usage in modern culture [9].

3.2   A Social Cloud Compute

A social cloud compute is structured to permit entree in flexible computable abilities offered by a fabric of cloud built over capitals of socially connected peer contribution. This cloud is a type of communal [10], as the requirements are possessed, offered as well expended through social community members. Also, with the help of this organization, customers are talented to implement plans on virtualized capitals that reveal (protected) admittance to contributed resources. In this cloud model, clients are able to run apps, possibly at the same time, on their assets on sandboxed lightweight virtual machines (VM) [9]. Although the notion of a social cloud computing can be useful to any kind of virtual atmosphere. Regarding in this context [11] hardwearing virtualized surroundings which are grounded on Xen, yet the time to generate as well as the visualize context of virtual machines can be understood in the following Fig. 1 which is more significant.

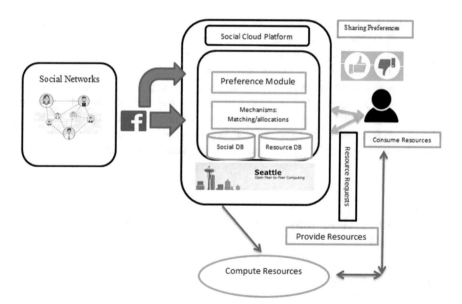

Fig. 1   A social cloud compute components

4 Social Compute to Cloud: Architecture

There are three areas examined in building social computing, namely a platform of social cloud, economic-social model, and technical socio adapters. A platform of social cloud [12] is the methodological execution for the building and simplification of the social cloud also essential middleware to allow resource allocation among "friends" of the Internet. Technical socio adapters are meant to perceive and understand social networks for the generation of distribution preferences. An economic-social model is to understand the design of a microeconomic public structure for the distribution of properties on the grounds of social networks concerning the client-specific zeal to consume the offered resources.

4.1 A Platform: Social Compute to Cloud

A social cloud to compute components can be understood as a model of the cloud which is displayed in Fig. 1 is a platform that is obligatory to facilitate and manage. Center in the system is a social clearing house in which all the information regarding clients includes the preferences allocation, stock resources, and requests. The house necessitates two of the databases aim to arrest user's social graph, also their allocation likings, and a resource leader to save track of its issues, obtainability, also provisions [9]. Middleware runs the elementary reserve materials, virtual resources, and tools for sandboxing, establishing also using capitals. It also describes the procedures looked-for for clients as well as requirements for entry and exits the system. Technical socio adapters can be understood in a situation as an application of Facebook, as well as required to offer admission to the essential features of socializing user's networks, and behaves similar to authentication resources. The resource allocation is enabled by user's sharing preferences which are the requirement of social clearinghouse [9]. Matching mechanisms govern suitable provisions of possessions through the user's allocation preferences crosswise their social network. Compute properties are the methodological legacy of clients that they find the money for to digest from the social cloud.

4.2 Economic-Socio Model

An economic-socio model defines a social-compute cloud that agrees with the preference matching type which is being used and executed. This method reflects non-profit sharing tools grounded on the preferences of the user. This sort of matching is positively functional in various types, counting the student's admittance to university and potential scholars in schools. Exact marketplace objectives depend upon algorithms that occur in figure matching problem-solutions for example: figuring out a principally reasonable explanation or existing high-user well-being [9].

4.3   Seattle: Research Platform

Seattle is one of the research platforms which are open source and educational which is crafted to construct a disseminated overlap network system on compute means. For example, servers, personal computers, and mobile telephones contributed to its clients [9]. It also sorts a simple simulated coating that runs on a donor's system and allows extra users to execute applications through diverse operating systems and architecture styles. Significantly, the virtual layer guarantees that apps are remotely sandboxed from further plans executing on a similar host. Seattle software is executed in the high-level programming language (Python) and Django framework is used to construct the clearinghouse.

5   Conclusion

Social computing uses social networks which is one of its services that is proposed in this work. Cloud computing evolving as a convenient paradigm due to its service provisioning availability. A social cloud is presented as a stage that assists the allocation of structure capitals among friends through digitally programmed socialize relations and networks. Programs on virtualized resources provided by their networks can be operated by the clients. Seattle is one of the open research access platforms to user's social network, which permits clients to draw preferences for sharing and apply similar algorithms to facilitate socially preference-based attentive resource distribution.

6   Future Work

In the future motivation towards the additional methods to identify them inevitably from their social network. Lastly, the user's usage is explored and communication of their families or friends is the main purpose as well as transport execution to a production equipped system.

References

1. Kumar D (2019) Review on task scheduling in ubiquitous clouds. J ISMAC 1(01):72–80
2. Cappos J, Beschastnikh I, Krishnamurthy A, Anderson T (2009) Seattle: a platform for educational cloud computing. ACM Sigcse Bull 41:111–115. https://doi.org/10.1145/1508865.150 8905
3. Gu K, Liu D, Wang K (2019) Social community detection scheme based on social-aware in mobile social networks. IEEE Access 7:173407–173418. https://doi.org/10.1109/ACCESS. 2019.2956149
4. King I, Li J, Chan KT (2006) A brief survey of computational approaches in social computing. In: Proceedings of the international joint conference on neural networks, Atlanta, GA, June 2009, pp 2699–2706. May P, Ehrlich HC, Steinke T (2006) ZIB structure prediction pipeline: composing a complex biological workflow through web services. In: Nagel WE, Walter WV, Lehner W (eds) Euro-Par 2006. LNCS, vol 4128. Springer, Heidelberg, pp 1148–1158
5. Kadushin C (2012) Understanding social networks: theories, concepts, and findings. Oxford University Press
6. Scott J (2006) Social network analysis: a handbook. Sage
7. Tavakolifard M, Almeroth KC, Barbara S (2012) Social computing: an intersection of recommender systems, trust/reputation systems, and social networks. IEEE Netw, 0890-8044/12/$25.00 © 2012 IEEE
8. Chard K, Bubendorfer K, Caton S, Rana O (2011) Social cloud computing: a vision for socially motivated resource sharing. IEEE Trans Serv Comput 5(4):551–563
9. Caton S, Haas C, Chard K, Bubendorfer K, Rana OF (2014) A social compute cloud: allocating and sharing infrastructure resources via social networks. IEEE Trans Serv Comput 7(3):359–372
10. Mell P, Grance T, The nist definition of cloud computing. Technical report 800-145. National Institute of Standards and Technology. https://csrc.nist
11. Thaufeeg AM, Bubendorfer K, Chard K (2011) Collaborative eResearch in a social cloud. In: Proceedings of the IEEE seventh international conference on E-science (e-science), Dec 2011, pp 224–231
12. Zhuang Y, Rafetseder A, Cappos J (2013) Experience with seattle: a community platform for research and education. In: Proceedings of the second GENI research and educational workshop

Internet of Things Based E-Hooding in a Smart City

Enugala Vishnu Priya Reddy and Polala Niranjan

Abstract The advertisement boards are acting as mediators nowadays for business development. In upcoming days, the smart hooding boards will lead the marketing smarter in the smart cities. The hooding boards are now becoming user-friendly by implementing the Internet of Things (IoT). Recently, at main centers/junction points, there is huge traffic so, the hooding boards are in high demand for marketing purposes. But the marketing for all businesses may not be flexible here in this IoT-based method, i.e., smart hooding boards it will more accessible and reduce the pollution. In the concept of smart cities, the administration is done by utilization of smart hooding boards which provide a smarter way of providing public services like municipal, electric bills and also receives complaints area wise accordingly. This concept of the smart hooding system will be beneficial to the politicians, film industry, sports events with the smart hooding system, mobile e-film facilities, fan park system, for conducting a political meeting, newly released films, e-medicine events, online sports matches.

Keywords E-hooding · Public service · Marketing · Smart city · User-friendly

1 Introduction

Unity is the power of a nation. It was proved in India that rural background village people had a great unity among themselves. Similarly, the joint family also has a great unity among the family members. The rapid growth of the cities and urbanization will affect the growth of an individual lifestyle. At the same pace due to the busy scheduled life, people come across religious festivals, cultural events, sport international events, national disaster times, war situations, etc., and these are only highlighted by the media but the actual concern is missing. It results in the unity of a nation comes to the out nook of the city. Management is required to administrate the city to make unity; this can be achieved by gathering all variety of people on to a single platform for

E. Vishnu Priya Reddy (✉) · P. Niranjan
Research scholar, Career Point University, Kota, Rajasthan, India
e-mail: vishnupriya.enugala@gmail.com

© The Author(s), under exclusive license to Springer Nature Singapore Pte Ltd. 2021   291
J. S. Raj et al. (eds.), *Innovative Data Communication Technologies and Application*,
Lecture Notes on Data Engineering and Communications Technologies 59,
https://doi.org/10.1007/978-981-15-9651-3_25

sharing or gathering the information. The administration system should be designed in combination with the implementation of IoT which facilities with smart hooding boards from the main city for marketing purposes, e-mobile film facilities, sports international events for entertainment, and the advertisement can be done at any place irrespective of the place demand. As the high demand location boards have huge usage, so, the location, which is not in demand, can also be used in an informative way with this entire city that will be connected with every place along with the unity of the administration.

In recent days, many mobile applications and televisions satellite channels are accessed directly by everyone but, in the end, the concept of those programs is not delivered correctly, and it misleads them. In an individual life, each spends most of the time lonely by watching television programs they react emotionally which makes some other individuals feel stress and pressure. In a busy scheduled life, people spend their free time with the mobile, online the daily televisions serials only this impact shows in real life. For instance, a TV serial dominates the family members at home, and they don't even communicate till it reaches the end of the program. This is due the individual doesn't match their age whereas at the IoT implementation smart hooding board, they feel free and open with their age group of people along improving the unity among themselves for a smart city.

2 Review Literature

The rapid growth of information on real-time devices made it depend on the Internet of Things (IoT) technologies. The relationship between smart city and IoT is mentioned by Arasteh et al. [1] Architecture details of smart cities are discussed, and the view is presented by Pellicer et al. [2], El-Baz and Bourgeois [3], Da Silva et al. [4], while security issue is identified by Ijaz et al. [5]. A gap between the Internet of Things and the Cloud of Things is discussed by Petroleo et al. [6]. The need for fog computing for the smart city applications is told by Perera et al. [7]. Electric vehicles charging is an application of smart city described by Shuai et al. [8]. As a smart city application along with the analysis by videos using distinctive in-depth learning algorithm by Wang and Sing [9].

In smart city's data-centric view of application and services, the usage of wireless sensor networks has been intrinsically diverse by Rashid and Rehmani [10]. Smart homes and buildings with advanced IoT technologies are reducing the consumption of resources (electricity, water) [3] and with sensors monitoring [11]. Due to pollution, city is becoming unhealthy to keep an eye monitoring, and alert of change in things has identified, and detection of crimes security issues can also be monitored in a smart city [12].

3 Purpose and Design of the Study

Social ethics and relationship are decreasing day by day when compared to the olden days. The development of urbanization with the latest technologies connects each corner of the place of smaller countries to make a large country. The bridges bind the gap between the countries. As the implementation of the IoT, infrastructures could enable various and massive opportunities in the highest research motivations which are revealed, and then, some useful applications are outlined. So IoT is the best platform for the public with the smart hooding board's usage. To be united with the upcoming latest technology, it is necessary to develop these smart hooding boards using IoT for a smart city, recently IPL fan park system, in Delhi mobile-based film theaters, mobile TV telecasting; some of the film theater telecasting live matches have been very popular and very successful.

At present, in cities, hooding boards are organized by the third-party agency to generate revenue to the city authorities. Due to this reason, good quality is not provided, and cities are not maintained properly which leads to accidents. A solution to this problem is the user-friendly service of the e-hooding board. A city with a smart and different type of hooding board facilitates people will watch and enjoy. IoT technology will be useful to control smarter boards organized every ever.

3.1 Design of the Concept

The concept was designed by the different types of the hooding board for the required location for various purposes as per the situation demand and along with service orientation of hooding is designed in this study, considering every corner of the city with IoT platform. There are three types of hooding available, namely commercial hooding, social service hooding, and user-friendly hooding.

4 Research Methodology for the Design

Searching and selecting different types of users are to connect the end user and provide a basic idea about the smart city's. The smart hooding board facilitates the different people in various locations with their lifestyle and provides end-to-end services to the users in the entire city.

4.1 Commercial Hooding

The commercial hooding generates major revenue in the entire smart city. Organizers can choose the type of hooding based on the following three categories of E-hooding, namely high demandable E-hooding, medium demand E-hooding, and low demand E-hooding.

In the earlier days, there were third-party agencies for advertisement agencies used to take lease wise for the overall amount. The total revenue income goes to the third-party agency. As the administrator for the advertisement of the hooding board, the revenue fluctuates depending on the demandable days and un-demandable days. Hence, the third-party agency is biding for long-term lease which might be manually manipulated and not a transpiration method and leads to more misusing of the system.

The implementation of smart hooding system biding is automatically generated by the online transpiration method. There are three types of situations in which demand for hooding is a customer need identifying and well in advance in the plan as they required those economically and situational demands of the requirement as per prefixed listing of demandable rate. The registered businessmen or required hooding board authorities get enrolled through BAR code/QR code for the commercial smart hooding boards. After registration, the detail of listing rates depending on the demand for the board gets updated daily. There is an option to trigger an online bid without watching updates every time.

4.2 Social Service Hooding Board

The social service hooding boards are established everywhere in the city at the street entries. This e-hooding board will display all types of the colony/street complaints registration, and OTP satisfies serves, pending serves, and satisfy all types of information can get at the e-hooding board. The local area citizens or concerned authority officer of smart city services like electricity, municipal taxes, pollutions complaints, sanitary sewer, accidents, emergency health care all types of government. The NGO service identifying and organizing through these e-hooding boards can be accessible at the street corner to the main high way central point, and all information and service results can be handled by IoT latest technologies from the top level to low level of organized authorities.

Consider, for example, a street of people with contact addresses and the communicating device that collects all the data at the point of the street area with e-hooding board provides privacy within the street family person having a major health problem like cardio arrest, the chance of theft problem, and so on like in all emergencies. The citizen of any other near citizen or whoever is near located people identify the emergency, and they can intimate the information to the nearest street e-hooding board with the help of communicating device like mobile or Internet, message, social media. So that the information is communicated to all the family members of the street by having a huge gathering. And also to inform the consent of government authorities regarding the emergency. People try to communicate and respond immediately to get escaped from the problem by the effects of e-hooding boards.

Smart city complaint hooding is the daily service that can be observed by the IoT support system so that if any problem or complaint faced by the citizen can be recorded using an OTP. Similarly, the concerned authorities receive the complaint with the help of OTP and solve the problem. All this information is appeared as compliant in the street e-hooding board. For these types of IoT method involve the top level to ground level officers of the smart city, the citizen can interact with the support of the IoT-based technology.

4.3 User-Friendly Hooding Boards

Presently, people are facing an individuality problem due to the social status maintenance of those people who maintain a high quality of equipment so the security reasons they only using this device. By the implementation of the e-hooding board in an area, at the time of special events like culture, vocational, and festivals, based on the events, the e-hooding board gives a lot of information which unites the area with like-minded people. In this method, so many entertainment hooding boards are currently running successfully. Other types of hooding are situational hooding, community-oriented hooding, daily events hooding, slum hooding boards, and so on.

The IPL fan park successfully telecast 36×26 ft of big-screen television for live cricket matches of 36 s biggest cities that are fortunately running and getting grand success in India per day more than 10,000 people who are watching on a single screen.

In Delhi, recently mobility film theaters are running more than 100 mobile theaters, and these can be occupied up to 300 people at a time; the situation of demand is based on the area and the interest of the people.

In some special situations like VIP person visiting a location of the smart city as due to the huge rush of people and heavy traffic situation message of the VIP person are not able to convey to the people, so the e-hooding board communicates the exact message by avoiding the problem faced through IoT platform.

The trendy situations like marriage and function can be telecasted to people living in different places in the world. By uploading the link and telecast to the people by communicating with the e-hooding board by the support of the IoT platform.

In the smart city, daily many events are conducted and important persons visit; this daily information can be displayed and spread the information through the e-hooding board easily to cover the programs.

The movie releasing events and heavy crowd situations demand the entertainment hooding board which very useful for the smart city environment people.

In slum area, hooding boards are very essential as a part of living style to develop and motivate about health care tips, educational help services, development of living styles; (to avoids superstition beliefs) all these will help the people in a slum area to educate by using of this e-hooding boards.

4.4 Implementation Work

E-hooding boards can be developed with the support of IoT elements from the smart city administration and citizens. In the smart city, people and organizing hooding administration boards are always connecting through Internet service by different types of communicating devices mode and connected to the cloud with respective departments.

Figure 1 depicts the architecture of smart hooding. Here, the administration of smart cities is always connected to the cloud through online system support but there is no need for the citizen, i.e., user people to connect to the Internet every time. They need an online facility for any special requirements, need to activate their device, and automatically communicate with the better service of higher authorities and also to

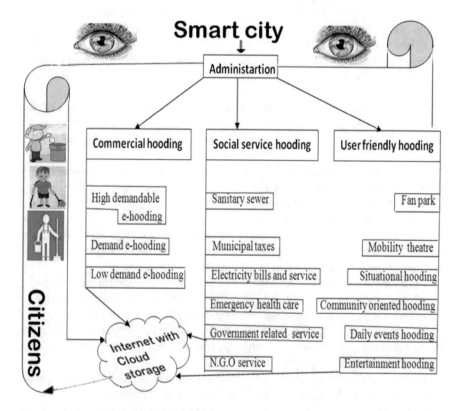

Fig. 1 The architecture of smart hooding

send feedback. All these are communicated through the Internet facility with cloud storage and retrieve the earlier data when required so that the administration can work automatically in a user-friendly manner. Smart hooding board activities are based on the earlier time demand, compared with the cloud storage data.

4.5 Statement of Ethical Approval with Sutiable Examples

The statements for ethical proof is mentioned with a similar few of the examples like a public corporation sector, devotional board, spots committee board in commercial, social service, entertainment events in India.

The Indian Railway Catering and Tourism Corporation (IRCTC) is a corporation that is the ethically approved one as a registered governing body. Considering in passenger point of view IRCTC e-ticketing is a similar concept of commercial hooding the passenger with the e-ticket gets the updates of the train through their registered contact details.

The Tirumala Tirupati Devasthanams (TTD) is the board of the organization for the devotional worship at Tirumala Tirupati which serves the devotee with the concept of social service hooding board in guiding them for traveling toward the worship place.

The Board of Control for Cricket in India (BCCI) is the largest committee for cricket gaming in India. In organizing the cricket matches, few of the matches were live telecasted named with as IPL fan parks, which are organized in some areas in India. The concept of IPL fan park is similar to the user-friendly hooding boards.

The above are the statements of the ethical approval concept for e-hooding boards in smart city.

4.6 Limitations

The study is limited to the area location with heavy rainfall and natural disaster to the remote location where the electricity supply is not sufficient. The implementation of the study utilizes a continuous power supply. This study is an exemption for physically handicapped. And also this study is an exemption for hackers and personal security.

5 Future Scope

In the upcoming days, there is a scope of implementing the e-hooding board as multi-user screens with connecting user mobile devices with the help of 5G technology, Wi-fi six networks. As people are getting busier, these e-hooding devices can also be used as mobile displays, and they also can be guided with indication e-hooding boards with the support and loading with the pre-scheduled data and alerting through the hooding board.

6 Conclusion

The most essential concept is to realize a smart city is with IoT. In a city, a data-oriented smart infrastructure is an eco-friendly system that is to be developed and established to maintain the urban environment in a smarter way. Nowadays, e-hooding boards became part of the development of the city. The increase in revenue from the e-hooding boards helps in the growth furthermore. The orientation of the smart hooding board service is to be user-friendly for the smart city people. The service-oriented e-hooding boards are the best, in educating people in the slum areas. Even from the entertainment point of view, these e-hooding boards will be helpful to reach throughout the city as mentioned in the above sections. Finally, for the better service to the smart city people, it is now very essential to develop e-hooding boards with the IoT technologies as they give 24/7 h of transparent services.

References

1. Arasteh H et al (2016) IoT-based smart cities: a survey. Accessed on Dec 2016
2. Pellicer S et al (2013) A global perspective of smart cities: a survey. In: Proceedings of the 7th international conference innovation mobile internet services ubiquitous computer. Taichung, Taiwan, July 2013, pp 439–444
3. El-Baz D, Bourgeois J (2015) Smart cities in Europe and the alma logistics project. ZTE Commun 13(4):10–15
4. da Silva WM et al (2013) Smart cities software architectures: a survey. In: Proceedings of the 28th annual ACM symptoms application computer. Coimbra, Portugal, 2013, pp. 1722–1727
5. Ijaz S, Shah MA, Khan A, Ahmed M (2016) Smart cities: a survey on security concerns. Int J Adv Comput Sci Appl 7(2):612–625
6. Petrolo R, Loscrì V, Mitton N (2017) Towards a smart city based on a cloud of things, a survey on the smart city vision and paradigms. Trans Emerg Telecommun Technol 28(1). Art. No. e2931
7. Perera C, Qin Y, Estrella JC, Reiff-Marganiec S, Vasilakos AV (2017) Fog computing for sustainable smart cities: a survey. arXiv preprint arXiv:1703.07079. Accessed on Apr 2017
8. Shuai W, Maille P, Pelov A (2016) Charging electric vehicles in the smart city: a survey of economy-driven approaches. IEEE Trans Intell Transp Syst 17(8):2089–2106
9. Wang L, Sing D (2016) Deep learning algorithms with applications to video analytics for a smart city: a survey. arXiv preprint arXiv:1512.03131. Accessed on Nov 2016
10. Rashid B, Rehmani MH (2016) Applications of wireless sensor networks for urban areas: a survey. J Netw Comput Appl 60:192–219
11. Hammi B, Khatoun R, Zeadally S, Fayad A, Khoukhi L (2018) IoT technologies for smart cities. IET Netw 7(1):1–13
12. Rathore MM, Ahmad A, Paul A, Rho S (2016) Urban planning and building smart cities based on the internet of things using big data analytics. Comput Netw 101:63–80

Content-Based Image Retrieval for Detection of Endometrial Cancer Using Radial Basis Function Network

K. Sujatha, B. Deepa lakshmi, Rajeswari Hari, A. Ganesan, and R. Shobarani

Abstract Endometrial cancer causes death among women community worldwide and that the content-based image retrieval (CBIR) is used to locate images in vast databases using machine vision techniques. Effective decision making is the only solution the physicians can offer to detect the occurrence of heart disease so that the patients can be relaxed and can undergo treatment at an appropriate time. The CBIR model on receipt of query on receipt of query, extracts the same set of features of a query which matches against indexed features index and retrieves similar images from the database. Further, these are facilitated by medical images stored in distributed and centralized servers are referred for knowledge, teaching, information, and diagnosis. Abnormal vaginal bleeding serves as the primary symptom of endometrial cancer. Thus, the system performance mainly depends on the features extracted from the computed tomography (CT) images that are adopted for indexing. Conversely, there is a shortage of ineffective scrutiny tools to identify the concealed relationship in the data pattern. This proposed technique intends to provide a study of existing techniques to form the knowledge base which will guide the gynecologists to take an effective decision. Nine features with radial basis function network were used to compute the formation of endometrial cancer at early stages. The feature selected must require lesser storage, retrieval time, cost of retrieval model and must support different classifier algorithms. Feature set adopted should support to increase the early diagnosis of the disease. This work summarizes the strength of local binary patterns (LBP) and

K. Sujatha (✉)
Department of EEE, Dr. MGR Educational and Research Institute, Chennai, India
e-mail: drksujatha23@gmail.com

B. Deepa lakshmi
Department of ECE, Ramco Institute of Technology, Rajapalayam, India

R. Hari
Department of IBT, Dr. MGR Educational and Research Institute, Chennai, India

A. Ganesan
Department of EEE, RRASE College of Engineering, Chennai, India

R. Shobarani
Department of CSE, Dr. MGR Educational and Research Institute, Chennai, India

© The Author(s), under exclusive license to Springer Nature Singapore Pte Ltd. 2021
J. S. Raj et al. (eds.), *Innovative Data Communication Technologies and Application*,
Lecture Notes on Data Engineering and Communications Technologies 59,
https://doi.org/10.1007/978-981-15-9651-3_26

its variants for indexing medical images. The efficacy of the LBP is verified using medical images from OASIS. The results prove good prospects of LBP and its variants which is due to the presence of unique binary patterns for normal and abnormal uterus conditions. The LBP features are used to train the intelligent classifier where feed-forward neural networks (FFNN) are trained with radial basis function network (RBFN). The RBFN facilitates the retrieval of images corresponding to normal and abnormal conditions.

Keywords Endometrial cancer · Magnetic resonance imaging · Local binary pattern · Radial basis function · Content-based image retrieval

1 Introduction

The image processing technology provides a non-invasive method for the diagnosis of endometrial cancer [1, 2]. The image processing algorithm monitors and facilitates a patient management system for drug treatment and offers an excellent cost-effective method [3, 4].

Nowadays, the printed copy of image formats is not used for the analysis of medical images. Since the printed copy of the images require more space, it would be better to use images in it's digitized or discrete form [5, 6].

The increase of medicinal information in digital libraries makes it harder to analyze them and diagnose the related disease. Since recovering text-based information belongs to a newly developed branch, an approach to solve this issue is required in indexing the metadata for the image. With these annotations, text-matching techniques that assess the similarity check between the required statement and the metadata are applied for retrieving images. Hence, this method is addressed as text-based or concept-based image retrieval [7, 8].

Generally to archive an image with an all-purpose vocabulary of therapeutic terms devours various reserves and requests broad cooperation attempts that are difficult to incorporate. It is reasonable to utilize instructive methodologies by beginning with more meticulous sets and effort simplification later. This strategy is utilized by Dean Bidgood in the composite SNOMED DICOM microglossary. Annotation of images by a human is a time consuming and cumbersome task and also leads to unrecoverable errors. Mark et al. have undergone the revise of therapeutic images by means of DICOM legends and exposed that 15% of explanation blunders arise on or after both creature and mechanism source. An additional considerable barrier in concept-based image retrieval systems is the fact that the inquiry does not allow the client to toggle and/or unite communication patterns during text deals [9, 10]. The initiative structure would mitigate the creature from the explanation task, by doing it routinely, and allowing image repossession by its content in its place of textual description. This framework is known as content-based image retrieval (CBIR) [11, 12].

CBIR provides a developing trend in therapeutic field publications. Comparative analysis of CBIR implementations in therapeutic imaging is presented by Long et al. Despite the growth of CBIR structures in medical imaging, the effectiveness of the structures in a number of medical functions is still very guarded. There exist only a couple of systems with relative success [13, 14]. To revision the uterine cervix cancer, the Cervigram Finder system was urbanized. It is a computer-aided structure where related images are retrieved from a database using local attributes from a client-distinct region in an image that is computed and using similarity measures. The Image Retrieval for Medical Applications (IRMA) structure is a mesh-based X-ray retrieval system. It allows the user to retrieve images from a database given an X-ray image as inquiry [15, 16]. CBIR for medical applications can be found in. Medical application is one of the priority areas where CBIR can meet more accomplishment due to population aging in developed countries.

2 Literature Survey

Endometrial cancer is a frequently occurring disease in women. Presently, Cervigram is used to diagnose uterus cancer. Some researchers have used a fully computerized automated scheme for the detection of endometrial cancer using image processing techniques so as avoid manual intervention by the physician. Effective image processing and intelligent algorithms facilitate a speedy and trustworthy detection of the reduced area of the uterus. Two phases are involved, which includes preprocessing and decision. The pre-processing phase involves feature extraction which is used by AI algorithms to make a decision regarding coronary heart disease. Hence, early detection of uterus cancer is achieved by segmentation and quantification [1].

Cervigram technique is used for the diagnosis of uterus cancer leading to various complications that cause death. Nayana Mohan et al. have projected an image handing out a method for identifying the uterus cancer. DICOM images of the heart are used for the revealing of endometrial cancer. The section of awareness is fragmented using enhancement diffusion filter and also usage of morphological operations like dilation and erosion. The section of the uterus is segmented along the medial line using fast marching based method. The diameter of the cancerous region is measured. This method is evaluated for a set of images with an accuracy of 86.67% for diagnosing endometrial cancer [2].

3 CBIR for Medical Applications

Comparative analysis of CBIR implementations in medical imaging is presented by Long et al. Despite the growth of CBIR frameworks in medical imaging, the utility of the frameworks is limited in the medical field. There exist two relatively successful schemes. Cervigram Finder scheme facilitates the detection of cancer in the uterus. The attributes extracted from the section of awareness of the uterus image are practiced by the algorithms for similarity measures, executed on a computer where such similar images are retrieved from the database and correlated with the abnormal condition. Medical applications that rely on image retrieval use an internet-based X-ray recovery system. The user can recover the images from the database with the help of a query for each X-ray image.

4 Methodology

CT images of the uterus for 141 patients were reviewed for identification of block in the heart. Firstly, 71 MRI scan images were analyzed to identify endometrial cancer and the remaining other 70 MRI scan images for validation. Randomly, the samples are selected with normal and abnormal images of the uterus to identify endometrial cancer (Fig. 1). Figure 2 depicts the block diagram for the detection and diagnosis of endometrial cancer. The MRI images shown in Fig. 3 correspond to various levels of endometrial cancer were obtained from the open-source database (https://radiop aedia.org/images). Initially, the MRI images of the uterus are pre-processed for noise removal, where the mean filter is found to get rid of the speckle blast present in the image. The signal to noise ratio is inferred to be 31db. The local binary features are extracted for all the MRI images of the uterus which helps to categorize and identify the normal and abnormal conditions. The LBP is used as normalized inputs to train and test the feed-forward multilayered architecture with RBFN.

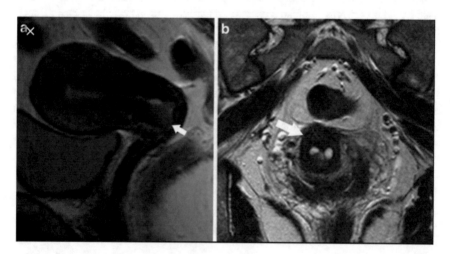

Fig. 1 MRI images with endometrial cancer

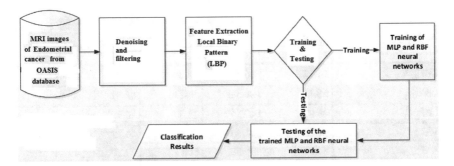

Fig. 2 Block diagram for diagnosis of endometrial cancer

Fig. 3 Training and validation of RBFN for diagnosis of endometrial cancer

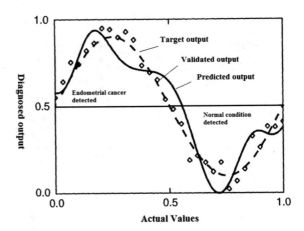

5 Results and Discussion

The purpose of experiments is to estimate and prove the capability of the LBP operator to represent the medical images mathematically. Experiments are performed separately to test the capacity of the operator to classify the images and to retrieve the images. A basic retrieval system that classifies and retrieves images using LBP whose binary pattern is unique for normal and abnormal uterus images to detect cancer (without the help of trained classifier) is tested and the results are shown in Fig. 3.

The algorithm presented here is capable of a new automated system for the detection of uterus cancer and quantification of the blocks. An acceptable PPV was obtained with slightly less sensitivity to quantitative evaluation. The user interaction and detection time can be improved by making small changes with respect to the parameters in the algorithm so that the diagnostic accuracy is maximized. Figure 3 shows the results for training and validation by RBFN for the identification of endometrial cancer (Table 1).

Table 1 Group-wise percentage of correctly retrieved images from OASIS MRI database

Method	Category				
	1	2	3	4	Average
$LBP_{8,1}^{riu2}$	54.12	36.50	38.89	41.86	42.04
$LBP_{16,2}^{riu2}$	58.33	39.02	36.11	55.81	46.59
$LBP_{24,3}^{riu2}$	48.67	46.34	38.89	48.83	44.31
$VAR_{8,1}$	58.28	41.46	41.66	44.18	45.45
$VAR_{16,2}$	54.34	36.58	36.11	37.20	40.34
$VAR_{24,3}$	52.51	36.58	38.88	34.88	39.77
$LBP_{8,1}^{riu2}/LBP_{16,2}^{riu2}$	68.79	53.66	58.33	53.49	56.82
$LBP_{8,1}^{riu2}/LBP_{24,3}^{riu2}$	72.21	63.41	63.89	55.14	61.93
$LBP_{16,2}^{riu2}/LBP_{24,3}^{riu2}$	68.79	60.98	61.11	51.16	58.52
$VAR_{8,1}/VAR_{16,2}$	62.09	46.34	47.22	48.84	50.00
$VAR_{8,1}/VAR_{24,3}$	54.44	43.90	44.44	46.51	46.02
$VAR_{16,2}/VAR_{24,3}$	62.06	58.54	58.33	51.16	55.68
$LBP_{8,1}^{riu2}/VAR_{8,1}$	55.35	48.78	52.776	48.83	51.73
$LBP_{8,1}^{riu2}/VAR_{16,2}$	62.21	51.21	50.00	44.18	50.01
$LBP_{8,1}^{riu2}/VAR_{24,3}$	60.32	48.78	55.00	44.18	49.43
$LBP_{16,2}^{riu2}/VAR_{8,1}$	70.54	63.41	63.89	51.16	60.23
$LBP_{16,2}^{riu2}/VAR_{16,2}$	64.43	51.21	58.33	51.16	54.44
$LBP_{16,2}^{riu2}/VAR_{24,3}$	58.87	48.78	47.22	44.18	48.86
$LBP_{24,3}^{riu2}/VAR_{8,1}$	68.98	58.54	63.89	48.84	57.95
$LBP_{24,3}^{riu2}/VAR_{16,2}$	62.19	43.90	44.44	55.81	50.57
$LBP_{24,3}^{riu2}/VAR_{24,3}$	60.22	48.78	52.77	48.83	51.13

6 Conclusion

An RBFN-based classification method is presented in this research to detect endometrial cancer from MRI scan images. It is inferred that an RBFN classifier is more
robust as compared with BPA, so as to reduce the false positive maintaining high
accuracy in the identification rate. At the first stage, the method discriminates against
the normal and abnormal conditions existing in the uterus followed by the identification of endometrial cancer. Thus, a classification-based approach has the potential to
ease the difficulty of the physicians by reducing the rate of manual interaction which
is eventually required to eliminate them completely.

References

1. Amant F, Moerman P, Neven P, Timmerman D, Van Limbergen E, Vergote I (2005) Endometrial cancer. Lancet 366(9484):491–505
2. Kitchener HC, Trimble EL (2009) Endometrial cancer state of the science meeting. Int J Gynecol Cancer 19(1):134–140
3. Cancer Research UK. CancerStats. Available via https://info.cancerresearchuk.org/cancer stats/. Accessed 4 July 2010.
4. Polin SA, Ascher SM (2008) The effect of tamoxifen on the genital tract. Cancer Imaging 8:135–145
5. Epplein M, Reed SD, Voigt LF, Newton KM, Holt VL, Weiss NS (2008) Risk of complex and atypical endometrial hyperplasia in relation to anthropometric measures and reproductive history. Am J Epidemiol 168(6):563–570
6. Minagawa Y, Sato S, Ito M, Onohara Y, Nakamoto S, Kigawa J (2005) Transvaginal ultra-sonography and endometrial cytology as a diagnostic schema for endometrial cancer. Gynecol Obstet Investig 59(3):149–154
7. Todo Y, Kato H, Kaneuchi M, Watari H, Takeda M, Sakuragi N (2010) Survival effect of para-aortic lymphadenectomy in endometrial cancer (SEPAL study): a retrospective cohort analysis. Lancet 375(9721):1165–1172
8. Sahdev A, Reznek RH (2008) Magnetic resonance imaging of endometrial and cervical cancer. Ann N Y Acad Sci 1138:214–232
9. Pecorelli S (2009) Revised FIGO staging for carcinoma of the vulva, cervix, and endometrium. Int J Gynaecol Obstet 105(2):103–104
10. Creasman W (2009) Revised FIGO staging for carcinoma of the endometrium. Int J Gynaecol Obstet 105(2):109
11. Dowdy SC, Mariani A (2010) Lymphadenectomy in endometrial cancer: when, not if. Lancet 375(9721):1138–1140
12. Tozzi R, Malur S, Koehler C, Schneider A (2005) Analysis of morbidity in patients with endometrial cancer: is there a commitment to offer laparoscopy? Gynecol Oncol 97(1):4–9
13. Kitchener H, Swart AM, Qian Q, Amos C, Parmar MK (2009) Efficacy of systematic pelvic lymphadenectomy in endometrial cancer (MRC ASTEC trial): a randomized study. Lancet 373(9658):125–136
14. Roland PY, Kelly FJ, Kulwicki CY, Blitzer P, Curcio M, Orr JW Jr (2004) The benefits of a gynecologic oncologist: a pattern of care study for endometrial cancer treatment. Gynecol Oncol 93(1):125–130
15. Benedetti PP, Basile S, Maneschi F, Alberto LA, Signorelli M, Scambia G et al (2008) System-atic pelvic lymphadenectomy vs. no lymphadenectomy in early-stage endometrial carcinoma: randomized clinical trial. J Natl Cancer Inst 100(23):1707–1716
16. Kong A, Powell M, Blake P (2008) The role of postoperative radiotherapy in carcinoma of the endometrium. Clin Oncol (R Coll Radiol) 20(6):457–462

Design and Development of a Mobile-Based Application for Automated SMS System Using Voice Conversion Technique for Visually Impaired People

S. Prasanna, K. Vani, and V. Venkateswari

Abstract Mobile phones have become an important and useful device for all people. Technology has made the user's life easy. Mobile phones are a flexible device to send a message by using a keypad on the phone. For visually impaired people, it is hard to send a message by using the keypad feature. To overcome the problem in the existing system, an android application has been developed for blind people. Android application development is a great platform for creating applications and games for all android users throughout the world. There are millions of blind peoples in the world with a huge drawback of not able to access mobile devices on their own. Disability of reading the text message and sending a text message has a big impact on their lives. In this research article, an architecture of the mobile application is suggested for the people who are visually challenged through which a user can able to send or receive SMS without the help of others. By using this android application, the visually impaired persons can convert their voice to text messages, and received text messages can be converted to voice. The Hidden Markov Model is described in which the process of conversion of text SMS to voice by changing the modulation of the signal and vice versa. The accelerometer sensor is used in this proposed work which keeps detecting the device movement such as shake, swing, or rotation and helps to enable/open the android application automatically. This application is a very flexible and most successful approach for visually impaired persons to send/receive speech recognized short message services.

Keywords SMS reader · Text to voice · Voice to text · And SMS sender · Hidden Markov Model · Accelerometer

S. Prasanna (✉) · V. Venkateswari
Department of Computer Science, Shri SSS Jain College for Women, Chennai, India
e-mail: s.prasana@shasuncollege.edu.in

V. Venkateswari
e-mail: v.venkateswari@shasuncollege.edu.in

K. Vani
Department of Computer Applications, Shri SSS Jain College for Women, Chennai, India
e-mail: k.vani@shasuncollege.edu.in

© The Author(s), under exclusive license to Springer Nature Singapore Pte Ltd. 2021
J. S. Raj et al. (eds.), *Innovative Data Communication Technologies and Application*,
Lecture Notes on Data Engineering and Communications Technologies 59,
https://doi.org/10.1007/978-981-15-9651-3_27

1 Introduction

In our daily life, the mobile phone plays an important role. SMS texting is an important action on the mobile phone and the usage of mobile device is very drastic worldwide. Many innovations and findings were being developed in mobile phones. Speech recognition and conversion plays an integral part of the innovation. For an ordinary human, entering a text for SMS is a simple task. However, for blind people, this task is too complicated. In the existing system, blind people cannot able to access messages on their own. Few methodologies were developed to overcome the difficulties for blind peoples. Braille keyboard is the technique being used for the visually impaired persons to read or to enter the text. It has domino dots using which one can recognize the text on their own. The people who are using mobile phones must memorize the keys and do the action which is difficult for inexperienced person. However, this device can do text-entry for blind people. But this device is too heavy and expensive. The alternative device for blind people to recognize text is the screen reader. The screen reader uses a text-to-speech engine which translates information into the speech. The problem in this device is that the user needs to remember the keyboard disposition. Only the experienced user can handle this type of technologies, inexperienced users have difficulties to operate these type devices. To overcome these kinds of problems and to make visually impaired people comfortable in using mobile phones, an application is proposed in this research article. This application can be operated very easily by blind people. This enables visually impaired persons to dial a number to make a call or send or receive messages in voice format.

2 Literature Review

An SMS communication system for blind people was developed which used various modules like global system for mobile communication (GSM), Braille keypad, loudspeaker, SD card, and buzzer. It worked in a communication channel for the blind people to communicate. With the help of AT commands operation, all the receiving SMS can be read by using the GSM module. This process is very helpful for blind people to know the content in the SMS. The blind person hears the message through the loudspeaker and also displays it on the LCD module. It also gives alarm using a buzzer to realize that the message is arrived [1]. To support the constructed application text-to-speech (TTS), recognition methods of handwritten English characters and synthesis of speech through the signal manipulation were developed which worked on handwritten character recognition. This application helped to visually impaired in government offices where they need to recognize the handwriting documents. The system recognizes the text and converts the received input text to the output as voice by signal modification process in which every character has converted into the signals. This project was based on OCR which read the characters from the scanned document or image, processes it, and extracts the feature and recognizing a pattern.

The image of the handwritten characters is captured in the three phases. This application uses the pre-process of an image as its initial phase. By following the initial phase, it includes feature extraction, segmentation, and recognition using OCR. The output is used to recognize characters in the form of a.txt file. In the last phase, all the characters are converted into voice by using a signal modulation method [2]. Implementation of an application using the conversion of characters into voice technique in which synthesized speech was extracted from the text. Here, the goal was to make a natural human speech to listen. The idea behind this work is to develop a PC-based text-to-speech conversion technique using MATLAB. The TTS conversion system was coded in MATLAB. The system takes a sentence, analyze each word, and finds out corresponding phonemes. It plays audio which is equivalent to the text input. It also displays a waveform and spectrum of the generated speech sound [3]. An Application for visually challenged people using the conversion of character into voice is architected by the method of TTS. İt was based on a finger reader-based character reading system for blind people. Raspberry-pi microcontroller is used to develop the conversion of TTS. The conversion is done by capturing the character with a camera, and the image is passed as an input to the TTS unit. Raspberry-pi microcontroller is the major one for installing the TTS unit. The receiving signal by the TTS unit is given to the speaker by using the audio amplified process.[4].

3   Accelerometer Sensor

The accelerometer sensor is a motion sensor that is activated when a movement is detected. It keeps verifying the device movement, such as shake, swing, or rotation. This sensor is used in this proposed work to open the application automatically. And also, it gets the shake input from the acceleration. The android service on Startcommand() is kept running continuously in the background in this application. The service and accelerometer sensor are used for opening the application without human intervention. Due to the gravitational effect on the earth, the accelerometer is measured as rest in 1 g. This sensor is used to measure 9.81 meters per second (mps) as its register. For feet measurement, it uses 32.185 feet per second (fps) as its register. A sensor force is measured as ms^{-2}. This is used in the model as three physical axes (x, y, and z) including the gravity force. The sensor manager initializes to bind the sensor to listen to the accelerometer and access the device sensor. Get is an instance of a class by calling context. getSystemService() argument with sensor service.

The x acceleration value in the graph will points positive only when the model is in a flat position by the motion to push it from left side to the right side. If the acceleration value possesses +9.81, then the model is in a flat position. This is closely similar to the acceleration value of the model (0 ms^{-2}) minus the gravity force (-9.81 ms^{-2}). If the acceleration value is exactly equal to +9.81 ms^{-2}, then the model is in flat position and the motion is forward to the sky. This is closely similar to minus the gravity force (-9.81 ms^{-2}).

Sensor type_accelerometer is used to detect the acceleration force along the axis x, y, and z including gravity. Unit of measure is ms^{-2}. This is a gravity-based sensor as the three-dimensional vector indicating the direction and magnitude. The coordinates are the same as used in the acceleration sensor. Linear acceleration is three dimensional which indicates the acceleration along each device axis excluding gravity. The output of the acceleration gravity and linear acceleration sensor follows the relation. Acceleration = gravity + linear acceleration.

4 User Selection Page

The user selection page is automatically opened by the accelerometer sensor. This page is used to select the call page or message page. A machine voice will direct the user. The voice informs the user about the page and says "please click in the center and say your dialog." Queue_flush synthesizes the speech from the text for a quick playback queue. An activity_selection method performs within a given time frame. It has a set of activities each marked by a start time and finish time. Once the button is clicked in the center, a dialog box appears for voice recognition. İt is easy to use an API by Google program for the conversion of voice into text which uses convolutional neural network (CNN) which was designed for automatic speech recognition (SPR). The API recognizes the languages. Through this technology, the speech is converted and doing the actions. Figure 1 shows the user selection page in which it is possible to select the call or message page by the voice input.

Fig. 1 User selection page

Fig. 2 Call page

5 Call Page

The calling page is shown in Fig. 2 is used to contact someone the user needs to call. Click the page to open a dialog box with the mic. Using the mic, user says the number to dial in voice. The voice is converted as the number, and the number is dialed. If the number is valid, it returns the successful call or it will say it is the wrong number. Through intent, a user can able to make a call. Intent callIntent = newIntent (Intent.ACTION_CALL). The action_call method will immediately call the number. The startActivity(callIntent) method is used to start the call action. Voice interaction is a kind of activity that is triggered by the user's voice.

6 Message Page

On the message page, messages can be sent and received. Once the button is clicked in the center a dialog box appears to receive the messages. Users should mention the number where the message should be sent and the content of the message. In android phones, the users plan to share information quickly. Intent is used to pass between components such as activities, services, broadcast receivers, etc. StartActivity() method is used to invoke activity, broadcast receiver, etc. StartActivity() method defines the intent is used to start an activity. To start a service, startservice(intent) method call is used. The intent must be constructed, and then action has to be specified. The ACTION_SEND is used to send data from one activity to another activity [5].

7 Voices to Text

Short message service (SMS) is a text message service that is a component of the mobile device. Mobile phone users can exchange SMS in a very easy and quick way. These are some features in the early mobile phone message system. Text-entry was difficult because of multiple key presses on a small keypad to generate letters. Messages are permitted only 160 characters. Nowadays, SMS has become a very easy task for the users to send a message. It has become much simple to type the text and there is no character limitation for sending the message. But this easy task is difficult for the visually impaired. On their own, they cannot send a short message and dial the number to call. To overcome the problem, an application is developed for blind people. Figure 3 depicts the flow chart for voice to text conversion. It is a voice recognition system in which the user will enter his/her voice speaking to the mic. It is converted to the text and is transmitted. A recognizer intent is created by setting the flag such as ACTION_RECOGNIZE_SPEECH. This recognizer takes the user input speech and returns to the same activity. LANGUAGE_MODEL_FREE_FORM-input is a free form of English. EXTRA_PROMPT helps in prompting the user to speak. Once the voice is transmitted as input to the function, Activity result is processed in the manner of responding to the input and the required action will be performed. Toast is a method used to display information in a short time. It makes the toast containing text and duration [6].

8 SMS and Call Sender

The voice is converted to text and it is transmitted to the receiver. During the call sending option, if the call is made to a valid number, then it will dial or send a message to the destination. Invalid numbers will be discarded. An important aim of this application The voice SMS allows users to send the input speech content and send voice messages as the text message. The user can manipulate text messages fast and easily without using a keyboard which minimizes the time and effort [7].

9 Object Recognition

This system is developed in the process of recognizing an object by just scanning the object by the camera through the user's mobile phone. Figure 4 depicts the image recognition scanning method using paintings. UPC database is used for collecting the information regarding the scanned image. SIFT algorithm is functioned for passing the scanned image as an input to the server-side and UPC database responds to the server. Finally, information is produced in the format of the text messages. Then, the test is converted into speech by using the TTS unit.

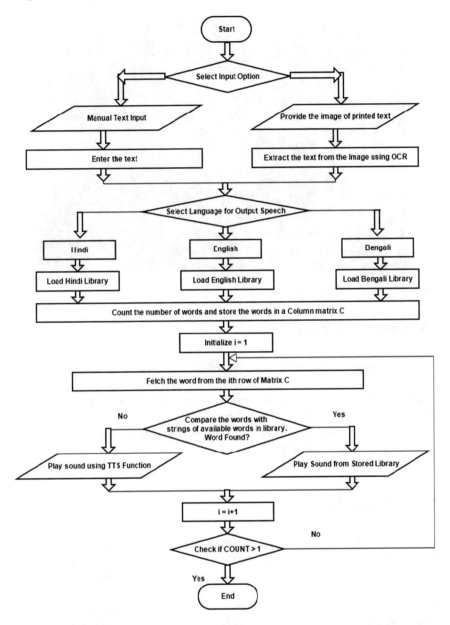

Fig. 3 Voice to text conversion

Fig. 4 Image recognition scans paintings

10 Navigation Recognition

This system is designed with the extraction of navigation recognition functions which is represented in Fig. 5. This app is used the augmented method which is a virtual reality model. This method derives from the geographic information system (GIS) concept which helps visually impaired people by providing the way from the start location to the destination as voice. This app also recognizes the shortest route by using a vector-based algorithm.

11 Working Principle

The proposed system works for the visually impaired people who can send SMS or make a CALL by speaking out the messages or numbers as shown in Fig. 6.

The application developed is targeted for blind humans, who are not able to use mobile phones for messaging or any other kinds of communicating devices, with the required comfort. The proposed application operates the message and call system, which work as the voice which is described in Fig. 7. The android application certainly listens to the messages and then responds. In this application, it converts the voice into text. To make a call people contact through voice and make a call. An accelerometer sensor is used in this project to open the application automatically. Voice- to- text technology is used to send and help them connected with people. Speech recognition technology is of particular interest due to the direct support of communications between humans and devices. Speech recognition is done by the

Fig. 5 Navigation screen of app

Internet, connecting to the Google server using the Hidden Markov Model algorithm. Android application is an open and free mobile platform. The software Eclipse IDE is the plug-in works along with the android development tools (ADT) which is the best kit for developing the powerful android application. The android SDK is also used in this work which operates beside programming language like Java and XML. Java technology is the programming language and platform independent [8, 9]. The Java application programming interface (Java API) provides a large number of software components to provide useful capability such as a graphical user interface (GUI).

12 Conclusion

An android application for smartphones is designed which is highly beneficial to the blind peoples. Using this application, the visually challenged people can able to send or receive messages in the format of voice and make calls by voice-enabled services. By using this application, those people can able to recognize object details just by scanning the object through the camera. This application also provides a map from the starting point to destination with the shortest route by navigating as a voice.

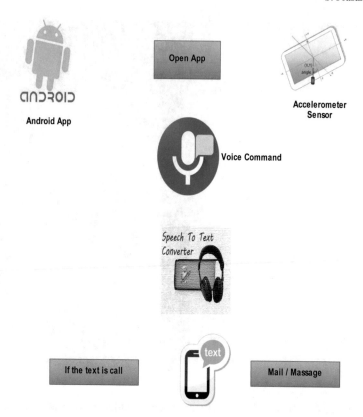

Fig. 6 Working flow of application

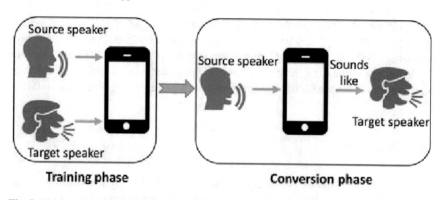

Fig. 7 Voice conversion application on mobile phones

References

1. Bentzen BL, Jackson RM, Peck AF (1981) Solutions for problems of visually impaired users of rail rapid transit. In: Improving communications with the visually impaired in rail rapid transit systems, vol 1. U.S. Department of Transportation, Urban Mass Transportation Administration, Washington, DC
2. Brabyn J (1985) A review of mobility aids and means of assessment. In: Warren DH, Strelow ER (eds) Electronic spatial sensing for the blind—contributions from perception, rehabilitation, and computer vision. Martinus Nijhoff, Boston, pp 13–27
3. Kirchner C, McBroom LW, Nelson KA, Graves WH (1992) Lifestyles of employed legally blind people: a study of expenditures and time use. Mississippi State University Rehabilitation Research and Training Center, Mississippi State
4. Jayant C, Ji H, White S, Bigham JP (2011) Supporting blind photography. In: Proceedings of the ASSETS 2011. ACM, pp 203–210
5. Makino H, Ishii I, Nakashizuka, M (1996) Development of navigation system for the blind using gps and mobile phone combination. In: Proceedings of the 18th annual international conference of the IEEE engineering in medicine and biology society, pp 506–507
6. Poppinga B, Magnusson C, Pielot M, Rassmus-Gröhn K (2011) TouchOver map: audio-tactile exploration of interactive maps. In: Proceedings of the MobileHCI 2011, pp 545–550
7. https://www.visioncue.com/MobileSpeak.html
8. Schlomer T, Poppinga B, Henze N, Boll S (2008) Gesture recognition with a Wii controller. In: Proceedings of the second international conference on tangible and embedded interaction, pp 11–14
9. Stepnowski A, Kamiński Ł, Demkowicz J (2011) Voice maps—the system for navigation of blind in urban area. In: Proceedings of the 10th WSEAS international conference on applied computer and applied computational science.CD-ROM, Venice, Italy

Wireless Sensor Networks Optimization Using Machine Learning to Increase the Network Lifetime

Sneha S. More and Dipti D. Patil

Abstract The wireless sensor networks optimization is the main approach for reducing energy consumption and for prolonging the network lifetime, by using the machine learning algorithms. The optimization of WSN achieves the clustering of nodes, the data aggregation of nodes, and reducing energy consumption. In this paper, more than 100 nodes are used to optimize networks with less energy consumption and less time requirement. This number of nodes gives the feasibility to the network. The energy consumption is to be reduced using the APTEEN protocol with threshold energy. The threshold value can be increased to 70%, and due to which the energy can also be reduced to 55%. The lifetime of the network happens when the dead node occurs in the system, where the minimum number of dead nodes exists to increase the network lifetime and to avoid packets loss. So, to reduce the dead nodes, the use of a machine learning algorithm is proposed. The results show the analysis of APTEEN protocol with a threshold and genetic machine learning algorithm which gives energy efficiency and improved network lifetime.

Keywords Wireless sensor networks (WSN) · Genetic machine learning algorithm (GMLA) · APTEEN protocol · Energy consumption · Network lifetime

1 Introduction

1.1 Wireless Sensor Networks

The wireless sensor networks could be known as the self-designed and less remote framework systems for screening physical or natural conditions, like sound, weight,

S. S. More (✉)
Savitribai Phule Pune University SPPU, Pune, India
e-mail: snehamore987@gmail.com

D. D. Patil
MKSSS's Cummins College of Engineering for Women, Pune, India
e-mail: diptivt@gmail.com

© The Author(s), under exclusive license to Springer Nature Singapore Pte Ltd. 2021   319
J. S. Raj et al. (eds.), *Innovative Data Communication Technologies and Application*,
Lecture Notes on Data Engineering and Communications Technologies 59,
https://doi.org/10.1007/978-981-15-9651-3_28

temperature, movement, vibration, contamination. The data found in the primary space sink with the system data, and it is observed.

The base station or sink acts as an associate interface among shoppers and also the framework. The specified knowledge from the system by infusing inquiries and social affair results from the sink will be recovered by someone. Ordinarily, a distant device prepares to contain many device nodes. The device nodes will impart among themselves utilizing radio sign. The wireless device gadgets likewise react to inquiries sent from the "control location" to perform express rules or grant identifying tests.

1.2   Genetic Algorithm

Genetic algorithms (GA) are flexible heuristic search algorithms that stay with the greater bit of evolutionary algorithms. Genetic algorithms rely upon the contemplations of typical assurance and genetic characteristics. These are shrewd maltreatment of subjective interest outfitted with recorded information to organize the hunt into the zone of better execution. They are normally used to make first-rate answers for enhancement issues and search issues.

Genetic algorithms reproduce the system of normal selection which suggests those species which can conform to changes in their condition can suffer and mirror the individuals. They reproduce "characteristic determination" among a person of consecutive age for handling an issue. Each age includes some masses of individuals, and everyone addresses a point in search space and possible course of action. Everyone is addressed as a string of characters/entire number or skim or bits. This string is practically comparable to the chromosome.

1.3   Machine Learning

Machine learning is that the field of concentrate that offers computers the capability to find out without any ambiguity changed. ML is known as one of the most energizing advancements that one would have ever run over. As it is apparent from the name, it gives the computer that makes it increasingly like people to have the capacity to learn. Machine learning is effectively being utilized today, maybe in a lot a greater number of spots than one would anticipate. Machine learning algorithms are utilized in a wide assortment of algorithms, for example, email separating and computer vision, where it is troublesome or infeasible to build up an evolutionary algorithm for viably playing out the assignment.

1.4　Energy Consumption in WSN

Wireless Sensor Network (WSN) sensors, when in doubt passed on in the non-accessible condition, are energized using little batteries nearby by techniques with adequate control, and superseding batteries are not another option. Contingent upon a battery restricts the sensor's lifetime just as makes a compelling arrangement and the leading group of WSNs a certified test.

1.5　Network Lifetime

Accordingly, network lifetime can on the other hand be characterized as the "time until the primary node or the base node dies." The most straightforward to catch pointer of this measurement is the greatest per-node load, where a node's heap compares to the number of packets sent from or directed through the given node. Obviously, the system arrangement that limits the greatest node load is the one that will guarantee the most extreme network lifetime.

This paper discusses the wireless sensor networks, machine learning, genetic algorithm with energy efficiency, energy consumption, and network lifetime for the optimization of data and improvement of packet loss to reduce the dead nodes.

2　Literature Review

The paper suggested by More and Nighot [1] discusses the proposal of power consumption, a lifetime of the network, and packet reduction; these are the major concerns in WSN. For minimizing power consumption and to prolong the lifetime of the network, the ACO algorithm is proposed. Lastly, comparative graphs of Anthocnet, Apteen, Eaar, and Apteen with threshold energy are clarified and examined during where it shows anticipated algorithmic program that gives attractive results over APTEEN protocol any place; power consumption has been reduced to 60% [1].

The authors Srilakshmi and Sangaiah [2] discuss the inner and outer factors; systems will rectify the progress that affects the localization of hubs, delays, steering instruments, geological inclusion, cross-layer style, the standard of connections, flaw identification, and nature of administration, among others. An applied math investigation of the data, the rationale in choosing explicit systems to deal with remote system issues, and a speedy dialog on the difficulties inalienable during this space examination region unit clarified [2].

A tale computation to push ahead uRLLWSNs execution by applying AI techniques and innate counts have been discussed by makers, for example, Chang et al. [3]. Using the K-means clustering computation for creating two-level orchestrate topology, the proposed structures check the fatal dataset, showed by individuals and make a grouping procedure of centrality change to maintain a strategic distance from over-trouble cluster heads. A multi-target enhancement show is characterized to simultaneously satisfy various improvement goals checking the longest compose the lifetime and the most imperative orchestrate the system and relentless quality [3].

Pal et al. [4] examine the hereditary equation-based cluster head decision that brings together agglomeration calculations to possess a much better burden adjusted system than the ordinary agglomeration calculation [4]. Recreation gives that the orchestrated objectives find the perfect cluster heads and join a drawn-out framework future than the standard agglomeration computations.

The creators Alsheikh et al. [7] depict that AI conjointly summons a couple of sensible plans that increase resource use and drag out the period of the framework. In this paper, the review over the year of 2002–2013 the AI systems modifies the location of the typical issues in remote sensor frameworks (WSNs) [7]. Advantages and downsides of each foreseen algorithmic program square measure evaluated against the contrasting disadvantage. They conjointly provide a comparative manual to help WSN planners in making fitting AI answers for their specific application challenges [7].

3   Proposed Work and System Information

3.1   *System Architecture*

The system architecture shown in Fig. 1 describes the detailed analysis of the system which gives the overall idea of the application for the optimization and increase

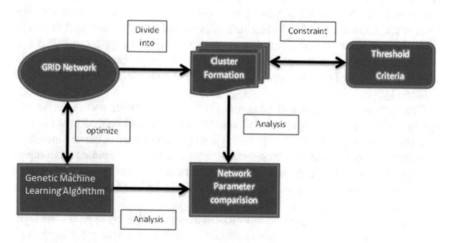

Fig.1 System architecture

in network lifetime. The system architecture is mainly divided into five modules, i.e., grid network, cluster formation, threshold criteria, genetic machine learning algorithm, and network parameter comparison.

The idea about grid network, cluster formation, and network parameter comparison has been illustrated from the paper [6]. The modules have good efficiency so that the proposed system architecture has been implemented with the same inputs from it. The proposed paper concentrates more on the threshold criteria and the algorithms.

The threshold criteria arc to be set here will be 70% due to which the energy reduces to 55%. The proposed system has the advantage of reducing the maximum number of packet loss, does not attain zero energies, and destroys dead nodes.

The genetic machine learning algorithms are the main part of the system where the network lifetime can be increased as compared to the previous paper [1]. This algorithm gives the best network lifetime and less energy consumption. The step by step procedure has been explained in the system architecture diagram.

Step 1: Pre-processes data.
Step 2: Generate offspring.
Step 3: Analyze gene effectiveness.
Step 4: Predictive model.

The above step by step procedure is the pseudo algorithm for genetic machine learning algorithm.

3.2 Proposed Work

The objective of the genetic machine learning algorithm is to minimize power consumption and increasing the network lifetime. Also, the proposed algorithm, i.e., apteen protocol with threshold energy by adding more range of nodes and giving more accurate data regarding nodes where it provides a minimum number of packet loss. During this proposed algorithm, threshold energy is set to 70 J due to which the energy consumption reduces to 55%. Below is the step by step procedure of increasing network lifetime by following the apteen_threshold approach:

(a) Firstly, set threshold energy to 70.00 J and all the variables.
(b) Calculate the energy of a considerable number of hubs, for example, inside the range and their neighboring hubs.
(c) Next is to contrast n, i.e., one hub's energy and neighboring hub's energy state nb on the off chance that v is more prominent than nb, at that point pick v.
(d) Now, think about this value and threshold energy, and on the off chance that this worth has a littler sum than the threshold energy, at that point pick v as cluster head.
(e) Next is to envision cluster head to be available in neighboring hops and not available then stop it. Else, send a cluster head to give the message.

The above steps give information about the Apteen protocol with threshold energy for the execution of increasing network lifetime.

4   Results and Analysis

The simulation results describe the system with all the required data, and various scenarios have been executed. The results of about clustering and routing are being provided for giving the idea about wireless sensor networks parameters. The graphs of the proposed algorithm and genetic machine learning algorithm with the existing protocol have been provided which includes inputs such as range of dead nodes, energy used per delivered packets, and range of packet loss dropped and delivered. Here, only the WSN parameter analysis is provided for the 250 nodes, in which the clustering of nodes and routing are the main parameters by which the network lifetime and energy consumption can be achieved.

4.1   Clustering of Nodes

Clustering engages the aggregation of the managing information and henceforward, supports the adaptability of steering calculations. Cluster-based controlling manages the problem of the hub heterogeneousness and coordinating overhead. The bundle enables the system of differed leveled directing inside which courses zone unit recorded between clusters move concerning enlarged course life expectancy and reduced administration overhead. Cluster head organizes the group individuals and their activities.

Cluster bottleneck problem develops because the cluster centers moreover around the congestion of its cluster, and this will maintain a strategic distance by tolerating the cluster approach. So, the exchanges inside the cluster are regularly between clusters or intra-cluster. The inter-cluster correspondence tends to the correspondence of center points inside the bundle; however, the intra-bunch correspondence tends to the correspondence between the groups through the passage centers.

4.2   Routing

The routing is a procedure to choose the appropriate way for the information to head out from source to goal. The procedure experiences a few troubles while choosing the course, which relies on, sort of system, channel qualities, and the exhibition measurements.

4.3 Graphs

The important intention of the framework is the analysis of all the systems that have enforced; the below graph Fig. 2 shows the comparison output of the higher system. The coordinate x-axis has interval values and another coordinate y-axis with a variety of dead nodes. To seek out the dead node who must sign on trace file once energy is zero and ensures time and provides correct output to the user. This task is finished by making the dead node awk script file which can offer the results. The projected system Apteen_threshold has results higher than the previous apteen; GMLA offers a minimum range of dead nodes at every interval.

The amount of energy used by all protocols is displayed in the below graph Fig. 3. As it is been mentioned each time that Adhoc network is a lot of power-hungry, thus the aim is accomplished at this time that GMLA has less power consumption which is able to have an effect on each network parameter and prove the good results.

In the previous system, the packet reduction rate is maximum as compared; it is overcome by threshold criteria as displayed in the graph Fig. 4. GMLA yields less

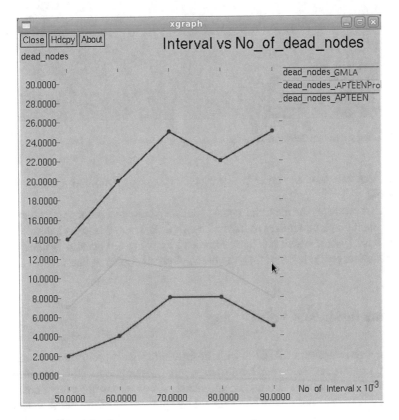

Fig. 2 Number of dead nodes

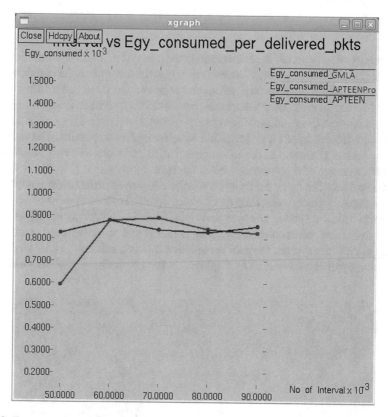

Fig. 3 Energy used per delivered packets

packet loss rate than the other two protocols that are apteen and apteen with the threshold.

GMLA produces the maximum packet delivery ratio than other applications. The base system graph maximizes as interval maximizes where this gives a disadvantage of the base system which has been improved by using a threshold system, where packet delivery ratio is better than the base system as shown in Fig. 5.

5 Conclusion and Future Scope

The proposed algorithm APTEEN with the threshold energy produces more accurate data, minimum packet loss, and minimum energy consumption. In this algorithm, the number of nodes used is 250 nodes, and which also shows better energy consumption. The network lifetime can be increased by applying a genetic machine learning algorithm for the dynamic networks with its predictive model where the data are

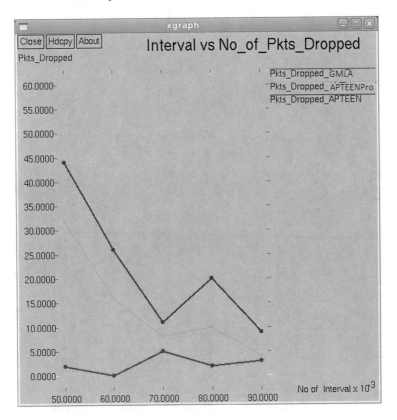

Fig. 4 Packets dropped graph

analyzed by its effectiveness and self-learning techniques. Due to which the occurrences of dead nodes become zero in the network. These two algorithms provide better results as compared to the existing system with better efficiency, less energy consumption, and increased network lifetime.

In the future, different advanced techniques can be used for increasing network lifetime and other approaches such as quality of service, time consumption, and so on. And also the number of nodes can be extended up to 500 nodes [5, 8–12].

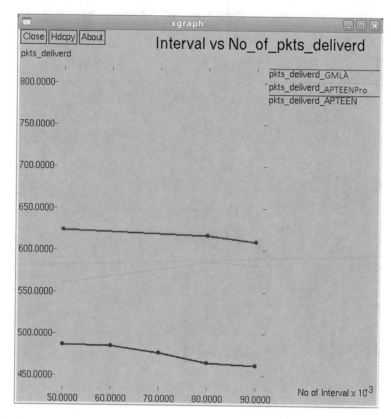

Fig. 5 Packet delivered graph

References

1. More SS, Nighot MK (2017) Optimization of wireless sensor networks using artificial intelligence and ant colony optimization for minimizing energy of network and increasingnetwork lifetime. IEEE
2. Srilakshmi N, Sangaiah AK (2019) Selection of machine learning techniques for network lifetime parameters and synchronization issues in wireless networks. J Inf Process Syst 15(4):833–852. https://doi.org/10.3745/JIPS.04.0125
3. Chang Y, Yuan X, Li B, Niyato D, Al-Dhahir N (2018) Machine-learning-based parallel geneticalgorithms for multi-objective optimizationin ultra-reliable low-latency WSNs. IEEE
4. Pal V, Yogita, Singh G, Yadav RP (2015) Cluster head selection optimization based on genetic algorithm toprolong lifetime of wireless sensor networks. In: Third international conference on recent trends in computing (ICRTC 2015)
5. Bayrakli S, Erdogan SZ, Genetic algorithm based energy efficient clusters (GABEEC) in wireless sensor networks. In: The 3rd international conference on ambient systems, networks and technologies (ANT)
6. More SS, Nighot MK (2017) Artificial intelligence and antcolony optimization based wireless sensor networks to minimize energy of network. Indian J Comput Sci Eng (IJCSE)
7. Alsheikh MA, Lin S, Niyato D, Tan H-P (2015) Machine learning in wireless sensor networks: algorithms, strategies, and applications. In: 2015 IEEE communications surveys & tutorials

8. Pinto AR, Camada M, Dantas MAR, Montez C, Portugal P, Vasques F (2019) Genetic machine learning algorithms in the optimization of communication efficiency in wireless sensor networks. IEEE
9. Townsend L (2018) Wireless sensor network clustering with machine learning. Nova Southeastern University Works
10. Kuila P, Gupta SK, Jana PK (2013) A novel evolutionary approach for load balanced clustering problem for wireless sensor networks. Swarm Evol Comput 12:4856
11. Vadlamudi R, Umar S (2013) A review of APTEEN in wireless sensor networks. IJCSET 9:306–311
12. Manjeshwar A, Agrawal DP (2002) APTEEN: a hybrid protocol for efficient routing and comprehensive information retrieval in wireless sensor networks. In: IEEE, proceedings of the international parallel anddistributed processing symposium (IPDPS.02), vol 02, pp1530–2075

Separable Convolution Neural Network for Abnormal Activity Detection in Surveillance Videos

S. S. Anju and K. V. Kavitha

Abstract Nowadays, the process of abnormal activity detection has gained a wide variety of applications in security and surveillance. Most of the public spaces are now installed with CCTV Cameras and the videos are continuously recorded. At present, the major part of surveillance cameras are tend to merely record the video for post-incident manual analysis. In this scenario, dependability on human monitors increase with a decrease in the reliability of the surveillance system. The security is at risk if the attention levels of the human monitors degenerate to unacceptable levels. The main aim is to enable surveillance cameras to understand activities in their environment instantly and report to the concerned users in case of events that need attention and to maximize the reliability of the surveillance system so that it could ensure maximum human benefit and security. The activities being recorded in a surveillance camera can be classified as normal and abnormal activities like accidents, theft, fire, crimes etc. This study proposes a separable convolution neural network which is trained and evaluated for finding better performance in the classification of activities based on surveillance videos. In particular, it examines the use of different optimizers namely Adam, RMSProp, and SGD in the proposed separable convolution neural network. The evaluation metrics used here are confusion matrix, accuracy, classification report with sensitivity, specificity, precision and f1-score. During evaluation the results showed that the proposed separable convolution neural network model trained with SGD optimizer has gained 94% accuracy with the test data.

Keywords Computer vision · Separable convolution neural network · Intelligent surveillance system · Activity detection · CCTV

S. S. Anju (✉) · K. V. Kavitha
Department of Computer Science and Engineering, Sree Chitra Thirunal College of Engineering, Trivandrum, Kerala, India
e-mail: anjuss1997@gmail.com

K. V. Kavitha
e-mail: kavitha@sctce.ac.in

© The Author(s), under exclusive license to Springer Nature Singapore Pte Ltd. 2021   331
J. S. Raj et al. (eds.), *Innovative Data Communication Technologies and Application*,
Lecture Notes on Data Engineering and Communications Technologies 59,
https://doi.org/10.1007/978-981-15-9651-3_29

1 Introduction

Abnormal activity detection in the field of surveillance and security are quite challenging, where there is a problem in classifying the sequences of video data recorded by CCTV cameras into normal or abnormal events like fire, accident, robbery etc. Over the last decade, real-time video analysis is observed as a rapidly improvising field, where it evident an extraordinary growth. The main goal of surveillance video processing and analysis is to find the abnormal events, which is very crucial to the public safety with a minimum manual labour. Intelligent video surveillance is a prominent area of research, which recognizes the events in the surveillance feed and classify the usual events into normal activities and unusual or suspicious activities events into abnormal activities [1].

For protecting offices, building, malls, houses and other infrastructures, government, public and private sector organizations are investing a large sum of money and according to the current market research corporations, in the upcoming years the trend of automatic security industry is going to grow exponentially. As of now, fire accidents and road accidents are increasing rapidly so it is equally important to detect abnormal events that can affect the peaceful life of a society. Intelligent surveillance [2] can solve this problem by analyzing surveillance data [3] and generating timely warnings.

Convolution neural networks (CNN's) can be used for solving the problem of abnormal activity detection. CNN's have the capability to learn features from image frames [4] and classify data with the assistance of learned features. Convolution is considered as the most important operations in artificial neural networks (ANN's). There are many variations in CNNs. One sort of CNNs are separable convolution neural networks. The separable convolution neural networks are widely used because of the following reasons:

- They can reduce over-fitting due to the benefit of having lesser number of parameters to adjust as compared to the standard CNN's.
- They are compatible for mobile vision applications because they are computationally cheaper with fewer no: of computations than standard CNN's.

A separable convolution layer [5] performs a depth-wise convolution that acts separately on channels followed by a point-wise convolution that mixes channels rather than convoluting together across all channels of an image, a separate 2D convolution on each channel is performed. This layer is preferred over usual convolution layers in many applications because it's an effective way to scale back the parametric complexity of early convolutions. Also, it can reduce no: of convolutions and hence speeds up training. MobileNet [5] and XceptionNet [6] proposed by Google are the main and most discussed applications of separable convolution neural networks.

The contributions of this work can be summarized as follows:

- A deep learning model is built for abnormal activity detection in surveillance videos.

- To address the issue of large no:of parameters and computational complexity, a separable convolution neural network architecture is proposed.
- To make the proposed method more efficient, three optimizers are cross checked for identifying best configuration in terms of accuracy.

The organization of this research work is summarized as: "Related works" section provides a brief history of some previous methods in abnormal activity detection. "Methodology'" section presents the details of the proposed system. "Results and Discussions" section interprets the experimental outcomes. Finally, the conclusion of the paper with future work is given in "Conclusion" section.

2 Related Works

The problem of abnormal activity detection in surveillance videos from a CCTV camera can be realized as a classification problem where from the image frames of each instance of the video is searched by the classifier for each of the activity a pattern. In recent years many machine learning and deep learning approaches were used for activity recognition from videos.

Abnormal activity detection is one of the most difficult problems in computer vision. Sultani [7] proposes a multiple instance learning solution to abnormal activity detection by considering only weakly labeled training videos. The major contribution of [7] is the introduction of a large-scale video anomaly detection dataset—UCFCrime. This new dataset serves as a benchmark for activity recognition on un-clipped videos due to large intra-class variations. The feature extractor used is 3D Convolutional Neural Network (C3D) [8].

C3D in [8] is proposed for video feature extraction. The spatial and temporal features from videos were successfully extracted using this network. But its use is very challenging for reasons like over-fitting of the weight filters. The C3D consists of 8 layers of 3D convolutions and 5 layers of pooling. The network was then trained on the Sports 1M dataset and was able to achieve remarkable performance compared to previous networks even though it was computationally complex.

Another approach for anomaly detection considered was to perform action recognition and then tune the network for detecting anomalies based on recognized actions [9]. However, it had some drawbacks to the specific use case are considering. It uses RGB-D data which contains the skeletal poses along with the normal video data. They are processed in two steps. The human skeletal data is passed through an RNN comprising of Bi-GRU units and the other and the normal video data is passed through a CNN with normal 2D convolutions and pooling operations. The output of the two networks is combined and the classification is done using a linear SVM. This framework has performed fairly well with action recognition tasks. Since CCTV data is only 2D video, the initial solution to the problem was to estimate the poses of people in the video and then give that as inputs to the RNN as in this architecture. This had drawbacks with poor performance in anomaly caused because of non-human type objects like cars, etc. Also, this will make the network computationally expensive.

In [10] the solution was to consider the poses of the people in the frames and find their variations across frames using an RNN network and then classify it using a linear SVM. For extracting the poses most efficiently, OpenPose is the best algorithm. This method performs fairly well even with some occlusion and clothing. Also, this offers excellent speed. However, the problem arises when only half the body is in the frame. So even though it responds well to multiple people in the frame this method would fail when it comes to anomaly detection. Also, this method does not consider the anomalies caused by other objects like cars, weapons etc. So this method will fail in giving good results for the problem.

3   Methodology

The design of Abnormal Activity Detection System for surveillance videos is shown in Fig. 1. It consists of mainly 4 blocks, which are Video (Input), Pre-processing, A separable convolution neural network and Activity detection (Output).

Video data collected by the CCTV is pre-processed to generate image frames. These image frames are transferred to the Deep Learning block for activity detection. The deep learning classifier used is a separable convolution neural network which is used to decide whether the streamed video frame has abnormal activities or not. If abnormal activity is identified then a notification is sent to the end user. The system recognizes the normal and abnormal activities i.e. fire, accident and robbery by classifying video frames. The detailed architecture of the proposed separable convolution neural network is given in Fig. 5.

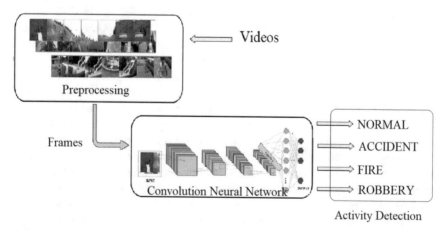

Fig. 1 Proposed abnormal activity detection system

3.1 Data Collection

In this step video footages are collected. UCFCrime[7] is a large-scale CCTV video dataset used for video anomaly detection, consisting of 1900 real-world videos of thirteen completely different abnormal events and normal activities captured by cameras. It's the most important dataset with quite fifteen times videos than existing anomaly datasets and includes a total of 128 h of videos. This publicly available dataset is collected for abnormal activity detection. It contains thirteen realistic anomalies together with Abuse, Arrest, Arson, Assault, Road Accident, Burglary, Explosion, Fighting, Robbery, Shooting, Stealing, Shoplifting, and Vandalism. These anomalies are selected by Sultani et al. [7] as they have a major impact on public safety.

The dataset as a whole is very large and had 13 anomaly classes. From this, a four class dataset of videos is selected where the chosen classes are Normal, Fire, Accident and Robbery.

The distribution of videos among the four classes in the dataset was not balanced. So some videos for these classes are downloaded from YouTube and added to the data. This is done because the system requires to search for maximizing the accuracy of all classes in a balanced way, such that the errors from a class have the same cost as those from the another class. The final distribution of videos among the four classes is expressed as a bar plot in Fig. 2. From this it is clear that the dataset is now balanced and have equal density distribution.

3.2 Data Pre-processing

In pre-processing step the selected data classes are subjected to video to frame conversion. Videos can be viewed as a series of frames or images which appears with no time lags. Hence for the process of abnormal activity detection, the detector can be applied to individual frames instead of applying in whole video. Also detecting

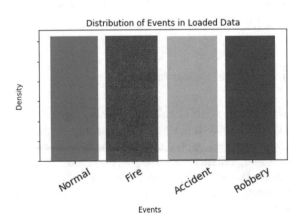

Fig. 2 The Distribution of videos in the dataset

Fig. 3 Visualizing an image in R, G and B channels

activity from individual frames can result in the requirement of low computation power than detecting activity from videos. It is a fact that for real-time detection of activities from videos, the detection process must look in every frame, in order to ensure that no event is going to be missed. Hence the videos in the dataset are converted into frames at a frame rate 30 fps. The frames in the resulted image dataset are analyzed for its properties like image height, width and no: of channels. By using image resizing techniques all the frames are converted to a specific dimension ($128 \times 128 \times 3$), which is the required input dimension for further steps.

The data is now frames (images) with height 128 pixels, width 128 pixels. The no: of channels is 3 which indicates that all images are in RGB format. Hence for denoting a single pixel, 3 values are necessary. Figure 3 visualizes an image in separate R (red), G (green) and B (blue) Channels.

3.3 Data Partitioning

In this step the dataset is partitioned into training data and testing data. As the result of data partition 75% of the data frames are in training data which is used for training the neural network and 25% of the data frames are in testing data which is kept for testing the trained network.

3.4 Data Augmentation

The performance of deep learning neural networks typically increases with the quantity of information available. The quantity of training data is increased by creating transformed versions of images in the training dataset that belong to the same class as the original image and the process of transforming images into new and different images is called image data augmentation [11]. Transformations include a range of operations from the field of image manipulation, such as shifts, flips, zooms, and far additional.

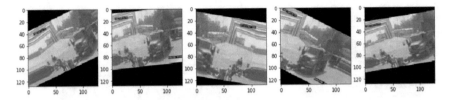

Fig. 4 Visualizing data augmentation in an image

The data augmentation output of a single image of the dataset is shown in Fig. 4. Image data augmentation is often solely applied to the training data, and not to the test data. Hence to boost the training results, all the images in training data is subjected to data augmentation. It will improve the size of training data as well as the model learned from this data is going to be ready to take care of distorted or shifted camera feeds also.

3.5 Separable Convolution Neural Network Architecture

For abnormal activity detection from surveillance videos, a separable convolution neural network is constructed [12–16]. At this context, the videos with no accident, no fire and no masked man with weapons like guns is treated as normal activity. On the contrary, if the video is having images of accident then it is classified as Accident, if the video is having images of fire then it is classified as Fire and if the video is having images of masked man with weapons like guns then it is classified as Robbery which all are the abnormal activities. After detecting the activity in the video, the communication is made to the respective persons about the abnormal activity via internet as an email or sms if the trained model and an internet communication is deployed in the camera module. If this model is deployed in the user side application then the videos coming through internet is processed and detection process takes place at the user side application and immediate response is given to the user as an alert.

Figure 5 shows the architecture of the proposed separable convolution neural network. The neural network accepts an input image of size $(128 \times 128 \times 3)$ and outputs one of the four class labels Normal, Fire, Accident and Robbery. It contains a total of 16 layers that together with 1 input layer, 4 separable convolution layers with RELU activation, 4 batch normalization layers, 3 maxpool layers, 1 flatten layer, 1 dense layer, 1 dropout layer with $dropout = 0.5$ i.e. 50% and the output dense layer with softmax activation.

The separable convolution has [17] two stages: depth-wise convolution and point-wise convolution. The depth-wise convolution focuses on the area information. The point-wise convolution is mainly involved with channel information. To some extent, separable convolution is considered as a convolution matrix decomposition [18]. The

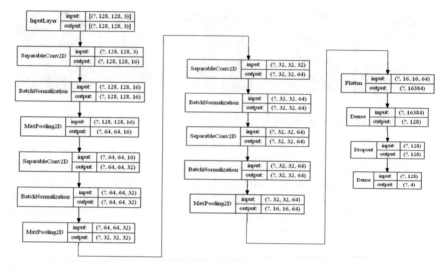

Fig. 5 The proposed separable convolution neural network architecture

use of separable convolution instead of using standard convolution makes the system more faster which is essential for activity detection from videos. In the proposed system, separable convolution layers are having a kernel size of (3 × 3). The primary layers learn high level features and the final layers are responsible for identifying activity from the output feature maps of previous layers.

The architecture Fig. 5 also gives the dimensions and specifications of each of the 16 layers. The use of batch normalization layer in between separable convolution layer and maxpool layer increases the stability of the neural network [19]. It supports higher learning rates because batch normalization makes sure that there's no activation that's gone really high or really low. Also this layer reduces over-fitting because it has a slight regularization effects.

4   Results and Discussions

The separable convolution neural network described in the methodology is trained using training data with the following configurations.

- Processor: Google COLAB GPU
- No: of Epochs: 100
- Batch Size: 64
- Loss: Sparse categorical cross entropy
- Optimizers: Adam, RMSProp, SGD.

For finding the best configuration, 3 different optimizers Adam, RMSProp and SGD are used for training. Optimizer is responsible for providing the most accurate results possible and to reduce the loss which is the convergence criteria. Adam (Adaptive Moment Estimation) works with momentum and is too fast and converges rapidly even though it is computationally costly. RMSprop (Root Mean Square Propagation) optimizer restricts the oscillations in the vertical direction. Therefore, it increases learning rate and converges faster. In SGD (Stochastic Gradient Descent) optimizer model parameters are more frequently updated, and have high variance and fluctuations in loss functions at different intensities. The learning rate is set as default as per optimizer. After training the 100 epochs the model is tested with the remaining testing data. To evaluate the proposed system, confusion matrix and classification report are produced. The learning curve of the neural network model at each time step is also analysed for identifying the variations in accuracy and loss at each epoch for the three optimizers.

The maximum accuracy obtained is 94% for the separable convolution neural network with SGD optimizer. While the same network trained with RMSProp and Adam optimizer yield 91% accuracy with the testing data.

4.1 Confusion Matrix

The confusion matrices [20] produced after the evaluation of the proposed neural network architecture with Adam, RMSProp and SGD optimizers is given in Figs. 6, 7 and 8 respectively. The classes are labeled as 0 (Normal), 1 (Accident), 2 (Fire) and 3 (Robbery). From the confusion matrices of Adam and RMSProp, it is observed that there is a slight confusion between class normal and class robbery. But for SGD this confusion is very small.

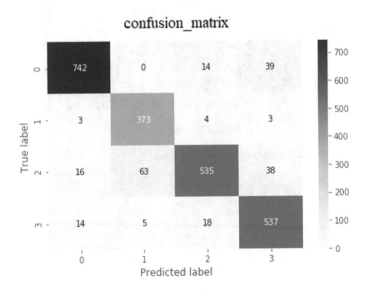

Fig. 6 Confusion matrix: network with Adam optimizer

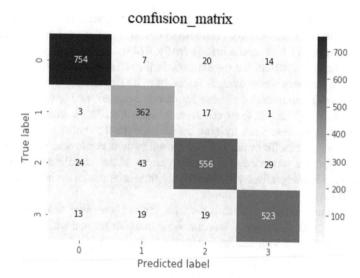

Fig. 7 Confusion matrix: network with RMSProp optimizer

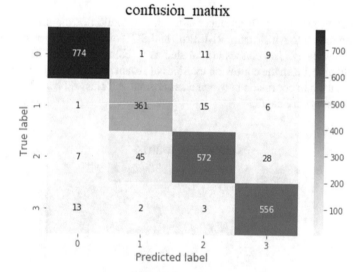

Fig. 8 Confusion matrix: network with SGD optimizer

4.2 Classification Report

The Classification reports [20] produced after the evaluation of the proposed neural network architecture with Adam, RMSProp, SGD optimizers is given in Figs. 9, 10 and 11 respectively. Maximum precision for prediction of all classes is obtained with SGD optimizer which are 0.97, 0.88, 0.95 and 0.93 for Normal, Fire, Accident and Robbery classes respectively. From this it is clear that the system identifies normal activities at a maximum precision of 0.97 than other classes.

```
                    classification_report

[INFO] evaluating network...
                precision    recall   f1-score    support

       Normal      0.96       0.93      0.95        795
         Fire      0.85       0.97      0.91        383
     Accident      0.94       0.82      0.87        652
      Robbery      0.87       0.94      0.90        574

     accuracy                           0.91       2404
    macro avg      0.90       0.92      0.91       2404
 weighted avg      0.91       0.91      0.91       2404
```

Fig. 9 Classification report: evaluating network with Adam optimizer

```
                    classification_report

[INFO] evaluating network...
                precision    recall   f1-score    support

       Normal      0.95       0.95      0.95        795
         Fire      0.84       0.95      0.89        383
     Accident      0.91       0.85      0.88        652
      Robbery      0.92       0.91      0.92        574

     accuracy                           0.91       2404
    macro avg      0.91       0.91      0.91       2404
 weighted avg      0.91       0.91      0.91       2404
```

Fig. 10 Classification report: evaluating network with RMSProp optimizer

```
classification_report

[INFO] evaluating network...
                precision    recall   f1-score    support

      Normal       0.97       0.97      0.97         795
        Fire       0.88       0.94      0.91         383
    Accident       0.95       0.88      0.91         652
     Robbery       0.93       0.97      0.95         574

    accuracy                            0.94        2404
   macro avg       0.93       0.94      0.94        2404
weighted avg       0.94       0.94      0.94        2404
```

Fig. 11 Classification report: evaluating network with SGD optimizer

4.3 *Learning Curves*

The proposed separable convolution network is trained with 3 different optimizers and the learning curves are drawn. The curves gives the variations of accuracy and loss at each epoch during the training. The learning curves of Adam, RMSProp, SGD optimizers is given in Figs. 12, 13 and 14 respectively.

4.4 *Comparison*

The learning process is carried out with three optimizers. For selecting the best among them for the proposed neural network, training time and the learning curves of accuracy and loss are compared. The comparisons of training time, accuracy curve and loss curve are given in Figs. 15, 16 and 17 respectively.

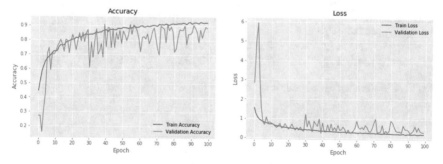

Fig. 12 Accuracy and loss variation during training: Adam

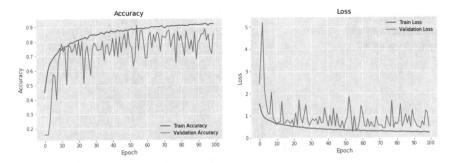

Fig. 13 Accuracy and loss variation during training: Rmsprop

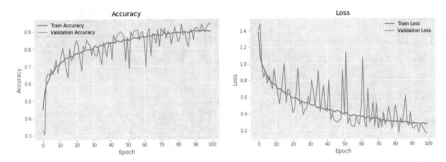

Fig. 14 Accuracy and loss variation during training: SGD

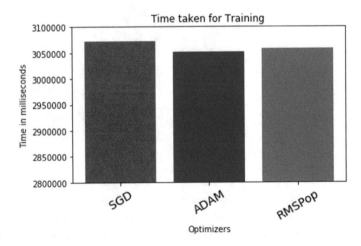

Fig. 15 Comparison of training time

Fig. 16 Comparison of training accuracy with different optimizers

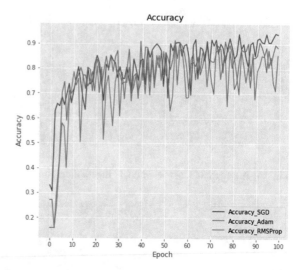

Fig. 17 Comparison of training loss with different optimizers

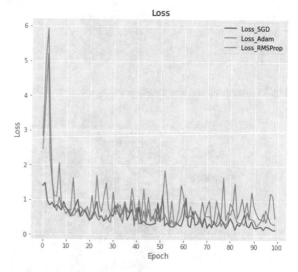

The accuracy of SGD optimizer is starting at a higher point than RMSProp and Adam optimizer. During further training also it keeps a high accuracy and from the Fig. 16 it is clear that variations on accuracy of SGD optimizer are much stable than others. Loss value implies how poorly or well a model behaves after each iteration of optimization. Loss can be seen as a distance between the true values of the problem and the values predicted by the model. Greater the loss is, more huge is the errors you made on the data. The loss of SGD optimizer is starting at a very low point than RMSProp and Adam optimizer. During further training also it keeps loss low and from Fig. 17 it is clear that variations on loss of SGD optimizer are much stable than others. Here a great accuracy with low loss means the system made low errors on a few data which is the best case. But when comparing the time taken for training in Fig. 15, SGD is taking more time and Adam consumes slightly lesser time than others.

Hence it is best to use SGD optimizer for the proposed abnormal activity detection system. The response time for one frame is also evaluated in Google Colab GPU environment and obtained an average of 25 ms with the test data which implies that the system can work in almost real-time.

5 Conclusion

Video surveillance is a prominent area of research which includes monitoring of real-time videos, recognition of human activities and categorization of events into normal or abnormal activities. For real-time detection and report of abnormal events recording in CCTVs an end-to-end deep learning system is proposed. Since deep networks need large amount of training data to achieve good performance data augmentation is performed to expand training data. A separable CNN model is built for video activity classification with four classes Normal, Fire, Accident, Robbery. The model is trained with 3 different optimizers which are Adam, RMSProp and SGD. The performance of these models are evaluated in terms of training accuracy and loss curves, confusion matrix, and classification report. On comparison the highest accuracy of 94% is gained with the model using SGD optimizer and precision 0.97, 0.88, 0.95 and 0.93 is obtained for Normal, Fire, Accident and Robbery classes respectively. From the evaluation of response time of individual video frame, it is clear that the system can detect abnormal activities in near real-time. As a future work, the no: of abnormal activity classes can be increased to improve the reliability of the system.

References

1. Chaudhary S, Khan MA, Bhatnagar C (2018) Multiple anomalous activity detection in videos. Proc Comput Sci 125:336–345
2. Ibrahim S (2016) A comprehensive review on intelligent surveillance systems. Commun Sci Technol 1, 7–14 (2016)
3. Chowdhry D, Paranjape R, Laforge P (2015) Smart home automation system for intrusion detection. In: 2015 IEEE 14th Canadian workshop on information theory (CWIT). St. John's, NL, pp 75–78 (2015), https://doi.org/10.1109/CWIT.2015.7255156
4. Cheong KH et al (2019) Practical automated video analytics for crowd monitoring and counting. IEEE Access 7:183252–183261. https://doi.org/10.1109/ACCESS.2019.2958255
5. Howard AG, Zhu M, Chen B, Kalenichenko D, Wang W, Weyand T, Andreetto M, Adam H (2017) MobileNets: efficient convolutional neural networks for mobile vision applications. arXiv:1704.04861. Available: https://arxiv.org/abs/1704.04861

6. Chollet F (2017) Xception: deep learning with depthwise separable convolutions. In: Proceedings of the IEEE conference on computer vision and pattern recognition (CVPR) June 2017, pp 1800–1807
7. Sultani W, Chen C, Shah M (2018) Real-world anomaly detection in surveillance videos. In: IEEE conference on computer vision and pattern recognition (CVPR)
8. Learning spatiotemporal features with 3d convolutional networks. In: ICCV 2015
9. Zhao R, Ali H, van der Smagt (2017) Two-stream RNN/CNN for action recognition in 3D videos. In: 2017 IEEE/RSJ international conference on intelligent robots and systems (IROS), Vancouver, BC, pp 4260–4267
10. Zhao R, Ali H, van der Smagt P (2017) Two-stream RNN/CNN for action recognition in 3D videos. In: 2017 IEEE/RSJ international conference on intelligent robots and systems (IROS), Vancouver, BC, pp 4260–4267
11. Mikolajczyk A, Grochowski M (2018) Data augmentation for improving deep learning in image classification problem. In: 2018 international interdisciplinary Ph.D. workshop (IIPhDW) . https://doi.org/10.1109/iiphdw.2018.8388338
12. Tripathi RK, Jalal AS, Agrawal SC (2018) Suspicious human activity recognition: a review. Artif Intell Rev 50:283–339
13. Chen Y, Peng F, Kang X, Wang ZJ (2019) Depthwise separable convolutional neural network for image forensics. In: 2019 IEEE visual communications and image processing (VCIP), Sydney, Australia, pp 1–4
14. Belyaev A, Kuzmina, VV, Bychkov AA, Yanakova ES, Khamukhin AV (2018) The hierarchical high-speed neural network image classification algorithm for video surveillance systems. In: 2018 IEEE conference of Russian young researchers in electrical and electronic engineering (EIConRus), Moscow, pp 1840–1843
15. Niklaus S, Mai L, Liu F (2017) Video frame interpolation via adaptive separable convolution. In: 2017 IEEE international conference on computer vision (ICCV), Venice, pp 261–270
16. Pfeuffer A, Dietmayer K (2019) Separable convolutional LSTMs for faster video segmentation. In: 2019 IEEE intelligent transportation systems conference (ITSC). Auckland, New Zealand, pp 1072–1078
17. Sironi A, Tekin B, Rigamonti R et al (2015) Learning separable filters. IEEE Trans Pattern Anal Mach Intell 37:94–106. https://doi.org/10.1109/tpami.2014.2343229
18. Mao Y, He Z, Ma Z et al (2019) Efficient convolution neural networks for object tracking using separable convolution and filter pruning. IEEE Access 7:106466–106474. https://doi.org/10.1109/access.2019.2932733
19. Ioffe S, Szegedy C (2015) Batch normalization: accelerating deep network training by reducing internal covariate shift. arXiv preprint arXiv:1502.03167
20. Anju SS, Kavitha KV (2019) Performance evaluation of various machine learning techniques for human activity recognition using smartphone. Int J Comput Sci Eng 7(8):316–319

Data Optimization-Based Security Enhancement in 5G Edge Deployments

S. Sree Lekshmi, Satwik S. Bandodkar, Vikas Vippalapalli, Alekhya Susarla, and Seshaiah Ponnekanti

Abstract Recently, edge computing and data analytics have been recognized as some of the key technologies to deliver low latency, high data rate services using the 5G networks. The 5G edge infrastructure is likely to cater to a mix of cellular and non-cellular devices mainly covering the Internet of things (IoT) and miniature sensor devices. The network vulnerability and threat mitigation at the edge become utmost important to protect the network from upstream attacks originating from the sensors. In mobile network deployments, the app data from multiple user devices can be collected at the mobile edge analytics server residing on the network side. These app data feeds can be processed at the edge server to dynamically build network quality maps. The quality analysis can be utilized to precisely monitor the experience of the users. Further, the serving cells or base stations serving those users within the edge can also be identified on the maps. This can help to pinpoint the real-time traffic variations per user in each cell. The user traffic trends, derived from the app-based quality analysis, can also help to identify security anomalies that may be taking place at the edge. In this paper, a novel framework to couple the traffic analysis and security monitoring at the mobile edge has been proposed with an example

---

S. Sree Lekshmi (✉) · V. Vippalapalli · S. Ponnekanti
Center for Wireless Networks and Applications (WNA), Amrita Vishwa Vidyapeetham, Amritapuri, India
e-mail: sslekshmi@am.amrita.edu

V. Vippalapalli
e-mail: vikasdec23@gmail.com

S. Ponnekanti
e-mail: seshaiahp@am.amrita.edu

S. S. Bandodkar
Amrita Center for Cyber Security Systems and Networks, Amrita Vishwa Vidyapeetham, Amritapuri, India
e-mail: satwiksb95@gmail.com

A. Susarla
Department of Electronics and Telecommunication Engineering, G. Narayanamma Institute of Technology and Science (For Women), Hyderabad, Telangana, India
e-mail: s.alekhya1503@gmail.com

J. S. Raj et al. (eds.), *Innovative Data Communication Technologies and Application*,
Lecture Notes on Data Engineering and Communications Technologies 59,
https://doi.org/10.1007/978-981-15-9651-3_30

347

manual case study. With the increasing use of artificial intelligence (AI)/machine learning (ML) in automatic traffic management and optimization in 5G, such app-based framework is likely to play a key role to harden the 5G edge fabric in the future deployments.

Keywords 5G networks · Security threats · Data optimization · Mobile edge

1 Introduction

Recently, there is a surge in the deployments of 5G networks to provide high data connections to mobile devices. 5G has been designed to support stringent requirements in terms of latency, capacity, throughput, availability, and reliability in the network. In the future, two key services related to a massive number of IoT and ultra-reliable low latency communications (uRLLC) [1] will be introduced in the 5G delivery. In such services, a large amount of data needs to be processed quickly (in real time) in the network from heterogeneous sources ranging from tiny sensors or IoT devices to massive machines, robots, and drones. With the 5G rollout, network architecture and security transformation are the key challenges for the service providers (SPs).

To meet the stringent requirements of 5G latency-sensitive services, the current research trends shift the analytics capability toward the edge of the network. The processing of massive amounts of data from different sources requires a data optimizer (DO) at the edge. The current network configuration is a mixture of wireless technologies (cellular/non-cellular) [2] which is used to support various IoT applications. The diverse set of IoT devices streaming the data (measurements) to the servers in the network presents security challenges at the edge. It is important to secure the edge for the upstream attacks from multiple sets of IoT devices that connect to the network through gateways. The continuous measurement data from various end devices can be collected at the edge servers to observe the real-time subscriber behavior analytics. The paper highlights the need for security screening through data optimization analytics at the edge. Understanding the 5G recent trends and threat landscape, this paper focuses on the design of a reference security framework coupled with DO to enhance 5G edge security.

The paper is organized as follows: Section 1 is the Introduction. Section 2 discusses the recent trends in 5G networks. Section 3 describes the 5G threat landscape. Section 4 discusses the reference design of a security framework together with DO to realize comprehensive edge security while maintaining the network performance. Section 5 discusses the experimental validation of the proposed edge security framework. Section 6 outlines the future scope and conclusions.

2 Recent Trends in 5G Infrastructure

5G network employs a set of innovations as listed below to cope with ultra-low latency, greater speed, and massive connectivity requirements:

- Software defined networking (SDN) and network function virtualization (NFV) [3] to create software-programmable networks
- Service-based architecture (SBA) [4] enabling control and user plane separation (CUPS) [5] design principles and foster innovation and placement of control elements in the cloud
- Network slicing [5] to cater needs of 5G applications in a wider set of verticals meeting the data rates and latency requirements
- Multi-access edge computing (MEC) [6] to provide rapid network services closer to the users.

2.1 Multi-access Edge Computing

The network elements in 5G architecture are softwarized to form network functions (NFs). 5G network comprises several distributed functions that can span data centers (DCs) covering edge and remote clouds. Table 1 shows the mapping between various 5G NFs, functionalities, and position in the network. In the 5G era, DCs should be scalable to support varying user traffic.

With the MEC concept, some of the core network (CN) functions can be placed at the edge DCs reducing the overall end-to-end latency. Edge computing improves the service experience by migrating the computational capabilities toward the network edge. By processing the information close to the users, the quality of

Table 1 Distribution of 5G network functions

Nodes	5G network function	Position
User plane function (UPF)	Data plane	Edge, regional and central cloud
Access mobility management function (AMF), session management function (SMF)	Packet control, security management	Central and edge location possible
Authentication server function (AUSF), Unified data management (UDM)	Subscriber management	Mostly central cloud
Security edge protection proxy (SEPP)	Roaming and interconnect signaling	Central cloud
Network slice selection function (NSSF), network repository function (NRF)	Resource management function	Central and edge location possible
Policy control function (PCF)	Controls policy rules	Central and edge cloud

service/experience improves and congestion is reduced on the mobile network as the bandwidth increases. The positioning of the NFs or the DCs is a key factor to be considered in an advanced network like 5G. Based on the service requirements, DCs can be placed at the edge (close to users) or aggregation points, or the central cloud.

2.2 ETSI-NFV MANO Framework

NFV decouples hardware and software resources, avoiding the need for dedicated network appliances for network operations. Network operations such as routing, firewall become software functions. With NFV, general-purpose hardware can instantiate network functions on-demand as per the service requirements. Network operations will be delivered by virtual machines (VM). NFV augment new capabilities to the network and demands novel management and orchestration principles to operate, administrate, maintain, and provision resources. Management and orchestration (MANO) [7] has been introduced by the European Telecommunication Standards Institute (ETSI). The MANO framework is an important enabler for the 5G infrastructure. The VNFs as well as the hardware resources together form the network function virtualization infrastructure (NFVI). The NFV MANO consists of three main functionalities:

- *NFV orchestrator (NFVO)*: The orchestrator in the NFV infrastructure involves the automation and management of NFVI. Orchestrator instantiates different VNFs to meet specific service demands. The end-to-end orchestration manages the service instances that will chain together to form the services.
- *Virtual network functions manager (VNFM)*: It manages the life cycle of VNFs. The fault-management configuration accounting performance and security (FCAPS) of VNFs will be managed by the VNFM.
- *Virtualized infrastructure manager (VIM)*: It is the management of NFVI which includes the physical resources, virtual resources, and software resources.

2.3 5G NFV Infrastructure NFVI Security

Traditional network security has strengthened the network at the core network nodes protecting the network perimeter with Gi-LAN hardening [8]. The current Gi-LAN implementation typically comprising load balancer (LB), firewall (FW), deep packet inspection (DPI) dynamically updates the virtual machines to provide security at the network level. This is supported by the Gi-LAN controller that instantiates the security functions depending on the traffic load at the perimeter of the network or on the Gi-LAN interface. In the edge services, however, IoT devices and gadgets are hardware constrained (low in memory, lightweight CPU, and limited bandwidth). Because of these device limitations, the tiny devices cannot host built-in security modules making them susceptible to vulnerabilities. To deliver holistic network security, the Gi-LAN security needs to be extended to the edge of the network which can protect the network from the upstream attacks of IoT devices that connect to the network. The next section describes the overview of the security threat landscape in 5G.

3 5G Security Threat Landscape

Cellular networks have been designed to reduce security threats at the air interface, DCs, and CN. The softwarization in the 5G networks results in a dynamic threat surface. Table 2 shows the categories of security threats [9] and the associated risk profile.

Table 2 Overview of threat landscape

Threats	Risk profile
Distributed denial of service (DDoS)	Service unavailable to a target user
Privacy breach	Accessing/releasing information without permission
Illegal intercept	Tapping of telephonic/internet-based services
Spoofing	Faking or imitating IP packets with a false address
IoT botnets	Infecting devices with malware to get control
Fake base station	To catch UE-IMSI, intercepting calls and location tracing
Fronthaul	Connects remote radio units at cell sites to distributed data centers
Backhaul	Connects edge/aggregate data centers to the core network
NFVI attack	The key challenge is to deploy standardized NFVI that can adapt to real-time threats, where slicing is hosted. It adapts to real-time threats
Application programming interface (API) attacks	Authorizing user credentials to protect sensitive data
Signaling storms	Hackers overloading the signaling traffic in the network (overloading bandwidth at signaling servers/DC with huge connections)
3rd party software	Low-quality components, poor coding/engineering process, lack of standard security functionalities
SIM hijacking	Convincing service provider and getting access to the original user account

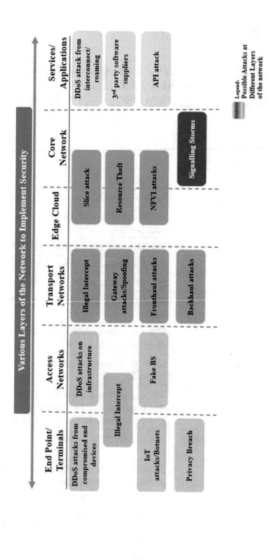

Fig. 1 Possible location for attacks in the network

Figure 1 includes the network locations and the type of possible threats encountered in mobile networks. The security should be enforced at various levels ranging from the end-user/terminal to the core network as explained.

- **Terminal security**: Secures the user layer consisting of devices ranging from sensors to massive machines/drones
- **Access network security**: Provides secure access to the network by users, ensuring security to signaling and data plane traffic
- **Transport network security**: Ensures secure transfer of data via different transport links such as front haul and backhaul networks
- **Edge security**: Implements security functions close to the users or at the network edge (Edge DCs)
- **Core network Security**: Utilizes security functions within and at the perimeter of the core network including interconnecting security
- **Application security**: Implements secure application programming interfaces (APIs) to protect the user data through the authorization.

Considering the above threat factors and understanding the network locations and types of possible attacks in the mobile network, it is necessary to avoid designing edge security strategies in isolation. In the 5G era, holistic security is important and can be achieved by considering the following steps:

- Design an enhanced security monitoring at the edge, based on the analytics within DO
- Automation of security monitoring with AI/ML techniques.

The following section explains the reference design of a security framework at the 5G edge utilizing the DO capability to enhance security monitoring function.

4 Reference Design Framework for 5G Edge Security

The edge data analytics and optimization can improve network performance by reducing network latency while contributing to the hardening of the security at the edge. This is normally facilitated by the design of the two modules which include DO [10] and security engine operating at the edge. In this section, the individual modules are outlined below followed by the combined edge security approach.

4.1 Data Optimizer

DO provides insights on different types of data in the cellular network from user equipment (UEs), core network, and access network. With the advent of app-related services in smartphones, a substantial amount of application-level data exist in the network. In the reference design framework, DO module has been designed to collect the network quality inference from the mobile apps (Network Performance Monitor Apps) placed on the UEs. The data from multiple devices can be collected from a centralized server on the network side to develop network quality maps. The quality of experience (QoE) map generation helps real-time data optimization. Further, in the 5G network, small access points or small cells are expected to be deployed densely throughout the network. The DO module can utilize the knowledge of the locations of the small cells and develop a dynamic cluster-based network quality map covering the following key performance indicators (KPIs):

- Throughput measurements in the network (downlink/uplink) which includes instantaneous, peak, average and cell throughputs
- Location of the users/devices in the network
- List of access points or base stations serving the users
- Types of traffic in the network (HTTP, video/voice, signaling messages)
- Radio frequency (RF) conditions indicating the signal quality such as received signal strength indicator (RSSI), received signal received power (RSRP), received signal received quality (RSRQ) and signal-to-interference noise ratio (SINR).

The quality maps feed into subscriber behavior analytics providing valuable insight into the subscriber's real-time as well as historical data consumption and contextual information. This information is expected to play a crucial role in the security monitoring at the edge as explained in the following section.

4.2   Security Monitoring at the Edge

The security monitoring at the edge can utilize the five key set comprising the source and destination IP, source and destination port, and protocol to identify security anomalies. The following applications and services can be categorized at the security monitoring stage.

- Encrypted or unencrypted traffic

Traffic types including websites/domains visited (HTTP, https, TCP, UDP, QUIC [11]), streaming services such as YouTube, Netflix, Hotstar, Amazon Prime, App traffic covering WhatsApp, Facebook, Twitter, Instagram, and background software updates.

Further, the security monitoring also interacts with the security controller connected to the security functions such as firewall, DPI and traffic detection function (TDF). By utilizing all the feeds, security monitoring can provide alerts in the event of any attacks taking place on the network.

4.3 Enhancing the Edge Security Utilizing DO Analytics

The scanning capability of security monitoring can be enhanced by coupling the information derived from the DO module described in the previous section. This handshaking between DO and the security monitoring function to exchange the subscriber analytics will enable effective protection at the 5G edge. The reference design security framework with DO is shown in Fig. 2. The reference model is based on distributed edge infrastructure constituting various types of end devices and DCs covering on-premises/aggregate points. The combined module of DO and security monitoring feeds the qualified alerts to undertake remedial measures in case of any network breaches.

By utilizing the interaction between the DO and security monitoring function, various enhancements can be provided in the overall security posture of the 5G edge components. For example, mitigation of *DDoS attack which typically results in a suspicious increases in the network traffic flow.* DO can alert the security engine about the vulnerabilities/threats, observing the sudden fluctuations in the traffic. Therefore, DO couple with a security engine can mitigate the risk early before a full-scale DDoS attempt.

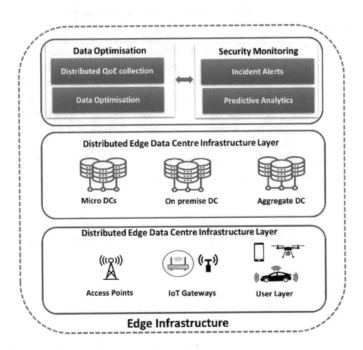

Fig. 2 DO-based edge security reference framework

To validate the above combined DO and security monitoring approach at the 5G edge, QoE-based detection of attack scenarios has been tested in long-term evolution (LTE) networks. The validation approach and results are outlined in the following section.

5 Experimental Validation of DO-Based Edge Security

This section elucidates a simplified case study work that has been undertaken to reproduce the effects of bidding down and DDoS attacks at 5G edge. The experiment has been carried out with the standalone network performance monitoring apps to observe and understand the effects of bidding down and DDoS attacks on mobile devices. In the validation process, the instantaneous app datasets from the UEs have been utilized to observe the network quality instead of the centralized server explained in Sect. 4.1.

5.1 Throughput Variation-Based Bidding Down Detection

In bidding down attack, the device capabilities were seized to reduce the data rate originally supported by the device. For example, in bidding down attack, a cellular device using LTE (4G) service can be downgraded to 3G/2G services. In the validation procedure, initially, a mobile device was reconfigured to different network types to emulate the bidding down attack. In the next step, the speed test app (Open Signal App) was utilized to understand the changes in uplink (UL)/downlink (DL) throughputs for different network types. With open signal speed test app, the instantaneous data has been collected from the live BSNL network by repeating the test several times. The data rates as shown in Table 3 vary depending on the network type

Table 3 UL/DL throughputs for different network type with speed test app

UL Throughput (Mbps)			DL Throughput (Mbps)		
2G	3G	4G	2G	3G	4G
0.01953	3.69	4.32	0.48632	8.29	9.63
0.00683	4.33	4.3	0.4541	8.92	10.4
0.02539	3.71	3.63	0.51074	8.51	7.54
0.06152	4.46	0.409	0.27539	10.7	9.14
0.01367	3.52	3.86	0.61035	6.97	7.24
0.00195	3.14	4.09	0.27929	9.4	9.34
0.02246	3.88	4.2	0.47558	0.92	0.89
0.01562	3.62	4.19	0.47753	0.584	2.33
0.02441	4.12	3.7	0.57324	0.189	0.631
0.00683	3.17	3.81	0.53906	1.4	1.33
0.04492	3.64	3.65	0.32031	1.49	0.77
0.02441	4.12	3.86	0.5732	0.7255	0.168

used for the testing purpose. The throughput results show that the network quality or throughput falls when the devices are downgraded in a bidding down the attack.

The mobile edge analytic server residing on the network side can continuously keep track of the throughput data from the app data feeds. It can be observed from Table 3 that uplink speeds for a user are in the range of 3.63–4.32 Mbps as long as the device has 4G connectivity. If an edge device has an uplink speed of 4.32 Mbps as highlighted in Table 3 for a long time and suddenly falls to 3.69 Mbps or 0.001953 Mbps, the mobile edge analytic server can identify this change by considering its previous traffic patterns. The edge server can then conclude that the user is affected by bidding down the attack.

5.2 Quality of Experience (QoE)-Based DDoS Detection

In denial of service (DoS) attacks, the attacker overloads the network with more traffic to destabilize it. To reproduce the effects of the DoS attacks at the edge of the 5G networks, the QoE has been measured on multiple devices (UEs/smartphones) in a selected cell in the LTE network. The throughput metric has been used to reflect the observed QoE in the edge devices. To emulate the increase in the network traffic in DoS attacks, a large file has been uploaded and the corresponding QoE has been measured at the same time in all edge devices using the Network Cell Info Lite App [12]. The experiment has been carried out at two different weak coverage locations. Prior to the throughput tests in these locations, certain test conditions have been fixed for the test. The description of the two throughput tests is explained below.

In throughput test 1, two smartphones/UEs connected to the same cell in the network have been considered with the test conditions as shown in Table 4.

Table 4 Conditions for throughput test 1

Test conditions	
Wi-Fi	OFF
Cell ID (both devices)	243
Coverage	weak
RSRP (UE 1)	-113 dBm
RSRP (UE 2)	-109 dBm
UE 1	iPhone XS
UE 2	iPhone 8 Plus
Network	LTE
File size	7.76 MB
File type	PDF
Upload location	Google Drive
Start time	10:23 PM

N.B. speed tests run at the same time on both handsets

With these UEs, three different cases have been tested to obtain the throughputs (download and upload speeds) as shown in Table 5. In case 1, UE 2 starts uploading a large file (7.76 MB). In case 2, the upload is 50% complete while in case 3, UE 2 is in the pre-completion stage of the uploading process. In all the test cases, UE 1 has been kept in a passive state. Figure 5 shows the upload speed for both UEs during the test for all the cases. It has been observed that the throughput for UE 1 falls as the uploading by UE 2 completes 50%. The upload speed for UE 2, however, improves to the initial conditions as it progresses through the uploading of the large file. This mimics the DoS attack scenario, where a sudden increase in the uplink data from the compromised device (UE 2) connected to a cell in the network causes throughput to fall for other edge devices (UE 1) in the same cell.

Table 5 Throughput test 1 results

Edge devices	Throughput during file upload (Kbps)		
	Case 1 initial conditions	Case 2 50% uploaded	Case 3 pre-completion
UE 1—download speed	4030	1880	4210
UE 1—upload speed	873.42	928.5	831.98
UE 2—download speed	8150	6500	7810
UE 2—upload speed	453.6	254.54	475.67

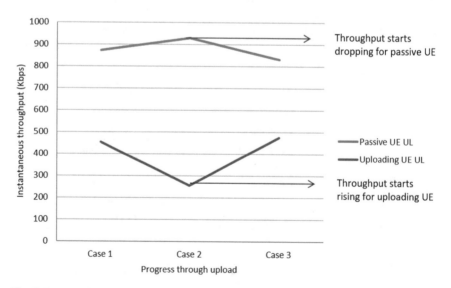

Fig. 5 Progress through upload speed in throughput test 1

In throughput test 2, two UEs connected to the same cell in the network have been considered with the test conditions as shown in Table 6. With these UEs, three different cases have been tested to obtain the throughputs as shown in Table 7. In case 1, UE 1 starts uploading 100 MB file. In case 2, the upload by UE 1 is 50% complete while in case 3, the upload has been completed. In all the test cases, UE 2 has been kept in a passive state.

Figure 6 shows the upload speed for both the UEs during the test for all cases. It has been observed that the throughput for UE 2 starts falling as UE 1 starts the uploading process. The upload speed for UE 2 improves as the upload has been completed by UE 1. The upload speed for UE 1, however, improves to the initial conditions as it progresses through the uploading of the large file. The effects for the DoS attack can be observed in UE 2, where the upload speed falls as the UE 1 starts uploading of the large file.

Table 6 Conditions for throughput test 2

Test conditions:	
Wi-Fi	OFF
Cell ID (both devices)	153
Coverage	Weak
RSRP (UE 1)	−106 dBm
RSRP (UE 2)	−105 dBm
UE 1	One plus 7
UE 2	Nokia 6.1 plus
Network	LTE
File size	100 MB
File type	MP4 video
Upload location	Google drive
Start time	02:21 PM

N.B. speed tests run at the same time on both handsets

Table 7 Throughput test 2 results

Edge devices	Throughput during file upload (Mbps)		
	Case 1	Case 2	Case 3
	Initial conditions	50% uploaded	Completion
UE 1—download speed	5.2	1.1	4.7
UE 1—upload speed	0.723	0.9342	0.7879
UE 2—download speed	2.3	0.414	0.483
UE 2—upload speed	0.938	0.414	0.738

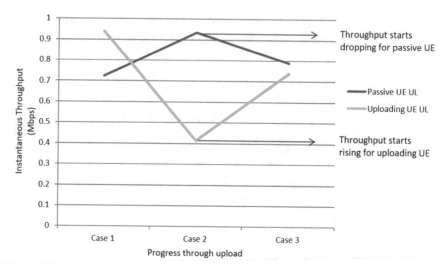

Fig. 6 Progress through upload speed in throughput test 2

6 Conclusions and Future Scope

In the context of 5G edge transformation, network security and performance optimization have been receiving significant attention in recent times. It is crucial to secure the 5G edge for upstream attacks from rapidly evolving cyber threats from a massive number of sensors/IoT endpoints. This requires an understanding of the real-time subscriber analytics to support data optimization and security monitoring at the edge. To highlight the novel approach of combined data optimization and security monitoring, the paper introduces a reference design edge security framework utilizing DO analytics. The paper also discusses the validation of the proposed framework with a simplified manual case study. The case study introduces effective test cases utilizing instantaneous app (network performance monitoring apps) data to observe the effects of bidding down and DDoS attacks in the mobile network. The test cases follow QoE-based attack detection in which throughput variations are considered as the key metric to observe the effects of the attack. The case study observes and analyzes the throughput variations in edge devices connected to the same cell, where one of the edge devices considered to be compromised tries to upload massive data. The quality analysis helps to understand the network performance and to identify any security anomalies in the network.

The next steps will focus on feeding all the data from the apps to a centralized server at the network side creating network quality maps. In this context, it is important to look into the latest MEC reference model to facilitate the implementation of the combined DO analytics and security monitoring approach proposed in the paper. The reference model has been recently published by ETSI [13]. Within the reference model, as shown in Fig. 7, network management or MEC orchestrator components

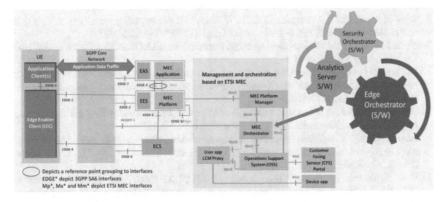

Fig. 7 MEC reference model coupled with DO analytics

implemented by software vendors need to be connected to the DO module to exchange information.

To realize effective edge security in the reference model, intelligent and automated security operations are required:

- *ML-based subscriber behavior detection system*: ML framework can understand the subscriber analytics and contextual information to support security monitoring at the edge. The framework interacts with security functions such as firewall, carrier grade network address translation (CGNAT), DPI to automate the monitoring process. With the application of AI/ML [14], security monitoring can raise real-time alerts to the operation center enabling predictive and proactive decisions.
- *Implementation of an intelligent Cyber Security Operation Center (CSoC) dashboard* which operates 24 × 7 in the backend to monitor and analyze network behavior/activities, reducing fatigue of the security analyst.

References

1. Popovski P et al (2019) Wireless access in ultra-reliable low-latency communication (URLLC). IEEE Trans Commun 67(8):5783–5801. https://doi.org/10.1109/TCOMM.2019.2914652
2. Raj D et al (2018) Enabling technologies to realise smart mall concept in 5G era. In: 2018 IEEE international conference on computational intelligence and computing research (ICCIC), Madurai, India, pp 1–6. https://doi.org/10.1109/ICCIC.2018.8782338
3. Bouras C, Kollia A, Papazois A (2017) SDN & NFV in 5G: advancements and challenges. In: 20th conference on innovations in clouds, internet and networks (ICIN), Paris, pp 107–111
4. Qiu B-J et al (2018) Service level virtualization (SLV): a preliminary implementation of 3GPP service based architecture (SBA). MobiCom
5. Lekshmi SS, Anjana MS, Nair BB, Raj D, Ponnekanti S (2019) Framework for generic design of massive IoT slice in 5G. In: 2019 international conference on wireless communications signal processing and networking (WiSPNET), Chennai, India, pp 523–529. https://doi.org/10.1109/WiSPNET45539.2019.9032878

6. Xiao Y, Jia Y, Liu C, Cheng X, Yu J, Lv W (2019) Edge computing security: state of the art and challenges. Proc IEEE, 1–24. https://doi.org/10.1109/JPROC.2019.2918437
7. Yousaf FZ et al (2019) MANOaaS: a multi-tenant NFV MANO for 5G network slices. IEEE Commun Mag 57(5):103–109
8. Raj D et al (2019) Effective Gi-LAN optimisation towards hardening the 5G service provider platform. In: 2019 international conference on wireless communications signal processing and networking (WiSPNET). IEEE
9. Ahmad I et al (2017) 5G security: analysis of threats and solutions. In: 2017 IEEE conference on standards for communications and networking (CSCN). IEEE
10. Sree S, Ponnekanti S (2019) Open RAN deployment using advanced radio link manager framework to support mission critical services in 5G. EAI Endor Trans Cloud Syst 5:162140. https://doi.org/10.4108/eai.15-3-2019.162140
11. Cook S, Mathieu B, Truong P, Hamchaoui I (2017) QUIC: better for what and for whom? In: 2017 IEEE international conference on communications (ICC), Paris, pp 1–6. https://doi.org/10.1109/ICC.2017.7997281
12. Network cell info lite v3.35, Mar 2017. Available at: https://play.google.com/store/apps/details?id=com.wilysis.cellinfolite&hl=cs. Accessed: 11 Sept 2017
13. Harmonizing standards for edge computing—a synergized architecture leveraging ETSI ISG MEC and 3GPP specification. https://www.etsi.org/
14. Lam J, Abbas R (2020) Machine learning based anomaly detection for 5G networks, 7 Mar 2020. arXiv:2003.03474v1 [cs.CR]

Automatic Detection of Crop Diseases and Smart Irrigation Using IoT and Image Processing

Anushree Janardhan Rao, Chaithra Bekal, Y. R. Manoj, R. Rakshitha, and N. Poornima

Abstract In India, wastage of water has become the most notable issue. It is imperative to decrease water misfortune in the fields because of different elements like funnel spillage or letting abundance water into the homesteads without proper knowledge. This undertaking is a mix of the savvy water system through cloud and recognition of yield ailments utilizing AI. Malady in crops on leaves decreases the nature of every item and the amount of agrarian item. Thus, picture preparing methods is being needed that will help in the precise and ideal discovery of illnesses and defeat the confinements of human vision. Creation of harvests can be expanded by distinguishing the ailment well in time. Programmed identification of plant infection helps in observing the harvest and mechanically distinguishes indications of the arrangement when they appear on plant leaves. Most plant sicknesses are achieved with the guide of growths, microscopic organisms and infections. Harvest assurance is performed particularly in enormous homesteads utilizing mechanized picture preparing innovation that can recognize infected leaf utilizing the shading data of leaves. These comprise of many advances information, like pictures, pre-preparing and extraction of components and grouping them on the stand-out premise.

Keywords Internet of things (IoT) · Machine learning · Node MCU · Image processing

1 Introduction

In India, farming assumes a significant job for advancement in food creation. Agribusiness is one of the fields where water is required in huge amount. Wastage

A. J. Rao (✉) · C. Bekal · Y. R. Manoj · R. Rakshitha · N. Poornima
Department of Computer Science and Engineering, Vidyavardhaka College of Engineering, Mysuru, Karnataka, India
e-mail: anushreerao23@gmail.com

N. Poornima
e-mail: poornima.cs@vvce.ac.in

© The Author(s), under exclusive license to Springer Nature Singapore Pte Ltd. 2021
J. S. Raj et al. (eds.), *Innovative Data Communication Technologies and Application*,
Lecture Notes on Data Engineering and Communications Technologies 59,
https://doi.org/10.1007/978-981-15-9651-3_31

of water is a serious issue in horticulture. Each time abundance of water is given to the fields, farming has brought about helpless yield when contrasted with populace.

Development. The water system is generally done utilizing trench frameworks in which water is siphoned into fields after normal time period with no input of water level in the field. This kind of water system influences crop wellbeing and produces a helpless yield since certain harvests are too touchy to even think about watering content in the soil. There are numerous strategies to spare or to control wastage of water in farming. Web of things (IoT) is an achievement in the advancement of innovation. IoT assumes a significant job in numerous fields, one of that is agriculture by which it can take care of billions of individuals on earth in future. The groundwater level is diminishing step by step because of impromptu utilization of water, shortage of land water and absence of downpours additionally causes the decrement in the volume of water on earth.

The main reason that illness identification in plants assumes an important job in the horticulture field is because the economy profoundly depends on horticultural profitability and infection in plants is become very common nowadays. Illness in plants prompts a critical decrease in both quality and amount. The appropriate consideration is not taken here, at that point it causes genuine impacts on plants and because of which influences the particular item quality, amount of profitability. Recognition of plant infection through some programmed method is gainful as it decreases an enormous work of observing in large ranches of yields, and at beginning phase itself, it distinguishes the manifestations of maladies for example at the point when they show up on plant leaves. It is exceptionally hard to screen plant illnesses physically. It requires an enormous measure of work, expertise in the plant maladies, and requires the extreme handling time. Henceforth, picture handling is utilized for the discovery of plant sicknesses. Ailment discovery includes the means like picture procurement, picture pre-handling, picture division, highlight extraction and characterization. This paper talked about the techniques used for the location of plant maladies using the picture of the leaves.

2 System Requirements and Specification

2.1 Software Requirement and Specification

An overview of the system requirements and its specification with purpose, scope, definitions, acronyms, abbreviation, reference provided by Software Requirements Specification (SRS) and also an overview gives a comprehensive description of the intended purpose and environment for software development. It describes what the software will be able to do, how well it performs the expected gestures and determine its accuracy. It also contains non-functional requirements and its imposing constraints the implementation or design. SRS establishes the basis for agreement between the suppliers and the customers on what the software product to do as well what not to do.

Some of the parameters of determining a good SRS are operating speed, availability, portability, response time, maintainability, security, footprint and speed of recovery from adverse events are evaluated.

2.2 Functional Requirements

A function of a software system or its components is defined by functional requirements such as computations, specialized subtleties, information control and other explicit usefulness details that characterize what a system should achieve. It holds all the information about each case wherein it includes the information of which systems uses which function requirement this is also called as quality requirements. This inflicts requirements on design and development. The hierarchy follows as: user/stakeholder request → feature → use case → business rule. A behavioural framework is illustrated by use case by one or more functional requirement. Though the analyst will derive the function requirements by validating a set of use cases, this will help the user in implementation. In our project, mage recognition of such interactive systems is focused on. In this sense, by analysing and training the image data from AlexNetModel.hdf5, a machine learning method would be desinged to classify the actual image fed into the system with high accuracy. This data is then used to derive the result. In this case, the result is the name of the disease that the crop has encountered.

2.3 Non-functional Requirements

In systems engineering and requirements engineering, the system will operate based on the standards described by the non-functional requirements. The system must perform as it is defined in the NFR. Non-functional requirements are not a specific function and it is of the form 'System shall be <requirement>'. The general properties of this system normally distinguish b/w success or failure based on the improvement of the project. A non-functional requirement is also called as 'quality attributes' of a system (Tables 1 and 2).

Table 1 Software requirements		
Required operating System	Any version of Windows
Required tools	Mathworks, Arduino
Runtime platform	.NET 4.5 and above
Architecture	64-bit
Server	ThingSpeak cloud server
DirectX	Version 9 and above

Table 2 Hardware requirements

Processor required	64-bit processor
Memory used	2 GB RAM
USB Support	USB 3.0
GPU	NVidia 600 s and above, AMD 5400 s and above
Sensor	Capacitive soil moisture, DHT11 sensors
Tools	NodeMCU, Motor, Pumps, Battery

3 System Analysis and Design

3.1 Existing System

The existing system for the smart irrigation system is to have a microcontroller system connected to the soil moisture sensors, etc. There are various systems for smart irrigation where there are certain disadvantages like less cost efficiency, complexity, etc. One of the existing systems includes picture of the soil near the crop's root part and estimates the water content optically. In this system, the sensors are embedded in a chamber under controlled illumination and buried at the root level of the plants. In the smartphone, an Android App was developed to operate directly on computing and connectivity components, such as the Wi-Fi network and the digital camera. Mobile App wakes up the smartphone by activating the device with user-defined parameters, then phone captures the pictures of the soil using the in-built camera through a non-reflective glass window and the estimated ratio between the dry and wet area of the image is achieved using RGB to grey process. Once the WI-Fi is enabled, to control the irrigation water pump, the ratio is transmitted to a gateway via a router. There exists a system, i.e. wireless sensor network-based automated irrigation system to optimize water use for agricultural purpose [1–3].

Temperature sensors and distributed wireless sensor network of soil moisture are placed in a crop field. With the help of Zigbee protocol, sensor information is handled [4]. The water quantity programming can also be controlled using an algorithm with threshold values of the sensors to a microcontroller [5]. The data inspection in the system is powered by cellular Internet interface and solar panel [6]. With the help of fixed wireless camera in crop field, disease area can be monitored using an image processing technique [7]. But these systems can be enhanced by using a cloud [8–10]. The system of checking for which disease a particular crop has is also necessary and the existing system has the image processing system to detect diseases for a particular crop [11–15]. Each crop will have to be tested in the places particularly for testing the crop diseases. There is no handy system where the farmer can get to know which disease is it [7, 10].

3.2 Proposed System

The proposed system includes having a website that will connect an IoT system and crop diseases detection system. The farmers with Internet access on phones can easily make use of this system. In the proposed system, both IoT-based irrigation system and machine learning-based crop diseases system have been included. The IoT-based system includes using the Node MCU and sensors like soil moisture sensor and DHT11 to sense the basic things like temperature and humidity. The values are uploaded in the cloud, and based on the threshold value set, the water pump will be controlled whether water has to be left to the fields or not. There would not be any human intervention necessary and the water is left automatically to the fields. This system can be used for various crops, as each crop sample can be accessed with our sensors. The user can set different threshold values for the different crop samples, and based on the data collected by these sensors, the relay will be triggered and water will be let into the fields.

The same system can have a crop diseases detection system where the image can be uploaded of the plant and the system will detect which disease, and based on that, the farmer can find remedies like pesticides or any medicines to cure that disease. This system uses a convolutional neural network algorithm, thereby increasing accuracy. It is used for the image processing technique. For the dataset, Alextnet model has been used which allows multiple-GPU.

3.3 System Architecture

System architecture for our project is as shown in Fig. 1 which shows the different elements and their relationships. The main webpage will have different components which further contains the different processes.

The system architecture can be consist of two different modules each interconnected. The user either directly or indirectly interacts with each of these modules in some way.

3.4 Control Flow Diagram

As soon as the webpage loads, can see that there are two options. One is for crop disease detection and the other is for smart irrigation. The smart irrigation system includes using the Node MCU and sensors like soil moisture sensor and DHT11 to sense the basic things like temperature and humidity. The values are uploaded in the cloud, and based on the threshold value set, the water pump will be controlled whether water has to be left to the fields or not. The same system has a crop diseases detection system where the image of the plant can be uploaded and the system will

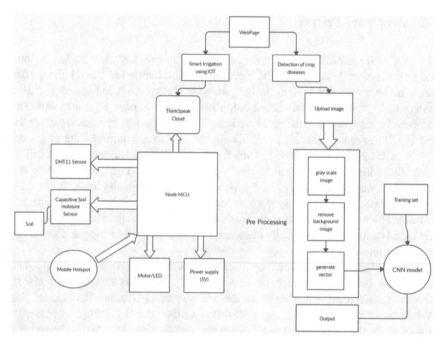

Fig. 1 System architecture

detect which disease. Once the system detects the disease, the user can then take the necessary measures to provide the specific pesticides or insecticides to the plants (Fig. 2).

4 Results and Analysis

4.1 System Testing

Testing plays a vital role in the development life cycle of the product where it detects the remaining errors from all the phases and it includes the testing methodologies like unit testing, integration testing, user acceptance testing and output testing.

4.2 Test Cases

A test case in software engineering defines a few collections of variables/conditions under which a tester can determine whether a software system is working correctly (Table 3).

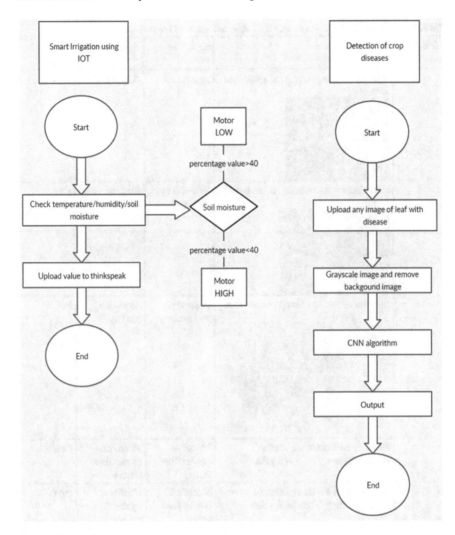

Fig. 2 Control flow diagram

4.3 Snapshots

The snapshots of the working project have been put in this section. Each of these represents a screen of the website and a suitable description of each snapshot has been given to explain each (Fig. 3).

This is the start page of the website. A user who is using the webpage for the first time can see the various options present on this webpage. Once decided, the user can choose either to check the crop disease part or can view the data from the smart irrigation using IoT option (Fig. 4).

Table 3 Some of the test cases during the implementation of the project

Test case no	Test cases	Expected output	Actual output	Pass/Fail
1	Provide an image of crop disease that is trained 	Apple rust	Apple rust	Pass
2	Provide an image of crop disease that is trained 	Strawberry leaf spot	Strawberry leaf spot	Pass
3	Provide an image of crop disease that is not trained 	Blueberry leaf healthy	Apple leaf healthy	Fail
4	Reading Soil Moisture sensor If curr. value < Threshold value	Motor on (water flow starts)	Motor on (water flow starts)	Pass
5	Reading Soil Moisture sensor If curr. value ≥ Threshold value	Motor off (water flow stops)	Motor off (water flow stops)	Pass

The above screenshot shows the IoT data collection page of the website. As soon as the user opens this page, the user can view the data collected by the sensors. The IoT module that haS been used can be started using the hotspot given by us. The sensors start measuring the values and collecting them. Those values or data can be seen on this page (Fig. 5).

This screenshot is of the crop disease detection page of the website. As soon as the user selects this option from the first page, he or she is directed to this page. As can be seen, there is an option to choose a file. So make sure to download the image to be detected. Once the image is downloaded, click on the 'choose file' option, and then choose the image that needs to be detected for a disease. Once the image has

Fig. 3 Start page

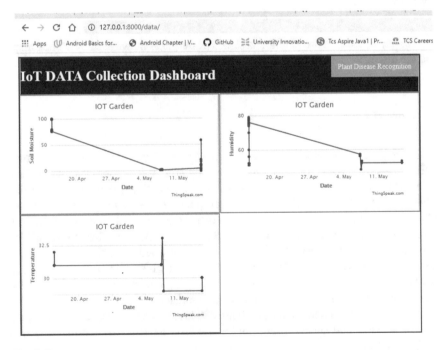

Fig. 4 Data page

Fig. 5 Disease detection page

been selected, click on the 'predict' button. The output is shown as above. The image and the disease that have been detected are shown.

5 Conclusion

In this project, a web application has been developed that assists farmers in two different aspects, one is to control the automatic flow of water into the agricultural fields depending on the factors like soil moisture, humidity, temperature and the other being able to detect which disease a crop has. The main intention behind this project is to make it easy for the farmers to control water flow into the fields by just connecting to the Internet and there would not be excess water loss to the fields. The amount of water necessary will be determined and left accordingly. The farmer need not take samples to the laboratories to find out which disease the crop has, it can be done by just clicking a picture and uploading it to the system and the CNN algorithm will use the image and find out which disease it is. This way the farmer can log in have two functionalities and this will lead to a smart irrigation technique. The TensorFlow and Django framework is used for detection of crop diseases and the IoT-based smart irrigation uses the Arduino IDE and NodeMCU. This was the intent behind the project and has been achieved to the best possible extent.

Acknowledgements If words are considered as the tokens of acknowledges, then the words play the heralding role of expressing our gratitude.

With proud gratitude, we thank God Almighty for all the blessings showered on us and for completing our project successfully.

We owe our gratitude to our beloved Principal, **Dr. B Sadashive Gowda** for his whole hearted support and for his kind permission to undergo the project.

We wish to express our deepest gratitude to **Dr. Ravi Kumar V**, Prof. and Head, Department of Computer Science and Engineering, VVCE, for his profound alacrity in our project and his valuable suggestions.

We wish to enunciate our special thanks to our paradigmatic and relevant project coordinators **Janhavi V** Associate Professor, **Usha C S** Assistant Professor, **Nithin Kumar** Assistant Professor, **Nagashree Nagaraj** Assistant Professor and our internal guide **Ms. Poornima N** Assistant Professor in Computer Science and Engineering, VVCE who gave us guidance throughout our project period with their valuable suggestions and for devoting their precious time in making this project a success.

In the end, we are anxious to offer our sincere thanks to our family members and friends for their valuable suggestions, encouragement and continuous support.

References

1. Gupta A, Mishra S, Bokde N, Kulat K (2016) Need of smart water systems in India. Int J Appl Eng Res 11(4):2216–2223
2. Al-Ghobari HM, Mohammad FS (2011) Intelligent irrigation performance: evaluation and quantifying its ability for conserving water in arid region. Appl Water Sci 1:73–83
3. Ganjegunte GK, Sheng Z, Clark JA (2012) Evaluating the accuracy of soil water sensors for irrigation scheduling to conserve freshwater. Appl Water Sci 2:119–125. Smith B. An approach to graphs of linear forms (Unpublished work style) (unpublished)
4. Gutiérrez J, Medina JFV, Garibay AN, Gándara MAP (2014) Automated irrigation system using a wireless sensor network and GPRS module. IEEE Trans Instrum Meas 63(1):1–11
5. Dursun M, Ozden S (2011) A wireless application of drip irrigation automation supported by soil moisture sensors. Sci Res Essays 6(7):1573–1582
6. Kim YJ, Evans RG, Iversen WM (2008) Remote sensing and control of an irrigation system using a distributed wireless sensor network. IEEE Trans Instrum Meas 57(7):13791387
7. Warne PP, Dr. Ganorkar SR (2015) Detection of diseases on cotton leaves using K-mean clustering method. Int Res J Eng Technol (IRJET) 02(04):425–431
8. Shergill D, Rana A, Singh H (2015) Extraction of rice disease using image processing. Int J Eng Sci Res Technol 135–143
9. Naikwadi S, Amoda N (2013) Advances in image processing for detection of plant diseases. Int J Appl Innov Eng Manage (IJAIEM) 2(11):168–175
10. Kamlapurtur Kar SR (2016) Detection of plant leaf disease using image processing approach. Department of ETE Education and Research in India, issue 2016-September. E-ISSN 2250-3153
11. Kaur Kra R (2018) Machine learning algorithms for disease classification in crop and plants. Int J Eng Res Technol 08-August-2018. e-ISSN 2455-2585
12. Kambale G (2007) Crop disease identification and classification using pattern recognition and digital image processing techniques. Professor of CSE MME College in India. P-ISSN 2278-8727
13. Selvaraj A, Shebiah N, Ananthi S, Varthini S (2013) Detection of unhealthy region of plant leaves and classification of plant leaf diseases using texture features. Agric Eng Int: CIGR J 15:211–217
14. Omrani E, Khoshnevisan B, Shamshirband S, Saboohi H, Anuar NB, Nasir MHNM (2014) Potential of radial basis function-based support vector regression for apple disease detection. Department of Biosystem Engineering, pp 2–19

15. Jhuria M, Kumar A, Borse R (2013) Image processing for smart farming: 'A digital image processing based algorithm which detect and recognize the plant diseases and various symptoms'. Institute of Electrical and Electronics Engineering

Refactoring OOPerl and Translating it to Python

V. C. Skanda⬤, Rachana Jayaram⬤, and N. S. Kumar⬤

Abstract The Perl Developer community is shrinking rapidly. Perl ruled the scripting world in late 90s and early 2000s. Perl 5 introduced Object orientation in a very unusual way in 1994. Companies have developed a vast amount of OOPerl code over the period of its popularity. As the scripting world has moved from Perl to Python, it has become difficult to find personnel to maintain legacy OOPerl code. This paper explores the program translation from Perl to Python for easier maintenance using compiler design techniques. Given a syntactically correct code written in Object Oriented Perl, the techniques developed here can be used to generate an equivalent code in Python. The Python code generated is logically identical to the input Perl code and given any input, produces an identical output.

Keywords Perl · Python · Program translation · Compiler design · Lex · Yacc · Object oriented programming

1 Introduction

Perl has turing-complete grammar. Its parsing can be affected by the run-time code executed during the compilation phase. For example, while evaluating the expression '2x5', Perl interprets 2 as a string since 'x' here is replication operator which replicates the string by the number specified in the second operand (5 in this case). Therefore, Perl cannot be parsed by a straight Lex/Yacc combination. Instead, the Perl interpreter implements its own lexer, which coordinates with a Yacc parser to resolve ambiguities [1].

V. C. Skanda · R. Jayaram (✉) · N. S. Kumar
Department of CSE, PES University, Bengaluru, India
e-mail: rachana.jayaram@gmail.com

V. C. Skanda
e-mail: skandavc18@gmail.com

N. S. Kumar
e-mail: nskumar@pes.edu

© The Author(s), under exclusive license to Springer Nature Singapore Pte Ltd. 2021, corrected publication 2021
J. S. Raj et al. (eds.), *Innovative Data Communication Technologies and Application*, Lecture Notes on Data Engineering and Communications Technologies 59, https://doi.org/10.1007/978-981-15-9651-3_32

It is said that *Only Perl can parse Perl* [2]. This is due to the fact that the Perl interpreter can simulate a Turing machine during its compiler phase. Therefore, it would need to resolve the halting problem in order to complete parsing in every case.

To understand the difficulty in translation, first the peculiarities of Perl should be taken into consideration. Namely, the concepts of blessing [3], special variables [4], confusing aspects of scoping (local, global and my) [5] can be challenging to translate as there is no 1 to 1 equivalent of aforementioned concepts in Python. However with intelligent parsing, perfect translation of these concepts were achieved.

2 Literature Survey

In his 1988 paper, R. C. Waters explores an abstraction-and-reimplementation based paradigm for program translation in which the source program is first analyzed in order to obtain a programming-language-independent abstract understanding of the computation performed by the program as a whole [6].

In his 1996 patent, Katsushige Hino introduces a programming language processing system in which a program described in a high level programming language is translated into another program written in lower level programming language, using term rewriting with pattern matching [7].

In 2018, Chen et al. presented a tree-to-tree neural networks approach towards translating legacy code. Our solution takes a grammar matching approach [8].

Perthon is an experimental Perl to Python translator that makes use of `Parse:`
`:RecDescent` for the parsing/lexing [9]. Our approach uses a 2-stage parser and Python Lex and Yacc (PLY) library [10] for building the translator.

3 Current Work

The translator developed here generates an error free, understandable and maintainable Python script that is equivalent to the input Perl source code. The Python Lex and Yacc (PLY) library was used to build a lexer and a 2-stage parser to achieve program translation. The lexer and parser developed handles the unusual cases previously mentioned.

For constructs with no direct Python equivalent, functions were implemented to mimic the behaviour of those constructs as in the cases of unary increment/decrement operators, print functions, some array operations and string-integer manipulations. In case of scalar context of arrays and special variables, 2 stage parsing was used to achieve translation.

The following constructs have been covered in the scope of this project.

1. Packages
2. Arithmetic, relational and logical expressions

3. Loops
4. Hashes
5. Subroutines
6. Constructors, Blessing
7. Scoping (local, global and my).

3.1 Architecture

Our architecture consists of 3 main stages as shown in Fig. 1:

1. Lexer which consists of regex rules for input code tokenization.
2. The sequence of tokens passed is matched to Perl grammar in the parsing stage. The parser built works in two stages.

   (a) In the first stage of parsing a type of look-up table is constructed. This look up table is designed to accommodate the intricacies of variable scope in Perl. Each variable entry in the lookup table is qualified with details about the function to which it belongs and whether the variable was defined locally or globally. This helps us in the second stage of parsing where lexical scoping of variables is of paramount importance.

   (b) In the second stage the parser utilizes the look up table generated previously along with production rules and mapping functions to map the tokens of each Perl statement to its equivalent Python statement. The generated python statement is subsequently added to an intermediate representation which is like an Abstract Syntax Tree (AST).

3 In the final step the generated Abstract Syntax tree is traversed using depth first traversal. Proper indentation is crucial for python; indentations are set to match the depth of the node containing the statement in the Abstract Syntax Tree.

The input to the program is a syntactically correct Perl code. The output is the equivalent Python code with comments intact for maintenance.

Fig. 1 System architecture

3.2 *Working*

In order to overcome the aforementioned difficulties in translating Perl, a lexer and parser was built from scratch which handled the ambiguous cases. The Perl parser was referred to for a better understanding on how to build out Perl to Python translator.

The lexer first tokenizes the Perl code. The tokens are recognized using regular expressions. The common tokens are name, string, number, reserved words, different operators, separators etc.

For the parsing stage a 2-stage parser is being used. In stage 1, the Perl code is parsed to know the scope of variables and also to know about the modules to be imported (if any). Using this information, a lookup table is built.

In stage 2, the Perl code is parsed again. This time, depending on the production rules and the scopes obtained from the lookup table, each Perl statement is mapped to its equivalent Python Statement. The generated python statements are fed to an AST (Abstract Syntax Tree). After the entire code has been parsed, the AST is traversed to obtain the equivalent python code.

Some Perl constructs supported in our tool are tabulated in Table 1 and listed below:

1. Packages—Package in python is a collection of functions whereas a package in Perl is a collection of statements which reside in their own name space. Classes in python are templates for creating objects and are a better match for packages in Perl. Thus packages in Perl are mapped to classes in python.
2. Sub routines and Methods - They are mapped to functions in python.
3. Blessing—In Perl, blessing is a way to make the Perl interpreter treat packages as object oriented classes and the blessed objects as the instances of those classes. The bless operator is generally used inside a method called new. This acts as a constructor. So the new method is mapped along with the bless operator to a constructor or __init__ in python.
4. Array—An array in Perl is an ordered list of values. Arrays are mapped to lists in python. There are many functions associated with arrays. They are mapped to their respective python list methods. So for example, `@arr = (1, 2, 'Hello);'` in Perl is translated to `arr=[1, 2, 'Hello];'` in python.
5. Hash—Hash is a collection of key value pairs in Perl. It is mapped to dict in python which is also a collection of key value pairs. So for example the hash `%data = ('adi', '32', 'Kumar' , 55);` is translated into the dict `data = 'adi':32, 'Kumar':55adi':32, 'Kumar':55` in Python.
6. Increment/Decrement operators—Like C, Perl has increment and decrement operators. But python does not support increment and decrement operators. They cannot be mapped to `var += 1` or `var -= 1` because in python this is a statement and not an expression. To simulate the effect of pre increment/decrement and post increment/decrement operators, then they are mapped to the functions. For example `++$x` is translated into `PreIncrement(x)` in Python.
7. Logical defined-or operator—It is similar to logical or operator in C. But the only difference is that instead of using the truth value of the left hand side expression,

Table 1 Perl constructs and their equivalent Python constructs

Perl construct	Python construct
Packages	Classes
Sub routines, methods	functions
Blessing	Constructor
Array	List
Hash	Dict

it checks whether the operand is defined [7]. Since this operator has no equivalent in Python, a function was developed. The function call replaces the operator in Python. So for example x//y in Perl is translated to dor(x,y) in Python.

All other operators are translated to their respective python equivalents. Scopes were handled using 2 stage parsing as mentioned earlier. Similarly conditional construct if-else if-else in Perl is translated to if-elif-else in Python, iterative constructs while, do-while, until in Perl is translated to while in Python, foreach to for in Python.

4 Results

The translating tool was run on several input Perl scripts encompassing all the constructs to be tested. Presented here are a few examples.

4.1 Example 1

This is an example printing the fibonacci series, demonstrating translation of loops. Here the for loop in Perl is translated to it's equivalent for loop in python (Table 2).
Translator Input: Perl Code

```perl
1   print "12 terms of the fibonacci series\n";
2   $a = 0 ;
3   $b = 1 ;
4   print "$a $b ";
5   for ( $i = 2; $i <= 12; $i++)
6   {
7       $c = $a + $b;
8       print "$c ";
9       $a = $b ;
10      $b = $c ;
11  }
12  print "\n";
```

Table 2 Output comparision for Example 1

Perl Code Output	Python Code Output
12 terms of the fibonacci series 0 1 1 2 3 5 8 13 21 34 55 89 144	12 terms of the fibonacci series 0 1 1 2 3 5 8 13 21 34 55 89 144

Translator Output: Python Code

```
1   str_ = lambda x: '' if x == None else str(x)
2
3   a = None
4   b = None
5   c = None
6
7   print( "12 terms of the fibonacci series\n", end='', sep='' )
8   a = 0
9   b = 1
10
11  print( str_(a) + " " + str_(b) + " ", end='', sep='' )
12  for i in range (2, 13, 1):
13      c = a + b
14      print( str_(c) + " ", end='', sep='' )
15      a = b
16      b = c
17
18  print( "\n", end='', sep='' )
```

4.2 Example 2

This is an example for getting the area of a rectangle. This demonstrates translation of packages, expressions, blessing and @INC [8]. There are 2 parts to this example. A server file which contains a package and its various functions and a client file which contains code which calls the functions defined in the server. They are translated into their equivalent constructs in python which is shown just below the Perl code (Tables 3 and 4).

4.2.1 Server Code

Translator Input: Perl Code - server.pm

```
1   package rect;
2   sub new
3   {
4       my $class = shift;
```

```
5      my $self =
6      {
7          length => shift,
8          breadth => shift
9      };
10     bless($self, $class);
11     return $self;
12  }
13  sub area
14  {
15     my $self = shift;
16     return $self->{length} * $self->{breadth};
17  }
18  sub perimeter
19  {
20     my $self = shift;
21     return $self->{length} * 2 + $self->{breadth} * 2;
22  }
23  1;
```

Translator Output: Python Code - server.py

```
1   class rect :
2       def __init__(self,*argv) :
3           arg_list = list(argv)[::-1]
4           self.__dict__.update(
5               {'length':arg_list.pop(),
6                'breadth':arg_list.pop()})
7
8       def area (self,*argv) :
9           arg_list = list(argv)[::-1]
10          return(self.length * self.breadth)
11
12      def perimeter (self,*argv) :
13          arg_list = list(argv)[::-1]
14          return(self.length * 2 + self.breadth * 2)
```

4.2.2   Client Code

Translator Input: Perl Code - client.pl

```
1   # !/usr/local/bin/perl -w -I E:\input
2   use lib 'E:\\input\\';
3   use server::rect;
4
5   $x=10;
6   $y=15;
7   print "Constructing a rectangle of length $x and breadth $y \n";
8
9   # Constructing a rectangle of length x and breadth y
10  $d = rect->new($x,$y);
11  $d->area();
```

```
12
13   # area and perimeter
14   $area = $d->area();
15   $perimeter = $d->perimeter();
16
17   print "A rectangle of length $x and breadth $y"
18   print "Has area = $area \t perimeter = $perimeter \n";
```

Translator Output: Python Code - client.py

```
1    str_ = lambda x: '' if x == None else str(x)
2
3    area = None
4    perimeter = None
5    x = None
6    y = None
7
8    # !/usr/local/bin/perl -w -I E:\input
9    import sys
10   sys.path.insert(0, 'E:\\input\\')
11   import server.rect as rect
12
13   x = 10
14   y = 15
15
16   print( "Constructing a rectangle of length " + str_(x) +
17          " and breadth " + str_(y) + " \n", end='', sep='' )
18   # Constructing a rectangle of length x and breadth y
19   d = rect.rect(x,y)
20   d.area()
21
22   # area and perimeter
23   area = d.area()
24   perimeter = d.perimeter()
25   print(
26       "A rectangle of length " + str_(x) + " and breadth " + str_(y))
27   print("Has area = " + str_(area) + " \t perimeter = " +
28         str_(perimeter) + " \n",
29         end='', sep='' )
```

4.3 Example 3

This is an example incorporating various types of variable scopes has been converted to equivalent Python code. It demonstrates local, global and my scopes in Perl and their translation to Python.

Translator Input: Perl Code

```
1    sub foo
2    {
3        print "deleted";
4    }
```

Table 3 Output comparision for Example 2

Perl Code Output
Constructing a rectangle of length 10 and breadth 15 A rectangle of length 10 and breadth 15 Has area = 150 perimeter = 50
Python Code Output
Constructing a rectangle of length 10 and breadth 15 A rectangle of length 10 and breadth 15 Has area = 150 perimeter = 50

```perl
5   sub foo
6   {
7       $z=shift;
8       my $y = 20;
9       local $z = 30;
10
11      # x: 10 y: 20 z: 30
12      print "foo before g : \n x : $x y : $y z : $z \n";
13
14      g();
15      # x: 2 y: 20 z: 3
16      print "foo after g : \n x : $x y : $y z : $z \n";
17
18      # y: 2
19      print "foo global : \n y : ", $main::y,"\n";
20  }
21  sub g
22  {
23      # x: 10 y: z: 30
24      print "g : \n x : $x y : $y z : $z \n";
25      $x = 2; $y = 2; $z = 3;
26  }
27
28  $x = 10;
29  # x: 10 y: z:
30  print "main before foo : \n x : $x y : $y z : $z \n";
31
32  foo($x);
33  # x: 2 y: 2 z: 10
34  print "main after foo : \n x : $x y : $y z : $z \n";
```

Translator Output: Python Code

```python
1   str_ = lambda x: ''
2   if x == None
```

Table 4 Output comparision for Example 3

Perl Code Output	Python Code Output
main before foo : x : 10 y : z : foo before g : x : 10 y : 20 z : 30 g : x : 10 y : z : 30 foo after g : x : 2 y : 20 z : 3 foo global : y : 2 main after foo : x : 2 y : 2 z : 10	main before foo : x : 10 y : z : foo before g : x : 10 y : 20 z : 30 g : x : 10 y : z : 30 foo after g : x : 2 y : 20 z : 3 foo global : y : 2 main after foo : x : 2 y : 2 z : 10

```python
3     else str(x)
4     x = None
5     y = None
6     z = None
7
8     def foo(x, y, z, * argv):
9         arg_list = list(argv)[::-1]
10        z = arg_list.pop()
11        my_y = 20
12        local_z = 30
13        # x: 10 y: 20 z: 30
14        print(
15          "foo before g : \n x : " + str_(x) +
16          " y : " + str_(my_y) + " z : " + str_(local_z) + " \n",
17          end = '', sep = '')
18        (x, y, local_z) = g(x, y, local_z)
19        #x: 2 y: 20 z: 3
20        print(
21          "foo after g : \n x : " + str_(x) +
22          " y : " + str_(my_y) + " z : " + str_(local_z) + " \n",
23          end = '', sep = '')
24        # y: 2
25        print("foo global : \n y : ", y, "\n", end = '', sep = '')
26        return (x, y, z)
27
28    def g(x, y, z, * argv):
29        arg_list = list(argv)[::-1]
30        # x: 10 y: z: 30
31        print(
32          "g : \n x : " + str_(x) +
33          " y : " + str_(y) + " z : " + str_(z) + " \n",
34          end = '', sep = '')
35        x = 2
```

```
36      y = 2
37      z = 3
38      return (x, y, z)
39
40   x = 10
41   # x: 10 y: z:
42     print(
43       "main before foo : \n x : " + str_(x) +
44       " y : " + str_(y) + " z : " + str_(z) + " \n",
45       end = '', sep = '')
46   (x, y, z) = foo(x, y, z, x)
47   # x: 2 y: 2 z: 10
48   print(
49     "main after foo : \n x : " + str_(x) +
50     " y : " + str_(y) + " z : " + str_(z) + " \n",
51     end = '', sep = '')
```

The translation tool successfully converts a Perl program to its equivalent python program. When the outputs for both the original Perl program and the python program are compared, they are exactly the same. A snippet of the tool UI can be seen in Fig. 2.

The peculiarities of Object Oriented Perl deters new programmers from picking it up leading to an overall shrinkage in the number of Perl developers. Despite this,

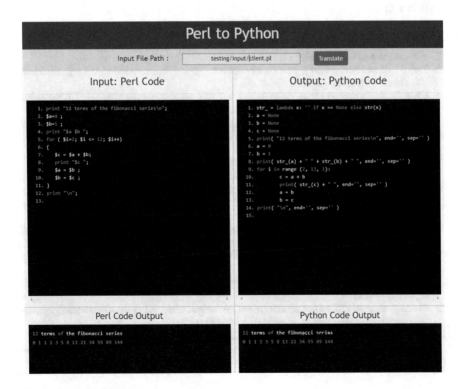

Fig. 2 System architecture

many companies still have a large amount of legacy Perl code in their code base whose maintenance is becoming tedious. Given that Python is has picked up in popularity and is inherently more maintainable than Perl, this tool can help upgrade legacy Perl code currently present in the industry to more maintainable Python code.

5 Future Work

Our tool currently can translate only syntactically correct Perl code to equivalent python code. There is a lot of work happening in the field of automated correction of syntax errors [11]. Recurrent Neural Nets have also been used to correct syntax errors [12]. These concepts can be integrated in order to provide syntactically correct python code from the Perl source code which can have syntax errors.

Also our tool currently does not support some Perl constructs like auto loaders, begin-end and value sub routines. Future work would involve encompassing all aspects of OOPerl.

References

1. Brown D, Levine J, Mason T (1995) Lex and Yacc. O'Reilly Media
2. McNamara J (2000) Spreadsheet::WriteExcel. Perl J 5(3)
3. Perl 5 Porters. Bless. Perl 5 version 32.0 documentation. https://perldoc.perl.org/functions/bless.html. Last accessed on 13 Aug 2020
4. Cross D. Perl's special variables. Perl.com. https://www.perl.com/pub/2004/06/18/variables.html. Last accessed on 13 Aug 2020
5. arturo. Scoping. https://www.perlmonks.org/?node_id=66677. Last accessed on 13 Aug 2020
6. Waters RC (1988) Program translation via abstraction and reimplementation. IEEE Trans. Software Engineering 14:1207–1228
7. Katsushige H (1996) Programming language processing system with program translation performed by term rewriting with pattern matching
8. Chen X, Liu C, Song D (2018) Tree-to-tree neural networks for program translation. In: Proceedings of the 32nd conference on neural information processing systems (NeurIPS 2018), pp 2–8
9. Manura D. Perthon–Python to Perl language translation. http://perthon.sourceforge.net. Last accessed on 13 Aug 2020
10. Beazley D. PLY (Python Lex-Yacc). https://www.dabeaz.com/ply. Last accessed on 13 Aug 2020
11. Lévy, J.: Automatic correction of syntax-errors in programming languages. In: Acta Informatica 4, pp. 271–292 (1975)
12. Bhatia S, Singh R (2016) Automated correction for syntax errors in programming assignments using recurrent neural networks. arXiv:1603.06129v1

Towards Determining and Delivering the Most Suitable Form of Diagrammatic Representation by Compressing Lengthy Texts

Varsha Kukreja and Aparna Halbe

Abstract Visualization of content is an essential component for thorough under-standing, interpretation and quicker and effective processing by the human brain, considering the abundance of data to which one is exposed daily. Determining the most appropriate form of a visual presentation can be time-consuming and chal-lenging. There is a need to reduce this manual effort in various domains and industries like marketing, education, travel and a business where presenting data is one of the crucial tasks. However, such textual content is absent to visual content converters that can be fed with structured, unstructured textual content and semi-textual content of various forms and recommend different suitable layouts and forms of visual repre-sentation. However, certain converters available in the market limit themselves in capacity to only structured content. Hence, developing a visual content generator that takes exhaustive textual content of various forms and recommends, a set of suit-able forms of diagrammatic representation to deliver the content is aimed at, thus saving time and enhancing the speed of processing data by the human brain by 60,000 times and expand the delivery of content by 4 times.

Keywords Visual content generator · Quantitative · Qualitative · Categorical · Automatic visual engine

V. Kukreja (✉) · A. Halbe
Department of Information Technology, Sardar Patel Institute of Technology, Mumbai, Maharashtra 400058, India
e-mail: Varsha.Kukreja@spit.ac.in

A. Halbe
e-mail: Aparna_Halbe@spit.ac.in

1   Introduction

1.1   Background

According to statistics available, 65% of learners are visual learners. Also visual content helps in a thorough understanding of data in much better form. It also helps in mining important patterns and observing trends. It is also easy to follow visual instructions with a rate of 323% better efficiency. As moving towards more digital content, there is a pool of legacy textual content that needs to be transformed and also the abundance of new data being created should be supported with a mechanism that automatically helps present this content in a better form which is quick and works with all forms of content including textual, semi-textual content, cases of structured, semi-structured and unstructured data effectively. Hence, aimed to create a visual content converter where users can feed in textual content and users are recommended with different suitable forms of visual presentation depending on data whether it is statistical, instructional, procedural or simply differentiating between two aspects. Users should also be able to choose between representations of different dimensions, shapes and characteristics and easily choose and embed them.

1.2   Motivation

The main motivation is to employ a better solution or implementation than the primary ones available in the market that works in low capacities with structured data and addressing a solution for this problem that works equally well with structured, semi-structured and unstructured data and identifies the most suitable form of a presentation not only for statistical data but also for procedural data as well.

- **Lack of tool for converting unstructured statistical data:**
  Although few tools are present, that convert structured statistical data into suitable statistical forms of presentations like line charts, histograms, bar graphs, etc. But statistical data present in unstructured form and semi-structured form is still a challenge for those tools.
- **Lack of tool for converting unstructured procedural data:**
  Apart from a wide pool of statistical data that can be easily represented in various mathematical forms, there still arises a need for tools that can identify different classes of such data and can represent them in various forms like trees, workflow diagrams, table of differences and Venn diagrams.

1.3 Objectives

Keeping the motivations in mind, the main objective of the project is to resolve the inadequacies in the present system. Having identified these, the individual solutions to these problems form the objectives of our system. These are succinctly stated as follows:

- **Create a robust technological framework to achieve a conversion of statistical, as well as non-statistical structured data:**
  The first step towards the ultimate goal of this project lies in the effective conversion of structured data into mathematical presentations of various forms and various dimensions and highlighting them to the user.
- **Enhancing the framework to handle the effective conversion of semi-structured and unstructured data:**
  Creating a model that could take unstructured data and could classify it into various categories as well as subcategories, such as paragraphs about sales distribution of company could be best represented by a pie chart, and a paragraph differentiating between two aspects could be best represented in tabular form. The model should also select the best suitable representation for data which could be workflow, pie chart, Venn diagram and hierarchical, mathematical forms using RNN model.

The rest of the paper is organized into a literature survey in Sect. 2, discussion of research methodology in Sect. 3 employed with a look at the system flow and approach, in Sect. 4 our prototype is presented and discussed results and limitations and conclude with Sect. 5.

2 Literature Survey

This paper [1] elaborates on how various forms of possible visualization for structured data sets can be derived and choosing the right one. It claims that every attribute can either be categorical or quantitative. Based on the available categorical and quantitative attributes, various possible finite numbers of mathematical forms can be represented by applying permutation and combination of these attributes.

The generation of arbitrary diagrams is focused on [2] as well and digs deep into constructing diagrams from basic elements. It follows a three-stage process that emphasizes on the construction of diagrams depending on the logical relations that can be extracted from the data and transforms the mapping of this relationship from mathematical form to graphical form and finally translating them into visual forms.

The paper [3] looks to solve the issues by the construction of knowledge graphs and uses BabelNet technologies at its centre. It also focuses on compressing and removing certain redundancies and mapping the concepts that could be relations from words into graphical relations and then delivering a visual form of the content.

The scope for improvement is identified in [4] which is a major challenge for users who are not familiar to a great extent with the world of statistics. It helps in quickly extracting the elements from data which could be mapped to a declarative language to create visualizations. This is done with the RNN model that helps in identifying such patterns.

3 Research Methodology

The components of the system have been discussed in detail in the following subsections.

3.1 Process Flow

To completely understand the complete complexity and context of our target domain, a comprehensive literature survey of research papers, articles and blogs is done on the topic of creating visual content from the textual content. The process drills down to three major steps which include generation of data set based on certain heuristics to get labelled corpuses with their appropriate form of visual content tagged. This data can be used to fine-tune the last layers of the RNN [5] model and can be used for effective training using transfer learning to better understand the language. The second step could be identifying key entities in the form of key-value pairs, where these keys depend on suitable forms predicted in the first step. A formal approach about necessity and type of each mathematical form was thoroughly found from the literature survey. This extraction of key entities would be supported by a POS [6] tagger and extracting relations through dependency parser trees. The third step helps in determining the most appropriate layout for the information presented to ensure a proper flow of visual content with textual content (Fig. 1).

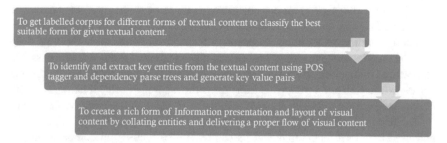

Fig. 1 Process flow of the system

Group A. TreeMap, PieChart - 1 numeric expressed as % , 1 category

Group B. Scatter Plot - 2 numeric (1X, 1Y) -(Optional :can include 2 categorical - 1 size & 1 color)

Group C. Linechart, BarChart- X (categorical variable) , Y(numeric variable)

Group D. Histogram - X, Y numeric - X in bins

Cases

1. 1 categorical variable - Count frequency of categorical variable (Group A, Group C)
2. 1 numeric variable - bin the vars and count frequency. (Group D)
3. 1 categorical v/s 1 numeric variable - Group by clause (Group A, Group C)
4. 1 numeric v/s 1 numeric variable - Group B
5. 1 categorical variable v/s 1 categorical variable - Not possible

Fig. 2 Cases for consideration for mathematical forms

3.1.1   For Structured Data

To realize the complexity of a system, begin by simplifying certain aspects of the first and second step of the process. Now, only structured statistical data is considered and determines the mathematical forms the data could be represented by, depending upon the characteristic of attributes. The attributes are now classified to be categorical if they account for less than 20% of uniqueness in the data set and based on other certain intuitive factors. The attributes which do not fall into this category can then be declared quantitative. Then based on the requirements of each mathematical model and permuting the available set of attributes for each type, those mathematical forms are tried to construct [7] (Fig. 2).

3.1.2   For Unstructured Data

For unstructured data, name entity recognition [8] is used that starts with few general entities, and with a few epochs of training, it drops out the general entities which do not appear in the training data and learn the entities of training data. It is the most predominant technique to extract entity [9] from the corpus and use them for classification or further analysis of documents. By extracting these entities, useful information is got and converts unstructured data into structured data (Fig. 3).

3.2   System Architecture

Initially labelled corpuses are collected, or data is generated in various formats consisting of statistical, numerical, procedural, sequential and instructional data and tagged with their proper format. By determining certain heuristics, an RNN [10]

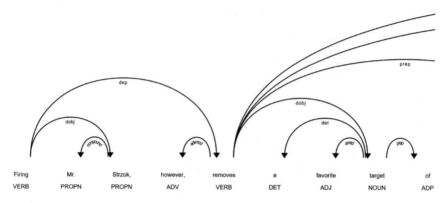

Fig. 3 Entity extraction

model is to be built that classifies data as statistical or procedural data. If the data belongs to the abovementioned two categories, it is further classified into subcategory best suitable for it, depending on certain heuristics. The key entities from data are extracted, and the proper layout of information presentation is constructed [11]. Using certain coverage and similarity metrics [12], the efficiency of conversion is measured, and the data is further used for training or is used to make modifications in the heuristics (Fig. 4).

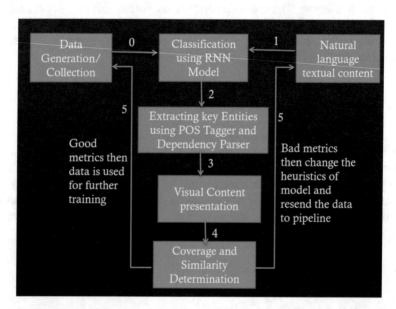

Fig. 4 Architecture diagram

4 Results and Discussions

In this section, our prototype, its scope, functionalities and limitations are discussed. Before discussing the functionalities, the assumptions made are stated as follows:

1. Under the present scope, assumed to deal only with statistical data.
2. The use case of structured statistical data available as csv, tsv and xlsx files is assumed.
3. Assumed that the user chooses at-most two attributes to get recommendations of suitable visual presentations.

4.1 Prototype

Our prototype is a web application that is fed with a structured statistical data file and is allowed to choose at the most two attributes and is recommended with suitable mathematical representations based on the cases discussed for each mathematical representation in research methodology. The algorithm identifies them to be categorical or quantitative and then does the desired permutation to achieve suitable results (Fig. 5).

The system is given an input csv file of Google Play Store ratings and reviews, and the user chooses to rate as one of the attributes. Since it is a categorical value, a group by clause is applied to find its categories, and it is declared to be categorical as its uniqueness is less than or equal to 20%.

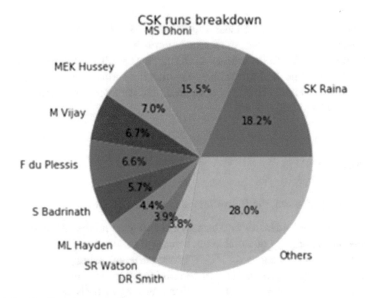

Fig. 5 Visualization of structured data

```
{'College Name': ['Govt PU College', 'Govt High School'],
 'Degree': ['BE in Engineering\nShridevi Institute\n'],
 'Designation': ['Software Engineer'],
 'Graduation Year': ['2017', '2013'],
 'Location': ['Bengaluru'],
 'Name': ['Software Engineer'],
 'Skills': ['HTML5 (2 years), JAVASCRIPT (2 years), MYSQL (2 yea
 'Languages: PHP, \n \nWeb Technologies: HTML5, CSS3, JavaScrip
```

Fig. 6 Extracting entities

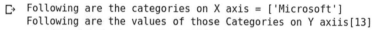

Following are the categories on X axis = ['Microsoft']
Following are the values of those Categories on Y axiis[13]

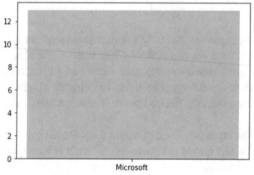

Fig. 7 Visualization of experience of a person

In the experiment, the model was trained with a sample of 250 annotated resumes. Using spacy and nltk libraries, the common entities were dropped out with a rate of 0.2, and the model learnt the entities from training data. Then the model was tested with a sample resume, and it was able to extract the entities, and graph of 'years of experience' was plotted. This is how unstructured data was converted to structured data and visualized (Figs. 6 and 7).

4.2   Results

Through the application, the following results are observed:

1. Classification of attributes into categorical and quantitative attributes.
2. Proper determination of suitable mathematical form.
3. Determination of categories for categorical attributes using group by clause.
4. The suitable mathematical form of recommendations for chosen attributes.
5. Representing unstructured data into visual form by using NER.

4.3 Limitations and Future Scope

The designed prototype cannot realize its full potential if it is not robust enough to handle all forms of data. This prototype is aimed at simplifying the three steps of the process. But a robust model will need a classification of data to the most suitable form for various dimensions and combinations of attributes and key entity extraction which will depend on the output from the first stage of the process and making a proper form of mapping for the third stage of the process.

5 Conclusion

In this literature, the unique problem of converting a wide range of data formats is presented including unstructured statistical formats, other non-statistical formats of structured and unstructured data, and proposed an entirely new process flow and its implementation to resolve these issues. Relevant objectives are designed which aligned with the framing of our problem definition. A prototype to that effect with several practical use cases implemented is then presented, which achieved the objectives to the scope decided to address. It was found that the system can be suitably adapted to various ecosystems with several tangible benefits. Lastly, assumptions, limitations and future scope of our research are discussed.

References

1. Hu KZ (2015) Towards DIVE: a platform for automatic visualization and analysis of structured datasets (2015)
2. Bateman J, Kleinz J, Kamps T, Reichenberger K (2001) Towards constructive text, diagram, and layout generation for information presentation
3. Reichenberger K, Golovchinsky G, Kamps T (1995) Towards a generative theory of diagram design
4. Rashidghalam H, Taherkhani M, Mahmoudi F (2016) Text summarization using concept graph and BabelNet knowledge base. IRANOPEN
5. Golovchinsky G, Kamps T, Reichenberger K (2006) Subverting structure: data-driven diagram generation
6. Dibia V Demiralp C (2018) Data2Vis: automatic generation of data visualizations using sequence to sequence recurrent neural networks
7. Reategui E, Klemann M, Finco MD (2012) Using a text mining tool to support text summarization. In: IEEE 12th conference on advanced learning technologies
8. Ramesh A, Srinivasa KG, Pramod N (2014) SentenceRank—A graph based approach to summarize text. ICADIWT
9. Schmitt X, Kubler S, Robert J, Papadakis M, LeTraon Y (2019) A replicable comparison study of NER Software: StanfordNLP, NLTK, OpenNLP, SpaCy, Gate. In: 2019 Sixth international conference on social networks analysis, management and security (SNAMS), Granada, Spain
10. Dawar K, Samuel AJ, Alvarado R (2019) Comparing topic modeling and named entity recognition techniques for the semantic indexing of a landscape architecture textbook. In: 2019 Systems and information engineering design symposium (SIEDS), Charlottesville VA, USA

11. Vychegzhanin S, Kotelnikov E (2019) Comparison of named entity recognition tools applied to news articles. In: 2019 Ivannikov Ispras open conference (ISPRAS), Moscow, Russia
12. Keretna S, Lim CP, Creighton D (2015) Enhancement of medical named entity recognition using graph-based features. In: 2015 IEEE international conference on systems, man, and cybernetics, Kowloon

Fraud Detection in NFC-Enabled Mobile Payments: A Comparative Analysis

Pinki Prakash Vishwakarma, Amiya Kumar Tripathy, and Srikanth Vemuru

Abstract Near field communication (NFC) is a short-range wireless communication technology that bestows contactless payments. Mobile Commerce is an amplification of electronic commerce and makes use of a mobile device to conduct a payment transaction. The growing popularity of mobile banking has led to an increase in fraudulent payment transactions thus resulting in financial losses. Therefore, the state-of-the-art solutions is required to detect and prevent fraud in mobile payments. This paper presents a comparative analysis of various fraud detection techniques for NFC-enabled mobile payments and evaluates each technique based on certain parameters.

Keywords Near field communication (NFC) · Mobile payments · Fraud detection

1 Introduction

Near field communication technology aids mobile payments with the help of mobile phones to conduct contactless payment transactions [1, 2]. In NFC-enabled mobile payments, the consumer needs to bring their NFC-based mobile phone in front of the point of sale (POS) to make a payment. For consumers making payment of goods purchased using a mobile phone is an easy and convenient way. The growing popularity of mobile banking has led to an increase in fraudulent payment transactions; thus, resulting in financial losses. Therefore, the state-of-the-art solutions is required

P. P. Vishwakarma (✉)
Department of Computer Science and Engineering, Koneru Lakshmaiah Education Foundation, Vaddeswaram, AP, India
e-mail: vishwakarmapp@gmail.com

A. K. Tripathy
Department of Computer Engineering, Don Bosco Institute of Technology, Mumbai, India
e-mail: amiya@dbit.in

S. Vemuru
School of Science, Edith Cowan University, Perth, Australia
e-mail: vsrikanth@kluniversity.in

© The Author(s), under exclusive license to Springer Nature Singapore Pte Ltd. 2021
J. S. Raj et al. (eds.), *Innovative Data Communication Technologies and Application*,
Lecture Notes on Data Engineering and Communications Technologies 59,
https://doi.org/10.1007/978-981-15-9651-3_34

to detect and prevent fraud in mobile payments. Fraud is treated as a cyber-crime eliciting colossal financial losses [3]. Fraud is booming desperately with furtherance in the mobile commerce industry. As a consequence, fighting fraud has evolved into an important concern to investigate. The fast growth regarding mobile payment manifests fraud to increase in the future.

Fraud detection and prevention identify fraudulent transaction accomplished through security flaws in mobile payment system at the same time preserving the nominal number of false alarms [4]. The fraud detection system based only on business rule may require more human reassessment which in turn increases the entire transaction processing time. Today, in the time m-commerce is growing rapidly, merchants have limited time to detect fraudsters. Furthermore, fraudsters are using cutting-edge technology to project new attacks in the system. Although automation in fraud detection offers much for high risk transaction, however, manual review (OTP/SMS/Email) residue foremost. In the time, m-commerce facilitates facund growth, payment fraud has come forth as one of the tremendous threats for the payment industry [5].

Consumer behavior analytics is perceived as a viable pow in fraud detection and prevention [6]. The consumer behavior profile is generated by capturing their behavior patterns and hidden/insensible facet of consumer behavior. The consumer behavior is used to justify the user each time a payment transaction is initiated. Further gives you a better understanding of normal user as to the abnormal user. The consumer behavior analytics is not only ethical about normal user behavior but endorse wherewith the fraudsters behave too. A fraud analytic system that avows layered approach and consumer behavior profiling across the mobile channel in real-time can competently remit many of the emanate fraud trends in the m-commerce area [7, 8]. It also helps in identifying flawlessly fraudulent patterns, allow fraud analytic engine to audit every transaction in real time, allowing it to detect intricate, cross-field referencing fraud.

Anomaly-based fraud detection lean on behavior profile and is payment modeled based on each behavior pattern. Fraud detection system monitors the behavior of the consumer for any change in the normal pattern. The banks are constantly down with stress to maintain a low false positive rate still nurturing immense consistency of fraud protection. It is imperative to understand the habitual behavior pattern of consumers to reduce false positive rate [7, 9].

The genuine consumer payment transaction can be declined when the purchase is made outside the normal geolocation. In such case, the transaction performed in another country would flag as a change in behavioral pattern. If this change is suspicious, then based on only one data attribute geolocation, the genuine consumer should not be considered as fraudster; therefore, cross-field referencing is important. In case of suspicious transaction, manual review (OTP, SMS and Email) plays an important role. However, this will amend the perceptiveness of the consumer and facilitate the banks to foresee new behavior of the consumer. The consumer understanding can also be improved through manual review in case of a suspicious transaction as there will be an interaction between the consumer and the bank to applaud the transaction in real time.

The remainder of this paper is organized as follows: Sect. 2 is the related work description. Then various fraud detection techniques comparison is described in Sect. 3 and finally, conclusion is concluded in Sect. 4.

2 Related Work

The most commonly used fraud detection methods are rule-based classifier, neural network, support vector machines, logistic regression, k-nearest neighbor, hidden Markov model, etc. [10–14]. These methods can be used individually or conjointly to classify the payment transactions in mobile payments. There are very few papers concerning fraud detection in mobile payments and while most papers are about credit card fraud detection. Prior fraud detection methods review data mining techniques, statistical models, machine learning, and artificial intelligence.

In supervised learning, there are known class labels in the given data, while in unsupervised learning, the algorithms try to find patterns in the given data. To identify fraudulent transactions when the training dataset is available, then supervised learning techniques are used whereas unsupervised learning techniques are used when there is no training dataset available. The dataset used for fraud detection is categorical data. Some algorithms make use of numerical data for fraud detection and so by using basic pre-processing techniques, the categorical data can be transmuted into numerical data [11]. Due to unavailability of mobile payments dataset for fraudulent/legitimate transactions, the academic research is confined.

In statistical model, fraud detection is not real-time based; therefore, fraudulent transactions are detected after the actual fraud happens. Moreover, manual fraud detection techniques are also not feasible due to the large volume of data [12]. Anomaly detection techniques can be used to find fraudulent transactions, for example, payment transactions carried out at odd hours or a sudden increase in the payment transaction amount.

The fraudsters do not follow single technique to commit fraud; they keep on improving their techniques; therefore, the behavior of fraudster is dynamic which leads to imbalanced datasets. Moreover, consumer behavior also changes over time making it difficult to differentiate between normal and abnormal transactions [2, 15]. In real-time-based fraud detection system, human involvement is important to detect whether a flagged payment transaction is fraudulent or legitimate. Therefore, multifactor authentication can be deployed to prevent fraud. Multifactor authentication plays an important role in fraud detection and prevention [2, 7, 15, 16]. In fraud prevention, it is imperative to integrate the prevention system with two-factor authentication.

In the score-based evaluation method [2], each payment transaction is scored based on the input parameters used for analyzing fraud detection. The input parameters are the consumer details, device fingerprint, consumer behavior, geolocation, and payment transaction details. If the transaction score is zero, then it is a legitimate transaction whereas if the transaction score is negative, then it is a fraudulent transaction. As machine learning algorithms are dependent on the input data, if irrelevant data is applied to the algorithm, then it may learn false things thus making it difficult to identify fraudulent transactions. Data mining techniques undergo overfitting and class imbalanced problem [17] whereas the score-based method for fraud detection in mobile payments has high accuracy with reduced false alarms [2].

3 Comparison of Various Fraud Detection Techniques

A comparative analysis is performed on NFC-enabled mobile payment transactions using Weka tool. To compare various fraud detection techniques, evaluation parameters such as precision, recall, F_1 score, true positive rate (TPR), and false positive rate (FPR) are used.

3.1 Dataset Description

The dataset is a real dataset from NFC-enabled mobile payment system [2] used for fraud detection. The input parameters used to detect fraud detection are the consumer details, device fingerprint, consumer behavior, geolocation, and payment transaction details. The dataset consists of total of 580 payment transactions for evaluation. Based on the transaction score, the target class is labeled, i.e., zero score yields a legitimate transaction while a negative score yields a fraudulent transaction.

3.2 Evaluation Approach and Metrics

Weka tool is used to evaluate the performance of the various fraud detection techniques in NFC-enabled mobile payment system. The fraud detection techniques used for comparative analysis with the score-based method are Naïve Bayes, decision tree (J48), random forest, bagging, and logistic regression. To determine the goodness of the various fraud detection techniques metrics like precision, recall, F_1 score, true positive rate (TPR) and false positive rate (FPR) have been considered. The metrics result is for 70:30 data distribution, i.e., 70% of the dataset is used for training and 30% of the dataset is used for testing purpose.

Table 1 Performance evaluation of different fraud detection techniques

Metrics	1: Naive Bayes	2: Decision tree (J48)	3: Random forest	4: Bagging	5: Logistic regression	6: Score based
Precision	0.976	0.976	0.967	0.846	0.975	0.992
Recall	0.966	0.966	0.966	0.92	0.971	0.969
FPR	0.003	0.003	0.394	0.92	0.068	0.069
F_1 score	0.968	0.968	0.961	0.881	0.973	0.9805
TPR	0.966	0.966	0.966	0.92	0.971	0.969

3.3 Comparison Results

In this subsection, the comparative results of various fraud detection techniques with score-based method are presented. The parameters used for comparison of various fraud detection techniques are precision, recall, F_1 score, true positive rate (TPR), and false positive rate (FPR) are as shown in Table 1.

The comparison results show that score-based method is leading the other fraud detection techniques with a high precision value of 0.992 and bagging has the lowest precision value of 0.846. Naive Bayes and decision tree (j48) have a low false positive rate of 0.03, logistic regression and score-based method have a moderate false positive rate of 0.068 and bagging has a high false positive rate of 0.92. Logistic regression has a high recall value of 0.971 followed by score-based method with a value of 0,969 and then Naive Bayes, decision tree, and random forest with a similar value of 0.966, and bagging method has the lowest recall value of 0.920.

The F_1 score is the weighted median of recall and precision is high for the score-based method with a value of 0.980 as compared to other fraud detection techniques as mentioned in Table 1.

The recall and precision performance of the various fraud detection techniques are presented in Fig. 1. The true positive rate (TPR) and false positive rate (FPR) performance of the various fraud detection techniques are presented in Fig. 2. Bagging method has a similar value of 0.92 for TPR and FPR.

4   Conclusion

The evaluation of various fraud detection techniques in NFC-enabled mobile payments is based on their precision, recall, F_1 score, true positive rate (TPR), and false positive rate (FPR). A substantial point to feature is the excellent precision scored with score-based method with a value of 99.21% showing that this method is good for fraud detection in NFC-enabled mobile payments. The score-based method for fraud detection in mobile payments has high accuracy with reduced false alarms.

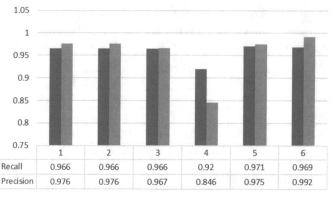

Fig. 1 Recall and precision performance of all the fraud detection techniques

Fig. 2 TPR and FPR performance of all the fraud detection techniques

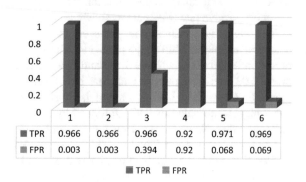

References

1. Coskun V, Ozdenizci B, Kerem Ok (2013) A survey on Near Field Communication (NFC) technology. Wirel Pers Commun 71(3):2259–2294
2. Vishwakarma PP, Tripathy AK, Vemuru S (2019) An empiric path towards fraud detection and protection for NFC-enabled mobile payment system. In: TELKOMNIKA, vol 17, No 15, Oct 2019, pp 2313–2320
3. Abdallah A, Maarof MA, Zainal A (2016) Fraud detection system: a survey. J Netw Comput Appl 68:90–113
4. Edge ME, Sampaio PRF (2009) A survey of signature based methods for financial fraud detection, Comput Secur 28(6):381–394. ISSN 0167-4048, https://doi.org/10.1016/j.cose.2009.02.001
5. Ter Chian F, Tan ZG, Cahalane M, Cheng D (2016) Developing business analytic capabilities for combating e-commerce identity fraud: A study of Trustev's digital verification solution. Inf Manage 53(7):878–891
6. Biometric technology underpins innovation at borders (2015). Biometric Technology Today, vol 2015, Issue 11, p 1
7. Vishwakarma PP, Tripathy AK, Vemuru S (2018) A layered approach to fraud analytics for NFC-enabled mobile payment system. In: Negi A, Bhatnagar R, Parida L (eds) Distributed

Computing and internet technology. ICDCIT 2018. Lecture notes in computer science, vol 10722. Springer, Cham, pp 127–131

8. Ash D (2011) The UK fraud landscape for financial services. Comput Fraud Secur 2011(4):16–18

9. Zoldi S (2015) Using anti-fraud technology to improve the customer experience. Comput Fraud Secur 2015(7):18–20

10. Kumar P, Iqbal F (2019) Credit card fraud identification using machine learning approaches. In: 1st International conference on innovations in information and communication technology (ICIICT), 25th and 26th Apr 2019. IEEE, pp 1–4

11. Thennakoon A, Bhagyani C, Premadasa S, Mihiranga S, Kuruwitaarachchi N (2019) Real-time credit card fraud detection using machine learning. In: 9th International conference on cloud computing, data science & engineering (confluence), 10–11 Jan 2019. IEEE, Noida, India

12. Amarasinghe T, Aponso AC, Krishnarajah N (2018) Critical analysis of machine learning based approaches for fraud detection in financial transactions. In: ICMLT'18, 19–21 May 2018. ACM, Jinan, China

13. Huang TU, Huang C (2019) Fraud payment research: payment through credit car. In: 10th International conference on e-business, management and economics, ICEME 2019, 15–17 July 2019. ACM, Bejing, China, pp 189–194

14. Tae C, Hung P (2019) Comparing ML algorithms on financial fraud detection. In: DIST 19, 19–21 July 19–21. ACM, Seoul, Republic of Korea, pp 25–29

15. Carminati M, Polino M, Continella A, Lanzi A, Maggi F, Zanero S (2018) Security evaluation of a banking fraud analysis system. ACM Trans Privacy Secur 21(3), Article No. 11:1–31

16. Vishwakarma PP, Tripathy AK, Vemuru S (2018) The fact-finding security examination in NFC-enabled mobile payment system. Int J Electr Comput Eng (IJECE). 8(3):1774–1780

17. Wei W, Li J, Cao L, Ou Y, Chen J (2013) Effective detection of sophisticated online banking fraud on extremely imbalanced data. World Wide Web 16:449 (Springer)

Feature Extraction of Unstructured Product Reviews

Ankur Ratmele and Ramesh Thakur

Abstract Today is the world of Internet and online shopping plays a very important role. There are many e-commerce websites from where user can buy a product. But it is very difficult for the user to select which feature of the product is good and which feature of the product is bad because there are thousands of reviews on an e-commerce website. With a huge number of reviews, it is very cumbersome for the user to read all the reviews and take a decision whether which feature is important for this product. It is a very stressful job to determine the important feature of the product from thousands of reviews. A review is a sequence of texts containing the opinion of customer based on their experience. There is no fixed format for the customer to write the reviews. The solution to this problem is to develop a technique that extracts and identifies important features from the unstructured reviews. In this paper, the proposed technique will extract the feature from unstructured reviews and matrix will be constructed to identify the importance of that features is known as Important Feature Review (IFR).

Keywords Important Feature Review (IFR) · Product features · Feature Extraction Algorithm (FEA) · Part of speech

1 Introduction

With Web 2.0 which is gaining popularity nowadays, opinion mining is one of the emerging research areas for the researchers because of the huge size of the reviews available in various online platforms such as social network, e-commerce websites

A. Ratmele (✉)
STME-NMIMS, Indore, M.P., India
e-mail: ankur22ratmele@gmail.com

IET-DAVV, Indore, M.P., India

R. Thakur
IIPS-DAVV, Indore, M.P., India
e-mail: r_thakur@rediffmail.com

and many more forums. When there were no such online portals, e-commerce exists then during that time if one wants to purchase the product, then there is need to do the survey or to ask many people regarding the product or feature of that product, i.e., how are the product and its features. But now are days there is an e-commerce website, different types of an online portal from where the reviews about the product and its features can be read.

There are huge amounts of reviews available for product with the growth of online shopping, i.e., via e-commerce website. Potential customers therefore find it very difficult to read each and every review to find the features and to identify significant features that are important among many identified features required for the customer. Feature selection seems to be the method of eliminating data set features that are unnecessary in regards to the task to be performed.

A technique needs to be developed which extracts features from unstructured reviews and identifies their importance. In this proposed system, Stanford parser, i.e., Stanford dependency parser is used to identify the dependencies in reviews because the reviews are unstructured and parts of speech tagging (POS) is used to identify verb, noun, adverb, adjective, etc.

In the next section, we would like to share related works on this field, proposed technique, experiment and evaluation followed by the conclusion.

2   Related Work

There are various approaches used to identify product feature. In [1], the implicit features were extracted in the sense of the implied features according to the opinion words and the similarity between the product features. They also create a matrix to demonstrate the relationship between words of opinion and features of the product and then use a new algorithm to filter the noises in the matrix. The implied features are defined by the meaning and the terms of thought.

In [2], a new pruning approach based on a grouping of features was proposed to prevent unnecessary pruning of the results. The method for calculating the similitudes based on semantics is abandoned in the grouping process. Rather, they find the feature terms as strings and then measure the similarities between them. By counting the feature words in reviews, authors consider that the feature words that describe the same objects and the feature words that describe the different objects are identical. The approach used in this paper does not guarantee that the fragments are segmented into one category with semantic relationships.

In [3, 4], the authors proposed a framework for the identification of product features using the conditional random fields (CRFs) model. In [5], the authors concentrate mainly on categorizing the product features the customers commented on. An unsupervised product-based twice-clustering features method of categorization is introduced. They proposed a method of categorization with the semantic association of two product-based features.

In [6], high-frequency aspects are listed and the corresponding opinion terms are extracted for the aspects to classify the polarity of each aspect. Naivè Bayes model used to build an aspect ranking algorithm that ultimately ranks the products where the weight of the features is determined from reviews.

In [7], the proposed method uses the dictionary of part speech and the corresponding opinions to identify the implicit features. The system considers not only the phrase noun/noun but also the phrase verb/verb as the features.

Hu and Liu [8] make the most significant contribution in the field of opinion mining, using the NLProcessor linguistic parser to analyze each review to split a text into phrases and generate the part of speech tag for each word.

Authors in [9, 10] focus on association rule mining which makes great use of the Apriori algorithm to recognize highly frequent aspects of the product. In [11], they use a semantic-based approach to extract feature and developed an algorithm. Recursive deep model is used to determine the sentiment of the reviews sentences. In [12], authors very well mentioned various techniques and datasets that are available in the research area of opinion mining and sentiment analysis.

3   Proposed Technique

The proposed technique is different from previous techniques as in the traditional techniques, the feature is either selected manually or noun is selected as features but in our technique, features are identified from the reviews which are noun as well as other than the noun. The overall architecture of the proposed technique is shown in Fig. 1. In this proposed technique, the reviews are extracted from any e-commerce website. In this work, the reviews are taken from the Amazon website. In this technique, reviews of mobile phone are taken from the Amazon website. The reviews are not in the fixed format they are unstructured reviews. Following are the main steps of the technique:

3.1   Preprocessing of Reviews

Preprocessing is one of the important steps whenever the reviews are unstructured. In our proposed technique, unstructured reviews are used, so preprocessing of reviews has to be done. In this step stop words which are not required in further processing are removed because stop words are of no use in our technique. Stemming and spelling correction in reviews is also performed in preprocessing. Spell correction is required because there can be spelling mistakes in reviews.

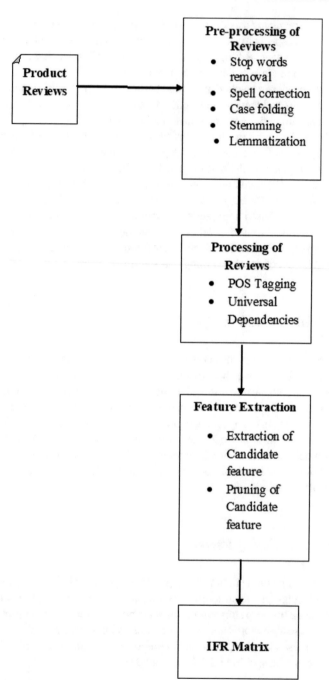

Fig.1 Proposed architecture

3.2 Feature Extraction

This is a very important step in the proposed technique. In this feature extraction, part of speech plays a very important role. Stanford POS tagger is used to find universal dependencies and tagged reviews which are input to the proposed algorithm. Suppose sample review is *battery and camera of this phone is good*. The tagged review is battery/NN and/CC camera/NN of/IN this/DT phone/NN is/VBZ good/JJ and universal dependencies are nsubj(good-8, Battery-1), cc(camera-3, and-2), conj(Battery-1, camera-3), case(phone-6, of-4), det(phone-6, this-5), nmod(Battery-1, phone-6), cop(good-8, is-7), root(ROOT-0, good-8). This input is given to an algorithm (FEA), and with the help of this algorithm, different features are extracted from the reviews. After the execution of FEA, the feat consists of all the features that are identified.

Feature Extraction Algorithm (FEA)

Input
Set of tagged reviews (TR_i),
Set of universal dependencies (UD_i)
$TR_i=TR_1, TR_2, TR_3,\ldots\ldots TR_n$
$UD_i=UD_1, UD_2, UD_3,\ldots\ldots UD_n$
Output:
Set of Feature Set (Feat)
Feat=($feat_1$, $feat_2$,$\ldots\ldots\ldots$, $feat_n$)

1. Begin
2.   for each TR_i in TR
3.   for each UD_i in TR_i
4.   if(UD_i = 'nsubj').
5.   for remaining UD_i in TR_i
6.   if (UD_i = 'conj') and arg.UD_i = 'noun.'
7.   Extract feat from conj, nn
8.   Feat.append(feat)
9.   else if (UD_i= 'Compound') and arg. UD_i = noun && $UD_i \neq$ 'dobj.'
10.   Extract feat From compound,nn,nnp
11.   Feat.append (feat).
12.   else if(UD_i='compund' && $UD_{i=}$ dobj && arg.$UD_{i=}$ noun).
13.   Extract feat from nn, compound, dobj
14.   Feat.append (feat)
15.   else if(UD_i = nmod)
16.   Extract feat from nmod, nn.
17.   Feat.append(feat)
18.   else
19.   Extract feat from nn, amod
20.   Feat.append (feat)

21.   Return Feat ////Integrate Feat = {feat₁,feat₂....featₙ}

3.3   Important Feature Review (IFR Matrix)

With the help of IFR Matrix, important features are identified from the set of features that are extracted from the above algorithm. The algorithm for constructing the IFR Matrix is given below:

IFR Matrix()
Input:
Feat (Features)
RS (Reviews)
Output: Matrix[Feat]][Review]

IFR(Feat, RS)

1.   Begin
2.   For each RS in RS_k
3.   For each Feat in $Feat_l$
4.   If $Feat_l$ present in RS_k then Matrix [Feat]][Review] = 1
5.   Else
6.   [Feat]][Review] = 0
7.   Return Matrix[Feat]][Review]

With the help of the above algorithm, the matrix is constructed as shown in Table 1 in which row consists of feature and column consists of reviews. For example, IFR Matrix is shown below:

In this, important feature is evaluated based on the calculation of total numbers of 1's in a row that is a feature present in the reviews.

$$\alpha = \frac{\text{Total numer of 1's in row}}{\text{Total number of RS}} * 100 \tag{1}$$

Table 1 IFR Matrix

	RS_1	RS_2	...	RS_n	Total
$Feat_1$	1	0	...	1	$\sum Feat_1$
$Feat_2$	1	1	...	1	$\sum Feat_2$
...	...	...	...	...	...
$Feat_m$	1	0	...	0	$\sum Feat_m$
$Feat_l$	1	0	...	1	$\sum Feat_1$

If the value of $\alpha \geq 15$, then product feature is important, otherwise product feature is not important. This values of α depend upon the features that have to be included in the experiment.

4 Experiment and Evaluation

The proposed system is implemented in Python. In this experiment, data set that contains reviews of mobile phone is used. With the use of Stanford POS, dependencies are identified which is used to determine product features by the FEA. The evaluation has been done on the basis of effectiveness of the feature extracted and importance of extracted features. Features that are extracted by the FEA are shown in Fig. 2. The important feature that is identified by the IFR Matrix is shown in Fig. 3.

The performance of the proposed algorithm is evaluated based on accuracy. The accuracy is the total number of correctly identified product feature (CIPF) divided by the total number of product feature (PF) shown in Eq. 2. Accuracy calculated by Eq. 2 for different important product features is shown in Table 2.

$$\text{Accuracy} = \frac{\text{CIPF}}{\text{PF}} * 100 \tag{2}$$

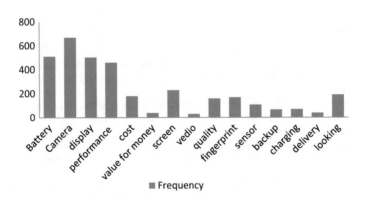

Fig. 2 Feature extracted by FEA

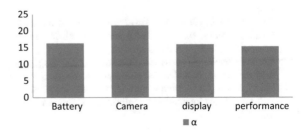

Fig. 3 Important feature identified IFR

Table 2 Accuracy of important product feature

Product feature	Accuracy (%)
Battery	82.6
Camera	89.5
Display	79
Performance	74

5 Conclusion

In this research, a technique is proposed which is used to extract product features and also identify important features of the product. This technique can be used to perform feature-based opinion mining or sentiment analysis. Analyzing the reviews is a very useful task for the customer as well as for the manufacturer. Customer can be used to analyze reviews to purchase a product while manufacture analyzes reviews to know what they want to improve in their product. This technique solves the problem of analyzing online reviews which are facing by the users while purchasing an online product. This technique can also be used by the manufacturer to identify the important features of the product which can be used to enhance the productivity of the product.

References

1. Sun L, Li S, Li J, Lv J (2014) A novel context-based implicit feature extracting method. In: 2014 International conference on data science and advanced analytics (DSAA), Shanghai, pp 420–424
2. Pan Y, Wang Y (2011) Mining product features and opinions based on pattern matching. In: Proceedings of 2011 international conference on computer science and network technology, Harbin, pp 1901–1905
3. Xu B, Zhao T, Zheng D, Wang S (2010) Product features mining based on conditional random fields model. In: 2010 International conference on machine learning and cybernetics, Qingdao, pp 3353–3357
4. Huang S, Liu X, Peng X, Niu Z (2012) Fine-grained product features extraction and categorization in reviews opinion mining. In: 2012 IEEE 12th international conference on data mining workshops, Brussels, pp 680–686
5. Jia W, Zhang S, Xia Y, Zhang J, Yu H (2010) A novel product features categorize method based on twice-clustering. In: 2010 International conference on web information systems and mining, Sanya, pp 281–284
6. Alengadan BB, Khan SS (2018) Modified aspect/feature based opinion mining for a product ranking system. In: 2018 IEEE international conference on current trends in advanced computing (ICCTAC), Bangalore, pp 1–5
7. Liu L, Lv Z, Wang H (2012) Opinion mining based on feature-level. In: 2012 5th International congress on image and signal processing, Chongqing, pp 1596–1600
8. Hu M, Liu B (2004) Mining and summarizing customer reviews. In: Proceedings of the tenth ACM SIGKDD international conference on knowledge discovery and data mining. ACM, pp 168–177

9. Hu M, Liu B (2004) Mining opinion features in customer reviews. In: AAAI, vol 4, no 4, pp 755–760
10. Somprasertsri G, Lalitrojwong P (2008) Automatic product feature extraction from online product reviews using maximum entropy with lexical and syntactic features. In: IEEE international conference on information reuse and integration, IRI 2008. IEEE, pp 250–255
11. Devasia N, Sheik R (2016) Feature extracted sentiment analysis of customer product reviews. In: 2016 International conference on emerging technological trends (ICETT), Kollam, pp 1–6. https://doi.org/10.1109/ICETT.2016.7873646
12. Ratmele A, Thakur R (2019) Statistical analysis & survey of research work in opinion mining. SSRN Electron J 6. https://doi.org/10.2139/ssrn.3366306

When Security Meets Velocity: Modeling Continuous Security for Cloud Applications using DevSecOps

Rakesh Kumar and **Rinkaj Goyal**

Abstract In the quest of velocity in time-to-market for the cloud applications, often, the security requirements are overlooked. It is mainly due to the preconceived notion that building security capabilities is very time-intensive and affects delivery timeline. However, the lack of required security capabilities in the applications have cascading impact on business objectives. The proposed research work addresses this issue by enabling automation in security activities. This work extends our previous work to analyze security challenges in cloud-enabled, cloud-native, cloud edge IoT, and cloud big data analytics applications and conceptualize a continuous security model for cloud applications using DevSecOps. The model proposes four inter-woven components: (1) *principles* to follow, (2) *application lifecycle stages* to implement, (3) *practices* to adopt, and (4) *automation ecosystem* to use. These components provide security elements to consider during cloud application development, deployment, and operation phases to bring velocity in building security capabilities and providing the continuous security assurance.

Keywords Cloud security requirements · Threats · Vulnerabilities and solution · Cloud security model

1 Introduction

The DevOps methodology has enabled agility and velocity in the cloud application delivery to achieve time-to-market business objectives to stay relevant in the competition [10, 13]. However, in the race against the delivery time, most of the applications are delivered with an insufficient level of required security capabilities. Security, mostly, is seen as the last thing to do in the application lifecycle. The lack of necessary security capabilities in the application leads to post-deployment secu-

R. Kumar · R. Goyal (✉)
University School of Information, Communication and Technology (USIC&T), Guru Gobind Singh (GGS) Indraprastha University, New Delhi 110078, India
e-mail: rinkajgoyal@gmail.com

© The Author(s), under exclusive license to Springer Nature Singapore Pte Ltd. 2021   415
J. S. Raj et al. (eds.), *Innovative Data Communication Technologies and Application*,
Lecture Notes on Data Engineering and Communications Technologies 59,
https://doi.org/10.1007/978-981-15-9651-3_36

rity challenges in the production environment, which is much costlier to manage and rectify. This can be addressed by building and validating the required security capabilities in early and every stage of the application lifecycle [19]. However, often it does not happen in practice, mainly, because building security capabilities is a time-intensive activity that can potentially delay the delivery timeline. This can be achieved by infusing automated security tasks along the DevOps continuous integration and continuous delivery (CI/CD) pipeline [9, 19]. The inclusion of security tasks along the DevOps practices is institutionalized as DevSecOps [4, 25]. DevSecOps extends the DevOps methodology to embed automated security tasks within the DevOps activities [2, 12, 19, 39]. The appropriate automation tools are integrated within the CI/CD pipeline to automate the security tasks and deliver the security enriched cloud applications [19, 20]. The technology-specific automation tools are selected as per the business-context and the type of cloud application. This research work primarily discusses the cloud-enabled and cloud-native application types. The cloud-enabled applications are the legacy applications that are originally designed for the static, monolithic servers that are ported to a dynamic cloud environment with some architectural and design changes to capitalize on the cost-effective cloud platform and infrastructure services. The cloud-enabled applications are technologically transformed into service-oriented architecture (SOA) to make them deployable to the cloud. However, the service-oriented approach does not capitalize on the full potential of the cloud. This led to the evolution of the cloud-native applications using the containerization of the applications. A cloud-native application uses a microservices architecture that enables features to be developed and packaged independently as containers, and these containers are dynamically orchestrated [7]. The containers capitalize on cloud attributes to provide automatic scaling, provisioning, upgrade, and redundancy. The cloud-native approach has further revolutionized other cloud applications, like cloud-edge IoT and cloud big data analytics applications [6, 21]. However, all these cloud applications have their own specific security challenges, including the common cloud-centric security challenges.

This work extends our previous published work [19] in which a continuous security model is proposed to capitalize DevSecOps methodology. It describes principles, workflow, practices, and tools for the design and implementation of security controls along the CI/CD pipeline for continuous security assurance of cloud application and service delivery. In this work, the security challenges that are specific to the cloud-enabled, cloud-native, cloud edge IoT and cloud big data analytics applications are analyzed to devise a continuous security model to build and validate security capabilities along the development, deployment, and operation phases of these applications using the DevSecOps methodology. In rest of the paper, Sect. 2 provides an overview of the related works. Security challenges to cloud-enabled, cloud-native, cloud edge IoT, cloud big data analytics applications are analyzed in Sect. 3. Section 4 provides a continuous security model to address the challenges, followed by illustration of the proposed model in Sect. 5. Section 6 provides conclusion and future research direction in related areas.

2 Related Work

The continuous security is an active research area for cloud application delivery. Jabbari et al. [13] outlined DevOps practices mapped to the software knowledge area. Myrbakken and Colomo-Palacios [25] defined DevSecOps with the challenges and benefits of its adoption. Kuusela [20] discussed different security testing methods in the continuous integration process and the corresponding tools for vulnerability scanning, static code analysis, configuration checking, security verification, and dependency verification. Koskinen [15] identified balancing security with the delivery pipeline's speed, securing the delivery pipeline, and insider threat as three major areas of challenges in building security in cloud applications. Ullah et al. [37] emphasized securing the major components of the CI/CD pipeline, like repository, CI server, deployment server, etc., using the appropriate level of access control and secured communication. Blake [3] recommended automated vulnerability scanning and remediation, along with policy enforcement and infrastructure monitoring, to contain the security risk associated with cloud applications. Parnin et al. [30] identified ten adages for implementing a CI/CD pipeline for practical consideration. In their case study, Paule et al. [31] found the DevOps team working on CI/CD pipeline is aware of the security attributes; however, it is not their priority to secure the CI/CD pipeline. It poses a risk to the business-critical development infrastructure. Rahman et al. [38] established that automation of the commonly used DevOps activities, like monitoring, testing, and deployment of applications, is useful for system's security. Shahin et al. [33] categorized different challenges, approaches, tools, and practices for adopting continuous practices. Mohan et al. [24] found access control, roles separation, manual execution of security tests, audit, security guideline enforcement, security issue management, and security team participation as hurdles for deployment process automation. They also recommended strong collaboration, communication, logging, process automation, separation of roles, and others as the best practices to follow for transformation to DevSecOps. In our other work [19] a continuous security model is proposed for automated DevSecOps using open-source software over cloud.

In brief, significant works are done in the area of continuous delivery and deployment, emphasizing the automation of CI/CD pipeline along with challenges and recommended methods, tools, and technologies. The proposed research work extends our previously published work [19] to analyze the security challenges in the cloud applications, specifically the cloud-enabled, cloud-native, cloud edge IoT, and cloud big data analytics applications, to devise a continuous security model using the DevSecOps methodology. The model provides principles to follow, application lifecycle stages to implement, practices to adopt, and automated ecosystem to use for automation of security tasks in the development, deployment, and operation phases of the cloud applications.

3   Analyzing Security Challenges to Cloud Applications

The security challenges to cloud applications are multi-fold. First, applications are not designed and developed from the required level of the security capability perspective. Second, cloud development and testing environments have their own inherent security challenges. Often, they are not equipped with the right set of tools to provide the desired level of security development and testing support at speed. Third, the production environments are not sufficiently equipped to provide run-time security protection from the ever-evolving threat spectrum.

3.1   Security Challenges in Application Design and Development

The choice of architecture, design, and technologies to use for cloud applications pose different security challenges. Lack of security awareness and seriousness to secure coding guidelines in developers can further aggravate these challenges. Nevertheless, the core security challenges remain centered around the assurance of data confidentiality, integrity, availability, accountability, and privacy by applying and managing user identity, authentication, and authorization policies and encryption techniques.

1. **Security Challenges with Cloud-Enabled Application**: Initially, the cloud was seen as a means of optimizing the computing resource utilization to drive the cost-efficiency. The applications originally developed and deployed over traditional data center's static server environment, were minimally re-engineered, termed as cloud-enabled, to run in the dynamic cloud environment. The application features were broken into services to adopt service-oriented architecture (SOA). These services deployed over the cloud environment were communicating through message passing and accessible to the users using browsers over the Internet [16]. However, the browser and SOA based services led to multiple security challenges [1, 11, 18, 36], like vulnerabilities in web services, web clients, unsecured APIs, encryption schemes, key management, and identity and access management, etc. The Open Web Application Security Project (OWASP) highlighted the top ten web application security issues, which shall be addressed during the application design and development [29]. These challenges are aggravated in the absence of sufficient security testing. The proposed continuous security model aims to address these challenges by introducing automated security checks and testing at the early stages of the application lifecycle. Further, the manual update of the inter-dependent web services component affects service availability. The automated framework to update the application components minimizes the application downtime during the updates.

2. **Security Challenges with Cloud-Native Application**: The cloud-enabled applications could not capitalize on the full potential of the cloud due to its own

inherent limitations, like, web services inter-dependencies for a complete feature, release to the production, scaling, etc. The velocity of digital transformation demands secured and efficient data processing, leading to cloud-native applications [16]. A cloud-native application leverages microservices architecture, developed and packaged as containers, and the containers are dynamically orchestrated [7]. The cloud-native applications harness the power of the cloud to enable faster time-to-market and increase operational efficiency by providing auto-scaling, auto-provisioning, auto-upgrade, and auto-redundancy capabilities.

However, data security and privacy challenges are more in cloud-native than monolithic or cloud-enabled applications, mainly due to the distributed nature of the microservices. The microservices-based container increases the attack surface as the number of entry and exit points increases with the large numbers of container deployment. The cloud-native applications require API authentication as microservices-based architecture relies heavily on APIs to interact with different entities, like, IoT devices, machine-to-machine communications, event-driven web applications, etc. The diversified technologies, including open-source components, used in microservices development complicate vulnerability detection. The challenges arise in logging and monitoring cloud-native applications due to the frequent orchestration of containers leading to changes in IP-addresses, port, and service endpoints [5, 14].

The cloud-native applications are designed for rapid development and large scale deployment in minutes with all the required security capabilities. It can be achieved only through an automated continuous security assurance framework aligned with the continuous development, integration, and deployment pipeline. A continuous security model is required to detect fast, contain locally, and recover instantly for any deviation from the desired behavior for a highly dynamic cloud-native application environment.

3. **Security Challenges with IoT and Edge Computing Applications**: Edge computing enables data processing at the location of data creation, addressing the latency and bandwidth issues, to make decisions faster by the smart IoT devices. It lessens the overhead of huge data transfer from the IoT devices to a centralized cloud environment for decision making with reduced cost for data storage. However, a part of filtered data is sent to the cloud environment for advance analytics for strategic business decisions and IoT device management enhancements [21, 43]. From security perspective, there is a positive aspect of IoT edge computing, as it does local data processing, so, security controls can be enforced at data originating source to contain the threats locally. However, the cloud edge IoT application poses new challenges apart from the inherent web interface vulnerabilities. Based on a centralized filtering policy, the fragmentation of data for sending to the cloud has to deal with the data security and privacy challenges [42].The northward scaling of edge devices require robust monitoring and maintenance to ensure their availability for continuity of critical decision making by the IoT devices. Access control, data encryption, and the use of virtual private networks are critical elements to protect the edge computing system [21]. Con-

sidering the scale of the IoT device that a cloud edge environment will have to manage, it requires an automated environment for configuration, scheduling, deployment, suspend, resume, and shutdown for a large number of edge applications and devices. An IoT edge cloud computing environment require an integrated continuous development, deployment, and operation environment with continuous security testing for the IoT edge based applications. The automated firmware and application patches are key for keeping these devices up-to-date.

4 **Security Challenges with Big Data and Analytics Applications**: Big Data refers to a large volume of data that inundates a business on a daily basis. It is nearly impossible to process the big data using traditional methods due to the large volume, speed of generation, and complexity of the generated data. Cloud is a more practical solution for big data analytic applications as it provides the computing and storage resources (platform-as-a-service, storage-as-a-service, etc.) with high availability, reliability, and wider accessibility in a cost-effective way [6]. Big Data can be structured, unstructured, or semi-structured. It is characterized by eleven key dimensions: volume, velocity, variety, veracity, validity, volatility, value, variability, visualization, valence, and vulnerability [35]. Each of these dimensions has a direct or indirect impact on the security and privacy for the big data as the data go through the three phases, acquisition, storage, and analytics, of its lifecycle. Further, the IoT revolution is expected to affect the big data acquisition, storage, and analytics methodology due to the continuous streaming of the data and associated security and privacy challenges. In the data acquisition phase, weak encryption scheme, identification, authentication of data source endpoints, data falsifying, network bandwidth configuration, communication link security, man-in-the-middle attack, etc., are some of the security issues to address. Data integrity, access control, backup and recovery, dynamic scalability, etc. need to be addressed during the storage phase [8]. The intelligent analytics applications apply different machine learning and artificial intelligence techniques, during the analytics phase, to provide insight for business decisions. The analytics application can be injected with malicious instruction by the attackers. The analytic phase is more susceptible to advance persistent threats (APTs). Using the data provenance methods to detect the anomaly in the data acquisition, data encryption, access control, secure data mining techniques, blockchain technology, etc., are some of the strategies to protect the big data [8, 35]. It requires speed in security assurance activities for the big data analytics applications, which can be achieved through an automated framework.

3.2 Security Challenges in Application Development and Testing Environment

It has been experienced that the development team depends on the customer infrastructure for application testing [31]. The availability of application development and testing environment with the necessary automation framework is an area of concern to deliver applications with velocity and security. Using cloud infrastructure for application development and testing environment naturally addresses both the issues. However, the cloud has its own security issues and challenges. Considering the limited scope of this paper, the different security issues and challenges of using the cloud are not explained, as this has been extensively discussed in different studies and security solutions to address the same [1, 11, 17, 18, 27, 34, 36]. The security of the development and testing infrastructure pipeline must be ensured by the unified DevSecOps team [31, 37]. The cloud infrastructure and the cloud-based automation tools can provide an automated, integrated development, testing, and deployment environment for end-to-end security testing at the desired speed [19].

3.3 Security Challenges in Application Production Environment

The increasing technological complexities in cloud application architecture and evolving attack spectrum aggravate challenges to protect the application production environment. Logging, monitoring, alerting, and incident response are critical security tasks to perform. False-positive keeps challenging security intelligence in detecting the right incident or to take proactive measures. With the cloud-native microservices-oriented technology, the end-to-end visibility of the application log, monitoring, and anomaly detection become more complex, especially with the frequent upgrades and deployment of the containers [5, 14]. For the edge IoT, it becomes even more challenging due to hundreds of thousand microcontroller devices in production. The logging and monitoring information gathered at different levels require a centralized analytic for runtime application behavior analysis to facilitate decision-making and provide dynamic protection to the application production environment. The Interactive Application Security Testing (IAST) and Runtime Application Security Protection (RASP) mechanism require advance analytics. An integrated cloud application development, deployment, and operation environment is required for continuous security assurance.

4 Addressing the Challenges: A Continuous Security Model using DevSecOps

The analysis of cloud-enabled, cloud-native, cloud edge IoT, and cloud big data analytics applications in the previous Sect. 3 reflects the need for a continuous security delivery model to fulfill the agility, velocity and security requirements of these applications as per the business context and objectives. Figure 1 depicts a continuous security model based on DevSecOps methodology having four inter-woven components: (1) *principles*, (2) *application lifecycle stages*, (3) *practices*, and (4) *automation ecosystem*.

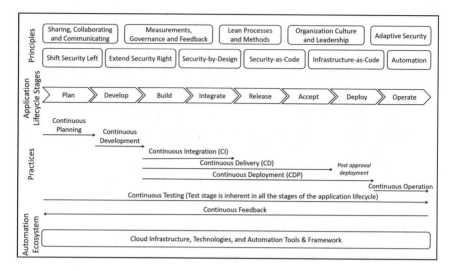

Fig. 1 Continuous security model for cloud applications using DevSecOps

4.1 Principles to Follow

Table 1 outlines the *principles* to follow for continuous security assurance of the cloud applications. The model extends the CALMS (Culture, Automation, Lean, Measurements, and Sharing) principles of DevOps [40]. The proposed principles are aligned with the DevSecOps methodology and applied across the application lifecycle stages.

4.2 Application Lifecycle Stages to Implement

The model (Fig. 1) proposes a ten stage application lifecycle (Table 2) that a cloud application spans through development, deployment, and operation phases [2, 12, 39]. These lifecycle stages provide an end-to-end logical workflow path of application delivery. These stages can be augmented and restructured as per the application business context. Every stages have different security assurance tasks, defined and aligned with the adopted continuous security practices (Table 4), that are executed as the application lifecycle progresses through these stages. The model imbibes the *Test* stage inherent to all the stages of the application lifecycle for continuous security assurance throughout the development, deployment, and operation phases.

Table 1 Principles to follow in the proposed continuous security model [2, 19, 20, 26, 28, 32, 40, 41]

DevSecOps principles	Description
Shift security left	Shift security left emphasizes incorporating security controls in the early stages of the application development. The objective is to mitigate the risk of discovering security issues in the later stages of application lifecycle where it costs more to fix and impacts on quality and time-to-deliver.
Extend security right	Extend security right advocates for incorporating security controls at all the stages of the application lifecycle. The objective is to practice continuous security testing to identify security issues in the application at the earlier stages and prevent their propagation to the later stages.
Security-by-Design (SbD)	SbD exploits a business context's threat analysis to provide design specifications of the security controls for every stages of the application lifecycle. SbD advocates using security design constructs like, authentication, access control, least privileges, securing secrets, securing the weakest link, fail securely, defense-in-depth, reusability, logging, monitoring, and auditing [26, 28].
Security-as-Code (SaC)	SaC implements security design constructs as code, complementing the "shift security left", "extend security right", and "security-by-design" principles. Like any other application features, the security controls, policy, checks, tests, platform configuration, logging, monitoring etc., are implemented in code to automate the security tasks of the adopted practices.
Infrastructure-as-Code (IaC)	IaC treats application infrastructure ecosystem, like, physical servers, devices, communication links, the operating system, physical and virtual resources, etc., as a programmable unit and applies software development methodology to develop and use version-controlled codes for their provisioning and configuration. With the software-defined approach, the resources can be scaled up and down as per need using standardized templates within defined timelines [2].
Compliance-as-Code (CaC)	CaC implements compliance, for security requirements of regulatory authority and organization security standards, as code to collect and provide the required data set for verification and validation by the designated authority [2].
Adaptive security	It advocates for continuous risk assessment for evolving threats and adapts proactive, detective, corrective, and predictive security measures to minimize the probability and impact of an attack.
Sharing, collaborating and communicating	The development, operation, and security experts share, collaborate and communicate information to progress along the application lifecycle. The team-specific processes and tools are replaced with a common set of lean processes and inter-working tools.
Measurements, governance and feedback	The unified DevSecOps team works for a common set of business goals and key performance indicators (KPIs). The measurements and analysis of the KPIs are automated and feedback to all the stakeholders to take necessary actions for any needed improvisation. An empowered governing body shall be in place to ensure the functioning of the application life cycle as per expectation.
Lean processes and methods	Being lean is to eliminate waste, bottlenecks, and bureaucracy in existing processes and methods by engaging the stakeholders to improvise them from the learning of previous release cycles. It will facilitate more effective automation and improvisation in delivery capabilities.
Automation	Automation complements all the above principles and is key for the effective and efficient adoption of DevSecOps to bring velocity and security together for cloud application delivery. An appropriate inter-working tool-suite and security linters shall be identified and integrated to automate the different security tasks of the adopted practices across the lifecycle stages [19, 20, 32, 41].
Culture and leadership	DevSecOps adoption is a techno-cultural transformation and shall be well-supported by the organization's leadership. It requires continuous awareness and training to the people on DevSecOps adoption philosophy and the value it brings to the business by enabling agility, velocity, and security in the application delivery in a cost-effective way [40].

Table 2 A cloud application lifecycle stages in the proposed model [2, 12, 19, 39]

Stages	Description
Plan	An end-to-end planning is made for all the activities encompassing all the stages of application lifecycle for development, deployment, and operations. The expert analyzes the application of functional and non-functional requirements. An adaptive security architecture is adopted as per the application threat modeling to implement incremental security controls with every delivery cycle.
Develop	The application features are developed as per security guidelines and are enforced using security linters integrated as plugins in the integrated development environment. The different security constructs, infrastructure provisioning, and configuration, and compliance requirements are developed as per IaC and CaC designs. The codes are committed to the source version control repository.
Build	This stage is triggered with every commit to source code repository to compile the changed code and resolve dependencies using source composition analysis (SCA) and static application security testing (SAST).
Integrate	The integrate stage prepares and validates the application as a single system comprises different software components. The system-level integration testing is performed to identify any interface or feature break. Infrastructure configuration and provisioning codes are also integrated and tested for automated delivery to staging and deployment to the production environment.
Accept	The application is tested for expected functional and non-functional features by the users and the quality assurance team at the staging environment. The automated installation, provisioning, and configuration testing is executed at the production replica staging environment. The load, performance, and regression testing is performed to validate the expected behavior. The different security tests like fuzz test, pentest, exploit test, DAST, etc., are performed, and security gaps are rectified.
Deploy	The accepted application binaries are deployed, either manually or automated, on the production environment followed by smoke testing. Any observed deviations are addressed immediately.
Operate	The production environment and the deployed applications are continuously monitored for malicious activities and expected performance levels using appropriate tools. The objective is to prevent, detect, respond, and predict security incidents and take proactive and reactive measures. The automated patch deployment and infrastructure orchestration tools provide speed and accuracy.
Adapt	The auto-scaling of infrastructure and auto replacement of compromised components are aimed in adapt stage using the right set of tools.
Test	It is inherent to all the stages to provide continuous security assurance by performing different security tests along with functional and non-functional tests. It employs source code review, SCA, SAST, DAST, IAST, RASP, pentest, vulnerability testing, exploits testing, etc., to identify security gaps and trigger their rectification.

4.3   Practices to Adopt

The DevOps and DevSecOps practices, suggested by the researchers and practitioners, aim to streamline the cloud application development, deployment, and operation to achieve time-to-market with velocity and security [2, 9, 12, 39]. The set of practices to adopt depends on the business context. The model proposes for eight interconnected practices aligned with application lifecycle stages (Table 3). For effective and efficient implementation of different tasks of the practices, specifically, the continuous security assurance tasks, the model proposes to use an automation ecosystem consisting of the cloud platform and appropriate tool-suite (Table 4).

Table 3 Continuous practices to adopt in the proposed model [2, 9, 12, 19, 39]

Practices to adopt	Description
Continuous planning	It identifies and plans for all the different tasks across the lifecycle stages of application development, deployment and operations phases. The functional and non-functional requirements, including infrastructure configuration, provisioning and security requirements, are analyzed by experts and finalized activities to fulfill these requirements are prioritized as per business objectives.
Continuous development	The DevSecOps team adopts an agile methodology to design and deliver working application units in iteration so that immediate feedback is available for any required improvement and enhancement in coming iterations. They adopt an adaptive security architecture for application design using appropriate threat modeling tools (Table 4) adhering to Security-by-Design principles. They identify security use, misuse, and abuse cases to develop appropriate security test cases. The team shares a version control system for the common codebase. The security plugins check security guidelines while the developer is coding. It also involves the development of IaC, SaC, and CaC codes for automation of infrastructure configuration, provisioning, and security compliance checks.
Continuous Integration (CI)	It enables developers to integrate their code into a shared repository multiple times in a day to get immediate feedback to rectify any unexpected integration deviation. An automated build triggers on every code check-ins. During the build, SAST and SCA resolve dependencies and identify any known vulnerabilities in the components. Automated continuous testing tasks perform the necessary unit, integration, and security tests to create a tested build package.
Continuous Delivery (CD)	It includes all the tasks of CI along with automated packaging and release of the build on a pre-production environment. Continuous testing extends to perform user acceptance, regression, load, performance, and DAST. The DAST involves the vulnerability scanning and pentesting of the released package. The tested build package is stored in an artifact repository. After necessary approval, the tested package is manually deployed over the production environment.
Continuous Deployment (CDP)	Continuous deployment is continuous delivery with a tested build package with auto-deployment using a suitable tool (Table 4). After that, continuous testing triggers to perform smoke tests, vulnerability scan and penetration tests on the production environment for smooth transitioning to operations.
Continuous operation	It involves automated continuous logging, scanning, monitoring, and analyzing system and application events. The graphical trending and analysis provide insights to identify anomalous events and notify to the DevSecOps team to initiate necessary remediation actions. In addition to analyzing external events for intrusion detection and prevention, the RASP mechanism provides an inside view of input data processing to identify application-level attacks. It has automated infrastructure orchestration and scaling capability to satisfy elastic demand and replacement of malicious components.
Continuous testing	The embedded testing tasks run concurrently with tasks of other continuous practices in the corresponding application lifecycle stage. This advocate fail-early to rectify-early to gain overall delivery performance. In addition to the traditional testing activities, it focuses on application security testing using different tools (Table 4). It employs SCA, SAST, DAST, and IAST security testing techniques to use appropriate tools as the application progresses through its lifecycle stages.
Continuous feedback	A data-driven automated feedback loop enables the sharing of relevant observations and the findings of continuous integration, testing, and operations. They are communicated and shared transparently to all stakeholders through a common platform. The continuous feedback from the integrated tool-suite used for reactive, corrective, preventive, and predictive actions as continuous improvement.

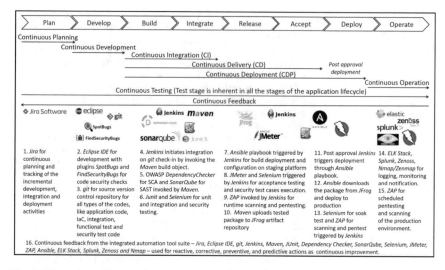

Fig. 2 Illustration of the proposed continuous security model using DevSecOps

4.4   Automation Ecosystem to Use

The automation ecosystem, the key component of the proposed model, intends to automate the different tasks of the adopted practices. The cloud platform becomes the natural choice for setting up the desired automation ecosystem for development, integration, testing, deployment, and operation activities. Table 4 provides a proposed set of widely used tool-suite for creating an automation ecosystem for delivering security enriched cloud applications [32, 41]. These tools are aligned with the continuous practices and the application lifecycle stages. However, it is not the only set of tools for setting up an automation ecosystem. The choice of the tools will depend on the business context for which the continuous security model is to be implemented. The functional analysis of the mentioned tools (Table 4) is not in the scope of this work, however, the inter-working of a subset of the tools are illustrated.

5   Illustration of the Proposed Continuous Security Model

Figure 2 illustrates usage of the proposed continuous security model (Fig. 1) as described in the Sect. 4. It depicts continuous security implementation for a cloud-enabled Java IT application using a set of inter-working tools selected from the Table 4. *Jira* is used for the continuous planning and tracking for development, integration, testing and deployment activities. The planning includes the infrastructure availability for the automation ecosystem, staging and production environment with their configuration and provisioning requirements. *Eclipse IDE* is selected for

Table 4 Security activities and automation tool-suite in the proposed model [19, 32, 41]

Continuous practices	Lifecycle stages	Security controls and related activities	Suggested tools
Continuous testing	All stages	Code review	Gerrit, Review board, Phabricator
		Software composition analysis	DependencyCheck, Synk, FOSSA, WhiteSource
		Static application security testing	Anchore, DockerBench, Clair, FindSecurityBugs, SonarQube, SpotBugs, DevSkim, PMD, Brakeman, ESLint, SCSS Lint, Actuary, ansible-lint, puppet-lint, git-hound, Checkmarx, Veracode, Fortify, Dagda
		Unit and integration testing	TestNG, JUnit, xUnit Framework, Rspec, Mocha
		Acceptance, load, performance, regression and smoke testing	JMeter, Selenium, Cucumber, ZAP Baseline Scan, FitNesse, Gatling, Locust, SoapUI, MicroFocus
		Dynamic application security testing (Vulnerability Scan, FuzzTest, PenTest, ExploitTest)	OpenVAS, OpenSCAP, Zed Attack Proxy, Wfuzz, Wapiti, Nmap, SQLMap, Vuls, Sslyze, Metasploit, Vega, Arachni, Acunetix, CloudSploit, Nessus
		Interactive application security testing	Hdiv, Contrast Security
		Infrastructure hardening, configuration and security testing	Cloud Custodian, InSpec, ChaosMonkey, Oscap, Ansible-Lint, Puppet-Lint, Foodcritic, Serverspecs, Dev-Sec.io, Test Kitchen, grsecurity, DockerBench
Continuous planning	All stages	Integrated project planning, execution and monitoring	Jira, Confluence, Phabricator, Redmine, Tuleap, OpenProject, Trello, Stride, Pivotal Tracker, Axosoft
Continuous development	Develop	Threat modeling, security requirement analysis, and adaptive architecture	ThreatDragon, CAIRIS, SeaSponge
		Development environment	NetBeans, Kdevelop, Eclipse, Eclipse Che
		Security use, misuse and abuse testcase	gauntlt, BDD-Security, Mittn
		Source code version control repository	Subversion, Git, Mercurial, GitLab, GitHub, Nexus
Continuous integration	Build, integrate	Integration automation	Concourse CI, Jenkins, GitLab, TeamCity, Travis CI, Bamboo, AWS CloudBuild
		Build automation	Maven, Gradle, Ant.

(continued)

Table 4 (continued)

Continuous practices	Lifecycle stages	Security controls and related activities	Suggested tools
		Artifact repository	Archiva, Nexus OSS
Continuous delivery	Build, integrate, release, accept	Staging infrastructure configuration and provisioning	Puppet, Chef, Saltstack, Ansible, Terraform, CFEngine, BCFG2, AWS CloudFormation, Azure Resource Manager, Cloud Deployment Manager
		Delivery automation	Spinnaker, GoCD, Docker, GitLab
Continuous deployment	Build, integrate, release, accept, deploy	Deployment automation	Drone, Foreman, Capistrano, Fabric, GitLab
		Production infrastructure configuration and provisioning	Puppet, Chef, Saltstack, Ansible, Terraform, CFEngine, BCFG2, AWS CloudFormation, Azure Resource Manager, Cloud deployment manager
		Secrets management	Blackbox, Harshicorp Vault, AWS KMS, Azure Key Vault, Ansible Vault, chef-vault, Docker Secrets
Continuous operation	Operate, adapt	Logging, Analysis, Visualization and notification	Elasticsearch, Logstash, Kibana, ElastAlert, Grafana, Graphite, Seyren, Graylog2, Rsyslog, Splunk
		Continuous monitoring and incident handling	CloudWatch, CloudTrail, reddalert, Sysdig Falco, Prometheus, Nagios Core, Zabbix, Munin, Cacti, Icinga, CollectD, OpenNMS, Gangila, Logz.io
		IDS, IPS, Security information and event management, and threat intelligence	Fail2Ban, Snort, Suricata, OSSIM, OSSEC, Structured threat information expression, Trusted Automated exchange of intelligence information
		Runtime application security protection	Hdiv, AppSensor, Contrast Security
		Infrastructure configuration, provisisoning, and orchestration	Kubernetes, OpenStack, Mesos, Swarm, Cloudify, Vagrant

(continued)

Table 4 (continued)

Continuous practices	Lifecycle stages	Security controls and related activities	Suggested tools
Continuous feedback	All stages	Collaboration and communication environment	Phabricator, Mattermost, Tuleap, Rocket.Chat, Jira, ServiceNow, CollabNet, Slack, Microsoft teams
		Quality and performance measurements, analytics, trending, and alerting	Elasticsearch, Logstash, Kibana, ElastAlert, Grafana, Graphite, seyren, Prometheus, Icinga, LimeSurvey

the Java based IT application development. The *SpotBugs* and *FindSecurityBugs* plugins integrated with *Eclipse IDE* for security coding guideline checks. The *git* used as shared source code repository for the multi-location development team. The integration automation is achieved through *Jenkins*. *Maven*, integrated with *Jenkins*, triggers the build every time a developer checks-in her code in *git*. At the build time, the *Dependency Checker* and *SonarQube* trigger the source code composition analysis (SCA) and static vulnerability scanning of the application source code, including identifying any open source dependencies. The objective of performing SCA and SAST is to address any security issues in the early stage of the development to save the time and money as compare to discovering it at the later stage. *SonarQube* also helps to identify and resolve license conflicts. *JUnit* and *Selenium* provides automated test framework for unit and integration testing and *ZAP* for security fuzz and penetration testing. *Ansible* provides an agent-less framework to create and execute a playbook that uses "ssh" for pushing the release configuration and deployment packages. *JMeter*, *Selenium*, and *ZAP* are used for load, performance, acceptance, security, vulnerability scanning and pentesting at the staging server. *JFrog* used as artifact repository for deployable release package. *ELK stack* (*Elasticsearch, logstash, Kibana*), provides a strong inter-working logging, monitoring, alerting, analytics capabilities to define rules on transaction logs for detecting the deviations and feedback to the team for relevant action. *Zenoss, NMap* and *ZAP* enable continuous scanning and monitoring of infrastructure, network, and application with alert notification and analytic report to feedback the DevSecOps team for the required actions. *Splunk* provides artificial intelligence based security information and event management (SIEM) functionality for threat intelligence. The integrated inter-working illustrated tools provide continuous feedback to the DevSecOps team for the tasks executed in the corresponding practices for reactive, corrective, preventive, and predictive actions as continuous improvement.

The illustrated implementation of the continuous security model (Fig. 2) can be extended for the cloud-native applications, for example, by integrating *Jenkins* and *Ansible* with *DockerHub* and *kubernetes* for building the docker container images and for their deployment and orchestration. In this, *Jenkins* pulls the source code from the *git* repository and builds the docker image. The integrated docker image is pushed to *DockerHub* for *Ansible* to deliver it to the staging server. Post testing, *Ansible* delivers the images to *kubernetes* from where it is deployed and orchestrated to the production environment. *docker-secret* can be used for managing the sensitive data.

Further, this illustration can be extended for the cloud edge IoT application and cloud big data analytic application for providing the continuous security. In case of the cloud edge IoT applications there could be tens to hundreds thousand devices to manage securely from deployment and operation perspective. Due to the very nature of the applications, selection of the tools for continuous security task automation shall be done appropriately. For example, the *git*, *Jenkins*, *Ansible*, *Docker*, etc., can provide the integrated continuous security pipeline for docker image controlled IoT devices like Raspberry, Orange Pi etc., wherein the updated docker images can be deployed securely over the IoT devices. Also, one can build the cloud edge IoT CI/CD pipeline using the Microsoft Azure IoT Edge [22], AWS IoT [23], etc.

6 Conclusion and Future Work

DevSecOps methodology is now increasingly being used to drive velocity and security together for cloud applications. The nature of security challenges varies for the cloud-enabled, cloud-native, cloud edge IoT, and cloud big data analytics applications. The automation ecosystem also varies to support the required security testing. The continuous security model proposed in this work provides the principles to follow in implementing the cloud application lifecycle by adopting continuous practices and using an appropriate automation ecosystem to automate the different security tasks. The model can be used for continuous security assurance for cloud-enabled, cloud-native, cloud edge IoT and cloud big data analytics applications. The model can also be useful for other applications based on the cloud-native technologies by integrating the appropriate automation tools in automation ecosystem. The comparative performance analysis of the different automation tools for a given cloud application context can be considered for future research work. The Interactive Application Security Testing and Runtime Application Security Protection mechanism is evolving with artificial intelligence and advanced data analytics techniques. This can be another area of future work to explore for building the additional runtime protection layer. The ultimate goal is to establish the trust and confidence of the users in cloud applications by providing continuous security and privacy assurance of their data.

References

1. Ardagna CA, Asal R, Damiani E, Vu QH (2015) From security to assurance in the cloud: a survey. ACM Comput Survey 48(1):1–50. https://doi.org/10.1145/2767005

2. Bird J (2016) DevOpsSec. O'Reilly Media, Inc. https://www.oreilly.com/library/view/devopssec/9781491971413/
3. Blake C (2020) Reducing risk with end-to-end application security automation. Netw Secur 2020(2):6–8. https://doi.org/10.1016/S1353-4858(20)30019-2
4. Carter K (2017) Francois Raynaud on DevSecOps. IEEE Software 34(5):93–96. https://doi.org/10.1109/MS.2017.3571578
5. Chandrasekaran A (2019) Best practices for running containers and kubernetes in production. https://www.gartner.com/en/documents/3902966/best-practices-for-running-containers-and-kubernetes-in-
6. Cloud Standards Customer Council (2017) Cloud customer architecture for big data and analytics v2.0 https://www.omg.org/cloud/deliverables/CSCC-Cloud-Customer-Architecture-for-Big-Data-and-Analytics.pdf
7. CNCF (2020) Cloud native computing foundation. https://www.cncf.io/
8. CSA (2013) Expanded top ten big data security and privacy challenges. Technical report. Cloud Security Alliance. https://cloudsecurityalliance.org/artifacts/expanded-top-ten-big-data-security-and-privacy-challenges/
9. Davis J, Daniels K (2016) Effective DevOps: building a culture of collaboration, affinity, and tooling at scale, 1st edn. O'Reilly Media, Inc.
10. Ebert C, Gallardo G, Hernantes J, Serrano N (2016) DevOps. IEEE Software 33(3):94–100. https://doi.org/10.1109/MS.2016.68
11. Fernandes DAB, Soares LFB, Gomes JV, Freire MM, Inácio PRM (2014) Security issues in cloud environments: a survey. Int J Inf Secur 13(2):113–170. https://doi.org/10.1007/s10207-013-0208-7
12. Gill NS (2018) DevSecOps tools and continuous security. https://www.xenonstack.com/blog/continuous-security-automation-devsecops/
13. Jabbari R, Bin Ali N, Petersen K, Tanveer B (2016) What is DevOps? A systematic mapping study on definitions and practices. In: Proceedings of the scientific workshop, proceedings of XP2016. ACM, New York, NY, USA, pp pp. 12:1–12:11. https://doi.org/10.1145/2962695.2962707
14. Jamshidi P, Pahl C, Mendonça NC, Lewis J, Tilkov S (2018) Microservices: the journey so far and challenges ahead. IEEE Softw 35(3):24–35. https://doi.org/10.1109/MS.2018.2141039
15. Koskinen A (2019) DevSecOps: building security into the core of DevOps. https://jyx.jyu.fi/handle/123456789/67345
16. Kratzke N (2018) A brief history of cloud application architectures. Appl Sci 8(8). https://doi.org/10.3390/app8081368
17. Kumar R, Goyal R (2019) Assurance of data security and privacy in the cloud: a three-dimensional perspective. Softw Qual Prof 21
18. Kumar R, Goyal R (2019) On cloud security requirements, threats, vulnerabilities and countermeasures: a survey. Comput Sci Rev 33:1–48. https://doi.org/10.1016/j.cosrev.2019.05.002
19. Kumar R, Goyal R (2020) Modeling continuous security: a conceptual model for automated devsecops using open-source software over cloud (ADOC). Comput Secur 97:1–28. https://doi.org/10.1016/j.cose.2020.101967
20. Kuusela J (2017) Security testing in continuous integration processes.https://pdfs.semanticscholar.org/4d55/0be7c00fb51a5594b40ca92234b0a538685d.pdf
21. Lin J, Yu W, Zhang N, Yang X, Zhang H, Zhao W (2017) A survey on Internet of Things: architecture, enabling technologies, security and privacy, and applications. IEEE Internet Things J 4(5):1125–1142. https://doi.org/10.1109/JIOT.2017.2683200
22. Microsoft Azure (2019) Continuous integration and continuous deployment to Azure IoT Edge. https://docs.microsoft.com/en-us/azure/iot-edge/how-to-ci-cd
23. Mobarak M (2019) Implementing a CI/CD pipeline for AWS IoT Greengrass projects. https://aws.amazon.com/blogs/iot/implementing-a-ci-cd-pipeline-for-aws-iot-greengrass-projects/
24. Mohan V, ben Othmane L, Kres A (2018) BP: Security concerns and best practices for automation of software deployment processes: an industrial case study. In: 2018 IEEE cybersecurity development (SecDev), pp 21–28. https://doi.org/10.1109/SecDev.2018.00011

25. Myrbakken H, Colomo-Palacios R (2017) DevSecOps: a multivocal literature review. In: Mas A, Mesquida A, Connor RV, Rout T, Dorling A (eds) Software process improvement and capability determination. Springer, pp 17–29. https://doi.org/10.1007/978-3-319-67383-7_2
26. NCSC (2019) Cyber security design principles. Technical report. National Cyber Security Center). https://www.ncsc.gov.uk/collection/cyber-security-design-principles
27. Neelaveni R (2019) Performance enhancement and security assistance for VANET using cloud computing. J Trends Comput Sci Smart Technol 01:39–50 https://doi.org/10.36548/jtcsst.2019.1.004
28. OWASP (2016) Security by design principles. https://www.owasp.org/index.php/Security_by_Design_Principles
29. OWASP (2017) OWASP top 10 application security risks—2017. https://www.owasp.org/index.php/Top_10-2017_Top_10 (2017)
30. Parnin C, Helms E, Atlee C, Boughton H, Ghattas M, Glover A, Holman J, Micco J, Murphy B, Savor T, Stumm M, Whitaker S, Williams L (2017) The top 10 adages in continuous deployment. IEEE Softw 34(3):86–95. https://doi.org/10.1109/MS.2017.86
31. Paule C, Düllmann TF, Van Hoorn A (2019) Vulnerabilities in continuous delivery pipelines? A case study. In: 2019 IEEE international conference on software architecture companion (ICSA-C). pp 102–108. https://doi.org/10.1109/ICSA-C.2019.00026
32. SANS (2020) Securing web application technologies (SWAT) and secure DevOps practices. https://www.sans.org/security-resources/posters/secure-devops-practices/175/download
33. Shahin M, Ali Babar M, Zhu L (2017) Continuous integration, delivery and deployment: a systematic review on approaches, tools, challenges and practices. IEEE Access 5:3909–3943. https://doi.org/10.1109/ACCESS.2017.2685629
34. Shakya S (2019) An efficient security framework for data migration in a cloud computing environment. J Artif Intell Capsule Netw 01:45–53 https://doi.org/10.36548/jaicn.2019.1.006
35. Sitalakshmi Venkatraman RV (2019) Big data security challenges and strategies. AIMS Math 4:860–879. https://doi.org/10.3934/math.2019.3.860
36. Tabrizchi H, Kuchaki Rafsanjani M (2020) A survey on security challenges in cloud computing: issues, threats, and solutions. J Supercomput. https://doi.org/10.1007/s11227-020-03213-1
37. Ullah F, Raft AJ, Shahin M, Zahedi M, Baba, MA (2017) Security support in continuous deployment pipeline. In: Proceedings of 12th international conference on evaluation of novel approaches to software engineering. https://arxiv.org/ftp/arxiv/papers/1703/1703.04277.pdf
38. Ur Rahman AA, Williams L (2016) Software security in DevOps: synthesizing practitioners' perceptions and practices. In: Proceedings of the international workshop on continuous software evolution and delivery, pp 70–76 (2016). https://doi.org/10.1145/2896941.2896946
39. Weeks DE (2019) DevOps reference architectures 2019. Technical report. Sonatype. https://www.sonatype.com/devsecops-reference-architecture-2019
40. Willis J (2012) DevOps culture (part 1). http://itrevolution.com/devops-culture-part-1
41. XebiaLabs (2020) Periodic table of DevOps tools (v3). https://xebialabs.com/periodic-table-of-devops-tools/
42. Xiao Y, Jia Y, Liu C, Cheng X, Yu J, Lv W (2019) Edge computing security: state of the art and challenges. Proc IEEE 107(8):1608–1631. https://doi.org/10.1109/JPROC.2019.2918437
43. Zhang J, Chen B, Zhao Y, Cheng X, Hu F (2018) Data security and privacy-preserving in edge computing paradigm: survey and open issues. IEEE Access 6:18209–18237. https://doi.org/10.1109/ACCESS.2018.2820162

Efficient Two-Layer Image Protection with Wavelet Transform Compression

M. Vaneeta, V. Sangeetha, and S. Swapna Kumar

Abstract The encoding complexity of an image format is a vigorously updating area of study in the field of two-layer protection with wavelet transform compression. In the proposed method, hybrid 2D-FDCT watermarking and RSA encryption for multispectral images predicted an efficient system. This approach satisfies the encryption security, robustness and classification accuracy retention of an algorithm. The two-layer protection of encrypted and embedded watermark image followed by wavelet transform compression minimizes the file size in the exhaustive process for encoding. An important merit is that encoding time is very much reduced in contrast to other security and compression mechanisms. The enhanced value of PSNR as well as trade-off of MES, normalized cross-correlation, the average difference and structural content improves the storage large file size medical image and improves bandwidth to an acceptable level.

Keywords Compression · DCT · Encryption · Image processing · Watermarking

1 Introduction

The Internet of things (IoT) is considered as the interconnection of computing devices such as in factory machinery, medical equipment or domestic appliances, enabling

M. Vaneeta (✉)
Department of Computer Science & Engineering, K S Institute of Technology, Visvesvaraya Technological University, Bengaluru, Belagavi, Karnataka, India
e-mail: vaneetam@ksit.edu.in

V. Sangeetha
Department of Information Science & Engineering, Sai Vidya Institute of Technology, Visvesvaraya Technological University, Belagavi, Bengaluru, Karnataka, India
e-mail: sangeetha.v@saividya.ac.in

S. Swapna Kumar
Department of Electronics & Communication Engineering, Vidya Academy of Science and Technology, APJ Abdul Kalam Technological University, Thrissur, Kerala, India
e-mail: swapnakumar.s@vidyaacademy.ac.in

© The Author(s), under exclusive license to Springer Nature Singapore Pte Ltd. 2021
J. S. Raj et al. (eds.), *Innovative Data Communication Technologies and Application*,
Lecture Notes on Data Engineering and Communications Technologies 59,
https://doi.org/10.1007/978-981-15-9651-3_37

them to send and receive data through the Internet. It is an emerging technology that has gained power in recent years [1]. Further, in modern world, multimedia communication and information security are two active components in a wide application such as academic, IoT applications and industry. The demand for fusion trend between them allows a secure transfer of multimedia files as per ITU-T, Rec. X.800 and IETF RFC 2828 protocols [2]. The data can be secured by applying authentication, confidentiality and protection against unauthorized disclosure. To meet such limitations, several methods such as watermarking and cryptography have been proposed in the literature.

Digital watermarking is a data hiding technology that is suitable for different multimedia applications [3]. It operates on different methods to cover objects, such as text content, media images, audio or video. Watermarking techniques are appropriate for ownership protection by hiding information. On the other hand, cryptography is the mechanism that encrypts the messages to render the information not understandable to any unauthorized entity. The conception of commutative watermarking-encryption (CWE) was first conversed in with a special prominence on watermarking in the encrypted domain [4]. To maintain the integrity of the security system, the data is encrypted watermarked data.

The wavelet compression method under lossy compression ensures a better compression rate and accuracy. After encryption, watermarking and compression, the image is aligned for efficient transmission. At the receiving end, the data decompressed and decrypted with the watermark extraction to recover the original image [5].

The remaining part of paper is planned as follows: Sect. 2 refers to the literature survey work. Section 3 discussed design methodology. Section 4 refers to the proposed method. Section 5 mentions the simulation practice. Section 6 shows result and analysis. Finally, the concluding Sect. 7 makes remarks.

2 Literature Survey

A grayscale image is just a two-dimensional matrix, whereas RGB image is a three-dimensional matrix. As per the requirement, the image is chosen for efficient utilization in the security aspects. Mohanty and Kougianos [6] presented the 'secure better portable graphics' (SPGB) for trustworthy media communication. The BPG image format used in IoT is suitable for the smart healthcare industry.

A unique approach adopted by Kougianos et al. [7] for a secure digital camera provides a double-layer protection on watermarking and encryption. It is compatible with different multi-media and system-on-a-chip (SoC) technology model. Mohanty [8] proposed 'secure digital camera (SDC)' architecture. This provides a two-layer protection with unified integration of watermarking and data encryption capabilities.

Tian et al. [9] introduced an innovative system in secure images captured by digital camera using twofold procedure that combines semi-brittle and vigorous blind-type watermarks. Ullah et al. [10] discussed the signcryption paradigm in public-key cryptography on digital signature and public-key encryption. This method has better bandwidth consumption and effective resource usage. Swapna Kumar et al. [11] expressed online signature verification system on touch interface-based devices. The dynamic time warping (DTW) algorithm is used as function-based technique to align processed signature storage and for feature extraction. Zebbiche et al. [12] proposed multiplicative watermark decoding. The watermark is here embedded with the DCT and the DWT. Tan and Zhao [13] proposed the hybrid of digital watermarking image and encryption algorithm. It uses the transform of the watermarking image in the time domain and uses the symmetrically encrypted image compression method, to reduce image size. This is done by reducing the root-mean-square error of the digital watermarking image.

Zhou and Tang [14] proposed a scheme of RSA algorithm for an image. It indicates the importance and shortcoming of RSA mathematical algorithms on information security issues. Yang and Li [15] adopted the discrete cosine transform (DCT) algorithm in digital watermarking by MATLAB and simulated using the Simulink made the image processing easier. Thereby, efficient algorithm on watermarking improved the data integrity. Chandorkar [16] discussed discrete wavelet transform (DWT) for SSD camera design, hardware capable of an imperceptible watermark embedding with the LeGall5/3 using VLSI architecture. The algorithm is suitable to assess with JPEG compression.

3 Design Methodology

Image coding is extensively used in a wide range of application. In this proposed method, the hardware and software implementations on two-dimensional forward discrete cosine transform (2D-FDCT) and two-dimensional inverse discrete cosine transform (2D-IDCT) are adopted. These approaches are normally used for image coding and decoding with specific control techniques to speed up the processing. The basic block diagram is shown in Fig. 1.

The 2D-FDCT and 2D-IDCT coding and decoding are considered to be a high-processing real-time operation. 2D-FDCT is used for pixel merging approach and resizes function to obtain an image of the desired size from the input image [17].

Fig. 1 Basic block diagram

3.1 Proposed Lemma Proof

Correlation is between two-layer protection and its inversion using 2D-FDCT.

3.1.1 Watermarking and Encryption

Consider a d-dimensional space of an original feature

$$\alpha = (\alpha_1, \ldots, \alpha_d) \tag{1}$$

The encryption function selected is

$$f_E(\alpha) = \alpha_R = \left(\alpha_{R(1)}, \ldots, \alpha_{R(d)}\right) \tag{2}$$

The hidden watermark of an image feature is

$$\beta = (\beta_1 c_1, \ldots, \beta_d c_d) \tag{3}$$

where R is considered to be RSA secret key combination, and c is DCT coefficient. The embedding function is defined by

$$f_\beta(\alpha, \beta) = \alpha + \beta \tag{4}$$

The encryption of the watermark image is represented as:

$$\gamma = f_\beta(f_E(\alpha), (f_E(\beta)) = f_E(\alpha) + (f_E(\beta) = \alpha_R + \beta_R \tag{5}$$

The decrypted watermarked image feature is shown as:

$$\delta = f_E(f_\beta(\alpha, \beta)) = f_E(\alpha + \beta) = f_E(\alpha) + f_E(\beta)$$
$$= \alpha_R + \beta_R = f_\beta(f_E(\alpha), (f_E(\beta)) = \gamma \tag{6}$$

The encrypted watermarked and decrypted watermarked transformation dimensional space remains the same. So, it is proved the prediction of watermark embedding, and ciphering is proposed based on this lemma.

3.1.2 Watermark Embedding and Extraction

Consider the host image α which is interleaved using M by scrambling the watermark. Implementing the 2D-FDCT on spatial data sequence $c(i, j), 0 \le i, j \le 7$, in the image matrix α of 8×8.

The corresponding 2D-IDCT sequence $f(\alpha, \beta), 0 \leq \alpha, \beta \leq 7$, in the frequency domain of matrix F, is defined as:

$$
f(\alpha, \beta) = \frac{1}{4} C(\alpha) C(\beta) \sum_{i=0}^{7} \sum_{j=0}^{7} c(i, j)
$$
$$
\cos\left(\frac{(2i + 1)\alpha\pi}{16}\right) \cdot \cos\left(\frac{(2j + 1)\beta\pi}{16}\right) \tag{7}
$$

The inverse transformation is represented by

$$
C', C' = \{c'(i, j)\}, \tag{8}
$$

$$
c'(i, j) = \frac{1}{4} \sum_{\alpha=0}^{7} \sum_{\beta=0}^{7} C(\alpha) C(\beta) f(\alpha, \beta)
$$
$$
\cos\left(\frac{(2i + 1)\alpha\pi}{16}\right) \cos\left(\frac{(2j + 1)\beta\pi}{16}\right) \tag{9}
$$

where

$$
C(\alpha) C(\beta) = \begin{cases} 1/\sqrt{2} & \text{if} (\alpha, \beta) = 0 \\ 1 & \text{for others} \end{cases}
$$

The matrix $M = \{m(\alpha, \beta)\}$, where $m(\alpha, \beta)$ represents the matrix element in the αth row and βth column.

$$
m(\alpha, \beta) = \begin{cases} 1/\sqrt{2}, & \alpha = 0 \cup 0 \leq \beta \leq 7 \\ \frac{1}{2} \cos\frac{(2\beta + 1)\alpha\pi}{16}, & 1 \leq \alpha \leq 7 \cup 0 \leq \beta \leq 7 \end{cases} \tag{10}
$$

The 2D-DCT image data matrix D and its inverse matrix C' are given as:

$$
D = MCM^{\mathrm{T}} \tag{11}
$$

$$
C' = M^{\mathrm{T}}DM \tag{12}
$$

The image watermarking is applied to 1D-FDCT and 1D-IDCT that are used for coding and decoding. The 2D-FDCT and 2D-IDCT reduce the complexity of image watermarking that reduces high computation performance and its easy extraction.

4   Proposed Methods

The proposed two-layer method block diagram efficiently and securely embeds the original image into a cover image referred to as a secret image which is shown in Fig. 2.

In this approach after the watermarked image, RSA encryption is applied to provide more security to an image. The image is compressed using wavelet transform for effective compression for an image to transmit over the network. At the receiver section, decompressed image is decrypted which is further decoded to obtain the original image from watermarked image.

Fig. 3 describes the decoding of the proposed method. The encrypted and watermarked compressed image can be decrypted to retrieve the original data without watermark being affected and can retrieve the original image without a watermark.

The image is compressed using DCT wavelet transform for effective compression for image to transmit over the network. At the receiver section, decompressed image

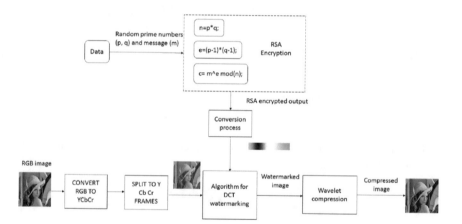

Fig. 2 Block diagram of processed encoding method

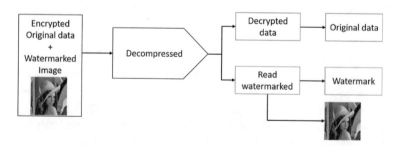

Fig. 3 Block diagram of processed decoding method

is decrypted which is further decoded to obtain the original image from watermarked image.

4.1 Flow Diagram

The flow diagram of the whole proposed process is shown in Fig. 4. Initially, data is encrypted with modified RSA encryption converted to cipher image. The image to be watermarked is converted from RGB to YCbCr. If it is RGB, then it has to be first converted to YCbCr format. Then each of the YCbCr frames is splitted, and watermarking is done on the Y-frame. If the image is gray scale, the watermarking can be directly performed.

After watermarking is done, the last step is compression. Performance analysis is done on the images obtained as a result. At the end, these steps are repeated for various formats.

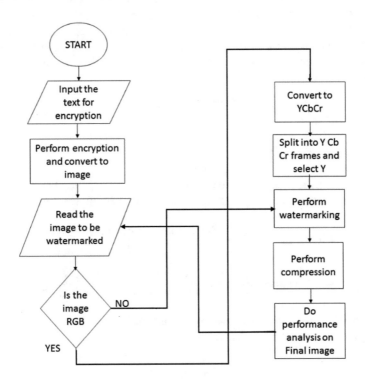

Fig. 4 Flow diagram

5 Simulation

Experiments are simulated using MATLAB15 Intel Inspiron 6400, 1.7 GHz processor. As proposed in the final results, a set of images after watermarking are compared with the original image to obtain the image. The obtained image after compression is compared with the preliminary image. After comparing, the performance analysis is done. Here, the parameters, PNSR, SR and BPP are calculated. The images of type JPEG, BPG and PNG of size 512 × 512 are used for testing.

5.1 Quality Assessment

The performance analysis measures the quality index of the images. Classical image quality metrics are used to measure the various parameters that cause the noise when the image is compressed [18–22]. The human visual system (HSV) model is not considered for objective or correlation of the performance assessment. A higher value of PSNR is considered for measuring the similarity between the original image and the compressed image. The mathematical expression which is shown in Eqs. (13) and (14), respectively, validates the PSNR and mean square error parameters.

$$\text{PSNR} = 10 \log_{10} \frac{255^2}{\text{MSE}} \tag{13}$$

$$\text{MSE} = \sum \sum \frac{(X(i, j) - Y(i, j))^2}{n \times n} \tag{14}$$

The mathematical expression which is shown in Eqs. (15) and (16) determines and validates compression ratio, encoding time, BPP and SR, respectively.

$$\text{Bits per pixel(BPP)} = \frac{\text{No. of bits to represent image vector}}{\text{No. of pixel of original image}} \tag{15}$$

$$\text{Speed up ratio (SR)} = \frac{\text{Compression time of WT}}{\text{Compression time of proposed WT}} \tag{16}$$

Normalized cross-correlation measurement is the comparison of two time series, but using a different scoring result of Eq. (17).

$$\text{NCC} = \sum_{i=1}^{m} \sum_{i=1}^{n} \frac{A_{ij} \times B_{ij}}{A_{ij}^2} \tag{17}$$

The measurement of an average of the difference between the original signal and the test image as well as structural contents is determined by Eqs. (18) and (19).

$$AD = \frac{1}{mn} \sum_{i=1}^{m} \sum_{i=1}^{n} [A(i, j) - B(i, j)] \tag{18}$$

$$SC = \frac{\sum_{i=1}^{m} \sum_{i=1}^{n} (A_{ij})^2}{\sum_{i=1}^{m} \sum_{i=1}^{n} (B_{ij})^2} \tag{19}$$

These important assessments of a particular component of difference metric depend upon the intended use of simulation results.

6 Results and Analysis

6.1 Analysis-I

First, the process of watermarking is carried out for grayscale images. After watermarking, the results are tabulated as shown in Table 1.

Table 1 shows the comparison of image size before and after watermarking on image of size 512 x 512. It is observed that RSA encrypted message that is watermarked provides two layer security and reduces the file size of an image in different format [23, 24]. For JPEG, the size has become 192 KB from the original size of 64 KB, for PNG, it is decreased to 35 KB, and for BPG, it has become 30 KB. In the case of RGB images, an intermediate step of converting from RGB to YCbCr is followed.

Figures 5 and 6 indicate the result obtained after encryption, watermarking and compression of the grayscale image. It may note that there is no much visible change in the image quality even after compressing the images. This follows the measurement technique [25] principles.

The comparison results of original and watermarked RGB images are shown in Figs. 7 and 8.

	Images	Original size	Proposed watermarked size
Table 1 Comparison of image size before and after watermarked image of size 512 × 512	JPG	64	192
	PNG	203	35
	BPG	31	30

a b c

Fig. 5 Original grayscale images: **a** JPG, **b** PNG, **c** BPG

a b c

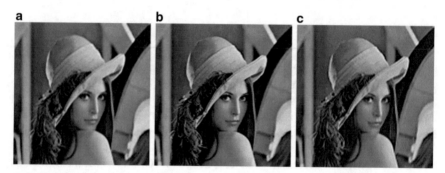

Fig. 6 Comparison result of original and watermarked gray images: **a** JPG, **b**: PNG, **c** BPG

a b c

Fig. 7 Original RGB images: **a** JPG, **b**: PNG, **c** BPG

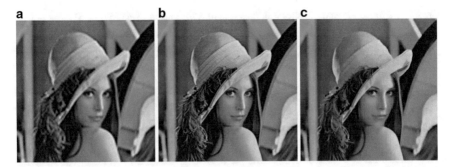

Fig. 8 Comparison result of original and watermarked RGB images: **a** JPG, **b**: PNG, **c** BPG

6.2 Analysis-II

The derived quality assessment of grayscale image based on the LENA image of 64, 203 and 310 KB size is tabulated in Table 2. It shows the performance analysis of grayscale JPG, PNG and BPG images using RSA encryption, 2D-FDCT watermarking and DWT compression [26].

From the assessment, it shows that the file size of JPG is far less than PNG and BPG image has the least file size among the three image formats. After using the watermarking algorithm, the PNG format file size is lesser than MSE and higher than PSNR. Also, the normalized cross-correlation is one, and the average difference is zero with the structural content which is close to one. BPG image is comparable to PSNR with PNG format. The average of differences between the original image and the watermarked image is lesser for BPG format.

After compression, from the table, it can be inferred that BPG format even though its values lie between that of PNG and JPG has higher performance in terms of both quality and file size. But as BPG is not wholly supported by every device, PNG can opt for applications in which size does not matter.

Table 2 Performance analysis after watermarking of grayscale images of size 512 × 512

Image	Format	Size (KB)	MSE	PSNR	Normalize cross-correlation	Average difference	Structural content
LENA (64 KB)	JPG	30	19.69	35.1871	0.9997	−5.80E−03	0.9994
LENA (203 KB)	PNG	49	4.70	41.4069	1	0	0.9997
LENA (31 KB)	BPG converted to JPG	30	10.78	37.8034	1.0002	−0.0107	0.9989

Table 3 Performance analysis after watermarking of RGB images of size 512 × 512

Image	Format	Size (KB)	MSE	PSNR	Normalize cross-correlation	Average difference	Structural content
LENA (67.5 KB)	JPG	38.3	20.1299	35.0924	0.9994	−1.77E −02	1.0001
LENA (462 KB)	PNG	472	6.6675	39.8912	0.9991	0.0094	0.9998
LENA (31 KB)	BPG converted to JPG	35	11.5709	37.4971	0.9999	0.013	0.9996

Table 4 Performance analysis after compression of RGB images of size 512 × 512

Image	Format	Size (KB)	MSE	PSNR	Normalize cross-correlation	Average difference	Structural content
LENA (68 KB)	JPG	38.3	20.8886	34.9317	0.9992	−1.77E−02	1.0004
LENA (462 KB)	PNG	439	7.3741	39.4537	0.9996	−0.0044	1.0003
LENA (31 KB)	BPG converted to JPG	24	35.5573	32.6215	0.9962	0.0222	1.0056

For other application where size has limitations, JPG can be used. RGB images have higher file size as compared to grayscale images. After watermarking, the PSNR of each of the three formats is comparable listed in Tables 2, 3 and 4.

After compression, RGB images after compression are listed in Table 4.

After compression, the PNG file size is reduced from 472 to 439 kb, and the PSNR value is still 39.4537 db which is just 0.437 db less than the watermarked one. The average difference is much less for PNG images.

6.3 Analysis-III

The statistics of the MSE of compressed images is shown in Fig. 9. MSE is maximum for BPG images and minimum for PNG images. Hence, PNG has better performance under this wavelet compression scheme.

In Fig. 10, the PSNR of compressed images is shown. Image quality is better for PNG images. But the file size after compression seems improved for BPG.

BPG image format has an only slight difference in terms of quality as compared to JPG while its file size is much reduced as compared to JPG.

Fig. 9 MSE comparison of
three compressed image
formats

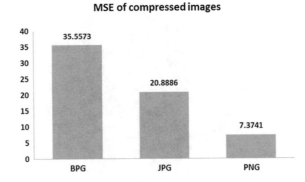

Fig. 10 PSNR comparison
of three compressed image
formats

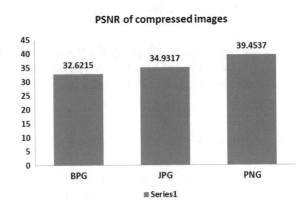

6.4 Analysis-IV

From Fig. 11, the result comparison on PSNR value obtained for BPG format image
with reference paper [6] shows there is 23.7% reduction from the original BPG

Fig. 11 Comparison of
proposed with reference
paper

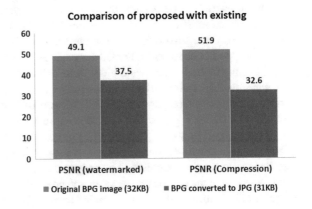

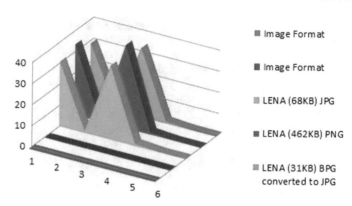

Fig. 12 Comparison of the existing and proposed method

image after watermarking. Similarly, for the compressed image, there is a reduction of 37.18%. But the image taken for analysis in [6] is an original BPG format which is not supported in most of today's available systems.

6.5 Analysis-V

See Analysis-II for Table 4 results that show the performance of proposed method is having significant improvement as compared to existing method.

From Fig. 12, it shows that there is a maximum peak on LENA (462 KB) PNG image performance in terms of PSNR and MSE index.

7 Conclusion

In today's modern communication, there are many image formats available with some particular characteristics. Hence, a scheme is developed to find the better formats that show improved performance and more secured. The proposed scheme aims at finding out the better format intended for each user through analyzing the various formats. This method addresses a high-performance two-layer security of encryption, watermarking followed by compression for analysis. Experimental results infer performance analysis of grayscale JPG, PNG and BPG images using RSA encryption, 2D-FDCT, DCT watermarking and compression and 2D-IDFT conversation. The good tradeoff between MSE, PSNR, normalized cross-correlation, the average difference and structural content of the reconstructed image can help to store huge size images in medical archives. The proposal is tested against many parameters, and it is seen that the size can be reduced considerably without compromising much in the PSNR. Future work can be extended to multimedia real-time applications by further reducing the encoding complexity and to improve PSNR value. This can even be extended to video formats for better compression of other images.

References

1. Catarinucci L, De Donno D, Tarricone L (2015) An IoT-aware architecture for smart healthcare systems. IEEE Internet Things J 2(6):515–526
2. Samanta S, Dutta S, Sanyal G (2011) Digital watermarking through embedding of encrypted and arithmetically compressed data into image using variable-length key. Int J Netw Secur Appl (IJNSA) 3(2)
3. Cancellaro M, Battisti F, Carli M, Boato G, De Natale FGB, Neri A (2008) A joint digital watermarking and encryption method. In: Proceedings vol 6819, security, forensics, steganography, and watermarking of multimedia contents X
4. Schmitz R, Gruber J (2017) Commutative watermarking-encryption of audio data with minimum knowledge verification. Hindawi Adva Multimedia 2017(Article ID 5879257)
5. Tseng K-K, He X, Kung W-M, Chen S-T, Liao M, Huang H-N (2014) Wavelet-based watermarking and compression for ECG signals with verification evaluation. Sensors. ISSN 1424-8220. https://doi.org/10.3390/s140203721
6. Mohanty SP, Kougianos E (2018) SBPG: secure better portable graphics for trustworthy media communications in the IoT. IEEE Access
7. Kougianos E, Mohanty SP, Cai W, Ratnani M (2009) VLSI Architectures of perceptual based video watermarking for real-time. In: Proceedings of the 10th international symposium on quality electronic design (ISQED), p 527
8. Mohanty SP (2009): A secure digital camera (SDC) architecture for integrated real-time digital rights management. Elsevier J Syst Archit
9. Ullah S, Rinner B, Marcenaro L (2017) Smart cameras with onboard signcryption for securing IoT applications. In: GIoTS
10. Tian L, Tai HM (2006) Secure images captured by digital camera. In: Ullah S, Rinner B, Marcenaro L (eds) Smart cameras with onboard signcryption for securing IoT applications. ICCE 2006
11. Swapna Kumar S, Divakaran S, Suresh D (2016) Touch interface based online signature verification system. In: 2nd IEEE international conference on engineering and technology (ICETECH), Coimbatore, TN, India, 17th & 18th Mar 2016
12. Zebbiche K, Khelifi F, Bouridane A (2008) An efficient watermarking technique for the protection of fingerprint images. EURASIP J Inf Secur 2008(Article number: 918601)
13. Tan Y, Zhao Y (2019) Digital watermarking image compression method based on symmetric encryption algorithms. Symmetry 11(12):1505
14. Zhou X, Tang X (2011) Research and implementation of RSA algorithm for encryption and decryption. In: IEEE 2011
15. Yang Y, Li H (2015) The application of DCT algorithm in digital watermarking by MATLAB and simulation. In: IEEE 2015
16. Chandorkar AN (2010) VLSI architecture of DWT based watermark encoder for secure still digital camera design. In: ICETET 2010
17. Al-Ani MS (2018) Hardware implementation of digital image coding using FDCT. J Theor Applied Inf Technol 96(6)
18. Bala Anand M, Karthikeyan N, Karthik S (2018) Designing a framework for communal software: based on the assessment using relation modeling. Int J Parallel Prog
19. Zhao F, Huang Q, Gao W (2006) Image matching by normalized cross correlation. In: 2006 IEEE international conference on, ICASSP 2006 proceedings, vol 2
20. Sahasrabudhe N, West JE Machiraju R, Janus M (2009) Structured spatial domain image and data comparison metrics. In: Proceedings visualization'99 (Cat. No.99CB37067). IEEE Xplore. Print ISSN 1070-2385

21. Hore A, Ziou D (2010) Image quality metrics: PSNR versus SSIM. In: 20th International conference on pattern recognition. IEEE Xplore
22. Santoso AK, Dr. Nugroho LE, Dr. Suparta GB, Dr. Hidayat R (2011) Compression ratio and peak signal to noise ratio in grayscale image compression using wavelet. IJCST 2(2)
23. Santhi V, Arulmozhivarman P (2013) Hadamard transform based adaptive visible/invisible watermarking scheme for digital images. J Inf Secur Appl 167–179 (Elsevier)
24. Viadya P, Mouli C (2015) Adaptive digital watermarking for copyright protection of digital images in wavelet domain. Elsevier Procedia Comput Sci 233–240
25. Choong KK (2018) Use of mathematical measurement in improving the accuracy (reliability) & meaningfulness of performance measurement in businesses & organizations. Measurement (Elsevier)
26. Vinothkanna MR (2019) A secure steganography creation algorithm for multiple file formats. J Innov Image Process (JIIP) 1(01):20–30

An Intelligent Text Encryption System Using Musical Notes

Reman Krishnan, Ryan Thomas, D. S. Akshay, Varsha Sunil Krishna, and R. S. Divya

Abstract Musical cryptograms have always been a matter of fascination for musicians and cryptologists across the world. In such cryptograms, each alphabet could be mapped to musical notes in numerous methods, which makes their detection difficult. The proposed system uses musical passwords to access the files and musical notes to encrypt and decrypt the text file data. When a user accesses a file, a musical keyboard appears and the user is prompted to play the musical password. Users define their own password as a sequence of musical notes of varying length. The musical password further serves as the key during encryption. The output would be a set of musical notes which represents the corresponding text in the file. The encrypted text file is then played as music. The proposed model is a symmetric model, where the same musical password used to encrypt should be used to decrypt data. The system ensures that a given plaintext with a given key generates different encrypted output each time due to usage of random mapping tables.

Keywords Musical cryptography · Encryption · Decryption · Symmetric key cryptography · Musical passwords · Musical notes

R. Krishnan (✉) · R. Thomas · D. S. Akshay · V. S. Krishna · R. S. Divya
Department of Computer Science and Engineering, Mar Baselios College of Engineering and Technology, Trivandrum, Kerala, India
e-mail: reman@gmail.com

R. Thomas
e-mail: rzj.ryan@gmail.com

D. S. Akshay
e-mail: akshaydez01@gmail.com

V. S. Krishna
e-mail: varshanshyam@gmail.com

R. S. Divya
e-mail: divya.rs@mbcet.ac.in

© The Author(s), under exclusive license to Springer Nature Singapore Pte Ltd. 2021   449
J. S. Raj et al. (eds.), *Innovative Data Communication Technologies and Application*,
Lecture Notes on Data Engineering and Communications Technologies 59,
https://doi.org/10.1007/978-981-15-9651-3_38

1   Introduction

Cryptography and network security as a branch of computer science have evolved over time, to offer techniques to increase the security of data across networks. Modern cryptographic systems rely on the intractability of certain mathematical problems to ensure security.

Music and its attributes were used in cryptography since olden days in the form of musical cryptograms. A musical cryptogram refers to a musical puzzle, which when compared to normal numeric puzzles, needs more time and skill to be decoded. They have been widely used by many music composers in their music compositions to protect them from being forged.

Music as a tool for cryptography provides some interesting properties, which makes its usage in cryptographic applications a useful one. Some of these interesting features could be listed as below:

- Uniqueness—Unless and otherwise there exists a deliberate attempt, no two tunes usually sound similar completely. This property of uniqueness of music could be used to generate unique musical tunes associated with data for cryptographic purposes.
- Undecipherable—Just by hearing music or by getting the music notes of a tune alone would not be enough to retrieve the data that was encrypted in music. There exist millions of possibilities, which make the system even more secure against a brute-force attack.

In the proposed system, music is used as a tool to encrypt text data and to access files. The system intends to make use of the above-mentioned properties of music to provide better security for text files and its contents.

2   Related Works

Musical cryptography has been a relatively unexplored field within the cryptographic domain. Though musical cryptograms were used by musicians to hide messages within their compositions since olden days, there has not been significant progress. With the advent of modern cryptographic systems like DES, AES, Blowfish, etc., much research has not been done about using musical notes for encryption purposes.

Sams [1] provided insights about the usage of music for cryptographic purposes and how they have been used in historic times. An extensive study about how musical notes were used by various musicians was discussed in his paper. Most researches done on this field emphasized the use of simple substitution techniques that map the English characters to musical notes. Dutta et al. [2] proposed a symmetric cipher that mapped the characters to musical notes with the help of mapping tables generated. The characters were mapped to three octaves of music notes. Dutta et al. [3] proposed a technique to hide messages using musical notes by using 26 alphabets and 10 digits from 0 to 9, where they were randomly assigned to musical notes. One of the major issues in designing a musical cryptographic model is the difficulty in finding a suitable substitution for characters. Kumar et al. [4] proposed the use of genetic algorithms to generate an optimal sequence of musical notes as encrypted output, which generated more soothing music as output instead of mere random sequences of notes. Yamuna et al. [5] used graph theoretic concepts along with the use of musical notes to encode a binary string. In [5], a CBC mode block cipher was proposed that took musical notes as an initialization vector (IV) and encoded binary strings to musical notes. A graph was constructed with these musical notes and the cryptographic key was derived from the degree of vertices.

Hindustani music and carnatic music, which are two main genres of Indian classical music, have been proposed to be used in cryptography. Rao et al. [6] proposed a cryptographic algorithm that uses carnatic music concepts for encryption. In [6], the characters were mapped to music notes, and the mapping was different for each scale (raga). The raga was proposed to be a part of the key in the proposed system, where the sender and receiver need to agree upon a particular scale(raga) for encryption. Pranav et al. [7] used a Hindustani music scale called Yaman Kalyan for encryption purposes. A semi-natural composition process was proposed in [7] to generate random musical notes based on the raga. Dutta et al. [8] proposed a cryptographic system entirely working on the concept of raga, with no symmetric or asymmetric keys used for encryption and decryption purposes. Dutta et al. [9] proposed a fuzzy logic-based approach to generate a symmetric cipher, which produces music sequences from the candidate notes to encode every character.

Kumar et al. [10] proposed a music cipher using a hybrid of Polybius and Playfair ciphers to encrypt a text and generate a musical output after encrypting. In [10], the key matrix for Playfair cipher was generated using Blum-Blum Shub Generator, a pseudorandom generator and the use of multiple ciphers contributed to the enhanced security. Maity [11] proposed a music cipher using modified Polybius cipher. In [11], Polybius cipher was modified using a 6×6 magic square and musical notes. The 6×6 magic square could accommodate 26 alphabets as well as 10 digits, and the musical notes were used for indexing the rows and columns of the magic square.

Passwords are widely used for user authentication. Usually, passwords are alphanumeric. A wide variety of attacks have been conducted on alphanumeric passwords, which exposed their vulnerability. As an alternative to alphanumeric passwords, graphical passwords were proposed. Suo et al. [12] provide a comprehensive survey regarding graphical passwords. The survey compares various graphical password methods available and their enhanced security has been compared to normal alphanumeric ones. Musical passwords, as well as graphical passwords, have been rarely studied and implemented.

Prakash et al. [13] proposed a voice biometric system, which authenticates a user through the musical notes sung by the user. In [13], the password was taken as the musical pattern sung by the user. Use of musical notes to create a password has been proposed by Kumar [14] in his research paper. The proposed system took musical inputs from users through a virtual keyboard and stored them in a database after hashing them with the MD-5 hash function. More research needs to be conducted regarding the viability of using musical passwords.

3 Proposed Methodology

The proposed system has been divided into three modules—encryption phase, decryption phase and database management. This section of the paper explains the design and working of each module. The working of each module has been explained and discussed.

3.1 Encryption Phase

Multi-stage encryption has been proposed to increase security—two-phase encryption is proposed. The encryption engine takes the file contents at once as input and generates corresponding encrypted output in the form of musical notes. The encryption module is designed to exhibit properties of a product cipher—by using multiple substitutions and transposition ciphers.

The conversion of plain text, which is made up of English characters, is converted to musical notes with the aid of multiple mapping tables. All characters are converted to discrete numeric values during the first phase of encryption and the numeric value obtained as output from the first phase is converted to musical notes during the second phase of encryption.

Mapping tables are randomly generated for each file. This ensures that different files are encrypted using different mapping tables, which eliminates reuse of mapping tables and eventually increasing security. In-built pseudorandom number generators (PRNG) in programming languages is used for random number generation.

First phase encryption works in two steps. The input contents are reversed, and then each character is mapped to discrete unique numeric values with the help of randomly generated mapping table values. The output of the first phase encryption would be a sequence of numbers (Fig. 1).

Second phase encryption also proceeds in three steps:

- The numeric value sequence obtained as output from the previous stage is XOR ed with the file password user sets.
- The password is converted to its ASCII values and then XOR ed with the numeric sequence.
- A nonlinear function, namely squaring function, is then applied to this XOR output. This alters the length of the numeric sequence. The squaring function takes each two digits of a numeric sequence and replaces them with the square of that two digit number, which would be four digits.

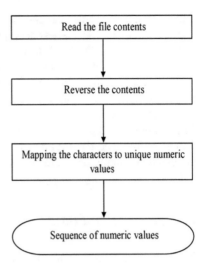

Fig. 1 First phase encryption

This increases the length of cipher text and makes it look different from plain text. The output of the squaring function is then mapped to musical notes with the help of another randomly generated mapping table. The generated music notes are then played (Fig. 2).

Music 21 [15] is a module available in Python language that aids in music generation. This module is imported and used for playing musical notes generated by the encryption engine. Parameters like BPM, duration of musical notes are set according to the user's preference. Once encryption is done, the output file generated is played in media player. This music file is saved at the same location where the encrypted

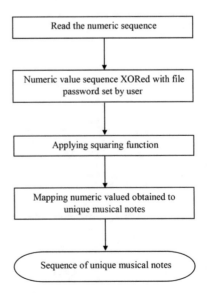

Fig. 2 Second phase encryption

file resides. This enables the user to hear the melody any number of times according to their wish until the file gets decrypted.

The use of XOR operation in second phase encryption changes the numeric sequence in accordance with the ASCII values of the password. XOR operation is done as an intermediate step before the squaring function is applied to the numeric sequence. This step works like a stream cipher—a numeric sequence which acts the plaintext gets XORed with the ASCII values of password, which plays the role of keystream in a standard stream cipher. The insecurity involved in using XOR alone has been eliminated by the application of squaring function, which alters the length of encrypted output before mapping them to musical notes.

3.2 Decryption Phase

The proposed system is a symmetric model, and the decryption is done exactly in reverse order of how encryption was performed. The musical password used to encrypt the file acts as the symmetric key for the proposed system and is used for decryption too. The first phase of decryption proceeds in three steps, as follows (Fig. 3):

- The musical notes are converted back to numeric values, by performing reverse lookup using the same mapping table used for encryption.
- Square root function is then applied, which takes each four digit number in the sequence and replaces it with a square root of that number, which would be a two digit number.

Fig. 3 First phase decryption

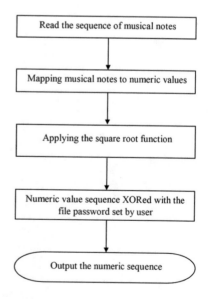

Fig. 4 Second phase decryption

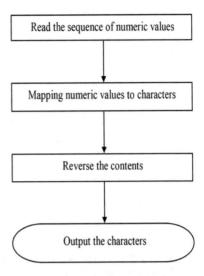

- The numeric sequence obtained is then XOR ed with the file password the user had set. Similar to that in the encryption phase, ASCII values of password characters are used in the decryption phase. This means that the decryption phase cannot decrypt accurately unless the user provides the exact password he/she used for encryption.

The second and final phase of decryption deals with converting the numeric sequence back to the original text. This phase follows two steps (Fig. 4):

- The numeric values are converted back to English characters by performing reverse lookup using the mapping table used for encryption.
- These characters are then reversed to obtain the original file contents back.

3.3 Database Management

To store the credentials associated with each user, a database needs to be maintained. Two database tables are proposed—one table exclusively for storing user's credentials, and others solely for storing information about the files that are presently encrypted.

User's credentials include username, password and their email id. Username and password are required for the user to login to the application. The typed credentials are matched against the database entries for user authorization. A first time user is required to sign up by providing credentials, which would then be stored in its database table. Email id of each user is asked during the sign-up phase, which enables the users to reset their credentials when needed. In such situations, an email would be sent to the user's mail id which instructs steps to reset credentials.

When a user encrypts a file by assigning a musical password, the entry for the corresponding file is created in this database table. Once a file is decrypted successfully, its entry needs to be removed from the database table. This ensures that the table keeps track of presently encrypted files only and also helps in reducing its size.

In certain situations, where the users reset credentials, it needs to be updated in database tables. If a user encrypts a file and then changes their credentials, the password changes. As the file was encrypted using the previous password, the files presently encrypted by the user need to be decrypted, and the file entry should be removed. This ensures consistency of data across database tables.

4 Security Analysis

1. **Brute-force attack on passwords**: Compared to use of alphanumeric passwords, musical passwords are expected to offer a greater degree of security to the proposed system. The proposed system uses a 60 key virtual musical keyboard which takes the musical input from the user or musical password. The password characters are mapped to these 60 musical notes—which cover five octaves. The probability for guessing a music note accurately in the musical password becomes $1/(60!)^n$, where n denotes the length of a password. Guessing becomes even more difficult when more music notes are included in the password, thus generating a highly secure musical password. Since the users can choose any melody of random length, the passwords are extremely difficult to guess accurately.

2. **Database attacks to yield password**: Users create passwords for each file they lock and encrypt. These passwords are stored in a database table. The security of the database tables should not be compromised to a slight degree. To secure passwords, the SHA-2 hash function is being used. The passwords are encrypted using SHA-2 hash, and the digest value is obtained. The security of passwords is further enhanced by converting the message digest obtained to musical notes. The conversion of digest value to musical notes, using the same mapping table which was used in encryption of text contents. This can increase the efforts for cryptanalysis drastically, as the cryptanalyst needs to first decrypt the musical password to obtain hash value.

3. **Frequency analysis of cipher text**: Encryption phase of the proposed system does the job of a substitution cipher. Substitution ciphers have been found vulnerable to frequency analysis attacks. In the proposed system, the vulnerability tends to reduced by ensuring that the same symbols in plain text get encoded to different cipher text. The squaring function applied during the encryption phase ensures that the cipher text and plain text are of different lengths. This enhances security, by eliminating frequency analysis attacks and helps in reducing statistical relationships between plain text and cipher text.

4. **Randomness in encrypted output**: For converting English text to musical notes, two-level mapping is done with the help of two mapping tables. These mapping tables are unique for each file—the probability of two files having the same

mapping table is nearly zero due to randomness involved in its generation. The uniqueness of the mapping table ensures the diversity in the generation of musical notes in cipher text. Mapping tables are concealed from access of users to avoid deletion of mapping tables from the system and to prevent unauthorized access, which otherwise could make the encrypted file vulnerable to cryptanalysis easily, as the cryptanalyst gets access to mapping tables.

Reuse of mapping tables is, as dangerous as a collision in hash functions and attacks could be easily done if an attacker has access to two files with the same mapping tables under any circumstances.

5. **Immunity against standard attacks**: The randomness in encrypted output helps in making the proposed system immune against standard cryptanalytic attacks, which include known plaintext attack, chosen plaintext attack and chosen cipher text attack. The reuse of mapping tables is extremely rare to occur due to pseudo-randomness present, and this means that an attacker needs to attack each text file individually, to get access to each mapping table associated with that particular file.

5 Results and Discussion

The proposed system has been implemented as a Windows-based application, with Python 3.6 used to program encryption and decryption modules. The whole application was built in Microsoft Visual Studio, and the encryption and decryption modules were linked to the application.

Python 3.6 supports random number generation using the *random* module available. Two mapping tables were created, with the aid of in-built random number generator. Ninety-five musical notes were used for encryption purposes, spanning about eight octaves. The characters were mapped to numbers, and these numbers were further mapped to musical notes, randomly. These mappings were used in the first phase as well as in the second phase of encryption. Mapping files were saved and concealed from external access in the system, for the decryption module to use them later. As discussed earlier in the implementation section, Music 21 [15] module has been used for playing the musical notes generated by the file.

To accept musical notes as a password from the user, a musical keyboard was designed as a Windows form, with 60 keys, spanning five octaves. This Windows form was made a part of the whole application. The played musical notes were concatenated as a single string and used as password. This password was sent to encryption and decryption modules. Use of squaring function in the second phase of encryption meant that the encrypted text contains more characters than that in the original file, effectively increasing its size.

Notepad files having size up to 2500 bytes were taken for experimental study and their execution time and output file size were studied. These text files contained English characters as well as digits and special characters. A graph was plotted

Fig. 5 Encrypted file size versus input file size

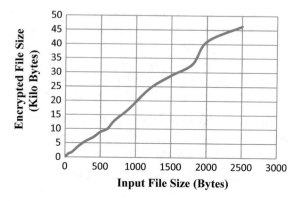

Fig. 6 Execution time versus input file size

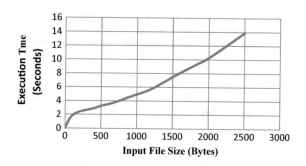

between input file size (in bytes) versus encrypted file size (in kilobytes). Clearly from the graph (Fig. 5), it can be concluded that the encrypted file size increases largely as the input file size is increased (Fig. 5).

The execution time of the proposed encryption algorithm has been found to increase as the input file size increases. The execution time includes the time taken for encryption as well as the time taken to generate the melody according to the user's parameters. The plot between execution time (seconds) versus input file size (bytes) has been given (Fig. 6). The execution time has been found nearly linear to input file size.

6 Conclusion and Future Scope

The primary goal of encrypting contents of a text file to musical notes and playing the contents as the music was achieved. The difficulty involved in detecting ciphers is enhanced when musical notes are being used for encryption. The accurate guessing of actual characters back from musical notes seems impractical due to randomness involved. Using musical passwords further enhances the security of the proposed system owing to the difficulty involved in exactly guessing the length of musical password and notes present in password unless a deliberate attempt occurs from the side of the user.

The methodology of using a medium like music, which has been veiled as entertainment and leisure and using it to produce an extensive encryption algorithm which is mathematically secure, highly robust against attacks and theoretically elementary in the realms of cryptography, is atypical.

Music generated by the proposed system lacks an esthetic appeal as the mapping was done randomly. Better algorithms could be incorporated in the proposed system in future which can generate more realistic and esthetic music sequences from the file contents. Further enhancements could be done to the proposed system to make them compatible with the PDF format of files, which are widely used. The proposed system could be extended to support texts in foreign languages other than English.

References

1. Sams E (1979) Musical cryptography. Cryptologia 3(4):193–201
2. Dutta S, Kumar C, Chakraborty S (2013) A symmetric key algorithm for cryptography using music. Int J Eng Technol 5(3):3109–3115
3. Dutta S, Kumar C, Chakraborty S (2010) A novel method for hiding message using musical notes. Int J Comput Appl 1(16):76–79
4. Kumar C, Dutta S, Chakraborty S (2014) Musical cryptography using genetic algorithm. In: 2014 International conference on circuits, power and computing technologies. IEEE, pp 1742–1747
5. Yamuna M, Sankar A, Ravichandran S, Harish V (2013) Encryption of a binary string using music notes and graph theory. Int J Eng Technol 5(3):2920–2925
6. Rao D, Koolagudi S (2019) Music cryptography based on carnatic music. Int J Eng Adv Technol 9(1):5107–5114
7. Pranav P, Chakraborty S, Dutta S (2019) A new cipher system using semi-natural composition in Indian raga. Soft Comput 24(2): 1529–1537 (Springer)
8. Dutta S, Chakraborty S, Mahanti NC (2011) Using Raga as a cryptographic tool. In: International conference on network security and applications 2011. Springer, Berlin, Heidelberg, pp 178–183
9. Dutta S, Kumar C, Chakraborty S (2015) Hidden messages using musical notes: a fuzzy logic approach. Int J Secur Appl 9(1):237–248
10. Kumar C, Dutta S, Chakraborty S (2015) A hybrid polybius-playfair music cipher. Int J Multimedia Ubiqut Eng 10(8):187–198
11. Maity M (2014) A modified version of polybius cipher using magic square and western music notes. Int J Technol Res Eng 1(10):1117–1119
12. Suo X, Zhu Y, Owen GS (2005) Graphical passwords: a survey. In: 21st Annual computer security applications conference (ACSAC'05). IEEE, p 10
13. Prakash R, Kumar S, Kumar C, Mishra KK (2016) Music password based biometric authentication. In: International conference on computing, communication and automation (ICCA2016). IEEE, pp 1016–1019
14. Kumar N (2012) User authentication using musical password. Int J Comput Appl 59(9):1–4
15. Music 21 documentation. https://web.mit.edu/music21/doc/index.html

Trumer: Rumor Detection on Twitter Using Machine Learning and NLP Techniques

Grusha Dharod, Harshal Dedhia, Jaineel Shah, Smita Sankhe, and Sneh Chitalia

Abstract A rumor/rumor is a form of a statement whose legitimacy is not yet confirmed. Rumor gives birth to the spread of misinformation in public. These days got most of the information about what is happening around the world from various online social media websites. Twitter is one of the most popular platforms, where people tweet about various events. One of the major problems all face on these social media platforms is "RUMORS." There is no authenticity regarding these tweets. In this work, an application "Trumer" is proposed that allows the user to input a tweet and calculates the probability of a tweet being a rumor in real-time. This application uses various parameters such as a retweet count, user's follower's count, the following count, verified status, related tweets, tense of the given tweet to calculate the probability of a tweet being a rumor. This application will allow users to determine the authenticity of the tweet and thus restricting the spread of misinformation.

Keywords Trumer · Rumor · Rumor detection · Twitter · NLP · Natural language processing · Hashtags

G. Dharod · H. Dedhia · J. Shah · S. Sankhe · S. Chitalia (✉)
Department of Computer Engineering, K. J. Somaiya College of Engineering, Vidyavihar, Mumbai, India
e-mail: csneh4@gmail.com

G. Dharod
e-mail: grushadharod@gmail.com

H. Dedhia
e-mail: harshaldedhia11@gmail.com

J. Shah
e-mail: jaineelchamp@gmail.com

S. Sankhe
e-mail: smitasankhe@somaiya.edu

© The Author(s), under exclusive license to Springer Nature Singapore Pte Ltd. 2021
J. S. Raj et al. (eds.), *Innovative Data Communication Technologies and Application*,
Lecture Notes on Data Engineering and Communications Technologies 59,
https://doi.org/10.1007/978-981-15-9651-3_39

1   Introduction

A rumor is defined as a statement, report, or a story whose truth is uncertain or doubtful. They circulate from person to person rapidly causing unwanted confusion among people. Rumors can be spread deliberately as a part of propaganda or can be spread due to misunderstandings as well.

Social media has gained a lot of popularity in the last few years. Many social media applications are emerging day by day. Every day, people share millions of posts. But, since everyone has the power to post whatever they want to, the social media platforms have thus become originators of rumors and spread of misinformation.

Twitter is one of the most popular social media platforms today. Approximately, 500 million tweets are tweeted every day [1]. Additionally, over 85% of all trending topics are news [2]. This makes Twitter one of the major sources to spread the news and also for people to voice their opinions. However, this also paves the way for spreading rumors. These rumors could lead to chaotic situations or even lead to undue losses for a person or an organization. In recent times, there have been many instances of rumors appearing on various social media platforms including Twitter.

A famous example is the 2013 rumor that the white house has been bombed, and then, US President Barack Obama is injured. This led to fear among people and also badly affected the stock market, leading to unfavorable economic situations. Another instance was a rumor that nine Indian banks are going to be shut permanently by the RBI. This rumor led to fear and chaos among the customers of these banks, with many rushing to withdraw all their hard-earned money and shifting to some other bank. Apart from the fear and chaos among common people, it also led to undue losses for these banks.

Therefore, the detection of such misinformation has become the need of the hour. This paper proposes a solution to determine the probability of a tweet being a rumor based on various parameters and present the results in a presentable manner.

2   Related Work

Recently, there has been a lot of attention toward detection of rumors, and there have been a lot of methods proposed to do the same. Some methods rely on human efforts while others rely on complex computations to get the desired outcome. Examples of using human effort for detecting rumors are websites like snopes.com and factcheck.org. These sites have teams of people who investigate the veracity of the rumors by looking up facts and evidence from various sources. Though these methods are very accurate, they are unable to investigate queries from every person on earth. This is where computation-intensive methods come in.

Earlier methods have treated rumor detection as a binary classification problem. One of the examples for rumor being a binary problem is shown in [3]. Slowly, there was a shift toward determining the probability instead of classifying into two buckets. For instance, in [4], the authors proposed such a rumor detection approach. They identified tweets that contain enquiry patterns (the signal tweets), made clusters based on content, created a statement that summarizes the cluster, and then used that statement to pull back in the rest of the non-signal tweets that discuss that same statement. Finally, the candidate rumor clusters were ranked based on various statistical features, in decreasing order of the likelihood of a statement being a rumor.

The approach followed in [5] was a parameter-driven approach, i.e., the detection is dependent on some predefined parameters to better detect different kinds of rumor spread. The authors identified characteristics of rumors on Twitter by examining linguistic style used to express rumors, characteristics of people involved in propagating information, and the network propagation dynamics. Then, they noted the key differences in each of these characteristics for the spread of rumor and non-rumor. Finally, these differences were used by a Hidden Markov Model to predict the veracity of real-time tweets.

For keyword extraction, hashtag analysis method was used which outperformed most of the traditional methods. TextRank [6] in which the sentence was tokenized and classified into POS tags. The drawback of this method was only some of the mentioned POS tags could be used as keywords. TF-IDF [7] in which generated keywords based on the frequency of their occurrence. Not all significant keywords have high occurrences in a tweet. SVM rank [8] is the combination of TextRank [6] and TF-IDF [7], however, combining them does not necessarily overcome their respective disadvantages. Word embedding and clustering [9] consider semantic relationships for keywords, however, it requires a significant amount of pre-processing to be done before the keywords that can be extracted.

[10, 11] are both based on stance detection. Both papers have classified stance on replies and not related tweets. Classes used in these papers are support, deny, comment, and query. The major issue was class imbalance which occurred due to misclassification of most tweets as "Comment." Furthermore, their models had difficulty in detecting the stance "Deny." Rumorlens [8] have not considered many of the parameters such as tense of a tweet, verified users in related tweets which is the deciding factor, for classification of the tweet as a rumor.

3 Dataset

3.1 For Stance Detection

In this step, two datasets were used. The first one was the Sentiment140 dataset [12].

As for the second dataset, it was created manually by taking some controversial tweets and finding related tweets using the Twitter API. It also included the majority of tweets from the PHEME dataset [13]. These tweets were divided into three categories and labeled as FOR, AGAINST, and NONE. This accounted for a total of 1980 tweets for the stance detection model (Fig. 1).

Fig. 1 Class distribution of dataset

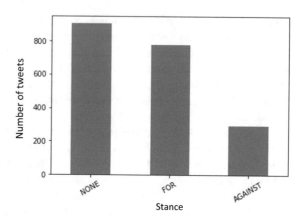

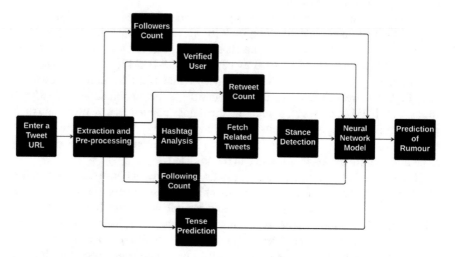

Fig. 2 Trumer architecture

3.2 For Final Model

Data from the PHEME dataset [13] was extracted. Then, for each tweet, the stance was determined using the stance detection model. Dataset consisting of all parameters required for the final model was then formed. This dataset consisted of around 4670 tweets.

4 Architecture

The proposed architecture of our system is shown in Fig. 2. The user will input a tweet URL in the system. From the input URL, various parameters such as the text of the tweet, hashtags in the tweet, retweet count of the tweet, followers and following count of the user, tense of the tweet, and whether the user is verified or not will be extracted.

Using the hashtags extracted, related tweets will be fetched, and then, the stance of each tweet will be predicted to the source tweet. There will be three stances viz. for, against, and none.

After this, the parameters extracted earlier along with the stance count is given to a neural network which has been trained to calculate the probability of a tweet being a rumor. Each of these steps is explained in detail in Sect. 5.

5 Implementation

The proposed methodology can be divided into four stages.

5.1 Pre-processing

1. **Extracting the basic parameters like followers count, verified_status, retweet_count, etc., from the tweet URL.**

The input as the URL of the tweet is taken as an input. The first step is to extract the tweet id from the URL. Then, using the Twitter API, the following parameters for the tweet are extracted.

- **Tweet text**: This will be useful further for tense detection and stance detection.
- **Verified status**: This will provide a sense of additional validity to the content of the tweet. If the tweeting user is a verified news source or a well-known personality in general, then it is less likely that the tweet is just a rumor, not backed by any solid evidence.
- **Retweet count**: Usually, retweeting a tweet implies that the user agrees with the post and possibly believes that it is true. Hence, a higher retweet count might reduce the chances of the tweet being a rumor.
- **Followers count**: The followers count helps to analyze the reach of a user. Generally, rumors are not started by famous users as it might reduce their credibility if it is found to be a rumor. Hence, a higher followers count would reduce the chances of the tweet being a rumor.

2. Finding tweets related to the source tweet from Twitter.

To perform stance detection in stage 5, the first step is to find a way to fetch related tweets. The most important task here is to extract the keywords, which can be used to search Twitter API to fetch other related tweets.

For this, two methods are used

(a) Hashtag and Proper noun Analysis Method:

A single tweet might not have many hashtags or proper nouns. To counter this, an algorithm is designed which is based on a dual-search pattern. Here, related tweets are searched for twice. For the first time, the search is for extracting the major hashtags and proper nouns and the second time to find related tweets using that.

A hashtag is a great tool to easily identify the tweets belonging to a particular topic. As for proper nouns, almost all of the tweets would be concerning some person, country, event, and so on. Hence, the proper noun will help us to further focus our search with respect to that entity.

A hashtag is an identifier which is written to easily separate content based on a theme/topic. It includes a "#" character followed by one or more words, with no spaces in between. Hashtags are generally used on social media platforms like Twitter, Facebook, Instagram, and so on.

- Searching for a particular hashtag helps one discover information on a particular topic, easily.
- Hashtags can also be used to direct information to some target audience.
- They are increasingly used by more and more platforms, thereby indirectly connecting the different sources of information.

This makes hashtags one of the major parameters to find related tweets. Hashtags and proper nouns are extracted from the tweet text. After getting a pool of extracted hashtags and proper nouns, the Twitter API is used to fetch related tweets using this pool. As a result, there would be a bigger collection of tweets to search for the hashtags and proper nouns. Finally, after extracting all the hashtags from this expanded collection of tweets, they are ranked in decreasing order of occurrence.

Lastly, the most frequent hashtags are considered to fetch related tweets using the Twitter API (Fig. 3).

(b) MonkeyLearn API [14]

In the rare cases where the tweet has neither of hashtags or proper nouns, this method is used to get keywords. On sending the tweet text, the API responds by providing the keywords along with their confidence scores. The keywords with confidence scores greater than 0.5 are considered. Considering these keywords, the Twitter API is used to fetch the related tweets. One of the major advantages of monkey learn is the identification of key phrases consisting of two or more words. Furthermore, it also provides a confidence level for each keyword. Also, there is no need to pre-process the sentence given for keyword extraction.

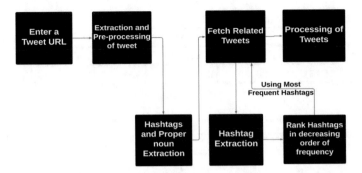

Fig. 3 Flow diagram of how the final set of related tweets are being obtained using hashtags and proper nouns

3. **Filtering out the tweets based on how much they are related to the source tweet.**

Even after fetching tweets using appropriate keywords, there is a possibility that some unrelated tweets might get through. To tackle this problem, cosine similarity [15] is used to further determine the level of relatedness of the source tweet with the fetched tweet. So, among the fetched tweets, after removing the stopwords from the tweet, the cosine similarity of each tweet is calculated with source tweet. From those, all the tweets are considered which are even very little similar to the source tweet. All those tweets which are completely different are ignored and have cosine similarity of 0. This ensures that completely irrelevant tweets are ignored, and tweets having very less similarity are not missing from the consideration for stance detection, thereby yielding better results.

5.2 Identifying the Tense of the Tweet

In the second stage, the tense of the tweet is determined. To achieve this, first, the structure of the sentence is found. For this task, the BLLIP Parser (also known as the Charniak-Johnson parser or Brown Reranking Parser) [4] and the WSJ-Penn TreeBank-3 (WSJ-PTB3) [16] model are used. BLLIP Parser not only has an f-score of 0.91 but also structures sentences as a tree, which allows it to find the tense of multiple sentences with great accuracy [4]. After getting the structure of the sentence, breadth-first search is used to find the main verb of the sentence. Finally, the POS-tag of this main verb is used to classify the tweet into present, past, or future tense.

The theory behind this step is that tweets posted in future tense have a higher chance of a tweet being a rumor due to the uncertainty concerning it and also less availability of any evidence as proof of the event. Now, one might ask that rumors can also be in the past tense, how can one ignore that? Yes, the rumor might be in the past tense as well. But, there is a way to counter that. As it is a past event, there will

Table 1 Results for stance detection model

FOR			AGAINST			Accuracy
Precision	Recall	F_1-score	Precision	Recall	F_1-score	
0.72	0.63	0.68	0.44	0.32	0.37	64.89%

be some evidence supporting or denying it. Some people will know about these facts and can voice their stance by tweeting for/against it. These stances are analyzed in stage 5, thereby handling the past tense problem.

5.3 Detecting the Stance of the Related Tweets and Classifying Them into one of fOR, AGAINST, NONE

Next step is to determine the stance of each related tweet with respect to the original tweet. For this task, the ULMFiT [17] algorithm is used for training the model. To train a model for stance detection on tweets, a considerably large dataset would be required. Since the ULMFiT algorithm was used, it was possible for us to achieve considerable results with a relatively smaller dataset. The algorithm uses transfer learning for text classification. Hence, a large generalized dataset is used to learn and understand the general structure of the language, followed by a smaller dataset focusing on the specific task of stance detection. The datasets used are mentioned in Sect. 3.1.

The results are shown in Table 1 which are obtained after applying the ULMFiT algorithm for the stance detection task. Table contains precision, recall, F_1-Score values for "FOR" and "AGAINST" categories, along with the overall accuracy of the model.

5.4 Final Model

To determine the probability of the tweet is a rumor, a model was developed which will take followers count, a user is verified or not, retweet count, number of related tweets in support, number of related tweets against the input tweet, number of related tweets which are neutral, number of tweets which are tweeted by verified user and are in support, number of tweets which are tweeted by verified user and are against the input tweet and tense of input tweet as input and give final value for the tweet being a rumor. All the input parameters were normalized. The dataset used to train this model is explained in detail in Sect. 3.2 (Fig. 4).

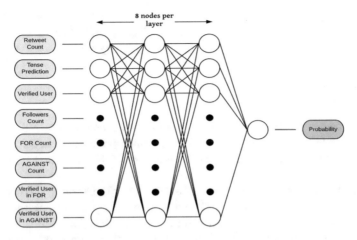

Fig. 4 Neural network model architecture

Initially, a basic neural network model was created with three hidden layers. Sigmoid was used as the activation function as the output value had to be between 0 and 1 for the probability. The same number of input and output dimensions was kept throughout the network. The loss was set to "mean_squared_error" and optimizer as "adam." In this scenario, an accuracy of 46% was achieved for ten epochs which increased to 53.51% for 1000 epochs. To further increase the accuracy, more layers were added; activation function was changed to "ReLU" for all the layers except the final output layer. By doing this an accuracy of 62.48% was achieved. Looking at the data again observation was made that the number of related tweets which are neutral that is making the dataset biased, so that parameter was dropped. By doing this an accuracy of 74.97% was achieved which was the best. However, one of the issues with the neural network models was the concentration of probability values close to 0.5. Even for the most obvious instances, the probabilities would rarely cross 0.6. So, instead of a neural network, logistic regression was tried, as the probability values were more linearly spread. As input features, normalized values were taken for followers count, retweet count, verified status, supporting tweets, denying tweets, verified supporting tweets, verified denying tweets, neutral tweets, and the tense of the tweet. The logistic regression model was trained, and an accuracy of 61.24% was achieved. Then, similar to the neural network model, neutral tweets input feature was removed. The accuracy jumped up to 74.84%. Thus, the accuracy was comparable to the neural network model, while having more linear values for probability. To increase the accuracy even further the SMOTE technique was tried to deal with imbalanced classes in the dataset. However, the results weren't as expected and the accuracy dipped to 68.59%. Then, the SMOTE technique was used on the neural network. Thus, an accuracy of 77.99% was achieved.

The results of the final model are shown in Table 2. Table contains precision, recall, F_1-Score, and the overall accuracy of the model.

Table 2 Results of the final model

Precision	Recall	F_1-score	Accuracy
0.7294	0.8988	0.8038	77.99%

6 Results

This method has achieved substantially better results in comparison to SEM-EVAL 2017 results [18, 19]. There were five teams which participated in the event and achieved scores 0.393, 0.464, 0.286, 0.536, 0.536, respectively, with a baseline score of 0.571.

Our application is tested on a few URLs. On tweet "The #pentagon is confirming the existence of UFOs. I knew some alien related shit was gonna come up soon. 2020 is so unbelievable that I'm just expecting the unexpected now" [20]. It shows the rumor probability of 0.6808393, i.e., 68%. It also displays various other parameters such as the followers count, retweet count, if the user is verified or not.

Figure 5a shows the output of our system. It displays the tweet text, rumor probability, i.e., 0.6808393 along with tense of the tweet, i.e., present tense and retweet count, i.e., 131.

Figure 5b shows the other parameters of the result such as followers count, following the count, verified account along with stance detection parameters such as for, against, and neutral.

7 Formulae

The formulae used are as follows:

$$P = \frac{(\mathbf{TP}_{C1} + \mathbf{TP}_{C2} + \cdots + \mathbf{TP}_{Cn})}{((\mathbf{TP}_{C1} + \mathbf{TP}_{C2} + \cdots + \mathbf{TP}_{Cn}) + (\mathbf{FP}_{C1} + \mathbf{FP}_{C2} + \cdots + \mathbf{FP}_{Cn}))}$$

$$R = \frac{(\mathbf{TP}_{C1} + \mathbf{TP}_{C2} + \cdots + \mathbf{TP}_{Cn})}{((\mathbf{TP}_{C1} + \mathbf{TP}_{C2} + \cdots + \mathbf{TP}_{Cn}) + (\mathbf{FN}_{C1} + \mathbf{FN}_{C2} + \cdots + \mathbf{FN}_{Cn}))}$$

$$A = \frac{(\mathbf{TP}_{C1} + \mathbf{TP}_{C2} + \cdots + \mathbf{TP}_{Cn}) + (\mathbf{TN}_{C1} + \mathbf{TN}_{C2} + \cdots + \mathbf{TN}_{Cn})}{\text{Total}}$$

where

P	is precision
R	is recall
A	is accuracy
$\mathbf{TP}_{Cn}$	is true positive of nth class
$\mathbf{FP}_{Cn}$	is false positive of nth class
$\mathbf{FN}_{Cn}$	is false negative of nth class1
Total	is the sum of true positive, true negative, false positive, and false negatives.

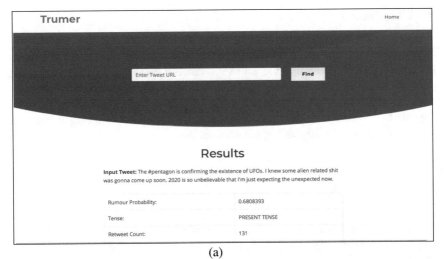

(a)

(b)

Fig. 5 **a** Test results—1, **b** Test results—2

$$\text{cosine similarity}(A, B) = \cos \theta = \frac{A \cdot B}{\|A\| * \|B\|}$$

8 Conclusion

Rumor Detection has become the need of the hour in today's world. In this paper, have been able to determine the rumor probability of a tweet with considerable accuracy using various parameters such as stance of related tweets, verified status of the Twitter user, retweet count, and so on. Hence, with this application, a way for the user is provided to check the veracity of the tweet before believing it and hope to alleviate the issue of spreading misinformation.

9 Future Work

In the future, this method can be further extended to other social media platforms where certain Twitter-specific parameters like retweet count, verification status, etc., might not be available. Secondly, our system only checks the veracity of the tweets which are input by a user. It can be expanded to look for rumors on the entire platform, thereby also detecting rumorous tweets which have not been checked by the users. Finally, our work is only considering English language tweets. This could be implemented for the major languages of the world.

To include comments for processing and determining the probability of a tweet being a rumor is being planned. Along with comments to implement image and video processing algorithms is also planned. The optimal ways to explore various external links are also thinking which are generally mentioned in the tweets.

References

1. Twitter usage statistics. https://www.internetlivestats.com/twitter-statistics/
2. Kwak H, Lee C, Park H, Moon S (2010) What is Twitter, a social network or a news media? In: International conference on world wide web, Raleigh, USA, pp 591–600
3. Qazvinian V, Rosengren E, Radev DR, Mei Q. Rumor has it: identifying misinformation in microblogs
4. Charniak E, Johnson M. Coarse-to-fine n-best parsing and MaxEnt discriminative reranking
5. Vosoughi S, 'Neo' Mohsenvand M, Roy D. Rumor gauge: predicting the veracity of rumors on twitter
6. Wongchaisuwat P (2019) Automatic keyword extraction using textrank (2019 IEEE)
7. Lee S, Kim H-J. Automatic keyword extraction from news articles using TF-IDF model
8. Cai X, Cao S. A keyword extraction method based on learning to rank
9. Zeng P, Yan Y, Xu J (2017) Automatic keyword extraction using word embedding and clustering (ICCSEC-2017)
10. Pamungkas EW, Basile V, Patti V. Stance classification for rumour analysis in twitter: exploiting affective information and conversation structure
11. Kochkina E, Liakata M, Augenstein I. Turing at SemEval-2017 Task 8: sequential approach to rumour stance classification with Branch-LSTM
12. Sentiment140 Dataset link. https://www.kaggle.com/kazanova/sentiment140

13. Pheme Dataset link. https://figshare.com/articles/PHEME_dataset_for_Rumour_Detection_and_Veracity_Classification/6392078
14. MonkeyLearn API. https://monkeylearn.com/
15. Cosine Similarity link. https://medium.com/@sumn2u/cosine-similarity-between-two-sentences-8f6630b0ebb7
16. Marcus MP, Santorini B, Marcinkiewicz MA, Taylor A. Treebank-3
17. Howard J, Ruder S. Universal language model fine-tuning for text classification
18. Gorrell G, Bontcheva K, Derczynski L, Kochkina E, Liakata M, Zubiaga A. RumourEval 2019: determining rumour veracity and support for rumours
19. Derczynski L, Bontcheva K, Liakata M, Procter R, Hoi GWS, Zubiaga A. SemEval-2017 Task 8: RumourEval: determining rumour veracity and support for rumours
20. Test tweet URL https://twitter.com/Adderalcoholic/status/1254937695111073793

Remote Surveillance and Communication Using Television White Space

E. Kaliappan, P. R. Madhevan, A. S. Aswath Ram, B. Ponkarthika, and M. R. Abinesh

Abstract Modern communication systems have numerous lags and redundancies because of the inherent noise and different technical disturbances that cause a tremendous deal of problem in transmission and reception. The range of FM and other radio waves are prevalently utilized for transmitting the voice information and has been old hat to that by itself. Utilizing the TVWS (Television White Space) any data type can be changed into audio i.e. implanting the information into an audio and disentangling it utilizing a basic radio receiving equipment. The communication channel has another advantage in terms of security; the data is encoded and decoded using user defined algorithms and this is difficult to crack. With these many advantages, the transmission also has certain difficulties in it. The data file contains errors or redundancy as the data is sent in a secured radio channel. This redundancy or error correction is done using Reed-Solomon coding, which is a famous coding technique that is employed for reducing the redundancy. The TVWS is widely used in research and other R&D in communication systems for developing the new 5G transmission in various countries. The TVWS has its advantage of high range and faster transmission speed compared to other communication channels. This is employed for the novel surveillance technique where motion is detected in a remote area where wired connection establishment is difficult. The motion thus detected is sent as an image to the owner of the property who can be several kilometers away.

E. Kaliappan (✉) · P. R. Madhevan · A. S. Aswath Ram · B. Ponkarthika · M. R. Abinesh
Electrical & Electronics Engineering, Easwari Engineering College, Chennai, India
e-mail: hod.eee@eec.srmrmp.edu.in

P. R. Madhevan
e-mail: madehavn1998@gmail.com

A. S. Aswath Ram
e-mail: aswathram3@gmail.com

B. Ponkarthika
e-mail: Ponkarthika.b@eec.srmrmp.edu.in

M. R. Abinesh
e-mail: abineshr1523996@gmail.com

© The Author(s), under exclusive license to Springer Nature Singapore Pte Ltd. 2021
J. S. Raj et al. (eds.), *Innovative Data Communication Technologies and Application*,
Lecture Notes on Data Engineering and Communications Technologies 59,
https://doi.org/10.1007/978-981-15-9651-3_40

Keywords Television Whitespace (TVWS) · Pulse Width Modulation (PWM) · Minimum Output Sum of Squared Error (MOSSE) · Convolution · Reed Solomon coding · Buffer

1 Introduction

With the current increase in communication techniques and modern surveillance methods the combination of exactly the correct communication channel with a comprehensive surveillance technique is needed for the hour for many remote places. Think of the border where a soldier has to be on guard to prevent trespassing and this place is in a remote place, where he can be easily overpowered by other forces like temperature, mental pressure, loneliness etc. Yes, he is very much capable of handling those but isn't it better to use computers for surveillance rather than wasting valuable human hours.

There are various surveillance techniques like drones and satellite imaging, which can be accurate enough to show the time on the watch worn by a person in the field of view. The major question of why this surveillance system that is proposed here should be used comes in mind. Think of a secure channel where anyone who interprets it can hear just noise but with the correct decryption algorithm, the transmitted data can be observed. So, the NOVEL idea in this surveillance system is the built communication channel. This Cyber Physical-System (CPS) is developed with the novel communication protocol where encryption, decryption and even the data frame format is user defined and can be tweaked based on the liking of the user. This secure communication is done using Television White Space (TVWS) whose frequency ranges from 470 MHz to 700 MHz [2]. This spectrum is unallocated, as all the television and broadcasting has switched from analog to digital, making this the perfect spectrum for communication [15]. The advantages of this spectrum are broadly classified for research purposes and establishing new communication protocols using it [2, 14, 15], [16]. Many developed countries have started working on 5G using this very spectrum, so the idea of "why this spectrum?" can be evaded easily with this explanation. The major advantages of TVWS over other communication channels and technologies are briefly explained in the upcoming chapters. The white space usage and the encoding and decoding techniques are user defined and can be changed at any instant in accordance to the data frame format [7]. The data frame format or data packet are being implemented using the user defined frames and rates [7, 13]. Because of the increasing requirement of radio frequency spectrum and bandwidth necessary to provide internet access and to sustain the congested network and the current network load, many people have come up with different ideas of communication protocols in order to overcome this challenge [7, 15]. These idle wasted frequencies are referred to as "white space". This process of 'recycling' spectrum is referred to as dynamic spectrum access, and poses a promising, less costly solution to increasing the available radio frequency spectrum [11, 15]. This serves as a very crucial advantage over other communication protocols wherein you have to account for the blocking of data transfer due to vegetation, obstacles in general but TVWS can flow through any of these obstacles in theory [1].

The system to be employed needs to capture the motion or trespassing in a remote place and send the image of the event to the user through the TVWS channel and display the motion detected.

2 Literature Survey

The range and the flexibility of TVWS network was clearly stated and because of the range, this wide spread network is used and cost of transmission was even more astonishing to be looked at [2, 10]. This chapter presents a case study extensively on TVWS for wide range network and communication. It deals with the practical difficulty over setting up a communication channel and also the advantages of using TVWS over other profoundly known technologies [10]. This chapter is specifically to roundabout the advantages of using TVWS in IoT and other novel upcoming M2M fields. The major disadvantage being the range is easily overcome by the usage of TVWS [2]. This paper presents the error-correcting codes that achieve the information that has been theoretically best and develops possible tradeoff between the rate and error-correction radius [3, 7, 12]. Concatenating the folded RS codes with suitable inner codes, the suitable binary codes are obtained to remain efficiently decoded up to twice the radius achieved by the standard GMD decoding[7, 12]. After the communication channel, next big setup problem is the motion detection algorithm. Something that is fast enough for tracking motion and also light weight so that the serial processing of the setup can be efficient. The basic algorithm for motion detection by using background subtraction is achieved from this. The motion detection algorithm used in this project uses a similar background subtraction technique for implementing a light weight motion detection so as to not harm the efficiency of the computing available [13]. Light weight algorithm is required since the transmission uses a complex software DMA and this takes up a lot of computing resources[9, 13].This paper gave us the exact knowledge on how to subtract motion frames from the video planes and use movement detection in a video surveillance technique[13]. The motion detection algorithm uses background subtraction and in order to improve the performance of the system [9, 13]. The paper also had a few insights on the lossless compression technique [9].This narrowband analysis paper gave an insight on how the transmission occurs and the design of data frame format for the communication system developed here [8, 12]. The paper had a basic analysis and also a in depth performance analysis as in range, susceptibility to noise etc between the TVWS and the normal GSM and other techniques[8].This paper discusses about various data protocols that are being used in the internet of things technology. Some of data protocols discussed in the paper are XMPP, MQTT, CoAP and DDS. The paper also differentiates these protocols based on quality of service (QoS). The categories for comparing these protocols were security, resource discovery, network packet loss, latency and bandwidth consumption [5, 6]. In order to fully understand the efficiency of the protocol designed here, there arises a need to fully understand the previous protocols and also their performance analysis. This paper has given greater insights about the difference in performance between the different protocols that is predominantly used for data sharing or communication between devices in modern times [5]. The paper deals with the image compression techniques and the compression was able to attain a 40.6% compression ratio by using a prediction-based algorithm [4]. This idea was also tested for the image compression technique. The compression techniques used was a deep surveyed study of multiple image compression techniques that were used in this paper for testing [3, 4]. The extensive literature survey helped us understand the complexity in the setup and the difficulties that might arise during the fabrication of the setup. Compression and then re-scaling issue still hold an unsolvable mystery in the image and video processing field [14].

3 Methodology

The methodology of this setup is divided into 2 parts

1. Transmitter
2. Receiver.

The outflow of the setup is such that the transmitter contains the motion detection unit built with Raspberry Pi 4b and the receiver contains a software defined Radio (SDR) for fine tuning the signal that is being received.

3.1 Transmitter

The Raspberry Pi has a practical hardware constraint on the transmission part because it is required to transmit in sub Giga Hertz band. In order to overcome it, a software PLL is used to adjust the base or core frequency to at least 10 times the normal frequency. And also, the transmitter has to get a sample for every 45 μs. Because of this constraint, the PWM clock frequency should be adjusted in such a way it improves the sub-sampling quality of our transmission.

With the PWM, it requires both high and low frequency band also because of the hardware constraint or the bottleneck in data transfer rate that is present on the raspberry Pi that is required to employ the DMA for transmitting the signal in sub Gigahertz band. This can attain the speed of 1X in 10 minutes of time. Here again there is a small difference from normal transmission setups. A stereo transmission i.e. Left + Right band is used, whereas the normal transmission uses the mono method.

Amplitude modulation is done only on right band and this gives us a sub-carrier in the band which is also a Double Side band suppressed carrier wave (DSB-S). The Frequency modulated (FM) wave is converted to suit the carrier and sent. Here is where most of the transmission starts. Sampling is done and these samples are

Fig. 1 Transmitter workflow

collected in a buffer in raspberry pi's memory, the clock-controlled register. Here the DMA is used to get higher frequency up and operating. These samples are sent to the input of a FIR software filter (FIR Bandpass) to limit only the maximum frequency to go through. The Transmitter workflow is explained in the illustration in Fig. 1.

3.2 Receiver

The wav file obtained from the receiver SDR kit is stored in the system and the following progress are done. The received file is demodulated using the demodulation program and the demodulated audio clipping is now converted to its corresponding BIN file or Binary file. This binary file is used for doing all the computation and file reconstruction. The binary file is then decoded using the user defined Reed Solomon decoding techniques and decoded data is logged in the binary file. The reconstruction of the image takes place and then the reconstructed image is stored in a temporary buffer. The temporary buffer is where the image processing and regeneration takes place.

After the image reconstruction the image is then pushed to the system memory by the receiver program. The forthcoming programs does necessary computation to improve the image dynamics in order to make it human readable and interpretable. The Receiver workflow is explained in an illustration in Fig. 2.

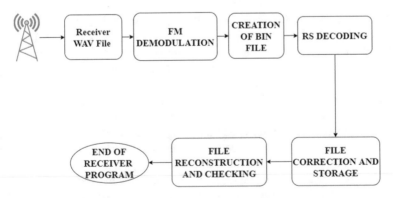

Fig. 2 Receiver workflow

4 Overall Workflow

First the camera module in the raspberry pi takes images about 14 frames per second on average and processes it, the first image of the frames is considered as background image and the subsequent frames are compared with the background frame for motion. Once an object of area above the defined limit comes into the frame, the object is detected and the object tracking algorithm moose comes into place. Minimum Output Sum of Squared Error (MOSSE) uses adaptive correlation for object tracking which produces stable correlation filters when initialized using a single frame. MOSSE tracker is robust to variations in lighting, scale, pose, and non-rigid deformations. It also detects occlusion based upon the peak-to-side lobe ratio, which enables the tracker to pause and resume where it left off, when the object reappears. There are two main components to the algorithm, they are initialization and tracking. For initialization purposes, the object contour created while object detection is sent to the moose filter.

The initialized filter is then correlated with a tracking window in the video to find the new location of the object and the tracker updates the filter as it tracks the template obtained after the pre-processing step is converted to the Fourier domain. A synthetic output is generated to initialize and update the filter. The synthetic output is also converted to the Fourier domain, and the filter is computed in the Fourier domain. The output of the correlation is converted back to the spatial domain to find the new position of the object in the next frames of the video during tracking. The correlation is performed in the Fourier domain to make the computations faster. Each tracked image is carefully analyzed for blurriness, since it's a moving object and the image can be blurry to avoid transmission of blurred images, various techniques are used to eject blur. The first method to consider would be computing the Fast Fourier Transform of the image and then examining the distribution of low and high frequencies, if there are a low amount of high frequencies, then the image can be considered blurry. However, defining when it is a low number of high frequencies and when it is a high number of high frequencies can be quite problematic, often leading to sub-par results. Some of other methods are, using just basic grayscale pixel intensity statistics, others are more advanced and feature-based, evaluating the Local Binary Patterns of an image. But the simplest method is calculating variance (standard deviation squared) of the Laplacian of the frame, if this is higher than a threshold the image can be considered as blurry.

After the image passes a blur test and has a difference metric higher than a defined threshold from the background to avoid transmitting images where the objects are in the end of frame, where only the final image will be ready. This image is then subjected to our own homemade compression algorithm to reduce the size before its transmitted. Before transmission the image is made to go through various processing to be able to convert to our transmission format. First the file is converted into symbols of eighth bit each and these symbols are attached to a leader and trailer tunes. The symbols are encoded in a two-level reed Solomon coding scheme. The symbols are then modulated by differential phase modulation with 8 different carrier frequencies generating the final tune, the tune generated is then converted to wav format for easier processing. Then the input tune is analyzed by the raspberry pi to generate the appropriate radio signals. First the bits are identified and then the data is transferred in chunks using to the PWM FIFO for excitation (the modulation technique used in Frequency modulation) since the FM transmits a DPSK modulated signal which is encoded by block reed Solomon coding its redundancy is still high. To improve performance and high frequency operation DMA is used to force data into PWM FIFO. Although the frequency output from the raspberry pi in square waves, a suitable filter would attenuate the unwanted frequencies and the desired frequency alone can be transmitted. The output from the pi alone can transmit at about 5 W which will be adequate with a reasonably good antenna to transmit signal around few 100 meters, with a good amplifier, filter arrangement the device can transmit very long ranges about kms.

Once the data is received the fm modulation is demodulated to retrieve the original signal. Then the original signal undergoes pre-processing to retrieve source clock frequency from the leader tune to properly decode the signal, then FFT of the signal is taken and the signals are sampled and quantized in software in order to get back the original bits, then the md5 checksum is done in order to calculate the correctness of data and the faulty blocks are found and reconstructed using reed Solomon decoder algorithm. Finally, the decompression by deconvolution of image with filter is done to retrieve back the original image.

As mentioned in earlier chapters and topics this setup uses a Reed Solomon coding for Forward Error Correction (FEC) and also for encryption and decryption it helps the system a lot. The light weight of the RS coding has been proved compelling after testing it multiple times and the computational expenditure by using this coding has once again proved irrefutable. The data frame format of the system that is developed is explained in Fig. 3.

The categorization of Data Frame Fromat is as follows:

- Lead modulator clock (74 symbols)
- Id1 initial mode select(mode data) (1 symbol)
- Id2 initial mode select(mode data) (1 symbol)
- Dummy frames for delimiting (4 symbol)

Lead Modulator Clock 74	Id1 1	Id2 1	Delimiting Frame 4	Starting of Information 1	Length of File Name 1	Length of File 1	File Name n	Start of Data 1	Data bit x	Mode data II 2	Trailer Bits 37

1 Symbol = 8 bits
overall 124 + n + x
symbols

Fig. 3 Data frame format

- Start of information header (1 symbol)
- Length of file name in binary(1 symbol)
- Length of file in binary big endian(1 symbol)
- File name (n symbols)
- Start of data (1 symbol)
- Data after every block 4 frames inserted for delimiter (x)
- Mode data II (2 symbol)
- Trailer bits (2 symbol).

The lead modulator clock carries the information about transmission and it is by far occupies the longest symbol space in the data frame format. This is done in order to reduce the frequency shifting and doppler shifting and other interference phenomenon at the receiver side. Id1 and Id2 are mode data that is pitched in as an extra information for the receiver when it decodes. Delimiting frames are added in order to provide a buffer space for the program to avoid errors and also marks the end of initial information frames. The start of information header is the first information that is provided to the receiver to start the decoder program up and running. After that the similar frame format used in most of the common communication applications is used. At the data bit after every data block 4 frames of delimiter are added in order to give a buffer space for avoiding errors. This data frame occupies symbols based on the size of the data. Here again dummy frames will be added to fill until the last block as required by the frame format. The second mode data or mode data II is to increase the redundancy and at the end trailer bits are added.

The LEADER is used to convey:

1. Tuning offset
2. Initial estimate of Tx clock rate, with respect to Rx clock rate
3. Inner coding scheme
4. Start of message.

The id1 and id2 are used to denote the mode of transmission and dfa is used to convey the outer coding scheme. The TRAILER is used to convey the end of message.

5 Final Deployment

The motion detection system starts working and the motion image is taken in. the image thus taken is then compressed. Once the image is compressed the transmission post processing kicks in, first the image files is converted to binary data of 0s and 1s. Then a symbol file is generated with leader bits, mode information, file name, file size and the binary data from the image followed by the mode information and trailer bits, all these data are then converted to symbol block as per the RS encoding scheme. After the two-layer RS encoding is done each of 8 symbols are used simultaneously by eight different differential phase shift keying modulators to generate the final

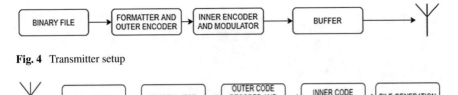

Fig. 4 Transmitter setup

Fig. 5 Receiver setup

audio file, this audio file is then sampled and stored in the memory of raspberry pi. The contents in the memory are then populated to the PWM FIFO rapidly and the PWM clock oscillator frequencies are controlled to produce desired frequencies according to the binary data received in the pin 14 of the raspberry pi. This signal is then radiated through an antenna. The receiving end consists of an antenna setup and RTL SDR, the antenna used is a dipole antenna with suitable size being or of the wavelength of radio used for maximum reception and impedance matching is ensured, the RTL SDR receives the data from antennae and converts into in phase and quadrature component signals and send it to the generate purpose processor of the pc it is connected to which samples the *i/Q* signals in very high sample rate in order to lock onto multiple frequencies, the RTLSDR SDK is used to lock onto the particular transmission frequency and receive the audio file, this audio file is then sent to demodulator program, the demodulation program first samples the input audio signals and determines the transmitter lock using the leader signals , then the new clock is used to identify the bits using 8channel DPSK demodulator, this bits are then arranged in symbols order received from the mode selected data at start of the audio file. The bits are then checked for errors and if any corrected using the reed Solomon decoder algorithm and then finally the original transmitted file is reconstructed and stored in the system. The exact workflow of the system transmitter is depicted in Fig. 4, and the receiver is elucidated in Fig. 5. This is the exact setup on the whole.

6 Experimental Results

The connections between the transmitter Pi and the receiver Pi are set up and the camera module is turned on in the transmitter pi computer. The SDR is set to run on the transmitter side for receiving. Once the framework is setup the camera is used to track motion in a confined place. Once the image is received for transmitting the image is compressed on the transmitter side.

As in Fig. 6, the compression has shown good results with a SSIM (Structural Similarity Index Mean) of 98 % and considerable reduction on file size. The output image and the original image is compared in Fig. 6 where Fig. 6a being the orig-

Fig. 6 (Left) Original Image and (Right) compression results

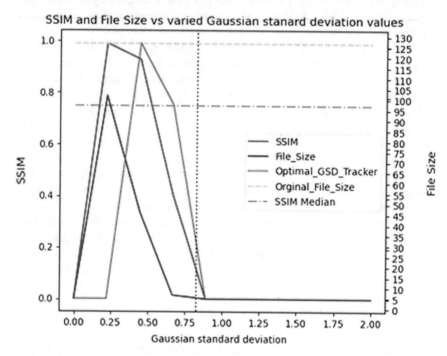

Fig. 7 Optimal tracker results

inal Image and (b) being the result of the compressed image. The metrics of the compression are represented in Fig. 7.

After compression the image data is converted into .wav format for transmitting and the transmission starts. This is depicted in Fig. 8.

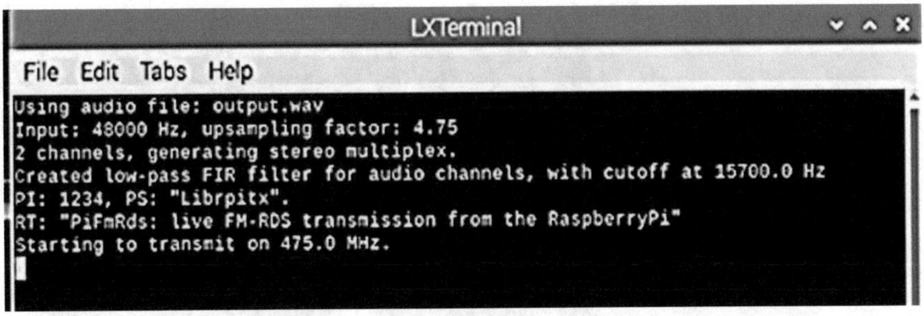

Fig. 8 Starting of transmission

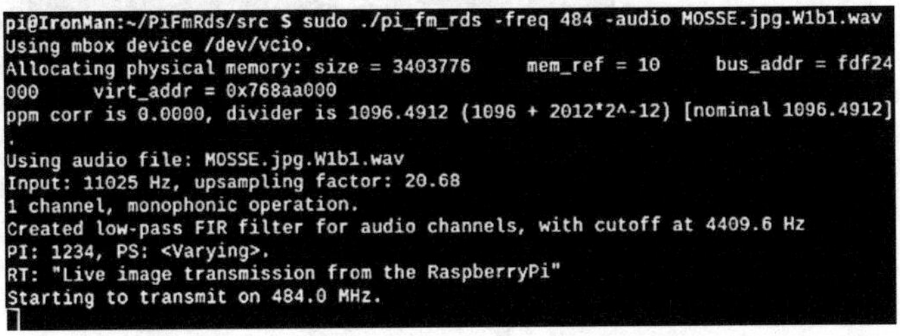

Fig. 9 Memory allocation

Once the transmission begins the memory allocation and data frame format is set for the output .wav file as depicted in Fig. 9.

The receiver side detects the start of transmission as the channel is programmed to do so and the receiver gets the preliminary start bit to start the receiving and demodulation protocol.

The image received is converted from binary to image file using the decoding algorithm and then the file can be opened by the concerned person who is monitoring. The test was done for a distance of approximately 50–100 m and the time taken for transmission of a 5 kb file in that average distance is 30 s. The image being averaged at 65 kb the time taken was approximately 390 s (including the decoding time and file creation). The file thus decoded is depicted in Fig. 10.

7 Conclusion and Future Scope

The developed model allowed surveillance i.e. Motion detection in a place where it should be critically used and without the use of Internet the captured image is thus transferred to a nearby place through wireless channel of TVWS spectrum. The transmission time is also considerably less and the loss of data in transfer is very low that stellar output images could be obtained. the downside of the findings is the transmission time taken for big files or chunky images and videos. Later

Fig. 10 Decoded image

technology advancements in this field of TVWS can facilitate even faster and can also transfer data to a distance of several kilometers altogether. Better antennae design and amplifiers can make this process easier and development of low-cost portable amplifiers can make it even better. If these API's can be made available for devices that are locally available making the setup an even better substitute for LORA or other communication protocols. This transfer is not just constricted to image but also to transfer of audio signals, messages, sensory data and if the technology is upped even could be a substitute for the currently used GSM communications because of the range flexibility and also the availability of spectrum channel.

Acknowledgements This work was supported by DST-FIST ProgrammeNo.SR/FST/College-110/2017, Government of India.

References

1. Aji LS, Wibisono G, Gunawan D (2017) The adoption of TV white space technology as a rural telecommunication solution in Indonesia. In: QiR 2017—2017 15th International conference on quality in research (QiR): international symposium on Eeectrical and computer engineering. vol 2017-Dec. Institute of Electrical and Electronics Engineers Inc., pp 479–484
2. Askhedkar A, Chaudhari B, Zennaro M, Pietrosemoli E (2020) TV white spaces for low-power wide-area networks. In: LPWAN technologies for IoT and M2M applications. Elsevier, pp 167–179 https://doi.org/10.1016/b978-0-12-818880-4.00009-0
3. Cao S, Wu CY, Krähenbühl P (2020) Lossless image compression through super-resolution. http://arxiv.org/abs/2004.02872

4. Centenaro M, Vangelista L, Zanella A, Zorzi M (2016) Long-range communications in unlicensed bands: the rising stars in the IoT and smart city scenarios. IEEE Wirel Commun 23(5):60–67
5. Development. IoT using data protocols: issues and performance. https://www.academia.edu/35593368/IoT_using_Data_Protocols_Issues_and_Performance
6. Guruswami V (2006) Algorithmic results in list decoding. Found Trends R Theor Comput Sci 2(2):107–195
7. Guruswami V, Rudra A (2005) Explicit codes achieving list decoding capacity: error-correction with optimal redundancy http://arxiv.org/abs/cs/0511072
8. Hassan DMA, Corral-De-Witt D, Ahmed S, Tepe K (2019) Narrowband data transmission in TV white space: an experimental performance analysis. In: 2018 IEEE international symposium on signal processing and information technology, ISSPIT 2018. Institute of Electrical and Electronics Engineers Inc., pp 192–196. https://doi.org/10.1109/ISSPIT.2018.8642786
9. Hernandez J, Morita H, Nakano-Miytake M, Perez-Meana H (2009) Movement detection and tracking using video frames. In: Lecture notes in computer science (including subseries lecture notes in artificial intelligence and lecture notes in bioinformatics), vol 5856 LNCS . Springer, Berlin, Heidelberg, pp 1054–1061. https://link.springer.com/chapter/10.1007/978-3-642-10268-4_123
10. Okeke C, Eng M. A comparative study between Hamming code and Reed-Solomon code in byte error detection and correction. www.ijraset.com
11. Rahman M, Saifullah A (2018) A comprehensive survey on networking over TV white spaces. Pervasive Mob Comput 59. arxiv.org/abs/1810.07120, https://doi.org/10.1016/j.pmcj.2019.101072
12. Reed IS, Solomon G (1960) POLYNOMIAL CODES OVER CERTAIN FINITE FIELDS* A code is a mapping from a vector space of dimension m over a finite field K (denoted by V,(K)) into a vector space of higher. Technical report 2. http://www.siam.org/journals/ojsa.php
13. Siddharth R, Aghila G (2020) A light weight background subtraction algorithm for motion detection in fog computing. IEEE Lett Comput Soc 3(1):17–20. https://doi.org/10.1109/locs.2020.2974703
14. Weinberger MJ, Seroussi G, Sapiro G (1996) LOCO-I: a low complexity, context-based, lossless image compression algorithm. In: Data compression conference proceedings. IEEE, pp 140–149 (1996)
15. Zhang W, Yang J, Guanglin Z, Yang L, Yeo CK (2018) TV white space and its applications in future wireless networks and communications: a survey

Classification of Fetal Heart Ultrasound Images for the Detection of CHD

T. V. Sushma, N. Sriraam, P. Megha Arakeri, and S. Suresh

Abstract Congenital Heart Disease (CHD) is the major cause of infant mortality accounting to about 28% of all congenital defects, thereby emphasizing the need for its early detection. Ultrasound (US) imaging modality is widely used in prenatal screening for monitoring the growth of the fetus. Clinically determining the abnormality is tedious and time consuming and depends on the expertise of the radiologist. Automated recognition of CHD from ultrasonic 2D imaging is proposed in this study which makes use of statistical features and pattern classifier such as support vector machine. Cineloop sequences with different abnormalities have been used for training. Features are classified using fine Gaussian and medium Gaussian kernels of SVM classifier with tenfold cross-validation resulting in an accuracy of 90% and 91.25%, respectively. The sensitivity obtained during the simulation is 90% and 92.5% for fine Gaussian and medium Gaussian, respectively, while the specificity is 90% for both the classifiers.

Keywords Congenital heart disease · 2D echocardiograph images · GLCM · SVM classifier

T. V. Sushma (✉) · P. Megha Arakeri
Center for Imaging Technologies, MSRIT, Bangalore, India
e-mail: sushtv@gmail.com

T. V. Sushma
S.J.C Institute of Technology, Chickaballapur, India

N. Sriraam
Center for Medical Electronics and Computing, MSRIT, Bangalore, India
e-mail: sriraam@msrit.edu

P. Megha Arakeri
Department of Information Science, MSRIT, Bangalore, India

S. Suresh
Mediscan Systems Pvt. Ltd., Chennai, India

© The Author(s), under exclusive license to Springer Nature Singapore Pte Ltd. 2021
J. S. Raj et al. (eds.), *Innovative Data Communication Technologies and Application*,
Lecture Notes on Data Engineering and Communications Technologies 59,
https://doi.org/10.1007/978-981-15-9651-3_41

1 Introduction

Congenital heart disease refers to the occurrence of heart defects at birth in the form of structural or functional abnormality. Early detection of CHD is of atmost importance since it is reported to be one of the major causes of fetal mortality [1–3]. About 43,000 children were reported to be suffering from serious defects in 2018 with only one-fourth of them receiving optimal care [4]. Early detection of CHD allows taking corrective measures either in utero or helps in planning the delivery appropriately increasing the chances of survival. Undetected cases result in live births or result in serious complications within a few years of birth. Many studies showing the importance of prenatal screening conducted over different population have been reported in the literature [4–6]. Ultrasound imaging modality has been widely used for monitoring the growth of the fetus and detecting any abnormalities. It is a preferred tool for prenatal screening as it is non-invasive, radiation free, and cost effective. Although tremendous advancement has been made in ultrasound such as 3D/4D ultrasound, Doppler ultrasound, 2D ultrasound are still widely being used as a prenatal screening tool because of its reduced computational complexity.

However, interpretation of the ultrasound images is a challenging task for radiologists since the ultrasound images are characterized by speckle noise resulting in low SNR value of the images. The effects of shadowing also make it difficult to delineate the boundaries of the underlying organs. Hence, identifying the organs and detecting any abnormalities rely greatly on the expertise of the radiologist. This has prompted a great focus on developing computer aided diagnostic systems and automated techniques that assist the doctors in diagnosis minimizing the chances of false predictions.

2 Related Background

Vijayalakshmi et al. [7], have proposed an automated technique for extracting the fetal heart chamber by using a region mask. Deng et al. [8] have proposed an automated technique for the detection of fetal structures in ultrasound images. Balaji et al. [9] have proposed a method for detecting damage to the heart muscles by automated analysis of the video. Work on automated detection of the plane of the view has also been reported. The recent trend of automated analysis in cardiac imaging using different imaging modalities has been discussed by Slomkaa et al. [10]. With the development of robust machine learning technologies, many researchers have proposed the development of computer aided diagnostic system to assist doctors for detection of abnormality in ultrasound images. Huang et al. [11] have presented a detailed survey of the use of machine learning in CAD systems developed for ultrasound images. Vidhi et al. have presented a review of the automated techniques for the detection of fetal abnormalities [10]. Automated analysis of fetal heart video which is an important initial step in the detection of CHD has been demonstrated by Bridge et al. [11] Availability of a larger database has prompted the use of deep learning such as Convolution Neural Networks (CNN) which is noted to provide promising results. Identifying the position of the fetal heart and the plane of view is a challenging task since a slight change in the position of the probe or relative movement between the fetus and the probe alters the view altogether. Many researchers have proposed automated view identification techniques using CNN [12–17]. The use of CNN for the automated analysis of ultrasound images has been demonstrated by Sridar et al. [18]. They have demonstrated the automated classification of B mode fetal ultrasound images from images of other regions and other modes of ultrasound images. Further, they have developed fetal biometry that serves as an input for further clinical research. The accuracy values for the different cases have been listed in Table 1. The use of deep learning model that demands the availability of large dataset can be overcome by the use of transfer learning as demonstrated by Gao et al. [19]. Table 1 lists some of the work carried out using machine learning algorithms on ultrasound images for the detection of abnormality.

3   Material and Methods

2D ultrasound cineloop sequences of fetal heart with gestation age ranging from 21 to 27 weeks with a frame rate of 25 fps and an average duration of 1 min were used. Four chamber apical view of fetal heart was used for analysis. The cineloop sequences were provided by Mediscan Pvt. Ltd., Chennai. 80 echocardiographic frames with 40 normal and 40 abnormal frames were used for the study. The normal videos had distinct four chambers while the abnormal videos included different fetal cardiac defects, such as the presence of only two chambers, additional growth in the right ventricle, septal defect, and the presence of only three chambers.

The cineloop sequences are converted into frames. The frames converted were stored in.png format. Ten frames of each cineloop sequence in different stages of the cardiac cycle were used for the study. Figure 1 shows the proposed CHD recognition using pattern classifier.

3.1   Preprocessing

The frames extracted were initially converted to grayscale. Since the fetal heart occupies only a small portion of the frame, the labels corresponding to patient information was removed, and the region occupying the four chambers of the heart was obtained by manually cropping the frames. The frames were then resized to a dimension of 256×256. Figure 2 summarizes the steps carried out in preprocessing.

Ultrasound images are in general characterized by speckle noise, and preprocessing plays an important role in delineating the chambers accurately. The anisotropic diffusion filter is applied to the resized frames followed by Laplacian

Table 1 Details of related work

Author	Dataset	Classifier	Accuracy	Observations
Sridar et al. [20]	4074 2D ultrasound images	CNN	B mode image classification Accuracy = 97.05%	Fetal structures are localized based on classification outcome
Narula et al. [21]	77 physiological hypertrophy in athletes, 66 hypertrophic cardiomyopathy	Ensemble model—ANN, SVM, RF)	Sensitivity—87% Specificity—82%	Images of only specific clinical characteristics were used
Lao et al. [22]	3381 with 78 cases of CHD	(i) Support vector machine (SVM) (ii) Random forests (RF) (iii) Logistic regression (LR)	SVM—94.76% RF—92.89% LR—98.06%	Classification of CHD has been performed on survey data collected
Balaji et al. [9]	Dataset of 60 patients consisting of both normal hearts and abnormal hearts	(i) PCA + BPNN (ii) PCA + KNN (iii) PCA + SVM (iv) DCT + BPNN (v) DCT + KNN (vi) DCT + SVM	(i) 86.66% (ii) 83.33% (iii) 80% (iv) 80% (v) 82% (vi)77.14%	Accuracy of the classification depends on the segmentation of LV region
Yue et al. [23]	800 ultrasound images	CNN	Poor segmentation quality has improved to 2.39% from 3.48%	The proposed method aids in segmentation using CNN

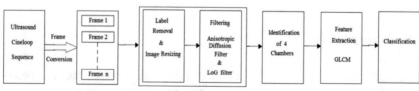

Fig. 1 Block diagram of the proposed classification technique

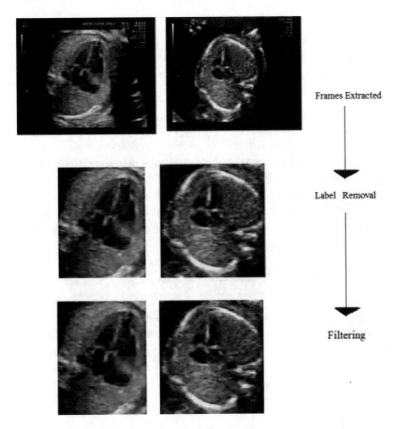

Frames Extracted

Label Removal

Filtering

Fig. 2 Pre-processing procedure

of Gaussian (LoG) filter which highlights the edges. This resulted in better delineation of the chamber boundaries visually. The application of the filtering operation results in the PSNR value 26.87 dB.

3.2 Feature Extraction

Tremendous research has been carried out toward image pattern classification with very good classification accuracies. Feature extraction plays a very important role in the image classification [24, 25]. Gray Level Co-occurrence Matrix (GLCM) [26, 27] is one of the well-known texture descriptors that are used for extracting second-order statistical texture features of images. It gives a measure of a variation in the intensity relative to each pixel. The features are determined by using the co-occurrence matrix which gives the frequency of occurrence of intensity values of pairs of pixels that are separated by a specific distance.

The following are the features extracted.

Contrast—It indicates the variation of the intensity present in the image. A value of zero indicates that an image is a uniform while a high value indicates the presence of edges and noise in the image. It is given by the following equation

$$\text{Contrast} = \sum_{i=0}^{G-1}\sum_{j=0}^{G-1}(i-j)^2 P_{i,j} \tag{1}$$

where G is the number of gray levels in the image represented with indices (i, j), and $P_{i,j}$ is the element i, j of the normalized GLCM matrix.

Correlation—It gives a measure dependency of a pixel to its neighboring pixels, i.e., it indicates how correlated the pixels are with each other. It is given by the following equation. A high value occurs if the pixels are uniformly distributed

$$\text{Correlation} = \sum_{i=1}^{G}\sum_{j=1}^{g}\frac{(ij)P(i, j)-\left(\mu_x \times \mu_y\right)}{\sigma_x \times \sigma_y} \tag{2}$$

where μ_x, μ_y represent the mean and the standard deviation of the row and the column sums of the GLCM matrix, respectively, and are given by the following equations.

$$\mu_x = \sum_{i=0}^{G-1} i P_x(i) \tag{3}$$

$$\mu_y = \sum_{j=0}^{G-1} j P_y(j) \tag{4}$$

σ_x, σ_y represent the mean and the standard deviation of the row and the column sums of the GLCM matrix, respectively,

$$\sigma_x^2 = \sum_{i=0}^{G-1}(P_x(i) - \mu_x(i))^2 \tag{5}$$

$$\sigma_y^2 = \sum_{j=0}^{G-1}\left(P_y(j) - \mu_y(j)\right)^2 \tag{6}$$

Entropy—It represents the amount of disorder in pixels of the image and depends inversely related to the uniformity of the image

$$\text{Entropy} = -\sum_{i=0}^{G-1}\sum_{j=0}^{G-1} P(i, j)\log(P(i, j)) \tag{7}$$

Homogenity—It is a measure of the smoothness of the image and varies inversely as the contrast

$$\text{Homogenity} = \sum_{i=1}^{G}\sum_{j=1}^{G} \frac{P(i, j)}{1 + (i - j)^2} \tag{8}$$

Sum Average—It is equal to the mean of the sum of the gray levels distributed in the image and is given by the following equation.

$$\text{Sum Average} = \sum_{i=0}^{2G-2} i p_{x+y}(i) \tag{9}$$

where

$$p_{x+y}(k) = \sum_{i=1}^{G} \Sigma_{j=1}^{G} p(i, j), \quad k = 2, 3, 4, \ldots 2G \tag{10}$$

Sum Entropy—It indicates the disorder related to the distribution of the gray level sum in the image.

$$\text{Sum Entropy} = - \sum_{i=2}^{2G} p_{x+y}(i)\log\{p_{x+y}(i)\} \tag{11}$$

Autocorrelation—This property indicates the repetitive pattern in the texture of the image, i.e., indicates the fineness or roughness of the image

$$\text{Autocorrelation} = \sum_{i=0}^{G-1}\sum_{j=0}^{G-1}(ij)p(i, j) \tag{12}$$

Cluster prominence and **cluster shade** are a measure of the skewness of the matrix with a high value indicating that the image is asymmetric

$$\text{Cluster Prominence} = \sum_{i=0}^{G-1}\sum_{i=0}^{G-1}(i + j - \mu_x - \mu_y)^4 . P(i, j) \tag{13}$$

$$\text{Cluster Shade} = \sum_{i=0}^{G-1}\sum_{j=0}^{G-1}(i + j - \mu_x - \mu_y)^3 . P(i, j) \tag{14}$$

where μ_x and μ_y are given by Eqs. (3) and (4).

4   Results and Observations

The statistical features derived using GLCM for normal and abnormal video frames are displayed using box plots as shown below. It is observed that the values of the features mostly overlap for both normal and abnormal images, however, the range being different with variation in the mean value.

Figure 3 shows that the mean value of contrast is higher for normal images as compared to abnormal images while the range is higher for abnormal frames.

Figure 4 indicates that the range and the mean value of correlation for abnormal images are higher compared to normal images

The mean entropy value as shown in Fig. 5 between normal and abnormal images is close to each other while the range is higher for abnormal images. Figure 6 indicates that mean value of the homogenity is higher for abnormal images as compared to normal images.

The sum average and sum entropy are represented by Figs. 7 and 8, respectively. The mean value of the abnormal images is found to be close to that of the normal images in both the cases while the range is larger for abnormal as compared to normal images.

The autocorrelation between normal and abnormal images as shown in Fig. 9 have a mean value close to each other indicating that both normal and abnormal images have a repetitive pattern. For abnormal images, this depends on the nature of the abnormality considered.

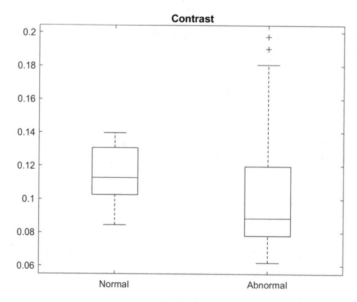

Fig. 3 Contrast between normal and abnormal frames

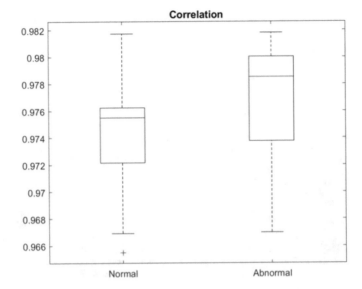

Fig. 4 Correlation between normal and abnormal frames

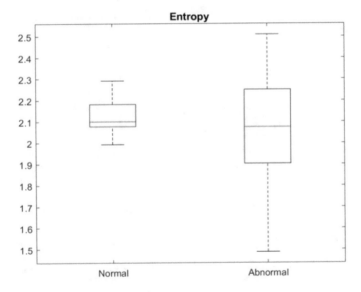

Fig. 5 Entropy between normal and abnormal frames

The cluster prominence and cluster shade are represented by Figs. 10 and 11, respectively. The mean value of cluster prominence is close to each other for normal and abnormal images. In case of cluster shade, as shown in Fig. 11, the mean value for normal images is more compared to as abnormal images; however, the range of

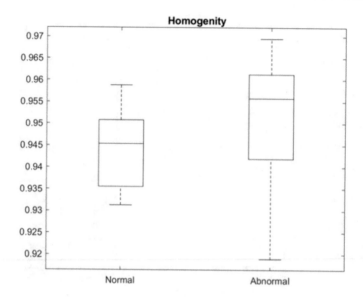

Fig. 6 Homogenity between normal and abnormal frames

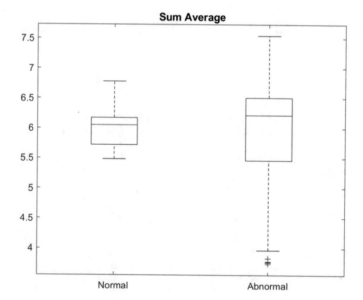

Fig. 7 Sum average between normal and abnormal frames

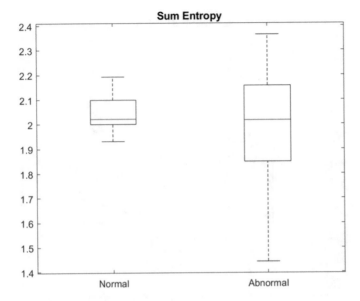

Fig. 8 Sum entropy between normal and abnormal frames

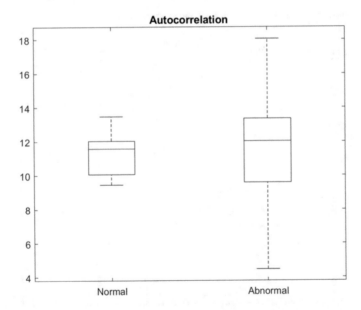

Fig. 9 Autocorrelation between normal and abnormal frames

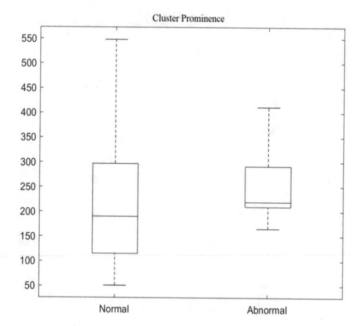

Fig. 10 Cluster prominence between normal and abnormal frames

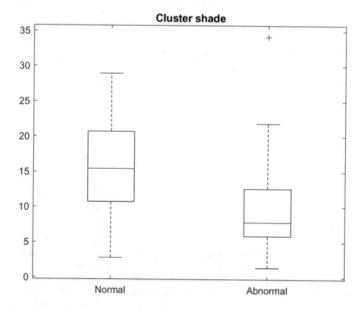

Fig. 11 Cluster shade between normal and abnormal frames

values is higher for normal images as compared to abnormal images for both the features.

4.1 Performance Metric

The Support Vector Machine (SVM) classifier is one of the widely used machine learning algorithms used for data classification [28, 29]. It determines discriminating hyperplanes between the different classes such that there is maximum separability between the classes. There are different types of SVM classifiers based on the hyperplane which separates the classes. The Gaussian kernel in SVM classifier is also known as the Radial Basis Function (RBF) kernel usually used to solve non-linear classification problems. In the present case, different kernels of SVM classifier, fine Gaussian kernel, and medium Gaussian kernel are used.

The classification is performed by using tenfold cross-validation in which the dataset is divided into groups of ten, and in each iteration, onefold is used for training, and the other folds are used for testing. The same process is repeated for all the folds, and thereby, all the observations are used for training as well as testing.

The confusion matrix gives a visual a performance of the classifier. The four quadrants of the confusion matrix are True Positive (TP), True Negative (TN), False Positive (FP), and False Negative (FN). The confusion matrix obtained using the two classifiers are shown in Tables 2 and 3.

The performance of the classifier can be computed using the confusion matrix using the following equations [30]

$$\text{Sensitivity} = \frac{TP}{TP + FN} \tag{15}$$

$$\text{Specificity} = \frac{TN}{TN + FP} \tag{16}$$

Table 2 Confusion matrix obtained using medium gaussian SVM classifier

True class	Predicted class	
	Normal	Abnormal
Normal	36 TP	4 FN
Abnormal	4 FP	36 TN

Table 3 Confusion matrix obtained using fine Gaussian SVM classifier

True class	Predicted class	
	Normal	Abnormal
Normal	37 TP	3 FN
Abnormal	4 FP	36 TN

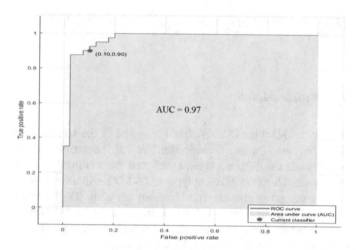

Fig. 12 AUC-ROC of fine Gaussian SVM classifier

$$\text{Accuracy} = \frac{\text{TN} + \text{TP}}{\text{TN} + \text{FN} + \text{TP} + \text{FP}} \tag{17}$$

$$\text{Error Rate} = \frac{\text{FN} + \text{FP}}{\text{TN} + \text{FN} + \text{TP} + \text{FP}} \tag{18}$$

The performance metrics of the two classifiers are as tabulated in Table 4.

The area under the curve—Receiver Operator Characteristics (ROC) curve is an important performance metric of the classifier. It is a plot of the True Positive Rate v/s the False Positive Rate and indicates the measure of separability of the classifier. In other words, it indicates how well the classifier can classify normal and abnormal cases. A good classifier has an AUC close to 1. Both the classifiers have AUC close to 1, with 0.97 for fine Gaussian kernel and 0.96 for medium Gaussian kernel SVM classifier as shown in Figs. 12 and 13, respectively.

Table 4 Confusion matrix of medium Gaussian SVM classifier

Performance metric	Fine Gaussian SVM classifier (%)	Medium Gaussian SVM classifier (%)
Sensitivity	90	92.5
Specificity	90	90
Accuracy	90	91.25

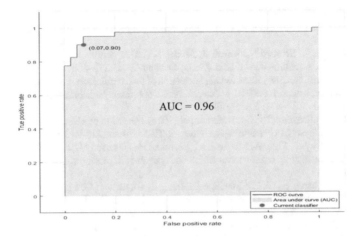

Fig. 13 AUC-ROC of medium Gaussian SVM classifier

5 Conclusion

The proposed study presents the use of statistical features extracted using GLCM for the classifying between normal and abnormal fetal heart. A dataset of 80 frames with gestation age between 21 and 27 weeks was used. This included 40 normal frames and 40 abnormal frames. Although there is overlap in the range of the feature values of normal and abnormal images, the fine Gaussian and medium Gaussian kernels of SVM classifiers, give appreciable accuracy values equal to 90% and 91.25%, respectively. As future work, the classification can be done by considering a larger and more heterogeneous dataset using different classifiers. Further, the effect on classification using different pre-processing techniques can be tested.

Acknowledgements We would like to thank Dr. Suresh, Mediscan Systems Pvt. Ltd., Chennai for providing the required ultrasound dataset for carrying out the work.

References

1. Aoki-Kamiya, C.: Congenital Heart Disease. Maternal and Fetal Cardiovascular Disease, pp. 163–177, Springer (2019)
2. Bravo-valenzula NJ, Peixoto AB, Junior EA (2018) Prenatal diagnosis of congenital heart disease: a review of current knowledge. Indian Heart J. 70:150–164
3. Chitra N, Vijayalakshmi IB (2017) Fetal echocardiography for early detection of congenital heart diseases. J. Echocardiogr. 15(1):13–17
4. Saxena A (2018) Congenital heart disease in India: a status report. Indian Pediatr. 55:1075–1082
5. Crispín Milart PH, Prieto-Egido I, Díaz Molina CA, Martinez-Fernandez A (2019) Detection of high-risk pregnancies in low-resource settings: a case study in Guatemala. Reprod. Health 16:80

6. Van Velzen, C.L., Clur, S.A., Rijlaarsdam, M.E., Bax, C.J., et. al.: Prenatal detection of congenital heart disease-results of a national screening programme. BJOG. **123**(3), 400–407 (2016)
7. Vijayalakshmi S, Sriraam N, Suresh S, Muttan S (2013) Automated region mask for four-chamber fetal heart biometry. J. Clin. Monit. Comput. 27(2):205–209
8. Deng, Y., Wang, Y., Chen, P.: Automated detection of fetal cardiac structure from first-trimester ultrasound sequences. In: 2010 3rd International Conference on Biomedical Engineering and Informatics, pp. 127–131. Yantai (2010)
9. Balaji GN, Subhashini TS, Chidambaram N (2015) Detection of heart muscle damage from automated analysis of echocardiogram video. IETE J. Res. 61:236–243
10. Slomka PJ, Dey D, Sitek A, Motwani M, Berman DS, Germano G (2017) Cardiac imaging: working towards fully-automated machine analysis & interpretation. Expert Rev. Med Dev. 14(3):197–212
11. Huang, Q., Zhang, F., Li, X.: Machine Learning in Ultrasound Computer-Aided Diagnostic Systems: A Survey. Biomed Research International Hindwai (2018)
12. Rawat, V., Jain, A., Shrimali, V.: Automated Techniques for the Interpretation of Fetal Abnormalities: A Review Applied Bionics and Biomechanics. Hindwai (2018)
13. Bridge CP, Ioannou C, Noble JA (2017) Automated annotation and quantitative description of ultrasound videos of the fetal heart. Med Image Anal. 36:147–161
14. Chen, H., Dou, Q., Ni, D.,Cheng, J.Z., Qin, J., Li, S., Heng, P.A.: Automatic fetal ultrasound standard plane detection using knowledge transferred recurrent neural networks. In: Medical Image Computing and Computer-Assisted Intervention—MICCAI 2015, Lecture Notes in Computer Science, vol 9349. Springer (2015)
15. Baumgartner, C.F., Kamnitsas, K., Matthew, J., Fletcher, T.P., Smith, S., Koch, L.M., Kainz, B., Rueckert, D.: SonoNet: Real-time detection and localisation of fetal standard scan planes in freehand ultrasound. IEEE Trans. Med. Imaging **36**(11), 2204–2215 (2017)
16. Kumar, A., Sridar, P., Quinton, A., Kumar, R.K., Feng, D., Nanan, R, Kim, J.: Plane identification in fetal ultrasound images using saliency maps and convolutional neural networks. In: 2016 Proceedings of IEEE 13th International Symposium Biomedical Imaging (2016)
17. Sundaresan, V., Bridge, C.P., Ioannou, C., Noble, J.A.: Automated characterization of the fetal heart in ultrasound images using fully convolutional neural networks. In: 2017 IEEE 14th International Symposium on Biomedical Imaging (2017)
18. Sridar, P., Kumar, A., Quinton, A.E., Kumar, R.K., Feng, D., Nanan, R, Kim, J.: Automatic identification of multiple planes of a fetal organ from 2D ultrasound images. In: MICCAI Workshop, Preterm Paediatric Image Analysis, pp. 1–10 (2016)
19. Gao, Y., Maraci, M., Noble, J.A.: Describing ultrasound video content using deep convolutional neural networks. In: 2016 IEEE International Symposium Biomedical Imaging , pp. 787–790. Prague (2016)
20. Sridar P, Kumar A, Quinton A, Nanan R, Kim J, Kumar RK (2019) Decision fusion-based fetal ultrasound image plane classification using convolutional neural networks. Ultrsound Med Biol 45(5):1259–1273
21. Narula S, Shameer K, Omar S, Dudley JT, Sengupta PP (2016) Machine-learning algorithms to automate morphological and functional assessments in 2D echocardiography. J. Am. Coll. Cardiol. 68(21):2287–2295
22. Lao Y, Li Z, Guo H, Cao H, Song C, Guo X, Zhang Y (2017) Predicting congenital heart defects: a comparison of three data mining methods. PLoS ONE 12(5)
23. Yue, Z., Li, W., Jing, J., Yu., J.,Yi, S.: Automatic Segmentation of the Epicardium and Endocardium using convolutional neural network. In: 2016 IEEE 13th International Conference on Signal Processing (2016)
24. Popescu, M.C., Sasu, L.M.: Feature extraction, feature selection and machine learning for image classification: a case study. In: 2014 International Conference on Optimization of Electrical and Electronic Equipment (2014)
25. Guo, J., Liu, L., Song, W., Du, C., Zhao, X.:The study of feature extraction and classification. In: 2017 International Conference on Progress in Informatics and Computing (2017)

26. Singh, S., Srivastava, D., Agarwal, S.: GLCM and its application in pattern recognition. In: 2017 5th International Symposium on Computational and Business Intelligence (2017)
27. Humeau-Heutier A (2019) Texture feature extraction methods: a survey. IEEE Access 7:8975–9000
28. Evgeniou, T., Pontil, M.: Support Vector Machines: Theory and applications In Machine Learning and Its Applications. Advanced Lectures (2001)
29. Cervantes, J., Garcia-Lamont, F., Rodriguez-Mazahua, L., Lopez, A.: A comprehensive survey on support vector machine classification: Applications, challenges and trends. Neurocomputing (2020)
30. Baratloo A, Hosseini M, Negida A, Ashal GE (2015) Part 1: simple definition and calculation of accuracy. sensitivity and specificity. Emergency 3(2):48–49

COVID-19 Risk Management: A Mobile Application to Notify Alerts and Provide Real-Time COVID-19 Updates

Jannaikode Yashwanth Kumar and Nenavath Srinivas Naik

Abstract The sudden outbreak of coronavirus has left people unanswered till date. COVID-19 virus cases are drastically increasing day by day. The only way to protect ourselves from COVID-19 is by maintaining social distancing and taking preventive measures. A mobile application-based solution is proposed for risk management of COVID-19. The objective of the proposed model is to help people with the status of cases in their region. The proposed model notifies the user regularly about the COVID-19 cases in nearby locations which help him/her to be cautious. Android studio and Google Firebase are used to develop the mobile application. COVID-19 information and state-wise cases are updated in the database regularly, and users are notified. The results are good in terms of response time and delivery of static alerts.

Keywords COVID-19 · Android studio · Geo-fencing · Google Firebase · Mobile application

1 Introduction

Coronavirus outbreak has a drastic effect on the lives of people to date [1, 2]. The recent COVID-19 pandemic poses a major threat to our society and way of life. To date, hundreds of thousands of people have died, and millions infected [3]. People with high immunity are least affected by this disease. People aged below 10 and above 60 are reported to be more vulnerable to disease from recorded statistics. WHO has declared the COVID-19 pandemic as a global health emergency [4]. Cleanliness and nutritious diet that provide immunity help to prevent the individual from the attack

J. Y. Kumar
Department of Information Technology, CVR College of Engineering, Hyderabad, India
e-mail: yashwanthkumar.jannaikode@gmail.com

N. S. Naik (✉)
Department of Computer Science and Engineering, IIIT Naya Raipur, Naya Raipur, Chhattisgarh, India
e-mail: srinu@iiitnr.edu.in

© The Author(s), under exclusive license to Springer Nature Singapore Pte Ltd. 2021
J. S. Raj et al. (eds.), *Innovative Data Communication Technologies and Application*,
Lecture Notes on Data Engineering and Communications Technologies 59,
https://doi.org/10.1007/978-981-15-9651-3_42

of the disease [5]. People nowadays are very much familiar with mobile applications. Mobile applications play a vital role in making things aware. Mobile apps are user-friendly, compared to web-based applications. People can access the desired information in a short time. Mobile applications come handy and are time-saving [6].

The mobile application helps to track registered devices so that users can know the vulnerable areas [7]. Likewise, mobile applications provide people to have a clear knowledge of the pandemic outbreak around them. As the number of cases is rapidly increasing throughout India, a mobile application that helps the user to be cautious which will surely help in a great way by alerting the users. Applications are suggested as mandatory resources by few governments which allow better contact tracing to ensure the safety of people [8].

Information on daily happenings would make people be more cautious and make them prepared for further consequences. Mobile applications provide easier access to required information on a topic in less time, unlike web-based applications. The proposed model alerts the user on entering and exiting the contained areas thus making people more cautious. There is a misconception in many people stating that the virus spreads through waves/mobile networks. Viruses cannot travel on radio waves/mobile networks. COVID-19 is spreading in many countries that do not have 5G mobile networks [9].

The rest of the paper is organized as follows. In Sect. 2, the literature survey of existing applications, and their limitations are described. The proposed mobile-based application for COVID-19 is described in Sect. 3. Section 4 presents the results and analysis of the proposed model. Section 5 concludes this paper with the future scope of the proposed model.

2 Literature Survey

Currently, there are mobile apps in the market that do provide information on COVID-19 to make people aware of this disease. Many states in India have come up with their state mobile applications to provide updates and information on disease. Tamil Nadu, Kerala, Punjab, Maharashtra, Mizoram states released mobile apps to ensure the safety of people during this pandemic [10]. Goa has come up with 'Test Yourself Goa' which helps the user to check whether he/she is prone to disease based on self-assessment tests. Besides giving data manually by handling a database, there are few API that provides COVID-19 updates [11]. The API's provide worldwide data and state-wise data of countries all over the world.

Existing applications have features like updates of newly occurred cases, state-wise cases reports, and information related to COVID-19 symptoms and preventive measures. Few applications are designed for quarantined people to follow a proper diet, they are accessible only to those who got registered with the government on testing positive. States like Maharashtra, Kerala, Karnataka have launched an application for people who are tested positive. The government has registered their mobile number of positively tested to provide health care facilities [12].

Besides applications from different states in India, the central government has launched a mobile application named 'Arogya Setu'. The application provides the status of COVID-19 and also displays the number of COVID-19 cases in their vicinity based on tests taken by the user. This application works with the help of Bluetooth. It tells how many COVID-19 positive cases are likely in a radius of 500 m, 1, 2, 5 and 10 km from the user [13]. The main drawback of this is the user may or may not give correct data in the self-assessment tests. 'Aarogya Setu' application is made mandatory by the government in the upcoming version of mobiles. People are asked to download the application by the government to know their health status by taking the self-assessment test and also to get the alerts of nearby vulnerable places.

Not only in India, but many countries all over the world have also come up with solutions to face COVID-19 pandemic efficiently. Many applications are, especially based on self-assessment and preventive measures. There is an issue of data privacy at stake; most of the app developers are now trying to put mechanisms in place to safeguard user privacy [14]. Privacy concerns have been raised, especially about systems that are based on tracking the geographical location of application users [15]. Most of the applications are intended to trace the location of people which denies the privacy of the user. The data tracked may be used for different purposes that relate to the privacy of the application user. Providing privacy has to be a primary attribute in an application. Besides mobile applications, there are websites to display information about the disease.

Web-based applications are not user-friendly for searching the data related to COVID-19. Mobile apps are better as they go with the notion of 'one click and you are there'. Mobile applications have received great demand during this pandemic to ensure the safety of people. China, South Korea and Singapore have led the way in developing contact tracing systems [16]. Mobile apps help people to access data in less time and are more convenient. Hence, a mobile application is developed to address user problems and provide an efficient platform in accessing data about disease and safeguarding their health. A software engineer and his team from Telangana had developed a web-based application which informs potential risk and symptoms through web application contact tracing and equips people with self-assessment tools [17].

3 Proposed Model

A novel model is proposed to face the pandemic efficiently by safeguarding the privacy of a user and providing real-time updates and alerts. The proposed mobile-based application ensures the privacy of the user as it does not take location details of the user into the database. It notifies the user when the user enters into the contained area and exits the contained area. Containment area or contained region is a vulnerable place where more cases are registered also called the disease-prone area. This area has the probability of spreading viruses compared to normal regions. The proposed mobile application provides the following data

Basic information about COVID-19
Cases in India and state-wise status
Media bulletin and government orders
Health facilities address
Contained regions on Google maps
Emergency contacts.

3.1 Approach Towards Problem

Firebase Database and Firebase Authentication are used to store data and authenticate users, respectively. With the help of Google drive, the latest versions of data are kept available to the user. This helps users to get information in one click as there are not many steps to avail it. Easy access makes the user comfortable to get through the details.

Initially, the user registers by entering basic details and verifies his/her mail. The user has to sign in to avail of the data. On updating the data in the database, the user gets a notification alert in background. Also, have the option of in-app messaging which alerts users while using the application. The status of cases is updated regularly in Firebase. Control room contacts and helpline numbers of states are made available, and one click can make the user call the required department. Static notifications are triggered manually on updating data by the admin, and dynamic notifications are triggered on entering and leaving the containment zones. Static notifications are triggered using Firebase cloud messaging, and dynamic notifications are triggered with the help of notification manager in android studio.

3.2 Geo-fencing

The feature geo-fencing is used to alert the people on entering contained areas. This works on opening the respective fragment in the application. Whenever the user enters the contained areas, notification manager immediately triggers, and the user gets notification stating 'You are in containment area', thus making the user more cautious. The containment regions are updated using Firebase by entering latitude and longitude of the location in the database. Google maps API is used to get the accurate location of the user. The applications of geo-fencing are many depending upon the project domain on which developer works [18]. Besides coding for geo-fencing, there are few API available that costs us as per requirements. One of them is TOMTOM, which is intended to define virtual barriers in real geographical locations. Together with the location of an object, whether that object is located within can be determined, outside, or close to a predefined geographical area. Customer, admin, requester, object are the roles created in this API [19].

3.3 Implementation

In Fig. 1, the user registers in the application and verifies his/her email which is mandatory. User requests for the required data, and the application retrieves the data from the database and displays on the application screen. Notifications are triggered by entering the containment areas.

Figure 2 shows the block diagram of the functioning of the application. The feature of Firebase in-app messaging is used which notifies the user on opening the application. In-app messaging enables one to alert the user on updates as well as navigate to a particular website in one click. In-app messaging can used to take the impressions of a particular activity, it can be pre-scheduled and it can be used in different layouts like a card, banner, etc. [20].

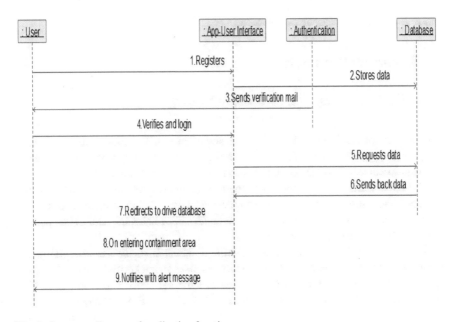

Fig. 1 Sequence diagram of application functions

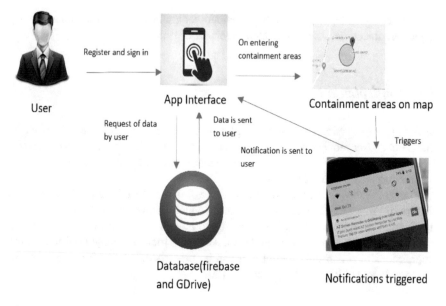

User

App Interface

Containment areas on map

Database(firebase and GDrive)

Notifications triggered

Fig. 2 Block diagram of application functionality

4 Results and Analysis

4.1 Application Design and Their Functionality

The mobile application user interface is as shown in Fig. 3. It shows the login screen (A), register screen (B) which take registration details. The homepage (C) has basic information about COVID-19, its symptoms and preventive measures.

Dynamic alerts are triggered when the user enters the containment zones. When the user enters into the red zone, he/she receives a notification as illustrated 'C' part of Fig. 4. Users can check other zones as well (B). This helps the user to be cautious if he/she travels through those areas.

In Fig. 5, A shows the scrolling real-time updates; B shows the emergency and important contacts; this screen also avails the user to make direct calls from the application on clicking on the number. C shows the manually triggered notification on the update of data in the app. This is done with the help of Firebase cloud messaging service [21]. On clicking the notification, user is directed to the application home screen if the user is already logged in.

Figure 6 shows different fragments of the mobile application; A shows drawer navigator which can be availed by clicking a button on the toolbar or swiping towards the right. In Fig. 6, B shows the health facility address and contacts, the status of COVID-19 cases in India and Telangana state can be seen in C. List of state-wise COVID-19 cases can be accessed from that screen (C). There are a few more fragments to facilitate all the information to the user. The feedback screen is made

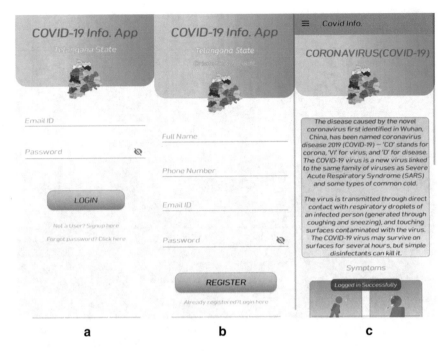

a   b   c

Fig. 3 Screenshots of the application interface

available to take suggestions and ratings from the user. Storage, download and load analytics of project can be acquired from Firebase console [22].

4.2 Analytic Reports

Firebase generates analytics based on user activity on the application with the help of Firebase authentication. Reports generated in Firebase can be accessed by the developer to monitor the performance of the application.

Figure 7 shows the downloads of data by the user, i.e. fetching of data from the database while using the application. Another graph shows the current storage of the database. The reports also depict the increase in statistics week over week.

The administrator knows the user's impression on notification; this helps in knowing the response from people which enables the developer to increase the quality and content of notifications as shown in Fig. 8.

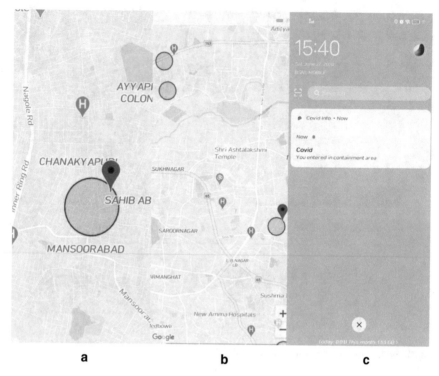

a   b   c

Fig. 4 Screenshots of containment areas and dynamic notification

5 Conclusions and Future Work

The proposed model helps the user to be attentive during this pandemic by sending regular alerts and real-time updates which help users to take preventive measures and ensures safety. Features availed in the application help the user to access information as well as contact the concerned department in case of an emergency. Dynamically triggered alerts keep the user notified of the contained places in which he/she is travelling. The proposed model has shown good results in fetching data from the database and authenticating the user during sign in. It has good scalability.

The current work is planned to be extended by facilitating users to receive alerts when the user is closer to the region. Features like alarm system to take medicine and sanitize regularly can be implemented in future. The scalability can be increased and additional features from Firebase can be enabled by purchasing the plans. Google maps API can be purchased for additional requests per day as the basic plan provides only 2500 requests per day. This makes the proposed model work smoothly even if the load is high. API's are also used to give out more information to the user. The geo-fencing API can be used for additional features and better results.

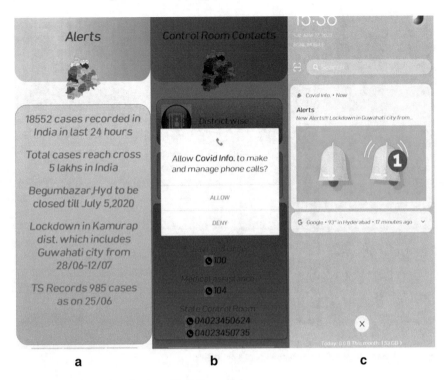

a   b   c

Fig. 5 Screenshots of application with different features

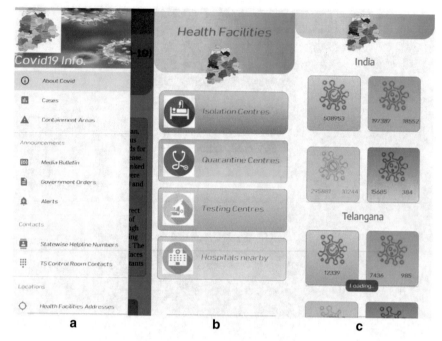

Fig. 6 Screenshots of different fragments of application

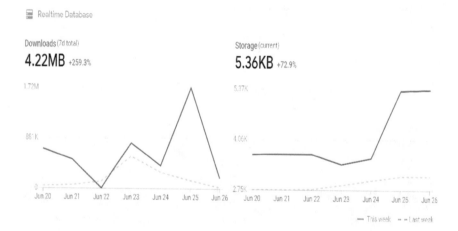

Fig. 7 Reports showing the storage and downloads of Firebase real-time database

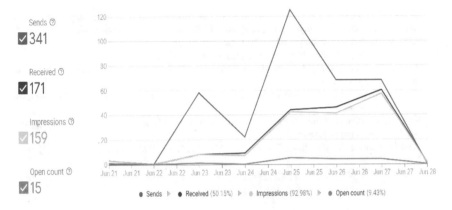

Fig. 8 Statistics of various activities in Firebase cloud messaging

References

1. BBC News.: WHO head: 'Our key message is: test, test, test. https://www.bbc.co.uk/news/av/world51916707/who-head-our-key-message-is-test-test-test
2. Pethick, A.: Developing antibody tests for SARS-CoV-2. Lancet **395** (10230)
3. Hernández-Orallo E, Manzoni P, Calafate CT, Cano J (2020) Evaluating how smartphone contact tracing technology can reduce the spread of infectious diseases: the case of COVID-19. IEEE Access 8:99083–99097. https://doi.org/10.1109/ACCESS.2020.2998042
4. COVID-19 outbreak: Migration, effects on society, global environment and prevention, https://www.sciencedirect.com/science/article/pii/S0048969720323998
5. How Reliable and Effective Are the Mobile Apps Being Used to Fight COVID-19?, https://thewire.in/tech/COVID-19-mobile-apps-india
6. A flood of coronavirus apps are tracking us. Now it's time to keep track of them, https://www.technologyreview.com/2020/05/07/1000961/launching-mittr-COVID-19-tracing-tracker/
7. Now, a mobile app predicts COVID-19 incidence days in advance, https://www.thehindu.com/sci-tech/science/now-a-mobile-app-predicts-COVID-19-incidence-days-in-advance/article31544706.ece
8. COVID-19 apps around the world, https://techerati.com/features-hub/opinions/COVID-19-apps-around-the-world/
9. 5G mobiles do not spread COVID-19, https://www.who.int/emergencies/diseases/novel-coronavirus-2019/advice-for-public/myth-busters?gclid=Cj0KCQjwudb3BRC9ARIsAEa-vUvdlwAV59nTqxfJ1xp7nKMD9TZsiT4mksqnq11xrTuO37kL9m1qwwaAj_tEALw_wcB#5g
10. Coronavirus Apps: Every App the Central Government And States Have Deployed to Track COVID-19. https://gadgets.ndtv.com/apps/features/central-state-governments-launch-coronavirus-mobile-app-list-2204286
11. Corona virus app API, https://coronavirus.app/map?compared=US,DE,FR,GB,ES
12. Mizoram launches coronavirus app to disseminate authoritative info to the public. https://www.deccanchronicle.com/technology/in-other-news/040420/mizoram-launches-coronavirus-app-to-disseminate-authoritative-info-to.html
13. Aarogya Setu- Features and tools https://en.wikipedia.org/wiki/Aarogya_Setu
14. Top-10 smart phone apps to track COVID-19,https://www.geospatialworld.net/blogs/popular-apps-COVID-19/
15. COVID-19 Apps. https://en.wikipedia.org/wiki/COVID-19_apps
16. COVID-19: The world embraces contact-tracing technology to fight the virus. https://www.livemint.com/news/world/COVID-19-the-world-embraces-contact-tracing-technology-to-fight-the-virus-11588237777342.html

17. CheckCOVID-19Now: A web app to spot coronavirus cases in Telangana, https://www.new indianexpress.com/cities/hyderabad/2020/apr/18/checkCOVID-19now-a-web-app-to-spot-coronavirus-cases-in-telangana-2131600.html
18. Geo-Fence ,Applications, https://en.wikipedia.org/wiki/Geo-fence
19. TOMTOM Geo Fencing API documentation, https://developer.tomtom.com/geofencing-api/ geofencing-api-documentation
20. Firebase in-app messaging, https://firebase.google.com/docs/in-app-messaging/compose-campaign?authuser=2
21. Sending messages to multiple devices, https://firebase.google.com/docs/cloud-messaging/android/send-multiple?authuser=2
22. Firebase console database https://firebase.google.com/docs/database/usage/monitor-usage?authuser=2

Visualization of Data Structures with Animation of Code

P. Sparsha, P. B. Harish, and N. S. Kumar⬤

Abstract Data structures play a major role in software development. However, data structures can be very tricky to understand. The best way to understand data structures is to incorporate them into action. Commonly, the data structures are learned by implementing in 'C'. Pointers in 'C' create another hurdle in learning. The proposed research work has developed a tool that helps in the interactive visualization of a user input code. Here, this paper proposes a web-based tool that takes 'C' program as the input from the user and does computations on it and finally creates animation of the respective data structures as well as the code. This helps the user to execute the one line code at a time and view the corresponding changes to the data structures. It provides features such as the animation of simple user programs in 'C' consisting of pointers, user-defined structures, and operations such as traversal, insertion and deletion in linked lists. This research work proposes to show the synchronized animation of code and schematic diagram for the linked list as a proof-of-concept.

Keywords Data structures · Visualization · Code animation · PLY · Pointers · Linked lists

1 Introduction

Data structure is a means by which the data is stored and arranged in the computer memory, in a way that can be easily retrieved and manipulated later. It forms the basis for most of the software. However, understanding the flow of execution can be more complex. Studies show that the human brain processes the visual information

P. Sparsha · P. B. Harish (✉) · N. S. Kumar
Department of CSE, PES University, Bengaluru, India
e-mail: harish.pb26@gmail.com

P. Sparsha
e-mail: sparshaprashanth3@gmail.com

N. S. Kumar
e-mail: nskumar@pes.edu

more efficiently than audio and text [1]. A visualization tool helps one to see data structures in action and understand them better.

This tool allows users to input their own 'C' program and experiment them to understand better. It helps the users visualize the working of their program and review various operations by generating the animation. This helps in interpreting the semantics behind the program in a easier way. The step-by-step execution of the program helps to spot the logical errors and debug them accordingly. This tool makes learning more effective and interesting. It offers an interactive learning session for the users. The proposed model provides a platform, where one can learn through experimentation and also have fun doing so.

2   Product Survey

To perform the specific set of operations upon encountering a line in the code, where lexer and parser tools are used. PLY (Python Lex-Yacc) [2] consists of Lex and Yacc modules. The Lex module is used to break the input text into a collection of tokens based on regular expressions. The Yacc module is used to recognize language syntax that has been specified in the form of context-free grammar [3].

With the development of technology, there are various platforms for animation of data structures. Data Structure Visualization [4] is one such tool. It provides an interactive interface for performing various operations on different data structures such as stacks, queues, lists, trees and many more. It provides a limited number of operations which can be performed on a particular data structure such as push and pop in stacks, enqueue and dequeue in queues. It uses a text box to provide any input required for the operation and buttons to select the operation to be performed. The animation in this tool is not code-driven. In addition to that, it provides animation of algorithms such as sorting and searching. The animation does not show the minute steps involved in performing these operations.

Algorithms and Data Structures Animations for the Liang Java [5] is a tool that provides the same features as [4] like visualizing the operations that can be performed on data structures. It provides the animation of certain algorithms as well. VisuAlgo [6] is another tool that provides animations of different data structures and algorithms based on menu driven user input. It also provides an online quiz to test one's knowledge. Data Structures and Algorithms animations using Processing.js [7] is a webpage that contains set of animations for data structures like stack, queue, cyclic queue and singly linked list. It provides buttons to take user input. Algorithm animations and visualizations [8] is a collection of algorithms animations and visualizations. It provides JavaScript and Flash animations for different data structures and algorithms. Algostructure [9] is a site featuring animations for different algorithms such as sorting and searching but does not take user code as input. Thus, these tools provide a limited set of instructions and are restricted to only those operations since they display visualization for existing code but not for user defined code. It does not relate the animation with the code.

This research work provides a tool that draws the animation for a user 'C' code. This has a wider scope as the user can input the code which can contain lines performing various types of operations. The animation is driven by the code and thus is not restricted to a set of operations. The proposed model has incorporated the visualization library from [4] in our source code to generate the animations. It contains various functions for drawing the animation.

3 Architecture

Figure 1 depicts the system architecture where a 'C' code is fed to the tool as input and the data structure visualization with code animation is shown.

The visualization tool consists of various parts each performing a specific function and making modifications to the stored data. The components of the visualization tool are shown in Fig. 2.

Fig. 1 System architecture

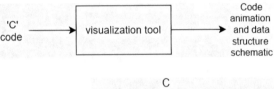

Fig. 2 Components of the tool

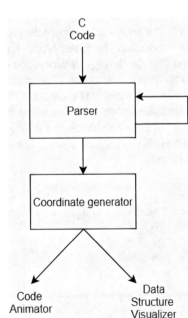

4 Implementation

Our implementation is a web based client server application. Flask [10] is a popular Python micro web framework. The back end uses Flask as the server. The server talks to the PLY engine for parsing. The implementation uses the static structure of a program to show the dynamic nature. To support the static nature, a piece of code is provided as an input to deliver a continuous structure.

4.1 Input

The interface consists of a text box where the user can input the 'C' code that needs to be visualized. Then, upon clicking the submit button the code is sent to the Parser. It should be a syntactically correct block of 'C' code. As the code is processed statically, all values should be known at compile time.

Figure 3 shows a snapshot of the editor where the user can insert the C code. Here is an example of a small 'C' code.

4.2 Parser

PLY is a pure Python implementation of popular compiler construction tools like Lex and Yacc. The Parser breaks the received single line into a collection of tokens specified by the regular expressions. Then, the syntax is checked using the productions specified as context free grammar. The Parser returns a list of dictionaries in the form of a JSON [11] object to the coordinate generator. The dictionaries contain information about each line in the form of key value pairs. In the case of conditional and iterative statements, the condition and the block statements are processed by recursively calling the Parser. The operator precedence and associativity have been handled.

Home Page

```
int p = 12;
int r = 5;
int q = p * r - 1;
```

Submit

Fig. 3 Editor to input code

Table 1 Output of parser and coordinate generator for code in Fig. 3

Line number	Output of parser	Output of coordinate generator
1	[{'p': ['int', 12]}]	[{'p': ['int', 12, [840,60], 0]}]
2	[{'p': ['int', 12]} , {'p': ['int', 12], 'r': ['int', 5]}]	[{'p': ['int', 12, [840,60], 0]} , {'p': ['int', 12, [840,60], 0] , 'r': ['int', 5, [840,185], 2]}]
3	[{'p': ['int', 12]} , {'p': ['int', 12], 'r': ['int', 5]} , {'p': ['int', 12], 'r': ['int', 5] 'q': ['int', 59]}]	[{'p': ['int', 12, [840,60], 0]} , {'p': ['int', 12, [840,60], 0] , 'r': ['int', 5, [840,185], 2]} {'p': ['int', 12, [840,60], 0] , 'r': ['int', 5, [840,185], 2] , 'q': ['int', 59, [840,310], 4]}]

In Table 1, the output of parser shows a list of dictionaries that store data gathered from parsing the code. Each dictionary in the list corresponds to a line in the code. Inside each dictionary exists key value pairs. The key is the name of the object and the value is the list containing details like datatype, value.

4.3 Coordinate Generator

The Coordinate Generator receives the list of dictionaries from the Parser and computes the coordinates and assigns a unique id to each object. Then, the updated list of dictionaries is sent to the code animator and data structure visualizer in the form of a JSON object using XMLHttpRequest (XHR) [12]. This enables us to retrieve data from the URL without having to refresh the page.

Table 1 depicts the output of the parser and coordinate generator after the execution of each line of code shown in Fig. 3. The output of the Parser shows the list of dictionaries. The first dictionary is associated with the first line of the code. For example, after the execution of the first line of code, the first dictionary contains the variable 'p' as the key and the corresponding value as '[int,12]' which is the datatype and the value of the variable 'p' respectively. Similarly, after the execution of the second line of code, the second dictionary contains the data collected by the execution until line 2.

The output of the coordinate generator shows the list (output of Parser) updated with coordinates and ids. For example, in the first dictionary, for the variable 'p', the coordinates '[840,860]' and id '0' are added to its value.

4.4   Code Animator and Data Structure Visualizer

It uses JavaScript to draw the animation. The coordinates and ids are used to position the objects on the canvas. As each line is processed, the corresponding line in the code is highlighted. The animation is implemented using the help of helper functions obtained from the Data Structure Visualization [4].

Some examples of these functions are:

(1) To Draw a Rectangle

$$\text{CreateRectange}(ID, label, width, height, initial_x, initial_y, xJustify, \\ yJustify, backgroundColor, foregroundColor) \tag{1}$$

(2) To draw an arrow from one object to another

$$\text{Connect}(fromID, toID, linkColor, curve, directed, label, anchorPosition) \tag{2}$$

Figures 4, 5 and 6 show the code animation and data structure visualization for the code in Fig. 3.

Figures 4 and 5 show the creation of a new variable 'p' and 'r' on the stack. In Fig. 6, a new variable 'q' is initialized and allocated memory on the stack.

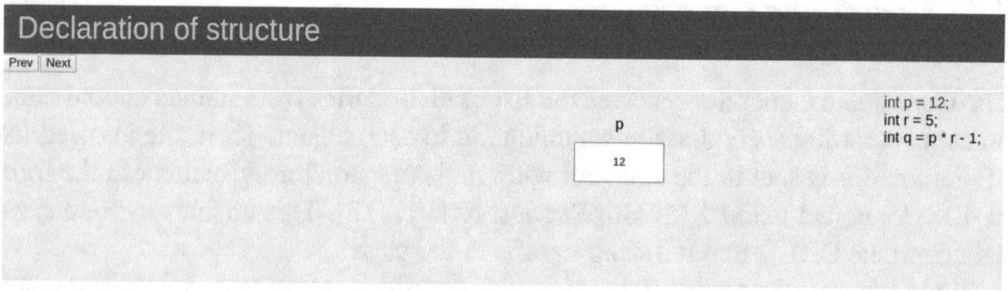

Fig. 4   Visualization after the execution of the first line

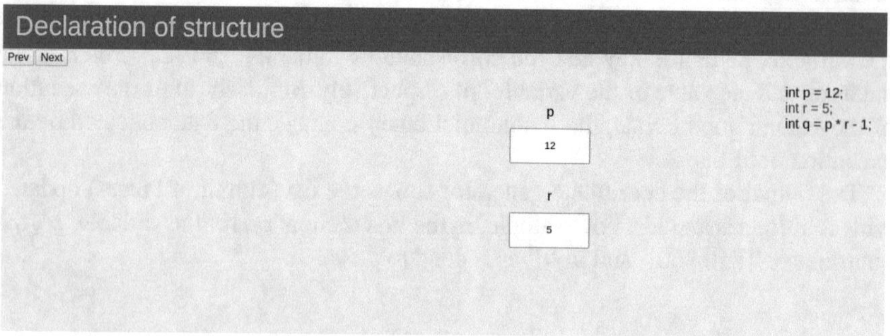

Fig. 5   Visualization after the execution of the second line

Declaration of structure

Prev Next

p

12

r

5

q

59

```
int p = 12;
int r = 5;
int q = p * r - 1;
```

Fig. 6 Visualization after the execution of the last line

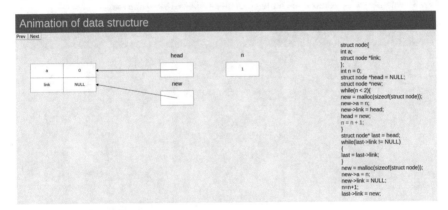

Fig. 7 Variable '*n*' is incremented during first iteration of the while loop

4.5 Examples

The line currently being executed is highlighted in red and the corresponding visualization of data structures is shown alongside.

Example 1 Figures 7, 8, 9 and 10 demonstrates the insertion of node in a linked list.

Figure 7 is a snapshot after the variable '*n*' is incremented during the first iteration. At this stage, the first node is added to the front of the linked list.

Figure 8 is a snapshot after variable '*n*' is incremented during the second iteration. At this stage, the second node is added to the front of the linked list. So the linked list now contains two nodes.

In Fig. 9 a new node is created and allocated memory. This node is yet to be added to the linked list.

Figure 10 is a snapshot after the execution of the last line in the code. Now, the third node is added to the end of the linked list.

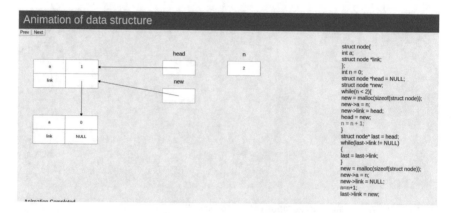

Fig. 8 Variable '*n*' is incremented during second iteration of the while loop

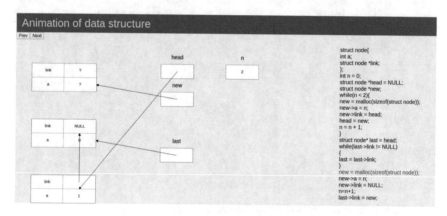

Fig. 9 New node is created

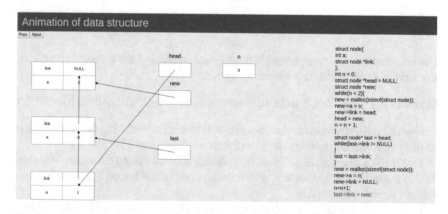

Fig. 10 Change the pointer of node 'last'

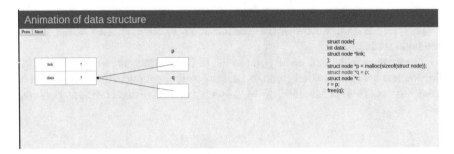

Fig. 11 New node '*p*' is allocated

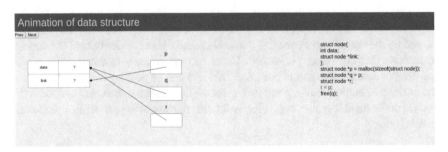

Fig. 12 A new node '*q*' is created

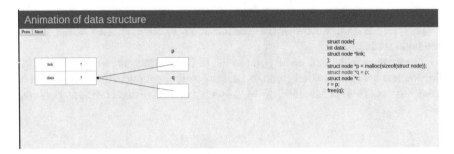

Fig. 13 A new node '*r*' is created

This example depicts the insertion of three nodes to a linked list. The first two nodes are added at the front of the linked list and the next node is added at the end. The animation and the corresponding line being executed are shown at each stage.

Example 2 Figures 11, 12, 13 and 14 demonstrates an example of a dangling pointer in case the of a linked list.

In Fig. 11, the new node '*p*' is allocated memory on heap.

Figure 12 is a snapshot after the creation of a new node '*q*'. This node points to the same address as '*p*'.

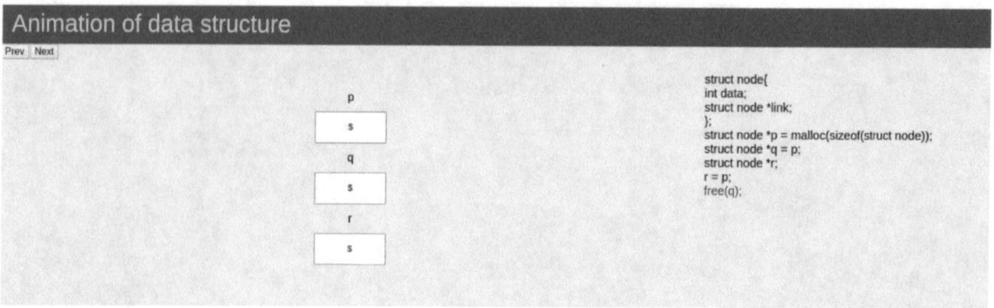

Fig. 14 Node '*q*' is freed

Figure 13 is a snapshot after the creation of a new node '*r*'. This node points to the same address as '*q*'.

The three pointers *p*, *q* and *r* point to the same location. So, when it is free (*q*), the pointers become dangling pointers i.e. the pointer pointing to the location that no longer exists. This is shown in Fig. 14. $ symbol is used to depict dangling pointer.

5 Results

This paper has explained the working of the tool that helps the users visualize data structures with the animation of the input 'C' code. They can view the changes caused by the execution of each line and understand the semantics behind the various operations that are being performed. Users can get a better understanding of the working of data structures and pointers. This tool can be used by teachers to explain the intricacies behind pointers and linked lists. The students can experiment with the code to understand these concepts clearly. These are some of the practical applications of this tool.

6 Future Work

The proposed research work has implemented certain functionalities that provide 'C' programming language. The proposed tool currently supports the linked list data structure. The proposed research work can be extended to more complex data structures like trees, graphs. This can be done by adding the necessary grammar rules in the parser. The extended research work can support dynamic features as well. Another possible improvement could be the user interface and better visualization.

References

1. The fidelity of visual and auditory memory (2019). https://doi.org/10.3758/s13423-019-01597-7
2. PLY (Python Lex-Yacc) Beazley DM, https://www.dabeaz.com/ply
3. Ahom AV, Lam MS, Sethi R, Ullman JD (2007) Compilers principles, techniques tools, 2nd edn. Pearson Addison. Wesley
4. Data structure visualizations. https://www.cs.usfca.edu/~galles/visualization/Algorithms.html
5. Algorithms and data structures animations for the Liang Java, C++, and Python Books. http://www.cs.armstrong.edu/liang/animation/animation.html
6. VisuAlgo—visualising data structures and algorithms through animation. https://visualgo.net/en
7. Data structures and algorithms animations using processing.js. http://littlesvr.ca/dsa-html5-animations/
8. Algorithm animations and visualizations. http://www.algoanim.ide.sk/
9. Algostructure—algorithms and data structures explained and animated for a better understanding of common problems. http://www.algostructure.com/
10. Welcome to flask—flask documentation (1.1.x). https://flask.palletsprojects.com/en/1.1.x/
11. Python JSON. https://www.w3schools.com/python/python_json.asp
12. XMLHttpRequest—Web APIs | MDN. www.developer.mozilla.org/en-US/docs/Web/API/XMLHttpRequest

A Machine Learning Enabled IoT Device to Combat Elephant Mortality on Railway Tracks

N. Krishna Chythanya, K. Madhavi, and G. Ramesh

Abstract Nature is for everyone, all the resources are to be shared by all living beings, but the great social animal-Man, spend all his mental and physical strength to monopolize the resources and create havoc in the ecological balance. Indian Railways is one among the top railway networks in the world with a wide coverage of approximately 115,000 km of the length of track and it almost equal to 1.5 times the earth circle. It is observed that almost 20 out of 101 elephant corridors in India has a passing Railway track, and the problem would only be augmented with more track being spread every year. It can be observed that the 2018 year was fatal for Elephants with 26 being killed by train accidents and announced officially. In this paper, the authors review the situation of Elephant mortality and suggest a mechanism to combat the same using Machine Learning algorithms and IoT-based device.

Keywords Railway track · Elephant mortality · Machine Learning · IoT device

1 Introduction

Under the sun, every element of nature is a gift to all living beings and every living being has a right to live peacefully in its own territory or habitat despite this knowledge, the great learned animal-The Man spends all his mental and physical strength to monopolize the resources and create havoc in the ecological balance. The irony of today is that mortality of wildlife does not create great news nor it creates hashtags in social media networking platforms. Least, the basic principle of live and let live should not be forgotten.

N. Krishna Chythanya (✉) · K. Madhavi · G. Ramesh
G.R.I.E.T., Hyderabad, India
e-mail: kcn_be@rediffmail.com

K. Madhavi
e-mail: madhaviranjan@grict.ac.in

G. Ramesh
e-mail: ramesh680@gmail.com

© The Author(s), under exclusive license to Springer Nature Singapore Pte Ltd. 2021   531
J. S. Raj et al. (eds.), *Innovative Data Communication Technologies and Application*,
Lecture Notes on Data Engineering and Communications Technologies 59,
https://doi.org/10.1007/978-981-15-9651-3_44

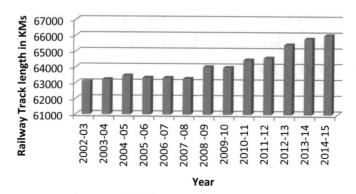

Fig. 1 Growth of railway track length in between 2002 and 2015 for a base idea

Indian Railways is one among the top railway networks in the world with a wide coverage of approximately 115,000 km of the length of track and it almost equal to 1.5 times the earth circle. Figure 1 gives an idea of growing railway track length up to 2015. The fastest transport for a huge number of industrially civilized human beings is creating a death trap for wild animals with its track spread through forest belt and near to habitats. The need for population estimation and safeguarding the ecosystem is increasing rapidly as wildlife-train collision is having a strong effect on the mortality rate of big animals like Elephants and other species. The impact on population dynamics can be understood only with proper mortality estimates considering the local scale studies to landscape-scale studies. Railways alone are having a bad impact on habitat loss for wildlife besides the noise pollution they create. It is not just one or two kilometers spread but the railway network on a global level, when considered is estimated at a distance exceeding over one million kilometers and rapidly growing at a pace of nearly 5% every year as a world bank statistics project in the past half-decade.

It is observed that almost 20 out of 101 elephant corridors in India have a passing railway track, and the problem would only be augmented with more track being spread every year as per the above statistics.

As compared to roads, the railways have a lesser impact on the following point basis shown in Table 1.

As observed from literature, there is an urgent need to do extensive research in mortality record mechanisms on train tracks as it is completely different from roads, and the patterns of roads can not be extrapolated (Fig. 2).

2 Literature Survey

In the multispecies surveys [1] conducted in the past, the following percentage of mammals were found dead on railway tracks Netherlands-26%, USA-36%, Spain-

Table 1 Comparison of roadways to railways on different parameters effecting the life around

Factors considered	Roadways	Railways
Traffic flow	Heavy	Less
Free intervals of traffic	Less	High
Wildlife mortality	High	Comparatively less
Width of corridors	Wide	Narrow
Chemical pollution	High	Low as electrified lines
Fencing	No	Yes in some high-speed routes

Fig. 2 Elephant Mortality in India between 1987 and 2007 owing to train hit

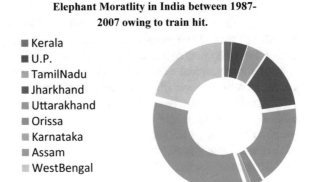

Elephant Moratlity in India between 1987-2007 owing to train hit.

- Kerala
- U.P.
- TamilNadu
- Jharkhand
- Uttarakhand
- Orissa
- Karnataka
- Assam
- WestBengal

38%. The sizes of species being killed on railway tracks due to train collision range from very small to very large could be small insectivores, small carnivores or large carnivores, and even bigger mammals like elephants.

The authors [1] have done an exhaustive survey on the research papers using zoological record databases, Google scholar for PhD dissertations, published peer-reviewed journal papers from ISI Web Of Science, to identify the work done on railway ecology considering multilingual research works spreading over languages like English, Spanish, French, etc. The authors pointed at the need for research in the following areas based on their literature review on the above sources they identified. Mortality, mitigating railway impacts, avoiding unsafe crossings, promoting safe crossings, habitat management, impact of fencing on the detecting, and controlling of wildlife mortality on railways.

Though often mortality is attributed to train collisions on railway tracks, yet the electrocution or collision with bridges or wires is also contributing to wildlife mortality, maybe in lesser numbers. Effect of uncapped catenary poles where hole-nesting birds breed but cannot escape once they are into it with electrified wires supporting is recently focused by a few researchers. Many times this non-train collision death of wildlife may be owing to grain search by the animals like goose or bears looking for sweet content in the shrubs or trees grew along the railway tracks. These also need to be considered in overall wildlife safety plans in railway tracks.

In general, most of the studies pertaining to wildlife mortality can be done on large mammals as the carcass would be lying for a long time and owing to its big size, it is easily identifiable. Many times large mammals with their charismatic looks attract public concentration with commercials involved in their anatomy parts. The following pie chart in Fig depicts elephant mortality between 1987 and 2007 due to train hit.

India has a widespread railway system with railway tracks spreading across length and breadth of its geographical area, and the train collisions with elephants have become a conservation concern in the recent past in India. The trends show that around 18 elephants, accounting to almost 45% of death of Elephants near Rajaji National Park were due to train collision. The numbers shown by Indian forest department was also alarming at a count of around 35 Elephants being killed during a one and half-decade period between 1990 and 2006. One of the factors effecting moment of these large mammals was the water scarcity during the high summer period and with thirstiness to quench their thirst, the large mammals in groups come out of their habitat and eventually in the wee hours of day particularly near to blind curves on railway track fell prey to train collisions.

In the recent LokSabha sessions for an unstarred question No.1902, on the topic of animals killed in train accidents, raised by one of the respected member of parliament, posed to Minister of Railways [2], the facts and figures about increasing numbers of animals, especially Elephants getting killed in rail accidents in past three years, and the present year was asked for and also the effective plan of action of concerned departments or government to control such mishaps was questioned. The following statistics were submitted by honorable railway minister of India as an answer to the above questions. The details of Elephants killed in train accidents in the past three years considering up to June 2019 is shown below with year-wise and zone wise (Fig. 3).

From Table 2 shown at the end of this paper, it can be observed that the 2018 year was fatal for Elephants with 26 being killed by train accidents and announced officially, mounting to approximately 40% of deaths over past 4 years with 2017, a bit better of having 23% and 2019 yet to complete is at a soothing figure of only 5

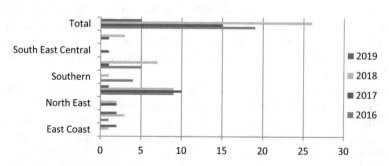

Fig. 3 Chart showing the number of elephants killed by train accidents in the past three years and up to June 2019

Table 2 Table showing many Elephants killed by train accidents in the past three years and up to June 2019 year-wise and zone wise

Year	Zone								
	East coast	Northern	North East	North East frontier	Southern	South Eastern	South East Central	South Western	Total
2016	0	1	0	9	4	5	0	0	19
2017	0	0	2	10	0	1	1	1	15
2018	1	3	2	9	1	7	0	3	26
2019	2	2	0	1	0	0	0	0	5

Elephants being killed making 0.07. The authors of this work also hope the count does not increase further.

Whereas the contents of Table 3 [3] shown at the end of this paper were the statistics are given on the animals other than Elephants mortality on railway tracks in past four years up to June 2019. Table 4 [2] in the last page is showing the percentage of animal death out of total death on the railway track in a zone per year from 2016 to June 2019.

It is clearly evident that for the habitat of animals in West-central and Northcentral railway zones of India, tracks are proving to be death beds. It is clearly visible that on average, the mortality rate percentage of animals in North Central is 31.377% over the past four years, and whereas for West Central it is 35.99%. These alarming numbers raise the question for safety of railway tracks for animals that have equal right as human beings to live on this earth.

The authors of this work would like to emphasis on this point as well that care should be taken to safeguard all animals besides Elephants on the railway track. The following action plan was mentioned by the Minister of Railways in his answer to curb the mortality of animals on railway track [2].

1. The imposition of speed restrictions in identified locations,
2. Provision of signage board,
3. Sensitization of train crew and station masters regularly,
4. Need-based clearance of vegetation on the sides of the track within railway land,
5. Construction of underpasses and ramps for the movement of elephants at identified locations
6. Provision of fencing at isolated locations, and
7. Forest department staff deputed in railway control offices to liaison with railway and Elephant trackers engaged by forest department for timely action by alerting station master and loco pilots.

The following actions were recommended by the researchers who worked out on conserving wildlife across railway tracks in Delhi-Dehradun track, where they could achieve zero mortality record in seven years.

- Sensitization of the train drivers
- Use of signage
- Improving visibility for the train drivers
- Joint patrolling
- Reducing the speed of trains in critical areas
- Making the sides of railway track elephant-friendly.

It is observed that often Elephants are hesitant to climb down after they climb up the track, this may be owing to a steeper gradient of the side of tracks. This can be handled by reducing steepness on two sides of tracks in concerned Elephant corridors. The factors that lead to high risk of train collisions with Elephants were observed to be as

Table 3 Statistics given on the animals other than Elephants mortality on railway tracks in the past four years up to June 2019

Years	Zone									Total
	Northern	North Central	North Eastern	North Western	South Central	South East Central	South Western	Western Railway	West Central	
2016	457	2283	649	1080	353	125	0	24	2974	7945
2017	445	3818	932	1222	413	225	0	39	4589	11,683
2018	351	4826	1146	1476	460	393	2	45	3926	12,625
2019	34	902	442	603	110	118	4	7	1259	3479

Table 4 Percentage of animal death out of total death on a railway track in a zone per year from 2016 to June 2019

Year	Zone									
	Northern	North Central	North Eastern	North Western	South Central	South East Central	South Western	Western Railway	West Central	Total
2016	5	28.7	8.1	13.59	4.44	1.57	0	0.3	37.43	7945
2017	3.80	32.67	7.9	10.45	3.5	1.92	0	0.3	39.27	11,683
2018	2.78	38.22	0.9	11.69	3.6	3.11	0.01	0.35	31.09	12,625
2019	0.97	25.92	12.7	17.33	3.16	3.39	0.11	0.2	36.18	3479

- Sharp turnings/blind curves
- High speed of the train (above 70 kmph)
- Low visibility for drivers because of fog
- It was also observed many trains—Elephant collisions occur during night hours between 6 pm and 6 am.

Elephant mortality in Assam due to hit by rail between 1990 and 2006 was very high and around 33 Elephants were killed of which 18 were female, 13 male, and sex of 2 was unidentified.

Indian Railway system makes use of different kinds of signal mechanisms. The prominent ones being as shown in the list below:

- Semaphore Signals
- Color Light Signals
- Position Light Signals
- Disk Signals
- Target Signals.

There was a proposal from professors of IIT-Delhi for a high tech device that makes use of sensors, cameras and lasers to identify the movement of elephants near the railway track, and the device shall send the signal to drivers within a range of 3–4 km so that trains can be stopped. This is apart from speakers being placed in the elephant moving areas, that emit a sound of bee's buzzing, that eventually deter elephants.

The researchers interested to get in depth overview of research and contributions done by others in this area can go through references [4–12] mentioned in the reference section at end of this paper.

3 Proposed Idea

Software that identifies wildlife is created and trained using Machine Learning algorithms that can give a very high accuracy of above 90% using convolutional neural networks learning techniques. The schematic diagram of the proposed idea is depicted in Fig. 4. Place a sensor on the head of locomotive as well as on top of signal post beside tracks in the most accident-prone areas and such route. As soon as a wild animal comes on to the track and stays there for more than a small fraction of time, the sensor detects and the application calculates the time required for the fast-approaching train if any. If so, either it will change the signal to orange or inform the signal station and IoT-based device on the locomotive. The device on locomotive would first honk for a while and slows down the speed by 50% so that immediately drivers can apply breaks and save the death of a wild animal. It is learnt based on the discussions held with a couple of station masters of different railway stations that a minimum distance of 25 m is expected when a sudden brake is applied to a fast-moving train to get it dead stop.

In case the wild animal is not near to a signal post, the IoT enabled locomotive sensor would capture the image at a specific distance and with Machine Learning algorithm help identifies whether it is a wild animal or not. If wild animal it honks and reduces speed to 50% for the manual appliance of breaks. In case of confirmation

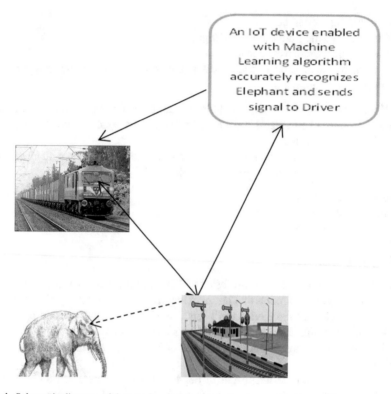

Fig. 4 Schematic diagram of the proposed idea

with a wild animal and the breaks are not applied by the driver while train reaches a specific distance from the wild animal on track still, the IoT enabled device controlled by the controller should apply complete brakes.

4 Expected Results and Experiment Set-Up

It is planned to make use of deep neural networks in image detection of Elephants. The convolution neural network type ResNet is the one wished to use and as very deep neural networks can face problems of vanishing gradients, this issue is efficiently handled using ResNet model of CNN and expected accuracy of above current acceptable works in the area, in identifying Elephant image correctly. Also, the IoT device with a signal delay of less than one second is expected to reduce the train collision with Elephants by a percentage of at least 80 on an average.

To implement the above experiment, imagenet Elephant data set as training data would be made use of, and Python programming language with packages including TensorFlow, plot lib, pandas, and keras would be used.

5 Conclusion

Earth is not only for us, It is for all living beings on it. The lives of wild animals should be respected and has to stop getting into their territory for our convenience. Hence, every care should be taken by humans to protect animals from killing by train hit. In this paper, a mechanism is proposed for combating Elephant mortality on railway tracks using Machine Learning and IoT device.

References

1. Barrientos R, Ascensão F et al (2019) Railway ecology vs. road ecology: similarities and differences. Eur J Wildlife Res 65:12 (Springer-Verlag GmbH Germany, part of Springer Nature 2019)
2. https://164.100.24.220/loksabhaquestions/annex/171/AU1902.pdf
3. https://www.railway-technology.com/features/featurethe-underdog-preventing-animal-casualties-on-railways-4532957/
4. Santos SM, Carvalho F, Mira A, Current knowledge on wildlife mortality in railways, Chapter 2, pp 11–22. https://doi.org/10.1007/978-3-319-57496-7_2. https://www.springer.com/978-3-319-57495-0
5. Sarma UK, Easa PS, Menon V, An occasional report of the conflict mitigation division of the
6. Wild Species Programme of the Wildlife Trust of India in partnership with the International Fund for Animal Welfare October 2006
7. Rangarajan M, Desai A, Sukumar R et al (2010) The report of the elephant task force , Ministry of Environment and Forests. 31 Aug 2010
8. NanajiWasnik R, Analysis of railway fatalities in Central India. J Indian Acad Forensic Med 32(4):311–314. ISSN 0971-0973
9. Menon V, Tiwari SK, Ramkumar K, Kyarong S, Ganguly U, Sukumar R (eds) Right of passage: elephant corridors of India, 2nd edn. Conservation Reference Series No. 3. WIldlife Trust of India, New Delhi
10. Roy M, Sukumar R, Railways and wildlife: a case study of train-elephant collisions in Northern West Bengal, India, Chapter 10, pp 157–176
11. https://www.railway- technology.com/features/timeline-165-years-history-indian-railways/
12. A scientific approach to understanding and mitigating elephant mortality due to train hits in Assam

Security Model for Internet of Things Based on Blockchain

J. T. Kruthik, K. Ramakrishnan, R. Sunitha, and B. Prasad Honnavalli

Abstract The Internet of Things has emerged as an area of incredible potential and growth, complemented by the advent of smart devices with increased computational power. However, these complex systems have large attack surfaces making them susceptible to various security attacks. Most common among these are data forgery and identity theft. Further, once the IoT is compromised, the smart devices may be controlled remotely by the attacker and used as botnets in Distributed Denial of Service attacks. The size and scale of the complex system result in technical challenges in management, synchronization, and security. In this paper, a blockchain model is proposed and implemented to secure the IoT system. Using Ethereum, a private blockchain model is developed, verified its functionality in a simulated IoT environment, and analyzed the performance of the model. Further, a web-based interface to the IoT system is developed to demonstrate the working of the blockchain.

Keywords Internet of Things · Security model · Network security · Blockchain · Ethereum · SHA256 · Cryptography · Hashing · Web-interface

1 Introduction

The world took a major leap when different computers were connected for the first time in 1969, marking the birth of the internet. The interconnection of computers and networks leads to faster development in the modern world. Then in 1999, the

J. T. Kruthik · K. Ramakrishnan · R. Sunitha (✉) · B. Prasad Honnavalli
PES University, Bengaluru, India
e-mail: sunithar@pes.edu

J. T. Kruthik
e-mail: jt.kruthik@gmail.com

K. Ramakrishnan
e-mail: ramakrishnanramakrishnan8@gmail.com

B. Prasad Honnavalli
e-mail: prasadhb@pes.edu

© The Author(s), under exclusive license to Springer Nature Singapore Pte Ltd. 2021   543
J. S. Raj et al. (eds.), *Innovative Data Communication Technologies and Application*,
Lecture Notes on Data Engineering and Communications Technologies 59,
https://doi.org/10.1007/978-981-15-9651-3_45

term Internet of Things was coined by Kevin Ashton as an interconnection of objects in the real world. Currently, the Internet of Things has grown into a key technology that enables faster interaction with the real world. Similar to the internet, IoT has also seen exponential growth in recent years. This growth is complemented by the development of smart devices. The increase in power of computing systems these days have enabled machines to learn and find patterns from the data. This has led to further improvement in the data processing and storage capabilities of the devices.

A smart device is an electronic device capable of collecting data via sensors and communicating with one another with minimal human interaction. These devices can take action on the environment based on this data. The data they collect from the environment is also shared among other smart devices using a peer-to-peer network. Data collected is of prime importance. The data processing may be done at the device itself or may be moved to the cloud. Further, this data is crucial for training, testing, developing, and maintaining the AI and ML models. The smart devices are a part of a peer-to-peer network since a lot of simultaneous communication is needed, which is not feasible in a client–server network. A peer-to-peer network means that each device may be able to communicate with other devices in the network and there is no fixed client and server. In most IoT systems involving an end-user, an interface is provided. Since few IoT systems use the internet, a web-based interface is convenient. Also, the interface must be able to monitor the devices in the network and notify the user in case of device crash, emergency, security breach, etc.

IoT systems use the network connection extensively for their processing. This means that existing security threats on the internet can affect the systems. Some of these security breaches in IoT are relatively new. They show the need for securing the network and hence its devices from security attacks. The infamous "Mirai" [1] IoT botnet attack is a Distributed Denial of Service attack carried out using IoT devices that run on ARC processors. The malware Mirai infects the IoT devices, by exploiting the default username and passwords, turning them into remotely controlled bots. This network of bots (botnet) was used to carry out a DDoS attack. By November 2016, the Mirai malware had infected over 6 million IoT devices. These devices were then used to bring down a cybersecurity blog. Since there are many devices, it is a challenge to find the attacker. Apart from DDoS attacks, IoT is also vulnerable to man in the middle attacks. The attacker secretly relays or alters the messages in the communication. These attacks could lead to severe problems such as identity theft and ransomware. Attacks like these bring down the users' trust in the system and hinder the growth of IoT. These attacks demonstrate the need for a strong and convincing solution for IoT security.

In 2008, Satoshi Nakamoto invented bitcoin [2]. This cryptocurrency has a distributed architecture and runs on a peer-to-peer network. Blockchain is a new technology introduced as the public ledger to track and validate the transactions done using bitcoin. It works using cryptographic hashing. Nowadays, there is a lot of research on extending blockchain for other applications as well, such as in the supply chain and smart contracts. Further, the growth of blockchain has brought more attention to its development. Ethereum is one such platform [3] that helps in developing new blockchain systems using solidity as the programming language.

In this paper, an IoT security model using blockchain is proposed and implemented. This paper is divided into 7 main parts and started with a literature review to understand the current progress in the field of IoT, blockchain, and IoT security. To demonstrate the model, a simple IoT network is built and simulate the transfer of data through the network. A front-end (end-user application) is developed for enhanced access to the IoT network. This simulation of the IoT system and the workflow is illustrated in the sub-section under the proposed system. Moving on, the basics of a blockchain is discussed and how to implement one. The architecture of the model is described and the various components of the system and conclude with the future scope of the project.

2 Literature Survey

A complete understanding of the working of an IoT system is crucial to identify the attack surface of the system. Mrs. Soumyalatha and Ms. Shruti provide an introduction to IoT systems and architecture [4]. The IoT system provides smart objects under it to connect to the internet and communicate with each other with minimum human interaction. In their paper, Soumyalatha, Shruti discuss IoT and its architecture. Further, they elucidate its relationship with wireless sensor networks (WSN). Through this paper, many applications of IoT are explained in brief along with the current tools, advantages, disadvantages, and challenges of IoT. This paper also proposes an idea for using IoT in the domain of Indian architecture.

There are different kinds of security attacks possible on an IoT system. To secure the system, it is important to know the existing security measures. Also, knowledge of various ways an attack can occur helps in finding the attack surface of the system. Once the attack surface is identified, then a security model can be developed. Minhaj and Khaled present and survey the security issues for IoT by reviewing and categorizing them [5]. They tabulate and map the security problems with the existing solutions. They discuss how blockchain can be a key enabler to solving many security problems. They identify open research problems and challenges for IoT security.

A blockchain model consists of various components with independent functionality. To implement a blockchain model, it is crucial to know the internals of a blockchain, how it works, and how it can be used in other systems as well. Christy Varghese gives a basic introduction to IoT technology and its architecture through the paper [6]. This paper also proposes building IoT systems using blockchain platforms. Also, a system is implemented using the Ethereum platform for IoT. The benefits of such a system are enlisted and discussed in brief.

Wang et al. focus on the research of model construction and performance evaluation [7]. Theoretical and data support is surveyed through literature. An IoT security model is established based on blockchain and InterPlanetary File system (IPFS). The security risks and system performance are analyzed for the exposed system. Then, the average latency and throughput of the system are analyzed. These analyses and tests demonstrate the effectiveness of the blockchain-based security model.

Jayavardhana, Rajkumar et. al. present a cloud-centric vision for worldwide implementation of IoT [8]. Various existing IoT tools such as RFID are discussed here. Further, the internal working of IoT in a cloud-based system is explained. The cloud implementation is done using Aneka. It is based on the interaction between public and private clouds. The paper concludes its IoT vision by expanding on the need for convergence of WSN, the Internet, and distributed computing directed at the technological research community.

Sathish and Dhiren introduce IoT and explain its capabilities to connect real-world objects into a unified system [9]. It discusses the serious concerns raised regarding personal information. The concern in IoT security about device and individual privacy is also discussed. They conclude the survey by summarizing the privacy concerns and security threats of IoT.

Atzori et al. address IoT and the integration of several technologies and communication solutions [10]. It explains how IoT is a system built from several other domains such as telecommunications, electronics, informatics, and social science. This survey reviews the visions of IoT paradigms and their enabling technologies. The paper also discusses the major issues faced by the research community in detail.

Nakamoto introduces cryptocurrency and a means to enable transactions in a peer-to-peer network [2]. This paper aims at enabling transactions among peers without a centralized system. This paper proposes a solution to the double-spending problem using a peer-to-peer network. It discusses various technologies such as hashing, proof of work, and immutable records. This paper explains the working of a blockchain and how it is used in the cryptocurrency bitcoin.

Seyoung, Sangrae et. al. discuss the management of keys using RSA public-key cryptosystems [11]. They use Ethereum to store the public keys and run the smart contracts since it provides easier management of the IoT devices. The proof of concept is illustrated with the help of few IoT devices. In this paper, an IoT system is built and is managed with the help of blockchain. This paper can be extended to manage a larger IoT system using blockchain. The size of the IoT system will add to the complexity of the model built.

3 Proposed System

3.1 Blockchain

A blockchain [12] is a ledger used to track and manage transactions in a network. It is maintained by a community of decentralized peer-to-peer networks. This means that the data is not stored in a central server, but is written to a ledger. Also, the nodes of the network connect to the peer-to-peer networks, unlike the traditional client–server architecture where multiple clients connect to a single server. Further, a decentralized system, in general, enables easier management of the transactions and quick decision making. Every node in the blockchain network has a copy of the ledger. This ensures transparency in the system, since each node can access the data without releasing on a particular node, but the entire network. Blockchain ensures that any transaction in the network is immutable. This guarantees that the data stored in the network is not modified at a later point in time, thereby preventing data tampering [13]. However, the data to be written must be verified before insertion. A blockchain can be viewed logically as a collection of 3 major components:

- Data: The transaction that needs to be stored in the chain. This data is generally hashed to store into a fixed-length field. The most common hashing used is the SHA256 hashing which ensures output size is 256 bits or 32 bytes.
- Previous information in the network: The existing information in the chain is stored as a hash of the data. The previous hash is used to track changes in the chain.
- Nonce (number only used once): This is a random number generated by each node in the network. This number is used to generate required hash values when appending the block to the blockchain and prevent replay attacks.

3.2 Working

When a transaction is initiated in the network, a new block is created with the transaction details hashed into the data field. To ensure that the transaction is valid, the transaction must be digitally signed by the user or node. This is done using public–private key pairs generated by the user for the first-ever transaction. The data is signed using the private key of the user. This private key must not be shared. The transaction signed with the private key must be broadcast along with the public key. The public key is used to validate the source as well as the transaction. Once the transaction is validated and accepted, the transaction can be added to the block. This append must be done by every other member in the blockchain holding a copy of the block. A block can accommodate a certain amount of transactions in it. Once the limit is reached, the block must be attached to the blockchain. This process of adding a block to the blockchain is accomplished after solving a cryptographic puzzle of generating a smaller hash value than the previous hash. The process of adding the block to the blockchain is called mining and the nodes that do this are called miners. While solving the puzzle, a lot of hash values need to be generated making this process computationally expensive.

Every block maintains a link to the previous block in the blockchain. This is how the chain is maintained among the blocks, and it is achieved by storing the hash value of the previous block in the current block. A hash is generated using a function called the hash function. The most commonly used hash function is SHA256. A hash is generated by applying the hash function on the data, in this case, the entire block. SHA256 hash function generates a 256-bit value for the input. Since the probability of a collision in the hash is negligible, it is safe to assume that different inputs will always generate different hash values. Further, a single input will always generate the same hash value. A unique property of the hash value is that it is not possible to generate the input string using its hash value. This ensures that the data is secure and can not be deciphered from its hash value. This can be viewed as a method of storing the information of the input. So, by tracking the hash of the previous block, its information is passed on to the current block. The hash of the current block will incorporate its information as well as the information of the previous blocks (Fig. 1).

The hash value generated by the miner must satisfy some requirements to append the block to the blockchain. The first requirement is that the hash generated must be lower than the target hash. The value of the target hash changes based on the difficulty of mining such hash values. The second requirement is that the hash generated must start with a specific number of leading zeros which is set by the blockchain. Since the data and previous hash cannot be altered when adding to the blockchain, the only way to generate different hash values is by concatenating a new number called the nonce. This number helps to satisfy the second requirement. So, each miner has to generate nonces, compute the resulting hash value, and repeat until the hash generated is less than the target hash. The nonce value is random and is best found using trial and error. Once the feasible nonce is found, the miner can append the block to the blockchain. This method is called proof of work. The proof of work mechanism makes the blockchain more secure. It deters security attacks like denial of service and spam since some computation must be performed by the sender (Fig. 2).

3.3 Advantages of Blockchain Over a Client–Server Model

The distributed nature and the proof of work mechanism implemented in blockchain provide certain important advantages over the traditional client–server model:

- **No single authority**: The blockchain is available to all in the network. The blocks can be added by any miner. This means that no single authority can take control over the blockchain network and the blockchain itself. This transparency of the network brings in consensus in the decision taken.
- **Consensus Algorithm**: Blockchain uses a peer-to-peer network and complex mechanisms such as proof of work and consensus algorithm. These components ensure that each node in the network participates in the decision being taken and that the decision is not always favorable to a particular node in the network.
- **Immutability**: The data once written in the blockchain cannot be easily modified or removed at a later time. This mechanism ensures that only valid blocks are added to the blockchain and encourages more trust among the users of the system.
- **Validation**: Any user in the network can verify the data written to the public blockchain using the right tools. This ensures that blocks in the network are valid.

3.4 Blockchain for IoT

It is noticeable that IoT and blockchain both work on peer-to-peer networks. Now, IoT is vulnerable to attacks such as data forgery and identity theft. If blockchain is incorporated within an IoT system, the security of the IoT systems can be improved. Also, this model may be easier to develop since both technologies work on peer-to-peer networks. The users of the IoT system can be assured of data safety since blockchain prevents updates in the data. Unlike the existing client–server model where the server or client could be targeted to attacks, the decentralized model of blockchain does not suffer from such vulnerabilities. However, this also means that each node must be protected equally well. In most cases, the blockchain being used will be a private blockchain, for instance in smart homes. This will reduce the chances of a remote attack from outside the network since only devices in the same network will be able to access the blockchain. Even in case of a compromise, the data cannot be altered since no single authority can control the chain. Further, any activity of the attacker can be found and tracked. Blockchain only provides support for operations such as read and write. It prevents any updates or deletes on the data stored in the network. These mechanisms can help in improving the security of IoT systems using the advantages provided by the blockchain. To use blockchain for IoT, it is necessary to track the parameters that need to be stored in the blockchain. Since the data flows from one device to another within the peer-to-peer network, few parameters that can be tracked include transaction id, timestamp, data being communicated, device id, and device name. These parameters can help in tracking the compromised devices and the history of transactions from or to the system, thereby helping in identifying the attacker.

3.5 Ethereum

Ethereum is an open-source public blockchain platform that was first proposed by Vatalik Buterin in 2013 and introduced in 2015. Ethereum also has a cryptocurrency called Ether which is the second-largest cryptocurrency in the world after bitcoin. Ethereum has spanned further in the blockchain domain. It also provides a distributed computing platform, unlike bitcoin. This platform enables users to build distributed applications (d-apps). A major component of these d-apps includes blockchain. These apps use blockchain as the background in their working model. Since Ethereum is open-source, it also allows the development and testing of the d-apps and blockchain models built. The access to the private is provided using programming languages such as solidity, serpent, etc. These languages are similar to python and are supported by rich documentation and an active development community. The platform allows compilation, execution, and simulation of the blockchain. The public blockchain in Ethereum is open to the internet. Once registered, miners can contribute to the network by computing the nonce of a block before it is added to the chain. This task requires high amounts of computational power and time. The miners are awarded for the same using the ether. Ethereum also allows users who generate the transactions to set an extra amount in cryptocurrency that is transferred to add the transaction to the blockchain. This extra incentive can attract more miners. Ethereum can be thought of as a transaction-state machine, which means that when a transaction is executed on the system, it changes its state and stores the same. Ethereum provides smart contracts. These are immutable rules or protocols which cannot be changed. These rules provide the interface to access the blockchain and must be followed by the users.

3.6 Smart Contracts

The smart contracts are digital contracts that enforce and control the access to the blockchain being built. These rules are similar to protocols and must be followed in transactions and access. Ethereum provides immutable smart contracts that can not be changed once compiled and deployed on the blockchain. A smart contract can be understood as a contract set between two parties in a deal. The contract would be a list of specifications, conditions, and terms. Both parties of the transactions will proceed only when the contract is valid. When the contract specifications or conditions are met, the contract holds and is said to be valid. Necessary actions are taken only when the contract is valid. Generally, this action would be a transaction among the parties. In case the deals are not abiding by the rules of the contract, the transaction can be penalized or considered invalid and dropped, and no further action is taken. Similarly, a smart contract is a set of rules written which are applied on the blockchain when a developer calls it. A user needs to communicate with the blockchain using smart contracts, be it reading data from the blockchain or writing data to the blockchain. Ethereum allows developers to create unlimited smart contracts. The advantage of a smart contract is the elimination of a middle man or a mediator for the transactions to be executed: Transactions are executed only if the rules of the contract are met. The smart contract works automatically. Developers can leverage this feature and eliminate middlemen and mediators and ensure a transparent transaction between the two parties. This can improve the trust in the system and hence elevate the user experience.

4 Implementation Module

4.1 Scenario

For this project, three raspberry pis have been chosen to mimic the actions of a camera, HVAC system, and voice assistant(Alexa). Each raspberry pi will act as nodes of the peer-to-peer network and will use the Ethereum platform to access the blockchain. Storing data on the blockchain is an expensive voyage to carry out because blockchain is best suited for transactions and cannot store large amounts of raw data. Instead, the hash value of the data is stored on the blockchain and the raw data and transaction details regarding the transaction on secondary cloud storage.

Unlike the client–server model, where there is a single point of failure, Ethereum is a distributed platform, where every node present in the blockchain has the copy of the chain and the transactions involved in it. Consensus algorithm helps execute and store the transactions into the network, making it highly difficult for attackers to tamper with the data, using these characteristics an IoT system is built, which can stand against many attacks like denial of service attacks.

Data is stored on the blockchain using a smart contract and every node in the blockchain can use the same smart contract to add the hash of the data to the blockchain provided they are members of the network. The data stored on the blockchain is the hash of the actual value generated by the SHA256 algorithm. This ensures that nobody on the network can view the transaction details. Storing the hash value instead of the actual data also reduces the amount of data stored on the blockchain drastically. Data is retrieved from the blockchain and compared with the data stored on the cloud storage; if the hash of the raw data matches the value retrieved from the blockchain, then it can be said for sure that the data is not tampered with. This data is later shown to the user on a web interface.

4.2 Ethereum Model

Launched in 2015, it is the world's programmable blockchain. Like other blockchains, it has a cryptocurrency called Ether (ETH). Unlike the client–server model, Ethereum is a distributed computing platform in which data is distributed to all the nodes storing the blockchain, therefore even if one copy of the blockchain is corrupted, other copies can be used. Ethereum allows developers to build new kinds of applications. These decentralized applications (d-apps) gain the advantages of blockchain technology. They are reliable and predictable, It will always run as programmed when uploaded to the Ethereum network. Because blockchain is decentralized and the data on it cannot be tampered DDoS attacks and forgery attacks can be avoided. In this project, the Ethereum blockchain platform is used to hold the SHA256 value of the data generated by the IoT devices.

4.3 Smart Contract

Smart contracts are computer programs that dictate the way nodes interact with the blockchain. In this project, having one smart contract using which the participating nodes(camera, HVAC, voice assistant) read or write data to the blockchain. The smart contract implemented mainly has two functions one is to write to a transaction, the function takes as input id, and the SHA256 output of the data as input. The provided id and SHA256 output of the data are stored in the blockchain; the id becomes crucial in storing data in a blockchain, without the id retrieving this data is highly difficult.

The second function is used to read data from the blockchain or precisely the transaction data or the SHA256 value of the data, the function takes the id as input and then searches the blockchain for the data stored corresponding to this id and returns it.

4.4 Architecture

The entire system consists of three raspberry pis which act as IoT devices, the camera stores its video, the HVAC system stores sensor data, and voice assistant stores the voice commands. Blockchain is used as an authentication and verification platform and not as a data storage platform because storing large data on the blockchain is not feasible as the blockchain is held by every node on the chain. The user must register all the nodes in the network to the blockchain by making a transaction. with the device id. Therefore, the device IDs of all the devices registered by the user are present on the blockchain. When the devices have to store the data, it is first checked with the blockchain if it is a registered device by invoking the read function using the device id. If the device id is already present on the blockchain, it is allowed to go ahead with the store transaction. The above method ensures the authenticity of devices using blockchain.

Once the authentication is completed, the SHA256 output of the data is stored in the blockchain with a user-defined id via the smart contract. The raw data is written to the cloud storage with the id defined by the user and the transaction hash. This id is highly crucial to retrieve the data stored on the blockchain. When the owner wants to view the output of the devices through the web interface, he/she requests for the data, the respective data is extracted from the database. A request is now sent to the blockchain to retrieve the data of the id, and if present data is read from the blockchain, the data is in SHA256 format. Therefore, the raw data extracted from the database is converted to SHA256 value. This value is compared with the data retrieved from the blockchain if a match occurs the owner can rest assured that the data is genuine. Therefore, to sum it up only devices registered to the IoT system can access the blockchain, and data tampering can be detected in case of any.

5   Results

Through this project, some of the major security issues in IoT generated data have been able to address. Prime among them are those related to protecting the data generated by the IoT devices from tampering and altering or deleting. Data confidentiality, data integrity, and tamper detection of the database are also addressed. The data requested by the user is validated before servicing this request.

Through the course of the project, found that the major reason for such security breaches is that the network is not secure. Any new node in the network must prove its authenticity before being able to participate in the transactions of the network. The authenticity is ensured by allowing only devices with registered Ethereum accounts to add or read from the blockchain.

The blockchain's distributed ledger is used to store our transactions, this ledger system makes it hard to alter the data because to do so, a hacker must take control of more than 50% of the network and must alter the hash of all the blocks which is computationally very expensive. Therefore can be ensured that data tampering is prevented.

A major shortcoming of a blockchain network is that storing the data is an expensive process. The use of a SHA256 hash function to generate the hash values, which are then stored in the blockchain, is a good workaround to address this. The data present in the blockchain can be viewed by every node in the network, but the data is encrypted using the SHA256 algorithm and then store it in the blockchain ensuring confidentiality.

When a user requests for data from the database, the data is retrieved and run the SHA256 algorithm over the transaction value of that particular transaction and then match this to the data present in the blockchain using the transaction ID. The user can stay rest assured about the data in the blockchain, but by matching the data present in the database to the data in the blockchain can be detected if the data is genuine in the database from the time it was stored in the blockchain. Even though the project protects data up to a certain extent, access control is necessary to display the data to the concerned user.

A web interface is built to demonstrate the functioning of the IoT system and blockchain. The web interface allows the user to view the live readings from the device network. In the case of a camera, the web interface displays the live feed onto the web page. For this purpose, flask is used. The web application uses the database and makes requests to the blockchain. In the case of sensory devices like HVAC, the history of readings collected is displayed. Sometimes the processed data is preferred over the raw input. For instance, the voice assistant converts the voice requests into text for further processing. The web interface allows access to this history as well. The following are the screenshots of the application.

Figure 3 shows the live feed from the camera which can be viewed by the client from a remote location. Similarly, Fig. 4 shows the temperature readings and humidity readings obtained from the HVAC system. Figure 5 shows the history of voice commands made to the voice assistant. The above diagrams of the interface show how the client can view and monitor the devices.

6 Future Scope

This project currently runs on a public blockchain. It can be restricted to a private blockchain to enforce strict access control. The project can be extended to use a more secure storage mechanism such as a distributed file system. The blockchain model does not handle all the security vulnerabilities of the IoT network, and hence must be incorporated along with the existing models. The future scope of this project lies in the fact that blockchain can maintain a secure accounting of the transactions in a peer-to-peer network. It can also help in identifying security breaches in the network and securing against the breach in the future.

7 Conclusion

IoT technology is increasing at a rapid pace. Its advantages are immense. Securing the data generated by the IoT network is crucial to ensure the growth and development of IoT. Further, authentication and authorization must be ensured on the IoT system. Securing the data can establish data privacy and eliminate data forgery and data tampering. These measures will improve the user experience and trust in the IoT system.

Blockchain provides this solution with its decentralized and immutable nature. In this project, these features are used to address the major issue of data security. Access control is guaranteed by ensuring that only registered devices can access the blockchain. Further, a decentralized and immutable nature enforces data availability and security.

Data forgery is prevented by the use of SHA256 hashing to encrypt the data before storing it in the blockchain. Data tampering is also avoided using authentication, authorization, and verification steps. Other attacks such as man in the middle are avoided by the use of the nonce field in the blockchain. Thus, using blockchain for IoT devices provides a good blend between the technologies and can avoid DDoS attacks, forgery attacks. The blockchain is used along with the existing IoT security models guaranteeing enhanced security in the system.

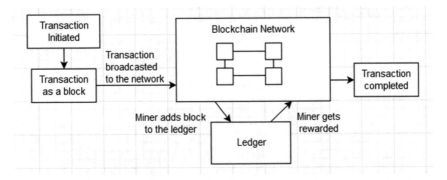

Fig. 1 Working of a blockchain

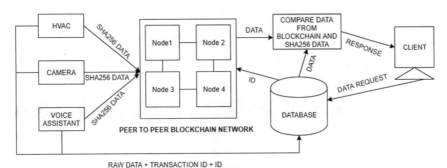

Fig. 2 Block architecture of the proposed model

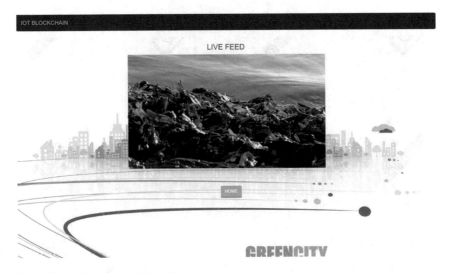

Fig. 3 Screenshot of the web interface—1

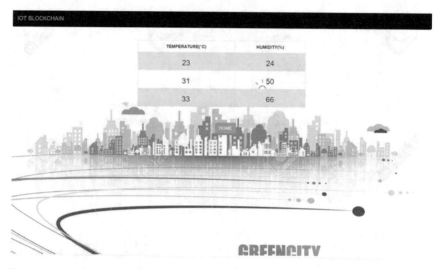

Fig. 4 Screenshot of the web interface—2

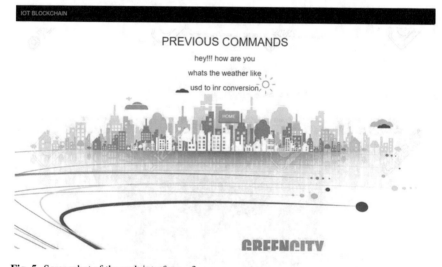

Fig. 5 Screenshot of the web interface—3

Reference

1. Antonakakis M, April et al (2017) Understanding the Mirai Botnet. In: 26th USENIX security symposium
2. Nakamoto S (2008) Bitcoin: a peer-to-peer electronic cash system. In: White paper
3. Wood G (2014) Ethereum: a secure decentralised generalized transaction ledger
4. Soumyalatha S (2016) Study of IoT: understanding IoT architecture, applications, issues, and challenges. Int J Adv Netw Appl (IJANA)

5. Ahmad MK, Salah K (2018) IoT security: Review, blockchain solutions, and open challenges. Future Gener Comput Syst 82:395–411
6. Varghese C (2019) IoT device management using blockchain. Int J Sci Eng Technol Res (IJSETR) 8
7. Wang Z, Dong X, Li Y, Fang L, Chen P (2018) IoT security model and performance evaluation: a blockchain approach. In: International conference on network infrastructure and digital content (IC-NIDC), pp 260–264
8. Gubbi J, Buyyab R, Marusic S, Palaniswami M (2013) Internet of things (IoT): a vision, architectural elements, and future directions. In: Future generation computer systems
9. Kumar JS, Patel DR (2014) A survey on internet of things: security and privacy issues. Int J Comput Appl 90
10. Atzori L, Iera A, Morabito G (2010) The internet of things: a survey. Computer networks
11. Huh S, Cho S, Kim S (2017) Managing IoT devices using blockchainplatform. In: 19th international conference on advanced communication technology (ICACT), pp 464–467
12. Crosby M, Nachiappan Pattanayak P, Verma S, Kalyanaraman V (2016) Blockchain technology: beyond bitcoin
13. Hofmann F, Wurster S, Ron E, Böhmecke-Schwafert M et al (2017) The immutability concept of blockchains and benefits of early standardization. In: 2017 ITU Kaleidoscope: challenges for a data-driven Society (ITU K), Nanjing, pp 1–8

Varying Definitions of LMS to Promote and Enhance Transactional Commitments on Account of Widespread Pandemic

Surekha Lanka, Narinder Singh Bagh, and Shrimoyee M. Sen

Abstract This paper is going to focus on concepts of space-based architecture (Blackboard system) to evaluate how well the components of the system function using the space paradigm and how it could be further implemented to the benefit of the students. The purpose of it is to increase space and performance in the data grid, which is composed of units that are independent of each other to attain high-performance application since application workloads are amplifying regularly. Especially in the current scenario of the current pandemic, every educational institution is required to continue their classes in an online delivery mode. The blackboard system is assimilated from a set of processing units. Blackboard has gained its popularity since the early stages, it is used by many educational managements including some of the health care services to teach, provide, and associate with one another.

Keywords Learning management system (LMS) · Tuple space paradigm · Space-based architecture · Dynamic scaling · Knowledge resources

1 Introduction

Many virtual training simulations have been introduced in the human-in-the-loop type. Many computers are needed on either side to use the system effectively and improve performance. To recollect, space-based architecture (SBA) is used to attain high-performance application through linear scalability and increasing simulation usage and its effectiveness. So, defining this problem is to integrate with the autonomous intelligent simulation model called AISM [1]. Which contain views that are used to describe a particular system, in this case, blackboard system. These views

S. Lanka (✉) · N. S. Bagh
Business and Technology Program, Stamford International University, Bangkok, Thailand
e-mail: surekha.Lnaka@stamford.edu

S. M. Sen
Faculty of Business Management and Marketing, Stamford International University, Bangkok, Thailand
e-mail: shrimoyeemukherjee@gmail.com

come from the stakeholder who is project managers, developers, system-engineer, and end-user. The aim is to obtain a well-built system which can qualify users interacting with multiple processes at the same time in addition to having a space that can maintain, transfer, and implement large data in the memory. To handle multiple tasks, the system would require space which is obtained by tuple space paradigm, the definition is distributed shared memory, for example, when a user enters the data into the system and tries to retrieve some information back, the data that was sent is partitioned in a certain pattern that the information that is sent back to the user is returned as the same pattern. This explanation is like a blackboard metaphor. It also involves parallel computing such as running accessed data or information concurrently. Blackboard systems consist of knowledge sources (KS) and are similar to hierarchical systems which have their highest priority at the top such as a top-down pyramid. System requirements constantly change. Therefore, software architectures should be able to manage agile processes and need to have the ability to meet future changes and system needs [2]. As there are many platforms available to teach and educate students, blackboard collaboration still proves to deliver effective online education. The two tools that are very similar in terms of functionality are zoom and google classroom [3]. Zoom has been engaged for online meeting, conferencing, and live chat, whereas google classroom is an easy option for schools to integrate into the system. However, blackboard would be more convenient as google classroom does not allow access from multiple domains, there are no automated updates and there are limited integration options. Blackboard collaboration could be more preferable as it is free while the zoom is only free for 45-min meetings along with having some secured issues where hackers can enter the session gaining sensitive information [4].

2 Background

In response to an ongoing pandemic of Covid19, there has been significant demand for school closures in 143 countries. With this sudden shift of not being able to show up at schools and universities, students will still be able to keep up their education and development by implementing the adoption of online learning. With many classes adopting virtual education colleges and universities have expressed significant interest to offer degrees and courses via the internet. According to Cruz et al. "The most collaborative experience is experienced since the beginning of the World Wide Web" In regards to technology challenges, online learners will be provided with guidelines on how to access and operational is the variety of the features and resources offered via blackboard collaboration. In addition, there will be peer-to-peer and student-to-faculty synchronous interaction [5]. This report will be focused on the concepts of space-based architecture (Blackboard system). Where the topic of leveraging the secrets of space-based will be looked up for more scalable API messaging, the benefits of in-memory computing, integration of middleware layers that are very much in need of the power, processing, and capabilities that the memory can provide. Space works together to store large amounts of data in memory and then archive that for high performance, dynamic scaling, and failsafe redundancy but they are also able to replicate and partition across all machines in the data grid, providing a very high degree of availability even when the machine or software fails. This idea of being able to use this persistence of information across many machines is very exciting. Now, the API processes can execute as they enter the system meaning there is no delay, as they enter the system, they can execute. This reduces many network hops and increases the security of the entire application environment through straightforward scalability and maintenance.

3   Framework Representation

Some designs such as application program interface, components of the system, implementation view, logical view, deployment view, data view, and the flowchart. To see what benefits it can provide to the students, how all the components work together, and how information is being inputted in the system.

1. Process view: The main concept is communication, how the students access their virtual classes will be looked at using the link that is provided, how the link was created, how the moderator invites the student to join the class. The diagram Figs. 1 and 2 will help system integrators understand how all the process runs together in parallel time and what activities are taking place.

   Target audience—Students in this particular section will be able to log in into online sessions where the link will be provided by the instructor but the intended audience would pertain to a system integrator.

   Area—Here, the distribution of functionality will be looked at and the synchronization between components which will be part of the non-functional requirement.

   Related Artifact—Not a specific artifact.

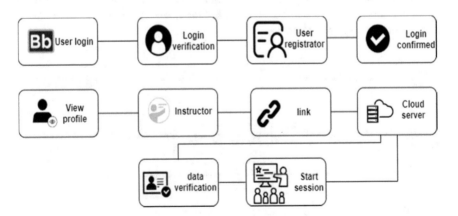

Fig. 1 Illustrates the API used by the students to gain access to virtual sessions

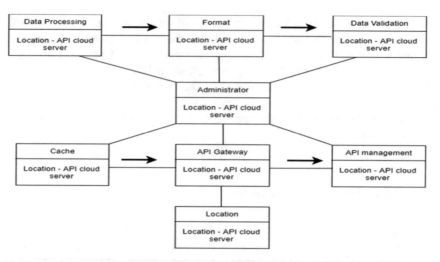

Fig. 2 Several processes running together in API cloud server

From Fig. 2 displays the process view of the blackboard system which will help to better understand how virtual classes are created, how the link is provided to the students, and how the students join the session at any location.

2. Implementation view: Blackboard system is a knowledge source which includes hierarchy concepts which have their highest priority at the top. This view will help the system programmers to understand the components and the modules of the system such as data grid, processing, and certain hypertext preprocessors used. Our space-based architecture has been used to distribute components from a service creation framework to dynamic environments [6]. The diagram Fig. 3 will display the scheduled design of the blackboard system.

Target audience—Students in this particular segment can set up their task by using scheduler connected to a database control but the intended audience would pertain to programmers, system engineers.

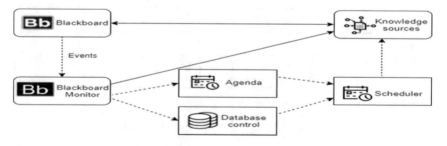

Fig. 3 Illustrates each knowledge source that needs to be executed

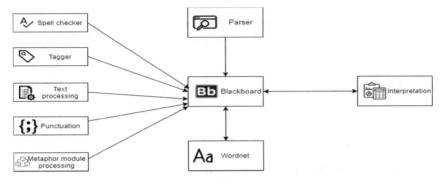

Fig. 4 Illustrates some tools and software required before the final execution of a task

Area—This area displays the components of the blackboard system including the interaction between different knowledge sources such as feature learning, classifiers, and decision engines.

Related Artifact—Implementation model, components.

3. Deployment view: Representation of architecture as a whole system, the kind of modules, components, hardware, and software used to deploy high design applications. A deployment will also consist of nodes which will describe the tools used in the system, these tools can always be implemented before the final execution state of the system. Figure 4 will display various modules and components, servers, hardware, and software used in the blackboard system.

Every data that is entered into the system will go through a spell checker, tagger, text processing, punctuation, and metaphor processing for matching contents with external contents. Parser means to resolve something into components which would describe the meaning, for example, the code integrated into the system is non-readable, therefore, enabling the code into human-readable form. The purpose of the parser is to analyze the source of the information and interpret the result.

Target audience—The diagram above displays the tools that are implemented in the system for students to make use of during the process of accomplishing the task, the intended audience here would be system deployment manager, who ensures that the hardware and software are fully functioning.

Area—Here, the modules will be looked at, components, hardware, and software used in the development environment.

Related Artifact—Deployment model.

4 Blackboard Architectural Pattern

There this platform aims at providing logical online information sharing between entities of software and user devices. So, utilizing this platform makes it feasible to

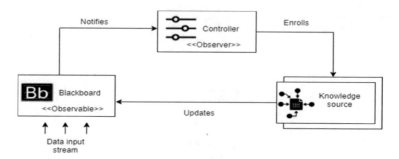

Fig. 5 Structure of the blackboard data input stream

further system development, which includes new sources of information and services, and to make the system highly scalable [7, 8]. The architectural design pattern is top to down approach as a hierarchy design which has their highest priority at the top, this design pattern helps to solve current problems that occur inside the system, using predictions that are backed up by some recommendation. Different algorithms are used to solve a particular problem and these algorithms are based on the result of another algorithm. Blackboard pattern includes words and segments from the intelligence community for a closed solution in a particular domain. The blackboard pattern includes its contents such as applications, implementation such as identification, speech identification, and consequences where quality attributes are met such as modifiability and reusability. The structure of a blackboard pattern involves three factors, these factors are class, collaborators, and responsibility which will be illustrated in Fig. 5 that will display the structure of updated information on the blackboard.

5 Framework Style

The layers of architectural style will be looked at, there are many layers such as resource-oriented layer and service-oriented layer, and web-application layered will be focused on and service-oriented layers which meet the functional and non-functional requirements that will be further discussed in the report. The PERA architecture that is modeled in multiple layers will also be looked at and stages including some of the architectural structures. Figure 6, there are five layers. The first layer is the user, who utilizes the system. Next is the environment which is the web that will be executed through blackboard collaboration and evaluator in the third layer. The common service layer provides some functionalities that are carried out by the user such as authentication, chat, and poll. The fifth layer is the communication middleware, which is run through the tuple space server, where several processes will be run simultaneously [9].

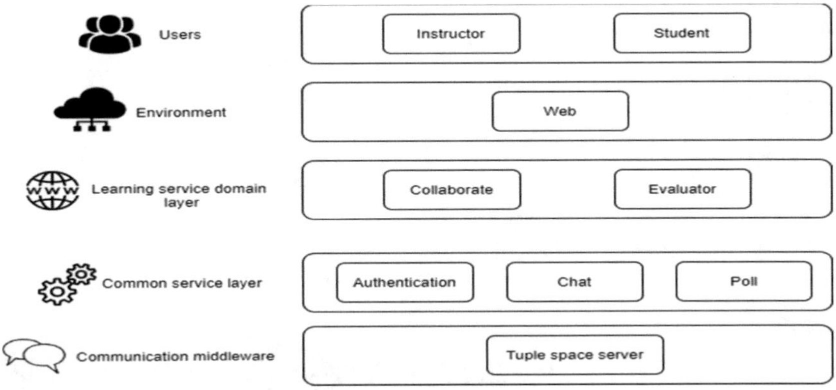

Fig. 6 Service-oriented layers of blackboard system during execution

6 Functional Flowchart

The space-based system technology aims at the seamless incorporation of various devices by developing pervasive computing environments, where various administrations can share information, perform operations, and interact with each other for joint task solving [10–12].

Pictorial representation explains the workflow of the blackboard system, like how students and instructors are logging in? And how they are accessing files? The flowchart below will illustrate how courses are being added, removed, and modified [13] which will be referred to as Fig. 7 of the report.

The aim here is to understand the key concept of how certain steps are required before a particular task is executed. A lot of time seem to enjoy the process of how the system works very well, this is because the system integrator, system designer, and system analyst makes sure that the performance of the system is high but behind every particular task that is executed can always be an error, therefore, the flowchart helps them understand how every step if necessary to go on to the next step, if there is a single step that is missing, then the entire system could be disastrous.

7 Testing

Usually goes through the validation of the system and its functionality to get the expected result, which is carried out before a program and its services are launched in the market, this is done to ensure that the bugs in the system have been removed which involves certain steps that will be looked into further in the report. There are different procedures used for testing such as quality assurance personnel and penetration testing to ensure the quality and accuracy of the product.

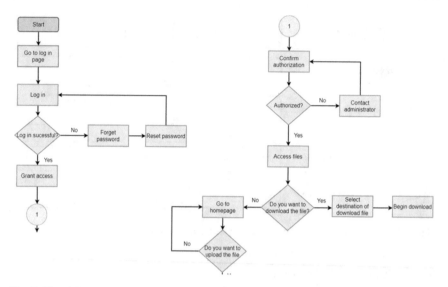

Fig. 7 Pictorial representation of blackboard files download and its accessibility

1. System's safety: The system will check for any possible attack or risk that can be harmful to the system. This usually involves many ways of dealing with the risks such as risk mitigation, risk acceptance, and risk exploitation. The system will always make sure that penetration testing is done to reveal all the vulnerability and reduce the risk of hackers attacking the system. The attacking can be of two types which as passive attack and active attack, a passive attack is where the user does not know the attack, whereas in active attack the user knows that someone is trying to gain access to the system. These terms are very important in security testing because if the system cannot analyze the threat, then the system cannot be protected. This is why security testing is done to ensure the threats that are minimized and no unauthorized user can access the activity taking place in the system. Table 1 will display a scenario as to what happens when a hacker tries to gain access to the system; this is to test the security of the system. In the below table which clearly shows a security scenario while an unauthorized user is trying to access the system [14].

2. Performance test: Is all about testing the speed, effectiveness, and the time taken for the system to respond to the user. No matter what kind of project is in the place, the performance should always be high. The performance of the system is tested within a certain set of the prescribed time; if the system takes longer than prescribed time, then the system developers and system analyst must go through every component of the system and document each of them in the form of documentation or video to check which component will be more effective and the ones that need to be changed. In the table below, a scenario during the performance testing of the system will be displayed [15].

Table 1 The testing output of security and performance outline

Portion of scenario	Possible values (security)	Possible values (performances)
Source	Hacker	System components
Stimulus	Attempt to log in to the system using malicious tools	Periodic events, a transaction of the system
Artifact	Cloud database	Cloud database
Environment	Normal operation or unusual operation	During average and overload use
Response	Results for response: – The system blocks the access of unauthorized IP – The system stores the information for further search – Administrator changes the security protocols	Results for response: – The system works fast during average use – Takes 28.2 s for the system to update information during average use – Takes 48.6 s or up to a minute to update information during overload use
Response measure	Any information downloaded will self-destruct with 35.7 s, ensuring the confidentiality and integrity of the system	Non-repudiation takes place if the debt is not removed from the system

8 Analysis of Existing Virtual Learning Methods

Learning management systems requires comprehensive toolkit and reporting tools for visualizing and analyzing valuable data in the area of online courses [16]. Few of the features that LMS must be ensured before using it are course status, last access by user, number of students enrolled in a course, number of students passing a course, course completion, total time spent by the students, and performance grade. Depending on the level of service and the budget, university is looking for, still, the quality of service must be ensured [17]. Some of the existing virtual learning models are Electa and LearnCube. Both of these models are used for virtual classrooms for anyone, anywhere. The drawback of using these models is the design, which is not very convenient for most people to use and the session capacities are 25 or 100 attendees along with only 14-days free trial. To have effective LMS, the system must be able to generate real-time reports, maximize training efficiency, record success metrics, and analyze the cost-effectiveness of the system [18]. Now, let's look at Fig. 8 which will illustrate what an effective LMS should contain.

Limitation of the system: Since every system has limitations even the blackboard system has some limitations as well, let's discuss the pros and cons of the LMS [19].

Another view of the limitations: In the blackboard system tools, mathematical equations are found to be impossible to solve and there are very few symbols in the tool. Students are not able to customize the appearance of the blackboard website to their comfort. For example, moving the announcement from top to bottom, only administrators are allowed to make changes. The system does not recommend any

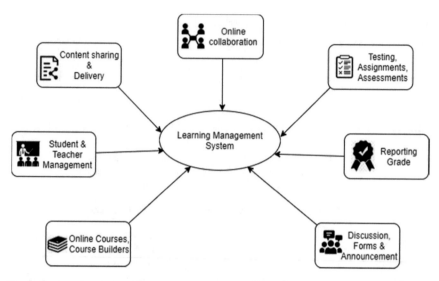

Fig. 8 Contentment of learning management system (LMS)

links or tools regarding the course which can be beneficial to students, only instructors of the particular course can provide the link [3]. This means the system does not have a recommender interface.

9 Conclusion

Every project experiences some drawback at some degree but can be concluded that this paper has met all the specifications and requirements which will successfully satisfy the end-user. When selecting an architecture, have to keep in mind that the result of the decision must be to handle the situation the right way, risk will always be there, and the risk can be simply reduced. There is no question that space is necessary for most of the system, without space the system cannot be implemented. With more time, the functionality can be developed and improved, add more features to the blackboard system for better user experience, making it more effective, and increasing its value concerning space-based architecture. In this paper, to eliminate the system from crashing is clearly understood and to reduce the risk at the same time, it always increases its value by adding more features and functionality to it. However, many universities have unique portals to share the courses materials and communication with instructors but they cannot handle all the tasks auto attendance, interaction with students, polling, which are similar to face to face classes. But the space-based architecture can handle multiple tasks and also available as a mobile application to attend the classes via mobile, user can simply share, download, and submit the assignments as well.

References

1. Lee SK, Yoo CG, Park JC, Park JH, Kang MG (2010) Design of the autonomous intelligent simulation model (AISM) using computer generated. In: Force"ICOSSSE'10: proceedings of the 9th WSEAS international conference on system science and simulation in engineering, October 2010, pp 302–307, available online at https://www.wseas.us/
2. Alokluk J (2018) The effectiveness of blackboard system, uses and limitations in information management. Intell Inf Manag 10:133–149. https://doi.org/10.4236/iim.2018.106012
3. Zoom is now free for K-12 schools! What you need to know (Internet). DGI communications. 2020 [cited 2020Aug15]. Available from: https://www.dgicommunications.com/zoom-is-now-free-for-k-12-schools-what-you-and-your-students-need-to-know/
4. Pappas C (2020) Google classroom review: pros and cons of using google classroom in eLearning (Internet). eLearning Industry. 2020 (cited 2020Aug15). Available from: https://elearningindustry.com/google-classroom-review-pros-and-cons-of-using-google-classroom-in-elearning
5. Politis J, Politis D (2016) The relationship between an online synchronous learning environment and knowledge acquisition skills and traits: the blackboard collaborate experience. Electron J e-Learn 14(3):196–222. Available online at www.ejel.org
6. Engelhardtsen FB, Gagnes T (2002) Using JavaSpaces to create adaptive distributed systems
7. Honkola J, Laine H, Brown R, Tyrkko O (2010) Smart-M3 information sharing platform. In: Proceedings of ISCC 2010 IEEE comp society 2010, pp 1041–1046
8. Smirnov A, Kashevnik A, Shilov N, Gusikhin O (2014) Context-driven on-board information support: Smart space-based architecture. In: 2014 6th international congress on ultra-modern telecommunications and control systems and workshops (ICUMT), St Petersburg 2014, pp 99–104. 10.1109/ICUMT.2014.7002086
9. Gopal R, Kay S, Lykins J, Ravishankar C, Vasavada Y (2018) System and architecture for space-based and mobile terrestrial sensor vehicles, and end-to-end network for aggregation and processing of sensor data. US Patent 10,063,311
10. Kiljander J, Ylisaukko-oja A, Takalo-Mattila J, Eteläperä M, Soininen J-P (2012) Enabling semantic technology empowered smart spaces. Comput Netw Commun
11. Balandin S, Waris H (2009) Key properties in the development of smart spaces. In: Proceedings of 5th international conference universal access in human-computer interaction, Springer, pp 3–12
12. Korzun D, Balandin S, Gurtov A (2013) Deployment of smart spaces in internet of things: overview of the design challenges Internet of Things. Smart spaces, and next generation networking. Springer, LNCS vol 8121, pp 48–59
13. Richards M, Software architecture patterns. O'Reilly Online Learning, O'Reilly Media,Inc.,ww.oreilly.com/library/view/software-architecture-patterns/9781491971437/ch05.html
14. Support, Blackboard, and More contact information. System requirements. Blackboard Student Support, https://ualr.edu/blackboard/welcome/system-requirements/. Accessed 29 July 2020
15. Cao S et al (2019) Space-based cloud-fog computing architecture and its applications. In: 2019 IEEE world congress on services (Services) Milan, Italy, pp 166–171. 10.1109/SERVICES.2019.00048
16. Gautam A (2019) 5 key reasons to choose an LMS with built-in reporting & analytics (Internet). eLearning industry. 2019 (cited 15 Aug 2020). Available from: https://elearningindustry.com/lms-reporting-key-reasons-choose-built-in-analytics
17. LMS reporting: why do we need data in digital learning process? (Internet). What LMS reporting tools do you need? (cited 15 Aug 2020). Available from: https://raccoongang.com/blog/lms-reporting-why-do-we-need-data-digital-learning-process/
18. Communications E (2020) Virtual classroom software (Internet). Virtual classroom software, Online training software, Electa LMS (cited 15 Aug 2020). Available from: https://www.e-lecta.com/

19. Hamad MM (2017) Pros & cons of using blackboard collaborate for blended learning on students' learning outcomes. High Educ Stud 7(2):7–16

A Survey on Recent Advances in Cyber Assault Detection Using Machine Learning and Deep Learning

Piyusha S. Pakhare, Shoba Krishnan, and Nadir N. Charniya

Abstract Cyber attacks hit companies, businesses, and common people every day. Cybercrime is increasing year by year as criminals that are trying to benefit from vulnerable sources. Software attacks are very difficult to detect as it hides in a very sophisticated way on the network. This survey paper gives a review of various machine learning (ML) methods used to detect different attacks. Several methods/architectures developed by researchers to detect cybercrimes using deep learning and machine learning techniques of classification are also discussed. It can be seen that machine learning and deep learning models are efficient in detecting cybercrimes with high accuracy when proper training is given.

Keywords Cyber attacks · Supervised learning · Unsupervised learning · Deep learning · Machine learning

1 Introduction

Cyber security is a very popular term today. Nowadays, almost every person uses computers and the Internet. People spend a long time on the Internet for conducting their banking transactions, shopping, communicating using e-mails, and other social media platforms. Due to this dependency, illegal computer activities are growing and changing like any type of crime. Any crime carried out using both computers and the Internet is called cybercrime or cyber attacks. Simply, cyber attacks are the attacks launched using one computer or more than one computer on some other computer/computers or a completely different network.

P. S. Pakhare (✉) · S. Krishnan · N. N. Charniya
V.E.S Institute of Technology, Mumbai, India
e-mail: 2018.piyusha.pakhare@ves.ac.in

S. Krishnan
e-mail: shoba.krishnan@ves.ac.in

N. N. Charniya
e-mail: nadir.charniya@ves.ac.in

© The Author(s), under exclusive license to Springer Nature Singapore Pte Ltd. 2021
J. S. Raj et al. (eds.), *Innovative Data Communication Technologies and Application*,
Lecture Notes on Data Engineering and Communications Technologies 59,
https://doi.org/10.1007/978-981-15-9651-3_47

Cybercriminals use malicious code to change or modify computer codes, logic, and even data to get access to confidential data, trade secrets, or sometimes just for identity theft. With an increase in communication and sharing the information (personal/public) on the Internet, as a result, "WWW" has become the main location for criminals to launch attacks and cause trouble.

Artificial intelligence is playing an active role in detecting cybercrimes. The new AI algorithms are unique in identifying data traffic, access, and transfer, and any distortions in "outliers" or data patterns. If something unusual happens, AI programs may need to dig deeper into the data to determine if the system has a security breach. Another way AI is used to protect against cyber attacks which are "supervised learning." The motivation for writing this survey paper is to civilize people about cyber attacks and prevent them from becoming victims of these attacks, as well as acquaint them with various machine learning and deep learning methods that have been developed to detect the attacks in the fastest way possible.

2 Literature Review

2.1 URL Interpretation Attack

Cybercriminals use URLs and change a certain part of it to redirect the victim to any fraud/fake website. Kumar et al. [1] have developed a multi-layer filtering model to detect bad URLs [2].

The first filter used is black and white. Whitelist allows schemes, domains, URLs, and blacklist blocks specific elements of your allowed locations. The second filter used is a naïve Bayesian filter. Alpha N-Bayes threshold training decides whether the URL is beneficial data for a naïve Bayesian model or not.

$$\alpha_{nbayes} = \max\left(\frac{P_1}{P_2}, \frac{P_2}{P_1}\right) \tag{1}$$

(where P_1 and P_2 represent malicious URL and benign URL, respectively; probability value [1] is calculated by the naïve Bayesian formula model). Considering some threshold value, if the value of a Bayes is greater than the threshold value, then that data is considered to be good for the naïve Bayesian model. The third filter used is the CART decision tree filter.

$$\alpha_{nbayes} = \max\left(\frac{n}{m}, \frac{m}{n}\right) \tag{2}$$

(where n makes up the number of leaf nodes for malicious URL; m makes up the number of leaf nodes for normal URL [1]).

If $\alpha_{nBayes} >$ threshold, then the URL is considered as benign; if not, then the classification result is recorded, and the URL is passed on to another filter. The last filter used is the SVM filter. This filter records its results and then combines it with naïve Bayesian filter and CART decision tree filter to collectively find the classification of the URL. In this way, the researchers have developed a model based on machine learning.

2.2 Distributed Denial of Service Attack

The major aim of the DDoS attack is to restrict access to legitimate users [3]. He et al. [4] have proposed a machine learning-based source side DDoS detection system. The architecture consists of virtual machine manager (VMM), a machine learning engine. VMM monitors the virtual machine (VM) and collects network traffic information and passes it to the machine learning engine which detects if any suspicious activity is present [4]. If any suspicious behavior is found in a single VM, then that VM is terminated [5]. If multiple VM shows suspicious behavior, then those networks are cast-off to defend against DDoS attack.

- Precision shows which part of the alarms is true alarms.

$$\text{Precision} = \frac{\text{TP}}{\text{TP} + \text{FP}} \tag{3}$$

- Recall shows the part of the attacks that are detected.

$$\text{Recall} = \frac{\text{TP}}{\text{TP} + \text{FN}} \tag{4}$$

- F_1 score is a frequently used standard to balance FP and FN.

$$F_1 = 2\frac{\text{Precision*Recall}}{\text{Precision} + \text{Recall}} \tag{5}$$

- A higher F_1 score indicates the good performance of the algorithm. It was found that supervised machine learning algorithms detect with good accuracy and recall as compared to the unsupervised machine learning techniques. Random forest supervised machine learning algorithm performs with the best accuracy, F_1 score and has the highest recall rate.

2.3 Malicious PDFs

PDFs can execute codes in the system. Criminals use this to infest the system with malware. Laskov and Šrndic [6] have developed a tool called PJSCAN which reliably detects PDF attacks with working false-positive rates. PJSCAN detects browser-based JavaScript attacks. The main indicator of JavaScript in the code is the keyword/JS. /JavaScript and /Rendition.

The architecture of PJSCAN proposed by the authors consists of two sections viz, feature extraction, and learning. Feature extraction part looks for JavaScript code in the PDF and does a lexical analysis on it. The resultant output is given as an input to the learning part. The model is then trained for detecting JavaScript in malicious PDF using an algorithm known as one-class support vector machine.

2.4 Brute—Force Attack

A brute-force attack is like trying to crack a password, username, or find a hidden web page or key used for encrypting a message, using a trial and error approach, and hoping, eventually, to guess correctly. Najafabadi et al. [7] have developed a machine learning algorithm for detecting SSH brute-force attack with the use of aggregated Netflow data at the network level. To extract the discriminating features, brute force Netflow is compared with the normal Netflow. By analyzing the features, it is found that, as compared to normal data, upload, download, and successful logins, number of bytes and number of packets in brute-force attack are very less. To build a predictive model, four classification algorithms used are 5-nearest neighbor (5-NN), naïve Bayes (NB), and two forms of C4.5 decision trees (C4.5D and C4.5N). Among the above models, the decision tree model has better performance and was easy to interpret.

2.5 Phishing Websites

Phishing is a type of social engineering attack that steals the user's data. The attacker masquerades as a trusted entity and dupes the victim into opening an email, instant message, or text message. Alswailem et al. [8] have used random forest machine learning classification technique to detect various phishing websites. Phishing websites have a definite pattern that can be considered as features. A total of 36 features of phishing websites was considered and categorized into three categories, namely features extracted from URLs, features extracted from page rank and page content. This study considered all the potential combinations of the named features. The final number of combinations was found using the formula:

$$\sum_{k=1}^{36} \frac{n!}{k!(n-k)!} \qquad (6)$$

(where k is the number of features that start from 1 to 36, and n is the number of all features which is 36).

To differentiate whether the website is phishing or legitimate, an algorithm is developed based on the random forest classification technique. This algorithm extracts the features using URL and DOM, i.e., document object model. URL extracts the URLs and page rank's features while the page content's features are extracted from DOM. These features are then sent to a classifier to differentiate between phishing and legitimate websites.

2.6 Man-in-The-Middle Attack

It is an active earwig where the hacker intercepts the connection between two legitimate hosts. He/she pretend to be the intended sender or receiver by intercepting the conversation through IP spoofing, HTTPS spoofing, wi-fi eavesdropping, and more. Dong et al. [9] have aimed a method to detect and find the location of Man-in-the-Middle (MITM) attack by examining the Round-Trip Time (RTT) in a fixed wireless network and measuring the received signal strength (RSS) from wi-fi access point. Using ARP cache poisoning, a testbed was deployed to simulate the MITM attack. The testbed environment included a client, a server, and a MITM attacker some distances apart. MITM attack was carried out using Ettercap (an open-source tool for launching a MITM attack on LAN) and Arpspoof (for launching ARP poisoning).

The proposed method to detect the MITM attack is to detect the variance in delay between normal transmission and intercepted transmission between two nodes. The client application measures the RTT from the client to the server. RTT is measured by taking the difference between the sent and receive time of the message. To calculate RSS, multiple wi-fi access points are in an indoor environment equal distances apart. Signal strength is found using the link budget equation. To predict the place of the attacker node, three machine learning algorithms viz. K-nearest neighbor (KNN), support vector machines (SVM), and Gaussian naive Bayes (GNB) are evaluated. From this, it was concluded that when MITM attack is present there is a huge delay and large standard deviation in measured RTT as compared to normal transmission. The measured RSS indicates the place of the server and the MITM attacker. Table 1 shows the performance accuracy of the ML models used to detect the above-mentioned attacks.

3 Machine Learning Algorithms Used for Cyber Attack Detection

3.1 Naïve Bayes Algorithm

Naïve Bayes algorithm is derived from the Bayes theorem. Naive Bayes classification algorithm is one of the most efficient text classification methods [10]. Bayes theorem finds the probability when other probabilities are known. The formula for naive Bayes is given below [11]:

$$P\left(\frac{A}{B}\right) = \frac{P(A)P\left(\frac{B}{A}\right)}{P(B)} \tag{7}$$

where $P(A/B)$ is the posterior probability of class A given predictor B.
$P(A)$ is the prior probability of class A
$P(B)$ is the prior probability of predictor B
$P(B/A)$ is a likelihood, the probability of predictor B given class A.
In real-world troubles, there are multiple A and B variables. In the case of cyber attacks, the naïve Bayes algorithm computes the probability of attacks as follows:
Step 1: Calculate the prior probability for a given class.
Step 2: Calculate conditional probability using each attribute for every class.
Step 3: Multiply the same class conditional probability.
step 4: Multiply prior probability with step 3 probability.
The major disadvantage of naive Bayes is that it assumes the predictors to be independent. It assumes that the attributes are mutually independent. But, in the practical world, getting a set of completely independent predictors is not possible.

3.2   Decision Tree Algorithm

The decision tree algorithm is a part of the supervised learning algorithm. This algorithm is used for solving the regression and classification [12] problems. The decision tree's structure is similar to a flowchart [13] as shown in Fig. 1.
Root node: It comprises of the entire population which is further divided into multiple homogeneous sets.
Leaf/Terminal node: Nodes that do not divide further are called leaf/terminal nodes.

Decision tree begins with a single node (root node) that represents all the records of the training dataset. If the records are in the same category, then that node becomes leaf node and is assigned a category to which it belongs. If not, then this

Fig. 1 The basic structure of a decision tree [13]

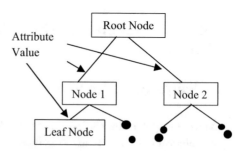

algorithm will use a method based on entropy, which is also called information gain and chooses the attribute which gives the best classification. This attribute becomes the test attribute of that node. For every test attribute's known value, a branch is created, and the dataset is divided into every branch in the same way. To create a sample decision tree for every division, this algorithm follows the same procedure recursively. This recursive division stops only when the below given conditions are true:

All the sample data of the given node belong to the same class.

After the split, certain branches do not have any sample records.

The major limitation of the decision tree is the overfitting problem.

3.3 Random Forest Algorithm

Random forest algorithm handles regression and classification [12] problems. A forest is a group of trees whereas random forest is a group of classification trees [14]. Random forest is constructed by merging the predictions of various trees; each one of them gets treated in isolation [15]. Regression trees, also called decision trees, are a tree that consists of the members of the decision/predictor variable on its leaf nodes and the entities of other dependent variables live on the intermediate nodes. For example, spam/no spam for emails, 0–9 for handwritten digits in pattern recognition, etc. Each tree gives its decision as output, and that output is its vote/solution for that problem. The last class is that class which has received most votes from the trees present in the forest. Working on this algorithm is in the image shown in Fig. 2.

1. Start by selecting random samples from the given dataset.
2. This algorithm will build a decision tree for every sample and will predict the output from every tree.

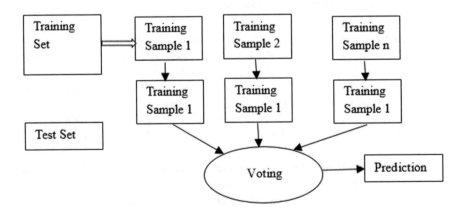

Fig. 2 Random forest algorithms steps [16]

Fig. 3 Support vector
machine [2]

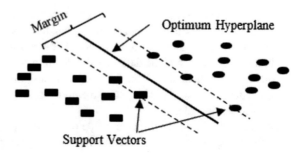

3. Voting takes place for every predicted output.
4. The output with most votes is the last output of the algorithm.

Constructing a random forest is very difficult and less intuitive when having a large accumulation of decision trees.

3.4 Support Vector Machine (SVM) Algorithm

The support vector machine belongs to the supervised machine learning algorithm. It is mainly considered for classification problems. It is reached from the principle of margin calculation [2]; i.e., it draws a margin between the classes as shown in Fig. 3. The margins are placed in such a way that the distance between the margin and the classes is long and minimizes the classification error. Large datasets need more training time. If the data consists of noise, in this case, target classes and overlapping, then SVM is a poor choice.

3.5 K-Nearest Neighbor (KNN) Algorithm

K-NN algorithm assorts the objects based on the closest training examples for pattern recognition inside the feature space. Instead of learning from the data, this algorithm memorizes it [17]. In this algorithm, each sample is given a class depending on its surrounding samples. This algorithm will predict the class in which the unclassified sample will belong by considering the class of its nearest samples.

Figure 4a shows the K-NN for $k = 1$; here, the new sample is classified with the help of only one known sample. Similarly, in Fig. 4b, the value of k is 4, which means the new sample is classified considering four of its closest/nearest neighbors. In this case, three neighbors are in the same class and one in a different class. Now, the class of unknown sample is predicted depending on the class in which its majority of the neighbors lie. The classification performance of the algorithm is dependent on the selection of K.

Fig. 4. **a** 1-NN decision rule, **b** k-NN [17]

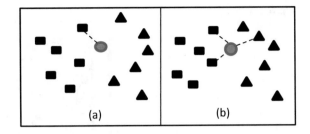

(a)   (b)

(If K is very small, then the data is mislabelled, and if K is very large, then many outliers from its neighborhood are involved that creates a problem. In the case of regression problems, the average of the K-NN output is the final prediction:

$$y = \frac{1}{k} \sum_{i=1}^{k} y_i \qquad (8)$$

(where y_i is the ith case of the samples, and y is the outcome of the query point).

4 Deep Learning Algorithm Used for Cyber attack Detection

4.1 Generative Adversarial Networks (GAN)

GAN is a combination of two neural networks viz, generator network and discriminator network. Figure 5 shows the working of GAN.

Generator network generates data samples, and the discriminator decides whether the data is newly generated or is a part of the original sample. This is done by using binary classification, where the output is in range 0–1. After training the model, the generator creates data that is difficult to differentiate from the original data. GANs solves the problem of creating data when do not have enough data. Also, it does not need human supervision. The best example is the Google brain project where researchers made use of GANs to develop techniques

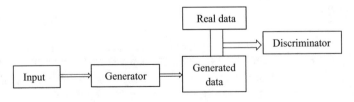

Fig. 5 Working of GAN [18]

Fig. 6 LSTM cell [20]

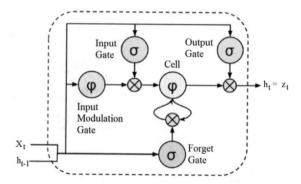

to encrypt the communication. Handling GAN is difficult, if the generator and discriminator are out of balance. The model will give unacceptable results.

4.2 Long Short-Term Memory (LSTM)

LSTM retains the information for a longer period which is a drawback of RNN. As a result, LSTM is widely used in name entity recognition, fraud detection, credit card frauds [19], etc. The cell structure of LSTM is shown in Fig. 6:

Forget gate: This gate removes the information which is not required by the cell. The result of this gate informs the cell state to forget particular information by multiplying that particular place by 0 in the matrix. Otherwise, if the result is 1, then the information is kept.

Input gate: This gate adds useful information to the cell. This gate is a sigmoid function in the range between [0,1]. Therefore, it only adds information and not forget/remove the data. To drop/delete the data, the tanh activation function is applied.

Output gate: This gate extracts utile information from the current state and gives it as an output. For this, first, a vector is created by applying the tanh function. Then, that information is modulated using a sigmoid function and filters the values to be recalled. Finally, the vector values are multiplied with the regulated values and send as output, i.e., input to the next cell.

5 Conclusion

People have become vulnerable to cyber attacks because of the easy availability of the Internet. Attackers come up with different ways to do fraud. With an increase in various types of cybercrime, there is a growing need of having a technology that can detect those crimes. Recently, machine learning algorithms are in increasing demand for detection and classification of malicious attacks. Attackers launch various

Table 1 Performance accuracy [1, 4, 6–8]

Authors	Year	Attacks	ML algorithms	Accuracy (%)
Kumar et al.	2017	Malicious URLs	Multi-layer filtering model	79.55
He et al.	2017	DDos	Random forest	94.96
Laskov and Šrndic	2011	Malicious PDFs	PJS$_{CAN}$	85
Najafabadi et al.	2015	Brute-force attack	5-NN	AUC—99.65
Alswailem et al.	2019	Phishing websites	Random forest	98.8

attacks to harm the victim, organization, companies. For the classification of these attacks, various machine learning and deep learning techniques are reviewed in this paper. Different architectures developed by the researchers related to various attack is depicted in this paper. Various machine learning algorithms like decision trees, naïve Bayes algorithm, random forest algorithm, PJSCAN, support vector machine, and deep learning algorithms like convolutional neural network, recurrent neural network (RNN), long short-term memory (LSTM), etc., are used by the researchers. All these algorithms use separate approaches to classify the data taken from the various datasets into benign or malicious. Some researchers have combined the algorithms to give a more efficient solution. From these classifications, the legitimate user will be alerted about the web page that he/she is opening which is genuine or not so that the user can take precautions to not fall in the traps of the cybercriminals. The future scope of this survey involves using ML algorithms (single or combined) for detecting various cybercrimes, calculating various parameters, and a comparative study of the new algorithms used.

References

1. Kumar R, Zhang X, Tariq HA, Khan RU (2017) Malicious URL detection using multi-layer filtering model. In: 14th international computer conference on wavelet active media technology and information processing. Chengdu, pp 97–100
2. Dey A (2016) Machine learning algorithms: a review. Int J Comput Sci Inf Technol 7(3):1174–1179
3. Smys S (2019) DDoS attack detection in telecommunication network using machine learning. J Ubiquit Comput Commun Technol 01(01):33–44
4. He Z, Zhang T, Lee RB (2017) Machine learning based DDoS attack detection from source side in cloud. In: IEEE 4th international conference on cyber security and cloud computing. New York, pp 114–120
5. Sadasivam GK, Hota C, Anand B (2016) Classification of SSH attacks using machine learning algorithms. In: 6th international conference on IT convergence and security. Prague, pp 1–6
6. Laskov P, Šrndic N (2011) Static detection of malicious javascript-bearing PDF documents. In: 27th annual computer security applications conference. Orlando, pp 373–382
7. Najafabadi MM, Khoshgoftaar TM, Calvert C, Kemp C (2015) Detection of SSH brute force attacks using aggregated netflow data. In: IEEE 14th international conference on machine learning and applications. Miami, pp 283–288

8. Alswailem A, Alabdullah B, Alrumayh N, Alsedrani A (2019) Detecting phishing websites using machine learning. In: 2nd international conference on computer applications & information security. Riyadh, Saudi Arabia, pp 1–6

9. Dong Z, Espejo R, Wan Zhuang YW (2015) Detecting and locating man-in-the-middle attacks in fixed wireless networks. J Comput Inf Technol, pp 283–293

10. Huang Y, Li L (2011) Naive Bayes classification algorithm based on small sample set. In: IEEE international conference on cloud computing and intelligence systems. Beijing, pp 34–39

11. Naïve Bayes for Machine Learning, https://machinelearningmastery.com/naive-bayes-for-mac hine-learning/

12. Classification and regression, https://www.geeksforgeeks.org/ml-classification-vs-regression/

13. Zhong Y (2016) The analysis of cases based on decision tree. In: 7th IEEE international conference on software engineering and service science. Beijing, pp 142–147

14. Jaiswal JK, Samikannu R (2017) Application of random forest algorithm on feature subset selection and classification and regression. In: World congress on computing and communication technologies. Tiruchirappalli, pp 65–68

15. Denil M, Matheson D, Nando de Freitas (2014) Narrowing the gap: random forests in theory and in practice. In: Proceedings of the 31st international conference on machine learning, vol 32 (2014)

16. Working of Random Forest, https://www.tutorialspoint.com/machine_learning_with_python/machine_learning_with_python_classification_algorithms_random_forest.htm

17. Imandoust SB, Bolandraftar M (2013) Application of K-nearest neighbor (KNN) approach for predicting economic events: theoretical background. Int J Eng Res Appl 3(5):605–610

18. Berman DS, Buczak AL, Chavis JS, Corbett CL (2019) A survey of deep learning methods for cyber security. MDPI

19. Wiese B, Omlin C (2009) Credit card transactions, fraud detection, and machine learning: modelling time with LSTM recurrent neural networks. In: Innovations in neural information paradigms and applications. Studies in computational intelligence, vol 247. Springer, Berlin, Heidelberg, pp 231–268

20. LSTM, https://medium.com/@kangeugine/long-short-term-memory-lstm-concept-cb3283 934359

Visual Attendance Recording System for Classroom Using Facial Features

Gunjan Bharadwaj and Pooja Pathak

Abstract Student attendance in schools, colleges, or universities plays a significant role in facilitating the learning process in a variety of ways. As there are a large number of students in a classroom, the main emphasis is given to detect and recognize all faces that are streamed in the camera. This process runs in parallel with the class lecture so that no extra time is required for calling out the names or roll numbers of the students to record the attendance. Face detection is used to locate the face in a image. There are two phases in automating the attendance process; first, the system is trained on the images of all students of the class. After this step, the system takes the live video of the class as input; it automatically detects and recognizes the face of each student present in the video and marks the presence of those students in a class. Those who are not found in the video will be marked absent automatically. The results show that the system performs well and saves nearly 5–10 min of a lecture.

Keywords Attendance automation · Face detection · Face recognition

1 Introduction

A large variety of biometric identification schemes are available for recording the identity of a person such as a fingerprint recognition [1], face recognition [2–4], iris detection [5], etc. Both fingerprint detection and iris detection schemes require one-to-one access of sophisticated hardware by each person, thereby consuming a lot of time when applied on a crowd of people. Face recognition system has proved its worth in many currently running online applications such as aadhar verification, passport authentication, criminal detection systems. It is about to be used in other applications like e-voting systems, individual identifications at the airports and online

G. Bharadwaj (✉)
Department of Computer Engineering and Applications, GLA University, Mathura 281406, India
e-mail: gunjan.bhartiya@gla.ac.in

P. Pathak
Department of Mathematics, GLA University, Mathura 281406, India
e-mail: pooja.pathak@gla.ac.in

© The Author(s), under exclusive license to Springer Nature Singapore Pte Ltd. 2021
J. S. Raj et al. (eds.), *Innovative Data Communication Technologies and Application*,
Lecture Notes on Data Engineering and Communications Technologies 59,
https://doi.org/10.1007/978-981-15-9651-3_48

face recognition-based attendance management systems. Facial recognition and other biometric features have been used by many researchers over time for attendance automation [5–8]. As a result, many organizations are already using face detection system in recording employee attendance. The proposed work emphasizes on using face recognition technique for recording student attendance in class. There are a large number of students in a class, and our main emphasis is to detect and recognize all faces which are streamed into the camera. This process runs in parallel with the class lecture so that no extra time is required for calling out the names or roll numbers of the students to record the attendance. A conventional method of recording attendance consumes at least 5–10 min. The proposed work replaces the conventional system by visually taking the input using a camera and marking attendance of students based on facial recognition. The overall purpose is to reduce the time and effort required to record the attendance as well as reduce the number of false attendance which may occur in the conventional system. The proposed system can easily replace the cumbersome task of maintaining a huge set of records of attendance.

Section 2 describes the proposed method; Sect. 3 describes the implementation and includes the results generated by the proposed method followed by a conclusion in Sect. 4.

2 Proposed Method

The proposed system is a process of recording attendance of a student using a camera and face recognition approach. A live video can be captured by a video camera. A face detection and face recognition algorithm will be made to run on the live video. Firstly, the face detection algorithm will extract the detected faces from the live video, and then, the face recognition algorithm will match these faces with the faces already stored in the database. If the match with the database is found, then the attendance will be marked. All other students in the class whose matching face is not found in the database will be marked absent for that class. The total number of detected faces will be matched with the total number presents marked in the database to verify that no student gets false present/absent. The flow diagram of the proposed method is as presented in Fig. 1.

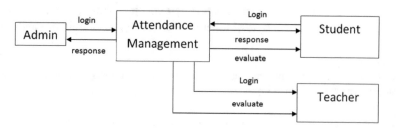

Fig. 1 Flow diagram of the automated attendance management system

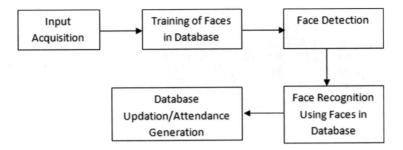

Fig. 2 Block diagram of automated attendance management system

The process consists of five steps as demonstrated in Fig. 2. Firstly, an input video will be captured using a video camera followed by training of the system on a set of faces. Training step will train the system on a set of known student's faces. In this step, the system learns to recognize a set of student's faces. Haar cascade classifier [9] is used for face detection. LBPH face recognizer has been used for facial recognition [10] since it uses both texture and shape information based on local binary patterns; histograms are extracted for small regions and converted into a single feature vector. Later, in step 3, the system will be tested on test data, where the system again captures live video and detects faces in the video. In step 4, the system recognizes the detected faces in step 3 using the database formed in step 2. Finally, in step 4, all the recognized faces will be marked present whereas all others will be marked absent. The system also performs a check that the total number of detected faces in step 3 should be equal to the number of presents in the database.

3 Implementation and Result Analysis

The proposed system has been built on the OpenCV framework using Python 3+ using SQLite3 for maintaining the record of attendance. Camera HD 720p Webcam Full HD 1080p have been used to scan the faces. Every part of the image is converted into a binary sparse format. The binary representation marks for their respective categories and maximum marked category decide the identity of the image. The proposed method works with accuracy. The implementation is carried out in the following five stages, and the results of the method are demonstrated in the figures.

3.1 Input Acquisition

The input is acquired using a video camera in a laptop. The students will be captured by the camera. The faces will be detected from the live video, and then, these cropped faces will be saved to the database for training purpose as shown in Fig. 3.

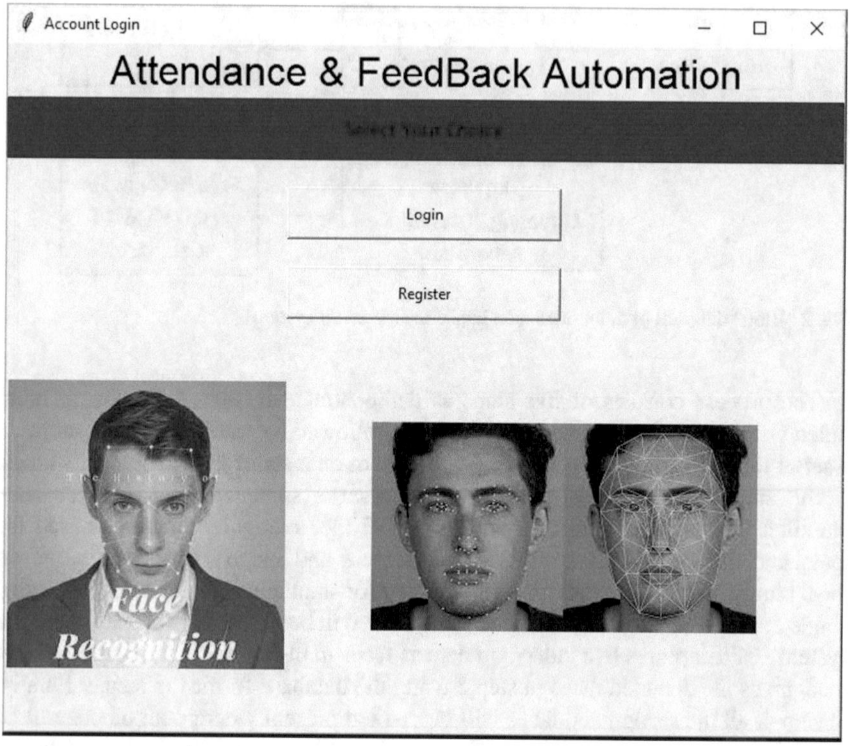

Fig. 3 Starting the GUI

3.2 Training

Training the proposed model with different faces is performed by taking the cropped faces from the first step of different angles and expressions (Fig. 4). Each student is allowed to sit in front of a camera at the time of training. Various pictures from different angles and expressions will be captured for each student by a camera and saved into the database (Fig. 5). These faces will be used by the database later for matching it with a detected face every time to mark attendance at the time of testing.

3.3 Face Detection and Face Recognition

After training, the proposed system is further tested with any of the trained faces. A new face will now be streamed live into the video camera. Face detection is done from live video camera using Haar classifier (Fig. 6). Now, the cropped face will be matched with that of the training data stored in the database using LBPH face

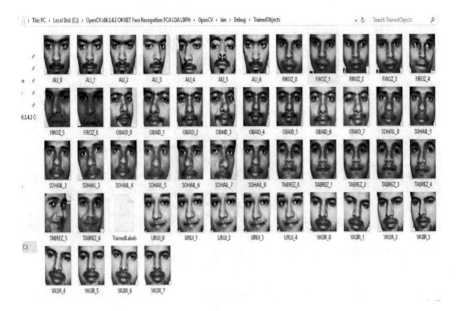

Fig. 4 Training data

Fig. 5 Training of the faces in database

Fig. 6 Face is detected

recognizer as explained in Sect. 2. The attendance updation process will then be carried out.

3.4 Database Updation and Attendance Sheet Generation

As soon as the student gets recognized by the system in the database, attendance will be turned into "Present" for that date and lecture period. The student whose attendance is not marked as present will all be turned to "Absent." Later, the system also counts the total number of presents and matches it with the total number of faces detected for that lecture period. The system works correctly if both the counts result as same. An attendance report can be generated if demanded by a system user (Fig. 7).

4 Conclusion

The proposed system requires only a video camera and permanent database storage that make it cost-efficient in terms of manpower and time as well as easily maintainable. The system prohibits false attendance as well as ensures that no student is left absent if he/she is present in the counting mechanism. As a part of modifications in future, the facial expressions can be extracted, and the machine can be trained to judge whether the student is good enough to give feedback, since many of the students do not even pay attention in the classes, and hence, the proposed model results in false feedback. The system can also be used to give statistics about the student's attendance. Also, our attendance recording system is intended to improve to give

Fig. 7 Database updated

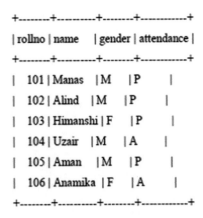

```
+--------+----------+--------+------------+
| rollno | name     | gender | attendance |
+--------+----------+--------+------------+
|  101   | Manas    | M      | P          |
|  102   | Alind    | M      | P          |
|  103   | Himanshi | F      | P          |
|  104   | Uzair    | M      | A          |
|  105   | Aman     | M      | P          |
|  106   | Anamika  | F      | A          |
+--------+----------+--------+------------+
```

good results in case of face occlusions and rotation by using improved algorithms for facial recognition.

References

1. Mohamed BK, Raghu C (2012) Fingerprint attendance system for classroom needs. In: India conference (INDICON) annual. IEEE, pp 433–438
2. Suma SL, Raga S (2018) Real time face recognition of human faces by using LBPH and Viola Jones algorithm. Int J Sci Res Comput Sci Eng 6(5):06–10
3. Cottrell GW, Fleming MK, Face recognition using unsupervised feature extraction. Submitted to INNC-90. Paris
4. Yambor WS (2002) Analyzing PCA-based face recognition algorithms: Eigenvector selection and distance measures. All Publications, p 39
5. Shoewu O, Idowu OA (2010) Development of attendance management system using biometrics. Pac J Sci Technol 2(1):8–14
6. Rahate M, Auti P, Kulkarni B, Mayande S (2015) A survey paper on automated attendance system. Int J Emerg Trends Technol Comput Sci 4(6)
7. Shubhobrata B, Gowtham SN, Prosenjit D, Aurobinda R (2018) Smart attendance monitoring system (SAMS): a face recognition based attendance system for classroom environment. In: 18th IEEE international conference on advanced learning technologies, pp 358–360
8. Selvi KS, Chitrakala P, Jenitha AA (2014) Face recognition based attendance marking system. Int J Comput Sci Mobile Comput 3(2)
9. Sander AS, Object detection using Haar-Cascade classifier. In: Institute of Computer Science, University of Tartu
10. Suma SL, Raga S (2018) Real time face recognition of human faces by using LBPH and Viola Jones Algorithm. Int J Sci Res Comput Sci Eng 6(5): 06–10

Retinal Image Diabetic Detection Using Multilayer Perceptron Classifier

B. Sadhana, B. Sahana, E. Deepthi, K. Pooja, Gayathri, and Prajna

Abstract The multi-layer perceptron classifier is adversely adapted to classify retinal images. The resultants may be normal or abnormal. Here, in the advanced scenario, diabetic retinopathy is a crucial issue in the medical field. So, the identification and cure of diabetic retinopathy are the major issues. To analyze the situation of processes, the retinal images for the detection of retinal image patterns, describing structural and disease patterns to appropriate elicitation of the image features, targeting diabetic retinopathy detection. Therefore, deploying suitable techniques to extract distinguishable features and application of relevant transformation and feature selection techniques to formulate an expressive feature vector and build efficient models to predict the patterns in the case of a retinal image using multilayer perceptron classifier. The major objectives in the case of diabetic retinopathy in human retinal images can detect and identify diabetic using multi-layer perceptron neural network (MLPNN) to classify retinal images as normal or abnormal. The classifier is used to find the best featured subset. To obtain highly accurate images for training and cross-validation datasets.

Keywords Neural network (NN) · Multi-layer perception neural network (MLPNN) · Retinal images database DIARETDB0 · Fundus photography

B. Sadhana (✉)
Department of Information Science and Engineering, CEC Bantwal, Mangalore, India
e-mail: sadhana26rai@gmail.com

B. Sahana
Department of Electronics Engineering Communication, RVCE, Bangalore, India
e-mail: sahanab@rvce.edu.in

E. Deepthi · K. Pooja · Gayathri · Prajna
Department of Information Science and Engineering, CEC Bantwal, VTU Belagavi, Belgaum, India

© The Author(s), under exclusive license to Springer Nature Singapore Pte Ltd. 2021
J. S. Raj et al. (eds.), *Innovative Data Communication Technologies and Application*,
Lecture Notes on Data Engineering and Communications Technologies 59,
https://doi.org/10.1007/978-981-15-9651-3_49

1 Introduction

Major causes of identification of diabetic retinopathy are restricted to detect diseases only for diabetic patients. Also to know the per cent of the diabetics and to provide privileges for doctors to update the details about the disease and its possible diagnosis. The undertaken project is mainly concentrated on detection and classification of diabetic retinopathy (DR), one of the most common reasons for blindness around the world. Diabetic retinopathy is difficult to identify in the premature and the method can be painstaking for technicians. Hence, a computer-aided analysis technique is built on deep learning which is proposed to detect and classify the diabetic retinopathy (DR) in the retinal image. A multi-layer perceptron is trained with a deep learning algorithm. The present work obtains retinal image as input and extracts the features which will be compared with trained features to predict and classify the diabetic level. A featured vector is established with contrasting framework namely mean and standard deviation, contrast, homogeneity, dissimilarity, etc. The dataset is trained N times to obtain the finest feature.

2 Literature Survey

Analog image processing and digital image processing are the major types of image processing concepts. In the case of hard copies, like photographs, printouts analog image processing technique can be adapted. A computer is used to modify digital images. To get the original information, it has to follow different phases of processing. Those various phases are like pre-processing phase, enhancement phase, and extraction of display information. Digital image processing has its own advantages over analog image processing. Medical image process is the method by which generates the different images of the human body for medical purposes and it helps to easily identify, observe, and detect the disease. Some patients may have a dark or dark blank spot in the eye. On detection of diabetic retinopathy from the digital fundus retinal images, the pre-processing methods are performed for the detection of diabetic, which is decisive in diabetic retinopathy grading. In all the recent works, it has been assumed that automatic retinal lesion identification for visual direction involves the development of automatic repeating screening which can automate the presence of any kind of abnormal images which are related to DR.

An ensemble-based framework is to improve multi-layer perception neural network detection [1]. The consistent MLP identification in digital fundus images concerned with the medical image processing is based on the collection framework that identifies the MLP. The collection framework is analyzed with the outputs of multiple classifiers and the combination of an internal mechanism of MLP detectors.

2.1 Automatic Detection of Diabetic Retinopathy

The major cause of blindness is diabetic. In an asymptomatic stage, which is effective for preventing blindness and reducing cost for health systems [2]. Most screening techniques acquire color photographs of the retina using digital color fundus cameras [3]. Photographs are then evaluated for the presence of lesions indicative of DR, including exudates (EXs) [4]. The application may reduce the workload and cost of analysis instead which has to be done by manual grading [5]. Many studies can be found in the literature regarding digital image processing for DR. Most algorithms comprise several steps. First, a pre-processing step is carried out to reduce the energy of the image, i.e., to attenuate image changes and variation by normalization of the original human retinal image [6]. Second, the segmentation step is carried out by using a histogram equalization method [7] Finally, only those extracted features of DR are retained for subsequent classification. This review gives an overview of the available algorithms for DR feature extraction and the automatic retinal image analysis systems based on the above-mentioned algorithms.

2.2 Detection Process of Exudates

The present methodology which uses a classifier called MLPNN for detecting and localizing the occurrence of exudates of low contrast digital images collected from patients [2]. Retinal fundus image is initially preprocessed and histogram algorithm method is adapted for segmentation of the exudates in the fundus images. This technology enables the ophthalmologists for detecting the appearance of exudates in less interval of time that helps in identifying the exudates and classify the diabetic level and thereby confirms the disease. The present back propagation neural network is the best choice to detect exudates. The Optic Disc is eliminated by using the RGB split method [1]. In order to obtain efficient detection of exudates and no exudates at the pixel level, binary thresholding, Gaussian filter, and equalization methods are used. This reduces the cost as well as the burden for the filtration of the normal images.

Major medical identification of dilation in blood vessels may be an effective technique while detecting retinal diabetic. Any other abnormalities in the retinal image may also cause diabetic in a patient along with the presence of exudates, lesions, etc. Fundus retinal images not only provide anatomical information of the retina but also the information of these pathological features [7, 8]. The healthy retinal image exhibits feature like a network of blood vessels, optic discs, and macula any change due to diabetic cause variations in features.

2.3 Feature Extraction of Fundus Image

Diabetic retinopathy identifies dilation in blood vessels, lesions or any other types of abnormalities, presence of exudates in the retinal image. Fundus image not only exhibits anatomical information but also provides information on pathological features. A network of blood vessels, macula, and optic discs are noticed in the healthy retina [9]. But in case of the diabetic patient retina, there may be variations in these features. MLPNN classifier can be adapted to detect diabetic by extracting the features using gray-level co-occurrence metrics method.

2.4 Algorithm Used for Extracting Fundus

After the final step, there may be a chance of high-intensity changes present in the retinal image and other additional features that are not visible. To reduce the intensity, histogram equalizations can be used which is uniform. HE is a technique which detects different intensity modifications in the given input image which improves the image global contrast. Finally, during the equalization process, both histogram equalization and contrast-limited adaptive histogram equalization are used.

But contrast limited adaptive histogram equalization gives a little better features than a simple one [10]. Basically, contrast-limited adaptive histogram exhibits smooth feature extractions than simple histogram equalization.

2.5 Proposed System

In the proposed system, features can be extracted using gray-level co-occurrence metrics and detection of diabetic retinopathy process is carried out. The retinal images were analyzed using specified algorithms. In the proposed system, it consists of different modules like retinal fundus image as input, preprocessing stage which does decolor and resizing of a given input image by using equalization and bitwise and median blur morphological closing to filter blood vessels; during feature extraction, the features are extracted by using GLCM method. Multi-layer perception (MLP) classifier classifies the retinal image.

Figure 1 represents the proposed system. For each input image. Initially, the image is pre-processed, next it enters into segmentation step, later features are extracted, and the extracted features will be fed into MLP classifier to compare the features which are stored on a trained dataset. MLPNN is adapted to test a feature vector of the retinal image which classifies images as normal or abnormal. Finally, the retinal image will be classified based on the diabetic level(class 1:very severe, class 2: moderate, class 3:normal).

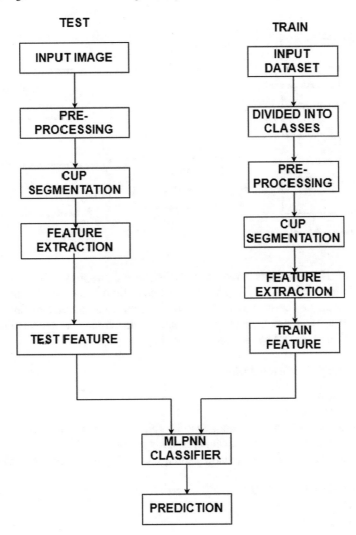

Fig. 1 Proposed system structure

3 Design

3.1 Abstract Design

Figure 2 shows the abstract design of the entire system. Initially, the input is fed into a preprocessing block. In pre-processing bitwise and median blur, a morphological closing method is used to remove blood vessels binary thresholding to extract

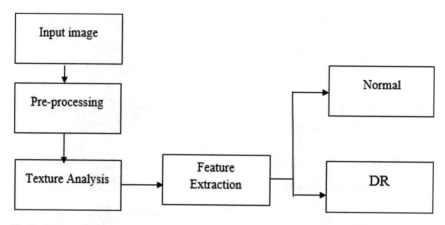

Fig. 2 Abstract design

retinal part. Next, it enters texture analysis where the pre-processed image undergoes histogram equalization to get a contrast and bright image and filtration takes
place by using the Gaussian filter. Next, the features are extracted by using GLCM
method. From the extracted features, one can decide whether the extracted features
are normal or abnormal by using MLP classifier.

Steps for Logic Implementation

- Input image
- Training the dataset
- Preprocessing of an input image (RGB split, Histogram equalization, binary
  thresholding, bitwise, and median blur morphological closing).
- Segmentation (RGB split, Gaussian filter, HE, taking mean and standard devia-
  tion).
- Feature extraction (GLCM)
- MLP classifier
- Final result.

4 Results and Discussion

Figure 3 shows the login page for admin. At first, admin should be registered with
the unique user name. Using the user name and password, admin must log in to the
application to perform further steps.

Figure 4 shows the user interface. The figure shows three main steps for diabetic
detection, i.e., preprocessing, edge detection, and analysis.

Figure 5 shows the preprocessing. Pre-processing stage removes the noise and
eliminates irrelevant information. Firstly, the retinal-colored image is preprocessed,
i.e., only green component is extracted as the intensity of the green channel is high.

Fig. 3 Login page for the admin

Canny edge detection can detect edges with suppressed noise which may smoothen the image using Gaussian filter.

Figure 6 shows the canny edge detection step. Canny edge detection can detect edges with suppressed noise which smoothens the image using Gaussian filter.

Figure 7 shows the analysis and diabetic retinopathy determination page. This page shows whether the image is diseased or not.

5 Conclusion

The proposed system enables users to detect diabetes through a desktop application. The system also ensures better accurate results when compared to other existing algorithms like ANN. Also ensures time saving when compared to a manual inspection which focuses on the detection of diabetic retinopathy using MLPNN. Retinal image classification results may be normal or abnormal. MLPNN can be used to find out the best and strong subset which holds all features to get highly accurate images for training as well as cross-validation of datasets. For future works, a device can be developed which is used to capture the image.

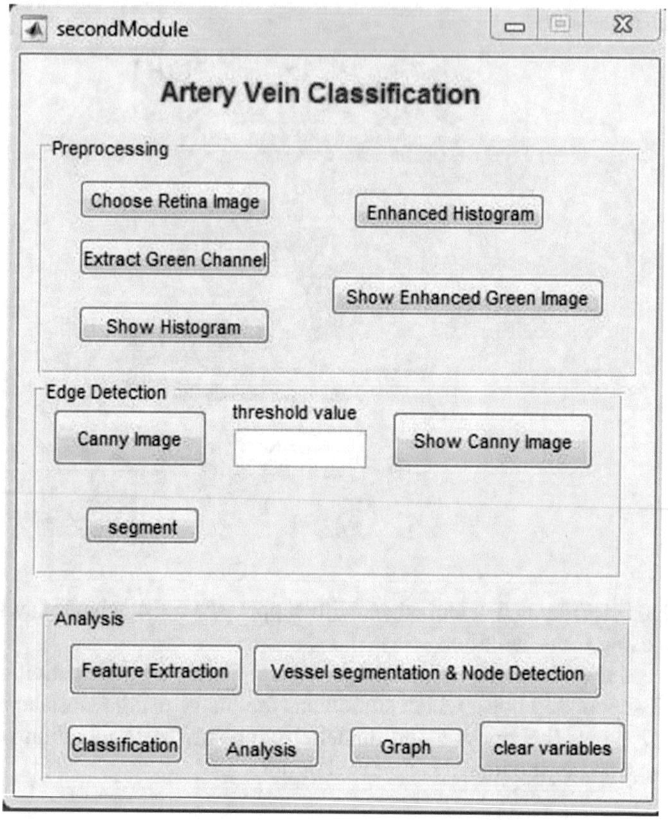

Fig. 4 User interface

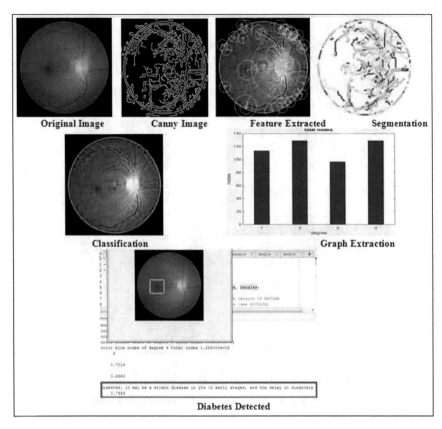

Fig. 5 Processing stage

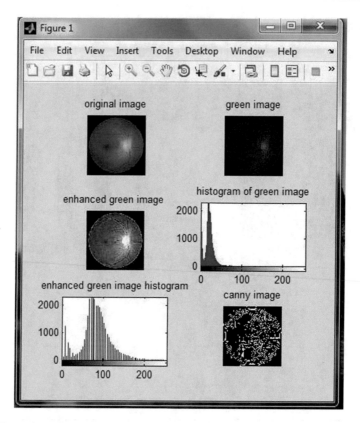

Fig. 6 Processing stage

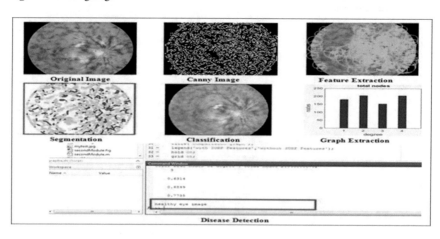

Fig. 7 Diabetic retinopathy determination page

References

1. García M, Valverde C (2013) Comparison of logistic regression and neural network classifiers in the detection of hard exudates in retinal images. In: 35th annual international conference of the IEEE EMBS Osaka, Japan, 3–7 July 2013
2. Kumar SJJ, Madheswaran M (2009) Extraction of blood vascular network for development of an automated diabetic retinopathy screening system. In: International conference on computer technology and development. IEEE https://doi.org/10.1109/Icctd.2009.212
3. Niemeyer M (2010) Retinopathy online challenge: automatic detection of microaneurysms in digital color fundus photographs. IEEE Trans Med Imaging 29(1)
4. Goldbaum M, Moezzi S, Taylor A, Chatterjee S, Boyd J
5. Hunter E, Jain R (1999) Automated diagnosis and image with object extraction, object classification and difference in retinal images. Br J Ophthalmol 83
6. Sinthanayothin C, Boyce J, Cook H, Williamson T (1999) Automated localisation of optic disc, fovea, and retinal blood vessels from digital color fundus images. Br J Ophthalmol 83
7. Jain AK (1996) Michigan State University, Jianchang Mao IBM Almaden Research Centre. Artificial neural networks: a tutorial
8. Klein D, Klein BE, Moss SE et al (1986) The Wisconsin epidemiologic study of diabetic retinopathy VII. Diabetic non-proliferative retinal lesions. Br J Ophthalmol 94
9. Kharat GU, Dudul SV (2008) Optimal neural network classifier for human emotion recognition from facial expression using singular value decomposition (SVD). Intern J Eng Res Indust Appl (IJERIA)1(4):155–166. ISSN-09704-1518
10. García M, Valverde C (2013) Comparison of logistic regression and neural network classifiers in the detection of hard exudates in retinal images. In: 35th annual international conference of the IEEE EMBS Osaka, Japan, 3–7 July 2013

A Quantum Machine Learning Classifier Model for Diabetes

Vasavi Kumbargeri, Nitisha Sinha, and Satyadhyan Chickerur

Abstract This paper demonstrates the advantages of quantum machine learning by using the quantum kitchen sinks (QKS) algorithm. Quantum machine learning (QML) is implemented along with the QKS to check the accuracy with the help of the hybrid algorithm. By optimizing the quantum learning, the accuracy of the algorithm is compared with the classical machine learning. The objective of the proposed work was to train a simple machine learning algorithm on a hybrid quantum computer/simulator to evaluate whether the hybrid quantum computer would be trained with higher accuracy compared to a conventional computer. The diabetics' dataset was used to train a logistic regression model as a binary classifier. Two qubits were used for implementing the QML algorithm on a quantum simulator. The Quantum kitchen sinks employ the quantum circuits which are used to nonlinearly tranform standard inputs into features that can be utilized by numerous machine learning algorithms. The features from the dataset were extracted using the quantum kitchen sinks (QKS), which is then fed into a classical logistic regression algorithm for training. This technique of extracting features from quantum techniques and using a classical computer for prediction empowers machine learning to obtain high accuracy.

Keywords Quantum computing · Machine learning · Quantum kitchen sinks · Diabetes

V. Kumbargeri · N. Sinha · S. Chickerur (✉)
School of Computer Science and Engineering, KLE Technological University, Hubballi, India
e-mail: chickerursr@kletech.ac.in

V. Kumbargeri
e-mail: vasavikumbargerik98@gmail.com

N. Sinha
e-mail: nitisha.sinha1997@gmail.com

S. Chickerur
Centre for High Performance Computing, KLE Technological University, Hubballi, India

© The Author(s), under exclusive license to Springer Nature Singapore Pte Ltd. 2021   603
J. S. Raj et al. (eds.), *Innovative Data Communication Technologies and Application*,
Lecture Notes on Data Engineering and Communications Technologies 59,
https://doi.org/10.1007/978-981-15-9651-3_50

1 Introduction

Machine learning focuses on designing computer programs that, when exposed to new data, can teach themselves to grow and change. The machine learning problems are solved by the quantum which enhances the ML along with the quantum algorithms, thus enhancing classical machine learning techniques and often accelerating them. To encode the classical dataset with the quantum, the computer process the available quantum information through an algorithm. Subsequently, routines for processing quantum information are implemented and by measuring the quantum system, the result of the quantum computation is readout.

The QML contributes to the intellectual advancements in quantum computing simultaneously with the applied research on machine learning. Although machine learning algorithms are employed to measure huge amounts of data, quantum machine learning intelligently enhances these abilities by generating opportunities for quantity state and system analysis [1]. One of these types of algorithms is called quantum kitchen sinks (QKS) [2].

Quantum kitchen sinks (QKS) is based on the idea of random kitchen sink algorithm, which is increasingly used for classifying the nonlinear distributed dataset. The conventional nonlinear kernel methods use the large nonlinear dataset for training the system. This hybrid quantum algorithm is used for supervised machine learning that shows that even a very small quantum computer can provide a very high-performance boost for classification tasks [3]. The QKS works with the quantum circuits to perform the nonlinearly transformed classical inputs within the specified features and further can be applied to the machine learning algorithms [4, 5]. In QKS, optimization of machine learning (ML) tasks can be greatly simplified using random nonlinear transformations. QKS' general idea is to randomly sample from a family of quantum circuits and use each circuit to perform a nonlinear transformation of input data into a measured bit string. Subsequently, a classical machine learning (ML) algorithm processes the concatenated results.

In classical machine learning, data is stored in a physical bit. A bit is in the binary state (either 0 or 1) and is mutually exhaustive. But in QML, the physical characteristics of smaller scale physics communicate among the molecules by adopting the quantum computations, so that quantum bits known as the qubits can further combine both the classical 0 and 1 state which leads to storing huge amount of data in a qubit form rather than the regular bit [6, 7]. This can be represented as shown in Fig. 1. This makes quantum computation much faster than classical approaches.

In [8–10], the authors propose to train a CNN model using a quantum algorithm for both the backward and forward passes. The proposed QCNN model enables the new frontiners in the image recognition domain for more complex convolutions such as larger kernels, high-depth input channels and high-dimensional inputs. The proposed method is based on the record-based dataset.

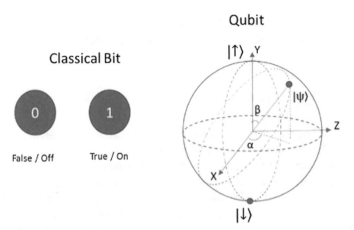

Fig. 1 Bit and qubit representation

2 Proposed Method

In this section, the proposed methodology is provided to implement the quantum kitchen sinks. The proposed block diagram is shown in Fig. 2. The approach proposed is a quantum circuit that is put between classical processes [11]. This variant of the quantum circuit will perform much better than complete quantum or complete classic algorithms.

As shown in the block diagram, the featurization is carried on quantum simulator/computer and the training for the model is carried out on a classical computer. This hybrid nature of QKS algorithm ensures in enhancing the accuracy of the machine learning model.The classical part of the block diagram is the necessary

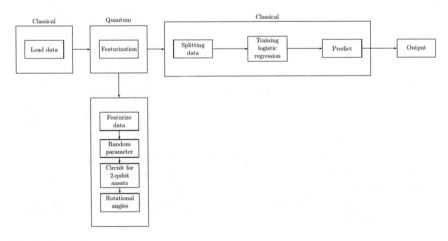

Fig. 2 Proposed approach

steps in traditional datascience. This paper focuses on the quantum aspect of the feature extraction and uses it to train a classical model for classification.

2.1 QKS Featurization

In this subsection, the QKS algorithm is explored to extract features using quantum computing. There is currently a lack of parallelism and performance in the means to extract features from datasets. Since revival and restoration require separate measurements of data points as bits and their properties, there is a loss in processing time when it cannot be parallelized or processed in a way that compares these bits from the start. The same limitation applies to extract features for a machine learning algorithm [12, 13]. Extracting the characteristics involves independently measuring each bit, its properties and translating them into feature vectors that can be fed into a classical machine learning model. It also involves an independent step to calculate the relationships between data points and/or their properties that can take several iterations. It is solved by quantum information processing algorithms via entanglement. Entanglement can be used to create non-local correlations of different, repeated geometric shapes between the qubits that store vertex locations. It is necessary to locate the triangle vertices and decide where each data point belong to in the collection of information and also about which particles reside in maximally entangled states. The local features are extracted based on quantum computers within a quantum feature extraction framework besides the quantum description. The featurization consists of four following methods:

- Random parameter generator
- Circuit for 2-qubit ansatz
- Rotation angles
- Featurize data.

2.1.1 Random Parameter Generator

Quantum theory's variational approach is a classic method of finding a quantum system's low-energy states. A test wave function is known as the ansatz defines the operation of some parameters and the value of the parameters can be obtained by this function which minimizes the energy's expectation value to certain parameters. The reduced acknowledgment is an estimation of the lowest energy-specific state, and the expected value tends to be the upper limit on the ground state's energy. A function is created which will draw all random variables needed. This is done on an ansatz (an educated guess) where two rotation angles for rotations on two separate qubits are needed. The same set of random parameters for all the data points in a dataset is extracted.

2.1.2 Circuit for 2-qubit Ansatz

A quantum gate or quantum logic gate is termed as the fundamental quantum circuit that functions on fewer qubits. They are analogous of classical logic gates for quantum computers. X and CNOT gate is performed on the qubit for rotation of the qubit. This ensures that the qubits are in a random excited state.

2.1.3 Rotation Angles

The quantum processor executes the electron spin for a qubit since it is probable to manage an electron spin. The electron spin connected with another electron spins with an extremely shorter time scale than a nuclear spin. Similarly, a nuclear spin is compatible with the quantum memory for a qubit as it is much longer connection time than an electron spin. In this step, two random rotation angles in the current loop are identified which is to be fed into the quantum circuit.

Pseudocode:

1. Identify the dimension of the feature space
2. A linear transformation matrix is structured according to a 2-qubit ansatz
3. Obtain the bias vector
4. Perform random transformation on data point to obtain rotation angles.

2.1.4 Featurize Data

The difference between the data aspects and features can be analyzed by the larger quantum state space which far better than the feature map automatically generated by the classical machine learning algorithm. The objective of utilizing quantum computers is to build classifiers by producing more complicated maps of data. This empowers the researchers to promote more powerful AI, for instance identifying the patterns in the data that are invisible to conventional computers. This step converts input dataset into a set of binary vectors according to the 2-qubit ansatz and draws all random parameters once and uses them for each data point.

The proposed QML program consists of the following steps:

1. Sample the random elements to use across all data points
2. Create and compile parametric circuit for 2q ansatz
3. Initialize empty list to hold 'random' features for each data point
4. For each data point:
- Create random features
- Identify rotation angles
- Collect result for the current value
- Collect random features for each data point in the list.

3 Experimental Data and Results

The dataset used for training is the diabetics dataset taken from Kaggle [14]. The National Institute of Diabetes and Digestive and Kidney Diseases delivers the dataset. Numerous constraints were laid on the selection of these occurrences from a longer database. Particularly, all the patients were at the age of a minimum of 21 years old of Pima Indian heritage.

The various attributes are selected from the dataset such as glucose, pregnancies, skin thickness, blood pressure, BMI, insulin, age, outcome and diabetes pedigree function.

The dataset consists of a total of 768 patient's record in which 650 records are used for training and 100 records are used for testing. The task for the algorithm is to obtain quantum features from the quantum computer and classify if the patient has diabetics or not using classical logistic regression.

The QML program was implemented using Rigetti's pyquil language and executed on Quantum Virtual Machine (QVM) using the QuilCompiler. Rigetti provides Quantum Cloud Services for users with a dedicated quantum machine image that is pre-packaged with the Forest SDK (Fig. 3).

Table 1 demonstrates that substantially good accuracy is achieved with fewer episodes which are not achieved in classical machine learning. The classical logistic regression (LR) is also used to classify the labels. This machine learning algorithm proved 64% effective. In comparison with just two qubits, the quantum kitchen sinks has given a precision boost of 66%. This increase points to a quantum computing benefit. Similarly, the confusion matrix for various other accuracies and classification report is as given in Figs. 4, 5, 6, 7, 8, 9, 10 and 11. Figure 5 indicates that the 64 class 0 instances are correctly classified with an $f1$-score of 0.78, whereas the classifier fails to unerringly predict instances of class 1 as a result $f1$-score is 0.10. This means that all class 0 instances are correctly labeled while some of the class 1 instances are considered as false negatives. Other classification reports may be understood in the same way.

4 Conclusion

The proposed work in progress approach shows that with just seven number of episodes, the hybrid quantum machine provides 66% improved accuracy with just two qubits as against 64% with classical logistic regression. This type of approach which involves both quantum computing and classical computing is anticipated that will help researchers to develop high-performance quantum computing application which can be used for various applications including quantum machine learning to obtain better and efficient performance and results.

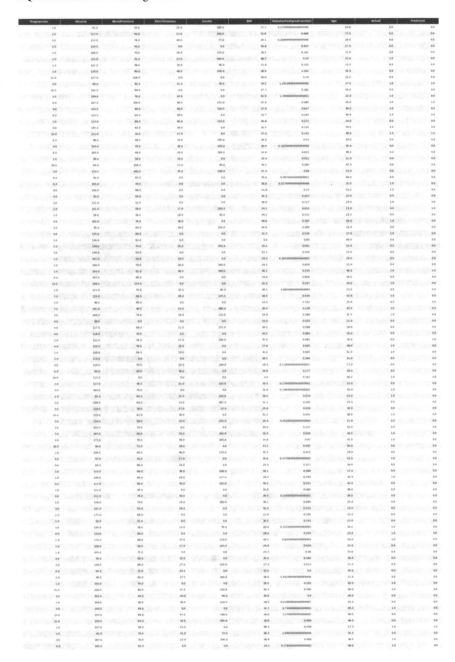

Fig. 3 Output Snapshot of the proposed approach

Table 1 Accuracy of quantum kitchen sinks

Number of episodes	Accuracy (%)
2	63
7	66
9	64
10	63
20	61
50	57
100	61
200	50

Fig. 4 Confusion matrix of the classifier with an accuracy of 66%

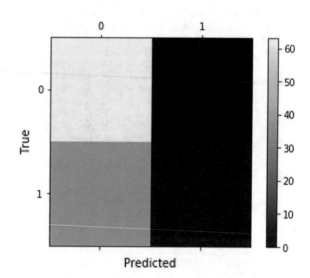

```
              precision    recall  f1-score   support

           0       0.64      1.00      0.78        63
           1       1.00      0.05      0.10        37

   micro avg       0.65      0.65      0.65       100
   macro avg       0.82      0.53      0.44       100
weighted avg       0.78      0.65      0.53       100
```

Fig. 5 Classification report with an accuracy of 66%

Fig. 6 Confusion matrix of the classifier with accuracy of 57%

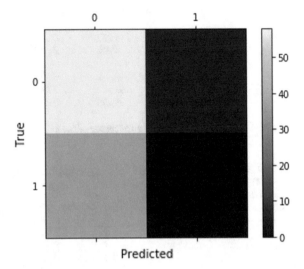

	precision	recall	f1-score	support
0	0.61	0.92	0.73	63
1	0.00	0.00	0.00	37
. micro avg	0.58	0.58	0.58	100
macro avg	0.31	0.46	0.37	100
weighted avg	0.38	0.58	0.46	100

Fig. 7 Classification report with an accuracy of 57%

Fig. 8 Confusion matrix of the classifier with an accuracy of 62%

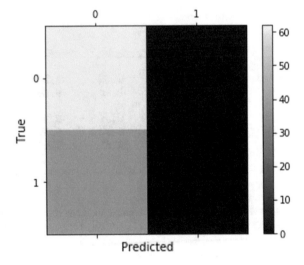

	precision	recall	f1-score	support
0	0.63	0.98	0.77	63
1	0.00	0.00	0.00	37
micro avg	0.62	0.62	0.62	100
macro avg	0.31	0.49	0.38	100
weighted avg	0.39	0.62	0.48	100

Fig. 9 Classification report with an accuracy of 62%

Fig. 10 Confusion matrix of classical logistic regression classifier with an accuracy of 64%

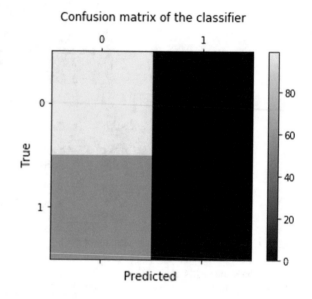

	precision	recall	f1-score	support
0	0.64	1.00	0.78	99
1	0.00	0.00	0.00	55
micro avg	0.64	0.64	0.64	154
macro avg	0.32	0.50	0.39	154
weighted avg	0.41	0.64	0.50	154

Fig. 11 Classification report of classical logistic regression classifier

Acknowledgements The authors prefer to recognize the support of Rigetti Computing and Management of KLE Technological University for supporting this research.

References

1. DeBenedictis EP (2018) A future with quantum machine learning. Computer 51(2):68–71
2. Wilson CM et al (2018) Quantum kitchen sinks: an algorithm for machine learning on near-term quantum computers. arXiv preprint arXiv:1806.08321
3. Gupta S et al (2017) Quantum machine learning-using quantum computation in artificial intelligence and deep neural networks: quantum computation and machine learning in artificial intelligence. In: 2017 8th annual industrial automation and electromechanical engineering conference (IEMECON). IEEE
4. Sergioli,G, Giuntini R, Freytes H (2019) A new quantum approach to binary classification. PloS one 14(5)
5. Tiwari P, Melucci M (2019) Towards a quantum-inspired binary classifier. IEEE Access 7:42354–42372
6. Jain S (2015) Quantum computer architectures: a survey. In: 2015 2nd international conference on computing for sustainable global development (INDIACom). IEEE
7. Narayanan A (1999) Quantum computing for beginners. In: Proceedings of the 1999 congress on evolutionary computation-CEC99 (Cat. No. 99TH8406), vol 3. IEEE
8. Roy A, Dutta D, Choudhury K (2013) Training artificial neural network using particle swarm optimization algorithm. Int J Adv Res Comput Sci Softw Eng 3(3)
9. Kerenidis I et al (2019) q-means: a quantum algorithm for unsupervised machine learning. Adv Neural Inf Process Syst
10. Kerenidis I, Landman J, Prakash A (2019) Quantum algorithms for deep convolutional neural networks. arXiv preprint arXiv:1911.01117
11. Mari A et al (2019) Transfer learning in hybrid classical-quantum neural networks. arXiv preprint arXiv:1912.08278
12. Agarwal T, Mittal H (2019) Performance comparison of deep neural networks on image datasets. In: 2019 twelfth international conference on contemporary computing (IC3). IEEE
13. Panigrahi S, Nanda A, Swarnkar T (2018) Deep learning approach for image classification. In: 2018 2nd international conference on data science and business analytics (ICDSBA). IEEE
14. Pima Indians diabetes dataset- https://www.kaggle.com/uciml/pima-indians-diabetes-database

An Enhanced Method for Review Mining Using N-Gram Approaches

Krishna Presannakumar and Anuj Mohamed

Abstract In the current scenario, when a person wants to purchase a product, they try to discover information concerning the product. For finding such information, they use reviews about the products on the Internet. This information always influences the purchase of products. The proposed work introduces a method for the classification of reviews using the relevant aspects of the product. The novelty of the work lies in exploiting the combination of higher-level N-approach for finding the relevant aspects from the product reviews. The most commonly used unigram approach fails to provide adequate accuracy when the dataset is very large. Proposed work uses a higher level of N-gram, a combination of unigram, bigram, and trigram for the extraction of relevant aspects from the reviews. The proposed method also calculates the polarity score of emoji in the review for the classification. The experimental analysis shows that the proposed method is better in terms of accuracy and execution time.

Keywords Review mining · Classification · N-grams · Aspect extraction · Opinion-based classification

1 Introduction

Sentiment analysis is an active research domain of computer science. Review mining is one of the significant areas of sentiment analysis. It is the computational analysis of the beliefs, opinions, behaviors, and feelings of the public toward different entities, individuals, issues, events, topics, and their attributes. The reviews extracted from social networking and online marketing sites help the business ventures to extract the good and bad traits of the products in a much easier way. The feedback

K. Presannakumar (✉) · A. Mohamed
School of Computer Sciences, Mahatma Gandhi University, Kottayam, Kerala, India
e-mail: krishna37941@gmail.com

A. Mohamed
e-mail: anujmohamed@mgu.ac.in

and reviews gathered from these platforms serve as an important source for future evaluation and improved product development. The research in review mining has recently become active because of massive volumes of review data available on the web [1]. The process of extracting opinions from the large volume of data is technically difficult but very useful. With social media rapidly changing on the Internet, consumers and business ventures are ever more using public opinions in these media for their assessment and future decision making. On the other hand, discovering and checking web opinion pages and extracting content confined to them remain a daunting task as each site usually includes a massive amount of opinionated information. The human data analyst will have trouble recognizing significant Web sites and summing up correctly the data and opinions expressed in them. Besides, human analysis of text knowledge is also considered to be less successful because of the substantial favouritism. As the amount of information to be evaluated rises, people also have trouble building reliable results. Automated opinion mining and summarization systems are thus important, as personal favouritism and unbiased opinion mining methods can solve the psychological limitations.

The objective of the proposed work is to make use of the highly unstructured text reviews about any product and classify these reviews into their respective class according to the sentiment polarity of the reviews. The reviews are classified into two classes, positive or negative according to the review title and the content. N-gram approach is gaining more attention from the researchers these days due to the wide range of applications that use this approach. Most of the research work in the area of review classification makes use of the unigram approach for classification. The combination of unigram, bigram, and trigram approaches is more suitable for classification because it contains more meaningful information. The proposed work uses a combination of unigram generated from the review title, bigram, and trigram created from the content to extract the relevant aspects of the review. The system accepts consumer reviews extracted from the Web sites as the input. The final output is the important aspects and opinions from the review along with the corresponding sentiment class. The existing works on the classification of reviews mostly depend on machine learning and lexicon-based approaches for the classification of reviews. When the dataset is very large, these methods do not provide accurate results. If the reviews are classified according to the aspects in the reviews, the classification is more accurate and the amount of time for the classification can be reduced. The work combines N-grams with aspect-opinion extraction to classify the reviews according to the opinion polarity. The proposed method is more accurate, and the use of N-grams provides a facility to handle the negations more efficiently. The polarity of the emoji is also used for the more effective polarity calculation of the review.

Through paying greater attention to the important aspects, consumers can make a smart purchasing decision accordingly, while businesses can focus on improving the quality of those aspects and thus effectively improve the product. However, manually recognizing the key aspects of the products from numerous reviews is infeasible for people. Thus, developing an automated system that can classify reviews using the relevant aspects is important. The rest of the paper is organized as follows: Sect. 2 provides a brief literature review about the works in the area of review mining.

Section 3 presents the proposed aspect extraction method which uses an N-gram approach. It focuses on how the aspects and relevant aspects are extracted from the review content and title. The preprocessing of the review is also explained. It also explains how the emoji in the reviews are handled and used for finding the polarity of the review. The proposed method for classification of reviews using relevant aspects is described in Sect. 4. The experimental results of the proposed work are detailed in Sect. 5. Section 6 provides concluding remarks and suggestions for future enhancement.

2   Related Works

There are numerous works associated with aspect-based opinion mining. Existing methods for recognizing aspects rely on lexicon approaches and supervised approaches. The lexicon-based approaches use a sentiment lexicon, consisting of a list of words, phrases, and sentimental idioms to determine the orientation of sentiment on each variable. The supervised methods use a set of data for training and testing. Some learning-based classification methods, such as support vector machine (SVM), Naive Bayes, and the maximum entropy (ME) model, are really popular [2].

Gobi et al. [3] proposed a new approach to identify the order preference based on the fuzzy technique and that approach called multi-criteria decision making (MCDM) perspective. It helps to extract the implicit and explicit feature models.

Bhumika et al. [4] discussed the sentiment analysis by using optimized support vector machine. Here, the parameters used for movie review and social tweets. In result analysis, comparison made between optimized support vector machine and Naïve Bayes classifier.

Hailong et al. [5] explained the process of dictionary-based word search approaches like WordNer and HowNet. It helps to detect the synonym, antonym and abbreviation of the wordset.

Fatehjeet et al. [6] introduced sentence classification using a dictionary-based approach for the Punjabi language. The developed system finds out the positive, negative, and neutral words from selected paragraphs.

Cruz et al. [7] suggested an approach called sentient analysis for dictionary-based approaches, and it uses the AcroDefMI3 and TrueSkill methods to obtain the correlation between the word polarities.

3   Aspect Extraction

The classification of the reviews based on the polarity of the aspect is very beneficial to the vendor as well as the consumer. The vendor can understand which areas of the product require improvement and which are the best aspects of the product. This knowledge can be used for improvisation and selling products by giving more focus to the good aspects. Figure 1 shows the architecture of the proposed system for review classification.

The working of the system can be divided into seven steps:

1. Extraction of data from the source: Reviews are collected from the Internet using the web crawling techniques
2. Preprocessing of the reviews: The unnecessary data, symbols, and characters are removed from the reviews
3. Generation of the N-grams: The review content and title are processed and divided into set of N-grams
4. Parts of Speech (POS) tagging: Each N-gram is associated with the corresponding POS tag
5. Extraction of aspect-opinion pairs: Relevant aspects and corresponding opinions are extracted from the reviews
6. Sentiment calculation of opinion features: The opinion score is calculated according to the polarity
7. Aspect-based classification: Reviews are classified into positive and negative classes according to the total score of the review.

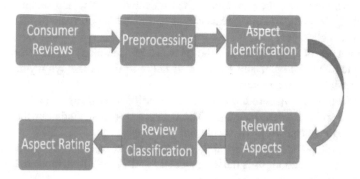

Fig. 1 Architecture of proposed review mining system

Fig. 2 Preprocessing

3.1 Dataset

The data used for the work is collected from www.amazon.in using Amazon web crawling. The consumer reviews can be directly fetched from the Amazon Web site using the Amazon Standard Identification Numbers (ASINs), which is the unique ID of the product. The consumer review of the unique ID will be extracted according to the number of pages given as the input to the system. Each data segment contains four attributes, rating of the review, title of the review, date, and review content. The extracted dataset is saved in the form of a comma-separated values (CSV) file.

Data preprocessing is a method of data mining that translates the raw data into a legible format. The extracted reviews are preprocessed to remove unwanted data. Figure 2 shows the steps followed in the preprocessing of reviews.

3.2 Aspect Extraction Using N-Gram

N-gram is a collection of the words that co-occur in a text. It is used to create features such as decision tree and Naïve Bayes and for other supervised machine learning models. Bigram contains a pair of words, and some times the first word can be a stop word, and usually stop word contains more data [7]. For example, the unigram "good" is a positive word, but the bigram "not good" carries negative significance. N-grams are used for a wide range of tasks. Aspect phrases are usually nouns and noun phrases, but they can also be verbs, verb phrases, adjectives, and adverbs [8].

Steps to create a new N-gram dictionary from a review are as follows:

1. For review, split the content into a set of word list
2. For the size of N-grams, type a number showing the size of the N-grams to be collected. For example, if the N-gram size is 2, bigrams will be created (Fig. 3).

N-grams are used, when creating a language model, to construct not only unigram models but also bigram and trigram models. The most important task is the extraction of aspect-opinion pairs. From bigram and trigram, the aspect-opinion pairs are extracted separately. Every aspect will have a corresponding opinion which conveys

Fig. 3 Aspect and opinion

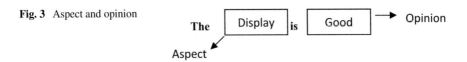

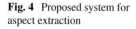

Fig. 4 Proposed system for
aspect extraction

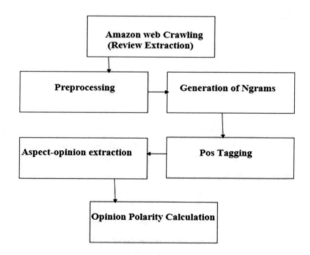

the sentiment toward the aspect. Figure 4 explains the important steps followed by
the proposed aspect extraction system.

In this work after the generation of trigrams, bigrams, and unigrams, only those
pairs of words are selected whose POS tag is either Adjective (JJ) or Adverb (RB)
or Noun (NN). For trigram, the combination of NN/JJ/RB is selected. For bigrams,
words with postag JJ/NN combination are selected. For the unigrams, only those
words with postag JJ are selected. The following algorithm explains the proposed
method used for extracting the aspects and the corresponding opinions.

Algorithm 1

The review title and content are processed separately. The reviews fetched from Web
sites are stored in the review dataset.

Input: Review Dataset
Output: Aspect-Opinion Pairs
Steps
1. Initialize review as Review Dataset
2. Initialize U as a set of unigrams #NULL at beginning
3. Initialize B as a set of bigrams #NULL at beginning
4. Initialize T as a set of trigrams #NULL at beginning
5. For each review
        {Pre-process review }
6. For each review in the training set
        {
        i. Find the title of the review
                For each adjacent word (w1) add the word into U
        ii. Find the content of the review
                For each pair of adjacent words in review content (w1, w2), add pair into B
                For each triplet of adjacent words (w1, w2, w3), add triplet into T
        }
7. Pos Tag every word in U, B, T
8. For each unigram u in U
        {
        If (postag is JJ), keep the unigram Else, remove u from U
        }
9. For each bigram b in B
        {
        If (postag is a combination of NN+JJ), keep the bigram
        Else, remove b from B
        }
10. For each trigram t in T
        {
        If (postag is a combination of NN+RB+JJ), keep the Trigram Else, remove t from T
        }

3.3 Aspect Score Calculation

The extraction of aspect-opinion pair will fetch every aspect and their corresponding opinions. But not every aspect needs to be relevant for the product [9]. So from a large number of aspects, only those aspects have to be selected which are most relevant. For finding the most relevant aspects from the list of aspects, Aspect_Value algorithm is proposed.

Aspect_Value Algorithm

Input: List of Aspect opinion pair
Output: List of Relevant Aspects Steps
1. Initialize Aspect_List=list of aspect from the aspect extraction algorithm
2. Aspect_score={}
3. Relevent_Aspect={}
4. For each aspect in Aspect_List
         Aspect_score[aspect]=+1
5. For each aspect in Aspect_List
         If Aspect_score[aspect]>threshold
         Add aspect to Relevent_Aspect

The algorithm begins with the list of all aspects which are generated using algorithm 1. It contains every aspect of the reviews. The score of each aspect represents the count of times the aspect is mentioned in the review. The relevance of the aspect is based on the value of the aspect. The aspects with the highest score are the most relevant aspect.

The relevant aspects are selected based on a threshold value which depends on the size of the dataset. Only those aspects which have aspect score greater than the threshold are selected as a relevant aspect.

3.4 Emoji Score Calculation

There is a growing use of a new generation of emoticons, called emoji, over mobile communications and social media. Emoji are visual representations of Unicode which are used as shorthand to convey concepts and ideas [10] . Over ten billion emoji have been used on social media in the last two years. The emoji also play an important role in the reviews. These emoji also contain the sentiment of consumer toward the product.

The emoji are extracted from the content of the reviews and then converted into corresponding Unicode and compare with the already created list of Unicodes for finding whether it is an emoji or not. The Unicode is used for further processing. A list of emoji and their corresponding polarity is saved in the form of a CSV file. The set of emoji and corresponding values are wrapped into a list, and it is used as a source for polarity detection. The emoji collected from the content of the review are examined using the list to find the corresponding polarity score. The score for each emoji in the review is calculated. This score is used along with a score of aspects, for finding the total polarity score of the review content.

4 Classification Using Opinion Ranking Algorithm

Classifications of the reviews are very important to know about the reaction of customers toward the product. The total polarity of the review is calculated by combing the polarity score of every opinion in the review. The score of the content and title is combined to calculate the total score.

For every review content

$$\text{BS} = \frac{\sum ap}{|a|} \text{Content Score,} \tag{1}$$

For every review title

$$\text{Title Score, TS} = \frac{\sum at}{|a|} \tag{2}$$

where the value of "a" is the total number of positive opinions; if the total number of positive is higher than the total negative, otherwise the value of a is equal to the total number of negative opinions.

p is the total polarity of the opinions in the review content, and t represents the total polarity of the opinions in the review title.

The review score is calculated using the review content score, review title score, and emoji score.

$$\text{Review score} = \frac{(w * \text{Title score}) + \text{Content Score} + \text{Emoji Score}}{(w + 1)} \tag{3}$$

The weight coefficient is represented using w. The weight coefficient is applied to the review title because a review title is always a very accurate assessment of the writer's mood and should be given extra consideration when determining scores of the review. For this reason, the score of a review can be measured as a linear composition of title score, body score, and emoji score as Eq. 3. The following algorithm describes the proposed method for the classification of the reviews using the relevant aspects.

Algorithm 3—Classification

Input: List of relevant aspects and opinions
Output: Positive class &Negative class for Reviews
Steps
For each review in Review list
1. Calculate polarity of opinions corresponding to relevant aspects
2. Calculate review content and title content polarity using Equation 1 & 2
3. Calculate total Review Score using Equation 3
        If Review Score >0
                Set class=Positive Else
Set class=Negative

5 Experimental Results

Experiments were conducted to compare the potential of the proposed method that uses the higher-level N-gram for review classification. The work includes two methods: The first method uses a combination of unigram, bigram, and trigram for finding relevant aspects from the product review and the second method is used for the classification of reviews which uses the relevant aspects extracted using the first method. The most commonly used comparison measure is the accuracy which measures the ratio of the number of correct classifications to the total number of assessed test cases. Other widely used evaluation measures are precision, F-score, etc. During the total score calculation of the review, a weight coefficient w is used for balancing the polarity of the review title and content. The weight coefficient is used with the review title for giving more weight to the title. The weight value is calculated using different experiments on the dataset. The weight value is fixed in such a way that the final accuracy of the system is at the peak.

Figure 4 shows the relationship between the weight value and evaluation measures. For each weight value, the accuracy and F-score of the dataset are calculated. The weight value w is fixed to 2 because the accuracy of the system is at a peak when the value of w is 2. The calculated weight coefficient is used for computing the total score polarity of the review (Fig. 5).

Fig. 5 Evaluation

Fig. 6 Comparison of methods

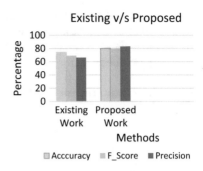

The existing works on the classification of reviews use methods like machine learning approach and lexicon-based approach for the classification purpose. These methods do not provide accurate classes when the data is relatively large. As the classification is based on the sentiments of the entire reviews, it may not provide the output that is being expected. The existing methods which use the N-gram approaches mainly focus on combining unigram and bigram. The proposed work combines unigram, bigram, and trigram to find the most useful aspect-opinion pairs. And thus, it provides much better results (Fig. 6).

The proposed work is using an N-gram approach for the extraction of the aspects of the review. As the number of the relevant aspect increases, the accuracy of the system also increases. The proposed method makes use of the N-grams for finding the most useful aspects and the corresponding opinions. These N-grams are used for classification purposes. The combination of N-grams provides exact and meaningful opinions about the aspect. This leads to more accurate classification of the review contents. A classifier with greater precision, accuracy, and F-score is considered more successful. The objective of the proposed method was to develop a system that can classify the product reviews more accurately in lesser execution time which was accomplished. After the quantitative and qualitative analysis of the result can be concluded that the proposed method is an efficient approach for product review classification. The evaluation of the proposed method shows that it has an accuracy of 78.2%.

6 Conclusion

In the field of sentiment analysis, aspect mining is relevant because every consumer in this competitive world tries to compare multiple products before buying. The company uses consumer feedback about their products to succeed and make improvements to the products. The proposed framework contains two main components, identification of aspects and classification of reviews based on these aspects. The work focuses mainly on extracting aspects from consumer reviews, which can be used to evaluate consumer feedback toward the product. Aspect extraction makes use of N-grams for finding the aspects along with the correct meaning. The classification of reviews is based on the score of the opinions regarding the aspect found within the review. The proposed method is more accurate, and the use of higher-level N-grams provides a facility to handle the negations more efficiently.

Acknowledgements The authors acknowledge the support extended by DST-PURSE (Phase II), Govt. of India.

References

1. Dey A, Jenamani M, Thakkar JJ (2018) Sentin-gram: an n-gram lexicon for sentiment analysis. Exp Syst with Appl J Elsevier
2. Kumar R, Sharan A, Yadav CS (2016) A framework for ranking reviews using ranked voting method. In: Proceedings of the second international conference on computer and communication technologies. Advances in intelligent systems and computing, Springer (2016)
3. Gopi N, Rathinavelu A (2018) Analyzing cloud based reviews for product ranking using feature based clustering algorithm. Clust Comput
4. Bhumika M, Vimalkumar B (2016) Sentiment analysis using support vector machine based on feature selection and semantic analysis. Int J Comput Appl
5. Zhang H, Gan W, Jiang B (2015) Machine learning and lexicon based methods for sentiment classification: a survey. In: 11th web information system and application conference, Tianjin (2015)
6. Kaur F, Bhatia R (2016) Sentiment analyzing by dictionary based approach. Int J Comput Appl
7. Cruz L, Ochoa J, Roche M, Poncelet P (2016) Dictionary-based sentiment analysis applied to specific domain using a web mining approach. Third annual international symposium (2016)
8. Awachate PB, Kshirsagar VP (2016) Improved twitter sentiment analysis using ngram feature selection and combinations. Int J Adv Res Comput Commun Eng
9. Mukherjee A, Liu B (2012) Aspect extraction through semi-supervised modeling. In: Proceedings of annual meeting of the association for computational linguistics (2012)
10. Lehman S, Felbo B, Mislove A, Sogaard A (2017) Using millions of emoji occurrences to learn any-domain representations for detecting sentiment, emotion and sarcasm. Conference on empirical methods in natural language processing

Alyce: An Artificial Intelligence Fine-Tuned Screenplay Writer

K. A. Eldhose, Charly Jose, S. Siddharth, Sreya Sarah Geejo, and Sneha Sreedevi

Abstract A good screenplay is one of the most important aspects of a great movie. Viewers should be able to relate and become a part of the emotional scenes in a movie. For a person to be able to write a good screenplay, they will need years of experience and immense talent. Those with the required experience and talents are not always available. Due to this fact, even the good stories can be viewed as an unstructured collection of information about an event. Stories do not expand the background details, and also it will describe the scenes vaguely. They do not paint the picture quite well for a movie. Stories just form the basis for a screenplay, but the actual details are further added or modified by the writers. In this digital era, with the recent advancement in technologies, this research work has proposed a new model, named "Alyce Garner Peterson" (or "Alyce"), which is an AI fine-tuned screenplay writer to generate the screenplay from a given story by incorporating the natural language processing and machine learning techniques. The proposed AI model Alyce outperforms the traditional screenplay writers and achieves what other writers strive to achieve in a short period of time.

Keywords Story · Screenplay · Alyce · Artificial intelligence · Machine learning · Natural language processing · Text-to-speech

K. A. Eldhose (✉) · C. Jose · S. Siddharth · S. S. Geejo · S. Sreedevi
Muthoot Institute of Technology and Science, Puthencruz 682308, India
e-mail: eldhoseka.abraham@gmail.com

C. Jose
e-mail: charlyjose@zoho.com

S. Siddharth
e-mail: siddharthajith97@gmail.com

S. S. Geejo
e-mail: sreyageejo@gmail.com

S. Sreedevi
e-mail: snehas@mgits.ac.in

1 Introduction

Data is available in large amounts from various sources, but extracting the required useful information from it is the existing research requirement. Text mining is the technique of analyzing and processing large amounts of unstructured textual data for the extraction of relevant information. It can be used to correlate data and come up with a solution to the problem in hand, which otherwise would be difficult to do with such vast amounts of data. The fundamental steps in text mining involve gathering the unstructured data to be given as input, followed by pre-processing and cleansing of this data to remove the anomalies. This is followed by the conversion of unstructured data to a structured format after which text mining tools are used to identify and extract the required pattern. Thus, it can be used for providing co-reference resolution of the characters and the words describing them in a given story.

Humans use natural language to communicate and convey ideas with each other while machine languages are used for communication between humans and machines. In traditional methods, generation of a script from a story takes a lot of time, effort and skill set. Also, for a better alignment and understanding the writer needs to have a proficiency in the language of the story and a vivid imagination. With machines significantly reducing the efforts taken by man in this modern era, the communication gap between humans and machines can be converged using natural language processing (NLP). NLP is a field of artificial intelligence (AI) that gives machines the ability to read, understand, process and analyze the natural languages used by humans. It can be used to extract the dialogues from the story for each of the characters. Natural language generation (NLG) is a technique used to convert data to its representation in natural languages like English. The theme, plot and narration of the various scenes can thus be generated. The plot generation process uses sentiment analysis of the story. Sentiment analysis is a method of text mining used to understand the sentiments and emotions portrayed by the textual input. The final script produced is given into a text-to-speech converter that reads it out with an emphasis on the emotions of each scene and dialogue.

Thus, the idea is a system to generate the script and audio transcript from a simple story by analyzing, identifying, extracting and processing sentences hence removing the need of a skilled and experienced screenplay writer when converting a story to a script for a movie or drama. This will help in decreasing the production cost and time for movie and theater adaptations of stories, especially world-famous ones for which the viewer's wait eagerly.

The graph in Fig. 1 shows the exponential increase in the amount of structured and unstructured information in digital sources. The amount of unstructured data is much higher than the amount of structured data. Stories are part of this unstructured data in the digital sources.

Fig. 1 Increase in data in
digital sources according to
https://seekingalpha.com/art
icle/4317825-elastic-big-
data-needseffective-search-
to-drive-value [18]

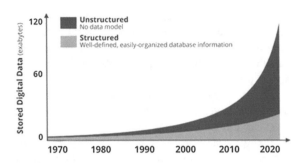

2   Related Works

As part of the literature survey conducted so far, some of the existing works related
to natural language processing (NLP) methods can be classified as identifying the
emotional arcs of the story, co-reference resolution, plot generation and text-to-
speech (TTS) conversion.

A story becomes successful when it has an emotional impact on the person reading
it. This emotional effect is based on how the story is shared with the reader and using
which emotional trajectory [1, 2] it is shared. Three tools are used for determining the
emotional arc of a story which are principal component analysis (PCA), clustering
and self-organizing maps (SOMs). Additionally, a hedonometer is used to find the
variations in happiness/pleasure. To construct the emotional arcs, the text of the
story is divided into segments of equal length and its size gives the fixed window
size. Sliding this fixed size window through the book, the sentiment scores can be
generated with the hedonometer, which comprise the emotional arcs. On applying
PCA to this, the six basic modes that will act as the core emotional arc types can be
obtained.

Every story can be mapped to one of these arcs. The six basic emotional arcs are:

"Rags to Riches" (rise)

"Tragedy" or "Riches to Rags" (fall)

"Man in a hole" (fall-rise)

"Icarus" (rise-fall)

"Cinderella" (rise-fall-rise)

"Oedipus" (fall-rise-fall).

Using hierarchical clustering, each of these books is mapped to that basic arc
from which has a least distance to its own arc. SOM is a type of dimensionality
reduction technique that is then used to find the collection of stories with most
common emotional arcs after removing noises from it.

Co-reference resolution is an important part in the processing of a text corpus
related to stories. It helps to resolve the possible references between characters. It

can be done using a deep learning method utilizing a suitable loss function. Recent research [3] has found that using a neural network-based model can improve the performance of co-reference resolution. It suggests a neural mention-ranking model for reference resolution which is based on assigning scores to pairs of the same entity and its possible co-reference.

The problem in identifying related mentions is another challenge. An improved deep learning system utilizing entity-level distributed representations [4] can be used to cluster-related mentions thereby identifying all the similar entities.

Text generation has seen many advances over the years which include the introduction of many technologies, for example, using generative adversarial networks (GANs) [5] like TreeGAN [6] and SeqGAN [7] or using conditional transformer language model like CTRL [8], etc. A recently released plug-and-play language model (PPLM) [9] combines a pre-trained language model (LM) with one or more simple attribute classifiers that guide text generation without any further training of the LM. It allows researchers to make use of the few pre-trained models out there: Rather than requiring everyone to train their own LM, PPLM lets users combine small attribute models with an LM to steer its generation. Attribute models can be 100,000 times smaller than the LM and still be effective in steering it. The tool can easily switch between sentiments and topics according to the given scenario. PPLM proceeds through a three-step process: forward pass, backward pass and vocabulary generation. In the first step, a forward pass is performed through the LM to compute the likelihood of the desired attribute. This is then followed by a backward pass that updates the internal representations of the LM based on gradients from the attribute model. In the final step, a new distribution of vocabulary is generated.

Tacotron [10] was designed to have an upper hand over an integrated end-to-end text-to-speech (TTS) system that can be trained on <text, audio> pairs with minimal human annotations. This kind of system allows for rich conditioning of various attributes or features like the sentiment. Adaptation to a new model will also become easier with this model, and also a single model is likely to be more robust than a multistage model where each component's error can compound. An improved system called Tacotron 2 [11, 12] was also proposed to improve the TTS.

For an end-to-end model, TTS is a difficult learning task. Unlike end-to-end speech recognition or machine translation, TTS outputs are continuous and output sequences are usually much longer than those of the input.

3 Methodology

A narrative story is fed into the model (Fig. 2). The phase splitter labels the narrative story and passes them in parallel through next two phases—dialogue processing and plot generation. The first parallel phase does dialogue processing along with sentiment analysis. The second parallel phase extracts the data required for generating the plot of the script and generates it. The output of two phases is then combined to produce an initial output, i.e., the screenplay. Text-to-speech synthesis is done to

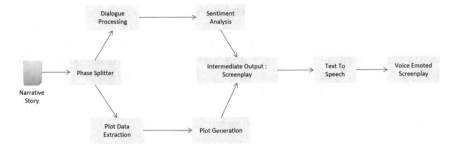

Fig. 2 Architecture diagram of Alyce

get an audio output which is then voice-modulated to get a final output of a voice emoted screenplay.

3.1 Phase Splitting

In this phase, the story is converted into JavaScript Object Notation (JSON) [13] object with each paragraph as an array of JSON objects which contains another array of JSON Objects having the sentence and their corresponding label as the content. This is then given as input into two sections—dialogues and narration. The identification of dialogues in the sentences can be done using simple regular expression matching of those words/sentences given within quotation marks. The remaining parts of the story can be labeled as non-dialogue/narration and can be used to generate the plot or scene settings. This conversion to JSON is important because it gives a standard structure to the data so that it can be sent between the different phases of the system.

3.2 Dialogue Processing and Plot Data Extraction

After the story is labeled, the dialogues identified are given as input to the dialogue processing phase. Here, the problem is to find the speaker of the corresponding line. This is because in most cases, the sentence containing the dialogue will only be having references to the speaker as "he", "she", etc. This can be resolved by correlating the reference to the corresponding speakers. Running the reference resolving code in the entire story will not be producing sensible results; hence, the sentences in quotes are pre-processed by replacing the actual text with a dummy text which will again be replaced after resolving the references. Replacing the references in the sentence containing the dialogue gives a modified story. Thus, to separate the dialogues and narration from this modified story, the speaker's name is identified from the dialogue containing the sentence using any NLP package and set up the dialogues in the format of a script.

This phase of Alyce involves the usage of NLP directly and indirectly. The co-reference resolution is done using NeuralCoref [14, 3]. NeuralCoref is built using SpaCy [15], an NLP package. This is the indirect method in which NLP is employed. However, after the resolution is done, to lay the content in the form of dialogues as in a script by identifying the speaker, NLP is used again. This is the direct use. The outputs are then sent to the corresponding sections.

3.3   Plot Generation

The plot plays a significant role in a screenplay. It helps in painting the context for a particular scene and helps the reader in visualizing the story in his mind. The structural and emotional aspects of dialogues are also dependent on the plot. Plot generation is the second phase of the screenplay writer. A lot of open-source text generation software exists in the market including image captioning systems. It identifies the properties of the images, and with a few of these main properties, it generates meaningful sentences that caption the images. However, these tools cannot generate complex and more humane texts as these are designed to write only a few words. On the other side, there are large transformer-based language models (LMs) that are trained on huge text corpora, designed specifically to generate texts. Even though they have powerful text generation capabilities, they cannot be controlled to switch between sentiments and topics. The idea is to incorporate PPLM [9] to produce more controllable texts.

The PPLM employs an attribute model to guide the generation of text by the LM. This attribute model can be as simple as a bag of words. Thus, in this phase, the keywords from story are taken along with keywords related to them to form a text file. This dynamically created text file will act as the attribute model for PPLM to guide the generation of plot data.

3.4   Text-to-Speech

Text-to-speech synthesis is the third phase of the screenplay generator. Usually, screenplay generation ends with the "writing" part, but with Alyce being more humane, text-to-speech is brought into consideration. Normally, when a word is synthesized by a speech engine, the pronunciation required for the word is gathered from different subword sounds and then glued together. For example, to synthesize the word "Impressive", it would be broken down into chunks like "imp", "pres" and "ive". These chunks would then be derived from other words. Imp's sound may be derived from "Important", Pres' sound would be derived from "President", and ive's sound would maybe from "Detective". These chunks, when glued together, will give the word "Impressive".

These techniques are intelligible up to a certain degree but are not natural. Such TTS engines are termed as concatenative TTS. The concatenative TTS combines high-quality audio clips to form speech. However, the generated speech will be emotionless. Parametric TTS engines are a statistical model that can be incorporated for emoting the modulations in the speech. A parametric TTS has two stages. First, extract linguistic features after text processing. Second, extract vocoder features that represent the corresponding speech signal.

To be more efficient, deep learning could be used to learn features from data, and it could be passed onto the vocoder features. The deep learning model should fine-tune the features, and the final output shall be more humane. Tools like Tacotron [10] and Mozilla TTS can be used to achieve this.

We are planning to use the Tacotron2 [16] implementation by Mozilla with WaveRNN [17] as the vocoder. The speech synthesized by Mozilla TTS is indeed natural but cannot sense the sentiment in the text. This is because of the lack of proper data sets.

In this phase, a new model is trained for generating the audio transcripts. Training is done on a whole new data set. To prepare the data set, first a particular celebrity's voice is chosen, say for example Scarlett Johansson. The dialogues from her movies are extracted and saved as an audio file. The data set is built having a format similar to <text, audio> pair. Once the training is done, all the dialogues are passed and corresponding audio transcripts will be generated.

4   Expected Results

On a system with Intel Xeon E5 2630 V4, nVIDIA 2 * 11 1080ti, 64 GB RAM with CUDA support along with the software packages NeuralCoref 4.0, SpaCy 2.1.0, PPLM, etc., installed we expect the following results.

On completion of the Identification phase of the proposed work, the references of characters in sentences are resolved and the dialogues of each character are identified and extracted from the text. The separation of other descriptions from the text is also done in this phase.

The outcome aimed to be produced at the end of the plot generation phase includes the script containing the dialogues and plot description generated from the details, keywords and sentiment of the story which was separated from the text in the previous phase.

On the successful completion of the text-to-speech phase, the result expected to be achieved is a fully functional screenplay transcripter that converts the textual script into an audio transcript and emotes the script while reading.

Consider the following story excerpt as the input to the model: The playground was filled with children that spring afternoon. John was waiting anxiously under the old apple tree for Ria to come. He takes out his phone again and dials her one last

Table 1 Expected outcome

Phase	Steps	Expected outcome	Expected accuracy
1	Dialogue masking	Story with masked dialogues	Above 90%
	Identifying characters	Possible characters identified	
	Resolving character references	Story with references substituted with corresponding characters	
	Dissecting dialogues and narration from story	Primitive form of movie script	
2	Extracting keywords from narration and sentiments	Possible keywords extracted for plot	Between 65 and 80%
	Plot generation	Plot description	
	Combining dialogues with generated plot	Finished screenplay	
3	Text-to-speech synthesis if narration and dialogues	Audio transcript generated	Between 50 and 90%

time out of frustration hoping that she would pick up the phone at least this time. He had to tell her the news as soon as possible before she came to know about it from anywhere else. Finally, she picks up the call. "Hi John.", she said in her usual cheerful voice. "Ria, where are you? I have been trying to call you for hours now", he replies in an unusually irritated voice. "Is everything OK?", she asks, and after a short pause, he replies "I'm sorry but Jane is no more (Table 1)".

The output expected for each phase for the above example is as follows:

Phase 1:

Identified Dialogues:

Ria: Hi, John.
John: Ria, where are you? I've been trying to call you for hours now.
Ria: Is everything OK?
John: I'm sorry but Jane is no more

Phase 2:

Extracted keywords: [playground, children, spring, afternoon, [John, anxious], old apple tree, [John, frustration], news, [Ria, cheerful], [John, irritated], [Ria, Jane, Sad]]

Possible plot developed:

It's a wonderful spring afternoon. It's 4:30 PM and John is sitting on a bench under the old apple tree anxiously waiting for his Ria to come. The playground is filled with children. Some of the kids are playing football while some others are playing Hide and Seek. John takes out his phone and dials Ria for the last time out of frustration. John hopes that she would pick up this time as he has to share an important news

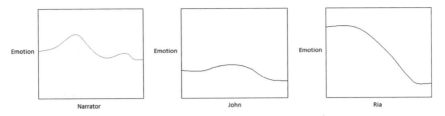

Fig. 3 Voice modulation expected

with her and it is important for him to inform her before she comes to know about it from anywhere else.

Combining phases 1 and 2:

[Narrator]

It's a wonderful spring afternoon. Its 4:30 PM and John is sitting on a bench under the old apple tree anxiously waiting for his Ria to come. The playground is filled with children. Some of the kids are playing football while some others are playing Hide and Seek. John takes out his phone and dials Ria for the last time out of frustration. John hopes that she would pick up this time as he has to share an important news with her and it is important for him to inform her before she comes to know about it from anywhere else (Fig. 3).

Ria: Hi John.
John: Ria, where are you? I've been trying to call you for hours now.
Ria: Is everything OK?
John: I'm sorry but Jane is no more

5   Conclusion

Video and animation are in the boom now but there is nothing that parallels the experience given by augmented reality (AR). Alyce in her full strength will be able to depict the generated screenplay in an AR format by using stick figures to give a more engaging and entertaining experience. Thus, in totality Alyce Garner Peterson will be a new NLP based prototype that aids the conversion of a story to screenplay. The solution simplifies the procedure by recognizing the characters in the scene and their dialogues. The summary system embeds the environmental data as the theme and generates the plot of that scene. The plot is further used to map emotions and dialogue delivery to the AR system. The future scope of the work can be extended to enable all genres of stories to be converted into screenplay and also to support the multiple languages.

References

1. Reagan AJ, Mitchell L, Kiley D, Danforth CM, Dodds PS (2016) The emotional arcs of stories are dominated by six basic shapes. EPJ Data Sci. https://doi.org/10.1140/epjds/s13688-016-0093-1
2. Reagan AJ, Tivnan BF, Williams JR, Danforth CM, Dodds PS (2015) Benchmarking sentiment analysis methods for large-scale texts: a case for using continuum-scored words and word shift graphs. CoRR abs/1512.00531. http://arxiv.org/abs/1512.00531
3. Clark K, Manning CD (2016) Deep reinforcement learning for mention-ranking coreference models. CoRR abs/1609.08667. http://arxiv.org/abs/1609.08667
4. Clark K, Manning CD (2016) Improving coreference resolution by learning entity-level distributed representations. CoRR abs/1606.01323 (2016), http://arxiv.org/abs/1606.01323
5. Goodfellow IJ, Pouget-Abadie J, Mirza M, Xu B, Warde-Farley D, Ozair S, Courville A, Bengio Y (2014) Generative adversarial networks
6. Liu X, Kong X, Liu L, Chiang K (2018) Treegan: Syntax-aware sequence generation with generative adversarial networks. CoRR abs/1808.07582 (2018). http://arxiv.org/abs/1808.07582
7. Yu L, Zhang W, Wang J, Yu Y (2016) Seqgan: Sequence generative adversarial nets with policy gradient. CoRR abs/1609.05473 (2016). http://arxiv.org/abs/1609.05473
8. Keskar NS, McCann B, Varshney LR, Xiong C, Socher R (2019) Ctrl: A conditional transformer language model for controllable generation (2019)
9. Dathathri S, Madotto A, Lan J, Hung J, Frank E, Molino P, Yosinski J, Liu R (2019) Plug and play language models: a simple approach to controlled text generation (2019)
10. Wang Y, Skerry-Ryan RJ, Stanton D, Wu Y, Weiss RJ, Jaitly N, Yang Z, Xiao Y, Chen Z, Bengio S, Le QV, Agiomyrgiannakis Y, Clark R, Saurous RA (2017) Tacotron: a fully end-to-end text-to-speech synthesis model. CoRR abs/1703.10135 (2017). http://arxiv.org/abs/1703.10135
11. Shen J, Pang R, Weiss RJ, Schuster M, Jaitly N, Yang Z, Chen Z, Zhang Y, Wang Y, Skerry-Ryan R, Saurous RA, Agiomyrgiannakis Y, Wu Y (2017) Natural tts synthesis by conditioning wavenet on mel spectrogram predictions (2017)
12. Tacotron 2: Generating human-like speech from text. https://ai.googleblog.com/2017/12/tacotron-2-generating-human-likespeech.html. Accessed on 20 Mar 2020
13. The json data interchange syntax. http://www.ecmainternational.org/publications/files/ECMA-ST/ECMA-404.pdf. Accessed on 20 Mar 2020
14. Neuralcoref 4.0: Coreference resolution in spacy with neural networks. https://github.com/huggingface/neuralcoref. Accessed on 21 Mar 2020
15. Spacy. https://spacy.io. Accessed on 21 Mar 2020
16. Shen J, Pang R, Weiss RJ, Schuster M, Jaitly N, Yang Z, Chen Z, Zhang Y, Wang Y, Skerry-Ryan RJ, Saurous RA, Agiomyrgiannakis Y, Wu Y (2017) Natural TTS synthesis by conditioning WaveNet on Mel spectrogram predictions. CoRR abs/1712.05884 (2017). http://arxiv.org/abs/1712.05884
17. Kalchbrenner N, Elsen E, Simonyan K, Noury S, Casagrande N, Lockhart E, Stimberg F, van den Oord A, Dieleman S, Kavukcuoglu K (2018) Efficient neural audio synthesis. CoRR abs/1802.08435 (2018). http://arxiv.org/abs/1802.08435
18. Seeking alpha. https://seekingalpha.com/article/4317825-elastic-big-data-needseffective-search-to-drive-value. Accessed on 27 Feb 2020

Multistage Model for Accurate Prediction of Missing Values Using Imputation Methods in Heart Disease Dataset

Pooja Rani, Rajneesh Kumar, and Anurag Jain

Abstract When machine learning is used for the design of a prediction model in medical science, then higher accuracy is essential. It becomes difficult to achieve higher accuracy due to unavailability of values in certain fields of data set. Therefore, it is necessary to deal with the issue of missing values effectively. This research work focuses on an efficient way to handle missing values. Authors have proposed a systematic methodology for the identification of missing value. Authors have used Cleveland Heart disease dataset from the UCI (University of California, Irvine) repository to test their experiments. Missing values are imparted using three different approaches, namely random, MISSHASH & MISSFIB. Four imputation methods k-nearest neighbor (KNN), multivariate imputation by chained equations (MICE), mean, and mode imputation were analyzed with the help of four classifiers Naive Bayes (NB), support vector machine (SVM), logistic regression (LR), and random forest (RF). Root mean square error (RMSE) of classifiers was compared to find the combination of the best imputation method. It has found that MICE imputation method has performed better related to other imputation methods. Moreover, its accuracy is independent of classifier and missing value distribution.

Keywords KNN imputation · Mean imputation · Mode imputation · Multivariate imputation by chained equations · Root mean square error

P. Rani (✉) · R. Kumar
Department of Computer Engineering, MMEC, Maharishi Markandeshwar (Deemed to be University), Mullana, Ambala, Haryana, India
e-mail: pooja.rani@mmumullana.org

R. Kumar
e-mail: drrajneeshgujral@mmumullana.org

A. Jain
Virtualization Department, School of Computer Science, University of Petroleum and Energy Studies, Dehradun, India
e-mail: anurag.jain@ddn.upes.ac.in

J. S. Raj et al. (eds.), *Innovative Data Communication Technologies and Application*,
Lecture Notes on Data Engineering and Communications Technologies 59,
https://doi.org/10.1007/978-981-15-9651-3_53

1 Introduction

Medical datasets are highly vulnerable to missing data that can affect predictions if not handled properly. Missing data may occur due to three mechanisms: missing completely at random (MCAR), missing at random (MAR), and missing not at random (MNAR). If the probability of missing data is not affected by observed values or missing values, then it is known as MCAR. If the probability of missing data is only affected by observed values, then it is known as MAR. If the probability of missing data depends upon missing values, then it is known as MNAR [1].

Missing data can be easily handled with complete case analysis method in which rows having missing values are not used for training. This approach is efficient only if the dataset contains a few missing values. If the training dataset contains a large number of samples that have missing values, it may reduce the size of the training data [2]. If the dataset has a large proportion of missing values then imputation is a better method, which fills missing values with some plausible value. Single imputation or multiple imputations can be used to impute values. In single imputation, missing values are replaced with values obtained using a certain rule producing a complete dataset. Although single imputation is better than complete case analysis, it cannot handle uncertainty in imputation. Multiple imputations produce multiple datasets by performing imputation multiple times. It can handle uncertainty in a better way as compared to single imputation. Several methods of imputations exist but which is preferable at a given situation is not clear. The focus of this research is to find an efficient imputation method.

Analyzing the impact of missing values on a model performance with a complete dataset having no missing data is possible through introducing missing values randomly and it may not provide an exact estimate. It is a major requirement in machine learning to analyze the impact of missing values because when the real data is collected, there may be a chance that some values may be missing. Authors have proposed a methodology which can be applied on any dataset to analyze the impact of missing values using different distribution and imputation methods. There are three distribution methods used in this research, so estimates provided by it are more accurate.

The remaining sections of this paper are organized as follows: Sect. 2 contains a description of different imputation methods used by different researchers. Section 3 contains methodology, classifiers, missing value distribution, and imputation methods used for performing experiments. Section 4 includes results and discussion. Section 5 contains the conclusion.

2 Related Work

Imputation method used by various researchers to handle missing values is given in Table 1.

Table 1 Imputation methods used by various researchers

S. No.	Authors	Datasets	Imputation methods
1.	Noor et al. [3]	Particulate matter dataset	Linear, quadratic, and cubic interpolation
2.	Kumar and Kumar [4]	Breast cancer dataset, hepatitis, and lung cancer datasets	Mean imputation, KNN, weighted KNN, and fuzzy KNN
3.	Sim et al. [5]	Iris dataset, wine dataset, glass dataset, liver disorder dataset, ionosphere dataset, and Statlog Shuttle dataset	Litwise deletion, KNN imputation, mean imputation, predictive mean imputation, group mean imputation, hot deck imputation, and k-means clustering imputation
4.	Nahato et al. [6]	Hepatitis dataset and breast cancer dataset	Most frequent imputation method
5.	Qin et al. [7]	Chronic kidney disease	KNN imputation method
6.	Venkatraman et al. [8]	DiabHealth dataset	Mean and mode imputation method
7.	AlMuhaideb and Menai [9]	25 medical datasets from UCI repository	Litwise deletion
8.	Kuppusamy and Paramasivam [10]	Nursery dataset	MissForest algorithm
9.	Hu et al. [11]	Surgical patients EHR dataset from University of Minnesota clinical data repository	Mean imputation, zero imputation, normal imputation, MICE imputation
10.	Sujatha et al. [12]	Cleveland heart disease dataset	Imputing missing values classifier (IMVC) method
11.	Karim et al. [13]	Solar radiation dataset	Spline interpolation method
12.	Mohan et al. [14]	Cleveland heart disease dataset	Litwise deletion
13.	Almansour et al. [15]	Chronic kidney disease dataset	Mean imputation
14.	Kim et al. [16]	Photovoltaic dataset	Linear interpolation, KNN imputation, mode imputation, MICE
15.	Stavseth et al. [17]	Questionnaire dataset	Complete case analysis, hot deck imputation, multiple correspondence analysis, MICE, latent case analysis, expectation maximization with bootstrapping
16.	Nikfalazar et al. [18]	Adult, Housing, Pima Autompg, Yeast and CMC	DIFC method combining decision tree and fuzzy logic
17.	Raja [19]	Dermatology, pima, yeast, and wisconsin datasets	Rough k-means centroid-based imputation

All the researchers had analyzed imputation methods with a dataset having missing data or with a complete dataset in which missing values were introduced randomly. A systematic methodology to analyze the impact of missing values on a completed dataset was not available. In addition, two new algorithms MISSHASH and MISSFIB for introducing missing values are also proposed.

3 Methodology, Missing Value Distribution, Imputation Methods, and Classifiers

3.1 Methodology

A four-stage model has proposed for finding the impact of missing values of a given dataset as shown in Fig. 1. In stage 1, a suitable value removal algorithm as per the missing value distribution requirement is selected. Imputation method is selected in stage 2. To predict the missing value, the classifier algorithm is selected in step 3. The parameter to estimate the accuracy of the predicted missing value is chosen in step 4. If the predicted value is sufficiently accurate, then the procedure is stopped; otherwise, the procedure is repeated with another combination of missing value

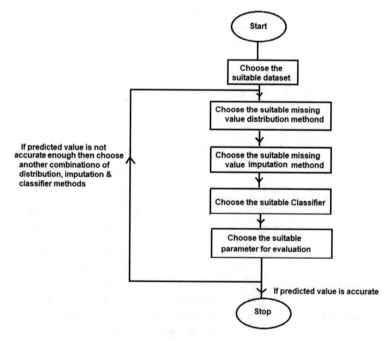

Fig. 1 A four-stage missing value prediction model

distribution, missing value imputation, and classifier. Proposed model was tested on cleveland heart disease dataset taken from the UCI repository [20]. Combination of missing value distribution, missing value imputation, and the classifier can be tested on different datasets. Different combinations of imputation method and classifier can be tested on input dataset to find out the best combination to reduce RMSE.

3.2 Missing Values Distribution Methods

Random Distribution of Missing Values

In this algorithm, missing values are distributed randomly in rows and columns of the dataset. A random function is used to generate row number and column number. After generating row number and column number, the value in that row and column is removed.

Algorithm   Random Distribution of missing values

Procedure: MISSRAND(DS,N)
Input:Dataset DS having R rows and C columns with no missing values
Variable N initialized with number of random values to be generated
Output:Dataset DS with missing values
1: $I = 1$
2: **while** $I \leq$ N **do**
3:     $INDEX1 \leftarrow$ Random Number in range of 0 to R-1
4:     $INDEX2 \leftarrow$ Random Number in range of 0 to C-1
5:     $DS[INDEX1][INDEX2] \leftarrow$ Null
6:     $I \leftarrow$ I+1
7: End While

Missing Values Distributed Using Fibonacci Series

Dataset is converted into a one-dimensional array and missing values are distributed in a pattern of Fibonacci series. Numbers in Fibonacci series are generated and values at those positions are removed. If the value at the index generated is already missing, then the next number of the series is generated. It is necessary to make sure that the index generated does not exceed the dimension of the array. If it happens, then a mod of the index with an appropriate number is calculated to generate the index within the range of the array. After introducing missing values, this array is again transformed into the dataset with rows and columns.

Algorithm Distribution of missing values using Fibonacci series

> Procedure: MISSFIB(DS,N)
>
> **Input:Dataset DS having R rows and C columns with no missing values**
> **Variable N initialized with number of random values to be generated**
> **Output:Dataset DS with missing values**

1: Convert dataset DS into one dimensional array A
2: $I = 1, N1=0, N2=1$
3: while $I \leq$ N do
4:     $N3=N1+N2$
5:     $INDEX=N3$
6:     $N1=N2$
7:     $N2=N3$
8:     if $INDEX \geq 4000$ then
9:       $INDEX \leftarrow$INDEX MOD 4000
10:     End if
11:     while INDEX+1 MOD 4000=0 OR A[INDEX]=Null do
12:       $N3=N1+N2$
13:       $INDEX=N3$
14:       $N1=N2$
15:       $N2=N3$
16:       if $INDEX \geq 4000$ then
17:         $INDEX \leftarrow$INDEX MOD 4000
18:       End if
19:     End While
20:     $A[INDEX] \leftarrow$Null
21:     $I \leftarrow$I+1
22: End While
23: Convert A into dataset DS

Missing Values Distributed Using Hashing Method

Dataset is converted into a one-dimensional array and missing values are distributed using a hash function. A nearest prime number M exceeding the size of the array is chosen. A random number R is generated within the range of the array. An index is selected by calculating the Mod of the R with M and value at that index is removed. If the selected index has already missing value, then the next index is calculated by adding the index with 2^i having $i = 0,1,2, \ldots N$. After introducing missing values, this array is again transformed into the dataset with rows and columns.

Algorithm Distribution of missing values using hashing method

---

    Procedure: MISSHASH(DS,N)
    Input:Dataset DS having R rows and C columns with no missing values
    Variable N initialized with number of random values to be generated
    Output:Dataset DS with missing values

1: Convert dataset DS into one dimensional array A
2: $M \leftarrow$ Nearest Prime Number greater than $(R*C)-1$
3: $I = 1$
4: **while** $I \leq N$ **do**
5:     $R \leftarrow$ Random Number in the range of 0 to $(R*C)-1$
6:     $INDEX = R*R$ MOD M
7:     $K = 0$
8:     **while** A[INDEX]=Null **do**
9:         $INDEX =$ INDEX+POWER(2,K)
10:         $K = K + 1$
11:     End While
12:     $A[INDEX] \leftarrow$ Null
13:     $I \leftarrow I+1$
14: End While
15: Convert A into dataset DS

3.3 Missing Value Imputation Methods

In missing value imputation, missing data is filled with an estimated value. This estimated value is inferred from the known values of the dataset [21].

Four imputations methods used in performing experiments are discussed in this section.

Mean Imputation
Missing values in each attribute of data are found and it is replaced with a mean of observed values of that attribute [5].

Algorithm Mean Imputation

    Procedure: MEANIMP(DS)
    Input:Dataset DS with missing values
    Output:Imputed Dataset DS with zero missing values

1: Find out the missing values in first attribute of dataset
2: Calculate mean of the attribute.
3: Replace all the missing values in the attribute with calculated mean
4: Repeat steps 1 to 3 for all the attributes in the dataset

Mode Imputation
Missing values in each attribute of data are found and it is replaced with the most frequent value in the observed values of that attribute [22].

Algorithm Mode Imputation

Procedure: MODEIMP(DS)
Input:Dataset DS with missing values
Output:Imputed Dataset DS with zero missing values

1: Find out the missing values in first attribute of dataset
2: Find out the value V which occurs most frequently in the attribute.
3: Replace all the missing values in the attribute with value V.
4: Repeat steps 1 to 3 for all the attributes in the dataset

KNN Imputation

In this method, k-nearest neighbors of the record with missing data is found and it is replaced with mean values of the attribute of these neighbors [23]

Algorithm KNN Imputation

Procedure: KNNIMP(DS)
Input:Dataset DS with missing values
Output:Imputed Dataset DS with zero missing values

1: Find out the record with missing values in any attribute of the dataset.
2: Select the K nearest neighbors of the record based upon remaining attributes of the record.
3: Calculate the mean value V of the corresponding attribute of the selected records.
4: Replace the missing value in the attribute with value V.
5: Repeat steps 1 to 4 for all the attributes in the dataset

MICE Imputation

MICE method performs imputation multiple times. It assumes the MAR characteristic of missing data. It means data is missing randomly. In this method, a regression model is used to predict the value of the missing attribute from the remaining attributes of the dataset [24].

Algorithm MICE Imputation

Procedure: MICEIMP(DS)
Input:Dataset DS with missing values
Output:Imputed Dataset DS with zero missing values

1: Perform mean imputation for missing values of all attributes.
2: Replace the mean imputation of one variable X with missing value.
3: Build a regression model to predict the value of x from other variables in the dataset.
4: Use the regression model to predict the value of x and replace the missing values in x with these predicted values.
5: Repeat steps 2 to 4 for all the variables in the dataset.

3.4 Classifiers

Machine learning classifiers used to analyze the impact of imputation methods are discussed in this section.

Naive Bayes Classifier
This classifier assumes independence among individual features of the data and calculates the probability of each class using these features. The Bayesian theorem is applied for calculating the probability [25]. Probability of occurring class i, if j occurs, is calculated using the formula given in Eq. 1. It is known as posterior probability.

$$P(i|j) = P(j|i)P(i) \div P(j) \tag{1}$$

$P(j|i)$ is the probability of occurring j with i which is known as likelihood. $P(j)$ known as prior attribute probability is the probability of occurring j. $P(i)$ is the probability of occurring class i which is known as prior class probability.

Logistic Regression
It is a suitable method for binary classification problems where each outcome variable y can have only two values 0 or 1. In it, a logistic function is used which is also named as sigmoid function. This function maps any real-valued number to a value in the range of 0 to 1. Probability of input X belonging to class 1 is predicted by combining input values X using weight using Eq. 2:

$$P(X) = \frac{\exp(\beta_0 + \beta_1 X)}{1 + \exp(\beta_0 + \beta_1 X)} \tag{2}$$

This equation is known as the logistic regression equation. Here, β_0 is intercept term and β_1 is a weight for input value X. Weight for each input X is learned from data provided for training. Probability calculated using Eq. 2 is compared to a threshold to find out the class of input data [26].

Support Vector Machine
SVM maps the input to the n-dimensional space to create a hyperplane with maximum margin. This hyperplane creates a boundary between two classes. Maximizing the margin ensures the data points will be accurately classified. Data points closest to the hyperplane are known as support vectors [27]. Hyperplane H_0 is defined as given in Eq. 3:

$$H_0 : \ w^T x + b = 0 \tag{3}$$

Two hyperplanes H_1 and H_2 are created in parallel to hyperplane H_0 as given in Eqs. 4 and 5.

$$H_1 : \quad w^T x + b = -1 \tag{4}$$

$$H_2 : \quad w^T x + b = 1 \tag{5}$$

Each input vector X_i should be satisfied by hyperplane with constraint given in Eqs. 6 and 7:

$$w X_i + b \geq +1 \text{ for } X_i \text{ having class 1} \tag{6}$$

and

$$w X_i + b \leq -1 \text{ for } X_i \text{ having class 1} \tag{7}$$

Hyperplane separating two classes in SVM is shown in Fig. 2.

Random Forest
It is an ensemble method, in which multiple decision trees are used to make the classification. Each decision tree participates in the classification process and final prediction is done using the majority vote. Decision tree suffers from the problem of overfitting, i.e., model provides good accuracy on training data only. This problem is solved by random forest. Samples for constructing decision trees are chosen randomly from training data. At each step of constructing trees subset of features is chosen randomly [28]. Following are major steps of this algorithm:

(1) Choose random samples from the dataset.
(2) Construct multiple decision trees from the chosen samples.
(3) Perform prediction using each decision tree.
(4) Make the final prediction by using a voting mechanism.

Fig. 2 Hyperplane separating two classes in SVM

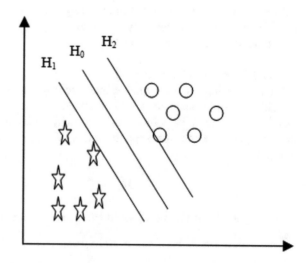

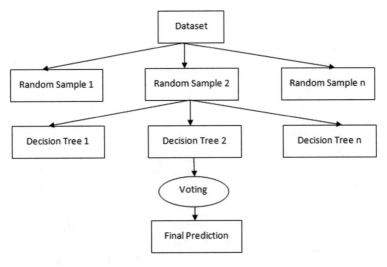

Fig. 3 Working of random forest

The working of a random forest is shown in Fig. 3.

Performance Parameter

Root mean square error (RMSE) was used as a performance parameter of the classifiers to evaluate the impact of imputation methods on classifier performance. Root mean square error (RMSE) calculates the deviation between the predicted and observed value by using Eq. 8 [29]:

$$\text{RMSE} = \sqrt{\frac{1}{N} \sum_{i=1}^{n} (P - O)^2} \tag{8}$$

Here, P indicates predicted value, O indicates the observed value and N indicates total samples.

The value of RMSE indicates the error in prediction so it should have a smaller value. If accuracy is low, RMSE will be high and vice versa [30].

4 Results and Discussions

4.1 Random Distribution of Missing Values (MISSRAND)

Missing values were randomly distributed across the dataset with a 10% to 30% missing ratio. After introducing missing values artificially, different imputation methods were used to impute these values and performance of NB, SVM, LR, and

Table 2 RMSE of classifiers with different imputation methods on a random distribution of missing values

S. No	Imputation method	Classifier	10%	15%	20%	25%	30%	Average RMSE
1	MICE	NB	0.4375	0.4260	0.4182	0.4667	0.4523	0.4401
2	KNN	NB	0.4486	0.4486	0.4412	0.4874	0.4559	0.4563
3	MEAN	NB	0.4375	0.4449	0.4523	0.4806	0.4908	0.4612
4	MODE	NB	0.4375	0.4595	0.4631	0.4806	0.4806	0.4642
5	MICE	SVM	0.4523	0.4595	0.4908	0.4908	0.4806	0.4748
6	KNN	SVM	0.4737	0.4908	0.5041	0.5233	0.4667	0.4917
7	MEAN	SVM	0.4559	0.4908	0.5138	0.5170	0.4840	0.4923
8	MODE	SVM	0.4840	0.4975	0.5073	0.5510	0.5233	0.5126
9	MICE	LR	0.4221	0.4182	0.4102	0.4667	0.3810	0.4196
10	KNN	LR	0.4260	0.4486	0.4182	0.4772	0.4021	0.4344
11	MEAN	LR	0.4299	0.4375	0.4102	0.4840	0.4142	0.4351
12	MODE	LR	0.4337	0.4486	0.4559	0.4806	0.4559	0.4549
13	MICE	RF	0.4182	0.4102	0.3980	0.4182	0.3723	0.4033
14	KNN	RF	0.3980	0.4412	0.4412	0.4702	0.4062	0.4313
15	MEAN	RF	0.4021	0.4375	0.4337	0.4874	0.4142	0.4349
16	MODE	RF	0.4375	0.4559	0.4102	0.4772	0.4375	0.4436

RF classifiers is analyzed in terms of RMSE using this imputed dataset to compare different imputation methods. Results are shown in Table 2.

MICE method provided the lowest RMSE with all the classifier. This has shown in Fig. 4.

4.2 Missing Values Distributed Using Fibonacci Series (MISSFIB)

Missing values were distributed across the dataset according to MISSFIB algorithm with a 10–30% missing ratio. After introducing missing values artificially, different imputation methods were used to impute these values and performance of NB, SVM, LR, and RF classifiers is analyzed in terms of RMSE using this imputed dataset to compare different imputation methods. Results are shown in Table 3.

MICE method provided the lowest RMSE with all the classifier. This has shown in Fig. 5.

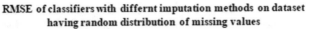

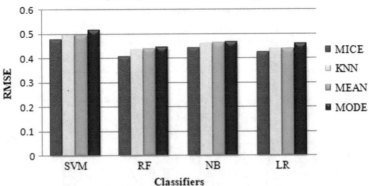

Fig. 4 RMSE of classifiers with different imputation methods on a random distribution of missing values

Table 3 RMSE of classifiers with different imputation methods on a dataset having missing values distributed using MISSFIB algorithm

S. No	Imputation method	Classifier	10%	15%	20%	25%	30%	Average RMSE
1	MICE	NB	0.4523	0.4412	0.4486	0.4631	0.4772	0.4564
2	KNN	NB	0.4523	0.4595	0.4772	0.4874	0.4908	0.4734
3	MEAN	NB	0.4412	0.4523	0.4772	0.4874	0.4941	0.4704
4	MODE	NB	0.4631	0.4874	0.4840	0.4772	0.5008	0.4825
5	MICE	SVM	0.4523	0.4631	0.4595	0.4449	0.4631	0.4565
6	KNN	SVM	0.4737	0.4840	0.4941	0.4737	0.5202	0.4891
7	MEAN	SVM	0.4595	0.4772	0.4840	0.4840	0.5008	0.4811
8	MODE	SVM	0.4941	0.4941	0.5008	0.4874	0.5265	0.5005
9	MICE	LR	0.4102	0.3980	0.4412	0.4299	0.4486	0.4255
10	KNN	LR	0.4260	0.4221	0.4559	0.4523	0.4840	0.4480
11	MEAN	LR	0.4102	0.4449	0.4559	0.4595	0.4908	0.4522
12	MODE	LR	0.4412	0.4559	0.4772	0.4840	0.4975	0.4711
13	MICE	RF	0.3938	0.4337	0.4260	0.4299	0.4299	0.4226
14	KNN	RF	0.4221	0.4375	0.4840	0.4412	0.4874	0.4544
15	MEAN	RF	0.4102	0.4449	0.4559	0.4595	0.4806	0.4502
16	MODE	RF	0.4260	0.4631	0.4631	0.4375	0.4702	0.4519

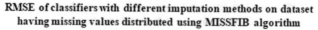

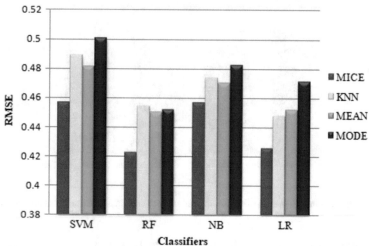

Fig. 5 RMSE of classifiers with different imputation methods on a dataset having missing values distributed using MISSFIB algorithm

4.3 Missing Values Distributed Using Hashing Method (MISSHASH)

Missing values were distributed across the dataset according to MISSHASH algorithm with a 10% to 30% missing ratio. After introducing missing values artificially, different imputation methods were used to impute these values and performance of NB, SVM, LR, and RF classifiers is analyzed in terms of RMSE using this imputed dataset to compare different imputation methods. Results are shown in Table 4.

MICE method provided the lowest RMSE with all the classifier. This has shown in Fig. 6.

From the analysis of Figs. 4, 5, and 6, it can be concluded that the MICE method is the best imputation method independent of missing value distribution and classifier used. In addition, it has shown better accuracy relative to other imputation methods. Results of Tables 2, 3 and 4 indicate that MICE reduced 9.08% RMSE as compared to MODE in MISSRAND distribution, 6.48% RMSE in MISSFIB distribution, and 16.38% RMSE in MISSHASH distribution.

5 Conclusion

In this research work, authors have systematically analyzed the different steps involved in the process of identification of missing value and given a methodology

Table 4 RMSE of classifiers with different imputation methods on a dataset having missing values distributed using MISSHASH algorithm

S. No	Imputation method	Classifier	10%	15%	20%	25%	30%	Average RMSE
1	MICE	NB	0.4299	0.4337	0.4337	0.4449	0.4221	0.4328
2	KNN	NB	0.4772	0.4595	0.4631	0.4667	0.4523	0.4637
3	MEAN	NB	0.4559	0.4486	0.4631	0.4840	0.4595	0.4622
4	MODE	NB	0.4772	0.4337	0.4631	0.4840	0.4595	0.4635
5	MICE	SVM	0.4559	0.4631	0.4486	0.4631	0.4667	0.4594
6	KNN	SVM	0.4737	0.4737	0.4874	0.5041	0.5265	0.4930
7	MEAN	SVM	0.4772	0.4772	0.5041	0.5202	0.5296	0.5016
8	MODE	SVM	0.4806	0.4702	0.5041	0.5450	0.5296	0.5059
9	MICE	LR	0.4299	0.4337	0.4375	0.4486	0.4182	0.4335
10	KNN	LR	0.4375	0.4449	0.4337	0.4806	0.4412	0.4475
11	MEAN	LR	0.4142	0.4631	0.4523	0.4806	0.4523	0.4525
12	MODE	LR	0.4523	0.4337	0.4299	0.5106	0.4975	0.4648
13	MICE	RF	0.4021	0.3980	0.4062	0.3587	0.3587	0.3847
14	KNN	RF	0.4337	0.4375	0.4299	0.4142	0.4806	0.4391
15	MEAN	RF	0.4486	0.4260	0.4062	0.4595	0.4449	0.4370
16	MODE	RF	0.4449	0.4182	0.4595	0.4908	0.4874	0.4601

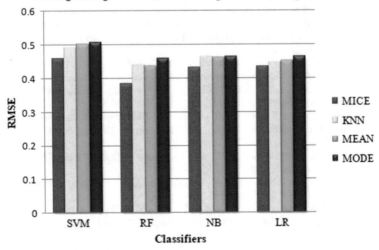

Fig. 6 RMSE of classifiers with different imputation methods on a dataset having missing values distributed using MISSHASH algorithm

for the same. To test the proposed methodology, authors have used three different algorithms for removal of values from the given dataset. After removal of values from the available dataset, authors have used four imputation methods MICE, KNN, MEAN, and MODE. To evaluate the performance of the imputation method, authors have used four classifiers Naive Bayes (NB), support vector machine (SVM), logistic regression (LR), and random forest (RF). Performance has evaluated on the scale of root mean square error (RMSE) parameter. From the experimental results, it can be concluded that MICE is the best method to estimate the missing value irrespective of the distribution of missing value and classifier method. MICE performed best with MISSHASH distribution method and RF classifier.

Research work related to feature selection from the same group of authors is already under review. After solving this missing value issue, now, authors have planned to use their work in the design of a decision support system in healthcare.

References

1. Jakobsen JC, Gluud C, Wetterslev J, Winkel P (2017) When and how should multiple imputation be used for handling missing data in randomised clinical trials–a practical guide with flowcharts. BMC Med Res Methodol 17(1):1–10
2. Bertsimas D, Pawlowski C, Zhuo YD (2017) From predictive methods to missing data imputation: an optimization approach. J Machine Learn Res 18(1):7133–7171
3. Noor MN, Yahaya AS, Ramli NA, Al Bakri AMM (2014) Filling missing data using interpolation methods: Study on the effect of fitting distribution. Key Eng Mater 594:889–895
4. Kumar RN, Kumar MA (2016) Enhanced fuzzy K-NN approach for handling missing values in medical data mining. Ind J Sci Technol 9(S1):1–7
5. Sim J, Lee JS, Kwon O (2015) Missing values and optimal selection of an imputation method and classification algorithm to improve the accuracy of ubiquitous computing applications. Math Prob Eng 2015:1–14
6. Nahato KB, Harichandran KN, Arputharaj K (2015) Knowledge mining from clinical datasets using rough sets and backpropagation neural network. Comput Math Methods Med 2015:1–13
7. Qin J, Chen L, Liu Y, Liu C, Feng C, Chen B (2019) A machine learning methodology for diagnosing chronic kidney disease. IEEE Access 8:20991–21002
8. Venkatraman S, Yatsko A, Stranieri A, Jelinek HF (2016) Missing data imputation for individualised CVD diagnostic and treatment. In: Computing in cardiology conference, vol 43, pp 349–352. IEEE
9. Al Muhaideb S, Menai MEB (2016) An individualized preprocessing for medical data classification. Procedia Comput Sci 82:35–42
10. Kuppusamy V, Paramasivam I (2016) A study of impact on missing categorical data—a qualitative review. Ind J Sci Technol 9(32):1–6
11. Hu Z, Melton GB, Arsoniadis EG, Wang Y, Kwaan MR, Simon GJ (2017) Strategies for handling missing clinical data for automated surgical site infection detection from the electronic health record. J Biomed Inform 68:112–120
12. Sujatha M, Anusha S, Bhavani G (2018) A study on performance of cleveland heart disease dataset for imputing missing values. Int J Pure Appl Math 120(6):7271–7280
13. Karim SAA, Ismail MT, Othman M, Abdullah MF, Hasan MK, Sulaiman J (2018) Rational cubic spline interpolation for missing solar data imputation. J Eng Appl Sci 13(9):2587–2592
14. Mohan S, Thirumalai C, Srivastava G (2019) Effective heart disease prediction using hybrid machine learning techniques. IEEE Access 7:81542–81554

15. Almansour NA, Syed HF, Khayat NR, Altheeb RK, Juri RE, Alhiyafi J, Alrashed S, Olatunji SO (2019) Neural network and support vector machine for the prediction of chronic kidney disease: a comparative study. Comput Biol Med 109:101–111

16. Kim T, Ko W, Kim J (2019) Analysis and impact evaluation of missing data imputation in day-ahead PV generation forecasting. Appl Sci 9(1):1–18

17. Stavseth MR, Clausen T, Roislien J (2019) How handling missing data may impact conclusions: a comparison of six different imputation methods for categorical questionnaire data. SAGE Open Med 7:1–12

18. Nikfalazar S, Yeh CH, Bedingfield S, Khorshidi HA (2019) Missing data imputation using decision trees and fuzzy clustering with iterative learning. Knowl Inf Syst 1–19

19. Raja PS, Thangavel K (2019) Missing value imputation using unsupervised machine learning techniques. Soft Comput 1–32

20. https://www.kaggle.com/ronitf/heart-disease-uci. Accessed on 01 Oct 2019

21. Saar-Tsechansky M, Provost F (2007) Handling missing values when applying classification models. J Mach Learn Res 8:1623–1657

22. Farhangfar A, Kurgan L, Dy J (2008) Impact of imputation of missing values on classification error for discrete data. Pattern Recogn 41(12):3692–3705

23. Thomas RM, Bruin W, Zhutovsky P, van Wingen G (2020) Dealing with missing data, small sample sizes, and heterogeneity in machine learning studies of brain disorders. In Machine learning, pp 249–266. Academic Press (2020)

24. Azur MJ, Stuart EA, Frangakis C, Leaf PJ (2011) Multiple imputation by chained equations: what is it and how does it work? Int J Methods Psychiatr Res 20(1):40–49

25. Dulhare UN (2018) Prediction system for heart disease using Naive Bayes and particle swarm optimization. Biomed Res 29(12):2646–2649

26. Musa AB (2013) Comparative study on classification performance between support vector machine and logistic regression. Int J Mach Learn Cybernet 4(1):13–24

27. Jain A, Kumar R, Mittal S, Rani P, Sharma R, Lamba R (2020) An optimized system for heart disease prediction by feature selection. Patent Application No. 202011004239A. Office of controller general of patents, Designs & TradeMarks, India. 8532 (2020)

28. Jabbar MA, Deekshatulu BL, Chandra P (2016) Prediction of heart disease using random forest and feature subset selection. In: Innovations in bio-inspired computing and applications. Springer, Cham, pp 187–196

29. Guo H, Yin J, Zhao J, Yao L, Xia X, Luo H (2015) An ensemble learning for predicting breakdown field strength of polyimide nanocomposite films. J Nanomater 2015:1–11

30. Ayilara OF, Zhang L, Sajobi TT, Sawatzky R, Bohm E, Lix LM (2019) Impact of missing data on bias and precision when estimating change in patient-reported outcomes from a clinical registry. Health Qual Life Outcomes 17(1):1–9

A Review on Natural Language Processing: Back to Basics

P. M. Dinesh, V. Sujitha, C. Salma, and B. Srijayapriya

Abstract Deep learning models have made incredible progress in tackling an assortment of natural language processing (NLP) issues. An ever-developing assortment of research, in any case, outlines the dependence of deep neural systems (DNNs) to ill-disposed models—inputs adjusted by acquainting little irritations with knowingly fooling an objective model into yielding mistaken outcomes. The powerlessness to aggressive models has gotten one of the fundamental obstacles blocking neural system organization into security basic conditions. This paper talks about the contemporary utilization of ill-disposed guides to thwart DNNs and presents an extensive audit of their utilization to improve the robustness of DNNs in NLP applications.

Keywords Natural language processing · Nlp toolkit · PROPOSEL · Probabilistic latent semantic analysis · SVM for spam categorization · Sentimental classification using machine learning techniques

1 Introduction

The arena of natural language processing (NLP) incorporates an assortment of subjects, which includes the computational processing and comprehension of humanoid tongues. Meanwhile, in the 80s, the arena has progressively depended on the information determined calculation including insights, likelihood, and AI [1, 2]. Late increments in computational force and processing data in parallel systems,

P. M. Dinesh (✉) · V. Sujitha · C. Salma · B. Srijayapriya
Department of ECE, Sona College Technology, Salem, India
e-mail: pmdineshece@live.com

V. Sujitha
e-mail: sujicara@gmail.com

C. Salma
e-mail: SK2336367@gmail.com

B. Srijayapriya
e-mail: Srijayapriyabharathi24@gmail.com

© The Author(s), under exclusive license to Springer Nature Singapore Pte Ltd. 2021 655
J. S. Raj et al. (eds.), *Innovative Data Communication Technologies and Application*,
Lecture Notes on Data Engineering and Communications Technologies 59,
https://doi.org/10.1007/978-981-15-9651-3_54

saddled by graphical processing units (Graphical Processing Units) [3] presently take into account "deep realizing," [4], which uses counterfeit neural systems (ANNs), in some cases through a very large number of training factors [5]. Likewise, the modern accessibility of huge informational collections, encouraged by modern information assortment forms, empowers the preparation of such deep structures [6–8]. As of late, scientists and experts in NLP have utilized the intensity of present-day ANNs with numerous favorable outcomes, starting in enormous portion [9] by means of the spearheading effort by Collobert et al. In the ongoing historical, the utilization of machine learning [10] has significantly upsurged. This has prompted huge developments together in central regions of NLP [11] and in sections in which it is straightforwardly used to accomplish down to earth and valuable targets. This article gives a concise prologue to both natural language processing and deep neural systems (DNNs) and afterwards presents a broad conversation on in what way deep learning presence is utilized to take care of recent issues in NLP [10]. Whereas a few different courses and records on the point have been distributed [12], none of them has widely secured the cutting edge in the same number of regions inside it. Besides, no other review has inspected not just the utilization of deep figuring out how to computational etymology yet, in addition, the fundamental hypothesis and conventional NLP errands. Notwithstanding the conversation of ongoing progressive advancements in the field, this article will be helpful to pursuers who need to acquaint themselves rapidly with the flow best in class before leaving upon additionally propelled research and practice.

In this survey, the methodologies used to make antagonistic content models to test and improving the unwavering quality of DNNs in NLP are quickly examined and make the accompanying commitments.

- A scientific categorization to order the methodologies is made to produce ill-disposed models in NLP.
- These methodologies are tested and order them utilizing our scientific categorization.
- Different sorts of cautious methodologies against antagonistic models in NLP is talked about.
- Primary difficulties are featured and future research bearings for ill-disposed models in NLP

2 English Dictionary for ML and NLP

Human-oriented PROsody and POS English lexicon for machine learning and NLP (PROPOSEL) [13] comprise of the pattern of stress and intonation in a language and part of speech. A language of users knowledge of words has a certain entity (104049) and is open-source incorporated with the natural language toolkit where text file can be accessed. The PROPOSEL extracts and combines the data available in databases and dictionaries this is especially done to interact with the corpus. It is a text-to-speech system from an NLP [13] module where the pronunciation, intonation is given by an individual of their own way. The automatic speech recognition, phonetics for given words can be known by using the CMU (Carnegie-Mellon) dictionary found in 1998 in Carnegie-Mellon University. The PROPOSEL can be implemented by python programming, the datatype called dictionary available in Python which maps the objects in key, value pair format. Here, the data type of the key is restricted whereas the value can be of any datatype.

ProPOSEL's product instruments are perfect with NLTK and empower clients to characterize and look through a subsection of the dictionary and admittance passages through term lesson, speech sounds interpretation, pronunciation check and musical organization. ProPOSEL additionally tended to a few progressively broad issues identifying with psychological parts of the vocabulary: the incomplete examples in the brain of a word reference client; the requirement for access and hunt by regular vibrations, mood, patterns of rhythms, and syntactic comparability; strong and normalized association of relating to the words sections from various sources; and simplicity of coordination into NLP applications.

3 Probabilistic Latent Semantic Analysis (PLSA)

PLSA [14] is beneficial and aimed at the examination of statistical techniques viz bi-mode and co-occurrence figures enhance the fields of natural language processing, deep learning, and machine learning for text analysis. The probabilistic latent semantic analysis is far superior to standard latent semantic analysis since it prefers mixture decomposition rather than single-valued decomposition. The machine learning system has to distinguish lexical level from the semantic level. Challenges faced in evidence recovery are related to the field of request-depended retrieval involving automatic indexing which uses the familiar technique of vector space model. The advantage of statistical models over the singular valued decomposition model is that it helps in interfacing different models. PLSA has subsequently to be believed as an auspicious original solo knowledge strategy by an extensive scope of utilizations in script knowledge and data recovery.

4 Support Vector Machines for Spam Categorization

The proliferation [15] of spam is solved by either of the two methods such as technical or regulatory. The technical process depends upon the address of the sender and header context which utilizes the filtering where even the valid messages are being blocked. The feature is analyzed by term occurrence and term occurrence reproduced by opposite article occurrence. The performance is based on the error rate, false alarm, miss rate, recall, and precision. The linear techniques are Rocchio and linear SVM

[11] and the nonlinear techniques take less execution time they are Boosting and Ripper.

The utilization of support vector machines (SVM's) in arranging email as junk or no junk by contrasting it with three other characterization calculations: Rocchio, boosting choice trees, and Ripper. All above calculations are tried on dual distinct informational collections: one informational index where the quantity of features was compelled to the thousand of finest features and additional informational index where the number of attributes was more than seven thousand. SVM's accomplished finest when utilizing double features. On behalf of the two informational collections, boosting trees and SVM's had adequate test execution as far as exactness and speed. In any case, SVMs had altogether less preparing time.

5 Emotional Analysis Using Artificial Intelligence

In sentimental analysis [7, 16], the review is rated as positive or negative. The three machine learning methods which exist and implemented are Naive Bayes, extreme entropy sorting, and support vector machines. By labeling the articles with their sentiment as positive or negative, it will be useful for customers to buy a product as well as a reader to buy a book or read. Sentimental cataloging could be supportive even in commercial brainpower application and replacing systems, where handler response whose feedback is summarized quickly. In the case of a survey, the comebacks given in natural language arrangement might be treated by sentiment categorization. The challenges faced in sentiment analysis is more understanding that is required in case of topic-based classification. Sentimental analysis is done for many purposes like a movie review, a survey about a product so that if the products are beneficial, one could buy such product or may decide not to by the product, this may also help for improving the features of a product of a company by simply customers answering the questions with positive or negative sentiments.

6 Hate Dialog Finding Using Artificial Intelligence

Hate speech [17] is the way of expressing a negative opinion on the basis of nationality, character, gender, creed, color, and so on by Nockleby [18]. The work on hate speech in 1997 was based on hostile messages and abusive messages but, nowadays, it refers to the cyberbullying. Later, the work was done on cursing or insults detection and these belong to the offensive language. The contents like cursing and vulgar language were focused by Xiang et al. [19]. For performing the task of text classification, the surface-level features are being taken into consideration that includes bags of words. The N-gram technique is beneficial because it contains the above-mentioned features. In the process of hate speech detection, predictive words are trained as well tested but often the problem of data sparsity occurs. The error can be rectified by word clustering. Brown clustering is the standard algorithm. The hate speech is related to sentimental analysis as it refers to the negative sentiment. The sentiment information includes a single-step and multi-step approaches. The number of positive, negative, neutral words occurrence can be analyzed by the supervised classifier.

7 Language Structural and Functional Features for Natural Language Processing

The motive of the linguistic typology [20] is to obtain the structural and semantic variations on the world languages. The NLP would provide guidance even for the languages that lack human labeled resources. The variation may occur in structures of real-world languages which may share at deep- and abstract-level features and this is limited to certain resource-rich languages. The unsupervised models do not depend on the manually annotated resources availability Snyder and Barzilay [21]. The world languages are compared systematically and documented which is based on the empirical observation Comrie [22], Croft and Lafferty [23]. The data and resources scarcity are the major challenges for multilingual NLP. The data scarcity can be overcome by transferring data to resource-poor languages from resource-rich languages whereas at earlier approaches they find their basis on parts of speech, dependency relations which are high-level delexicalized features and universal features (Table 1).

Table 1 Consolidation for machine learning abilities for NLP

No.	Review topic	Outcomes from review
1	English Dictionary for ML and NLP (PROPOSEL)	PROPOSEL is a language engineering resource for cognitive aspects of the lexicon
2	PLSI based text indexing	Probabilistic latent semantic analysis is used for unsupervised knowledge technique in an extensive range of applications of script knowledge and info recovery systems
3	SVM for junk cataloguing	Support vector machines are widely used for junk or non-junk classification in information analysis
4	Sentimental classification using machine learning techniques	Sentiment analysis is used for classification of text in movie reviews using machine learning techniques
5	Hate dialog findings using artificial intelligence	Hate speech detection and classifications are using support vector machines
6	Language structural and functional features for natural language processing	Use of information in existing typological material in the enlargement of NLP procedures

8 Conclusion

The neural networks and deep learning settle the vast majority of the issues acquired in NLP. The shrouded states between input word and yield vector structure concentrated system for careful and productive learning. This innovation can be utilized as the foundation of artificial intelligence. Future attempts to be done in this field by incorporating cross-language IR and machine-human exchange. Likewise, reinforcement learning is gaining dependable ground in gaming and spreading its application in different fields including natural language processing. It centers around higher long haul prize, for example, answer for pole balancing issue. The prize-driven programming is an indispensable piece of TensorFlow also.

9 Future Work

Through these extensive reviews, a system for transliteration of text from one language to another language using machine learning techniques has been planned to develop which will be helpful for most of the minor language community peoples.

References

1. Jones KS (1994) Natural language processing: a historical review. In: Current issues in computational linguistics: in honour of Don Walker. Dordrecht, The Netherlands. Springer, pp 3–16
2. Liddy ED (2001) Natural language processing. In: Encyclopedia of library and information science, 2nd edn. Marcel Decker Inc, New York
3. Coates A, Huval B, Wang T, Wu D, Catanzaro B, Andrew N (2013) Deep learning with cots HPC systems. In: Proceedings of ICML, pp 1337–1345
4. Raina R, Madhavan A, Ng AY (2009) Large-scale deep unsupervised learning using graphics processors. In: Proceedings of ICML, pp 873–880
5. Goodfellow I, Bengio Y, Courville A, Bengio Y (2016) Deep learning, vol 1. MIT Press, Cambridge
6. LeCun Y, Bengio Y, Hinton G (2015) Deep learning. Nature 521(7553):436–444
7. Schmidhuber J (2015) Deep learning in neural networks: an overview. Neural Netw 61:85–117
8. Ciresan DC et al (2011) Flexible, high performance convolutional neural networks for image classification. Proc IJCAI 22(1):1237
9. Collobert R, Weston J, Bottou L, Karlen M, Kavukcuoglu K, Kuksa P (2011) Natural language processing (almost) from scratch. J Mach Learn Res 12:2493–2537
10. Goldberg Y (2017) Neural network methods for natural language processing. Synth Lect Hum Lang Technol 10(1):1–309
11. Liu Y, Zhang M (2018) Neural network methods for natural language processing. Comput Linguist 44(1):193–195
12. Young T, Hazarika D, Poria S, Cambria E (2018) Recent trends in deep learning based natural language processing. IEEE Comput Intell Mag 13(3):55–75
13. Brierley C, Atwell E (2008) ProPOSEL: a human-oriented prosody and PoS English lexicon for machine-learning and NLP. In: Coling 2008: Proceedings of the workshop on cognitive aspects of the Lexicon (COGALEX 2008), pp 25–31

14. Hofmann T (2013) Probabilistic latent semantic analysis. arXiv preprint arXiv:1301.6705
15. Drucker H, Wu D, Vapnik VN (1999) Support vector machines for spam categorization. IEEE Trans Neural Netw 10(5):1048–1054
16. Tripathy A, Agrawal A, Rath SK (2015) Classification of sentimental reviews using machine learning techniques. Proc Comput Sci 57:821–829
17. Schmidt A, Wiegand M (2017) A survey on hate speech detection using natural language processing. In: Proceedings of the fifth international workshop on natural language processing for social media, pp 1–10
18. Nockleby JT (2000) Hate speech. Encycl Am const 3(2):1277–9
19. Xiang G, Fan B, Wang L, Hong J, Rose C (2012) Detecting offensive tweets via topical feature discovery over a large scale twitter corpus. In: Proceedings of the 21st ACM international conference on information and knowledge management Oct 29, pp 1980–1984
20. Ponti EM, O'horan H, Berzak Y, Vulić I, Reichart R, Poibeau R, Shutova T, Shutova E, Korhonen A (2019) Modeling language variation and universals: a survey on typological linguistics for natural language processing. Comput Linguist 45(3):559–601
21. Snyder B, Barzilay R (2008) Unsupervised multilingual learning for morphological segmentation. In: Proceedings of acl-08: hlt 2008 Jun, pp 737–745
22. Comrie B (1989, Jul 15) Language universals and linguistic typology: syntax and morphology. University of Chicago press
23. Croft B, Lafferty J (eds) (2003, May 31) Language modeling for information retrieval. Springer Science & Business Media

Road Extraction Techniques from Remote Sensing Images: A Review

Dhanashri Patil and Sangeeta Jadhav

Abstract Road detection from remotely sensed images is a fundamental task in the geographic information system. On account of applications like urban management, traffic control, and map updating, road extraction from remote sensing images has significant research importance in recent times. Road extraction from satellite images is a crucial task as these images are noisy and contain lots of information. So it becomes difficult to process large amount of data. The important parameters for road detection are road features and its corresponding classification methods. These parameters decide the performance accuracy of the road extraction system. The systematic analysis of existing road detection techniques is elaborated in three important sections: different features of road, supervised, and unsupervised classification techniques. The main objective of this comprehensive survey is to render the analysis of different classification methods like mathematical morphology, SVM, CNN, etc. By using multiple features of the road, the system performance can be improved.

Keywords Remotely sensed images · Mathematical morphology · SVM · CNN

1 Introduction

The first earth observation satellite was launched by America in 1978; this revolution further led the research in different remote sensing applications like weather forecasting, consistent global measurement of the earth including sea measurement, oceanic plant life, and ozone layer [1]. Remote sensing is defined as measuring physical parameters of objects at distance through emitted or reflected energy. In this field, information is collected by employing acoustic waves or electromagnetic

D. Patil (✉)
D. Y. Patil College of Engineering, Akurdi, Pune, India
e-mail: dpdhanashripatil@gmail.com

D. Patil · S. Jadhav
Army Institute of Technology, Pune, India
e-mail: djsangeeta@rediffmail.com

waves emitted by target [2]. Major remote sensing applications are agriculture, water resources, land map, biodiversity, and disaster management.

The road is a key element of transportation and backbone for the urbanization. The extraction of the road from RS image is having great significance in a geographical information system [14]. Information retrieval of the road can be done using manual and automatic extraction. Manual extraction needs human involvement, while efforts and time could be saved using automatic extraction techniques. Road detection is generally classified into two types: road detection and centerline extraction. In road detection techniques, road and non-road pixels are classified separately, whereas in the second method, pixels which belong to the centerline of the road are categorized.

In ancient years, runways were detected from an aerial image by using runway markings. Hypothesis and test paradigm were utilized for the same [12]. This technique was initiated using multiple cooperative methods, by employing road surface texture and edges [13]. Trajectory extrapolation, anomaly detection, and masked correlation were introduced to eliminate the problem of alignment [19]. Table 1 gives a summary of some ancient road detection techniques.

The challenges in the field of RS are mainly due to the calibration of sensors mounted on aircraft or satellite and distortion of colors. The shadow of buildings and trees on the roads also makes the object detection tangled. The satellite image contains huge information that requires lots of computation time and memory. To address these challenges, research in this domain has become crucial. The main objective of this survey is to highlight various supervised and unsupervised classification techniques for road extraction.

The image database is elaborated in Sect. 2. The road features are mentioned in Sect. 3. Preprocessing techniques for enhancement of an image are defined under Sect. 4. Different extraction methods are discussed in Sect. 5. The overall observations are concluded in the last section.

2 Image Database

RS technique has become an essential part of information technology companies and different government sectors, to interpret intended parameters from the scene [3]. Because of this, the large number of RS images that cover every corner of the earth has made available for different applications [3]. Google Earth, Ikonos, GeoEye, World View, and Quick Bird are a platform which dispenses high-resolution images [4].

Massachusetts road data set is also made available for research by Mnih. These images are available in tagged image file format. For supervised learning the training and testing data set are to be prepared separately. CASNET data set is also made available for research in road detection. This data set includes 224 images from Google Earth along with its road segmentation map and centerline map [9]. To improve the accuracy of networks, sometimes a large amount of training data samples are required. This can be done by data augmentation techniques, addressed in the

Table 1 Summary of ancient road detection methods

S. No.	Method	Features	Remark
1	Multisource knowledge integration technique [14]	Lines and edges	1. Graphs were generated using connectivity relationship between adjacent pixels 2. Multiple delineation algorithms were used to detect roads
2	Semiautomatic method [15]	High contrast ribbons	1. Ribbon center detection technique was adopted on grayscale images 2. Graph was constructed by considering ribbon pixels
3	Least square B-spline snakes [16]	Linear features	1. This is a semiautomatic technique where seed points are marked by observer 2. Dynamic programming and LSB snakes' technique is used to extract feature automatically
4	Multiresolution extraction [17]	Geometric, radiometric, topological, and contextual features	1. In low-resolution images, curvatures and vanishing gradients are used as line points which further connected into contours 2. In high-resolution images, polygonal approximation is used with edge extraction technique
5	Visual model [18]	Spectral, spatial, and geometric features	1. Small sections of road were detected using strip detector 2. With direction and distance proximity techniques, these strips were connected together

preprocessing section. The augmented data set becomes large, which can be stored in GPU memory.

3 Feature of Roads

Roads are lengthened in their characteristics, and this feature makes it interpretable from multiple objects which are present in RS image. Road features have been classified into four categories: geometric, radiometric, contextual, and topological [5]. The accuracy of the network is contingent on the number of features used. Some of the methods have used multiple features.

3.1 Geometric Feature

The road is continuous, and its width is much smaller compared to its length. The proportion among length and width is substantial [4]. Geometric feature inference technique is used to extract rural roads; in the first phase, parallel road lines were identified, and grouping and connection formation were done in the second phase [49]. Large area and long strip features were used for removing non-road areas with the help of tensor voting techniques [50].

3.2 Radiometric Feature

The street surface for most of the part is homogeneous. These are having decent stand out from the adjoining zones [6]. The radiometric feature has been used for segmentation of an original image. A homogeneous region was obtained using Geary's C method. To obtain the road information, the local homogeneity of gray value was calculated by Geary's C method [51]. Geary's C is a measure of spatial autocorrelation which calculates the correlation between adjacent observations. Roads were extracted in the suburban area by using radiometric features like edges, a standard deviation of intensity, and color. This method is classified into three categories: segmentation, grouping, and extraction [52].

3.3 Contextual Feature

Sometimes, there is an array of trees alongside roads or shadow projected by flyover. These both are considered contextual features [6]. A tensor voting technique for road detection used multiple features, in which contextual features are used to avert the possibility of removing small lines that are part of a road segment. This method implies small values and standard deviation of curvature of road segments [53].

3.4 Topological Features

Roads do not get interrupted anywhere. So the continuity of road pixels in an image is considered as topological features. The road intersects with each other and forms its network [4]. Topological information was used with multi-conditional generative adversarial network to acquire road information [54].

4 Preprocessing

The essential point of the preprocessing method is to lessen the dimensionality of the image which brings down the complexities of information prediction from large surroundings [7]. This technique makes use of top principle components by applying PCA to RGB image. Some of the preprocessing techniques which have been used are as follows:

4.1 Data Augmentation

When dealing with neural networks, a large number of data samples are required to train the network. In the RS field, the number of images can be increased using data augmentation techniques. One image is incrementally rotated by different angles [8]. This can be done by extracting fixed or random position patches from an image. Then the size of the data set can be boost by flipping these patches [9]. This technique is generally used in a deep neural network to avoid the problem of overfitting.

4.2 Image Enhancement

In mathematical morphology, image enhancement is carried out by dilation and erosion operations; dilation is used to make the desired object thicken in binary image, and erosion makes them shrink [10]. As the measure of RS image is huge, it winds up hard for storing and processing the tasks. So the original image is first divided into several sub-images with the size power of two to make use of beamlet transform [11].

4.3 Edge Detection

This is a crucial step in preparing labeled data samples for supervised methods. Boundaries at the road and non-road areas are considered as edges. So the area where intensity changes abruptly can be detected using operators like Canny, Sobel, or Prewitt. These operators work on edge gradients. The noise immunity can be improved by combining these operators with some filtering and thresholding techniques.

5 Road Extraction Methods

Different methods have been used by numerous scientists to detect roads from high-resolution images. Road extraction techniques are mainly classified into two categories: supervised techniques and unsupervised techniques. According to Fig. 1, different classification techniques of road detection are classified as follows.

5.1 Supervised Techniques

In the supervised model, training of the network is done by means of labeled samples. Some examples of the supervised methods are artificial neural network (ANN), Markov random field (MRF), and support vector machine (SVM). The overall accuracy of the system depends upon labeled samples.

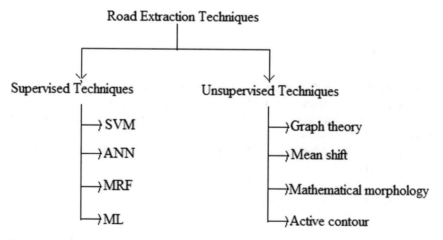

Fig. 1 Classification of road extraction methods

5.1.1   Support Vector Machine (SVM)

Support vector machines were introduced to nonlinearly map input vector to feature space, and this was done by utilizing polynomial input transformation [20]. It has classified hyperspectral RS image using SVM and has marked outstanding performance in comparison with neural network and K-NN classifiers [21]. Edge-based features like width, length, gradient, and intensity were used [22] to extract road from RS image; this experiment was conducted using SVM into the RAIL framework. SVM as a classifier was used to divide images into two categories, road group or non-road group, and it has been concluded that SVM achieves higher accuracy than Gaussian maximum likelihood (GML) [23]. Still, some discontinuous road segments were obtained due to shadows of trees. By making use of SVM along with segmentation and full lambda function, roads were detected from a high-resolution image with spectral, spatial, and texture features. It is seen that this strategy has less precision for the complex street database [24].

The SVM technique is efficient and easy to operate. But as the complexity of roads in an image increases, then accuracy degrades. It also falls in discontinuous road segmentation due to occlusion of shadows.

5.1.2   Artificial Neural Network

This system works similarly to the biological neural system. Artificial neurons are connected together, and this network can be trained by sending information repeatedly through connections between neurons. Roads were detected using spectral information with the assistance of a neural network classifier. This technique also made use of a gray level co-occurrence matrix to generate a road raster map using texture parameters [25]. A structured neural network was proposed by integrating three techniques: feature extraction, fusion network, and skeleton network [26]. They have made use of encoder and decoder to generate a feature map which is followed by a fusion network to eliminate the problem of discontinuity of road in the output of the feature map. At last, skeleton of the centerline of roads was generated for ease of vectorized result [26]. Modified ANN was used to improve the classification accuracy of road and non-road pixels by using 3×3 window of the road mean vector with spectral information of pixels [27].

ANN technique gives better accuracy of recognition, but it does require a large number of training images. For the less number of the data sample, this network gets over fitted. The time required for the training of a network is large.

5.1.3   Convolutional Neural Network (CNN)

CNN is a deep neural network that is a combination of an input layer, an output layer, and hidden layers. Unlike ANN, all the neurons in the first layer are not connected to neurons present in the second layer. So the overfitting problem can be addressed with the help of CNN. This method was used on high-resolution images to the categorized road and non-road information, and the obstacles present in the output were filtered using wavelet transform [28]. A road structure refined convolutional neural network was proposed to construct a structured road network by employing geometric information in cross-entropy loss [29]. Cascade end-to-end convolutional neural network (CASNET) is used to determine road and its centerline. The problem of non-homogeneity was addressed to some extent using CASNET. It consists of two convolutional networks cascaded in series: The first network is used to extract road, and the second network detects centerline using a feature map generated by the first network [9].

This method overcomes the limitation of time in context with ANN. It has been also observed that CNN is efficient to detect roads under occlusion of the shadow of a tree. Thus, it improves the accuracy of recognition for complex data sets [9]. However, CNN requires large data samples for training. To avoid the problem of overfitting, separate samples for training and testing data set are needed to be prepared. Roads under large occlusion cannot be detected using this technique.

5.1.4 Markov Random Field

MRF belongs to probability theory which describes random variable relation in a one-directional graph. Gaussian Markovian model was used to segment road targets in the image. The task of segmentation utilizes conditional probability and prior probability which leads to good segmentation results [30]. A hierarchical multiscale MRF was proposed for the detection of line, connected components, and road junctions. This technique made use of two MRF models that work on contextual information and structural knowledge [31]. Curvilinear features were used with MRF to detect road segments, and it is concluded that this technique provides robust results by using a watershed transform and region adjacency graph [32].

It has been observed that the accuracy of results can be improved by integrating MRF with a few other techniques. Similarly, by utilizing different source information (radar image and optical image), the recognition rate can be improved.

5.2   Unsupervised Techniques

This technique works on non-labeled data samples for training. The output is predicted by applying uncategorized data to various algorithms. Graph theory, mathematical morphology, mean shift, and active contour are the techniques based on this principle.

5.2.1 Graph Theory

It gives relationships in terms of vertices or nodes connected by edges which are also known as clusters. This field belongs to discrete mathematics. When it is used for image segmentation, pixels can be considered as vertices, and based on similarity or difference in weights, edges are applied. Graph theory was used for the formation of road after extracting road center pixels. This task was carried out utilizing three techniques: graph formation, parametric tracking, and refining of a graph. The results of road detection have been improved to a great extent due to the use of structural and parametric representation [41]. A graph-based technique was used for the segmentation of buildings and roads from very high-resolution satellite images. This technique deals with complex data and makes use of a global feature. Segmentation of image takes place using eigenvalue and eigenvector which sometimes leads to complex processing for a large amount of data [4].

Road detection results have been improved by using a parametric and structural form of network. But computations due to large data handling become complex with this technique.

5.2.2 Mean Shift

The mean shift technique has a great range of applications in the field of pattern recognition and image segmentation. This is an iterative method based on kernel estimation function to locate maxima. Road extraction was carried out using a mean shift algorithm, which first computes the mean shift-vector. This vector further translates kernel to converge near-zero gradient estimation [42]. The drawback of slow convergence has been eliminated by introducing the quasi-Newton method with the existing method. In this technique, gradient information is used to construct the Hessian matrix to reduce computational time [43]. A semiautomatic method for centerline extraction of road works on the geodesic method; it selects seed points that are linked together. But the accuracy gets affected due to non-precise seed points, and this drawback has been overcome using mean shift technique which calculates modes of kernel function [44] (Table 2).

5.2.3 Mathematical Morphology

This technique belongs to geometric information processing. Image transformation can be carried out by conducting different operations like erosion, dilation, etc. Road enhancement was carried out by erosion and dilation operators, and structuring elements were set to eliminate unwanted information [45]. Roads were selected from advanced land observing satellite images using the mathematical morphological algorithm and Euclidean distance technique [46]. This technique also gives the better result to find out different types of roads like linear, curvilinear, crossing road, and breakage road.

Table 2 Summary of road detection techniques

S. No.	Algorithm	Road features	Remark
1.	Cascade end-to-end convolutional neural network [9]	Pixel intensity	1. Two different networks are used for road detection and centerline extraction 2. Feature map of first network is applied as input to second network 3. This system has high accuracy in occlusion of trees and buildings 4. For continuous occluded area, roads cannot be detected
2.	Fusion of SVM and mathematical morphology [24]	Spatial and texture features	1. Fusion of SVM and mathematical morphology can overcome discontinuity 2. At corners and intersection area of road accuracy are less
3.	ANN [25]	Intensity, edge, spectrum	1. Roads are extracted using spectral information 2. Road map was computed using GLCM 3. Discontinuity, noisy, overfitting
4.	SVM [33]	Edge gradient, length, width	1. FCM was used for non-supervised clustering of RS images 2. SVM used as a classifier 3. Integration of SVM and FCM improves classification accuracy
5.	Knowledge-based method [34]	Intensity, edges	1. Hough transform used to detect small segment of road and to connect discontinuous segment of road 2. Limitations are oversegmentation, susceptible to occlusion and shadow

(continued)

This method generally has been used for restoring features and segmentation of an image. But still, the problem of homogeneity occurs in the output. Accuracy can be improved by combining more than one technique with mathematical morphology.

Table 2 (continued)

S. No.	Algorithm	Road features	Remark
6.	Mathematical morphology [35]	Geometric feature	1. Each pixel in this technique constitutes feature vector on which extraction is based 2. Discontinuity and susceptible to noise are the limitations
7.	Multistage framework [36]	Spectral features	1. This technique eliminates the problem of discontinuity using multistage post-processing approach 2. Road junction detection is the issue faced by this technique
8.	Object-level algorithm [37]	Single-pixel width	1. Structural features used for road width and perimeter calculation 2. Object-based approach used to extract multiscale information to reduced shadow on road 3. Producing small spurs around centerline is the drawback of this algorithm
9.	Regression-based algorithm [38]	Shape features and spectral features	1. This system was build using multivariate adaptive regression algorithm 2. It cannot extract area around road intersections
10.	Vectorization approach [39]	line and curve segments	1. Image is segmented to identify road network followed by decision making 2. It cannot perform well for complex road data set.
11.	Multiresolution analysis (MRA) [40]	Topological feature	1. MRA is used for two sequential steps that are street segment and street intersections 2. Discontinuous segmentation is limitation of this method

5.2.4 Active Contour

These are also known as snakes, which precisely portray the outline of an object in an image. This enables to locate continuous boundaries in sub-regions. In a semiautomatic approach, the median filter was first used for preprocessing. Then seed points related to the road were entered by users. Finally, using the active contour model, road segments were extracted [47]. The drawback of the earlier technique of detecting the non-road segment was addressed by combining snakes and multiresolution analysis. This also improves convergence [48].

6 Conclusion

This paper gives a summary on road extraction techniques that have been used to detect road from remotely sensed images. It describes different platforms, which can be referred to collect data samples. Features of road segment are defined in four important categories: geometric, radiometric, topological, and contextual. The preprocessing technique of road has been elaborated to enhance the intended parameters in an image by removing noise. It restores road features and increases the data set as per the requirement of various algorithms. Road extraction methods are described in supervised and unsupervised techniques. Various methods under these categories are highlighted along with their advantages and limitations.

Accuracy of road extraction from the remotely sensed image depends upon features selection and extraction techniques used. Few techniques have marked good results with multiple extraction methods. Most of the techniques have got higher efficiency, but still, research is required in some aspects. The main shortcoming that comes in the picture is the discontinuity of road. As roads are occluded due to buildings, parking lots, trees, etc., it becomes difficult to improve the accuracy of extraction. Many techniques have faced this issue, and non-road segments will be detected in the result. Accuracy is still remaining as an important challenge for the complex road data set. It has been observed that, by using a single road feature, it becomes difficult to achieve a good detection accuracy. So in the near future, there is a significant research scope to implement a road detection system using multiple features.

References

1. National Aeronautics and Space Administration. www.nasa.gov.in
2. Daptardar AA, Kesti VJ (2013) Introduction to remote sensors and image processing and its applications. IJLTET 2:107–114
3. Zhang L, Zhang L, Du B (2016) Deep learning for remote sensing data. IEEE Geosci Sens Mag 4:22–40

4. Wang W, Yang N, Zhang Y, Wang F, Cao T, Eklund P (2016) A review of road extraction from remote Sensing images. J Traf Transport Eng 3:271–282

5. Garnesson P, Crard Giraudon G, Montesinos P (1990) An image analysis system, application for aerial imagery interpretation. In: Proceedings 10th international conference on pattern recognition, vol 1, pp 210-212, Atlantic City, USA

6. Vosselman G, Knecht J (1995) Road tracing by profile matching and Kalman filtering. In: Gruen A, Baltsavias E, Henricsson O (eds) Workshop on automatic extraction of manmade objects from aerial and space images. Birkhauser, Berlin, pp 265–274

7. Mnih V, Hinton GE (2010) Learning to detect roads in high-resolution aerial images. In: Daniilidis K, Maragos P, Paragios N (eds) Computer vision—ECCV. Lecture Notes in Computer Science, vol 6316. Springer, Berlin

8. Panboonyuen T, Jitkajornwanich K, Lawawirojwong S, Srestasathiern P, Vateekul P (2017) Road segmentation of remotely-sensed images using deep convolutional neural networks with landscape metrics and conditional random fields. Remote Sens 9:1–19

9. Cheng G, Wang Y, Xu S, Wang H, Xiang S, Pan C (2017) Automatic road detection and centerline extraction via cascaded end-to-end convolutional neural network. IEEE Trans Geosci Remote Sens 55(6):3322–3337

10. Awad MM (2013) A morphological model for extracting road networks from high-resolution satellite images. J Eng 1–10

11. Sghaier M, Lepage R (2016) Road extraction from very high resolution remote sensing optical images based on texture analysis and beamlet transform. IEEE J Sel Topics Appl Earth Observ Remote Sens 9(5):1946–1958

12. Huertas A, Cole W, Nevatia R (1987) Detecting runways in aerial images. In: Proceedings of the 6th national conference on artificial intelligence, pp 712–717

13. McKeown DM, Denlinger L (1988) Cooperative methods for road tracking in aerial imagery. CVPR (1988)

14. Fischler MA, Tenenbaum JM, Wolf HC (1981) Detection of roads and linear structures in low-resolution aerial imagery using a multisource knowledge integration technique. Comput Gr Image Process 15(3):201–223

15. Hu J, Sakoda B, Pavlidis T (1992) Interactive road finding for aerial images. IEEE workshop applications of computer vision, pp 56–63

16. Gruen A, Li H (1997) Semi-automatic linear feature extraction by dynamic programming and LSB-snakes. Photogrammetr Eng Remote Sens 63:985–995

17. Baumgartner A, Steger C, Wiedemann C, Mayer H, Eckstein W, Ebner H (1996) Update of roads in GIS from aerial imagery: verification and multi-resolution extraction

18. Bajcsy R, Tavakoli M (1976) Computer recognition of roads from satellite pictures. IEEE Trans Syst Man Cybern SMC-6:623–637

19. Quam LH (1978) Road tracking and anomaly detection in aerial imagery. SRI Int (1978)

20. Cortes C, Vapnik V (1995) Support-vector networks. Mach Learn 20(3):273–297

21. Melgani F, Bruzzone L (2004) Classification of hyperspectral remote sensing images with support vector machines. IEEE Trans Geosci Remote Sens 42(8):1778–1790

22. Yager N, Sowmya A (2003) Support vector machines for road extraction from remotely sensed images. In: Petkov N, Westenberg MA (eds) Computer analysis of images and patterns. CAIP 2003. Lecture Notes in Computer Science, vol 2756. Springer, Berlin

23. Song M, Civco D (2004) Road extraction using SVM and image segmentation. Photogramm Eng Remote Sens 70(12):1365–1371

24. Bakhtiari HRR, lAbdollahi A, Rezaeian H (2017) Semi automatic road extraction from digital image. Egypt J Remot Sens Space Sci 20:117–123

25. Kirthika A, Mookambiga A (2011) Automated road network extraction using artificial neural network. In: Proceedings of IEEE international conference recent trends in information technology, pp 1061–1065

26. Cheng G, Wu C, Yu QH, Meng JS et al (2019) Recognizing road from satellite images by structured neural network. Neurocomputing 356:131–141

27. Mokhtarzade M, Valadan Zoej MJ (2007) Road detection from high-resolution satellite images using artificial neural networks. Int J Appl Earth Observ Geoinf 9:32–40

28. Jiang Y (2019) Research on road extraction of remote sensing image based on convolutional neural network. EURASIP J Image Video Process 1–11

29. Wei Y, Wang Z, Xu M (2017) Road structure refined CNN for road extraction in aerial image. IEEE Geosci Remote Sens Lett 14:1–5

30. Yong X, Shaoguang Z, Yuyue X (2008) Markov random field for road extraction applications in remote sensing images. Int Arch Photogramm Remote Sens 37:241–246

31. Perciano T, Tupin F, Hirata R, Cesar R (2011) A hierarchical Markov random field for road network extraction and its application with optical and SAR data. IGARSS, pp 1159–1162

32. Geraud T, Mouret J-B (2004) Fast road network extraction in satellite images using mathematical morphology and markov random fields. EURASIP J Appl Signal Process 16:2503–2514

33. Zhu DM, Wen X, Ling CL (2011) Road extraction based on the algorithms of MRF and hybrid model of SVM and FCM. In: International symposium on image and data fusion, pp 1–4

34. Hu J, Razdan A, Femiani JC et al (2007) Road network extraction and intersection detection from aerial images by tracking road footprints. IEEE Trans Geosci Remote Sens 45(12):4144–4157

35. Valero S, Chanussot J, Benediktsson JA, Talbot H, Waske B (2010) Advanced directional mathematical morphology for the detection of the road network in very high resolution remote sensing images. Pattern Recognit Lett 31(10):1120–1127

36. Das S, Mirnalinee TT, Varghese K (2011) Use of salient feature for the design of a multistage framework to extract roads from high-resolution multispectral satellite Images. IEEE Trans Geosci Remote Sens 49(10):3906–3931

37. Huang X, Zhang L (2009) Road centreline extraction from high resolution imagery based on multiscale structural features and support vector machines. Int J Remote Sens 30:1977–1987

38. Miao Z, Shi W, Zhang H et al (2013) Road centerline extraction from high-resolution imagery based on shape features and multivariate adaptive regression splines. IEEE Trans TGRS 10:583–587

39. Hormese J, Saravanan C (2015) Automated road extraction from high resolution satellite images. Int Conf Emerg Trends Eng Sci Technol 24:1460–1467

40. Peteri Renaud & Ranchin Thierry.: Multiresolution snakes for urban road extraction from Ikonos and QuickBird. In: Proceedings of the 23rd Symposium of the EARSeL, pp. 69–76(2004)

41. Unsalan C, Sirmacek B (2012) Road network detection using probabilistic and graph theoretical methods. IEEE Trans Geosci Remote Sens 50(11):4441–4453

42. Revathi M, Sharmila M (2013) Automatic road extraction using high resolution satellite images based on level set and mean shift methods. ICCCNT, pp 1–7

43. Yang C, Ramani D, De Menthon D, Larry D (2003) Mean-shift analysis using quasi-Newton methods. In: Proceedings of the international conference on image processing, vol 2, pp 447–450

44. Miao Z, Wang B, Shi W et al (2014) A semi-automatic method for road centerline extraction from VHR images. IEEE Geosci Remote Sens Lett 11(11):1856–1860

45. Awad MM (2013) A morphological model for extracting road networks from high-resolution satellite images. J Eng

46. Castro F, Centeno J, Road extraction from ALOS images using mathematical morphology. ISPRS TC VII Sympos 38:457–461

47. Anil PN, Natarajan S, A novel approach using active contour model for semi-automatic road extraction from high resolution satellite imagery. In: Second international conference on machine learning and computing, pp 263–266

48. Peteri R, Ranchin T (2003) Multi resolution snakes for urban road extraction from IKONOS and Quickbird images. In: Proceedings of 23rd EARSeL annual symposium remote sensing in transition, pp 141–147 (2003)

49. Liu J, Qin Q, Li J, Li Y, Rural road extraction from high-resolution remote sensing images based on geometric feature inference. ISPRS Int J Geo-Inf 6(10):314

50. Sun K, Zhang J, Zhang Y (2019) Roads and intersections extraction from high-resolution remote sensing imagery based on tensor voting under big data environment. Wirel Commun Mob Comput

51. Sun K, Zhang JP (2017) Road extraction from high resolution remote sensing images based on multi-features and multi-stages. J Inf Hiding Multimed Signal Proces 8

52. Grote A, Heipke C, Rottensteiner F (2012) Road network extraction in suburban areas. Photogramm Record 27(137):8–28

53. Maboudi M, Jalal A, Michael H, Saati M (2016) Road network extraction from VHR satellite images using context aware object feature integration and tensor voting. Remote Sens 8

54. Zhang Y et al (2019) Road topology refinement via a multi-conditional generative adversarial network. Sensors (Basel, Switzerland) 19(5):1162

The Advent of Deep Learning-Based Image Forgery Detection Techniques

R. Agarwal, O. P. Verma, A. Saini, A. Shaw, and R. Patel

Abstract Technological advancement has been playing a crucial role in the generation of digital images and processing them using various image manipulation software that are available in the market today. It has thus become essential to come up with an efficient image forgery detection technique to classify these images as either authentic or forged. The application of image forensics can be seen in national intelligence agencies, scientific publications, social networks, etc. In the past few years, deep learning-based techniques have achieved commendable results in the field of computer vision by involving applications such as object recognition, facial recognition, images classification, and many more, by making their use more evident day by day. The paper gives a comparison between various forgery detection techniques that are based on deep learning with techniques that do not use a neural network architecture for feature extraction.

Keywords Image forgery · Image forgery detection · Copy-move forgery detection · CNN · Deep learning

R. Agarwal (✉) · O. P. Verma · A. Saini · A. Shaw · R. Patel
Department of Information Technology, Delhi Technological University, Delhi, India
e-mail: ritu.jeea@gmail.com

O. P. Verma
e-mail: opverma.dce@gmail.com

A. Saini
e-mail: saini.akshay92@gmail.com

A. Shaw
e-mail: ashu.shaw09@gmail.com

R. Patel
e-mail: patel.ritik73@gmail.com

1 Introduction

Manipulations or imitations are not new to the world but have been in existence since the times when it was only limited to crafts and writings. In this modern era of digitization, with rapid technological advancements being made in the image processing software, the number of doctored images has grown to a large extent and there is a reason to question the legitimacy of the images in the present time. Doctored photographs can frequently be seen appearing in various places such as magazines, forensics, photographs that we receive in our e-mail inboxes, scientific studies, etc. As a result, many of the images in standard media and on the Internet are doctored and subject to credibility. This has called for the development of strategies or techniques to verify the authenticity of the image. Thus, many automatic forgery detection algorithms have been proposed for this purpose.

Various tools are available in the market that are free of cost along with some advanced software that are used for image tampering. Many social media companies such as Facebook are concerned about the doctored images that are created by these tools such as GNU GIMP and Adobe Photoshop. These images are the primary sources of spreading fake news in the market and also to incite the mob about some sensitive issues of the society.

Section 1.1 describes the classification of different image forgery detection (IFD) methods and then Section 1.2 presents the basic framework of IFD. The paper focuses on IFD techniques which involve copy-move forgeries in the images. An overview of copy-move forgery (CMF) is presented and existing pixel-based techniques of copy-move forgery detection (CMFD) are presented. The paper reviews various deep learning-based approaches that have been making their way in recent times in the research field of IFD.

Comparison between the deep learning-based approaches and the existing pixel-based approaches is given by analyzing them based on different performance measures. We hope that the paper helps fellow researchers to find new research problems in the field of IFD.

1.1 *Image Forgery Detection Classification*

Many IFD mechanisms and techniques have been decided over the past years in various research papers. These techniques can be roughly grouped into two categories: passive approach and active approach. Active approaches involve embedding particular information about the input image during the time of creation as a digital watermark. Active approach has a major drawback as most of the digital media sources generating these images do not tend to create these watermarks at the time of creation of images. Thus, we would only be limited to the images produced by digital cameras to apply these active approaches. Also specialized hardware is required for the application of these active approaches.

Passive or blind approaches on the other hand use the statistics of the image or the image content to verify the authenticity of the image, without using any sort of digital signature or a watermark. The assumption on which these approaches are based is the underlying value of the statistics of an image change, which can be studied to detect forgery.

The passive image forgery detection approaches can be grouped into the following five categories:

- Pixel-based IFD
- Pixel-based approaches basically concentrate on the pixels of a particular image for forgery detection. Some of the pixel-based approaches are CMFD, splicing, resampling, and statistical analysis. CMFD is the most popular method among all the techniques [1].
- Camera-based IFD
- Camera-based techniques are based on the fact that cameras follow a sequence of steps to preprocess the image, such as quantization, color correlation, white adjusting, gamma correlation, and JPEG compression, from capturing an image to saving it. Various cameras have different standards and these techniques work on these standards.
- Geometry-based IFD
- The following approaches are based on a principal point, which is the projection of camera's center on an image plane that makes object measurement in the world and its position relative to the camera. These can be further classified into categories: principal point and metric measurement.
- Format-based IFD
- Format-based IFD is another form of passive approach which is based on the format of the images, mostly the JPEG format. The lossy compression techniques produce statistical correlation which is then utilized to detect image forgery. Some of these techniques are Joint Photographic Expert Groups (JPEG) quantization, JPEG blocking, double JPEG. An image if present in a compressed form becomes difficult to find the forgery in image. However, these techniques can easily live up to this task.
- Physical environment-based IFD
- The following methods are based upon the 3D connections between the physical object, camera, ad light. Fake images are created by altering the lighting in the images and merging them together and also working upon matching the environments of both the images to most extent by contrast adjustment, etc. Some approaches following this mechanism are light direction (2d), etc.

1.2 Basic Framework of Image Forgery Detection

Image forgery detection is a binary classification task, with the objective of the method to classify the image as either forged or authentic. A general structure for blind or passive IFD is given below which involves the following major steps:

- Preprocessing: The step involves applying some operations on the image before the feature extraction process, such as the conversion of the image from RGB to grayscale, histogram equalization, smoothing, etc.
- Feature extraction: The features of each of the classes are extracted which are able to differentiate among them. Certain features are then selected which are more informative and sensitive to manipulations in the image.
- Feature matching: The step of feature matching involves the matching of the feature vectors of different regions, e.g., the rectangular regions in block-based methods, which are similar to each other.
- Filtering: Some feature vectors which might not be similar but still give a positive result in terms of similarity and thus have to be filtered out from the actual feature vector pairs which are similar.
- Classification: This step classifies the image as belonging to either of the class: forged or authentic image, using a trained classifier.
- Post-processing: Image forgery localization in many copy-move and splicing forgeries is one of the examples of post-processing operation which gives additional information about the forgery in the image after being classified as forged.

2 Types of Image Forgery

The alteration of images essentially involves adding one or more features, deleting or changing some of the important features from a photo without leaving any visible traces on the image. Taking into account the various methods that have been used for forging an image, they can be classified into three major categories: CMF, spliced images, and image resampling.

CMF involves copying a part of an image, irrespective of size and shape, and pasting in a different location in the same image in to hide a feature or add a new one to the image. This kind of forgery is hard to detect since the basic features of the forged part are not different from that of the authentic image such as the textures, brightness, etc., as they are generated from the same image.

Splicing images for making them forged involves cut and paste from many images to make a fake image. The borders of the spliced region can be undetectable when splicing is performed precisely. Splicing although disturbs the Fourier statistics of a high order, and hence, these characteristics can be utilized to detect image forgeries.

Image resampling creates some specific periodic correlations in the image. Many geometric transformations such as scaling, rotation, skewing, stretching, flipping, and so on can be performed that can be used to make a forged image.

Out of all these approaches, CMFD is the most widely studied, and hence, we will be focusing our attention to this kind of IFD in the future of this report. CMFD utilizes the concept that the illumination, color, noise, etc., of the source and target regions, are the same since they are derived from the same image. An experienced forger can also apply some post-processing techniques after CMF processes such as rotation, scaling, JPEG compression, etc., which makes the detection further difficult and complex. Hence, the foremost most important point in this detection method is the feature extraction which are invariant to these above post-processing operations.

CMFD methods can further be classified into two categories: Block-based and keypoint-based approaches. In block-based methods, the image is divided into some size of blocks and a feature vector is then generated for these blocks. The feature vectors which are similar are then matched to find the forged region. One of the most efficient block-based techniques that was introduced in 2010 was aimed at detecting the CMF with the help of Zernike moments which have properties like robustness to noise, rotational invariance, etc. In keypoint-based techniques, the input image is not subdivided into blocks, whereas the feature points are extracted from a particular region that has high entropy. The same processing steps of CMF are followed as described above. While we were creating a feature vector for each rectangular region, here, the feature point is extracted using different techniques like SIFT, SURF, MIFT, etc.

3 Traditional IFD Approaches

There are various techniques that have been developed and used for IFD as well as localizing the tampered area in the forged image.

Su et al. [2] provide an efficient algorithm to detect duplicated area which is observed mostly in video forgery. The method proposed uses the concept of exponential Fourier moments (EFM) features which is extracted from every block of the current frame of the video, where the block size can be varied from 8×8 to 24×24. The blocks are then subjected to a newly created block matching algorithm which uses Euclidean distance as a similarity measure to find out of multiple matching pairs with great computational efficiency. Post-processing method is applied to remove falsely matched pairs and find the tampered region in the forged frame.

Kaur et al. [3] proposed a method wherein the image is first converted to grayscale from colored images, the next step is the feature extraction step, using the GLCM algorithm and PCA, the features are extracted. The GLCM algorithm basically provides a co-occurrence matrix. The feature matching is done on the matrix using the similarity measure to sort the features lexicographically, to detect the forgery in the images.

Mayer et al. [4] introduced the concept of lateral chromatic aberration (LCA) to locate the area in which copy-move forgery has been carried out. The proposed method detects forgery by comparing both local and global estimates of LCA vectors and hence identifying the localized LCA inconsistencies. The method requires finding out keypoints around which local LCA can be calculated. The inconsistency of local LCA is comapared global LCA to detect the forged area and then posing forgery detection as a testing problem and finding out a parameter which is optimal when inconsistency is Gaussian and IID and plays a vital role in decision making.

Jaafar et al. [5] introduce a new method to detect CMF based on k-d tree algorithm. Initially, the image is converted to grayscale from colored image. The image is then divided according to a z parameter; now using SIFT, we extract the feature vectors. Feature descriptors thus obtained are used to create a k-d tree, indexing the higher dimensional features in the top. Feature matching is carried out in two steps; first features are mapped using the Euclidean similarity measure and then the 2NN method is applied. To improve the efficiency of the method, RANSAC is applied to remove any outliers or false feature matches.

Chen et al. [6] suggest a new approach to detect CMF and localize the tampered area in the image by considering Fractional Quaternion Zernike Moments (FrQZM) as feature descriptors and a state-of-the-art feature matching algorithm called Patch-Match algorithm. The proposed approach is carried out in three steps involving preprocessing, feature extraction, and feature matching. The methods start with dividing image into circular overlapping blocks with radius increasing by one pixel value. Quaternion representation is used to represent a pure quaternion matrix of each circular block. PatchMatch algorithm involves computing reflective offsets using random offsets created using random initialization and priority is assigned to every patch in accordance to this position and reflective offset of the patch. Random search is then used to update the mapping offset which can then be used as optimal offsets, else the matching algorithm is applied again using updated offsets as input.

Mahfoudi et al. [7] use the concept of color dissimilarity maps where, first using SIFT authors extracted the keypoints which is followed by matching of SIFT descriptors using two nearest neighbor (2NN). Finally, clustering is done to localize the forged area and to minimize the false rates; the results are further filtered using a local dissimilarity map (LDM).

Wo et al. [8] begin with extracting multi-radius PCET features for each pixel with different radii. For every circular region, PCET moment is calculated and the amplitude of all the computed PCET moments is considered as feature vector and resultant is a feature matrix which is a combination of all such feature vectors. To find out the tampered area, similarity between the feature vectors in the feature matrix must be computed. All matched pairs form a match pair set and mapping this set on the image gives a coarse detection result. A post-processing method called same affine transformation selection and spatial information is employed to filter out mismatched points and get an accurate detection result.

Chowdhury et al. [9] propose a GLCM-based texture analysis method to detect copy-move forgery. Keypoints are chosen and its SIFT features are computed. The keypoints are matched using the two nearest neighbor (2NN) and the matched features are clustered using an agglomerative clustering technique. To further improve the results, RANSAC is performed to remove the outliers, and further on the inliers, GLCM algorithm is applied to refine the results (Table 1).

Table 1 Comparative study of traditional IFD methods

Year	Technique	Method used	Pros/Cons	Datasets used	Performance
2020	IFD based on the demosaicing algorithm [10]	Color filter array detection	The method has problems with smaller images of size 700 * 700 pixels or less, in detecting the area with modifications	CASIA	Accuracy: 79%
2019	CMFD using Feature point matching [11]	SIFT features and RANSAC clustering algorithm	Robust against attacks such as scaling, noise addition, and rotation	MICC-F600	TPR: 97.50%
2019	IFD and localization using color dissimilarity maps [7]	SIFT features, 2NN matching algorithm, and local color dissimilarity Maps	Minimizes the false rates using LCDM	COVERAGE	F1-score: 100%
2019	CMFD using texture analysis [9]	SIFT keypoints and GLCM inliers	Finds out parameters which reduces false positives matches	MICC, COVERAGE	Accuracy: 70.07%
2019	Fast image forgery detection [5]	SIFT features, k-d tree and 2NN for feature matching	Uses RANSAC to remove outliers and reduce false feature matches	MICC-2000	Accuracy: 88%
2018	Keypoint-based CMFD technique [12]	SIFT, DBSCAN	Detects multiple forgeries successfully in images and is computationally efficient	MICC-F2000	TPR: 98.5%
2018	CMFD using blur and rotation invariant technique [13]	Fast Fourier transform	Method is robust to rotational and blurring attacks	MICC	Accuracy: 90–95%

(continued)

4 Limitations of Traditional Forgery Detection Methodologies

The major limitations in IFD are due to the various processes applied over the original image before pasting it over the same image to hide forgery. Various operations for

Table 1 (continued)

Year	Technique	Method used	Pros/Cons	Datasets used	Performance
2018	EFM features for video duplication region [2]	Exponential Fourier moments features	The concept of EFM employed for feature extraction of every block in the image greatly improves the efficiency of the entire algorithm	CASIA V2.0	Accuracy: 93.1%
2018	Forgery detection using LCA [4]	Global and local estimate of lateral chromatic aberration vector	Algorithm used greatly reduces the estimation time without incurring any errors	Sony DSC-H50 images	Accuracy: 84%
2018	Image forgery detection using FrQZM [6]	FrQZM feature descriptors and PatchMatch algorithm	This method uses FrQZM features which take advantage of all the four components of FrQZM coefficient hence considering both magnitude and phase information in the features	GRIP, FAU	F-measure: 0.8848
2017	Image forgery detection using PCET features [8]	Multi-radius PCET features	The proposed method can successfully detect forged areas with large scale rotation, smoothening, and noise degradation	Kodak images	Precision: 94.9% Recall: 74.5% F1-score: 83.2%

processing the image like scaling, rotation, bending, compression, and noise addition make detecting a forgery, a tedious task. The work discussed cannot be extended to other media sources such as videos, audios, etc.

The techniques discussed have a high computational cost compared to the existing deep learning methods and most of the techniques discussed are not robust to noise in the images, blurriness, and do not scale well with low resolution images. Another major issue is when the images are smaller or the number of images is less, the accuracy of these algorithms drops significantly.

Most of the traditional methods have a low detection rate compared to the existing deep learning-based methods. Another issue that frequently arises in the techniques is that the post-processing operations in the image like rotation, imprinting, additive noise, and contrast adjustment prove to be a challenge for these type of CMF techniques.

5 Image Forgery Detection Mechanisms Based on Deep Learning

Over the period of time, a sudden shift to methodologies using deep features extracted from an image or blocks of images using CNN=based architecture has been more profound than the existing techniques. A numerous block-based image forgery detection using deep learning has been proposed by different authors; all following the same basic structure as shown in Fig. 1.

Huang et al. [14] proposed a method that uses a nine-layer CNN where the first, i.e., input layer accepts an image of size 227 × 227 × 3, followed by five convolutional layers with feature fusion and max pooling for extracting meaningful features from an image, followed by two fully connected layers, and finally a SoftMax layer which uses probabilities to decide the nature of image under examination.

Thakur et al. [15] use a deep learning-based CNN is used to classify images as authentic and forged and it is followed by detecting and localization of forged area in the image using a machine learning-based saliency algorithm. The proposed DCNN model is composed of fifteen layers, an input layer, three convolutional layers as well as max pooling layers, four RELU followed by two fully connected, and finally an output layer to classify images as authentic and forged. After classification, a saliency-based algorithm detects the tampered area. Gaussian pyramids are created using preprocessed images which further helps in creating feature maps. Saliency map is obtained by computing mean of all the feature maps. These grayscale images obtained are analyzed for changes in pixels, color, intensity, and a suspected region is generated as output.

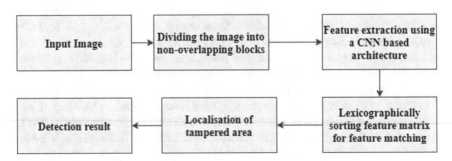

Fig. 1 Steps followed in block-based IFD

Zhong et al. [16] proposed a method that takes a color image as input for prepro-cessing via pyramid feature extractor which comprises of three feature extractor blocks to extract multi-scale and multi-dimensional deep features. The feature corre-lation matching blocks are responsible for obtaining candidate matching maps by learning coarse and fine details from hierarchical features extracted. Finally, the post-processing module analyzes the maps to accurately detect and localize the forged area in the image.

Qu et al. [17] introduced use of RGB and noise stream for forgery detection and localization. Using CNN network, RGB spatial features are extracted from the stream, and using an SRM filter layer, noise stream is obtained which is further used to obtain noise feature map. The ROI layer in the architecture selects spatial features from RGB and noise feature maps. A bilinear pooling layer is responsible for fusion of the features from two stream, and finally, a SoftMax layer detects and localizes the tampered area.

Zhang et al. [18] use a CNN-based architecture and trained it on pristine images. Patches are extracted from tampered images by moving a sliding window of size 32*32 and stride equal to 10. The RGB color space is to converted to YCrCb color space and the input to SCNN is a patch with only CrCb information. For every patch, a confidence score of being tampered region is computed and stored in the probability map. The sliding window covers the entire image and a complete probability map is obtained.

Cozzolino et al. [19] proposed a system which takes an input image and produces a same sized residual image as input image and contains natural spatial information and only camera model specific features. The two images are paired to be used in the training phase to compute Euclidean distance between them and back-propagate the error. The CNN is responsible of computing the distance and training process. An average of the residual of the pristine images is computed as a clean reference. This clean reference is compared with test image by using Euclidean distance a similarity measure. Pixel-wise distance is obtained on a heat map which can further be used as a decision map.

Kim et al. [20] use a CNN architecture composed of one high pass filter, two convolution layers, two fully connected layers, and one output layer. The method produces a manipulated image from a pristine image and creates a learning and testing set. The input image is given as an input to a high pass filter to extract deep and meaning features and reduces the size of image to $252 \times 252 \times 3$. The resulting image is then subjected to various convolution layer where operations such as filtering, pooling are carried out. The result is subjected to two fully connected layer with dropout 0.5 and finally an output layer for classification task.

Thakur et al. [21] instead of directly using the image to get the features, passed the residuals of the image through the CNN architecture based on VGGNet. Every image is converted into grayscale, then the residuals are calculated using SD_MFR, and on that LFR is applied to reduce the number of features. These residuals are fed to the CNN architecture which is composed of convolution layers, fully connected layer, and an output layer to predict whether image is forged or not.

Muzaffer et al. [22] utilized a CNN architecture based on AlexNet to extract the features, after which a feature matrix is formed on which feature matching is done. The feature matrix is lexicographically sorted and Euclidean distance used as a similarity measure to compute distance between two vectors and compare with a predetermined threshold. In the post-processing step, false matches are removed and the shift vectors are calculated; if the range of blocks with same shift vector crosses a predefined threshold (i.e., 32), a CMF is detected (Table 2).

6   Comparative Results and Discussion

Different authors use different measures such as accuracy, F1-score, true positive rate, and false positive rate for analyzing the performance of their respective proposed IFD methodologies.

The detection accuracy of the traditional IFD techniques [2, 4, 5, 8–10, 13] is 79, 70.07, 88, 90, 93.10, 84, and 74.50%, respectively, as shown in Fig. 2.

Figure 3 displays performance analysis of deep learning-based IFD techniques [14, 15, 18–21, 23–26], which have detection accuracy 94.26, 98.76, 94.80, 89.70, 99.97, 85.35%, 96.70, 95, 98.80, and 95%, respectively.

7   Conclusion

In this paper, various techniques for IFD and localizing the tampered area in a forged image have been surveyed and discussed. It has been observed in recent times that the CNN-based forgery techniques are favored over the existing approaches because of its strength in obtaining deep invariant features from an image that makes them a state-of-the-art forgery detection methodology. This paper has mentioned a few benefits and drawbacks of the deep learning-based forgery detection techniques and arises a need for the development of an effective and highly accurate methodology that can be robust against any type of attack on images, which intends to manipulate images for malicious activity.

Table 2 Comparative study of deep learning-based image forgery detection methods

Year	Technique	Method used	Pros/Cons	Datasets used	Performance
2020	Using Residual and CNN framework for forgery detection [21]	VGGNet	Using precalculated weights reduces computational time	CoMoFoD	Accuracy: 94.26%
2020	InceptionNet for image forgery detection [16]	Dense-InceptionNet	Better accuracy than most of the state-of-the-art methods	CASIA	F1-measure: 0.6429
2019	Deep visual features for splicing detection [23]	Inception Model and R-CNN	Highly accurate and reliable	CASIA V1.0	Accuracy: 98.76%
2019	CMF detection using AlexNet [22]	AlexNet	Low computational time	GRIP	F-measure: 0.93
2019	Using LSTM for image forgery detection [24]	Hybrid LSTM and encoder-decoder	Pixel level localization with high precision	NIST'16	Accuracy: 94.80%
2019	Using motion blur to detect digital image forgery [25]	Deep CNN for MBK estimation	MBK are sensitive to size and texture distribution and are robust to translation invariance	Natural camera images	Accuracy: 89.7%
2019	Object detection for image forgery [17]	R-CNN	Low reliability as compared to other methods	NIST'16, CASIA	F1-measue: 0.722
2018	Adversarial attacks for image forgery detection [26]	SPAM, Bayar2016, Xception	High detection results with images having adversarial noise	JPEG images	Accuracy: SPAM: 99.97 Bayar: 99.98 Xception: 98.81
2018	Forgery detection based on boundary detection [18]	Shallow CNN	Performs well on low resolution images	CASIA V2.0	Accuracy: 85.35%

(continued)

Table 2 (continued)

Year	Technique	Method used	Pros/Cons	Datasets used	Performance
2018	Camera-based forgery detection and localization [19]	PRNU noise pattern extraction using CNN	Provides very good results even when the PRNU is computed for small number of images	Dresden, VISION	Accuracy: 96.7%
2018	Imager forgery detection using deep features [14]	Nine-layer CNN	can provide quite accurate detection for three category of image forgery: splicing, re-touch, and recompress	CASIA	Accuracy: 95%
2018	Saliency algorithm for forgery detection and localization [15]	Deep CNN and Saliency-based algorithm	High detection and localization reliability	CASIA, DVMM, BSDS-300	Accuracy: 98.80%
2017	Forgery detection via integrating two CMFD techniques [27]	Statistical-based feature detector and any CNN-based feature detector	The use of tampering possibility maps reduces false positives and false negatives	IFS-TC images	F1-measue: 0.4925
2017	Image manipulation detection [20]	CNN and high pass filter	Possible to apply to other multi-media than images	Boss Base 1.01	Accuracy: 95%

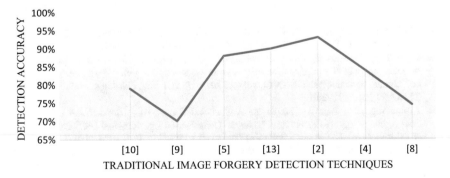

Fig. 2 Comparison of traditional IFD techniques in terms of accuracy

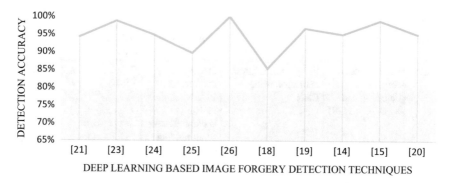

Fig. 3 Comparison of deep learning-based IFD techniques in terms of accuracy

References

1. Kashyap A, Parmar RS, Agarwal M, Gupta H (2017) An evaluation of digital image forgery detection approaches. Int J Appl Eng Res 12(15):4747–4758
2. Su L, Li C, Lai Y, Yang J (2018) A fast forgery detection algorithm based on exponential-Fourier moments for video region duplication. IEEE Trans Multimed 20(4):825–840. https://doi.org/10.1109/TMM.2017.2760098
3. Kaur T, Girdhar A, Gupta G (2018) A robust algorithm for the detection of cloning forgery. In: 2018 IEEE international conference on computational intelligence and computing research, pp 1–6
4. Mayer O, Stamm MC (2018) Accurate and efficient image forgery detection using lateral chromatic aberration. IEEE Trans Inf Forensics Secur 13(7):1762–1777. https://doi.org/10.1109/TIFS.2018.2799421
5. Jaafar RH, Rasool ZH, Alasadi AHH (2019) New copy-move forgery detection algorithm. In: Proceedings of 2019 International Russian Automation Conference RusAutoCon 2019, pp 1–5, 2019, doi: https://doi.org/10.1109/RUSAUTOCON.2019.8867813
6. Chen B, Yu M, Su Q, Shim HJ, Shi YQ (2018) Fractional quaternion zernike moments for robust color image copy-move forgery detection. IEEE Access 6:56637–56646. https://doi.org/10.1109/ACCESS.2018.2871952
7. Mahfoudi G, Morain-Nicollier F, Retraint F, Pic M (2019) Copy and move forgery detection using SIFT and local color dissimilarity maps. Glob. 2019—7th IEEE global conference on signal and information processing. Proceedings, pp 1–5, doi: https://doi.org/10.1109/GlobalSIP45357.2019.8969355
8. Wo Y, Yang K, Han G, Chen H, Wu W (2017) Copy-move forgery detection based on multiradius PCET. IET Image Process 11(2):99–108. https://doi.org/10.1049/iet-ipr.2016.0229
9. Chowdhury M, Shah H, Kotian T, Subbalakshmi N, David SS (2019) Copy-move forgery detection using SIFT and GLCM-based texture analysis. In: IEEE region 10 annual international conference, proceedings, TENCON, vol Oct 2019, pp 960–964. doi: https://doi.org/10.1109/TENCON.2019.8929276
10. Armas Vega EA, González Fernández E, Sandoval Orozco AL, García Villalba LJ (2020) Passive image forgery detection based on the Demosaicing algorithm and JPEG compression. IEEE Access 8:11815–11823. doi: https://doi.org/10.1109/ACCESS.2020.2964516
11. Li Y, Zhou J (2019) Fast and effective image copy-move forgery detection via hierarchical feature point matching. IEEE Trans Inf Forensics Secur 14(5):1307–1322. https://doi.org/10.1109/TIFS.2018.2876837

12. Soni B, Das PK, Thounaojam DM (2018) multiCMFD: fast and efficient system for multiple copy-move forgeries detection in image. ACM Int Conf Proc Ser 53–58. doi: https://doi.org/10.1145/3191442.3191465

13. Verma A, Kapoor V, Roy S (2018) Efficient copy-move forgery detection using blur and rotation invariant technique. ACM Int Conf Proc Ser 123–127. doi: https://doi.org/10.1145/3301326.3301327

14. Huang N, He J, Zhu N (2018) A novel method for detecting image forgery based on convolutional neural network. The 17th IEEE international conference on trust, security and privacy in computing and communications (IEEE TrustCom 2018). 12th IEEE international conference on big data science and engineering. Trust. 2018, pp 1702–1705. doi: https://doi.org/10.1109/TrustCom/BigDataSE.2018.00255

15. Thakur A, Jindal N (2018) Machine learning based saliency algorithm for image forgery classification and localization. ICSCCC 2018—1st international conference on computing, communications, and cyber-security, pp 451–456. doi: https://doi.org/10.1109/ICSCCC.2018.8703287

16. Zhong J, Pun C, Member S (2020) An End-to-End Dense-InceptionNet for image copy-move forgery detection. 15:2134–2146

17. Qu S (2019) An approach based on object detection for image forensics. In: 1st Int Conf Ind Artif Intell IAI 2019. doi: https://doi.org/10.1109/ICIAI.2019.8850791

18. Zhang Z, Zhang Y, Zhou Z, Luo J (2018) Boundary-based Image forgery detection by fast shallow CNN. In: Proceedings of international conference on pattern recognition, vol Aug 2018, pp 2658–2663. doi: https://doi.org/10.1109/ICPR.2018.8545074

19. Cozzolino D, Verdoliva L (2018) Camera-based image forgery localization using convolutional neural networks. In: *European signal processing conference*, vol Sept 2018, pp 1372–1376. doi: https://doi.org/10.23919/EUSIPCO.2018.8553581

20. Kim DH, Lee HY (2017) Image manipulation detection using convolutional neural network. Int J Appl Eng Res 12(21):11640–11646

21. Thakur R, Rohilla R (2020) Copy-move forgery detection using residuals and convolutional neural network framework: a novel approach, pp 561–564. doi: https://doi.org/10.1109/peeic47157.2019.8976868

22. Muzaffer G, Ulutas G (2019) A new deep learning-based method to detection of copy-move forgery in digital images. In: 2019 Scientific meeting on electrical-electronics & biomedical engineering and computer science. EBBT 2019, pp 1–4. doi: https://doi.org/10.1109/EBBT.2019.8741657

23. Saleem S, Dilawari A, Khan UG (2019) Multimedia forensic: an approach for splicing detection based on deep visual features . Int Conf Robot Autom Ind ICRAI 2019:1–6. https://doi.org/10.1109/ICRAI47710.2019.8967380

24. Bappy JH, Simons C, Nataraj L, Manjunath BS, Roy-Chowdhury AK (2019) Hybrid LSTM and encoder-decoder architecture for detection of image forgeries. IEEE Trans Image Process 28(7):3286–3300. https://doi.org/10.1109/TIP.2019.2895466

25. Song C, Zeng P, Wang Z, Li T, Qiao L, Shen L (2019) Image forgery detection based on motion blur estimated using convolutional neural network. IEEE Sens J 19(23):11601–11611. https://doi.org/10.1109/JSEN.2019.2928480

26. Gragnaniello D, Marra F, Poggi G, Verdoliva L (2018) Analysis of adversarial attacks against CNN-based image forgery detectors. In: European signal processing conference, vol Sept 2018, pp 967–971, doi: https://doi.org/10.23919/EUSIPCO.2018.8553560

27. Li H, Luo W, Qiu X, Huang J (2017) Image forgery localization via integrating tampering possibility maps. IEEE Trans Inf Forensics Secur 12(5):1240–1252. https://doi.org/10.1109/TIFS.2017.2656823

Low-Light Image Enhancement Using Deep Convolutional Network

R. Priyadarshini, Arvind Bharani, E. Rahimankhan, and N. Rajendran

Abstract Low-light image enhancement is generally regarded as a challenging task in image processing, especially for the images captured at nighttime or images taken in low-light conditions. It is because various factors of the image such as contrast, sharpness and color coordination should be handled simultaneously and effectively. To reduce the blurs or noises on the low-light images, many papers have contributed by proposing different techniques. One such technique addresses this problem using a pipeline neural network and a convolutional neural network. Due to the irregularity of intensity calculation in the working of the pipeline neural networks model, a hidden layer is added to the model which results in a decrease in irregularity. The CNN algorithm uses more number of hidden layers in identifying features of the image in the pixel level, and the process of repairing in the pixel level will increase the intensity of the image and thus increases the clarity of the image. The proposed model using DCNN increases the intensity accuracy percentage by 10–15%.

Keywords Low-light image · Deep CNN · Factors of image · Intensity of image · Neural network model · CNN · Nighttime image · Dark image · Improvisation of image

R. Priyadarshini (✉) · A. Bharani · E. Rahimankhan · N. Rajendran
Department of Information Technology, B.S. Abdur Rahman Crescent Institute of Science and Technology, Chennai, Tamil Nadu, India
e-mail: rspdarshini@gmail.com

A. Bharani
e-mail: arvindbharani00@gmail.com

E. Rahimankhan
e-mail: rahimankhan.khan@gmail.com

N. Rajendran
e-mail: rajendran.n81@gmail.com

1 Introduction

Low-light images are those type of images which are captured at very little light. Such kind of images is not clear. They lack a few essential features of the image which results in looking unclear or noisy. Some of the essential features which are being lost are the sharpness, brightness and color coordination of the image. Adjusting the pixel level is a simple way to enhance an input image. But this was proved to be an imperfect method to enhance an image completely. Then several models for image enhancement were found by scientists across the globe. Retinex theory [2], dehazing model and deep refinement networks are some of the algorithms used in low-light image enhancement so far. Histogram equalization [1] method was used to improve contrast in images. This is done by stretching out the intensity range of an image which allows for areas of low contrast to gain higher contrast. However, balancing dark region recovery and bright region preservation is a difficult task, and this model proved to have minimal image enhancement capabilities is preferred. Retinex theory (single and multi-scaled) is based on how the retinex in human eye works. But this method resulted in an image that was under or overly enhanced. The pipeline neural network model is one of the techniques used for enhancing a low-light image. The model uses the concept of denoising, discrete wavelet transform (DWT), logarithmic transformation, single image super-resolution convolution neural network (SRCNN), autoencoder [8] and blending function. In the proposed work model, a hidden layer called principal component analysis (PCA) [10] and combined with the output of PCA and autoencoder. The chances of losing some important features can be reduced in our proposed model. It was found that our model gives better results than the existing LLIE model [8].

2 Related Work

Several methods and techniques have been established for the enhancement of low-light images. Every image can be tonally distributed. The representation of this distribution is called a histogram. Initially, histogram equalization [1] was introduced for the betterment of contrast of an image by using tone transfer adjustment functions. It is plotted by taking intensity on X-axis and frequency or probability of those intensities on Y-axis. Further modifications led to contrast limited adaptive histogram equalization (CLAHE) [1], brightness error bi-histogram equalization (BBHE), dynamic histogram equalization (DHE) and quantization bi-histogram equalization (QBHE). It provides better results for either over-exposed or under-exposed images and is less efficient when nonlinear functions are involved. Linear correction helps in providing a better perception of an image. By using nonlinear functions, visible changes can be brought in the contrast of both dark and bright regions of an image.

One such function is the gamma correction or gamma transformation function, which is a logarithmic function that helps in the enhancement of the contrast by adjusting the gamma parameter. Gamma parameter gives the connection between the value of the pixel and the luminous intensity per unit area. Nonlinear filters such as block matching and 3D filtering (BM3D) [1] help in the extraction of noise from the image. It takes place in three steps, a grouping of fragments (block matching), filtering of fragments and aggregation of fragments which is the overlapping of macroblocks.

Finally, the introduction of autoencoders [9, 12], by the usage of concepts of the artificial neural network, became an icebreaker in the field of image processing, as it plays a major role in analysis, learning and understanding of complex concepts of deep convolution network [11].

The camera sensors are very much expensive; therefore, the quality of the image is improved by the LL-refinement network, and it outperforms the method in the work conducted by Lincheng et al. 2018. In their work, they have proved that Refinement Net is faster and better than the existing methods, and it reduces the noise and develops the image as a high-resolution image [7]. The future work of research paves the way to increase the number of hidden layers [4].

There is another work carried out from India [6], to enhance the visibility of low-light images using DNN, in their work two models have been trained, out of which one model is the RGB model in which each color component is trained through the channels, and the second model is training the raw image of short exposure with reference to long exposure of the image.

The work is carried from Iowa state university [5]. An autoencoder is designed to capture the signals from the image, and the main work of the proposed system is to over amplify the lighter part of the images and train the same. The proposed method can be used to enhance the images taken in poorly lighted places. It also overcomes the cost of the sensors used in the high-end cameras. The limitation of the system is indicated by the accuracy level achieved to enhance a high-resolution image.

Learning to see in the dark, the application of deep learning on enhancing the visibility of low-light images and the application of artificial intelligence in the image processing field has been elaborated.

2.1 Existing System

A pipeline and convolutional neural network model is designed for low-light image enhancement. The low-light image is sent through a denoising network. Firstly, they acquire the low-frequency part of the image using DWT. However, the size reduces into 4 times of the original size. And at the same, logarithmic transformation of the image is also acquired which will be used later. They separate the task of super-resolution reconstruction and build a small SRCNN consisting of only three convolution layers. The output of SRCNN is taken as an input by the autoencoder, which combines the deep and shallow features. Finally, the output of autoencoder and logarithmic transformation (sent through 1X1 CONV) is combined using a blending function.

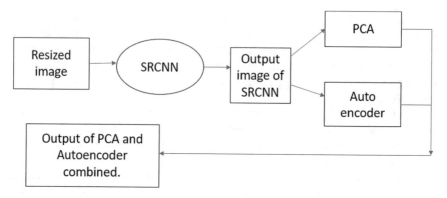

Fig. 1 Convolutional model with principal component analysis (PCA)

Pipeline neural network model is considered as a more efficient way than other techniques to enhance a low-light image. The model can be improved by adding hidden layers. Its efficiency is being elaborated by adding a hidden layer called principal component analysis (PCA) [7]. The autoencoder used in the pipeline neural network model does not work perfectly in a consistent way. This is a major disadvantage for autoencoders. Sometimes, it might lose the most important features of an image. This irregularity in the working of an autoencoder might lead to producing an unconventional result. Adding of PCA would manage the inconsistent working of autoencoders. PCA works more invariably compared to autoencoders. So, once the output of PCA and autoencoders is combined, even if the features had been lost by the autoencoders, PCA would preserve the same (Fig. 1).

3 DCNN-Based Low-Light Image Enhancer (DCNNLLIE) Model

The DCNNLLIE consists of five phases, in which the original image is preprocessed using denoising net and DWT. The resized image is given as input to the SRCNN, and the third phase is the combination of logarithmic function with PCA and autoencoder. The fourth phase is the training phase with the help of PCA. The last phase is introducing the blending function to give enhanced low-light image using DCNN (Fig. 2).

The low-light image is denoised first using a denoising net. The logarithmic transformation and discrete wavelet transform (DWT) of the denoised image are done simultaneously. The result of DWT on the image is 4 times smaller than the input image. To bring it back to its original size, the resizing of the image is carried out. It is enhanced into a higher-resolution format using SRCNN. The output of single image super-resolution convolutional neural network is sent into an autoencoder and a PCA [10] at the same time. The output from both the modules is combined. This output

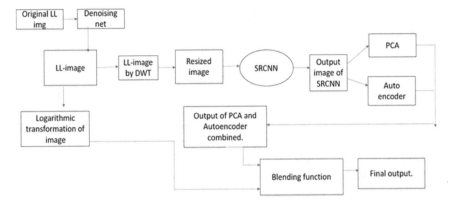

Fig. 2 DCNN-based low-light image enhancer

is collaborated with a logarithmically transformed image using blending function to get enhanced output.

3.1 Denoising Process of Images

The process of removing noise from an image is called as denoising. There are different types of noises in an image. Some of the types of noises are Gaussian noise, Poisson noise and speckle noise. The existing low-light image enhancement (LLIE) model has a denoising net which concentrates more on removing only Gaussian noise of the image. There is a method function to remove every type of noise. The level of noise been removed can be evaluated by comparing it with the original image. The denoising techniques are done in such a way that the sharp edges and textile features are maintained throughout the process. To accomplish a very high quality of denoising, the time taken for the process will be more. The first process that takes place in the pipeline neural networks model is denoising. In this model, only the Gaussian noise is removed from the low-light image.

3.2 Discrete Wavelet Transform (DWT)

Discrete wavelet transform is one of the efficient and popular techniques used for compression of images. In this technique, firstly the image pixels are being converted into wavelets, and then the compression of an image takes place using the wavelets. The images are compressed in such a way that the low-frequency parts of the image are brought to one point and the high-frequency parts of the image are brought to another point. Usually, the high-frequency parts contain all the unwanted details of the image. Hence, to omit those unwanted details, DWT is performed. The low-frequency part of the image contains all the important details of the image. After splitting the image into two parts, the low-frequency part is used for the enhancement process whereas the high-frequency part is completely omitted. The output of DWT consists of four forms of the input image. This can be done using Haar basis function. The four forms of the band show vertical, horizontal and diagonal information of the image, and they are as follows the LL band, LH band, HL band and HH band. The LL band consists of the low-frequency part of the image, and the remaining three bands have the high-frequency part. Usually, once DWT is performed, the image size will reduce to four times its original size. The reduced image is then brought back to its original size using the resize function. After denoising, the DWT process is done in the pipeline neural networks model.

3.3 Logarithmic Transformation

The logarithmic transformation of an image means every pixel value of the image is being converted into its logarithmic form. The formula for logarithmic transformation is

$$Pxio = C \log (Pxii + 1) \tag{1}$$

Pxoi—pixel value of the output image.

Pxii—pixel value of the input image c-constant.

   Plus one is done so that even when the pixel intensity is 0, there will be log (1) which shows one as the minimum value. If one is not added, then on the right-hand side, it will become log (0). The value of log (0) is infinity. As a result, it becomes difficult when you get the logarithmic transformation of a pixel value as infinity. During this process, the dark pixel values get expanded whereas all the high pixel values are converted into their log form. This leads to an output image which looks a little brighter than the input image. And finally, according to each pixel's intensity, different forms of output images will be produced. This process would help the model to store all the different levels of brightness possibilities of the low-light image.

3.4 Super-Resolution Convolutional Neural Network (SRCNN)

The main aim of super-resolution convolutional neural network (SRCNN) is to convert a low-resolution image to a high-resolution image. SRCNN performs four main operations. The first operation is the upscaling. The process of upscaling a low-resolution image to a desired high-resolution size is called upscaling. The second operation is called extraction of features. The process of extracting only the essential features of the low-resolution image is called feature extraction. These extracted features are highlighted in the high-resolution image which makes the image look clearer.

The third operation is mapping. All the feature maps representing the change from low-resolution patches to high-resolution patches are mapped. Finally, the reconstruction of the HR image from HR patches is done. The SRCNN gives a better scaling and color for the image. As a result, the image becomes more interpretable. All the output forms would look brighter than the original low-light image. The resized output of the DWT module is considered as the input for SRCNN.

These extracted features are highlighted in the high-resolution image which makes the image look clearer. The third operation is mapping. All the feature maps representing the change from low-resolution patches to high-resolution patches are mapped. Finally, the reconstruction of the HR image from HR patches is done. The SRCNN gives a better scaling and color for the image. As a result, the image becomes more interpretable. All the output forms would look brighter than the original low-light image. The resized output of the DWT module is considered as the input for SRCNN.

4   DCNN-Based Combined PCA and Autoencoder Output

4.1   *Principal Component Analysis (PCA)*

Principal component analysis (PCA) is a method of analyzing data by reducing the dimensions. It is done by identifying the main/principal components in data. PCA is used to identify the presence of any patterns in the data. This shows the similarities and differences in a set of data. The data is compressed (reduction in dimensionality) without any loss, which is its biggest advantage. It also reduces the noise present in the data. PCA allows redistributing the orientation of input data and view at a different angle [3]. The main pros of PCA are its way of reducing the dimensions without any loss and the way it works fast. But PCA does not characterize the nonlinear information present in the image.

In the LLIE pipelined network, autoencoders have been used. This led to some disadvantages like slow processing, inefficiency to work in every image, etc. So, in our project, autoencoders have been used and clubbed its output with the output of PCA in an aim to balance out the cons of the two methods. That is, linear features are recognized by PCA while the nonlinear features are recognized by autoencoders. Also, PCA works faster than autoencoders. In this way, it did not tend to lose any features of our dataset (images).

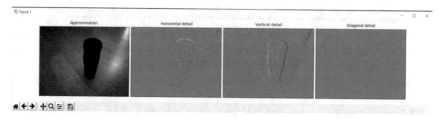

Fig. 3 Output after performing denoising and DWT

4.2 Combining PCA Output and Autoencoder Output

Finally, the outputs are combined. This process ensures that no important features of the image are being lost by the model during the enhancement process. As mentioned above, when an autoencoder misses some features, the PCA will act as a backup and prevent the model from producing an unconventional result. Likewise, PCA cannot characterize the nonlinear transformations of an image. Now the autoencoders act as a backup for PCA. It ensures that most of the required nonlinear transformations are extracted effectively. Generally, PCA works very fast. Hence, there would not be any increase in the time (caused by PCA) taken by the model for the enhancement process.

4.3 Blending Function

The process of blending two images according to the RGB values of the pixels is called a blending function. Basically, the color blending process takes place between two images. Blending function is a very simple process. It can be achieved by sending the images through a 1 * 1 convolution layer. Using a set of rules, the pixel values chosen from both the images are combined (Fig. 3).

4.4 Result Analysis

The desired output from the logarithmic transformation function is blended along with the combined result of PCA and autoencoder. By incorporating this function, the sharpness and brightness of the image can be acquired from logarithmic transformation and the color from the combined result of PCA and autoencoder.

The accuracy percentage of the DCNN method is increased from 10–15% as shown in the graph. The increase in accuracy of the image clarity is due to the combined method and DCNN hidden layer functionality. The increase in accuracy is shown in Fig. 4. The intensity of the image is increased, and the cumulative frequency

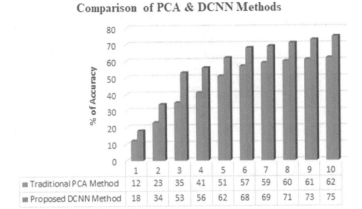

Fig. 4 Accuracy of DCNN method compared with the PCA method

is plotted as shown the low-light intensity is observed when the number of pixels is low. It is improved in the DCNN method to some extent as shown in Figs. 5 and 6.

The light intensity is measured for the raw original low-light image and the enhanced image. The number of pixels and its cumulative frequency is an essential factor for finding out the clarity of the images. The overall cumulative frequency and the intensity of a low-light image are shown in Fig. 6, whereas Fig. 5 shows the light intensity of the enhanced image when the number of pixels increases the intensity increases. The intensity is increased using the filter and training factor in blending function of DCNN.

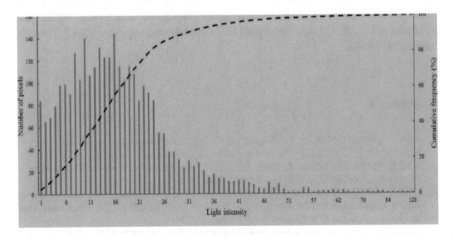

Fig. 5 Cumulative Intensity Graph I

Fig. 6 Cumulative intensity
of enhanced LLI Graph II

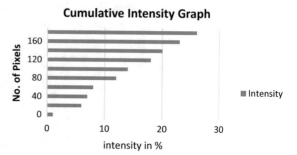

5   Conclusion

In the proposed system, it is found that the output is slightly enhanced than the existing system. This is due to the inclusion of PCA as a hidden layer. As mentioned above, the drawback of autoencoder is balanced by including PCA. The chances of losing important features have been comparatively low.

One of the biggest advantages of autoencoder is, and it can effectively identify the nonlinear transformations. Combining the two modules will ensure that no linear and nonlinear transformations of the image are lost. The limitations of our model are, adding an extra hidden layer would make the model a little bit more complex. The complexity in the system can be reduced by creating a single module that does the work of both PCA and autoencoder.

The future work of the proposed system can be improved by combining the DCNN with the restricted Boltzmann machine algorithm, where the pixel level amplification can be enhanced using the hidden layers; however, there may be a trade-off in time. In another direction of research, the low-light enhancement can be done by training the model using DCNN combined with RGB models and exposure time in terms of the original image.

References

1. Ancuti C, Ancuti CO, Haber T, Bekaert P (2012) Enhancing underwater images and videos by fusion. In: 2012 IEEE conference on computer vision and pattern recognition, pp 8188. IEEE
2. Iqbal K, Salam RA, Osman A, Talib AZ (2007) Underwater image enhancement using an integrated colour model. IAENG Int J Comput Sci 34(2)
3. Iqbal K, Odetayo M, James A, Salam RA, Talib AZH (2010) Enhancing the low quality images using unsupervised colour correction method. In: 2010 IEEE international conference on systems, man and cybernetics, pp 1703–1709. IEEE
4. Jiang L, Jing Y, Hu S, Ge B, Xiao W (2018) Deep refinement network for natural low-light image enhancement in symmetric pathways. Symmetry article (2018)
5. Lore KG, Akintayo A, Sarkar S (2016) LLNet: a deep autoencoder approach to natural low-light image enhancement, vol 3
6. Praveena M, Pavan Kumar V, Asha Deepika R, Sai Raghavendhar Ch, Rahul Sai Reddy J (2019) Enhancing visibility of low-light images using deep learning techniques. IJITEE, vol 8

7. Priyanka SA, Wang Y-K, Huang S-Y (2017) Low-light image enhancement by principal component analysis. IEEE transaction, image processing
8. Rana SB, Rana SB (2015) A review of medical image enhancement techniques for image processing. Int J Current Eng Technol 5(2):1282–1286
9. Rizzi A, Gatta C, Marini D (2003) A new algorithm for unsupervised global and local color correction. Pattern Recogn Lett 24(11):1663–1677
10. Sandbhor B, Kharat GU (2015) A review on underwater image enhancement techniques. Int J Adv Res Comput Sci Soft Eng 5(5):676–680
11. Sathya R, Bharathi M (2015) Enhancement of underwater images using wavelength compensation method. Int J Innov Res Comput Commun Eng 3(3):1829–1835
12. Talwar S, Kochher R (2015) Evaluation of underwater image dehazing techniques. Int J Sci Emerg Technol Latest Trend 21(1):11–13

Sign Language Recognition Using Convolutional Neural Network

Salakapuri Rakesh, Avinassh Bharadhwaj, and E. Sree Harsha

Abstract In today's world, communication is very important. A language is needed to communicate. Most of the specially abled people, use a different language for communication called sign language. This language helps them to communicate with other people with their hand expressions. These expressions will be different from country to country. In this paper, American sign language is used. This paper deals with helping specially abled people to communicate with people who don't know sign language by using the approaches of computer vision and deep learning. Our paper uses convolutional neural network to solve this problem. The first part of our paper focuses on capturing different hand expressions in the form of video by the person and translating them to text using a convolutional neural network. The other part focuses on the reverse of it, showing GIF upon converting text. Integrating these two parts will help in two-way communication.

Keywords American sign language · Computer vision · Communication · Convolutional neural network · Deep learning · Sign language

1 Introduction

The sign language is the basic and important communication tool used especially by the people who can't listen and talk. Sign language includes gestures used by hand fingers, i.e., each letter in the English alphabet has a distinct gesture. The deaf and dumb people communicate with each other by learning the sign language, i.e., the gestures used by particular sign language. These sign languages are classified into

S. Rakesh (✉) · A. Bharadhwaj · E. Sree Harsha
Department of IT, CBIT, Hyderabad, India
e-mail: srakesh_it@cbit.ac.in

A. Bharadhwaj
e-mail: avinasshbharadhwaj@gmail.com

E. Sree Harsha
e-mail: etukuri.sreeharsha@gmail.com

J. S. Raj et al. (eds.), *Innovative Data Communication Technologies and Application*,
Lecture Notes on Data Engineering and Communications Technologies 59,
https://doi.org/10.1007/978-981-15-9651-3_58

different categories like British sign language (BSL), French sign language (FSL), American sign language (ASL), Chinese sign language (CSL), Irish sign language (ISL), etc. In this paper, American sign language is chosen.

To live in this world, communication plays an important role. But normal people unable to communicate with specially abled people like deaf and dumb people because normal people don't know the sign language. So, this barrier has to be removed between specially abled persons and normal person and to create a two-way communication channel by applying concepts of computer vision and machine learning.

2 Related Work

A lot of research is being held on finding solutions to these sign languages predictions. But a good solution did not come across. Different techniques can be used to get the solutions but the powerful tool for getting the predictions accurately is convolutional neural networks. So, convolutional neural networks are used to get the predictions accurately. The examples of different techniques include image-based hand gesture recognition techniques [1], support vector machines [2], neural network technique for selfie camera [3] and convolutional neural network technique for selfie video [4], gesture recognition like hidden Markov model (HMM) [11], Eigenvalue-based [12], etc. The challenges in sign Language recognition are sign extraction, sign feature representation, video trimming, sign video background modeling and sign classification [5].

Fundamentally there are two techniques for sign language recognition, they are vision-based gesture recognition and sensor-based gesture recognition [6]. Lots of research has been done on sensor-based techniques like wires, helmets, gloves etc.[7–9]. However, because of the disadvantage of wearing it continuously is very difficult, consequently, further research work is concentrated on image-based methodologies [10].

The dataset created by Akash is used in this process and downloaded from Kaggle through the link https://www.kaggle.com/grassknoted/asl-alphabet. This dataset consists of 29 class labels where each class label has 3000 images which are generated using data augmentation technique, out of 29 classes had used 25 classes. Each image size is 200 × 200. A model is created using customized convolutional neural networks and trained the model with this dataset. This model works good only for the images present in the dataset, but if tried to work with real-time images, then only 6 classes were predicted accurately. So, the model has been trained in different ways by changing the number of layers present in our customized convolutional neural networks and activation functions used in different layers. Like this a lot of possibilities is tried but did not get the results accurately and have observed the reason for not getting the results correctly is, the clarity of images present in that dataset were poor, hence decided to work with a new dataset which consists of more clear images and that type of dataset didn't find on the Internet. So a new dataset with high-resolution images has been created with 25 classes where each class has 25 images with size 4000 × 3000.

3 Proposed System Overview

In our paper, a methodology is proposed as shown in Fig. 1 which contains two features 1. Speak 2. Listen. where the speak feature is used to create GIF based on the query. Listen to feature contains two sub-features where the user can upload an image or a video.

In our proposed methodology, a solution has to be generated which is helpful for two-way communication, i.e., given a recorded video consisting of gestures, the string or text is needed to get as output and vice versa. In the first process, i.e., generating the text from the given video, the process involved is to generate frames from the video. These frames go through the preprocessing techniques used in our algorithm and then given to the CNN model. The CNN model stores the responses and this process continue until the end of the video are reached. In this process, it may generate multiple frames for a single sign. To overcome this problem, the spell checker API is used at the end of generating all responses. The flowchart for this process is shown below in Fig. 2.

In the second process, i.e., generating a GIF from the given text, the first character of the text is taken as input and generate its equivalent image and store it in an array, and for second and third characters, equivalent images of all input characters are generated. Once the end of the text is reached then from the stored equivalent

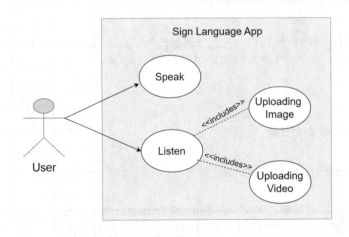

Fig. 1 Use case diagram for proposed methodology

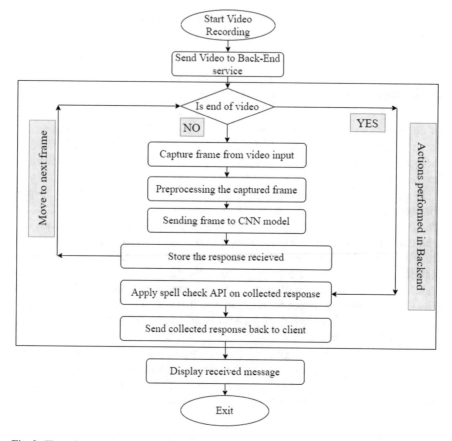

Fig. 2 Flow chart used to convert video to text

images, a GIF is needed to generate and give this generated GIF as output so that they can understand by looking at the GIF. The flowchart for this process is shown in Fig. 3.

4 Implementation

4.1 Creating New Dataset

A new dataset has been created with 25 classes where each class has 25 images which are shown in Fig. 4. These are performed by 5 different users from each user, five different sets have been taken, totaling $5 * 5 * 25 = 625$ sign images, each image size is 4000×3000.

Fig. 3 Flow chart used to generate GIF from given text

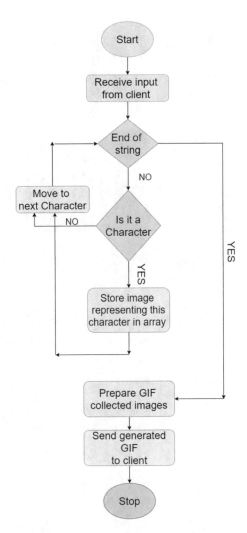

4.2 Implementation of Convolutional Neural Network Architecture

The proposed CNN architecture as shown in Fig. 5, uses a sequential model. This model consists of an input layer with an input size 256 × 256 and is followed by convolution layers with different window sizes followed by an activation function and max-pooling layers of pool size 3 × 3. The convolutional windows are having a size of 7 × 7, 7 × 7, 5 × 5, 5 × 5, respectively and numbers of filters in each layer are 32,32,64,64, respectively. Dense layers and SOFTMAX regression are used in the classification stage. The RMSPROP optimizer is used while compiling but did not get desired results so changed the optimizer to ADAM optimizer.

Fig. 4 The 25 classes images present in our dataset

4.3 Implementation of Listen Feature

This feature takes video as input. This video is divided into frames and then each frame will undergo pre-processing steps like resizing into 256 × 256 and this pre-processed frame will be given as input to the trained CNN model, CNN model gives a

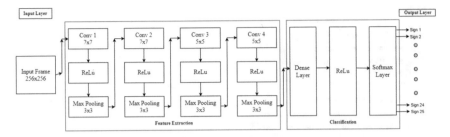

Fig. 5 CNN architecture

character as output. This character is stored in a list, as each gesture may be captured in multiple frames for the elimination of duplicate characters by comparing CNN generated output character with previously generated output character and if both are different it will be stored. After processing the whole video, an array of characters stored is converted to a string and given as input to spellchecker API which predicts the most possible replacement for a given string, this predicted string will be returned as output (Fig. 6).

```
SET spell to CALL SpellChecker
SET classifier to pretrained_model
class_labels : =['A', 'B', 'C', 'D', 'E', 'F', 'G', 'H', 'I', 'K', 'L', 'M', 'N', 'O', 'P', 'Q', 'R', 'S', 'T', 'U', 'V', 'W', 'X', 'Y',' ']
FUNCTION : main_func
      PASS IN      : spell , classifier , class_labels , video
            ans : = " "
            FOR each frame IN video:
                  img : = resize frame into (256,256)
                  img_array : = convert img into numpy array
                  Input_array :  = expand array dimension(i.e, convert 3D - 4D)
                  predictions : = classifier.predict(input_array)[0]
                  max_arg : = get argmax from predictions
                  predicted_char : = class_labels[max_arg]
                  IF last element in ans ≠ predicted_char THEN
                        ans : = ans+predicted_char
            words : = ans.split(" ")
            output : = ""
            FOR each word IN words:
                  corrected_word :  = spell.correction(word)
                  output : = output + corrected_word
                  output : = output + " "
            PASS OUT : output
END FUNCTION
```

Fig. 6 Listen feature pseudocode

```
FUNCTION : gifHelper
PASS IN      : input_text, gifs_folder
  filenames : = ['Images/start.jpg']
  input_text : = convert input_text to uppercase
  IF input_text exist in gifs_folder THEN
    PASS OUT gif
  Else THEN
    FOR each character IN input_text:
      uni_code := get unicode value of character
      IF uni_code is any alphabet or space THEN:
        append path of image having character gesture into filenames
    CALL generateGif PASS IN  filenames, "gifs/"+input_text+".gif"
    PASS OUT gif saved in gifs_folder
END FUNCTION

FUNCTION : generateGif
PASS IN      : filenames , gif_name
  images := assign a empty list
  FOR each file IN filenames:
    CALL imread from PACKAGE imageio PASS IN file THEN SET to image_read
    append image_read to list of images
  CALL mimsave from PACKAGE imageio PASS IN gif_name, images ,duration as 1 THEN SET to gifOutput
  PASS OUT gifOutput
END FUNCTION
```

Fig. 7 Speak feature pseudocode

4.4   Implementation of Speak Feature

This feature takes text as input. A folder is created which contains sign images and another folder which stores gifs that were generated previously. When a text is given as input it is converted it into the upper case then started searching for the file in a folder of gifs whether it exists or not, if it exists then send directly to the user else start considering alphabets in the given text, get images of the sign and store them in the same sequence order and convert it to gif, save it with the name of the text and send it to the user (Fig. 7).

4.5   Implementation as Microservices

Two features are implemented as microservices. As these microservices can be converted into a form of container and these containers can be managed by orchestration to provide continuous availability. As servers have larger computational power there is no issue of application getting crashed. Developers can connect to these microservices using the HTTP protocol which supports transmission of file size up to 2 GB and has tested them by creating an application that calls these API.

5 Results

Our model has been tested with many different test cases in both ways that are generating text from the video and generating GIF from the text and some of the test cases are shown below.

Figure 8 shows the video with "WHAT" string is given as input to the model and the model predicted it as "WHAT" and Spell Checker also predicted it as "WHAT". Figure 9 shows the video with "TEXT" string given as input to the model and the model predicted it as "TDXT" whereas Spell Checker predicted it as "TEXT". Figure 10 shows the video with "THUMB" string given as input to the model and the model predicted it as "THUAB" whereas Spell Checker predicted it as "THUMB". So, the advantage of using spell checker API is shown in both Figs. 9 and 10.

Figure 11 is the example test case of generating GIF from the given string. A string "TESTING" have been given as input and got the GIF which is showing the exact string as output. The generated GIF screenshots were taken and represented as Fig. 11.

Figure 12 shows the accuracy of our proposed methodology. For 30 epochs, the CNN model accuracy is 95.23% and Validation accuracy is 91.33%. Figure 13 shows the relationship between the number of epochs and model loss, as seen, a model loss is decreasing if the number of epochs is increased. If epochs are increased to more in number, then Neural Network starts to overfit which means it has started to learn from static noise data and which leads to a decrease in real-time accuracy and producing wrong results. And if the number of epochs is decreased to less than 20 it started to underfit and produces an accuracy of less than 80% based on the epochs used. For example, from Fig. 12, it is clearly understood that for 5 epochs accuracy is 12% and for 10 epochs it is 55% and for 15 epochs it is 75% approximately.

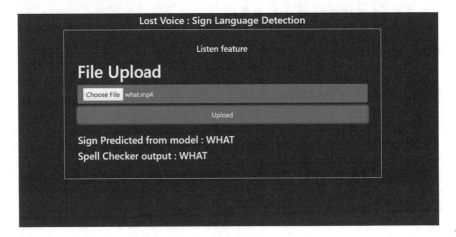

Fig. 8 Test case 1

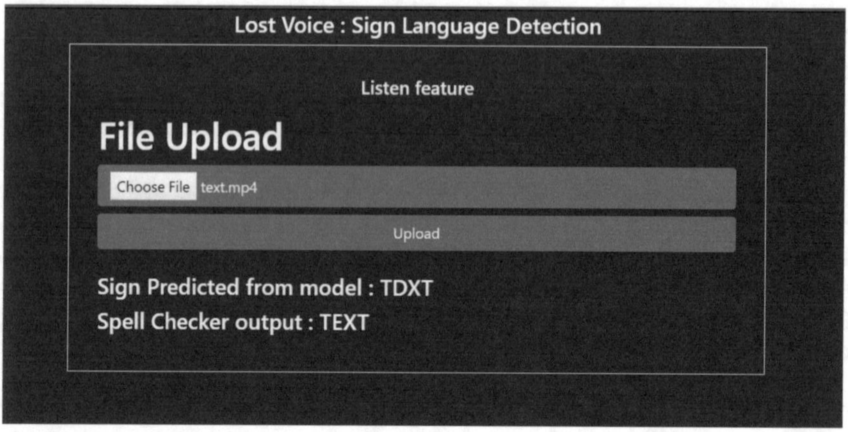

Fig. 9 Test case 2

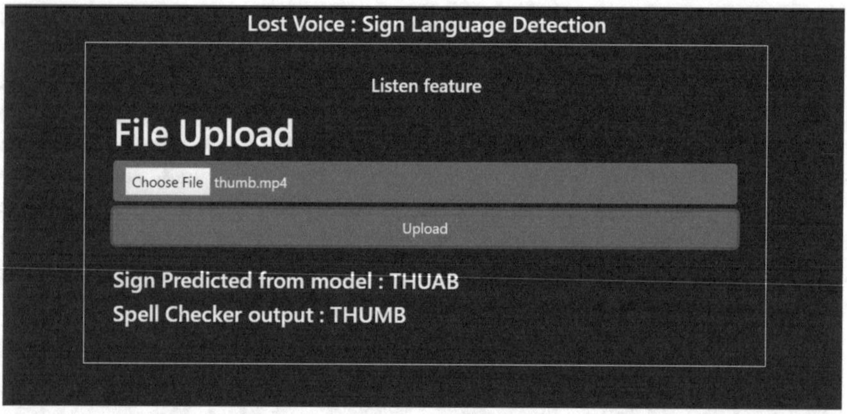

Fig. 10 Test case 3

6 Conclusion and Future Scope

In this paper, the hand sign language recognition system of American sign language using convolutional neural network have been discussed. Our proposed methodology and architecture for the neural network achieved an accuracy of 95.23% and validation achieved an accuracy of 91.33%. The microservices have been developed in which convolutional neural network was placed, and then these microservices were connected by front-end services and tested in the real world, where 18–20 signs were able to be recognized successfully out of 25 signs. Accuracy can be increased as the dataset is being increased by adding high-resolution images, and our future scope is to detect the remaining signs accurately.

Fig. 11 Test case 4

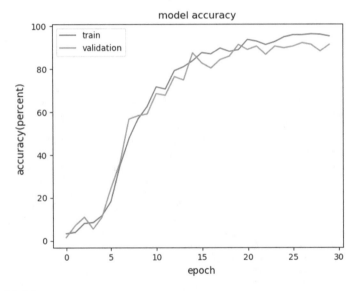

Fig. 12 Model accuracy

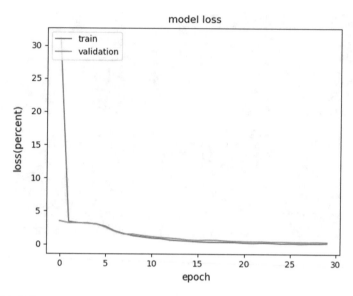

Fig. 13 Model loss

References

1. Nikam AS, Ambekar AG (2016) Sign language recognition using image based hand gesture recognition techniques. In: 2016 online international conference on green engineering and technologies (IC-GET)
2. Kumar A, Thankachan K, Dominic MM (2016) Sign language recognition. In: 3rd InCI conference on recent advances in information technology, RAIT-2016
3. Ananth Rao G, Kishore PVV (2016) Sign language recognition system simulated for video captured with smart phone front camera. In: Int J Electr Comput Eng 6(5):2176
4. Anantha Rao G, Kishore PVV, Anil Kumar D, Sastry ASCS (2017) Neural network classifier for continuous sign language recognition with selfie video. Far East J Electr Commun 17(1):49
5. Parton BS (2005) Sign language recognition and translation: a multi disciplined approach from the field of artificial intelligence. J Deaf Stud Deaf Educ 11(1):94–101
6. Chonbodeechalermroong A (2015) Dynamic contour matching for hand gesture recognition from monocular image. In: IEEE 12th international joint conference on computer science and software engineering. pp 47–51
7. Ramirez-Giraldo D (2012) Kernel based hand gesture recognition using kinect sensor. In: Image, signal processing, and artificial vision (STSJVA), 2012 XVII Symposium of IEEE
8. Zhu C, Sheng W (2011) Wearable sensor-based hand gesture and daily activity recognition for robot-assisted living. In: Part A: Systems and Humans, IEEE, Transactions on Systems, Man and Cybernetics, pp 569–573
9. Camastra F, De Felice D (2013) L VQ-based hand gesture recognition using a data glove. In: Neural nets and surroundings. Springer, Bertin, pp 159–168
10. Verma VK, Srivastava S, Kumar N (2015) A comprehensive review on automation of Indian sign language. In: IEEE international conference on advances in computer engineering and applications, pp 138–142

11. Lang S, Marco B, Raul R (2011) Sign language recognition using Kinect. In: Rutkowski LK, Marcin, Scherer RT, Zadeh R, Lotji Zurada J (eds) Springer, Berlin, pp 394–402
12. Singha J, Das K (2013) Indian sign language recognition using Eigen value weighted euclidean distance based classification technique. Int J Adv Comput Sci Appl 4(2)

An Optimization Based deep LSTM Predictive Analysis for Decision Making in Cricket

Arjun Nelikanti, G. Venkata Rami Reddy, and G. Karuna

Abstract This paper focuses on developing a system that provides support to the on-field umpire in cricket in order to make decisions like leg before wicket using two cameras placed at on-field umpire's position so that the cameras can capture video where the batsman attempting to play the ball. Initially, these videos are processed to obtain the ball movement data using background subtraction and spatial tracking. A new technique named as Spider-Squirrel Optimization-based Deep Long Short-Term Memory (SSO-based deep LSTM) is proposed to perform the path prediction after the batsman intercepts the ball. Finally, the result of the prediction path obtained using the first camera video and the second camera video are considered to analyze the leg-before-wicket event using the predictive confidence-based decision. The effectiveness of the proposed SSO-based deep LSTM is computed which revealed Mean Square Error (MSE) of 1.107.

Keywords Background Subtraction · Spatial Tracking · deep Long Short-Term Memory · Spider-Squirrel optimization · Mean Square Error

1 Introduction

Intelligent video surveillance (IVS) is broadly utilized in accident detection, patient monitoring, traffic control, and public security [1, 2]. Object tracking becomes very popular in computer vision and surveillance-related tasks. The tracker is utilized to search the continuous labels of moving objects over time. Trajectory path detection is

A. Nelikanti (✉)
Research Scholar, CSE, JNTUH, Hyderabad, India
e-mail: anelikanti@gmail.com

G. Venkata Rami Reddy
Professor and Additional Controller of Examinations, JNTUH (SIT), Hyderabad, India
e-mail: gvr_reddi@jntuh.ac.in

G. Karuna
Professor, CSE, GRIET, Hyderabad, India
e-mail: karunavenkatg@gmail.com

© The Author(s), under exclusive license to Springer Nature Singapore Pte Ltd. 2021
J. S. Raj et al. (eds.), *Innovative Data Communication Technologies and Application*,
Lecture Notes on Data Engineering and Communications Technologies 59,
https://doi.org/10.1007/978-981-15-9651-3_59

utilized in traffic management, automatic visual surveillance, sports video analysis, suspicious activity detection, and so on. The Mean-Shift tracking algorithm (MS) is utilized for selecting the probability distribution of the pixels in the target region as a tracking feature with low complexity and high execution efficiency [3]. In the Bayesian estimation theory, the KF has a good prediction probability of location in the next moment and fast convergence speed. The LSTM model [4, 5] associates LSTMs with the social pooling layer and predicts trajectory as the sequence generation.

In cricket, the high expense and number of technical requirements of tracking technologies restricted their usage in any matches, training academies, and competitions other than the matches played at an international level. For these applications, computer vision seems to be the best option, but there are very few computer vision algorithms and systems that operate at a low cost to enhance the experience of cricket.

One such high cost and infrastructure involved system in cricket analysis is Hawk-Eye [6] where six synchronized cameras are fixed around the ground field, as shown in Fig. 1. These cameras track the ball's moment while the bowler releases the ball from hand until the required location on the field. Two sets of three cameras are gen-locked which makes usage of six cameras. The Hawk-Eye system makes images into a 3D image which then calculates where the ball pitched, interception of batsman's pad is done manually, and the lateral movement is extended further off/beyond the wickets.

The third umpire with a set of many assisted technical persons will make use of this for making LBW event decisions with computers making processing of the video. Despite being successfully illustrated on TV, it has some problems: visibility, accuracy, and calibration of cameras, cost, robust communication, speed of computation, replacement in the case of a sudden fault, etc.

Fig. 1 Hawk-eye setup

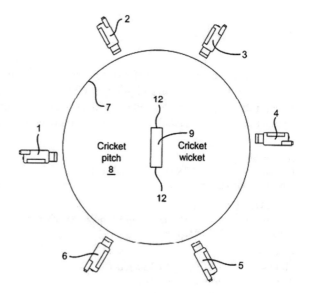

This research is focused to design the trajectory prediction by proposing an SSO-based deep LSTM classifier using two cameras. The proposed approach involves the following phases, namely frame extraction, moving object detection, ball object identification, spatial tracking of ball object, hitting point identification, and the path prediction for camera-1 and 2.

The major contribution of this research is SSO-based deep LSTM classifier for path prediction. An effective trajectory prediction mechanism is developed using the proposed SSO-based deep LSTM for predicting the trajectory from the input video. The path prediction process is performed on the basis of the deep LSTM classifier [7] in order to predict the path effectively such that the weights of the classifier are trained using the SSO algorithm.

2 Motivation

In years after the change in cricket format, new methods are included and umpires have more weight to carry apart from making decisions, i.e., carrying a camera fixed on their head cap. The purpose of using this is limited to provide entertainment given to the viewers of a cricket match. The idea of utilizing the information gathered from this facility motivated to carry out the research work to analyze the LBW event which also addresses a few drawbacks of Hawk-Eye, i.e., cost, time, and manpower.

3 Related Work

Arif Ahmed et al. [8] modeled a fusion-based approach for detecting the trajectory by selecting the trajectory which is accurate from the given trajectories. The trajectories are generated independently by using two tracking algorithms, namely L1APG and CT, and fused them using an unsupervised classification method to select a better trajectory.

Hu et al. [9] developed a video target tracking approach using the prediction and re-matching strategy under occlusion conditions. Initially, the mean-shift algorithm was combined dynamically with the Kalman filter to obtain tracking of the unoccluded target. After that, to estimate the position of the occluded target, the Kalman filter was integrated with target prior information. At last, the process of the occluded targets was re-matched using normalized cross-correlation for obtaining the optimal position of the target accurately and quickly.

Thomas et al. [10] presented an overview of computer vision applications for sports analysis and their challenges. The multi-camera ball tracking application provides detailed information for aiding the referee, coaching, and providing reviews to TV viewers. The applications for tracking players which are used for tactical analysis depend on semi-automated approaches.

Chalkley et al. [11] conducted a study to examine the ball flight prediction accuracy for determining LBW in cricket by using three cameras video-based decision-making tasks. Participants have viewed videos that represented the umpire's perspective and asked to estimate the final location of the ball at the stumps under temporary occlusion conditions by using a computer cursor. This study indicates that further research is required on perceptual-cognitive demands and other factors like removal of environmental influences like wind and pitch changing conditions, knowledge, and bowler's past history.

Baker [12] presented a method to determine the trajectories of cricket ball in the middle of bowler and batsman using the aerodynamic forces on cricket balls. This method uses the full trajectory equations developed to study the debris flight in severe windstorms. The presented method can calculate trajectories of all bowling speeds of different ball types and can also handle complex trajectories that involve late and reverse swing.

Mahon et al. [13] presented a study to address the factors that impact the decision making of a referee in a basketball game, where basketball officials were shown video clips and asked to detect infractions and fouls to test their ability in detection. In the proposed study, knowledge-priming and infraction-priming were given to the referees. This study explains the complexity in decision making of the referee in infraction-detection and indicates the necessity to consider key features such as positioning, time pressures, and low and high-frequency fouls while creating training and testing tools to affect processing information and decision making.

Southgate et al. [14] conducted a descriptive study to determine the accuracy of cricket umpire's ability to take leg-before-wicket decisions and to investigate whether the decision-making accuracy is affected when umpires monitor the feet of bowler's feet while delivering the ball. In the proposed study, four umpires have reported their judgments based on conditions specified in rules, i.e., observing the front foot of the bowler, observing back foot of the bowler before delivering the ball, no foot condition. To assess the accuracy of responses given by umpires, video recording aided by the use of superimposed wicket to wicket lines is utilized. The results indicate that the umpire's performance would be improved by relieving the umpires from judging no-ball deliveries.

Liang et al. [15] developed a deep learning approach for basketball detection. Here, the basketball detection model was trained using region-enabled FCN that considers the ResNet as a network backbone. Also, Soft-NMS, multi-scale training strategy, and OHEM were included for achieving higher detection accuracy. The method was faster and highly effective for basketball detection but failed to detect the basketball that is occupied entirely by the players.

Yoon et al. [16] developed an approach for recognizing the basketball players and their interaction with the athletes. Also, the Yolo was used for detecting and classifying the objects. However, the player tracking approach was introduced for performing better under a poor condition in which the camera angle changes and shifts dynamically.

Hui [17] presented the mean-shift algorithm for motion video tracking in the sports. Initially, the prediction method was introduced for locating the target location, and then the iterative calculation was made on the mean-shift algorithm for determining the true location of target. The method guarantees real-time tracking and reduces the computational complexity and time consumption, but the track objects with similar background color were not considered.

Kong et al. [18] developed a joint framework for action recognition and tracking of athletes. Here, the scaling and occlusion robust tracker, named SORT, was introduced for localizing the location of a specific athlete in every frame. Also, the scale refinement was achieved based on edge box (EB), and the occlusion recovery was performed using the candidate obstruction. Moreover, LRRCN was established to obtain tracking results.

Guan and Li [19] presented the Kalman filter for moving target tracking and the trajectory generation in the sports videos. Initially, the location and ten ID information on objects blob were received through image segmentation. Then, the Kalman filter and the centroid tracking algorithm were introduced for improving tracking precision.

Wu et al. [20] presented an approach for the pedestrian trajectory prediction. This framework used the encoder-decoder framework based on BiRNN. The major problem of integrating social interactions was addressed by BiRNN.

4 Proposed Methodology

4.1 System Description

The general system of our multi-camera tracking is illustrated in Fig. 2.

This section elaborates on the proposed method for trajectory prediction using the proposed SSO-based deep LSTM. Two cameras are located as specified in Fig. 3c which captures the ball moving toward the batsman. Initially, the video capturing the hit-ball by batsman or batsman body is taken from camera-1 (front view), and the selected video is subjected to the frame extraction phase, and then the extracting frames are given to the moving object detection phase.

Here, the detection of moving objects is performed based on background subtraction. After detecting the moving objects, the ball objects are identified based on shape, size, and color. Once the ball objects are identified, the spatial tracking of ball objects is performed. Finally, the path prediction is performed using deep LSTM, which is trained by the proposed optimization algorithm, named SSO. The proposed algorithm is inspired by two metaheuristic swarm intelligence algorithms, namely SMO [21] and SSA [22] which aims to solve optimization problems. Thus, the predicted path from camera-1 is obtained. On the other hand, the same steps are followed for the video captured by camera-2 (side view) to obtain the prediction path. The results of the prediction path obtained by the first video and second video are fed to the predictive confidence-based decision to obtain the final output.

Fig. 2 Overview of the
proposed system

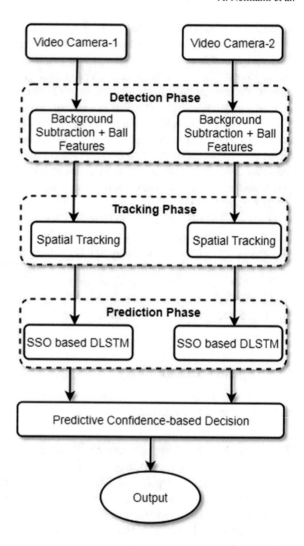

4.2 Object Detection Using Background Subtraction Model and Ball Features

Once the keyframes are extracted from the input video, the extracted keyframe is given to the object detection phase which uses background subtraction (to detect the moving objects). It is also utilized for localizing and obtaining the centroid of connected pixels moving on the foreground. This approach aims to detect the objects moving by calculating the difference between the present frame and the background image or the background model. The output obtained from the moving object detection is a batsman with bat and ball. After the detection of the moving

Fig. 3 a, b Cameras used, **c** field view with cameras alignment and **d** lenses and tripod used

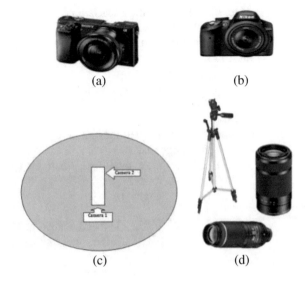

(a)   (b)

(c)   (d)

object, the ball object is identified based on size, shape, and, color. This step is very necessary for finding the ball object to predict the path effectively.

4.3 Object Tracking

This section presents the spatial tracking of the ball object. Here, the features of the object are taken to determine the location of the ball in the initial frame where the object appears. Based on the direction of spatial coordinates, i.e., (x, y) from the initial frame, the object is tracked in the next frame. Again, the location of the same object is tracked in the consecutive frame, and so on until the required keyframe. Now, these tracked locations are used to identify pitching and impact of the ball using the below Algorithm 1.

Algorithm 1: Sequential Decisions
Input: Ball Locations
Output: Pitching and impact locations
for 1 to size(ball locations)
    if change in Y
        pitching location
    end if
end for
for pitch_loc to size(ball locations)
    if change in Y
        after_pitching(k)=ball_loc (pitch_loc)
        k=k+1
        count=1
    else
        if count is 0 and change in X
            after_pitching(m)= ball_loc (pitch_loc)
            j=j+1
            m=m+1
        else
            break
        end if
    end if
    impactlocation=after_pitching(end)
end for

Algorithm 1 takes all the ball locations in terms of spatial coordinates (x, y) from the video and finds the first deflection, i.e., change in y's value (increasing to decreasing). Now, this deflection is considered pitching as the ball hits the ground and bounces. Now the impact location is found from the remaining ball locations by finding the next deflection of either or both x's and y's coordinates of ball location. This deflection location is considered as impact location. For the ball bowled full toss to the batsmen, the first deflection is considered as an impact location.

4.4 Path Prediction and Analysis

After identifying the impact point based on algorithm 1, the deep LSTM [7] classifier is utilized to predict the path. The conventional deep learning classifiers using sigmoidal activation units suffered from vanishing gradient problem. This may lead to information loss with time because of decaying gradient values through layers. This problem effectively utilized multiplicative input, forget gates, and output gates for preserving the state information. As deep LSTM holds several benefits than the other classifiers, it is highly effective for achieving path prediction using the memory cell of the classifier. Here, the memory cell is utilized to store the state information, and it acts as an accumulator.

Training of deep LSTM using spider-squirrel optimization algorithm: This section elaborates on the learning method used to estimate the weights of deep LSTM using training data. However, the objective function of the path prediction is optimized using the proposed SSO. The main aim of SSO is to train deep LSTM with the optimal weights. The SMO [21] is very efficient for training and is flexible in swarm intelligence-enabled algorithms and improved computing speed. The disadvantage of SMO is minimal convergence and is highly sensitive to the hyperparameters. On the other hand, SSA [22] is motivated by the dynamic searching behavior of squirrels, and it is an effective way for locomotion also known as gliding. The behavior of squirrels is formulated mathematically considering the features of food search. The SSA attains global optimal solutions with enhanced convergence behavior. The effectiveness of SSA is more precise and consistent and provides effective solutions for real-time issues. Here, the drawbacks of SMO are resolved by SSA that provides an optimal convergence rate. Thus, the integration of SSA and SMO enhances the overall system working of the algorithm.

The equation to update the position in SMO is as follows:

$$Y_{hp}^{u+1} = Y_{hp}^u + N(0, 1) \times \left(F_p - Y_{hp}^u\right) + N(-1, 1) \times \left(Y_{tp}^u - Y_{hp}^u\right) \tag{1}$$

The SSA [22] is employed for addressing the real-world issues, as they avoid local optima exploring the search space by exploring the global optimum. The SSA algorithm is highly efficient and offers better performance while evaluating the solutions, and there is less parameter needed for fine-tuning, and the SSA algorithm is employed due to its simpler algorithmic structure. As per the SSA algorithm, the solution update is expressed as,

$$Y_{hp}^{u+1} = Y_{hp}^u + O_k \times G_l\left(Y_g^u - Y_{hp}^u\right) \tag{2}$$

$$Y_{hp}^{u+1} = Y_{hp}^u + O_k G_l Y_g^u - O_k G_l Y_{hp}^u \tag{3}$$

$$Y_g^u = \frac{Y_{hp}^{u+1} - Y_{hp}^u + O_k G_l Y_{hp}^u}{O_k G_l} \tag{4}$$

After substituting Eq. (4) in Eq. (1), the obtained equation is given as,

$$Y_{hp}^{u+1} = Y_{hp}^u + N(0, 1) \times \left(\frac{Y_{hp}^{u+1} - Y_{hp}^u + O_k G_l Y_{hp}^u}{O_k G_l} - Y_{hp}^u\right)$$
$$+ N(-1, 1) \times \left(Y_{tp}^u - Y_{hp}^u\right) \tag{5}$$

$$Y_{hp}^{u+1} = Y_{hp}^u + \frac{N(0, 1)Y_{hp}^{u+1}}{O_k G_l} + N(0, 1) \times \left(\frac{O_k G_l Y_{hp}^u - Y_{hp}^u}{O_k G_l} - Y_{hp}^u\right)$$
$$+ N(-1, 1) \times \left(Y_{tp}^u - Y_{hp}^u\right) \tag{6}$$

$$Y_{hp}^{u+1} - \frac{N(0, 1)Y_{hp}^{u+1}}{O_k G_l} = Y_{hp}^u + N(0, 1) \times \left(\frac{O_k G_l Y_{hp}^u - Y_{hp}^u}{O_k G_l} - Y_{hp}^u \right)$$
$$+ N(-1, 1) \times \left(Y_{tp}^u - Y_{hp}^u \right) \qquad (7)$$

$$Y_{hp}^{u+1} \left(1 - \frac{N(0, 1)}{O_k G_l} \right) = Y_{hp}^u + N(0, 1) \times \left(\frac{O_k G_l Y_{hp}^u - Y_{hp}^u}{O_k G_l} - Y_{hp}^u \right)$$
$$+ N(-1, 1) \times \left(Y_{tp}^u - Y_{hp}^u \right) \qquad (8)$$

$$Y_{hp}^{u+1} \left(\frac{O_k G_l - N(0, 1)}{O_k G_l} \right) = Y_{hp}^u + N(0, 1) \times \left(\frac{O_k G_l Y_{hp}^u - Y_{hp}^u}{O_k G_l} - Y_{hp}^u \right)$$
$$+ N(-1, 1) \times \left(Y_{tp}^u - Y_{hp}^u \right) \qquad (9)$$

Thus, the final update equation of the proposed SSO algorithm is expressed as,

$$Y_{hp}^{u+1} = \frac{O_k G_l}{O_k G_l - N(0, 1)}$$
$$\left[Y_{hp}^u + N(0, 1) \times \left(\frac{O_k G_l Y_{hp}^u - Y_{hp}^u}{O_k G_l} - Y_{hp}^u \right) + N(-1, 1) \times \left(Y_{tp}^u - Y_{hp}^u \right) \right] \quad (10)$$

where,

Y_g^u: best location

Y_{hp}^u: current location

Y_{tp}^u: random location

F_p: mean of ball locations

Y_{hp}^{u+1}: next location

O_k: random gliding distance

G_l: Gliding constant and

N: random number.

By integrating the optimal features of SSA with SMO, the performance of path prediction can be increased with the global optimum solution. The output obtained from the proposed SSO-based deep LSTM from the input video is the predicted path.

The steps involved in the algorithm of proposed SSO are as follows:

Algorithm 2: Spider Squirrel Optimization(SSO)
Input: Gradients, Input data, response, epochs
Output: Updated Gradients
Initialize Population, Gliding constant (Gl), Gliding distance (Ok)
while Termination criteria is not satisfied
        for each gradient element
            for 1 to Max_pop
                Updated weights are calculated using eq (10)
                Calculate optimisation_loss
            end for
            if Isoptimisation_loss minimum
                Updated Gradients = Updated weights
            end if
        end for
end while

5 Implementation

The implementation of the developed method is done in MATLAB tool using a PC with the Windows 10 OS, 8 GB RAM, and Intel i3 core processor.

5.1 Camera Specifications and Position

We have used two DSLR cameras as shown in Fig. 3a, b using a standard tripod with the lenses shown in Fig. 3d.

The first camera used is Nikon D3200 shown in Fig. 3b, and size of the captured image is 1280 × 720 at the frame rate 60fps and used focal lengths—18–55 mm. The other is Sony Alpha a6000 shown in Fig. 3a which has 1920 × 1080 video resolution, and the number of frames per second or capture rate is 60fps and used focal lengths—55 to 210 mm. These two cameras are placed in two different positions to get the play of batsmen, i.e., side view and front view. The cameras are placed approximately 1 mtr high from the ground level and 22 yards apart as shown in Fig. 3c.

5.2 Data Collection

As far as we are aware, there does not exist off-the-shelf datasets for evaluating LBW decisions. Therefore we have to build a new dataset for this work. We designed our

Table 1 Dataset details

	Front view	Side view
Frame (pixels)	1280 × 720	1920 × 1080
Capture rate	60 fps	60 fps
Number of videos	80	80
Key features	Pitching and impact	Bounce and height

own cricket LBW situation videos. The details are shown in Table 1. We have captured around 80 videos at 60 fps in both views.

The object in our proposed datasets is a cricket ball with the specifications mentioned in [23]. The speed of the ball in the motion is approx 35–60 m/s used in taking the samples.

5.3 LBW Event Analysis

According to the cricket law 36 [24], the LBW out decision is analyzed by a sequence of events, i.e., pitching, impact, and hitting of wickets. Using the proposed methodology which followed the cricket law 36, the LBW event analysis is carried out in the following sequence.

Cricket ball detection and tracking: From two different views, we have successfully detected and tracked the cricket ball using background subtraction and spatial tracking for finding pitching and impact of the ball.

Here, front view video is considered for finding the pitching of the ball using algorithm 1, and the result is shown in Fig. 4a. If the pitching of the ball meets the condition, then we further analyze the video to find the impact of the ball using the same algorithm 1, and the result is shown in Fig. 4b.

Ball path prediction: After the sequential events, i.e., pitching and impact of the ball meet the conditions, then the path trajectory is estimated which is used to find whether or not the ball hits the wicket after the ball is intercepted by the batsman. Figure 5 shows the predicted path in both the views using the proposed method.

(a)   (b)

Fig. 4 **a** Pitching and **b** Impact

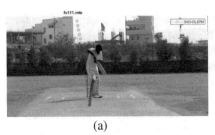
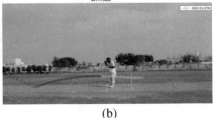

(a)   (b)

Fig. 5 Path prediction in **a** front view and **b** side view

6 Results and Discussions

In this section, we discuss the event analysis results. As pitching and impact can be known in front view video, so the person at the front view camera acted as the umpire to make decisions about pitching and impact which are recorded manually by him.

The correctness means the ratio of similar decisions made by the umpire and proposed method to the sum of similar and conflicting decisions made by the umpire and proposed method.

$$\text{Correctness} = \frac{\#\text{of similar decisions}}{\#\text{of similar} + \#\text{of conflicting decisions}} \tag{11}$$

Table 2 shows the decisions of pitching and impact by the umpire are 90.6 and 87.5% accurate and by the proposed method are 97.3 and 97.5% accurate.

After the decisions of pitching and impact, the decision of hitting wickets is taken by the two umpires positioned near the cameras on the field in both the views. The side view predicted path gives the height and movement of the ball beyond the impact point. Table 3 shows the result of hitting stumps decision by umpires and the proposed method in both views.

Table 2 Pitching and impact decisions in front view

Total	GT	Pitching decision		GT	Impact decision	
		Umpire	Proposed method		Umpire	Proposed method
80	75	68	73	80	70	78

Table 3 Hitting decision

View	Total	Hitting decision		
		Umpire	Proposed method	Correctness (%)
Front	80	63	69	91.3
Side	80	66	72	91.6

From the total videos, the correctness of the decision taken by umpires obtained is 91.3% in front and 91.6% in side views using Eq. (11).

In Fig. 6, the path predictions to find the hitting of wickets are shown which are obtained by running different algorithms like Kalman filter, DLSTM, and SSO-based DLSTM.

The comparative analysis of trajectory prediction using the MSE parameter of different methods is depicted in Fig. 7. If the path predicted is accurate, then the decisions made based on ball locations can be accurate. When the training data percentage is 40, the MSE computed by Kalman filter, deep LSTM, and the proposed model is 4.645, 3.434, and 2.142, respectively. For 80% training data, the MSE

(a)   (b)

Fig. 6 Comparing with other methods, **a** front view and **b** side view

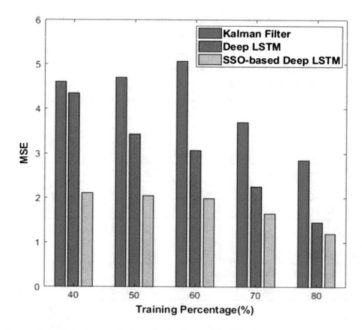

Fig. 7 Comparing with other methods, **a** front view and **b** side view

Table 4 Out decision

View	Total	Final decision (out or not out)		
		Umpire	Proposed method	Correctness (%)
Front	80	63	69	91.3
Side	80	66	72	91.6

computed by Kalman filter, deep LSTM, and the proposed model is 2.817, 1.317, and 1.107, respectively.

6.1 Analysis of Decision

After collecting the decisions from both the umpires and the proposed method, this decision data is filtered and analyzed by using predictive confidence-based decision matrix. This matrix is used to find out the correctness of the umpire using Eq. (11). The decision matrix uses simple if-else logic applied on data collected.

Table 4 shows final decisions taken by the umpire and proposed method about the LBW event whether the batsman is given out or not out. These decisions are taken on all videos in both the views, we have a total of 80 videos and out of which, 63 and 66 are correct decisions made by umpire in front and side views, respectively.

The experiment is carried out with the following limitation:

- As in updated cricket guidelines, we have red, white, and pink color balls, but we have experimented with red color balls only.

7 Conclusion

Mainly this paper addresses key issues of LBW decision making for umpires on the field. The straight umpire (front view) can take decisions about pitching and impact using spatial tracking of the ball and leg umpire (side view) can analyze the bounce and hitting of wickets. For this, an effective path prediction method named SSO-based deep LSTM, which aims to predict the trajectory path from the video frames, is proposed. The proposed method gives MSE value of 1.107 when compared with other methods, i.e., Kalman filter and deep LSTM. This work provides performance statistics about batsmen, bowler, and umpire. The results have shown that video visualization can provide cricket coaching with visually measurable and comparable summary records and is thus an economical means for evaluating skill levels and examining progress objectively and consistently. In the future, the system can be enhanced to automatically comment on the event occurred and generate sports news from live text commentary scripts.

Acknowledgements Authors thank Mr. Rameshwar, Mr. Vinay, Mr. Suheb, Mr. Javith, and Mr. Bharath for their help in database generation work.

References

1. Tripathi RK, Jalal AS, Agrawal SC (2018) Suspicious human activity recognition: a review. Artif Intell Rev 50(2):283–339
2. Yu H, Wang J, Sun X (2019) Surveillance video online prediction using multilayer ELM with object principal trajectory. In: Signal, image and video processing, pp 1–9
3. Comaniciu D, Meer P (2002) Mean shift: a robust approach toward feature space analysis. IEEE Trans Pattern Anal Mach Intell 5:603–619
4. Alahi A, Goel K, Ramanathan V, Robicquet A, Fei-Fei L, Savarese S (2016) Social lstm: human trajectory prediction in crowded spaces. In: Proceedings of the IEEE conference on computer vision and pattern recognition, pp 961–971
5. Pei Z, Qi X, Zhang Y, Ma M, Yang YH (2019) Human trajectory prediction in crowded scene using social-affinity Long Short-Term Memory. Pattern Recogn 93:273–282
6. https://www.hawkeyeinnovations.co.uk/. Accessed 06 June 2020
7. Chauhan S, Vig L (2015) Anomaly detection in ECG time signals via deep long short-term memory networks. In: Proceedings of IEEE international conference on data science and advanced analytics, pp 1–7
8. Ahmed SA, Dogra DP, Kar S, Roy PP (2018) Unsupervised classification of erroneous video object trajectories. Soft Comput 22(14):4703–4721
9. Hu ZT, Zhou L, Yang YN, Liu XX, Jin Y (2018) Anti-occlusion tracking algorithm of video target based on prediction and re-matching strategy. J Vis Commun Image Represent 57:176–182
10. Thomas G, Gade R, Moeslund TB, Carr P, Hilton A (2017) Computer vision for sports: current applications and research topics. Comput Vis Image Understand 159:3–18. ISSN 1077–3142. doi: https://doi.org/10.1016/j.cviu.2017.04.011
11. Chauhan, Chalkley, Daniel et al (2013) Predicting ball flight in cricket from an Umpire's perspective. Int J Sports Sci Coaching 8(3):445–454. doi:https://doi.org/10.1260/1747-9541.8.3.445
12. Baker J (2010) Calculation of cricket ball trajectories. Proc Inst Mech Eng Part C—J Mech Eng Sci 1:1–12. https://doi.org/10.1243/09544062JMES1973
13. Mac Mahon C et al (2007) Referee decision making in a video-based infraction detection task: application and training considerations. Int J Sports Sci Coach 2(3):257–265. doi: https://doi.org/10.1260/174795407782233164
14. Southgate David C, Neil B, Lyndall K (2008) The effect of three different visual monitoring strategies on the accuracy of leg before wicket decisions by cricket umpires. Clin Exp Optometr 91:385–393
15. Liang Q, Mei L, Wu W, Sun W, Wang Y, Zhang D (2019) Automatic basketball detection in sport video based on R-FCN and Soft-NMS. In Proceedings of the 2019 4th international conference on automation, control and robotics engineering, ACM, p 9
16. Yoon Y, Hwang H, Choi Y, Joo M, Oh H, Park I, Lee KH, Hwang JH (2019) Analyzing basketball movements and pass relationships using realtime object tracking techniques based on deep learning. IEEE Access 7:56564–56576
17. Hui Q (2019) Motion video tracking technology in sports training based on Mean-Shift algorithm. J Supercomput 1–17
18. Kong L, Huang D, Qin J, Wang Y (2019) A joint framework for athlete tracking and action recognition in sports videos. IEEE transactions on circuits and systems for video technology, 2019

19. Guan S, Li X (2019) Moving target tracking algorithm and trajectory generation based on Kalman filter in sports video. J Vis Commun Image Represent 102693
20. Wu J, Woo H, Tamura Y, Moro A, Massaroli S, Yamashita A, Asama H (2019) Pedestrian trajectory prediction using BiRNN encoder–decoder framework. Adv Robot 33(18):956–969
21. Bansal JC, Sharma H, Jadon SS, Clerc M (2014) Spider monkey optimization algorithm for numerical optimization. Memetic Comput 6(1):31–47
22. Jain M, Singh V, Rani A (2019) A novel nature-inspired algorithm for optimization: squirrel search algorithm. Swarm Evolut Comput 44:148–175
23. https://www.lords.org/mcc/laws/the-bal. Accessed 06 June 2020
24. https://www.icc-cricket.com/about/cricket/rules-and-regulations/playing-conditions. Accessed 06 June 2020

Classifying Automated Programming Contracts Using TanH2 Decision Tree Classifier

S. V. Gayetri Devi and C. Nalini

Abstract The process of drafting software coding contracts with only a slight inter-mediation by the developers remains pivotal in systematically coming up with optimized assertions (contracts) rendering extensive and enriched testing of software modules. The framework put forward begins with shaping up of the behavior and structural relevant dependency specifications as conditions or restraints on a decision tree and thereafter materializes them into automated coding contracts. Contract refining nuances are hence adopted in order to leverage on enhanced feasible assertions. This is exercised by customized K-means clustering algorithm to build accurate clusters of contracts analogous to each other. The proposed classification of isolated contracts is then made by adapting activation function of neural networks to compute entropy during decision tree classifier construction as a TanH2 entropy classifier tree method with improved memory and computation relevant performances and adequate effectiveness of faults exposure ability.

Keywords Contracts classification · K-TanH means clustering · Activation function · Optimization · Decision tree

1 Introduction

The operations absolutely necessary when developing software solutions integrate verification and validation. Verification strives in assuring that the software evolved is in consonance with system criteria, in seeking workarounds for development specific challenges, completeness and work as per requirements. Validation in contrast verbalizes on software comprehensiveness at every developmental position. Verification is along the lines of dynamic with static analyses. The assertions mandate on what is

S. V. Gayetri Devi (✉) · C. Nalini
Department of Computer Science and Engineering, Bharath Institute of Higher Education and Research, Chennai, India
e-mail: Gayetri.venkhatraman@gmail.com

C. Nalini
e-mail: drnalinichidambaram@gmail.com

© The Author(s), under exclusive license to Springer Nature Singapore Pte Ltd. 2021   739
J. S. Raj et al. (eds.), *Innovative Data Communication Technologies and Application*,
Lecture Notes on Data Engineering and Communications Technologies 59,
https://doi.org/10.1007/978-981-15-9651-3_60

expected from the arguments or parameters and what emerges out as resultant. In the essence of all run-time checks, the contracts figure out if the prerequisites are adequate on accessing a procedure and the postconditions are adhered on returning. Bottlenecks in either of these deviate the expected system behavior. Formal analysis tools during their usage have the contracts validated for fundamental conditions for adherence with function calls sought, and conformance of postconditions when operation is concluded.

To avoid any more difficulties while integrating several specifications and design-oriented influences because of enormous software intricacies with exponential growth of assertions, formation of a framework for contracts to be built from the code which is subjected to analysis and subsequently fetching contracts without higher degree of manual involvement by programmers is very much desired. To summarize, the contracts development technique must be improvised to handle only feasible contracts for time-bound software verifications.

2 Related Works

Tests accentuated by the directing standards of contracts are accurate in exposing the shortcomings in the software applications. Incorporating contracts in software coding is burdensome with rising number of assertions, and for this reason, one main demanding task is the intelligent choice of suitable contracts and allied test cases to pursue software testing.

Panhalkar and Doye [1] showcase about progressive approaches in decision tree classification and precise the improvement made and many methods utilized in decision tree. Methods such as clustering and height balanced based decision tree induction, support vector machine supervised classification, and evolutionary and fuzzy approaches to forming decision tree [2] are highlighted.

Adibi [3] presented a method to build a genetic algorithm-based optimal tree to choose efficient features using new operators for selection, mutation and crossover as continuous as well as discrete variables. The proposed method is compared with other established classification techniques [4] for test datasets and also real-world data showing enhanced performance of the decision tree.

Chabbouh et al. [5] projected a multiple objective evolutionary method to determine optimized oblique decision trees for imbalanced binary classification problems of data mining. The technique ODT-Θ-NSGA-III algorithm driven by the capability to identify the local optimum in the ODT search margins maximizes both recall and precision [6]. The algorithm shows better performance as against several existing classification methods on imbalanced benchmark datasets.

Pang et al. [7] showed test case on account of classification method with clustering characteristics for classifying them as detailed and inaccurate clusters to minimize cost for regression testing. The method [8] is formed with coverage details attained from earlier release(s) of software. Furthermore, the outcomes of the method are considered to evaluate coverage criteria and fault volume in the former developments.

Saettler [9] studied the issues to construct oblivious decision tree incurring maximum of k-classification faults and presented a randomized rounding algorithm with error probability of maximum $(n - 1)/2n2$ with best probable achievable logarithmic factor in the tree cost without breaching the error restraints [10]. The method is suitable for the problem's multiclass form with costs that are not uniform. Most of the prevailing assertions prioritization methods are assessed based on the efficiency but practically, this extent faces several real-world constraints [11]. For instance, various contracts might consume varied time of execution in the ranking process and the errors thereafter detected by these contracts might have diverse severity [12]. The efficacy of contracts prioritization is also interrelated with the verification time of all the contracts [13]. Bearing in mind, the objective of contracts prioritization, it is agonizing when the classification time is near to time expended on verification of all contracts. So, it is essential to assess the competence of prioritization methods allowing for their real prioritizing time and verification time [14]. The creation of contracts specifications in accordance with the classification tree technique is required to be done separately from implementation [15, 16]. This is in most cases taken place ahead of implementation phase and accomplished by somebody apart from the software programmer incurring more time being inevitable due to greater probability of faults being identified [17]. It is also formidable to estimate the effectiveness of assertions classification due to restricted testing resources. It is also vital to optimize the contracts execution because of the cost of individual contracts in place of the whole cost. Seemingly, such contract classification differs from the present classification problems, and hence, a totally different technique for this issue is required.

3 Methodology to Optimize and Classify Programming Contracts

The methodology brought forward presupposes the derivation of software code contracts in an automated arrangement followed by the application of an enhanced K-means clustering algorithm to rank the feasible and non-feasible assertions and the subsequent classification of those contracts. To derive the contracts and then have them optimized, the semantics and also the accompanying behavioral features are at first apprehended through valid static as well as dynamic analyses. These particulars are then constituted as norms in form of a decision tree to verify the test software. The criteria relevant to the system specifications are drawn up as contracts in order for the software being developed to be consistent with.

The estimated elements of the algorithm—particle swarm optimization (PSO)—are then modified to optimize the realizable contracts to be complied with. The method subsumes a congregation (also called as swarm) of programming contracts. Each destined contract is stipulated as a particle in the algorithm across the available examining boundaries. A distinct position of the code contract conveys its complete-ness and forms the lowest and the maximum degree of its compliance by the software, and this in our context is the fitness coordinates.

Next, the contracts that have similar aspects are sealed as clusters with the help of K-TanH means clustering—an improved variation of K-means clustering—and separated as feasible and infeasible contracts thus with K value set as 2 for the clus-tering work. When compared with K-means clustering, the introduction of activation function—TanH function—as modified form K-TanH clustering allows segregating contracts appropriately by accurately assessing feasibility of assertions grounded on behavioral related dependencies which are then converted into contracts. For each cluster, the centroid is calculated using the mean for all the assertions inside the relevant cluster. After this, the contracts are allocated to the reorganized centroid of the two clusters until the termination criteria are satisfied in accordance with the novelties of the test software and contracts, the complicatedness in the specifications, defect count during previous software verifications, alikeness in the faults located by the assertions, impact of the contract on the verification performance, time pertinent challenges, etc. The calculated centroids are put up as the fundamental centers of the two clusters for K-TanH means algorithm.

The enhanced entropy formula used for classification of the contracts is formu-lated by the introduction of tangent hyperbolic (activation) function while evaluating the entropy of the decision tree holding the dependencies of the classes of the source files as contracts in accordance with the specification. Entropy is an extent of uncer-tainty or information, and hence, the decision tree requires entropy to determine the homogeneity of the contract sample to partition the information as subsets with instances holding analogous values by inferring what should be added subsequently in the decision tree (Information gain) for maximizing decrease of the identified uncertainty.

$$\tan H \, E(s) = \tan H \left(\sum_{i=1}^{c} -P_i \log_2 P_i \right) \tag{1}$$

where

$E(s)$ is the entropy of the classification tree, also referred using letter 'H.'

P_i is the proportion of the contracts that belong to class c for a specific node.

TanH is tangent hyperbolic activation function used for entropy calculation.

Referring to the equation provided above, 'Pi' is basically the probability related to frequentism of a class 'i' in the work's data. There are two classes, a 'feasible' class and a 'non-feasible' class considered in the proposed work. Introducing activation function in the formula is to evaluate the degree of disorder in the contracts data at a quicker time. The calculated entropy value with hyperbolic tangent activation func-tion TanH along with Information gain is used to classify the contracts with classifier tree. The footsteps and the elementary approach for the TanH2 tree classification are illustrated in Fig. 1.

To summarize the steps of the above flow diagram, the phase after extracting the optimized contracts is the realization of clusters. Contracts that are alike are connoted as clusters. K-means clustering stands as one of the simplest techniques for clustering.

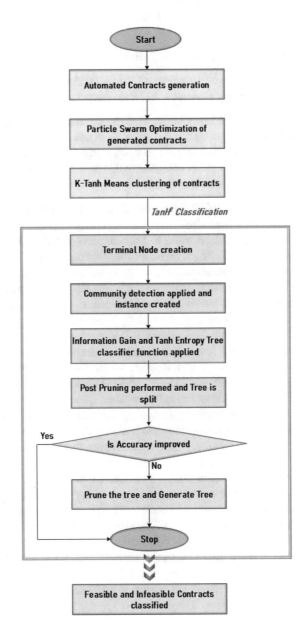

Fig. 1 Steps for TanH2 tree classification of contracts

In the naivest form, the dataset is divided into K-clusters. An objective function is then utilized to determine the eminence of partitions such that related objects fall in one cluster and unrelated objects fall in further groups.

In this technique, the centroid of one cluster is established to indicate a cluster. The centroid is actually considered as the center of the cluster that is computed as mean value of the points inside the cluster. Then, the clustering quality is assessed by estimating the Euclidean distance between the centers as well the point. This distance must be higher.

Step 1: Select a value for K wherein K denotes the count of clusters.

Step 2: Iterate every point and then allot the cluster with the most adjacent center to it. When every element is iterated, calculate the centroid of the entire clusters.

Step 3: Iterate every single element from dataset and compute Euclidean distance between the centroid and point of each and every cluster. When any point is existing in the cluster that is not most adjoining one to it, reallocate that specific point to nearest cluster. After completing this for all of the points inside the dataset, estimate again each cluster's centroid.

Step 4: Accomplish Step 3 till there exists no new allotment done between the two succeeding iterations.

In the proposed work, K-TanH means clustering, a modified form of the original K-means clustering is taken for the labeling of feasible contracts and also unfeasible ones. A contract is clustered as 'feasible,' if there exists valid behavioral dependencies (transformed later into contracts) between the methods of the source code classes based on software input specifications. A contract is grouped as 'non-feasible' or 'infeasible' if otherwise. Existing K-means clustering incorrectly clusters contracts from the source file as non-feasible ones in spite of valid behavioral dependencies in effect in the test software under verification. This can be evident in source file—contextListener.java for instance where K-means method inaccurately clusters three contracts as non-feasible even with the presence of valid dependencies. The enhanced K-TanH clustering fittingly considers these three contracts as feasible and isolates them appropriately by evaluating the contracts' feasibility on the account of those dependencies which are then adapted into contracts in due course. The incorrect isolation of assertions is one vital reason to go ahead with the proposed algorithm.

The count for clusters denoted as 'k,' in our work, is at first designated as 2, each one of which relates to infeasible and feasible assertions. The contracts are distributed to the neighboring cluster obvious based on the spacing among the one-to-one centroid as well as contracts. This positioning is assessed via Euclidean distance concerning two contracts. For every single cluster, the centroid is rationalized by computing mean of the contracts altogether in the associated cluster. Afterwards, the contracts isolation to the equivalent clusters is reiterated on the basis of the rearranged centroid of clusters till the explicit criteria for the termination are identified. These criteria are reliant on wide-ranging factors like contracts' newness as

Fig. 2 Split for
attribute—contracts

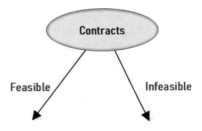

well as test modules, specifications intricacy, number of bugs known during previously performed verifications, likeness of the faults located by a contract, contract's consequence on the proficiency of verification and time pertinent constraints.

For avoiding difficulties during the fixation of centroid without manipulating the consequential clusters eminence, it has been initiated with a preliminaries stage of training—a perceptron model of neural networks to produce two basic cluster centers. The ensuing centroids are the straightforward cluster centers for the K-TanH means clustering. The TanH: Tangent hyperbolic function represents the activation function affianced in the network's hidden layer. This activation function speeds up the centroid calculation and hence tailors the K-means clustering algorithm to result in improved version named K-TanH means clustering. After the clusters are learned over the training group and the training assertions that are committed to every cluster can be reserved for training the specific classifier.

After clustering, the two clusters form the layers of the two decision trees. The attribute—'contracts' is actually split into two branches 'feasible' and 'infeasible', and the records are split at the node 'Contracts' with respect to the attribute's values, as shown in Fig. 2.

Thus, the contracts are first clustered using K-TanH means technique and then a decision tree is constructed for feasible and infeasible clusters, one each, respectively, and the work proceeds to classifying the assertions using the implemented TanH2 decision tree classifier. Classification thus works on predicting an unknown or the target contract from the provided known contracts assuming that the known contracts lead to a precise acknowledgment of the target contract class. The result of the entropy computation using TanH function is used to classify the clusters.

4 Implementation and Experimental Results

Being a segment in the development of a multi-layered tool for verifying multi-threaded Java software, the automated contracts are at first formed followed by application of K-TanH means clustering that is supplied with the extracted assertions (contracts) and the assimilated behavior-specific dependencies fetched from the code to test resulting in refined categorized contracts in the form of clusters. Then, classification is performed using decision tree that builds its own rules for splitting for classifying using the two measures—information gain and entropy. The usage of tangent hyperbolic function TanH for entropy calculation speeds up the formation of decision tree for quicker and accurate decision making. The code is implemented in Java version 1.8 as web application with JSP and servlet source code using Eclipse as the development platform and Windows 10—operating system. The source code under test is available on github: https://github.com/nardevar/Banking. Minimal spacing within the contracts inside individual clusters and a larger space between the two clusters are crucial for efficient clustering.

Table 1 depicts the number of contracts in each source file instance and the corresponding cluster to which a specific contract falls under, i.e., feasible or infeasible cluster.

When developing and then optimizing the contracts classification model, estimating how accurately it foretells the expected result gets crucial. The Metrics for Regression are at first calculated. Root mean squared error, mean absolute error, relative absolute error and root relative squared error are the metrics to detect errors variation in the set of prediction estimates being computed. The Kappa statistic measure is taken up to measure the concurrence between two categorizations sets of a specified dataset. It measures how our contracts classification outcomes using TanH2 classifier fare well as compared to the classification using entropy.

Table 2 portrays the evaluation of performance metrics for contracts classification using existing decision tree method and proposed TanH2 tree classifier.

Accuracy evaluation delivers a medium of assessing the performance of a model and hence may aid to clarify our comprehension of (predictive) indicative models facilitating the comparison of various algorithms or methods, in our work being TanH2 classifier as against decision tree classification using entropy. It is a ratio of

Table 1 Contracts classification using TanH2 tree classifier

#	Source file	Contracts count	Cluster name
1	EmployeeServlet.java	27	Feasible
2	MainControllerServlet.java	33	Feasible
3	Now.java	3	Feasible
4	ContextListener.java	2	Feasible
5	AccountNotFoundException.java	0	Non-feasible
6	AccountsNotFoundException.java	0	Non-feasible
7	BillPaymentException.java	0	Non-feasible
8	DateException.java	0	Non-feasible
9	PayeeNotFoundException.java	0	Non-feasible
10	TransferAccountNotFoundException.java	0	Non-feasible
11	TransferException.java	0	Non-feasible
12	UserNotFoundException.java	0	Non-feasible
13	BankDB.java	8	Feasible
14	DBConnection.java	0	Non-feasible

Table 2 Estimates of classifier's predictive performance

#	Description	Existing decision tree classifier	Proposed TanH2 classifier
1	Total number of instances of source files	14	14
2	Correctly classified instances of source files	11	14
3	Percentage of correctly classified instance of source files (accuracy)	78.5714%	100%
4	Incorrectly classified instances of source files	3	3
5	Percentage of incorrectly classified instance of source files	21.4286%	0%
6	Mean absolute error	0.2143	0
7	Root mean squared error	0.4367	0
8	Relative absolute error	71.8421%	0%
9	Root relative squared error	116.9072%	0%
10	Kappa statistic	−0.1053	1
11	Performance time	165 ms	154 ms

the number of correct classification of the contracts out of all the contracts classified. It holds good to use accuracy as a metric for small count of class values, for example, 2, as in our work, also referred to as a problem pertaining to binary classification.

Accuracy thus expresses the ratio of the total number of classifications of contracts that are correct. Accuracy = (instances of true classified source files with contracts)/(total number of source files) × 100.

Though accuracy is a vital metric, it is not at all times satisfactory to measure the performance of a model and is applicable in cases of no class imbalance (the percentage of instances of the entire classes being same), a case hardly existent in any real-world scenario. Hence, it requires other metrics that are briefly explained below.

Sensitivity or recall tells about completeness as to what percentage of correctly classified contracts did the classifier label as being correctly classified. The perfect best score is 1 which is achieved by the proposed classifier method. It is a suitable metric in situations wherein false negative outplays false positive. Sensitivity = TP/actual positives.

Precision refers to the exactness as what percentage of contracts that the classifier labeled as being correctly classified are actually correct. It is a valuable metric in circumstances wherein false positive is a greater worry than false negatives. Precision = TP/predicted positives.

Table 3 Detailed per class break down of existing decision tree classifier's prediction accuracy

Class	TP rate	FP rate	Precision	Recall	F-Score	ROC area
Non-feasible	0	0.083	0	0	0	0.833
Feasible	0.917	1	0.846	0. 917	0.88	0.833
Weighted average	0.786	0.869	0.725	0.786	0.754	0.833

F-measure or F-score or F_1-score is used to identify an optimal balance of precision as well as recall by combining both the metrics and representing the Harmonic mean of the two. F-score = 2 * (Precision * Recall)/(Precision + Recall).

Receiver operating characteristic—ROC curve is a foremost visualization method to compare the performance of the assertions classification algorithms. It encapsulates the compromises between the rates of true positive as well as false positive for the classification model via various probability edges. It is a graphical scheme that exemplifies the performance of a classifier model as its estimate threshold is diverse.

Weighted average is the weighted average of any of the metrics, namely precision, recall or F1-score.

Table 3 presents the prediction accuracy measures for conventional decision tree method as given below:

Table 4 depicts the metrics for predicting accuracy for proposed TanH2 tree classifier as given below:

A confusion or error matrix is typically employed as a quantitative way to portray the accuracy of classification. It refers to a table showing correlation between the results of the classification result as well as a reference result. In our work, it is taken to ascertain the performance of the classification problem where the contracts can fall in two classes—feasible and infeasible classes of assertions/contracts. It is a NXN matrix, where N holds for the number of target classes, in our work being a 2X2 matrix for the two classes. The overall accuracy is computed by the summation of the number of correctly classified instances of source files and dividing this by the total number of files.

In the proposed work, the classifier model predicts that a contract is classified correctly, the contract may or not be correctly classified in reality, and the confusion matrix helps to comprehend the model's reliability. There are four probable ways to understand the actual and predicted values:

Table 4 Detailed per class break down of measures for accuracy prediction—proposed TanH2 tree classifier

Class	TP rate	FP rate	Precision	Recall	F-score	ROC area
Non-feasible	1	0	1	1	1	1
Feasible	1	0	1	1	1	1
Weighted average	1	0	1	1	1	1

Table 5 Confusion matrix—existing decision tree classifier

Confusion matrix	Predicted		Accuracy = 78.5714%
	Non-feasible	Feasible	
Actual	0 (TP)	2 (FN)	
	1 (FN)	11 (TP)	

Table 6 Confusion matrix—proposed TanH2 classifier

Confusion matrix	Classified as		Accuracy = 100%
Actual	Feasible	Non-feasible	
	5 (TP)	0 (FN)	
	0 (FN)	9 (TP)	

- True positives—TP: Model projected that the specific contracts are classified rightly and those contracts were actually classified properly.
- True negatives—TN: Model has predicted that particular contracts were accurately classified and those contracts were really not classified exactly.
- False positives—FP: Model has estimated that certain contracts were not appropriately classified and those contracts were in reality classified precisely.
- False negatives—FN: Model expected that contracts were not classified truly and those contracts are not in fact classified faultlessly.

To deduce accuracy of the proposed model, the model is evaluated with values—actual as well as predicted ones:

Table 5 gives confusion matrix for existing decision tree classification as given below:

Table 6 has the confusion matrix for every class (feasible as well as infeasible classes) for proposed TanH2 decision tree classifier method as:

In contrast with traditional decision tree classifier (refer to Table 2), introducing the activation function: TanH function in the tailored form—K-TanH means clustering, largely groups the contracts more accurately through proper assessment of the contracts feasibility relative to behavior pertinent dependencies being shaped eventually as contracts. The algorithm iterates until centroid change is exhibited. The next step is the classification using proposed TanH2 decision tree classifier.

The contracts that are developed in an automated way after being optimized, clustered and classified using TanH2 tree classification have 14 input files for processing for contracts to be drawn and then grouped as two clusters: cluster1—feasible as well as cluster 2—infeasible and then classified with proposed method having acquired processing time of 74 ms, memory and CPU consumption of 33.53% and 28.37%, respectively.

By analyzing the dependencies in relation to the behavior in order to isolate dependencies as the infeasible and feasible ones, the accuracy of proposed classifier method is validated to confirm whether the contracts are classified correctly. This shortens the execution time by quicker exposure to errors with fine-tuned contracts.

Confusion matrix as discussed above is taken for validation with which parameters such as sensitivity or recall, specificity (same as that of recall but considers negative cases), precision, F-measure and ROC curve (refer Tables 3 and 4 for the values obtained) can be measured. Accuracy is also computed and found to be 100% for proposed work and 78.57% for prevailing classifier tree method.

ROC area (also referred to as AUC—area under ROC curve) denotes probability that a positive instance chosen at random from the test information is being ranked beyond a negative instance selected at random, on the basis of ranking created by the classifier.

Figures 3 and 4 show the ROC curve visualization and cluster visualization of contracts, respectively, classified with existing decision tree classifier.

Figures 5 and 6 portray the ROC curve visualization and clusters visualization of contracts, respectively, having classification with proposed TanH2 classifier tree.

The ROC curve delivers distinct information about the classifier's behavior. This curve is formed by the plotting of the true positive rate as against the false positive rate at varied threshold backgrounds. The best result is that the entire positive ones are ranked above the whole negative instances, and in this case, the AUC is 1. In the

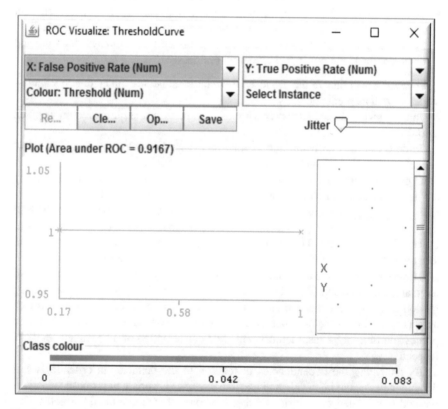

Fig. 3 ROC curve visualization of contracts with existing decision tree classification

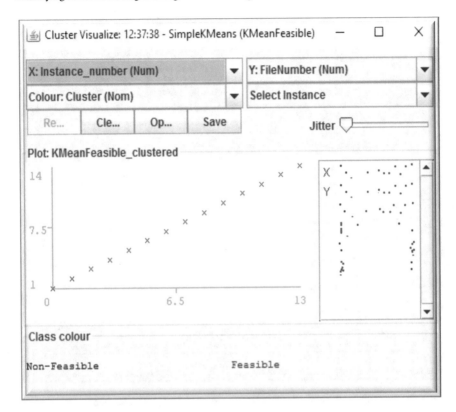

Fig. 4 Cluster visualization of contracts with existing decision tree classifier tree

worst case, it is 0. In the case where the ranking is essentially random, the AUC is 0.5, and if it is significantly less than this, the classifier has performed anti-learning. A good curve of ROC holds greater space under it due to the quicker shooting up of true positive rate to 100%. A bad curve has a very small area. A classification algorithm with perfect accuracy has an area of 1.0. From the experiments, it is evident that existing classification tree method has ROC area of 0.83 (Refer Table 3 and Fig. 3), while proposed classifier has an area of 1.0 (refer Table 4 and Fig. 5) and hence proposed model leading to better performance.

5 Conclusion and Future Work

For effective verification and validation of various software, it is very important to have vital focus on source code related dependencies analyses. This paper provides an inclusive scheme to arrive at contracts with minimal involvement of programmers. Having their derivation realized, the proposed work forms time impacted

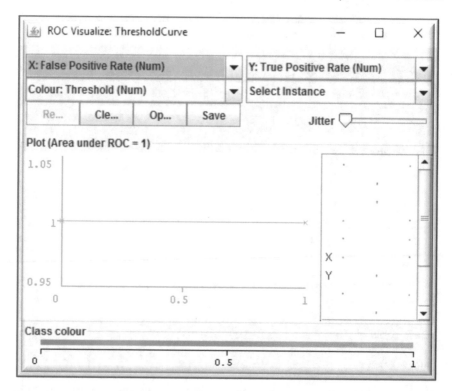

Fig. 5 ROC curve visualization of contracts with proposed TanH2 classifier tree

contracts by adequately tailored K-TanH means clustering along with TanH activation function based entropy determination for classifying the clustered contracts as an enhanced TanH2 entropy classifier. Experimental evaluation of memory, time and CPU consumption for the classification process is indicated. The scope for further research in this direction requires addressing the class imbalance problem through focus on synthetic dataset generation in order to balance the training dataset for better accuracy for different testing scenarios. Also in further lines, the usefulness of classified contracts can be validated by means of metaheuristic optimization algorithms to reinforce the software verification quality in due course.

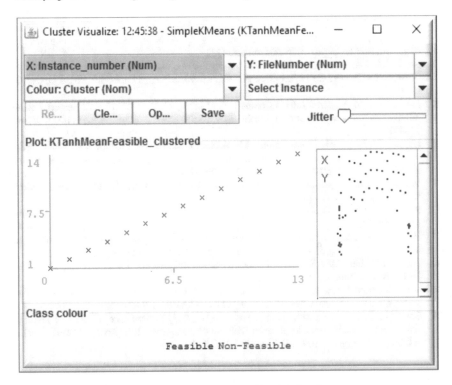

Fig. 6 Cluster visualization of contracts with proposed TanH2 classifier tree

References

1. Panhalkar A, Doye D (2017) An outlook in some aspects of hybrid decision tree classification approach: a survey. In: Satapathy S, Bhateja V, Joshi A (eds) Proceedings of the international conference on data engineering and communication technology. Advances in intelligent systems and computing, vol 469. Springer, Singapore. doi: https://doi.org/10.1007/978-981-10-1678-3_8
2. Gayetri Devi SV, Nalini C (2020) Optimization of automated software contracts generation by modified particle swarm optimization. Int J Fut Gener Commun Netw 13(1):629–637
3. Adibi MA (2019) Single and multiple outputs decision tree classification using bi-level discrete-continues genetic algorithm. Pattern Recogn Lett 128:190–196. ISSN 0167-8655. doi: https://doi.org/10.1016/j.patrec.2019.09.001
4. Gayetri Devi SV, Nalini C (2019) A systematic judgment to automated programming contracts generation. Int J Recent Technol Eng (IJRTE) 8(2). ISSN: 2277-3878
5. Chabbouh M, Bechikh S, Hung C-C, Said LB (2019) Multi-objective evolution of oblique decision trees for imbalanced data binary classification. Swarm Evolut Comput 49:1–22. ISSN 2210-6502. doi: https://doi.org/10.1016/j.swevo.2019.05.005
6. Gayetri Devi S, Chidambaram N, Narayanan K (2018) An efficient software verification using multi-layered software verification tool. Int J Eng Technol 7:454. https://doi.org/10.14419/ijet.v7i2.21.12465
7. Pang Y, Xue X, Namin AS (2017) A clustering-based test case classification technique for enhancing regression testing. J Softw 12:153–164. https://doi.org/10.17706/jsw.12.3.153-164

8. Gayetri Devi SV, Nalini C (2020) Prioritized automated generation of contracts with modified swarm optimization. Int J Adv Sci Technol 29(8s):2432–2439. Retrieved from https://sersc.org/journals/index.php/IJAST/article/view/14731

9. Saettler A, Laber E, Pereira FAM (2017) Decision tree classification with bounded number of errors. Inf Proces Lett 127:27–31. ISSN 0020-0190. doi: https://doi.org/10.1016/j.ipl.2017.06.011

10. Jha A, Dave M, Madan S (2019) Comparison of binary class and multi-class classifier using different data mining classification techniques. SSRN Electr J . https://doi.org/10.2139/ssrn.3464211

11. Hasnain M, Pasha MF, Ghani I, Imran M, Alzahrani MY, Budiarto R (2020) Evaluating trust prediction and confusion matrix measures for web services ranking. IEEE Access 8:90847–90861. https://doi.org/10.1109/ACCESS.2020.2994222

12. Visa S, Ramsay B, Ralescu A, Knaap E (2011) Confusion matrix-based feature selection. CEUR Workshop Proc 710:120–127

13. Saifan AA (2001) Test case reduction using data mining classifier techniques. J Softw 11(7):656–663

14. Singh A, Bhatia R, Singhrova A (2018) Object oriented coupling based test case prioritization. Int J Comput Sci Eng 6:747–754. https://doi.org/10.26438/ijcse/v6i9.747754

15. Suppriya M, Ilavarasi AK (2015) Test case selection and prioritization using multiple criteria. Int J Adv Res Comput Sci Softw Eng

16. Somasundaram A, Reddy US (2017) Modelling a stable classifier for handling large scale data with noise and imbalance. https://doi.org/10.1109/ICCIDS.2017.8272643

17. Bin Tariq O, Lazarescu MT, Iqbal J, Lavagno L (2017) Performance of machine learning classifiers for indoor person localization with capacitive sensors. IEEE Access 5:12913–12926. doi: https://doi.org/10.1109/ACCESS.2017.2721538

Automatic Dermatoglyphics Multiple Intelligence Test Based on Fingerprint Analysis Using Convolution Neural Network

Komal Darji, Sagar Darji, Sumit Nisar, and Abhijit Joshi

Abstract Fingerprints play a very crucial role in uniquely identifying a person, but it can only be able to derive the innate potentials, multiple intelligence and personality as well from fingerprints. Currently, the dermatoglyphics multiple intelligence test (DMIT) is used to generate reports by indicating various traits and potentials exhibited by a person. The existing systems have major drawback in deriving the fingerprint type and ridge count for all ten fingers that are identified manually by the DMIT experts. These two are essential parameters and given as input to the DMIT software to generate the report. This research work has been proposed to automate the manual work carried out by an expert. Image processing techniques and convolutional neural networks are used in the development of the proposed system. Image processing techniques are used to perform various operations on fingerprint images to remove the noise. Then the noise-free images are sent to convolutional neural network to correctly identify the fingerprint type out of the five classes. Further, the correct classification is used to categorize his/her personality, learning type, acquiring ability and the individual perception rate. The results obtained in this project are encouraging as the accuracy remains at the acceptable level by considering that the sample size used in the process is inadequate.

Keywords Inborn potential · Manual process · DMIT · Personality · Image processing · CNN · Automatic report generation · Minute extraction

K. Darji (✉) · S. Darji (✉) · S. Nisar (✉) · A. Joshi
Department of Information Technology Engineering, Dwarkadas J. Sanghvi College of
Engineering, Mumbai, Maharashtra 400056, India
e-mail: darjisagar7@gmail.com

S. Darji
e-mail: darjisagar7@gmail.com

S. Nisar
e-mail: sumitnisar786@gmail.com

A. Joshi
e-mail: abhijit.joshi@djsce.ac.in

© The Author(s), under exclusive license to Springer Nature Singapore Pte Ltd. 2021
J. S. Raj et al. (eds.), *Innovative Data Communication Technologies and Application*,
Lecture Notes on Data Engineering and Communications Technologies 59,
https://doi.org/10.1007/978-981-15-9651-3_61

1 Introduction

Every individual is unique and has a distinctive story to narrate. All human beings are creative, genius and need to succeed in their journey of life [1]. It is important to empower, enable, dream, believe themselves and create positive change in a person's life by understanding the person's inborn potential. It leads them to discover their passion and helps them to become successful in their life. Human behavior and understanding a fellow being have also been the most challenging task. How a person reacts to a situation? What are the strengths and weaknesses of a person? How does a person adapt to the surroundings? etc. have been a big question to get an answer to. What one should opt for? as his career, where does his interests lie? etc. are big question marks when a child reaches a certain age. This has been difficult even for a parent to understand his own child.

For the same reason, they undergo aptitude tests, IQ tests, etc. to find the potential of their child. However, the main drawback of these tests is that they gauge only acquired potentials, which are temporary and change along with time, surrounding, interests, etc. These are short-lived, and hence, one might not find pleasure in doing the same work after an interval. In a country, like India, where parents need to communicate and understand their child and his/her potential, these tests are of no use. It will create discontent in a child's mind [1]. Academic pressure and expectations from parents lead to wrong decisions. Every individual possesses a unique way of learning, grasping things from their surroundings and understanding the way they are. This is what is known as the Inborn Learning Potential. To know about these inborn potentials and using them for the learning is not an easy task. It is a huge challenge, especially for parents, to understand their child, his/her interests, strengths and inborn potentials, so much, so that it takes years for them to comprehend their child's character.

Therefore, to bridge the gap, fingerprints can be used to find one's innate potential. It is a unique and scientific test, devised using technology, which analyzes the ridges and patterns on the fingerprints (dermatoglyphics). Seven million fingerprint samples were documented and put together along with researches, biological facts, observations and psychological theories to verify the authenticity of the test. The test provides the inborn traits, personality and character of the person. Knowing this information, it helps parents to understand their child better and guides them in a way to groom their child accordingly. It suggests to them his/her areas of interests and strengths, which further help them wave his/her way to a good quality life.

Dermatoglyphics is the scientific study of the human fingerprint. Usually, fingerprints are formed during the 13th–19th week of an embryo and developed by the 24th week [2]. It relates to the development of the infant's brain. It precisely represents the brain cell's distribution among the various lobes of the brain. Dermatoglyphics analysis is an amalgam of neuroscience, psychology, genetics and medicine. It forms its basis on the relation of fingerprints and brain lobes along with its formation and the number of ridges.

The salient features of epidermal ridges are [2]:

- Fingerprint patterns are unique to every individual.
- The origin of the nervous system and the skin is from the ectoderm in the embryo; therefore, any change/upset during the formation period leads to a direct effect on both.
- This means that environmental factors can influence the development of epidermal ridges although it is a genetic process.

Using dermatoglyphics, a trained analyst can read a number of details about the individual one being examined, including personality traits, strengths (and dominance) of the left and right sides of the brain, chromosomal anomalies and even character traits, including inborn strengths and talents. The DMIT is a scientifically and medically proven method of revealing intelligence, inborn skills, talents and potential. This tool is useful for all age groups and provides solutions in different aspects of life like career path, personality, etc. With such knowledge, one can acquire great heights in his personal as well as social life.

Dermatoglyphics experts, who study the fingerprint patterns for each of the ten fingers, conduct DMITs. This study results in the two essential parameters (fingerprint type and ridge count), which are given as input to the DMIT software. Based on these inputs, the DMIT software generates the report. Here, the manual process does not only consume time but also requires more human hours. If one needs more reports, one has to deploy more workforce in the form of a dermatology expert. This is when technology can be used to make the process efficient, with pinpoint accuracy and more productive. In this project, the above concerns are addressed using suitable tools and technologies.

The rest of the paper is organized as follows: Sect. 2 provides review of traditional methods, existing systems and drawbacks of each along with the area useful in the development of the proposed system. Section 3 presents the proposed approach in detail. Section 4 explains the architecture of the proposed system with the functionality of each module. The experimental results obtained from our proposed system are discussed in Sect. 5. The paper ends with conclusion and future scope of the system.

2 Literature Review

This section covers a detailed survey on the existing systems, their drawbacks and all the factors that motivated us to propose our solution. Before that, the useful areas in the development of the proposed systems are discussed. Now, let us see these useful areas.

2.1 Overview of Various Areas

This section emphasizes the role of various areas and techniques, namely image processing, template matching and convolution neural network required in the course of building the system.

2.1.1 Image Processing

In order to examine the dermal ridges on one's hand, they first need to be taken onto the desktop. This is done using a high-quality fingerprint scanner that captures every ridge perfectly. An instruction that should be kept in mind while scanning the fingerprints is that we must carefully take the entire fingerprint from left to right. This ensures that we do not miss out on the main features—delta and center, required to find the type of the fingerprint pattern. These scanned images are basically an array or matrix of pixels, stored in JPEG format. Image processing is a broad category of techniques used to manipulate an image as desired. The technique to be applied on an image depends upon the output required. In this system, we require an enhanced, noise-free image fingerprint image. The ridges on the fingerprint image need to be made thin, for proper identification of the pattern. These requirements led us to use normalization, smoothing and thinning techniques in our system.

Normalization is a process that changes the range of pixel intensity value [3]. Smoothing is a key technology of enhancement which can remove noise in images. So, it is a necessary functional module in various image processing software. It improves the quality of images. There are multiple filters [4, 5] like mean filter, bilateral filter, median filter, low-pass filter, etc. Based on comparison, we decided that a bilateral filter is the best for our system because it preserved the edges of the image which is the basic requirement of our system, to preserve the ridges of fingerprint [3]. Thinning is a morphological operation, normally applied on binary images, that is used to remove selected foreground pixels. It outputs another binary image and is primarily used for skeletonization. In this mode, it cleans up the output of edge detectors by reducing all lines into a single pixel thickness. This process helps us to recognize the ridges of the fingerprint by creating a gap between two ridges, thus making it easier to find the type.

2.1.2 Template Matching

Template matching is a technique in digital image processing for finding small parts of an image which matches a template image [6]. Fingerprints consist of two main features that need to be identified in order to classify them [7]. The two features are the deltas and the center. The delta point is a pattern of a fingerprint that resembles the Greek letter delta. It is the point on a friction ridge at or nearest to the point of divergence of two type lines. Centers are also called core. In dermatology, core point means the center area of the fingerprint. One fingerprint image may have multiple core points. Core points can also be found in different shapes. For instance, in whorl pattern, it is found in the center of the spiral, and in loop pattern, it is found at the topmost region of the innermost ridge/loop. These can be used as template images that were to be found in the original image. But the problem that arose, in identifying the delta, was its position, shape and size in the fingerprint image. Using template image, it is not possible to identify deltas as they are not of a similar size or at the same position. In identifying the core points, the same problem also arose, along with an additional problem, such as fingerprints that can have different types of core points. Template matching is useful in this scenario wherein [6] a processed template image (of delta and center) is used to find in the original image. But due to the problems mentioned above, it was difficult to implement it for our system [8]. In MATLAB, the accuracy was about 10%, whereas in python, template matching did not show any results with almost 0% accuracy.

2.1.3 Convolutional Neural Network

The drawbacks of template matching led us to collect fingerprint images and develop a convolutional neural network model. Convolutional neural networks are the most widely used image classification model due to its ability to automatically learn from the large number of filters/images from the training dataset in parallel [9]. It creates a feature map based on the filter applied to the input and then uses it to find similar detected features. Convolutional layer forms the building block on the network and hence helps in predicting output for an image dataset. This process is useful to classify the fingerprint images into its correct label, which is the same as the work of a CNN. But due to lack of image dataset, we had to try template matching first. As it did not yield very good results, we gathered fingerprint images so that we can at least train our model for the common types of fingerprint. Therefore, we first trained the model to two subtypes, namely L and R that had the maximum data. After gaining about 70% accuracy, we increased our dataset and trained our model for the four main classes of fingerprint type. We have gained about 60% accuracy in the same, which can be further increased to almost 100% if we get more fingerprint images. Therefore, it was decided to use CNN to find the main class of fingerprint images.

Now, let us see the traditional methods generally used by parents to find the potentials in their children.

2.2 Traditional Methods

This section emphasizes the prominent traditional methods, namely aptitude test and palmistry that are used to identify the traits of an individual along with the pitfalls of these methods.

2.2.1 Aptitude Test

Aptitude means one's natural ability to do a particular type of work. If the person is naturally talented, then one can say that he/she has outstanding aptitude. However, some abilities are acquired, developed and understood over a period of time and hence are not considered innate. Innate ability is different from acquired skills or knowledge. These are merely achieved through years of training and learning. Therefore, in recent times, aptitude tests are used to check the knowledge that one has acquired over a certain period. This test is mostly used to make career decisions or employment decisions. It assesses learned skills or knowledge and is hence known as achievement tests. It is used in various examinations to check if a person is qualified or not. It is nowhere used to check a person's innate abilities, strengths or weaknesses.

2.2.2 Palmistry

Palmistry, or chiromancy, is the practice of characterization and predicting the future based on the study of the palm (chirology). Chiromancy involves the practice of anticipating a person's character or future life by reading the palm lines on the person's hand. Different lines on the palm indicate different meanings like heart line, lifeline, etc. These lines and bumps indicate interpretations due to their relative size, intersections between lines, etc. However, scientific literature typically regards palmistry as a pseudoscientific or superstitious belief. It is regarded as cold reading, meaning the practice of using high probability guessing and analyzing details based on the cues and signals given by the person. Palmists are therefore called psychics, who practice cold reading. It is a completely baseless prediction about one's life and personality.

2.2.3 Drawbacks of the Traditional Methods

The traditional methods used to identify an individual's potential were not up to the mark and had some issues. Palmistry or chiromancy was basically prediction. It provided information on baseless assumptions, and therefore, it was also referred to as cold reading. Such assumptions or predictions cannot be used to form a basis of an individual's life. One cannot make important life decisions such as a career path on the basis of prediction. Aptitude tests are taken in a certain set of environment and under a certain pressure over an individual wherein he/she is examined for his/her abilities. This pressure tends to have an effect on the individual's mind which impacts the answers. It is basically a test for knowledge and not for inborn potential. It tests for acquired potential, which varies with time and age, whereas inborn potentials are totally different from the acquired one. A person born with a specific trait or a talent cannot be tested using tests that give wrong output. These outputs mislead an individual into taking a wrong path. Therefore, these methods cannot be used to find innate abilities.

2.3   Existing System

It was in 1823 that the scientists discovered for the first time the relation between fingerprints and innate intelligence [10]. Through medical researches, it was claimed that the development of fingerprints and the neocortex happens during the same period. This was even verified by the prime genetics research sector in China [10]. Fingerprints are seen to be constant throughout the lifetime. This correlation has led scientists to integrate genetics, psychology, embryology, dermatoglyphics and neural science with the theory of multiple intelligences to deduce a person's inborn personality, talents and potentials. Hirsch and Schweicher discovered that even before the formation of fingerprints, the dermal nerves and blood vessels are arranged in an unusual fashion, which made them believe that both the nervous system and the vascular system help in determining the formation of fingerprints [11]. This further resulted in establishment of relation between fingerprint patterns and innate abilities. Jelovac et al. have even related fingerprint patterns with psychological conditions and different diseases [12].

Another study by the scientists at Barcelona University indicated that fingerprints of intellectually disabled people were very different from that of a normal person [10]. Intellectually disabled people had more arches and circular patterns on their fingerprint. This further proved the point that fingerprints are associated with intelligence. Medical experts have confirmed the relation of dermal ridges with one's multiple intelligence and inborn potentials.

Therefore, by determining the type of pattern on the fingers, one can deduce his/her innate abilities. The current DMIT software needs to be given two inputs, one being the ridge count and the second being the exact type of fingerprint pattern for each finger. This input needs to be found by the dermatoglyphic expert. They do so by using the scanned images. The fingerprints are first scanned using a highly accurate scanner. Then all the ten images of each person are stored separately and examined by the experts. Each fingerprint is made noise-free so that it can be examined properly, and the main pattern is found, which is called as the main class of the fingerprint type. After having found the main class, the ridge count for each finger is found manually by counting each line in the zoomed noise-free image. This is then further used to find the exact subtype out of the 31 types that have recently been introduced, thus making it a time-consuming process. After the two inputs have been found, the current system corresponds to the predefined classification in different fields like personality, multiple intelligence, learning abilities, etc. to generate a report. In this way, the current DMIT system works.

2.3.1   Drawback of Existing System

The current system requires a DMIT expert to manually find out the two important parameters: pattern type and ridge count, which are used as input for the DMIT software. This is a tedious and a time-consuming task. Each expert needs to examine the fingerprint images by zooming them and hence deducing the pattern type and ridge count. The entire time taken to generate a report using this system is approximately 2 h, which means that daily an expert can only make 3–4 reports. This hampers the productivity of the organization as well as the expert. Inadequate use of technology is a major drawback in the current manual system. Keeping in mind the productivity and other issues in the existing process, we decided to address them in our proposed system. The proposed system will use the technology to not only save time, but it will also improve productivity. Now, let us see the approach followed in the development of the proposed system.

3 Proposed System

The main purpose of developing the system is to eliminate the manual process of finding the pattern type in the fingerprint image with the automated process. By giving raw fingerprint images as input of all the ten fingers, our system would generate a report consisting of the personality, learning type, acquiring ability and the perception rate of the individual along with his/her personal details. In order to find the pattern, first, the fingerprint images will be preprocessed to remove the noise from them. During preprocessing, various image processing operations are carried out such as normalization, smoothing and thinning. These preprocessing operations are also performed on the image dataset (training data) so that they can be used to train the model and further useful to predict the pattern type. The CNN model will be trained first for five labeled classes mentioned below.

- Normal whorl is one of the main fingerprint classes. In this case, the fingerprint has two deltas. For ridge count calculation, if the main class is found to be a whorl, then the nearest distance out of the two deltas is considered as the final ridge count.
- Double loop whorl is another type of the main fingerprint class. In this case, the fingerprint has two deltas, and its center is a combination of two loops. For ridge count calculation, if the main class is found to be a whorl, then the nearest distance out of the two deltas is considered as the final ridge count.
- Loop is characterized by one delta present on the fingerprint, and the ridge count is defined as the number of lines between the delta and the center of the loop. Depending upon the flow of the ridges, i.e., whether it is toward the thumb or the little finger, there are two types of loops: radial and ulnar. Therefore, loop as a main class was not considered; rather, its two subclasses were considered in finding the results.
- Arch is the one which has no deltas. Here, the ridges are found by the waves created that develop a particular tent-like structure.

These five classes are encoded into a numeric array using one hot encoder, which uniquely identifies each. The array looked like [0 0 0 0 1] for normal whorl, [0 0 0 1 1] for double loop whorl and so on for the rest of the classes. The dataset was split into train and test set, after which the model was trained using the train set. The test set was used to check the prediction done by the model. The new fingerprint images were given to the model for predicting the pattern type of all the ten fingerprint images, classify them correctly and, hence, generate the report. Each pattern type denoted a unique meaning [13]. For instance, finding the personality of a person is determined using both thumbs. The left thumb is the primary and reveals the more dominating personality of the person, whereas the right thumb is considered secondary and reveals the less dominating personality of the person. The same way, acquiring style uses all the ten fingerprints to deduce out of the four: reflective, affective, cognitive and reverse thinkers. Percentages of arches, whorls, radial loop, ulnar loop and double loop whorls are also found. The class having the maximum percentage reveals the individual's acquiring style, e.g., if an individual shows 60% arch and 40% whorl, then his acquiring style will be 60% reflective and 40% cognitive, respectively. Arches correspond to reflective, radial loop corresponds to affective, whorls correspond to cognitive, and ulnar loop and double loop whorls correspond to reverse thinkers. This information is used to generate the report and output a PDF, which can be downloaded and stored in the database.

Getting an idea of the overall approach followed in the development of the system, now let us see the architecture of the system.

4 System Architecture

Figure 1 illustrates the detailed architecture of the proposed system.

The data flow architecture style is followed as data moves from one module to the other. Looking at Fig. 1, one can say that the architecture is roughly divided into three parts, namely image processor, class identifier and report generator. Now let us see the functionalities of these blocks.

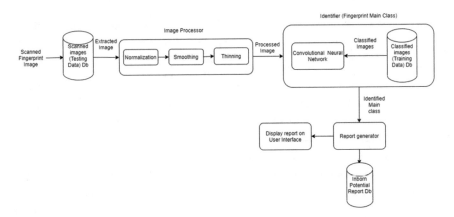

Fig. 1 System architecture

Fig. 2 Original image

(1) **Image Processor**: To work with any image, one must first make the image noise-free and gray scale for improving the accuracy of the result. This module is responsible for doing the same using various image processing techniques. Firstly, the fingerprint images are scanned and stored in scanned image(s) DB, which are used as a testing data. Now this testing data, i.e., the original image (see Fig. 2), is given as input to the image processor block. Various image processing techniques are applied. First, the image is normalized to increase the contrast in the image. As discussed in Sect. 2.1.1, adaptive Gaussian normalization technique is used, and the generated output (see Fig. 3) is given as input to the next process, smoothing.

Smoothing is responsible for removing the high frequency noise from the image. This is the stage where noise is completely eliminated from the image using a bilateral filter (see Fig. 4). It helps in preserving the edges which is much needed in fingerprint images. This output is then used by the thinning process, to reduce the thickness of the ridges in the fingerprint image (see Fig. 5). It helps in easing the task of next block CNN, which is used to identify the class. Thinning is the final process of this block, and the generated output image is given to the class identifier block.

(2) **Class Identifier**: As mentioned in the previous section, the fingerprint images have five classes, namely arch, radial loop, ulnar loop, normal whorl and double loop whorl. The role of this block is to classify all the ten fingerprints of a single person into its main classes. CNN is used for this purpose. Firstly, the CNN model is built by adding three different types of layers such as input layer, seven hidden layers (5 conv_2d layers + 2 fully connected layers) and an output layer. The functionality of three layers is as follow:

Fig. 3 Normalized image

Fig. 4 Smoothened image

Fig. 5 Thinned image

- **Input Layer**: This layer acts as an entry point or the start point for the images. It is at this layer that the processed fingerprint images are given as input to the neural network. It contains image data and represents the image pixels into matrices
- **Hidden Layer**: This layer comprises five convo layers and two fully connected layers. The convo layer is responsible for extracting features from the input layer. The fully connected layer is responsible to classify images into various labels. It contains weights, biases and neurons that help in training the model. This layer acts as a connection between the neurons of one layer and neurons of another layer.
- **Output Layer**: This layer outputs the main class of the given input fingerprint image. These classified image labels are input to report generating module.

(3) **Report Generator**: This block takes the class identified for ten fingerprints by class identifier as an input and finds correlation of each type. Each fingerprint pattern relates to a definite meaning. People having loops on their thumbs are said to be steady personality, arches indicate influential personality, whorls indicate dominant personality, and double loop whorls indicate compliant personality. These personalities further give details about their strengths, dislikes, etc. [14]. The same way, acquiring ability of a person is found by calculating the percentage of each pattern in the ten fingers. If arches have the highest percentage, the individual has reflective ability, for radial loop, it corresponds to affective, for whorls, it corresponds to cognitive, and for ulnar loop and double loop whorls, it corresponds to reverse thinkers. This correlation is put

in the HTML code which generates a report for the individual, displayed in the form of a PDF file. The report includes personal details of the individual and detailed information about his/her personality, learning type, acquiring ability and the perception rate of the individual.

5 Experimental Results:

In this section, the experimental results obtained from the proposed system are presented along with their analysis. Figure 6 depicts the predicted class by the system.

The predicted class by the system is ulnar loop (L) and radial loop (R) for the original image of type L and R, respectively. It is found that the results are overlapping. This is due to the inadequate amount of fingerprint images in the dataset to correctly identify the pattern. To check the accuracy of the system, a dataset of 200 images is used. The results of this test are presented in the form of a bar graph (see Fig. 5 and 6, respectively).

As can be seen in Fig. 7, normal whorl and arch are classified accurately. This happens because the dataset has a larger number of images of those patterns. Figure 7 also clearly depicts that classes: Radial loop, double loop whorl and ulnar loop are not classified accurately. This is due to two main reasons: lack of dataset and the similarity with other classes. The fingerprint patterns are minutely different from each other, and to accurately identify them, the dataset of fingerprint images would be required in lakhs. Another factor affecting the prediction is the size of the dataset. Here, the dataset used to train the model has a large number of fingerprint images for normal whorl and arch with 456 and 420 images, respectively, whereas the number of images for radial loop, double loop whorl and ulnar loop is 338, 236 and 324, respectively. Due to this, overfitting of the model takes place, and it identifies them

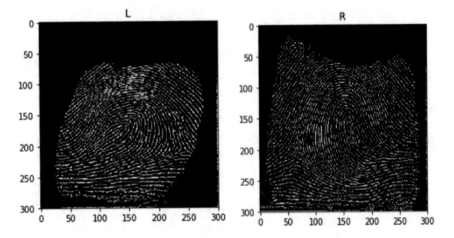

Fig. 6 Predicted main class (final output)

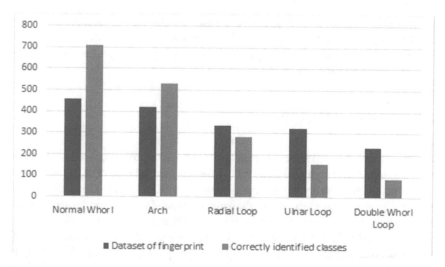

Fig. 7 Comparison between original number of images in dataset vs. classified output labels given by the model

incorrectly. This was noted in the correlation matrix generated by the system (see Table 1). Table 1 shows the correlation matrix for the input image, ulnar loop. In Table 1, W1 depicts normal whorl, X is arch, R is radial loop, L is for ulnar loop, and W2 is double loop whorl (the naming conventions are just used for the purpose of ease of readability). The numeric values corresponding to W1 and L are very close to each other. But due to the higher number for normal whorl, it overfits the model, thus incorrectly classifying ulnar loop as normal whorl.

Table 2 shows the confusion matrix including all the classes. It depicts the original and the predicted pattern class for the image, clearly mentioning whether they were

Table 1 Correlation matrix

	W1	X	R	L	W2
Correlation value	0.32859775	0.17849554	0.164336099	0.32235745	0.00618824

Table 2 Confusion Matrix

Predicted class	Actual class					
		W1	X	R	L	W2
	W1	416	20	60	82	132
	X	25	395	7	86	19
	R	6	2	260	13	3
	L	3	1	11	143	1
	W2	6	2	0	0	81

Table 3 Precision, recall and F1 score

	W1	X	R	L	W2
Precision	0.59	0.74	0.91	0.9	0.91
Recall	0.91	0.94	0.77	0.44	0.34
F1 score	0.72	0.83	0.83	0.59	0.5

classified correctly or not. For instance, from Table 2, we can conclude that 416 images for W1 were classified correctly. It classified 20 images with the actual class X as W1, i.e., incorrectly classified them. The same way, 60 images from class R, 82 images from class L and 132 images from class W2 were incorrectly classified as class W1.

Table 2 is further used in calculating the precision, recall and the F1 score for each class as shown in Table 3. The formula used for precision, recall and F1 score is:

Precision = true positives/(true positives + false positives).

Recall = true positives/(true positives + false negatives).

F1 score = (2 * precision * recall)/(precision + recall).

Precision is used as a metric to identify how many images are actually classified correctly with respect to the actual number of images present. Recall is used as a metric to give the total positively identified images, and F1 score gives us a balance between precision and recall. Table 3 presents the accuracy with respect to a particular class and not the entire dataset. From Table 3, it can be inferred that about 59% of the images from the class W1 are identified correctly, 91% of the images are classified as W1, and the overall accuracy for W1 is 72%. Similarly, this can be inferred for the rest of the classes.

6 Conclusion and Future Scope:

It can be concluded that the proposed system has overcome the productivity problems in the existing system by automatically finding out the personality, learning type, acquiring ability and the perception rate of the individual person by just getting raw fingerprint images as input. The proposed research work has used image processing techniques and CNN to obtain the desired results. The lack of dataset is the major hindrance in achieving acceptable accuracy. Due to which, overfitting is observed as seen in the previous section. Obtaining the images of rare fingerprint classes and in ample numbers to build a good dataset remains as one of the major challenges. Due to this, the CNN model could not be trained as per expectation. Though, this can be addressed by just increasing the fingerprint images of each type and training the CNN model. In this way, the manual task has been carried out by the expert to overcome through our system by creating more time for experts to indulge in other activities and make efficient use of the technology.

The system currently finds four personalities, learning type, acquiring ability and the perception rate of the individual giving fingerprints. Further scope of the proposed system is to overcome the lack of dataset and hence generate five different types of report, i.e., HR report, career report, spouse compatibility, genetic mapping and lifetime report. These reports are based on the fingerprint analysis performed by the system. For instance, an individual's strengths, weaknesses and personality could be considered by HR, while hiring or assigning a task. If a person has good leadership skills and can plan the task, it would lead to better results; to get these reports, the system needs to classify fingerprints, not only in the four main classes, but all the 31 types (which include the sub-classes) for obtaining an utmost accuracy. Based on different parameters, one can generate the desired report. This specialized report is beneficial to both user and counselor.

References

1. Thumbrule (2012) https://www.thumbrule.com.np/faq.html. Last accessed on 10 Jul 202.
2. Thumbrule (2012) https://www.thumbrule.com.np/introduction-of- dermatoglyphics.html. Last accessed on 10 Jul 202
3. Jiang X (2001) A study of fingerprint image filtering. In: Proceedings 2001 international conference on image processing (Cat. No.01CH37205), Thessaloniki, Greece, vol 3, pp 238–241. https://doi.org/10.1109/ICIP.2001.958095.
4. Han K, Wang Z, Chen Z (2018) Fingerprint image enhancement method based on adaptive median filter. In: 2018 24th Asia-Pacific conference on communications (APCC), Ningbo, China, pp 40–44. https://doi.org/10.1109/APCC.2018.8633498
5. Fahmy MF, Thabet MA (2013) A fingerprint segmentation technique based on Morphological processing. In: IEEE international symposium on signal processing and information technology, Athens, pp 000215–000220. https://doi.org/10.1109/ISSPIT.2013.6781882.
6. Li T, Arif A, Cheng C (2017) A dual-image template based fingerprint matching method. In: 2017 10th international congress on image and signal processing, bioMedical engineering and informatics (CISP-BMEI), Shanghai, pp 1–5. https://doi./org/10.1109/CISP-BMEI.2017.830 2039
7. Mehtre BM (1993) Fingerprint image analysis for automatic identification. Machine Vis Apps 6:124–139
8. Tian J, Peng Y (2012) Research of the Matlab application in the fingerprint identification system. In: 2012 international conference on image analysis and signal processing, Hangzhou, pp 1–5. https://doi.org/10.1109/IASP.2012.6425005
9. Pavithra R, Suresh KV (2019) Fingerprint image identification for crime detection. In: 2019 international conference on communication and signal processing (ICCSP), Chennai, India, pp 0797–0800. https://doi.org/10.1109/ICCSP.2019.8698014
10. Jelovac N, Milicić J, Milas M, Dodig G, Turek S, Ugrenović Z (1998) Dermatoglyphic analysis in borderline personality disorder and schizophrenia–results of a Croatian study. Coll Antropol. 22(1):141–148
11. Thumbrule (2012) https://www.thumbrule.com.np/scientific-date.html. Last accessed on 10 Jul 2020
12. Simon-Zorita D, Ortega-Garcia J, Cruz-Llanas S, Gonzalez-Rodriguez J (2001) Minutiae extraction scheme for fingerprint recognition systems. In: Proceedings 2001 international conference on image processing (Cat. No.01CH37205), Thessaloniki, Greece, vol 3, pp 254–257. https://doi.org/10.1109/ICIP.2001.958099.

13. Educareeroptions (2017)https://educareeroptions.com/discdope-personalities-traits-through-dmit-dermatoglyphics-multiple-intelligence-test/. Last accessed on 10-Jul 2020

14. Inborn Potentials (2015) https://inbornpotentials.wordpress.com/#:~:text=Every%20human%20possesses%20unique%20way,called%20as%20Inborn%20learning%20Potential.&text=Sometimes%20it%20takes%20them%20years,and%20inclinations%20of%20their%20child. Last accessed on 10-Jul 2020

Real-Time Monitoring of Crop in Agriculture Using Wireless Sensor Networks

Kirankumar Y. Bendigeri, Jayashree D. Mallapur, and Santosh B. Kumbalavati

Abstract The paper works on the issue of wireless sensor network (WSN) employed in the agriculture application. Most of research works in WSN are dedicated for various applications like military, medical, home automation, and so on, where the primary intention was to introduce a technology for agriculture with the evolving sensor technologies. Such technology can be considered as E-agriculture. Proposed work consideres sugarcane as a crop under monitoring, where the work mainly aims at monitoring humidity (H), temperature (T), pressure (P), and light intensity (LI) required for the intended crop and also on the land using LBR and MBR routing protocols. Simulation is first carried in Qualnet simulator and extended to real-time sensor. Simulation result shows that MBR routing protocol has achieved a better performance, when compared to LBR.

Keywords Wireless sensor network (WSN) · Agriculture · Sensenut nodes · Qualnet simulator

1 Introduction

Wireless sensor network (WSN) is a current trend in the communication with increase in various applications right from home automation, industry monitoring, and human body monitoring. Recently, WSN has also proven to be used for agriculture application at lower cost. Over the development of sensors, it was restricted for commercial

K. Y. Bendigeri (✉) · J. D. Mallapur
Department of Electronics and Communication, Basaveshwar Engineering College (A), Bagalkot, Karnataka 587103, India
e-mail: kiranbendigeri@gmail.com

J. D. Mallapur
e-mail: bdmallapur@yahoo.co.in

S. B. Kumbalavati
Department of Electronics and Instrumentation, Basaveshwar Engineering College (A), Bagalkot, Karnataka 587103, India
e-mail: sbkumbalavati@gmail.com

© The Author(s), under exclusive license to Springer Nature Singapore Pte Ltd. 2021
J. S. Raj et al. (eds.), *Innovative Data Communication Technologies and Application*,
Lecture Notes on Data Engineering and Communications Technologies 59,
https://doi.org/10.1007/978-981-15-9651-3_62

application and the recent trend has made these sensors to be used in agriculture application. The proposed research work mainly aims at agriculture community. Work is carried out for dveloping a remote control on crop growth using different sensors like temperature, humidity, pressure called as HTP sensor. Another such kind of sensor called TL is used to measure temperature and light intensity. Sensor nodes are introduced in the field to monitor the necessity of crop. The values obtained from these sensors can be used to make decision on the yield of crop. For example, humidity of soil has some predefined values as obtained by sensor when placed in soil, if the vapor content in soil decreases, corresponding change in values is obtained. Alert to farmer can be made so as to take neccessity action and help in better productivty of crop. Proposed work also considers use of routing protocols, as most of the most of researches considers monitoring of crop using hop-to-hop, i.e., each sensor sending data to gateway. Drawback in such methodology is sensors and gateway need to be placed at a very short distance, thus making it a very small communication area. Whereas routing protocols use nodes with sensors, further sensor data can be sent to nodes which act as neighbours till it reaches destination, thus increasing communication area. Proposed work uses two routing protocols MBR and LBR for monitoring of crop and land with sensors having antenna that can cover up to 1 km for communication between the nodes. Thus, our work helps in covering agricultural land for a larger area at lower cost.

2   Lierature Review

In this paper, [1] the sensor nodes have a sensors in particular soil pH, leaf wetness, soil dampness, atmospheric weight sensors joined to it. Based on the soil dampness, it sprinkles the needed water. In this paper [2], author has shown how an irrigation management can be done using WSN. Soil temperature and dampness are noted through GPRS. This paper [3] is proposing a complete solution for agriculture using farmer-based WSN and GSM technology. The data obtained on the field is sent to farmer for necessary action. This paper [4] shows a genuine way of sending data for WSN-based greenhouse monitoring. In this paper [5], author proposes a framework that can genuinely monitor farming natural data like temperature, moisture, and light intensity. This paper [6] shows how an irrigation management system (IMS) can be executed effectively using WSN. Remote monitoring is done using (GPRS) to report soil temperature, soil moisture. One of the recent implementation toward usage of optical satellite sensor toward agriculture-based application in WSN has seen in the work of Kussul et al. [7], the authors have presented a technique that analyses the pixel-based sensory data to perform classification. Chiang et al. [8] have used hardware-based approach for assessing the solar radiation toward crop harvesting. Gutierrez et al. [9] have used sensor network and mobile application for monitoring the irrigation process. The technique presents an irrigation sensor where the smart phone is used as an essential component of irrigation. Kone et al. [10] have used

MAC parameters of IEEE 802.15.4 standard to develop an analytical framework for monitoring extracted information about crop cultivation.

Qu et al. [11] have used WSN-based mechanism for aggregating the readings for performing validation of an aggregated data associated with crop leaf area index. A similar line of work was carried out by Bhanu et al. [12] where the authors have introduced a simple hardware-based technique using a microcontroller. Liqiang et al. [13] have constructed a hardware platform for extraction data for monitoring crop cultivation. The technique also uses networking protocol to carry out communication of extracted data to base station. The drawback that can be obtained from current research work is no papers are discussing on placement of nodes, cost-effective sensors, and strong software platform. Through proposed research work highlighting on the drawback mentioned and proposing solution for it as discussed in the following section.

3   Implementation

The work is carried in Qualnet simulation for test scenario. Nodes can be plced in random and uniform pattern. Qualnet simulation with random placement of nodes with sensors is shown in Fig. 1. Qualnet simulator can visualize behavior of different

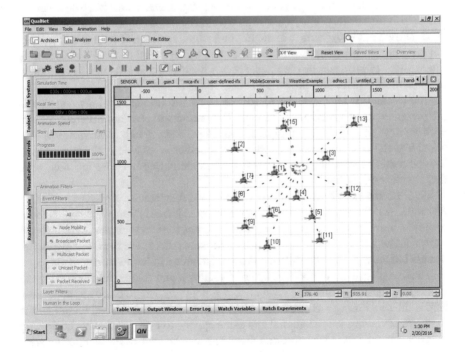

Fig. 1   Test scenario of WSN for agriculture application

Fig. 2 Different colored
sensor nodes for different
measuring parameters

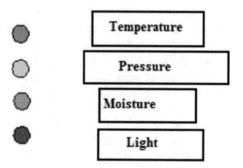

networks. For example, call connect, call terminate of mobile communication can be visualized. Therefore, it will set a base for actual implementation. Similarly, a sensor network with nodes and sensor can be placed in simulation area and using this simulator, the behavior of the network can be identified. This methodology can be extended to real-time scenario of sensors and nodes in agriculture area.

Presently, four sensor nodes are considered for interface of the network with programming as shown in Fig. 2. The readings of sensor obtained are more acceptable for the application utilized. Simulation results have distinctive results. Smaller area when considered irregular topology functions well. Larger area works well for regular topology. The implementation is conveyed in two regions, for small crop zone like fruit and vegetables and large zone like wheat, sugarcane, jowar, etc. These results will help in the territory of agriculture production by helping agriculture community. Work carried out helps farmer to continuously monitor crop. If crop conditions are not favorable, then farmer can be sent for action, and by doing so, manpower and high productivity of crop can be obtained. Device itself helps in effective development and growth of a plant thus providing benefit to the farmer.

The research work meets the disadvantage of current methodology and fulfills the present need of agriculture. The experimentation will consider different sensors with applications like, monitoring of soil moisture, water level indicator, atmosphere condition, the pest or disease observing all together accessible with every sensor. The placement can be in a manner that sensors are placed in entire recorded zone that are placed close to the crops.

Figure 2 shows different types of sensors used temperature, pressure, moisture, and light. As discussed, two deployments are considered. Figure 3 shows random deployment of nodes and few nodes with different sensors to monitor crop; Fig. 4 shows uniform placement of sensor nodes. Sensors colored are respectively placed. Timely base sensors can be shifted so that entire area can be covered.

Random placement of nodes indicated by number and colored line indicating the corresponsing sensors placed in that area. It is not necessary to place all the sensors across the monitoring area, because when moisture sensor is placed in soil and if there are deviations can be alerted and same applies to every corner of monitoring area considered, this is how cost-effective WSN can be applied for monitoring area. As shown in Fig. 3, if the node distribution is taken random, then few area will have

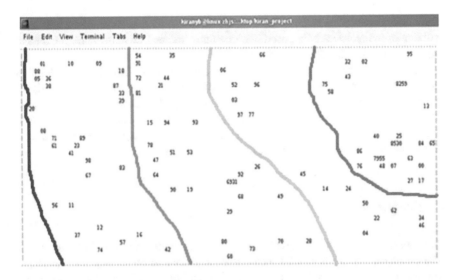

Fig. 3 Random placement of sensor nodes

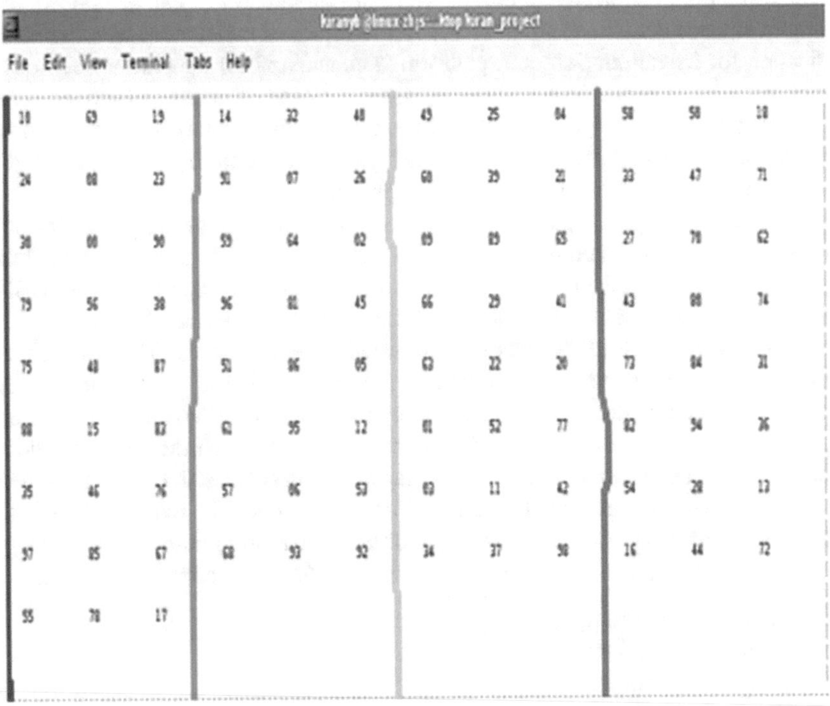

Fig. 4 Uniform placement of sensor nodes

many nodes and few areas will have no nodes to monitor, such scenario will have problem of many nodes being used in communication and few nodes to send data at larger distance leading to loss of energy.

Figure 4 showing uniform placement of nodes which will favor in a way that less number of nodes can cover the entire zone and every one of the nodes can effectively be utilized for measuring the parameters decreasing the burden on every sensors. Clustering method is also used in unifrom placement of nodes that reduces traffic of activity on all nodes. For carrying out the research work, we propose to utilize Sensenuts modular remote sensor equipment and programming.

4 Proposed Work

Considering the fact that uniform node placement is better in terms of saving energy of each node. Battery is crucial factor since sensors and nodes are in the field for live monitoring. Proposed work considers agriculture land with sugar cane crop. Near to the crop nodes, the sensors are placed. Data obtained from these sensors can be further communicated to farmer. Thus, any action needed can be noticed, for example, if any fluctuations are observed in the values of moisture sensor, then it is indication for irrigation. Sensenut platform is considered for simulation. Sensenut has both hardware and software platform, where hardware is nodes and software can be languages like C, C ++ , JAVA, Eclipse, etc. Routing protocols can be used to measure parameters related to plants like humidity, temperature, pressure, and light intensity, which is discussed in the following section.

Sensenuts development platform has a wide variety of sensor modules, which are compatible with radio module. They can be directly mounted on the radio module and the sensor-related data can be reported to the microcontroller. Sensenuts WSN platform is programmable to accomplish different tasks. Sensenuts brings the real-world scenario, where algorithms can be tested on real devices. The researchers write own algorithms and program the real motes according to the algorithm. The assumptions been made in the simulation environment are automatically overruled as the transfer of data between the devices takes place through the actual medium (wireless) with all available interferences and works on IEEE 802.15.4. Wide range of sensors can be connected with the motes depending upon the requirement of the application. Like environment monitoring, home automation, agriculture monitoring, industrial control, location tracing, etc. Some of the features about Sensenuts are:

- Flexible at MAC and physical layer
- Programming in C language
- Has API's to perform different tasks
- Easy to code routing algorithms and write applications.

Print Window: It allows the motes to display some custom messages on the GUI. At times, an application developer may be interested to integrate some third-party sensors or any custom sensor which may not be available as a part of Sensenut development platform at the moment. Extender module is suitable for such applications, which has the following features:

- This module actually consists of header pins which gives access to all the pins of the processor on the board
- External sensors can be connected to radio module using extender
- Extender may also be used for hardware debugging when there is a requirement of checking the voltage on some particular pin or view an output waveform from an output port on a oscilloscopes
- Connecting custom sensors
- Debugging hardware/checking output on number of ports.

In order to create a real-time behavior of a WSN, consider a gateway module as shown in Fig. 5. This gateway helps to exchange the data between a radio module and end user (computer in our simulation). As shown in Fig. 5, a gateway has a mini USB and computer has a USB port that is connected together. The gateway sends the data of radio module onto computer for the user reference. Similarly, a radio module shown in Fig. 6 and this radio module has a microcontroller, battery power, and antenna. The microcontroller is used to receive and send the instructions to be performed that also control the instruction of radio module. Battery is provided in terms of shells as of now, when implemented in real can be provided with rechargeable batteries so as to reuse it again. Antenna is provided to transfer the data wirelessly which is mounted onto radio module, for example, considering an open area, data can be transmitted for a distance of 200 m.

The implementation part consists of an agriculture area with a long crop that is grown, for example, a green house or sugar cane, that contributes for economy as it

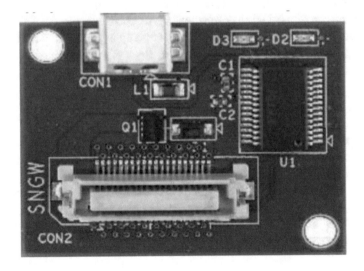

Fig. 5 PCB layout for gateway module

Fig. 6 Showing a radio module

yields on yearly basis with high product. From survey most of research papers that we have gone through have considered a simple hop communication- or unicast-based routing protocols. The proposed method considers routing protocols that enhances the efficiency of network by providing energy efficiency and network lifetime. Two routing protocols have been proposed, through which the simulation is carried out. Firstly, MAC-based routing (MBR) which can be used when the destination is always the PAN coordinator. As per IEEE 802.15.4, during network setup, each node gets associated with a coordinator. In MBR, routing is done in such a way that every node sends the packet to the coordinator to which it is associated. This process keeps on repeating till the time the packet is received by the destination node, i.e., PAN coordinator. If the packet gets dropped, it generates an exception and the node resets itself.

Secondly, we consider level-based routing (LBR) which is used in multihop scenario and can be used when destination is PAN coordinator.

The temperature measurement with respect to crop is given by the relation:

$$\Delta T = \frac{Q}{mc} \tag{1}$$

ΔT— Indicates change in temperature.
Q—Amount of heat absorbed or released.
C—Specific heat of the body.

Pressure that we do consider through sensor which actually measures with the crop is given by:

$$P = \rho g h \tag{2}$$

P—A water pressure.

ρ—Density of water.

h—Height of crop.

g—Acceleration due to gravity (approximately 9.81 m/s2 on Earth's surface).

Relation for humidity (H) for crop measuring real time data can be given by:

$$H = \frac{m_v}{v} \tag{3}$$

where m_v is the water vapor present per unit volume (v).

$LI = \frac{L}{4\pi d^2}$ Relation for Light Intensity *(LI)* related to electromagnetic energy emitted per unit time is given by:

L—Luminosity.

LI—Light intensity.

d—Distance between earths to star squared.

Consider the sensor nodes monitoring for a sugar cane shown with single radio module mounted with sensor as shown in figure in 7, the total of four sensors are made use, usually farmer has about half acre of land to an acre of land. For initial deployment with certain limitations in number of sensor nodes, consider monitoring of a single crop. One of the sensors capable of measuring light intensity and temperature is connected to sugar cane and famer being at a far distance, making use of two radio modules that actually carries the data forward to reach at the destination which has a radio module with gateway. The radio module at the destination mounted with gateway is used to receive data from neighboring nodes and bringing onto a computer or mobile. Similarly, the algorithm required to run a particular routing is sent from eclipse platform to radio modules through gateway; therefore, gateways act as two way communication device. After the setup is done and once the codes are brought onto radio modules through Sensenut, live readings are obtained for any sensors that are connected with the radio modules. An advantage of this are, the desired code based on the requirement can be designed and implemented with the radio modules and sensors respond accordingly. Thus, successfully able to achieve real deployment of sensors in a green house which has capability to measure the parameters relating to growth of a plant using LBR and MBR routing protocols.

Figure 7 shows a complete setup of WSN, where monitoring of crop sugar cane using LBR and MBR begins with live readings obtained on computer display for our reference. As discussed radio modules are battery powered which should be observed as most of chances are with the battery only. Thus, finding an unique approach to replace typical way of agriculture by introducing technology and calling it as e-agriculture.

5 Simulation

The temperature and light readings obtained for MBR protocol are as shown in Fig. 8. Basic purpose of doing this is as most sensors are meant for monitoring and sending

Fig. 7 Scenario of WSN setup for measuring various parameters of crop

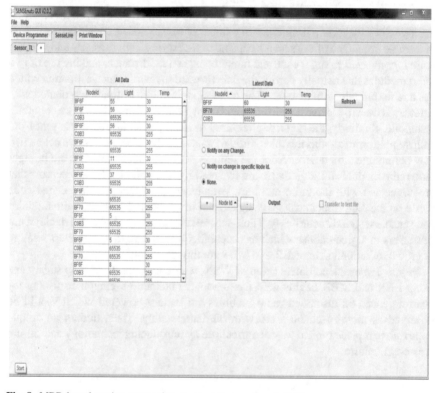

Fig. 8 MBR-based routing protocol to measure temperature & light

data through MAC based routing protocols that uses only hop-to-hop communication between two sensors thus reducing short range of communication between nodes having considered many nodes under different routing protocols where network is setup with source node, destine node, and intermediate nodes. The routing protocols like LBR, MBR, AODV, and LEACH can be made use, considering LBR and MBR for our simulation. There is definite increase in energy, power, and lifetime of network through use of these routing protocols. The readings are measured considering three sensor nodes. During simulation considering number of sensors varying from one to four that means nodes are taken as source and destination with two intermediate nodes. The same pattern of deployment is also used for LBR routing protocol, which also successfully senses the temperature and light intensity related to the crop. LBR has a better performance compared to MBR, as it is based on multipath routing algorithm. The simulation is designed such that, whenever any change with temperature or sunlight takes place its time is noted down in the simulation window.

The simulation is carried out using the real-time sensors. Real-time data for greenhouse was measured in terms of temperature, pressure, humidity, and light intensity. Two different routing protocols are considered for the same, which are LBR and MBR. Both the routing protocols are held for measuring all four parameters. The simulation inputs are number of real sensors are two, radio modules are four with one gateway, battery through 9 V cells are provided to each radio modules. Eclipse is simulation platform where desired routing protocols are simulated with help of Sensenut GUI. The nodes are able to send real-time data to target (laptop in our case). Sensor radio module is placed in greenhouse and at about 1 km away, we have laptop with radio module. We have successfully obtained data through other radio modules.

Figure 9 shows battery power consumption for MBR and LBR routing protocols. The simulation is held for measuring the parameters of light and temperature using sensor. Initially, we calculate the life of network considering a single radio module then extend it with four sensors. As the number of sensors increases, burden on individual sensors to handle and process data reduces, thus increasing network lifetime.

Figure 10 shows the measurement of temperature using sensors which are classified on different hours of one day. Readings are obtained on hourly basis for morning, noon, and evening hours. The device works well for all the temperature or light intensity conditions. Simulation uses two routing protocols, for which LBR is based on multi-hop routing protocol, which works well for carrying the information to the destination. Thus, exponential pattern of reading obtained and it is obvious, as temperature in the morning begins with lesser value, reaches peak value then decreases to a lower value.

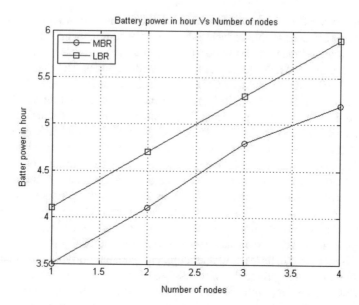

Fig. 9 Showing a battery consumption for LBR and MBR routing protocols

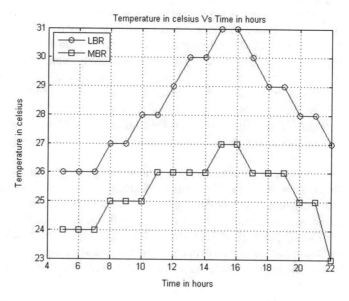

Fig. 10 Showing a temperature measurement for LBR and MBR routing protocols

6 Conclusion

The proposed research work is carried out to monitor the agriculture area using sensors. Most of the research work highlights on the sensor by directly communicating it to the gateway or cloud, which limits the communication range and it would be impossible to cover even 1 acre of land using plenty of sensors. The proposed work with the help of sensor sends the data to neighbors using routing protocols that will successfully cover 1 acre of land using four sensors to make it more cost effective. Future plan is to use different kinds of routing protocols like AODV, DSR to test the system performance.

References

1. Sakthipriya (2014) An effective method for crop monitoring using wireless sensor network. Middle-East J Sci Res 1127–132, Feb 2014
2. Million M, Marco Z, Antoine B (2012) Successful deployment of a wireless sensor network for precision agriculture in malawi-wipam. In: 3rd IEEE international conference on Networked Embedded Systems for Every Application, (NESEA-12), pp 1–6, Aug 2012
3. Deepti B, Reddy S (2013) WSN based closed loop automatic irrigation system. Int J Eng Sci Innov Technol 2(3): 229–237
4. Li X, Deng Y, Ding L (2008) Study on precision agriculture monitoring framework based on WSN. In: 2nd international conference on anti counterfeiting, security and identification (ASID), pp 182–185, 20–23 Aug 2008
5. Xiao L, Guo L (2010) The realization of precision agriculture monitoring system based on wireless sensor network. In: International conference on computer and communication technologies in agriculture engineering (CCTAE), vol 3, pp 89–92, 12–13 June 2010
6. Mafuta M, Zennaro M, Bagula A, Ault G, Gombachika H, Chadza T (2013) Successful deployment of a wireless sensor network for precision agriculture in Malawi. In: 3rd international conference on networked embedded systems for every application (NESEA), pp 1–7, 13–14 Dec 2012
7. Kussul N, Lemoine G, Gallego FJ, Skakun SV, Lavreniuk M, Shelestov AY (2016) Parcel-Based Crop Classification in Ukraine Using Landsat-8 Data and Sentinel-1A Data. IEEE J Selected Topics Appl Earth Observations Remote Sens 9(6):2500–2508
8. Chiang CT, Lin JX, Liu L (2015) Design of a CMOS Calibrated Monolithic Illumination Meter for Monitoring Solar Radiation of Tomato Crops. IEEE Sens J 15(9):5285–5290
9. Gutiérrez Jagüey J, Villa-Medina JF, López-Guzmán A, Porta-Gándara MÁ (2015) Smartphone Irrigation Sensor. In: IEEE Sens J 15(9):5122–5127
10. Kone CT, Hafid A, Boushaba M (2015) Performance Management of IEEE 802.15.4 Wireless Sensor Network for Precision Agriculture. IEEE Sens J 15(10):5734–5747
11. Qu Y, Zhu Y, Han W, Wang J, Ma M (2014) Crop leaf area index observations with a wireless sensor network and Its potential for validating remote sensing products. IEEE J Selected Topics Appl Earth Observ Remote Sens 7(2):431–444
12. Bhanu BB, Rao KR, Ramesh JVN, Hussain MA (2014) Agriculture field monitoring and analysis using wireless sensor networks for improving crop production. In: 2014 eleventh international conference on wireless and optical communications networks (WOCN), Vijayawada, pp 1–7
13. Liqiang Z, Shouyi Y, Leibo L, Zhen Z, Shaojun W (2011) A crop monitoring system based on wireless sensor network. Procedia Environ Sci 11:558–565

Cognitive Learning Environment and Classroom Analytics (CLECA): A Method Based on Dynamic Data Mining Techniques

Miftah Al Karim, Asif Karim, Sami Azam, Eshtiak Ahmed, Friso De Boer, Ashraful Islam, and Fernaz Narin Nur

Abstract With the advent of modern data analytics tools, understanding the bits and pieces of any environment with the abundance of relevant data has become a reality. Traditional post event analyses are evolving toward on-line and real-time processes. Along with versatile algorithms are being proposed to address the data types suitable for dynamic environments. This research would investigate different dynamic data mining methods that can be deployed into a modern classroom to assist both the teaching and learning atmosphere based on the past and present data. Time series data regarding student's attentiveness, academic history, content of the topic, demography of the classroom and human sentiment analysis would be fed into an algorithm suitable for dynamic operations to make the learning ambience smarter, resulting in better information being available to educators to take most appropriate

M. Al Karim
Department of Engineering, Computer and Math Sciences, Auckland University of Technology, Auckland, New Zealand
e-mail: miftah.alkarim@aut.ac.nz

A. Karim (✉) · S. Azam · F. De Boer
College of Engineering, IT and Environment, Charles Darwin University, Darwin, Australia
e-mail: asif.karim@cdu.edu.au

S. Azam
e-mail: sami.azam@cdu.edu.au

F. De Boer
e-mail: friso.deboer@cdu.edu.au

E. Ahmed
Faculty of Information Technology and Communication Sciences, Tampere University, Tampere, Finland
e-mail: eshtiak.ahmed@tuni.fi

A. Islam
School of Computing and Informatics, University of Louisiana At Lafayette, Lafayette, USA
e-mail: ashraful.islam1@louisiana.cdu

F. N. Nur
Department of Computer Science and Engineering, Notre Dame University, Dhaka, Bangladesh
e-mail: fernaznur@gmail.com

© The Author(s), under exclusive license to Springer Nature Singapore Pte Ltd. 2021
J. S. Raj et al. (eds.), *Innovative Data Communication Technologies and Application*,
Lecture Notes on Data Engineering and Communications Technologies 59,
https://doi.org/10.1007/978-981-15-9651-3_63

787

measures while teaching a topic. The research objective is to propose an algorithm that can later be implemented with proper hardware set-up.

Keywords Dynamic data mining · Time series analysis · Smart environment · Classroom behavior · Intelligent guidance

1 Introduction

Applications of intelligent technology in almost all the sectors of modern society have become particularly evident in the last five to six years. One of the key problematic issues arising out of today's in-class learning style is, in most cases, not being able to recognize student's degree of attention and the degree of mental involvement. The students may be sitting in the class, but it could very well be that their attention is dropping, or they are just not able to creatively engage with the lecturer. This seriously hampers their ability to learn, think, and remember the knowledge altogether. Cognitive learning is considered as a key method that encourages students to use their brains with increased efficiency as the level of engagement process remains high [1]. However, to develop a classroom setting that is conducive to cognitive learning requires high degree of digital involvement and data analytics to fully comprehend students' behavior, which dynamic in nature and varies with time and environment. In this research initiative, a framework has been proposed to collect a range of relevant data from a live classroom and through an innovative and effective data mining algorithm, valuable insights can be obtained that should boost the students' performance; and in fact will encourage them to learn how to learn and internalize concepts with ease. Students should also feel confident in connecting, explaining, and justifying their ideas. It has also been observed from previous research that learning analytics can in general can play a key role in enhancing and understanding of students' learning behaviors [2]. Additionally intelligent analysis of learning data proved useful for all parties alike such as policymakers, instructors, and learners [3].

1.1 Novel Contribution of This Research

This research proposes an effective and novel implementation of one or multiple algorithms suitably prepared to monitor, control, and provide suggestions to the audience as well as the instructor in a classroom. The dynamic platform would be capable of understanding different patterns in real time and make necessary adjustments or provide insights for a better management. The key outcome that our research attempt achieves;

(1) Analyze attentiveness of the audience as a function of time.
(2) Understand the dynamic nature of the student–content interaction.

(3) Understand the lecture delivery approaches based on the history of a student group.

(4) Make intelligent decisions in the forms of non-intrusive controls and suggestions that does not reveal confidential data of a student group.

The outcome of this research initiative would be to use simple classroom utilities such as microphone, close circuit cameras as data collection points and turn those data into valuable knowledge to control volume level, airflow, and timing of a content delivery and approach of a content delivery.

2 Related Works

This work is motivated from some of the previous attempts to introduce intelligence in a learning environment. One of those is embedded computer networking that makes it possible to implement new information paradigm called ambient intelligence (AmI). AmI can facilitate control schemes that takes full advantage of man–machine awareness. In [4], an AmI system is deployed based on speech recognition techniques, RFID technique to identify different role players, and a fuzzy approach for behavioral analysis. The purpose of implementing AmI is to control a test bed intelligent classroom. For understanding the learning approaches of students paper [5] uses the level of comprehension of a content. Based on this comprehension, a system is developed to scrape websites and present relevant data at teacher's disposal for better interaction with the students. In [6], a method has been proposed in a smart class room to detect the gesture of lecturer to strategically capture video and virtually control mouse pointer. The paper prepared a hybrid human model to understand the motion features and an algorithm called primitive-based-coupled hidden Markov model is presented for action recognition. In [7], performance of seven individuals has been analyzed based on different vocal behavior such as fundamental frequency, frequency range, jitter, shimmer, and words per minute. The idea is to search for a relation between teacher's vocal characteristics as well as being effective in the classroom environment to teach. In [8], researchers have deployed association rules mining and fuzzy representations to examine and analyze student learning, behaviors, and experience within a computer-aided classroom activities. Association rule mining [17] sheds light on how learners with various cognitive types interacted with a simulation for problem solving while to fuzzy representation was used to inductively gauge how students handle questionnaire data.

3 Proposed Methodology

The proposed method is based on classroom behavior of the audience (students) at different times of a session, at different contents of the session, and the past academic history of that student group.

The data input points are as mentioned earlier close circuit camera (if required multiple), strategically placed microphones, content of the lecture, and academic records. Cameras would be used to understand students' attentiveness toward the instructor. Simple image processing techniques such as eye-tracking and lip-tracking will be developed to understand different levels of attentiveness. At the same time, body movements or physical posture would be analyzed to predict whether students are enjoying the lecture or not. The microphones would be used to detect classroom noise. Noise-level detection would be carried out using signal-to-noise ratio. Here, the voice of the instructor is the signal and the humming generated due to the students talking to each other is the noise.

The signals generated from the instructor's voice would also be used to understand his/her psychological condition at real time. Instructors whether feeling happy to deliver a lecture or feeling a bit irritated toward a certain audience may produce signature in the signals. This research would use data mining techniques to investigate those emotion signatures. Application of data mining algorithm for developing intelligent system has been studied for quite sometimes now [18].

Based on the pattern generated in a classroom, finally, a decision-making process would be invoked to take intelligent decisions in form of non-intrusive suggestions or minor controls in classroom ambience such as volume control, airflow control, regulation of illumination, shedding, and the method of delivering the content. The overall project outline can be summarized as shown in Fig. 1.

3.1 Detailed Discussion on the Proposed Methodology

In this section, an overview of the research steps will be discussed. In a real classroom, multiple cameras would be proposed for the tracking purposes. Here, only one camera feed and one student have been chosen for demonstration purpose.

The first target data will be tracking of eyes. A student can either look toward the instructor or away based on his attention level. By tracking the cornea region, a trace of relative attentiveness can be done. The tracking will be carried out using eye detection followed by color thresholding and calculating the area of the bounding boxes. Similarly, by tracking the lips data on whether a student is paying attention, talking, or yawning can be aggregated. Figure 2 shows the method of tracking and some levels of identified patterns. These patterns are prepared to train a set of supervised dynamic data mining algorithms. For thresholds, the elbow method is the recommended one.

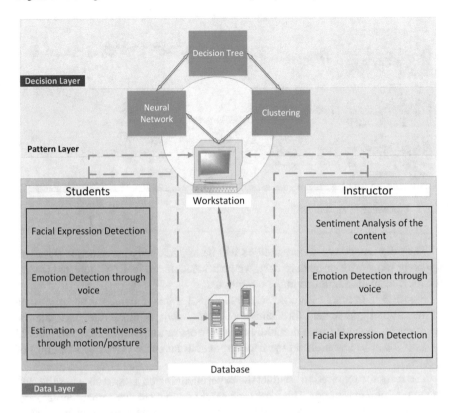

Fig. 1 Research work flow

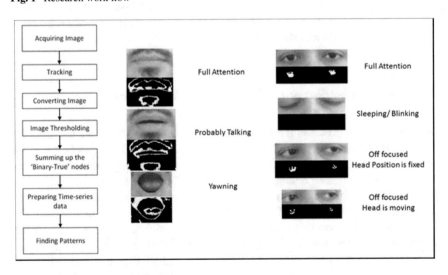

Fig. 2 Processing of eye and lip data

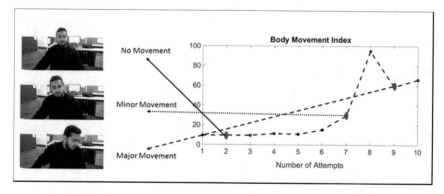

Fig. 3 Processing of movement data

The other image processing method will be to track body movements. Based on the body movement or posture data, whether students are feeling interested toward the topic or feeling fatigued may be estimated.

Figure 3 is used as reference where along with image data, a time series data of overall body movement for a small period is shown. In this figure, each attempt represents a second in time domain. In real-life operation, sample frequency would be a bit lower to comprehend the data throughout an entire period (40–60 min) of lecture.

The resource required to conduct the experiment inside a classroom is subjected to the size of a classroom. The ideal case would be to have one camera monitoring one student. However, depending on the aperture of the camera, the number can vary. For detecting the motion of the body, a distance-based analysis needs to be carried out. It is because the body movement from one meter versus the body movement detected from a ten meter distance would vary.

3.2 Data Collection and Mining the Obtained Data

Only using image processing may not provide accurate results all the time. As the data developed here is dynamic in nature, accuracy would be very important for any data mining platform to retrieve meaningful and reusable patterns [14]. That is why strategically placed microphones would also be used. The primary job of the microphones will be to collect voice signals both from the instructor and inattentive student group creating noises. Figure 4 shows three different types of voice signals. The first type is the composite signal that has both the voice of instructor and students' humming in the background. The second signal is the instructor's voice (authenticated voice signal can uniquely identify an individual with high degree of certainty [13]), and the third one is the filtered signal that has gone through a *butterworth high-pass filter* [16] in order to cut out the low humming sounds as much as possible.

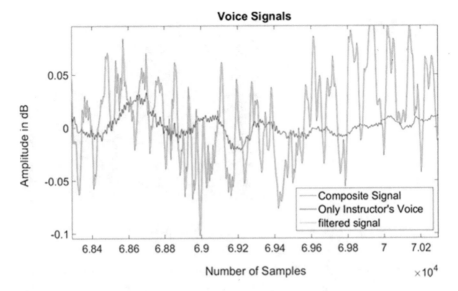

Fig. 4 Voice signals

Once the voice signals are captured and recorded, the signal-to-noise ratio would be used as a time series data to understand whether the students are attentive or talking to each other. Figure 5 shows the signal-to-noise ratio in terms of the voice signal of the instructor and the level of humming noises created by the inattentive students. The noise is calculated by simply finding out the differences between the composite signals and the filtered signals.

Along with these classroom data, other types of data such as student's academic background, previous reactions toward similar lectures, instructor's methods of delivering the lecture, and the characteristics of the content would also be considered. If found relevant classrooms geometry can also be taken account of, as at different season of the year, window side seats may have different comfort levels that may in turn put impact toward learning. Figure 6 shows the temperature distribution in a

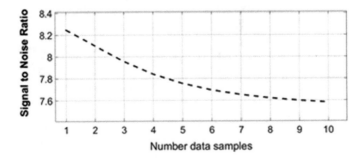

Fig. 5 Signal-to-noise ratio

Fig. 6 Classroom
temperature distribution

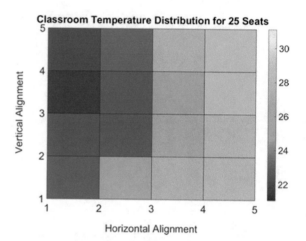

classroom at summer time. The data is randomly prepared to test the algorithm. The right side of the picture represents window side seats at a sunny summer day. The idea here is to find out the correlation between temperature distribution and irritation developed among the students.

Finally, difficulty level of the content and student's academic background can also be correlated to understand the student–topic interaction. In Fig. 7, a fictitious relation is shown between student's CGPA and mean attentiveness throughout a sixty minute session that has a correlation of '0.8387'. Based on different complexity level of the content, several of such analysis would be carried out. These later types of data would be used to train and prepare the unsupervised data mining algorithms.

Fig. 7 CGPA versus
attentiveness

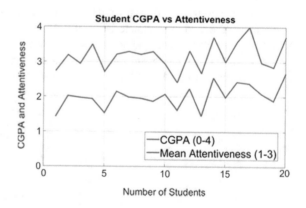

4 Proposed Algorithms and Expected Outcome

To understand different patterns in the data at first, a test set of different data types would be prepared. Below in Fig. 8, an example of the visualization of a set of data is given. The data is prepared considering that students are attending contents of same difficulty level on two different days. The circles in the plot show the cluster prepared from an algorithm called 'k-means' [9]. The other types of clusters would be based on the content levels of the topic and attentiveness of the students, etc. The recurrent neural network (RNN) would then be used to understand and predict the student's attentiveness based on the previous data. The RNN would be modified in real time based on per minute-based attentiveness of the students.

In a recurrent neural network (RNN), which is a type of artificial neural network (ANN) [12], connections between nodes usually generate a directed graph along a temporal sequence, giving it the edge in exhibiting temporal dynamic behavior [10].

In simplest of terms, (1) and (2) define how an RNN evolves over time:

$$O^t = f_1(h^t; \Phi) \tag{1}$$

$$h^t = f_2(h^{t-1}, k^t; \Phi) \tag{2}$$

where O^t is the output of the RNN at time t, k^t is the input to the RNN at time t, and h^t is the state of hidden layers at time t.

As RNNs work with a feedforward approach, RNNs can utilize their internal state, often simply called 'memory' to process variable length sequences of inputs. This makes them suitable for data processing where data features are seemingly unsegmented but connected in some way [10]. RNNs are well equipped to process arbitrary length sequence data.

Finally, based on these two, supervised and unsupervised learning methods [15], physical non-intrusive decision process would be carried out using decision tree

Fig. 8 Unsupervised clusters

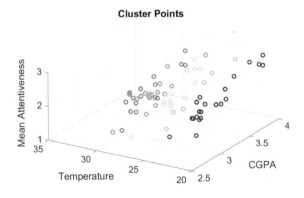

algorithm (DT). DT based 'Learning', in majority of the instances makes use of an upside-down tree-based progression method [11].

As the study requires significant effort in pattern mining, the three proposed machine learning algorithms have to work together. For example, to train the supervised algorithm NN and DT, a previously observed and agreed pattern has to be used. Once the system starts collecting data and predicting, based on the accuracy of the algorithm, a bidirectional relation has to be established. For the unsupervised machine learning activities, along with the traditional elbow method, an exploratory data analysis can be used to predefine the number of clusters.

The cluster index has been calculated based on the elbow method and the threshold is set at 5.

5 Conclusion

This research proposes a novel algorithms which can effectively be deployed to monitor, control a classroom and raise recommendations that can highly improve students' attentiveness and knowledge internalization ability as well as insights for a better management. The proposed platform works in real time and is able to adapt to the changes in a range of atmospheric and situational variables. Additionally, the system aids greatly in comprehending the lecture delivery approaches based on the history of a student group and producing intelligent decisions in the forms of non-intrusive controls and suggestions without revealing confidential data of a student group.

The proposed system will bring in the edges of digital technologies into the classroom and will boost the cognitive learning ability within the student cohort; thereby, benefitting not only the students' themselves but also the instructors and the academia in general. It is also believed that the advancement of technologies such as artificial intelligence and image recognition will dramatically increase the effectiveness of the system.

References

1. Prestine NA, Legrand BF (1991) Cognitive learning theory and the preparation of educational administrators: implications for practice and policy. Educ Administration Q 27:61–89
2. Wong A, Chong S (2018) Modelling adult learners' online engagement behaviour: proxy measures and its application. J Comput Education 5:463–479
3. Hwang GJ, Hung PH, Chen NS, Liu GZ (2014) Mindtool-assisted in-field learning (MAIL): An advanced ubiquitous learning project in Taiwan. Educational Technol Soc 17(2):4–16
4. Ramadan RA, Hagras H, Nawito M, Faham AE, Eldesouky B (2010) The intelligent classroom: towards an educational ambient intelligence testbed. In: 2010 Sixth international conference on intelligent environments

5. Talwar J, Ranjani S, Aras A, Bedekar M (2013) Intelligent classroom system for qualitative analysis of students' conceptual understanding. In: 2013 6th international conference on emerging trends in engineering and technology (ICETET), pp 25–29

6. Ren H, Xu G (2002) Human action recognition in smart classroom. In: Proceedings of fifth IEEE international conference on automatic face gesture recognition, pp 417–422

7. Schmidt CP, Andrews ML, Mccutcheon JW (1998) An acoustical and perceptual analysis of the vocal behavior of classroom teachers. J Voice 12:434–443

8. Angeli C, Howard SK, Ma J, Yang J, Kirschner PA (2017) Data mining in educational technology classroom research: can it make a contribution? Comput Educ 113:226–242

9. Karim A, Azam S, Shanmugam B, Kannoorpatti K, Alazab M (2019) A comprehensive survey for intelligent spam email detection. IEEE Access 7:168261–168295

10. Bang S, Tijus C (2018) Problem solving using recurrent neural network based on the effects of gestures. In: Proceedings of the 10th international joint conference on computational intelligence

11. Marouf AA, Ashrafi AF, Ahmed T, Emon T (2019) A Machine Learning based approach for mapping personality traits and perceived stress scale of undergraduate students. Int J Modern Educ Comput Sci 11:42–47

12. Mohamed A, Mazumder MDA (1999) A neural network approach to fault diagnosis in a distribution system. Int J Power Energy Syst 19(2):696–703

13. Kathed A, Azam S, Shanmugam B, Karim A, Yeo KC, Boer FD, Jonkman M (2019) An enhanced 3-Tier multimodal biometric authentication. In: 2019 international conference on computer communication and informatics (ICCCI)

14. Albayati MB, Altamimi AM (2019) An empirical study for detecting fake facebook profiles using supervised mining techniques. Informatica 43

15. Alloghani M, Al-Jumeily D, Mustafina J, Hussain A, Aljaaf AJ (2019) A systematic review on supervised and unsupervised machine learning algorithms for data science. Unsupervised and semi-supervised learning supervised and unsupervised learning for data science, pp 3–21

16. Dogra A, Bhalla P (2014) Image sharpening by Gaussian and Butterworth High Pass Filter. Biomed Pharmacol J 7:707–713

17. Chen F, Chen YPP (2014) Association rule mining (Frequent Itemset, Association Rule, Support, Confidence, Correlation Analysis). Dictionary of Bioinformatics and Computational Biology

18. Joseph SIT (2019) Survey of data mining algorithm's for intelligent computing system. J Trends Comput Sci Smart Technol (TCSST) 1:14–24

19. Sathesh A (2019) Enhanced soft computing approaches for intrusion detection schemes in social media networks. J Soft Comput Paradigm (JSCP) 1:69–79

Cloud Computing Adoption: Influencing Factors on Online Firms

Manjula Shanbhog

Abstract In recent years, cloud computing is remaining as a contemprory technology that adds feather to the informtion science paradigm. Cloud technology has several benefits that need to be considered while adopting it to overcome the existing challenges. This proposed research paper identifies significant factors for online firms to consider this technology and analyze their effects on online business firms by collecting data from various firms to examine and intrepret the rsesult.

Keywords Cost saving · Space saving · Cloud service provider reputation · Data control · Data security · Attitude

1 Introduction

Cloud computing provides resources for deploying information technology as services more attractively, reasonably, and easily to the world of business. Cloud computing and e-commerce [1] are like two faces of the same coin. The services provided by them can be regarded as time efficacious and cost-efficacious. Cloud computing helps in reducing the expenses that happens while setting the IT infrastructure of an organization [2], where e-commerce helps the online stores in establishing the business without using any physical setup.

Cloud computing helps the e-commerce owners to eliminate the concern and expenditure while investing the personal standalone desktops along with its maintainance and upgradation of hardware and software.

According to Svantesson and Clarke [3], cloud computing is referred as The technology, where user uses the computer/desktop at place, but its hardware and/or software are operated, monitored and maintained by the third party called Cloud service provider from another remote place.

M. Shanbhog (✉)
CET IILM Greater Noida, Noida, India
e-mail: manjulashanbhog@gmail.com

Cloud technology offers many models of services. XAAS can be denoted as the service taxonomy where X denotes "anything or everything," S denotes "Service," so XAAS stands for "Anything as a Service" [4].

2 Research Objectives

The objectives of this study are as follows.

(1) Analyze the significant aspects that affect the perception to accept cloud computing automation by online companies.
(2) Analyze the effect of these significant aspects affecting the perception to accept cloud automating services by online companies.

3 Literature Review

The most significant intention of any organization is to strengthen and upgrade it financially. To achieve this, one of the point to be considered is to decrease its possible expenditure; this can be achieved fairly using cloud computing automation. According to Marston et al. [5], managing fund and controlling the expense are the main diffculty that an organization faces, so the most important criteria for adopting cloud computing is the advantage of cost with regard to capital and operational expenditure. Cloud technology on the other hand reduces investment cost, reduces operation and maintenance cost, and reduces hardware and software cost. Rather than spending the money for implementing and maintaining software and hardware infrastructure, cloud computing offers services to businesses to access hardware and software on rental basis exterrnally [6].

Accordingly the Hypothesis formulated is.

Hypothesis 1: The Cost Saving Significantly Affect Online Firms Trust

Onsite infrastructure occupies space in the organization. Moreover, the bigger the organization, the more infrastructure is needed, and so more space will be occupied by these equipments. So the precious space will be unable to utilize for other purpose [7]. In a virtualized environment, computing resources can be dynamically created, enlarged, diminished, or moved as demand varies [8]. An organization can employ virtualization which gains greater elasticity while encouraging more effective use of IT resources. Any business company tend to have huge amount of data, they need to worry for running out of storage memory or require to buy more storage disk. Cloud computing gives the option of storing data on someone else's hardware and the user need not have to worry about the storage space [9]. From the above review of literature, the hypothesis formulated is.

Hypothesis 2: The Space Saving Significantly Affect Online Firms Trust

Privacy is an issue which hinders the usage of cloud technology. It is about the protection and careful use of the customers personal information. The cloud user provides the database with sensitive information of organization or individual, so privacy concerns will continue to grow [10]. Siani Pearson [11] argued that it is very important to take privacy into account when designing cloud services. From the above review of literature, the hypothesis formulated is.

Hypothesis 3: The Concern for Information Secreacy Significantly Affect Online Firms Trust

Security is another major issue that hinders the acceptance of cloud technology. According to Nariman Mirzaei [12], the cloud computing is going to happen but the most important factor need to be focused is security. Security issues such as phishing, data loss are some of the serious problems faced by the cloud consumer. So Security is the powerful preventing factor against the effective implementation and deployment of cloud computing services for individuals and organizations [13]. For the world wide popularity in accepting cloud technology, Security has encountered as the major obstacle [14]..

From the above review of literature, the hypothesis formulated is.

Hypothesis 4: The Concern for Information Safety Significantly Affect Online Firms Trust

The cloud service provider are managers also, i.e., they look after the physical infrastructure, the infrastructure includes various resources, whcih includes "data" as one of them, and the service user feels insecurity because the data is in others environment. Ali Khajeh-Hosseini et al. [15] stated that the cloud customer is worried and insecured about his data handing over to the cloud. In cloud, the customers data is spread all over the world. The term "control on resources" or "the ownership of resources" is directly proportional to the term "trust on managing system," as we have less amount of control on resources we trust less on the managing system [16]. The customer of cloud is the owner of his information, but he has lack of control over the data, lack of visibility over the data and resources, and he is concerned about the data loss also. The customer has to trust the service provider to maintain a practicable trust association. Service level agreement plays an important role between service providers and cloud service user and helps to develop confidence and faith with each other [17].

From the above review of literature, the hypothesis formulated is.

Hypothesis 5: The Concern for Data Control Significantly Affect Online Firms Trust

Cloud service users have little knowledge about the service provider and thus they are unsure to have trust on them [18]. Especially, users' are confused and worried about the potentials of cloud providers [19]. Selecting the type of cloud and the service is not so easy for a customer or an organization to choose, because there are a lot of providers who allege to offer the best possible service. It would be easier to rely on the service providers reputation. The reputation of cloud services or cloud service providers will certainly influence cloud users' options to select the cloud services [20].

From the above review of literature, the hypothesis formulated is .

Hypothesis 6: The Cloud Computing Service provider's Reputation Significantly Affect Online Firms Trust

One of the important features in human behavior is considered to be "Trust" [21]. When people or parties engage in transaction with a particular entity, if they rely on it, then the entity is regarded as a trustworthy entity [22]. Generally, we say that a person or an organization is reliable if it is trustable [22]. As the service user looses the access control of his resources, so there arises a sense of insecurity and which leads to dearth of trust on service provider. This reason has hampered the admittance of wonderful computing technology as cloud services [23]. Trust and security must be build by the cloud service providers to reduce the concern of a user to embrace cloud technology [23].

From the above review of literature, the hypothesis formulated is.

Hypothesis 7: The Online Firms Trust Will Significantly Affect the Perception to Accept Cloud Technology

According to planned behavior theory, attitude is human pscyhological attribute that states the feeling as good or bad toward a particular situation or thing. Organizations various factors are considered in deciding to adopt this contemproary technology. An organization's attitude influences a lot to adopt the new technology [24]. Attitude with other constraints were considered to be the significant factors of behavioral intentions to use the cloud services [25].

From the above review of literature, the hypothesis formulated is.

Hypothesis 8: The Online Firms Attitude Will Significantly Affect the Perception to Accept Cloud Technology

4 Research Methodology

From the literature review, various components were identified that affect the perception to accept this contemproray computing technology by online conpanies. To analyze research objective 2, exploratory method and descriptive method both are considered. This study used survey which helped to collect general view of people related to the topic. Non-probability, convenience and snow ball sampling method is used to study the above research objective. The questionnaries were formulated with close-ended structure and responses for the questions need to be choosed from the given set of options. The structured questionnaire along with the set of options were distributed to the respondents. Cronbach's coeffcient Alpha method is used to test the quentionnaire quality.

Karl Pearson's coefficient of correlation is used to analyze the hypothesis in this study. This is the most popularly used method to check the dependencies between two variables. Regression method is used in this research to measure the dependency level among the independent and dependent variables.

5 Result and Discussion

5.1 Reliability Test

The consistency or the quality of the constructed questionnaire measured through Cronbachs Alpha resulted in 0.784, which dipicts that the result obtained through the formulated questionnair will be almost the same and is independent of the time and places of the responses collected.

5.2 Sample and Response Rate

Among 500 individuals, only 75 of them responded to the survey. From a total number of 75 respondents, the responses included for the main study were only from those who were either full-time or part-time employee or who were self-employed or business owner, among these again the filter was applied and collected responses for main study were only from those who were either business owner or with overall responsibility for online communication or online retail, i.e., the information was collected only from those who were capable of taking online communication or online retail decision for their company. Again the filtration was applied and the responses were collected only from those who knew cloud computing technology. So, from 75 total responses, 21 responses were filtered out. Therefore, the total potential sample number decreased to 54.

5.3 Descriptive Statistics

Demographic Characteristics

The respondents demographic characteristics which this research used for analysis along with their response is given in table. The respondents were majorly classified on gender, occupation, and employment position (Table 1).

Table 1 Demographic characteristics

	Respondents	Count	Percentage
Gender	Female	24	33.2
	Male	51	66.8
Occupation	Employed fulltime	37	49.3
	Employed part time	17	22.7
	Self-employed/Business owner	14	18.7
	Home duties	2	2.7
	Student	4	5.3
	Looking for work	1	1.3
Employment position	Business organization employer/owner	22	32.4
	Online communication/Information technology head	38	55.9
	Head Person/Manager–who is reponsible for taking decision on daily basis for smooth running of the business	5	7.4
	Business organization employee	3	4.4

Correlation Analysis

As stated, the author has used correlation method to observe the dependencies among the vaious variables and coefficient of significance considered in this study is 0.05. If the significant value obtained for a hypothesis is less than 0.05, it indicates that the stated hypothesis is true and verified according to the data obtained. so accordingly hypothesis which supported are H1, H6, H7, and H8 (Table 2).

H1 supported implies online firms perception to accept cloud computing technology directly proportional to cost saving. Online firm believes that if they adopt the technology then cost saves.

Similarly, according to hypothesis H6, cloud service providers reputation plays a significant role on firms to accept the cloud technology for their business, i.e., online firm trust those service provider who are reputed.

According to hypothesis 7, Online firms trust on cloud technology motivates them to adopt it.

From hypothesis 8, it indicates that the attitude of online firms toward the technology is the decision factors for its adoption, i.e., if the firms have positive attitude toward the technology then firms go for adopting it.

Multiple Linear Regression Analysis

Regression Analysis 1

This analysis method is used to analyze the amount of dependencies that exit among the experimental variables. This linear model contains a variable "Trust" which is

Table 2 Summarized report for the above proved hypothesis

Hypothesis	Pearson correlation	Significant value	Result
H1	0.249	0.035	H1 is supported
H2:	0.155	0.131	H2 is not supported
H3:	−0.140	0.157	H3 is not supported
H4:	-0.149	0.142	H4 is not supported
H5:	−0.163	0.120	H5 is not supported
H6:	0.383	0.002	H6 is supported
H7:	0.848	0	H7 is supported
H8:	0.576	0	H8 is supported

dependent on 6 independent variables, they are "Cost Saving," "Space Saving," "Worriness regarding Information Secreacy," "Worriness regarding Information Safety," "Worries regarding Goverenence of data," and "reputation of cloud service provider."

The data of interest from the above table are correlation coefficient (R), goodness of fit (R square), and significance of coefficient (sig). The correlation coefficient value 0.469 indicates that direct relationship exists among the dependent and independent expiremental variables. The goodness of fit (R square) is 0.220 which shows that independent variables influences 22% on the dependent variable or it can be interpreted as 22% variation exits. The significant value obtained is 0.002, which is less than 0.05 coefficient of significance; this shows the considered model is statistically significant (Table 3).

From the above table, it makes us clear that the factor "Trust" majorly depends on CCSPR-Cloud computing service provider reputation. So we can conclude that online firms can build trust on this novel cloud technology if the service providers are reputed or reliable.

Table 3 Linear Regression-a

R	R^2	Adjusted R^2	Standard.error of the estimate	F Value	Significance
0.469	0.220	0.190	0.936	7.203	0.002

Table 4 Coefficient table

Coeff	Unstd coeff		Standardized coefficients	T value	Significance
	B values	Standard error	Beta value		
Constant Value	1.811	0.490		3.699	0.001
Cloud computing service provider's reputation(CCSPR)	0.461	0.128	0.465	3.600	0.001

Regression Analysis 2

For the remaining part of the model, the independent factors are "Trust" and "Attitude" and the concluding dependent factor considered is "Perception to accept cloud computing technology."

The significant value obtained is 0.000 which is less than 0.05 considered value and so the model obtained in significant. R^2 valuee is 0.745, which is quite high, indicates both the independent factors, i.e., "Trust" and "Attitude" influence 74.5% on the dependent variable "Perception to accept cloud computing technology."

Between Trust and Attitude, the factor "Trust" greatly affects the factor "Perception to accept cloud computing technology" when compared to the factor "Attitude". From table it shows that the independent variable "Trust" influences 77.9% on dependent variable "Perception to accept cloud computing technology." The another independent variable "Attitude" controls dependent variables "Perception to accept cloud computing technology" by 28.3%. The above model considered is statistically significant (Table 4).

6 Conclusion

This research paper anlyzes the acceptance of cloud technology by online companies. 8 hypothesis were formulated. To assess the hypothesis survey sampling methodology was conducted. Various dependent and independent factors were identified, among them independent factors identified are – cost saving, space spaving, worriness regarding information secreacy, worriness regarding information safety, woriness regarding governance of data and reputation of cloud computing service provider. The dependent variable considered is "Trust on cloud technology." i.e., the Trust on cloud technology was dependent on various independent factors. The hypothesis and model was tested using corelation and regression methods. The research conducted indicates that online companies Trust(T) majorly get influenced by the independent factors "cost saving(CS)" and "CCSPR-Cloud Computing service provider Reputation."

In the second stage, the independent variables "Trust on cloud computing technology" and "attitude of online firms" were considered while the final dependent

Table 5 Regression analysis 2

R	R^2	Adjusted R^2	Standard.error of the estimate	F value	Significance
0.863	0.745	0.735	0.565	70.285	0.000

Table 6 Co-efficient table

Coefficients					
Regression	Unstd coeffs		Std coefficients	T value	Sinificance
	B	Standard. error	Beta value		
(Constant)	-3.0×10^{-1}	4.31×10^{-1}		-6.98×10^{-1}	4.90×10^{-1}
Attitude	2.83×10^{-1}	1.27×10^{-1}	1.89×10^{-1}	22.07×10^{-1}	0.33×10^{-1}
Trust	7.79×10^{-1}	0.89×10^{-1}	7.51×10^{-1}	88.34×10^{-1}	0.00×10^{-1}

variable was "Perception to accept cloud technology." The hypothesis 7 and 8 were tested again using corelation and regression methods. We clearly got the result as "Perception to accept cloud technology" mainly influenced by Trust(T) and attitude of the Web-based-company toward cloud technology (Tables 5 and 6).

So, for online firms to adopt cloud technology, cloud service providers reputation plays a significant role. So the service provider has to develop and gain trust on online firms toward themselves. The trust is developed if the online firms get what ever they were promised by the service provider. To keep a check on the provided and received services with various parameters, a log need to be maintained, and this can be attained and maintained by the level and aggrements of the services(SLA) which is an official compromization database between the user and provider of cloud services.

References

1. Abdulkader SJ, Abualkishik AM (2013) Cloud computing and E-commerce in small and cloud computing and e-commerce in small and challenges. International Journal of Science and Research (IJSR) 2(12):285–288
2. Buyya R, Yeo CS, Venugopal S, Broberg J, Brandic I (2009) Cloud computing and emerging IT platforms: Vision, hype, and reality for delivering computing as the 5th utility. Future Generation Comput Syst 25(6):599–616
3. Svantesson D, Clarke R (2010) Privacy and consumer risks in cloud computing. Computer Law Secur Rev 26(4):391–397
4. Shanbhog M, Singh M (2015) Benefits and Challenges that Surrounds Around Adopting Cloud Computing in E Commerce. Int J Sci Res 4(8):1419–1423
5. Marston S, Li Z, Bandyopadhyay S, Zhang J, Ghalsasi A (2011) Cloud computing—the business perspective. Decis Support Syst 51:176–189
6. Cfo Research (2012) The business value of cloud computing. Cfo Publishing
7. Kell T (2015) Reduce your on-site computer hardware costs with cloud computing. Retrieved From Http://Blog.Marconet.Com

8. Sharma A, Vatta S (2013) Cloud Computing: Taxonomy And Architecture. Int J Adv Res Comput Sci Softw Eng 3(5):1410–1417
9. Miller M (2008) Cloud computing: web-based applications that change the way you work and collaborate online. Que
10. Wang J, Zhao Y, Jiang S, Le J (2010) Providing privacy preserving in cloud computing. IEEE
11. Pearson S (2009) Taking account of privacy when designing Cloud Computing services. In: Icse workshop on software engineering challenges of cloud computing, Vancouver (pp 44–52). Ieee
12. Mirzaei N (2009) Cloud computing. Independent Study Report , pp 2–4
13. Yeboah-Boateng EO, Essandoh KA (2014) Factors influencing the adoption of cloud computing by small and medium enterprises in developing economies. Int J Emerg Sci Eng (Ijese) 2(4):13–20
14. Chen Y, Paxson V, Katz RH (2010) What's new about cloud computing security?
15. Khajeh-Hosseini A, Sommerville I, Sriram I (2010) Research challenges for enterprise cloud computing, The 1st ACM symposium on cloud computing, SOCC
16. Global Netoptex Incorporated (2009) Demystifying the cloud important opportunities, crucial choices. Http://Www.Gni.Com. pp 4–14, 13 Dec 2009
17. Chandrahasan RK, Priya SS, Arockiam L (2012) Research Challenges and security issues in cloud computing. Int J Comput Intell Inf Sec 3(3)
18. Habib SM, Hauke S, Ries S, Muhlhauser M (2012) Trust as a facilitator in cloud computing:a survey. J Cloud Comput
19. Km K, Q, M. (2010) Establishing trust In cloud computing. It Professional 12:20–27
20. Huang J, Nicol DM (2013) Trust mechanisms for cloud computing. J Cloud Comput
21. Pearson S (2013) Privacy and security for cloud computing. (Pearson S, Yee G eds) Springer
22. Zissis D, Lekkas D (2012) Addressing cloud computing security issues. Future Gener Comput Syst 28(3):583–592
23. Hwang K, Li D (2010) Trusted cloud computing with secure resources and data coloring. IEEE Internet Computing 14–22
24. Bharadwaj SS, Lal P (2012) Exploring the impact of cloud computing adoption on organizational flexibility: a client perspective. In: Proceedings of 2012 international of cloud computing, technologies, applications and management, IEEE
25. Alotaibi MB (2014) Exploring users' attitudes and intentions toward the adoption of cloud computing in Saudi Arabia: an empirical investigation. J Comput Sci 10(11):2315–2329

Change Detection in Remote-Sensed Data by Particle Swarm Optimized Edge Detection Image Segmentation Technique

Snehlata, Neetu Mittal, and Alexander Gelbukh

Abstract Satellite images help in monitoring change detection as they are the big repository of information. An imperative task from the prospects of land development monitoring, disaster management, resource management, and environment evaluation is change detection. For change detection, segmentation of an image is being performed for locating the areas of interest. Nature-inspired particle swarm optimization is a metaheuristic algorithm that is simple, robust, and makes a fewer number of assumptions for the problem considered. This paper implements a particle swarm optimization (PSO) algorithm in MATLAB environment as edge detection segmentation technique for satellite images, which are being acquired from Google Earth. For qualitative analysis, the results are compared with the conventional edge detector operators such as Sobel, Canny, and Prewitt with the help of entropy values. It has been observed that PSO outperforms the conventional edge detection image segmentation methods, thereby giving better edges and clarity in images for change detection.

Keywords Particle swarm optimization · Artificial intelligence · Satellite images · Segmentation

1 Introduction

In image processing, the most basic operation is image segmentation for a better understanding of images and is considered to be the first phase in image processing.

Snehlata (✉) · N. Mittal
Amity University Uttar Pradesh, Noida, India
e-mail: snehsheoran312@gmail.com

N. Mittal
e-mail: nmittal1@amity.edu

A. Gelbukh
Instituto Politécnico Nacional Mexico, Mexico, Mexico
e-mail: gelbukh@cic.ipn.mx

Image processing helps in the visual appearance of an image along with preparing the images for analysis. Segmentation is about identifying the areas of interest as it subdivides an image into similar regions. Various practices find their application in the field of image segmentation. The major application of image processing covers medical images and satellite images. Satellite images are also a big storehouse of information. Processed satellite images have applications ranging from land use land cover monitoring, natural resource management, forest fires, natural disasters, change detection, and many more. Two major approaches for change detection in satellite images are object-based and low-level local approaches [1]. Remote-sensed data has been characterized by four different types of resolutions such as spectral, spatial, temporal, and radiometric. The extraction of edges from remotely sensed data is also very crucial as the edges will help us identifying changes. The changes are studied over a sometime for the identification of change detection for land, water, resources, natural disasters, and many more. The extraction of information from such a dataset is a taxing task and has invited the interest of many researchers. Nature-inspired algorithms such as particle swarm optimization, artificial bee colony algorithm, ant colony optimization also find application in processing remotely sensed data. These swarm optimization algorithms represent a subset of artificial intelligence and are gaining importance as these techniques are offering great flexibility and changeability [2]. Particle swarm optimization is considered to be a metaheuristics algorithm that makes no or fewer assumptions about the problem which needs to be optimized. It can also perform search operations where the candidate solutions are large in number. It is also robust and can show high efficiency in locating the global optimal solution. PSO is also considered to be a derivative-free technique and is also less sensitive to the kind of objective function [3].

The paper has the following structure. Section 2 discusses the related work, Sect. 3 offers the proposed methodology based on particle swarm intelligence, and the results are being discussed in Sect. 4 followed by a conclusion in Sect. 5.

2 Related Work

Bhandari et. al [4]. considered color satellite image segmentation by putting in use the cuckoo search algorithm. It has been supported by Tsallis entropy for multilevel thresholding and was evaluated against the wind has driven optimization, artificial bee colony (ABC), differential evolution (DE), and PSO, where the proposed method turned out to be more robust and effective. Bhandari et al. [5]. suggested the application of modified ABC-based segmentation for unearthing the optimum multi-level thresholds and the technique was evaluated against genetic algorithm (GA), PSO, and ABC algorithm by considering Kapur's, Otsu, and Tsallis objective functions.

Senthilnath et al. [6]. proposed the clustering approach using a novel bat algorithm for finding solutions in crop-type classification with the help of a multispectral satellite image and compared the results with other two meta-heuristic approaches such as particle swarm optimization and genetic algorithm. Muangkote et al [7]. enhanced the optimal multilevel thresholding effectively by implementing an improved moth-flame optimization algorithm on satellite images and comparison was done with five existing methods. Ryalat et al. [8]. investigated PSO, fractional-order Darwinian particle swarm optimization, and Darwinian particle swarm optimization for the segmentation of medical images, and it was observed that FOD-PSO is better with respect to speed, stability, and accuracy. Zhao et al. [9]. used 2D Kullback–Leibler divergence and modified PSO for presenting a segmentation algorithm, which in turn proved to be effective & robust.

Suresh and Lal [10]. presented an improved differential-PSO by using chaotic sequences that have replaced the random sequences and comparison was done with PSO, harmony search, cuckoo search, and differential evolution. PSNR, mean, MSE, STD of fitness values, SSIM, and total execution time parameters were considered for validating the experimental results. Kapoor et al. [11]. presented the Gray Wolf Optimizer Algorithm application for the segmentation of satellite images. The implemented application is efficient and accurate with respect to average intra & inert cluster distance & DB index. Naeini et al. [12]. used PSO in consort with minimum distance classifier for object-based classification in satellite images with very high spatial resolution and compared the performance with honey bee mating, artificial bee colony, and genetic algorithm, and it outperformed all the other techniques. Jasmine and Annadurai [13]. presented image enhancement of real-time video by making use of PSO along with adaptive cumulative distribution. Liu et al. [14] worked in the area of image segmentation by presenting an improved hybrid PSO. Gamshadzaei and Rahimzadegan [15]. provided a PSO-based index for water spread area detection by using satellite images. Chang et al. [16] presented the application of PSO for band selection of hyperspectral images. The proposed method indicated better classification accuracy and dimensionality reduction rate. Borjigin and Sahoo [17]. presented the use of the PSO algorithm for obtaining optimal threshold values for each component of an RGB image. Di Martino and Sessa [18]. used Chaotic Darwinian PSO for multi-level image thresholding. The method was applied to images that were initially compressed by using fuzzy transforms. Nandini and Leni [19]. presented a shadow detection algorithm with the help of PSO. The resulting accuracy was validated by using recall parameters and precision. Chakraborty et al. [20]. applied an improved PSO-based multi-level image segmentation technique. The technique used minimum cross entropy thresholding.

3 Proposed Methodology

3.1 Particle Swarm Intelligence

Kennedy and Eberhart pioneered the particle swarm optimization algorithm [21] and it is based on swarm social behavior. For obtaining the optimal solution, this algorithm performs the global best solution concept in use. Particles are generally referred to as agents and are responsible for performing optimal solution search. The trajectories of the agents are adjusted by deterministic and stochastic components. The particles move randomly and are also influenced by their best-achieved position and group position. The algorithm of PSO is represented as below [22].

1. Input: Randomly initial particles are generated
2. Output: Fitness value along with optimal particle
3. Start: Initialization of swarm
5. Evaluation of all particles using FF by using eq.
6. While not met = termination criteria, do
7. Velocity update
8. Each position update
9. Fitness function evaluation
10. Worst particle are replaced by best particle
11. LB and GB update
12. End while
13. Informative features subset returned.

PSO works on two main factors, represented in Eq. 1 as position update and Eq. 2 presents velocity update [22]. Here, LB_1 and GB_1 represent the current local and global solutions at 1st iteration, $rand_1$ and $rand_2$ represents random numbers and are in the range of [0,1], w is the inertia weight value and the constants are represented as c_1 and c_2.

$$X_{ij} = X_{ij} + v_{ij} \tag{1}$$

where

$$v_{i,j} = w * \times v_{ij} + c_1 \times rand_1 \times \left(LB_1 - x_{i,j}\right) + c_2 \times rand_2 \times \left(GB_1 - x_{i,j}\right) \tag{2}$$

This paper is to propose the application of particle swarm optimization-based edge detection for change detection in remotely sensed data. The process followed in the proposed methodology is depicted in Fig. 1. The first step is to read the input satellite image, followed by the initialization of parameters such as position and velocity along with the fitness function. Each particle keeps track of local best as well as global best solutions in the solution space. The final output image with edge detection is obtained and the entropy value for the same is computed and compared with the conventional methods.

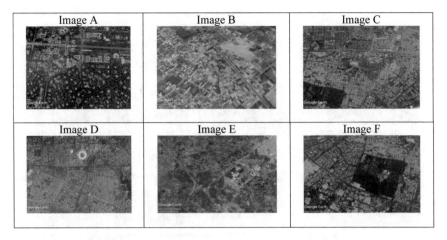

Fig. 1 Proposed methodology flowchart

4 Result and Discussion

The method was executed on 06 satellite images acquired from Google Earth is represented in Fig. 2. For every image, conventional edge detector operators such as Sobel, Canny, and Prewitt are being applied in the MATLAB environment. To reveal more edges thereby improving the quality, particle swarm optimization-based edge detection method is also implemented on the same set of images. The output images of Sobel, Canny, and Prewitt and PSO are depicted in Fig. 3. For evaluating the performance of all the methods, entropy value is computed and is being compared.

The entropy value can be calculated as per Eq. (3). Here, the total gray area is given by L and p is the probability distribution at each level:

Fig. 2 Original satellite images acquired from Google Earth

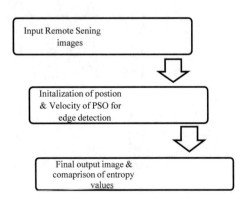

Input Remote Sening images

Initalization of postion & Velocity of PSO for edge detection

Final output image & comaprison of entropy values

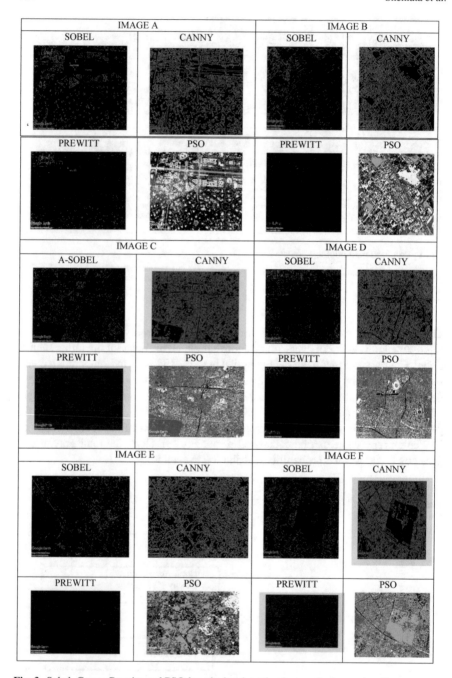

Fig. 3 Sobel, Canny, Prewitt, and PSO-based edge detection images for image A to F

Table 1 Entropy values for Sobel, Canny, and Prewitt & PSO

Image	Sobel	Canny	Prewitt	PSO
A	0.2235	0.5792	0.2228	0.9251
B	0.2025	0.6307	0.2007	0.8591
C	0.1872	0.7077	0.1841	0.8302
D	0.1908	0.7303	0.1899	0.8724
E	0.1754	0.6627	0.1726	0.8211
F	0.2116	0.6523	0.2091	0.8942

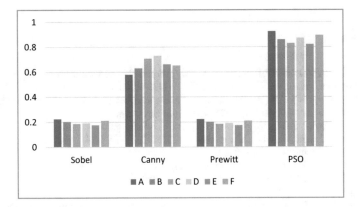

Fig. 4 Representation of entropy value for Sobel, Canny, and Prewitt & PSO

$$E = -\sum_{i=0}^{L} p_i.log_2 p_i$$

The entropy values for all the images are depicted in Table 1 and represented in Fig. 4. From the below-mentioned table, it has been observed that the entropy value for image A is 0.2235, 0.5792, 0.2228, and 0.9251 for Sobel, Canny, Prewitt, and PSO-based edge detector. The values for image B is 0.2025, 0.6307, 0.2007, and 0.8591. Similarly, the entropy values for images C, D, E, and F are represented. It has been observed that the highest entropy value is being provided by PSO-based edge detection. High entropy value represents more information in an image, leading to better image quality and better detection of edges. These edges will help in analyzing the changes occurring at a pace over some time and hold a great demand concerning land monitoring, resource management, forest fires, polar ice melting, and natural disasters.

5 Conclusion

The detection of edges in an image plays a significant role in the analysis of images. Satellite images are a great repository of information and, therefore, it is very essential to perform segmentation as this will helps in better understanding to change detection. In this paper, proposed PSO edge detection has been compared with Sobel, Canny, and Prewitt edge detection operators and it has been observed that PSO-based edge detection has higher entropy values. High entropy value represents an image of better quality, better detection of edges, hence supports better change detection.

References

1. Khan SH, He X, Porikli F, Bennamoun M (2017) Forest change detection in incomplete satellite images with deep neural networks. IEEE Trans Geosci Remote Sens 55(9):5407–5423
2. Chakraborty A, Kar AK (2017) Swarm intelligence: a review of algorithms. In: Nature-inspired computing and optimization (pp 475–494). Springer, Cham
3. Lee KY, Park J (2006) Application of particle swarm optimization to economic dispatch problem: advantages and disadvantages. In: 2006 IEEE PES power systems conference and exposition, Atlanta, GA, pp 188–192
4. Bhandari AK, Kumar A, Singh GK (2015a) Tsallis entropy based multilevel thresholding for colored satellite image segmentation using evolutionary algorithms. Expert Syst Appl 42(22):8707–8730
5. Bhandari AK, Kumar A, Singh GK (2015b) Modified artificial bee colony based computationally efficient multilevel thresholding for satellite image segmentation using Kapur's, Otsu and Tsallis functions. Expert Syst Appl 42(3):1573–1601
6. Senthilnath J, Kulkarni S, Benediktsson JA, Yang XS (2016) A novel approach for multispectral satellite image classification based on the bat algorithm. IEEE Geosci Remote Sens Lett 13(4):599–603
7. Muangkote N, Sunat K, Chiewchanwattana S (2016) Multilevel thresholding for satellite image segmentation with moth-flame based optimization. In: 2016 13th international joint conference on computer science and software engineering (JCSSE) (pp 1–6) IEEE
8. Ryalat MH, Emmens D, Hulse M, Bell D, Al-Rahamneh Z, Laycock S, Fisher M (2016) Evaluation of particle swarm optimisation for medical image segmentation. In: International conference on systems science (pp 61–72). Springer, Cham.
9. Zhao X, Turk M, Li W, Lien KC, Wang G (2016) A multilevel image thresholding segmentation algorithm based on two-dimensional K-L divergence and modified particle swarm optimization. Applied Soft Computing 48:151–159
10. Suresh S, Lal S (2017) Multilevel thresholding based on Chaotic Darwinian Particle Swarm Optimization for segmentation of satellite images. Applied Soft Computing 55:503–522
11. Kapoor S, Zeya I, Singhal C, Nanda SJ (2017) A grey wolf optimizer based automatic clustering algorithm for satellite image segmentation. Procedia Comput Sci 115:415–422
12. Naeini AA, Babadi M, Mirzadeh SMJ, Amini S (2018) Particle swarm optimization for object-based feature selection of VHSR satellite images. IEEE Geosci Remote Sens Lett 15(3):379–383
13. Jasmine J, Annadurai S (2019) Real time video image enhancement approach using particle swarm optimisation technique with adaptive cumulative distribution function based histogram equalization. Measurement 145:833–840
14. Liu S, Zhou K, Qi H, Liu J (2019) Improved hybrid particle swarm optimisation for image segmentation. Int J Parallel Emergent Distributed Syst 1–7

15. Gamshadzaei MH, Rahimzadegan M (2019) Particle swarm optimization based water index (PSOWI) for mapping the water extents from satellite images. Geocarto International, (just-accepted), pp 1–14

16. Chang Y L, Ayele AA, Huang M Y, Yuan H, Chang L, Chang WY (2019) Particle swarm optimization-based hotspot analysis and impurity function band prioritization using multiple attribute decision-making model for band selection of hyperspectral images. In: IGARSS 2019– 2019 IEEE international geoscience and remote sensing symposium (pp 3808–3811) IEEE

17. Borjigin S, Sahoo PK (2019) Color image segmentation based on multi-level Tsallis–Havrda– Charvát entropy and 2D histogram using PSO algorithms. Pattern Recogn 92:107–118

18. Di Martino F, Sessa S (2020) PSO image thresholding on images compressed via fuzzy transforms. Inf Sci 506:308–324

19. Nandini DU, Leni ES (2019) Efficient shadow detection by using PSO segmentation and region-based boundary detection technique. the Journal of Supercomputing 75(7):3522–3533

20. Chakraborty R, Sushil R, Garg ML (2019) An improved PSO-based multilevel image segmentation technique using minimum cross-entropy thresholding. Arabian Journal for Science and Engineering 44(4):3005–3020

21. Kennedy J, Eberhart R (1995) Particle swarm optimization. In: Proceedings of ICNN'95-international conference on neural networks (vol 4, pp 1942–1948), IEEE

22. Abualigah LM, Khader AT, Hanandeh ES (2018) A new feature selection method to improve the document clustering using particle swarm optimization algorithm. Journal of Computational Science 25:456–546

Artificial Intelligence-Based Scribe

V. G. Ganesh Bharathwaj, P. S. Gokul, M. V. Yaswanth, S. Sivasurya,
and A. Meena Kabilan

Abstract Scribe is used for helping physically challenged and short-term disabled students for instance, fracture. And also, the scribe is less efficient as they have major problems like lousy handwriting, and the student may have an introverted feeling due to their presence. These drawbacks of the manual scribe can be avoided with the help of an automated AI scribe. To build this AI scribe, this paper makes use of machine learning (To convert the text to their equivalent handwriting), Google speech to text API (Convert student speech to text), and Tobii Eye-tracking technology (To enable students to draw diagrams).

Keywords Physically challenged students · Short term disabled students · Google speech to text API · Tobii · Machine learning · Image processing

1 Introduction

There are more technological advancements available for the physically challenged persons and they may have a difference in opinion. Not many of the new technologies help differently-abled people as well as short-term disabled people in the field of education. There are many difficulties for such type of person, especially during examinations. The objective of the design is to make a proper AI system such that

V. G. Ganesh Bharathwaj (✉) · P. S. Gokul · M. V. Yaswanth · S. Sivasurya · A. Meena Kabilan
Department of CSE, Sri Sairam Engineering College, Chennai, India
e-mail: vgganesh99@gmail.com

P. S. Gokul
e-mail: 9gokul@gmail.com

M. V. Yaswanth
e-mail: yaswanthvmarneni@gmail.com

S. Sivasurya
e-mail: gamerganapathi@live.com

A. Meena Kabilan
e-mail: meenakabilan.cse@sairam.edu.in

© The Author(s), under exclusive license to Springer Nature Singapore Pte Ltd. 2021   819
J. S. Raj et al. (eds.), *Innovative Data Communication Technologies and Application*,
Lecture Notes on Data Engineering and Communications Technologies 59,
https://doi.org/10.1007/978-981-15-9651-3_66

even physically challenged and short-term disabled students will have a smooth interface with the system and provide a better way of writing examinations.

Initially, the student must answer the given questions verbally which is converted text with the help of Google API; then, the text is converted to the student handwriting with the help of machine learning and image processing (In case of permanent disability, the text is converted to standard handwriting specified by the institution) algorithms. If some answer requires a diagrammatic solution, it may be difficult for the student to specify the construction of the diagram to the scribe. To overcome this problem, Tobii Eye-tracking technology is used which enables the student to construct the diagram by tracking eye movements (with the help of sensors). This technology can also be used for specifying punctuations. To make it more efficient, when complicated words occur, students have to select the correct spelling from the option to remove the added advantage over normal candidates.

2 Literature Survey

For converting standard text to user text, a special type of Recurrent Neural Network (RNN) is used which is referenced from Sam Gueydan's. (Aug 21, 2016). Scribe helps for realistic handwriting with TensorFlow, using this idea alone need not be a complete solution for the issue discussed above. This technology should help students to attend all the given questions. In examination, students also face many diagrammatical questions, to help them answer this type of question, Tobii technology can be used. Tobii technology allows users to draw diagrams. AI scribe enables the user to draw a diagram using eye motion with the help of an eye-tracking sensor. Additionally, with the help of Tobii technology candidates can also select many punctuations and symbols (".", ",", " < ", " > ", "?," etc.).This technology also helps to select many special mathematical symbols that are not in the ASCII list such as integral, pi, gamma, alpha, etc.

3 Module Description

The proposed methodology is classified into three modules namely, speech to computer text, computer text to candidate's handwriting, and Tobii. Speech to text conversion can be done by Google's speech recognition. Differently-abled students can answer their question orally that oral answer will be converted to a text answer.

An important part of this project is to convert computer text to a candidate's handwriting. Intentions are very clear that the candidate who uses this software/system should not have any added advantages or drawbacks. Computer text looks very clean and neat this becomes an added advantage for the exam candidate so converting the text to human handwriting is required; this can be done with the help of machine learning algorithms. The conversion to user handwriting does not complete this

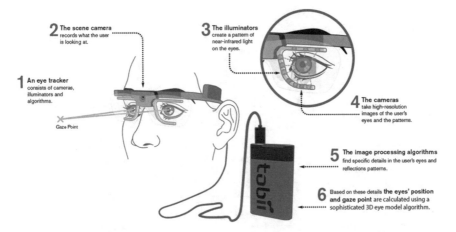

Fig. 1 Working principle of Tobii technology

design. Figure 1 depicts the working principle of Tobii technology. Sometimes candidate has to draw diagrams for a certain question this can be achieved by Tobii which is an eye-tracking technology that puts you in control of the device by using eyes. Using Tobii, a student can draw diagrams using his eye movements.

4 Module Implementation

(a) Speech to Computer Text

The first module converts an oral answer of a student to a text which is implemented using the python program by importing speech _ recognition and pyaudio (Python Audio). While running the program, voice is recorded as input to the device's microphone; then, the program reads this voice input and converts them to computer text. With this program, many languages can be converted to text so candidate those who choose different languages will also be benefited by this design. The workflow of the speech recognition system is shown in Fig. 2, and the code with the corresponding output is displayed in Fig. 3.

(b) Computer Text to Candidate's Handwriting

The second module is to convert text to human handwriting; this is implemented by importing matplotlib, TensorFlow this is just opposite to optical character recognition (OCR) in which each character of text is compared with student's handwritten character, and each character in text is replaced by his/her handwriting to achieve this Recurrent Neural Network (RNN) is used.

RNN is used here because it helps to store huge amount of data related to student's handwriting.

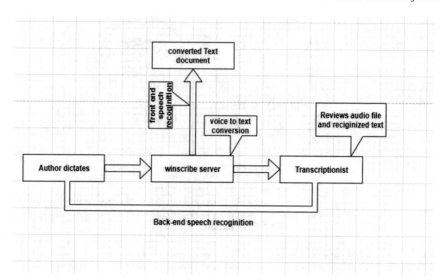

Fig. 2 Workflow of speech recognition system

Fig. 3 Code and output for speech recognition

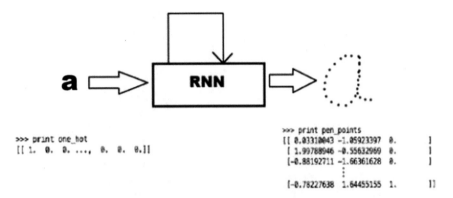

```
>>> print one_hot                    >>> print pen_points
[[ 1.  0.  0. ...,  0.  0.  0.]]     [[ 0.03310043 -1.05923397  0.        ]
                                      [ 1.99788946 -0.55632969  0.        ]
                                      [-0.88192711 -1.66361628  0.        ]
                                                      :
                                      [-0.78227638  1.64455155  1.        ]]
```

Fig. 4 Basic working principle of computer text to human handwriting using RNN

RNN is a type of neural network where the internal memory is used for storage purposes [1]. As the name suggests, RNN is recurrent which means it performs the same task n number of times. The output of the RNN not only depends on input but also on previous output. A special thing about RNN is its internal storage where it can process a sequence of inputs. This extraordinary quality of RNN helps in many applications such as unsegmented, connected handwriting recognition, or speech recognition. In other neural networks, all the inputs are independent of each other. But in RNN, all the inputs are related to each other.

The major advantages of Recurrent Neural Network (RNN) over other neural networks are RNN captures a series of inputs so it can predict the next output in the given input sequence and RNN performs computationally effective parameter sharing.

In Fig. 4, the standard letter "a" is converted to a human handwritten form of "a".

Before understanding the conversion first, understanding the basic concepts involved in the handwriting of humans is mandatory. Each character is written using continuous dots and every character has an End of Strokes (EOS). EOS is a point where the pen/pencil is lifted to stop writing or to write the next character.

As mentioned above, every character is made of continuous dots and each dot has x,y coordinates with this machine learning (ML) model can be trained to predict the future dots of the character. Along with the x,y coordinate, it is also important to know EOS of each character it is represented by 1 if EOS is reached else it is represented by 0. As shown in Fig. 4, the one-hot encoded value of letter "a" is sent as input to RNN labeled as A then output along with x,y coordinates vector of EOF.

To train the model, the following algorithms namely, The Long Short-Term Memory (LSTM) Cell, The Mixture Density Network (MDN), and The Attention Mechanism performs the specific set of tasks to be handled to generate the output which is portrayed in Fig. 5.

The Long Short Term Memory networks (LSTM) is a special kind of Recurrent Neural Network (RNN) that is used to remember the previous data and output for generating the next output [1]. When the entire gap between the relevant information

Fig. 5 Flowchart for
conversion of computer text
to candidate's handwriting

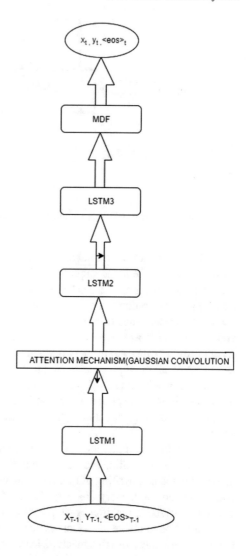

and the point where it is needed to become a very large LSTM is best suited. Each character is written using continuous dots, so LSTM stores the information about all the previous dots for the generation of correct data (character or word). Each character can be uniquely identified using the end of stroke property. In LSTM, previous data and important information are stored in cell state. There are three steps involved in LSTM. Initially, decisions should be made on which information should be thrown away using gates and based on the output produced by the gates decisions are made (if 1-Not thrown away and 0-Thrown). The next step is to decide which new information is going to store in the cell state this happens in two stages. The

Fig. 6 Graphical representation of MDN

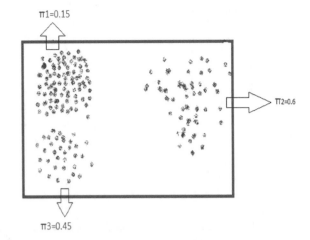

π1=0.15

π2=0.6

π3=0.45

further step involves the selection of value to be updated and create a new vector that is added to the state. Then, the filtered output is decided based on the cell state.

The Mixture Density Network (MDN) is a type of neural network which is used to predict the complex distribution of data. It gives an idea about the randomness present in the data. Figure 6 displays the graphical representation of MDN. It allows us to predict when the target variable will become noisy. They estimate a parameter π for each of these distributions. π is the probability that the output value was drawn from that particular component's distribution. In the handwriting model, the MDN learns about the messy and unpredictable parts of handwriting.

The attention mechanism is used to memorize long source sentences in neural machine translation (NMT) [1]. Suppose the model wants to write "Ganesh," Fig. 7 depicts the graphical representation of Attention Mechanism. First decoding takes place character by character to make up the word. For Attention, it makes use of gate recurrent unit (GRU) instead of LSTM. Since GRU is used, it is enough to consider only one state at a time which simplifies the process of reading a long sentence by dividing it into different modules. Each state generally represents a single word or character. So, to develop a sentence, a combination of n states is used for generation. Since all the parameters of this window are differentiable, the model learns to shift the window from character to character as it writes them.

Now consider the model, a student wants to write a long sentence "PSG is a good boy," it is known that only one state can be used at a time as gate recurrent unit(GRU) is used. So, each state represents a single word, for example, "PSG" represents one state, " is" represents the second state, and so on. Next, the score of each state is determined and the context vector is calculated. Then, the context vector is concatenated with the previous output which is then passed to a decoder which is described in Fig. 8. The corresponding output is represented in Fig. 9, and the ouput code is showcased in Fig. 10.

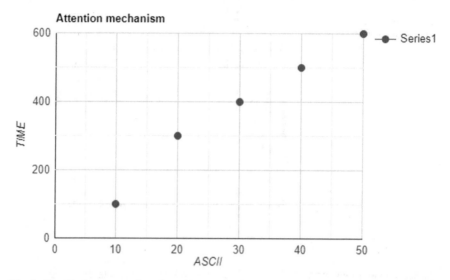

Fig. 7 Graphical representation of attention mechanism

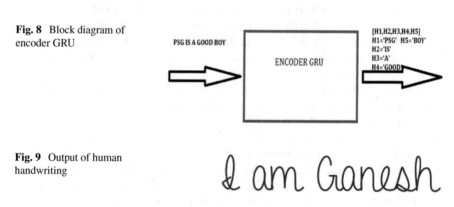

Fig. 8 Block diagram of encoder GRU

Fig. 9 Output of human handwriting

(c) Tobii Technology

Tobii is an Eye-tracking innovation. Eye-following is a sensor innovation that makes it feasible for a PC or other gadget to know where an individual is looking. An eye tracker can distinguish the nearness, consideration, and focal point of the client. Figure 11 displays the graphical representation of Tobii technology. In AI scribe, It is utilized to allow the client to give diagrammatically answer for a question. This strategy is more viable than a manual copyist as it is hard to clarify diagrams to a manual recorder. The fundamental idea is to utilize a light source to enlighten the eye making profoundly obvious reflections, and a camera catches a picture of the eye exhibiting these reflections. The picture caught by the camera is then used to distinguish the impression of the light source on the cornea and in the pupil. It will then be able to figure a vector framed by the edge between the cornea and student reflections the bearing of this vector, joined with other geometrical highlights of the reflections,

Fig. 10 Output with code

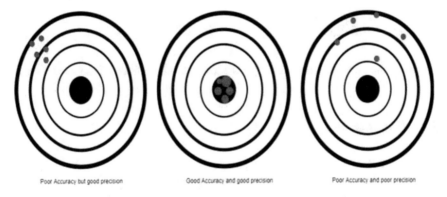

Poor Accuracy but good precision   Good Accuracy and good precision   Poor Accuracy and poor precision

Fig. 11 Graphical representation of Tobii technology

is then used to compute the look heading. Building up an appropriate baseline is basic to removing a solid pupil reaction. There are a few ways to deal with doing this, however, they, for the most part, include showing a plain stimulus and afterward set as a baseline, the pupil width in the few hundred milliseconds legitimately going before the key stimulus. The necessary degree of exactness and accuracy relies upon the idea of the eye-tracking study. Little vulnerabilities, the occurrence can be basic when the study looks information in understanding examinations or concentrates with a little stimulus. During information assortment, exactness, and accuracy are utilized as markers of the eye tracker information validity. A system with great exactness and accuracy will give progressively substantial information as it can honestly depict the area of an individual's look on a screen. Exactness is the normal distinction between the genuine boosts position and the deliberate look position.

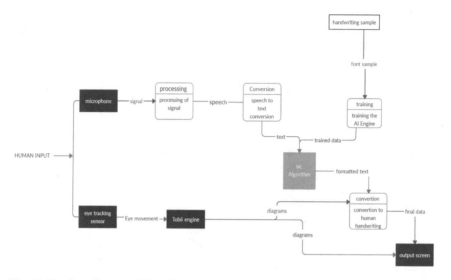

Fig. 12 Dataflow diagram of AI scribe

5 Data Flow Diagram

Figure 12 data flow diagram describes the inflows and outflows of the AI scribe.

5.1 Conclusion

Technology should help physically challenged people because they need it more than anybody. This paper is one such help for those people. And also helps the physically challenged students to present answer scripts like other students without the help of a human scribe. The model learns the handwriting of the particular student and writes answers given by the student orally in his/her handwriting as well as enabling to draw diagrams using Tobii technology. The proposed methods will certainly help those who are struggling to attend examinations.

6 Formulas

MDN cap predicts the pen's (x,y) coordinates by drawing them from a Gaussian distribution, distribution can be modified to make the handwriting cleaner or messier.
To calculate deviation

$$\sigma jt = \exp\left(\sigma^{(1+b)}\right) \tag{1}$$

where

σ—deviation.

b—bias.

To Calculate Probability

$$\pi jt = \exp\left(\pi^{(1+b)}\right) / \sum \exp(\pi^{(1+b)}) \tag{2}$$

where

π—probability.
    b—bias.

References

1. Greydanus S (2016) Scribe: realistic handwriting with TensorFlow, https://greydanus.github.io/2016/08/21/handwriting/
2. Graves A (2013) Generating sequences with recurrent neural networks. https://arxiv.org/abs/1308.0850
3. Opoka G (2018) https://github.com/Grzego/handwriting-generation
4. Vasques S (2018) https://github.com/sjvasquez/handwriting-synthesis
5. https://www.tobiipro.com/learn-and-support/learn/eye-tracking-essentials/how-do-tobii-eye-trackers-work
6. https://towardsdatascience.com/understanding-rnn-and-lstm-f7cdf6dfc14e
7. https://towardsdatascience.com/understanding-neural-networks-from-neuron-to-rnn-cnn-and-deep-learning-cd88e90e0a90
8. https://www.analyticsvidhya.com/blog/2020/02/cnn-vs-rnn-vs-mlp-analyzing-3-types-of-neural-networks-in-deep-learning
9. https://towardsdatascience.com/https-medium-com-rachelwiles-have-we-solved-the-problem-of-handwriting-recognition-712e279f373b
10. Bojja P, Velpuri NSST, Pandala GK, Lalitha Rao Sharma Polavarapu SD, Kumari PR (2019) Handwritten text recognition using machine learning techniques in application of NLP

A Hybrid Approach to Review Mining—Restaurant Data in Depth Analysis

P. Raghavendra Babu, S. Sreenivas, U. S. VinayVarma, and N. Neelima

Abstract The eatery is a growing market and along with it grows the competition. To stay on the top, one must have satisfied and happy customers and their reviews are significant for a successful business. Nowadays, restaurants need to take customer reviews into account to enhance the customer experience. In this paper, a hybrid methodology is proposed to overcome this problem, faced by the restaurants, using sentimental analysis on the reviews and differentiate the positive and negative aspects of the restaurant. This paper highlights the importance of machine learning algorithms and is used to find patterns in data that help to make wiser decisions and predictions. The sentiment of the reviews are classified into positive and negative, and the score of each sentiment is also measured. The proposed approach gives a classification accuracy of 84.76% which is better than the existing methods.

Keywords Data mining · Multi-dimensional sentiment · Sentiment score · Topic modelling · Sentiment analysis · Classification

1 Introduction

The food industry is quite diverse and in practise, satisfying the population requirements is more important than producing sufficient quantities, keeping in mind, the immense competitiveness in the marketing segment. Eateries are relatively easy to set up and the major initial costs go into marketing, but the rise in its economy can be achieved by the satisfaction of the consumer alone. Currently, the trend is to check the reviews and then go to a restaurant, as reviews are honest opinions. Sentimental analysis algorithms are used frequently to make critical decisions. These can be used to enhance the customer experience to improve the sales and their areas of improvement. The sentiment is a feeling which shows a person's attitude and thought towards something. Sentimental analysis (Opinion Mining) analyses the sentiment of people

P. Raghavendra Babu · S. Sreenivas (✉) · U. S. VinayVarma · N. Neelima
Department of Electronics and Communication Engineering, Amrita School of Engineering,
Amrita Vishwa Vidyapeetham, Bengaluru, India
e-mail: sreenivasjimmy10@hotmail.com

831

J. S. Raj et al. (eds.), *Innovative Data Communication Technologies and Application*,
Lecture Notes on Data Engineering and Communications Technologies 59,
https://doi.org/10.1007/978-981-15-9651-3_67

and categorises them a positive, negative or neutral. The internet is a vast ocean of information and a part of which is sentimental information.

Data mining is a practise used to find distinctive patterns in a dataset having vast raw data to fetch helpful details. Several data mining principles are applied to understand useful details in the information. This information helps in precise decision making at the right time by various restaurants. This is an important factor for whichever restaurant in the cut-throat competition of the market as it is not feasible for them to inspect and trace of all these comments.

2 Literature Review

The fundamental concern of the seller and researchers concerning consumer behaviour is customer satisfaction. In the case of hotel consumers, if they are satisfied with the service or food, they pass on the reviews to others by word of mouth [1]. Online customer reviews also known as "electronic word of mouth" furnish data on several aspects of a product or service. Online hotel reviews aid to the hotel operators to monitor the customers by analysing the reviews. Thus, online reviews would be a dependable source for the customers checking for the information about the products [2]. Also, the rating of the restaurants as per the industry standards are based reviews on food, service and decor/ambience [3].

Sentimental analysis is a computational identification or a study of opinions expressed in the form of text, which determines the consumer attitude towards the product is positive, negative or neutral [4]. Anastasia et.al expounded by considering the companies, GOJEK and Grab by collecting their twitter reviews on their service/product. Classifiers can be used in a wide range of applications such as early detection of cancer cells in [5] to surveillance applications using face recognition in [6]. Around 126,405 tweets were collected which contains GOJEK and Grab keyboards [7, 8]. Finding reviewer's sentiment is often very crucial because it tells us about the sentiment of a large dataset which is made from its integral components [9]. Different papers emphasise one point that text-based documents are better compared to the rating that is in numbers [10]. Analysis of text documents containing score rating is proven to improve the classifier's quality [11]. Yet another option had come into the picture with neutral scores, as it uses three classes [12]. The tweets on a particular topic were processed and were classified using three algorithms: SVM, decision tree machine and Naïve Bayes learning classifiers. The larger the data set, the better the performance the classifiers achieve. Based on the kind of dataset, the algorithm is chosen. For example, for the small size of dataset Naïve Bayes is preferred similarly for the large data set, SVM is preferred, and the decision tree algorithm is independent of the size of the dataset [4]. Naive Bayes computes future probability predictions from the information that have been given [1]. It predicts the probability of unknown sample by using the Eq. (1)

$$P\left(\frac{H_i}{y}\right) = \frac{\left(P\left(\frac{y}{H_i}\right)P(H_i)\right)}{P(y)} \tag{1}$$

where $P(\frac{H_i}{y})$ is the conditional probability of entropy H_i given y, P $(\frac{y}{H_i})$ is the probability function of y given H_i, P (Hi) is the prior probability, P(y) is the evidence. Support vector machine can categorise text and hypertexts which helps in reducing the usage of labelled data and one of the methods used subgradient descent.

In this approach, for certain implementations, the number of iteration does not depend on the number of data points. There are two types of decision trees known as classification trees which classify into yes or no type, and regression trees are used for predicting the class. The regression tree uses continuous target variables. They considered the few features of restaurants in blogging-based consumer comments. The opinion of these few features reflects the user's experiences. Suppose, if a user says "this restaurant food tastes good" (the taste of the food provided reaches the expectations) reflects customer's positive sentiments regarding the taste of the food of the restaurant [2]. Due to the enhancement of machine learning and its methodologies, sentiment analysis has attained huge popularity. Recent developments in sentiment analysis help to analyse the multi-dimensional sentiments and provides what one feel about food, service, etc., either positively or negatively.

3 Proposed Methodology

This paper implements a hybrid approach to find out the aspects that are needed to be improvised (or) require more attention. For business applications, determining the positive and negative sentiment is not enough. Determining the sentiment associated with the desired keyword will be more appropriate to improve the business. It requires an in-depth analysis to determine the positive and negative score associated with the desired keyword.

For example, the food here is delicious that is considered as "positive" senti-ment, and it is associated with food and categorise this as "food-positive". Similarly, the ambience is not great is considered as negative sentiment and is mapped as "ambience-negative". The proposed methodology is carried into two sections: data cleaning and sentiment classification. These are further scrutinised in the following sectors. To do opinion mining, it is essential to clean the data which is explained in the pre-processing stage. Then, the sentiment of the review is determined. The clas-sifier is trained with this sentiment to predict the sentiment of the unknown review. The general approach is explained in Fig. 1 (Figs. 2 and 3).

Fig. 1 The generalised approach of review classification

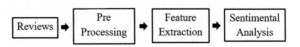

review	is_bad_review
0	We did not like the fact that breakfast was not included although you could pay extra for this The room we stayed in was lacking in space a bit and the bathroom was very badly lit to use the mirror in there which meant getting ready for evening a bit tricky but only a small negative comment	0
1	Rooms are nice but for elderly a bit difficult as most rooms are two story with narrow steps So ask for single level Inside the rooms are very very basic just tea coffee and boiler and no bar empty fridge	0
2	The room was big enough and the bed is good The breakfast food and service on the hotel is good outside the hotel there is a big park which is very good for walk in the morning and evening Many people are having picnics and do some bicycling	1
3	Rooms were stunningly decorated and really spacious in the top of the building Pictures are of room 300 The true beauty of the building has been kept but modernised brilliantly Also the bath was lovely and big and inviting Great more for couples Restaurant menu was a bit pricey but there were loads of little eatery places nearby within walking distance and the tram stop into the centre was about a 6 minute walk away and only about 3 or 4 stops from the centre of Amsterdam Would recommend this hotel to anyone it s unbelievably well priced too	1
4	No real complaints the hotel was great great location surroundings rooms amenities and service Two recommendations however firstly the staff upon check in are very confusing regarding deposit payments and the staff offer you upon checkout to refund your original payment and you can make a new one Bit confusing Secondly the on site restaurant is a bit lacking very well thought out and excellent quality food for anyone of a vegetarian or vegan background but even a wrap or toasted sandwich option would be great Aside from those minor minor things fantastic spot and will be back when i return to Amsterdam	1

Fig. 2 Reviews before pre-processing

```
POSITIVE reviews:[('bed',), ('bicycling',), ('breakfast',), ('evening',), ('having',), ('morning',), ('outside',), ('park',),
('people',), ('picnics',), ('building',), ('centre',), ('confusing',), ('minor',), ('staff',), ('away',), ('bath',), ('beaut
y',), ('brilliantly',), ('couples',), ('decorated',), ('distance',), ('eatery',), ('inviting',), ('kept',), ('little',), ('load
s',), ('lovely',), ('menu',), ('minute',), ('modernised',), ('nearby',), ('pictures',), ('places',), ('priced',), ('pricey',),
('recommend',), ('spacious',), ('stop',), ('stops',), ('stunningly',), ('tram',), ('true',), ('unbelievably',), ('walking',),
('amenities',), ('aside',), ('background',), ('check',), ('checkout',)]

NEGATIVE reviews:[('room',), ('booked',), ('booking',), ('day',), ('hotel',), ('big',), ('change',), ('check',), ('constant',),
('duplex',), ('high',), ('offered',), ('order',), ('placed',), ('turned',), ('way',), ('window',), ('able',), ('angry',), ('ann
oying',), ('arrival',), ('asked',), ('asleep',), ('attach',), ('audio',), ('available',), ('begin',), ('belongings',), ('bes
t',), ('bio',), ('broken',), ('btw',), ('cause',), ('ceiling',), ('ceilings',), ('city',), ('clock',), ('closed',), ('com',),
('contained',), ('costs',), ('determine',), ('disturbing',), ('double',), ('evening',), ('explaining',), ('fall',), ('floor',),
('fully',), ('garden',)]
```

Fig. 3 Reviews after pre-processing and TF-IDF vectorization

3.1 Data Pre-processing

Data pre-processing is critical in any opinion mining process, as they directly influence the success rate of the proposed methodology. The conversion of data to an understandable format for the computer is known as data pre-processing. This process reduces the dimensionality of the dataset under analysis by conversion of data to lowercase and by removing unnecessary characters like punctuations, integers, extra spaces and tags, (i.e., HTTP, www, etc.) as raw data is noisy and unclean. The major role of pre-processing is to filter out unnecessary data. Pre-processing incorporates several techniques like integration, cleaning, reduction and transformation.

These are two significant processes of the pre-processing module. The stemming process converts all the words to be inflected which are present in the text into a root form called a stem. For example, "automate", "automation" and "automatic" are converted into the stem "automat". Stemming speeds up the performance in applications where accuracy does not play a key role. There are mainly two errors in stemming:

- Over Stemming.
- Under Stemming.

Over stemming happens when two words are stemmed to the same root word that is of different stems. Under stemming occurs when two words are stemmed to the same root word that is not of different stems, whereas, lemmatization of a word includes its base form as well as the inflected form. For example, the words "playing", "played" and "plays" have "play" as their lemma. Therefore, lemmatization is considered to be more accurate.

3.2 Feature Extraction and Indexing

3.2.1 TF-IDF

TF-IDF called Term Frequency and Inverse Document Frequency, the score of TF-IDF means a score used in text mining and information revival. This quantity is a score used in calculating what importance that the term has in the document. TF (Term Frequency) which calculates how often a term repeats in the document. The term frequency is often divided by the size of the document in a way of normalisation is given in Eq. 2. Where "t" is the term to measure term frequency.

$$\text{TF}(t) = \frac{\text{(Number of times term } t \text{ appears in a document)}}{\text{(Total number of terms in the document)}} \tag{2}$$

IDF (Inverse Document Frequency) is a measure of how significant a word is. During the computation of Term Frequency, the terms are to be treated equally essential. There is a high chance for articles to repeat multiple times yet they are insignificant. Thus, one should decrease the most frequent words, to increase the less frequent ones, by using Eq. 3. Here, "t" is the term for which inverse document frequency is measured.

$$\text{IDF}(t) = \log_e\left(\frac{\text{Total number of documents}}{\text{Number of documents with term } t \text{ in it}}\right) \tag{3}$$

3.2.2 N-Grams

These are texts that are extensively used in natural language processing and text mining. These are a group of concurrent terms in a given space and during computation of the N-grams, you move single word forward (though one can move K number of terms forward in more advanced scenarios). If K = No of terms in a sentence X, then the number of N-grams for that sentence is given by Eq. 4.

$$\text{Ngrams}_x = K - (N - 1) \tag{4}$$

These are used for various such tasks. E.g., when building a model, these are utilised to build not only 1-g models but also 2-g and 3-g too. N-gram models can be utilised in numerous tasks like word breaking, text summarization and spelling correction. In this proposed methodology, unigram (i.e., N = 1) is used for the application of word breaking. Therefore, each word is separated and can be easily extracted.

3.2.3 LSI (Latent Semantic Indexing)

It is a method of extraction of contextual usage of words by the method of applied statistical computations on a large collection of text. It is used for retrieval of information, analysis and identification of pattern in the unconstructed text. It is an unsupervised method of uncovering synonyms.

3.3 Classification Methods

3.3.1 NB Classifier (Naïve Bayes)

This technique is a group of probabilistic techniques depending on the principle of theorem (Bayes theorem) with "naive" hunch of conditional independence model among all combination of features. Bayes theorem is used to calculate the probability $P(\frac{a}{b})$ and represents the set of outcomes where b represents the instance that needs to get classified, which represents particular characters. The Bayes theorem is based on the probability concepts given in Eq. 5.

$$P(a/b) = P(b/a) * P(a)/P(b) \tag{5}$$

where, $P(a/b)$ gives the probability of "a" being true given "b" is true, $P(b/a)$ gives the probability of "b" being true given "a" is true, $P(a)$ gives the probability of "a" is true and $P(b)$ gives the probability of "b" being true. Gaussian NB (Naïve Bayes) Classifier: In this classifier, each feature is related to continuous values that are distributed with regards to Gaussian distribution. This distribution is known as normal distribution. The Gaussian conditional probability is given by Eq. 6.

$$P(X_i/Y) = \frac{1}{\sqrt{2\pi\sigma^2_y}} \exp\left(-\frac{(x_i - \mu_y)^2}{2\sigma^2_y}\right) \tag{6}$$

where, $P(X_i/Y)$ gives the conditional probability of occurrence of Xi given Y has occurred, "σ" denotes the variance and "μ" denotes the mean.

3.3.2 Xgboost (EXtreme Gradient Boosting)

It implements grading boosting machines to force the computing powers to its limit in the boosted trees algorithms as it is built for performance enhancement. It is designed to use every bit of hardware resources and memory for boosting. It offers advanced features for model tuning and algorithm enhancement. It mainly performs three types of gradient boosting (Regularised GB, Gradient Boosting(GB) and StochasticGB).

$$\hat{y}_i = \sum_{k=1}^{k} f_k(x_i) \tag{7}$$

where "y_i" is the weighted samples considered, and "x_i" are the samples from the decision trees. The output of each training sample that is the sequential weighted samples are taken as average, and the output is determined as per Eq. 7.

4 Results

The dataset considered to evaluate the proposed method consists of 1500 reviews and is taken from Kaggle restaurant reviews dataset, which is open-source. Out of these 1500 reviews, 750 are positive and the remaining 750 are negative. The training and testing set are divided as 70 and 30% (i.e., 1050 reviews are used for training and 450 reviews are used for testing).

4.1 Sentiment Score

After TF-IDF, the most frequently occurred positive sentiment and negative sentiment words are plotted to their occurrences which are shown in Fig. 4. In Fig. 4a, the x-axis represents the positive words which were extracted from the reviews and y-axis represents the corresponding word frequencies in the reviews. Similarly, in Fig. 4b, the x-axis represents the most frequent negative words from the reviews and y-axis represents the corresponding frequencies.

4.2 Classification

After separating the positive and negative sentiment words, these features are used to train the classifier to classify the reviews. The proposed approach used XGboost algorithm to classify the given reviews into positive and negative sentiment. The

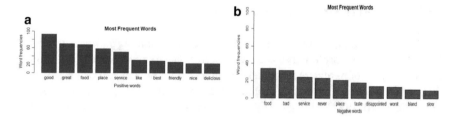

Fig. 4 **a** Most frequent positive words **b** Most frequent negative words

sentiment score for each review is calculated such as positive score and negative score. The classification of positive and negative reviews with the corresponding scores is shown in Fig. 5.

Several experiments were conducted by considering the reviews in the dataset as 500, 1000 and 1500. Also, the state-of-art methods such as Gaussian Naive Bayes and KNN were implemented to provide comparative analysis. It was observed that as the number of reviews is increased, the classification accuracy of Naive Bayes was decreased. The Naive Bayes classifier works better when the dataset is limited and it is most likely used for predicting the sentiment.

As the stop words are removed, the XGBoost algorithm works better under such a scenario. The corresponding experiments results were plotted in Fig. 6. To provide the performance evaluation, metrics such as accuracy, true positive rate and true negative rate are used. The performances of the proposed approach and existing approaches such as Naïve Bayes and K nearest neighbour algorithm (KNN) were recorded and given in Table 1. The number of correct predictions for positive tweets out of actual positive and negative tweets out of actual negatives is given by true positive rate and true negative rates. From Table 1, it is observed that the proposed approach had better classification accuracy for sentiment analysis.

Further, the proposed approach extends to find out the aspects where further improvement is needed. It provides a user-friendly platform, where the user can enter the aspects for which the sentiment need to be determined. The sentiment score

	tweet	sentiment_score	sentiment
0	Wow Loved this place	0.400000	positive
1	Crust is not good	-0.350000	negative
2	Not tasty and the texture was just nasty	-1.000000	negative
3	Stopped by during the late May bank holiday of...	0.200000	positive
4	The selection on the menu was great and so wer...	0.800000	positive
5	Now I am getting angry and I want my damn pho	-0.500000	negative
6	Honeslty it didnt taste THAT fresh	0.300000	positive

Fig. 5 The sentiment scores for some of the sample reviews from the dataset

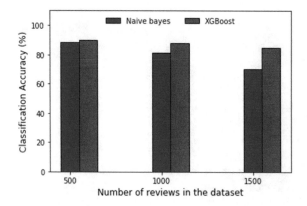

Fig. 6 Performance analysis of proposed and existed approaches

Table 1 Performance analysis of different classifiers for sentimental analysis

Methodology	Accuracy (%)	TP rate (%)	TN rate(%)
TF-IDF + Gaussian Naïve Bayes	70.24	68.16	66.87
TF-IDF + KNN	76.78	70.58	73.16
TF-IDF + XGboost (Proposed Approach)	84.76	78.58	82.59

associated with some of the user desired keywords is plotted in Fig. 7 and can easily identify the aspects where the user responses are positive and negative. This helps the business for further improvement.

The in-depth analysis is provided using LSI which provides the weight of each topic extracted from count vectorizer. These are plotted in Fig. 8.

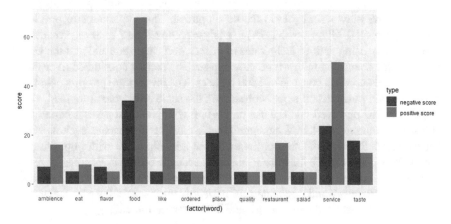

Fig. 7 Sentiment analysis with respect to the desired keyword

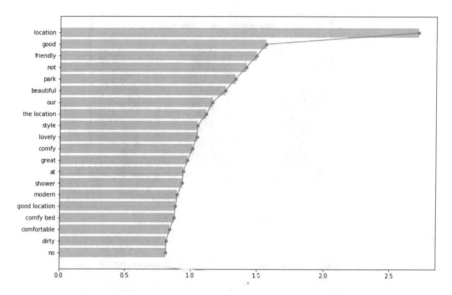

Fig. 8 In-depth aspect analysis of reviews

From the results, it is clear that the proposed hybrid approach provides better sentiment analysis towards business applications and helps to determine the aspects to be further improved.

5   Conclusion

The business applications such as mobile industries, shopping malls, hotels or restaurant's reviews play a vital role in the development. This paper aims to provide a hybrid approach (XGboost with LSA) to analyse feedbacks and reviews provided by consumers by using various techniques to extract, sort, segregate and plot the words based on the frequency with which they appear. The dataset considered to evaluate the proposed approach consists of 1500 reviews. The reviews are pre-processed and classified into positive and negative based on the sentiment score computed. From Table 1, it was observed that the use of TF-IDF with XGboost gives a better classification accuracy of 84.76% compared to the existing approaches such as Naïve Bayes and KNN. Classification of reviews does not help much with improvisation unless the aspect in which the improvement is needed to be found. For this, LSA was used and the corresponding weights related to each sentiment words are computed and plotted. Figures 7 and 8 provide a user-friendly interface for the end-user to find out the aspects that need to improvise and require attention, with ease.

References

1. Laksono RA, Sungkono KR, Sarno R, Wahyuni CS (2019) Sentiment analysis of restaurant customer reviews on tripAdvisor using Naïve Bayes. In: 12th international conference on information and communication technology and system (ICTS), Surabaya, Indonesia, pp 49–54
2. Chanwisitkul P, Shahgholian A, Mehandjiev N (2018) The reason behind the rating: text mining of online hotel reviews. In: proceedings of the 2018 IEEE 20th conference on business informatics (CBI), Vienna, Austria, 11–13 Jul 2018, pp 149–157
3. Gan Q, Yu Y (2015) Restaurant rating: industrial standard and Word-of-Mouth—a text mining and multi-dimensional sentiment analysis. In: 48th Hawaii international conference on system sciences, Kauai, HI, pp 1332–1340
4. Kaviya K, Roshini C, Vaidhehi V, Sweetlin JD (2017) Sentiment analysis for restaurant rating. In: IEEE international conference on smart technologies and management for computing, communication, controls, energy and materials (ICSTM), Chennai, pp 140–145
5. Madan K, Bhanu Anusha K, Pavan Kalyan P, Neelima N (2019) Research on different classifiers for early detection of lung nodules. Int J Recent Technol Eng 1037–104
6. Raju J VVSN, Rakesh P, N Neelima (2019) Driver drowsiness monitoring system. Smart Innov Syst Technol (SIST) 169:675–683
7. Alexander P, Patrick P (2010) Twitter as a corpus for sentiment analysis and opinion mining. In: Proceedings of LREC
8. Anastasia S, Budi I (2016) Twitter sentiment analysis of online transportation service providers. Int Conf Adv Comput Sci Inf Syst (ICACSIS), Malang, pp 359–365
9. Jean Y Wu, Pao Y Predicting sentiment from rotten tomatoes movie reviews
10. Gayatree G, Elhadad N, Marian A (2009) Beyond the stars: improving rating predictions using Review Text Content, WebDB, vol 9
11. Moontae L, Grafe R (2010) Multiclass sentiment analysis with restaurant reviews, Final Projects from CS N, vol 224
12. Jayashri K, Kinikar M (2013) Machine Learning algorithms for opinion mining and sentiment classification. Int J Sci Res Publ 3.6:1–6

Fresko Pisces: Fish Freshness Identification Using Deep Learning

Anandhu Suresh, Arathi Vinayachandran, Chinju Philip, Jithu George Velloor, and Anju Pratap

Abstract Fish freshness identification plays a prominent role in fishery industry applications. Recently, serious malpractices are observed in fishery industries, which is highly becoming as a socioeconomic issue. The recent development of convolutional neural networks has accomplished great outcomes in the field of image classification. This paper focuses on novel method of dealing with the advancement of fish freshness identification utilizing convolutional neural network. VGG-16 network through transfer learning is used as the classification algorithm. This will predict the percentage of freshness, remaining shelf life of a fish on giving its eye image, gill image, and skin discoloration like features as input. The data set was prepared by collecting more than 6000 real samples from various fish markets and shops of Kerala using cameras during a time period of 3 to 4 months. After data set collection, pre-processing was done in order to select the most suitable images and the model. The Final conclusion about freshness is formed after analyzing the condition of the eyes and the gills and some manual method of feature detection. The specialty of the strategy exists in the limit of the strategies to viably acclimate to new examples which could be an after effect of the comprehensive propriety of the specified parameters. The system shows up the exceedingly exact outcomes when contrasted with the ground truth.

A. Suresh (✉) · A. Vinayachandran · C. Philip · J. G. Velloor · A. Pratap
Department of Computer Science and Engineering, Saintgits College of Engineering, Kottayam, Kerala, India
e-mail: anandhusuresh321@gmail.com

A. Vinayachandran
e-mail: arathi.v1620@saintgits.org

C. Philip
e-mail: chinju.philip1620@saintgits.org

J. G. Velloor
e-mail: jithu.gv1620@saintgits.org

A. Pratap
e-mail: anju.pratap@saintgits.org

© The Author(s), under exclusive license to Springer Nature Singapore Pte Ltd. 2021
J. S. Raj et al. (eds.), *Innovative Data Communication Technologies and Application*,
Lecture Notes on Data Engineering and Communications Technologies 59,
https://doi.org/10.1007/978-981-15-9651-3_68

Keywords Deep learning · Convolutional neural network · VGG-16 · Quality index method · Transfer learning

1 Introduction

Fish is an imperative source of minerals such as calcium and phosphorus, zinc, iodine, magnesium, and potassium. Hence, the request for fish as an important means of food and its items are expanded in more numbers around the world. Fish provides nutrients that are essential to cognitive and physical development, especially in children, and are an important part of a healthy diet. As numerous nations does not develop the required fish of their demand so are imported from coastal countries [1]. Other than its nutritive esteem, fish is additionally a product of worldwide exchange and outside trade workers. The research for such an initiative started from the very beginning of mankind. In such scenarios, people were familiar with natural features of each species of fish and were able to determine the freshness manually. Later, such introspection became impractical and other means of process came into research. Automatic fish freshness identification is a demanded technique for many fishery applications, for example, marine ecology, automate logging of the catch in business and research fishing vessels, etc. Traditionally, the fish skin patterns as the entire shape have been utilized by shery analysts to recognize the fish. In any case, the lighting conditions, changes in skin shading and other factors impact the fish type. This makes it hard to physically group the fish species and effectively decipher the quality. The utilization of whole fish image as contributions for the preparation framework was not viable as neither memory use is ideal nor fish type is effectively recognizable. Fish skin images only can be fused in the mentioned fish framework as an enhancement and not as the important identification input [2]. Fish type distinguishing proof dependent on fish-eye images, gill images, skin pictures, and helpful parameters separated from these pictures is the approach embraced in the planned framework. The expanding purchaser enthusiasm for nourishment of quality and security which leverages the developing interest for responsive and fast logical advancements. The objective of freshness assessment of fish is to assist the maintenance or improvement of freshness by minimizing customer complaints about quality [3].

2 Literature Survey

The study related to this approach started from 2008 based on bio-genic amines and fish freshness assessment using a multi sensor system based on voltametric electrodes, then on 2009 meat and fish freshness inspection system based on odor sensing was proposed to use a metal oxide semiconductor-based electronic nose; the smell signature stored at room temperature is also measured. The rapid detection of fish freshness based on fish-eye image helps to choose the fresh fish-eyes images [4].

Next step was to figure out the strength of every image signal. Finally, the relation between the change of the image's energy and the value by modeling of partial least squares regression was analyzed. Non-destructive detection of fish freshness during its preservation by combining electronic nose and electronic tongue techniques in conjunction with chemometric analysis which total viable counts (TVC) of the fish were detected by the conventional method. E-nose and E-tongue data were analyzed by principal component analysis. A support vector machine regression model was applied to establish a relationship between the combined data from E-nose and E-tongue and from TVC values for quantitative determination. Fish freshness detection is done by using a computer screen photoassisted-based gas sensor array that uses chemical sensor array, where the optical features of layers of chemical, and sensitive to volatile compounds typical of spoilage processes in fish, which are interrogated by a very simple platform based on a computer screen and a webcam. An array of metalloporphyrins is used here to classify the fillets of thawed fish according to their storage days and to monitor the spoilage in a time period of 8 h. Evaluation of fish and fillet freshness quality was also done on 2010, which depicts that the texture and structure of fish muscle to verify the freshness and quality attributes that depend on several parameters such as hardness, cohesiveness, springiness, chewiness, resilience, and adhesiveness, as well as the internal cross-linking of connective tissue and the detachment. Fish freshness monitoring by using an E-tongue based on Polypyrrole modified screen-printed electrodes technique was implemented on 2011, where the method is used as an array of voltaic electrodes that are chemically modified based on screen-printed electrodes [5]. The pattern of responses are provided by the array that can be used to discriminate and evaluate the fish freshness state. Multicolor biosensor for fish freshness assessment with the naked eye was another technique, where hypoxanthine has been chosen as the index of fish freshness, the concentration of which is directly determined to the color of the sensing system. The color of the sensing system has a good relationship with hypoxanthine concentration in the concentration range of 0.05 mm {0.63 mm}, and the latest one was on 2019, which is a computer vision-based technique using fish skin tissue (Fig. 1).

3 Objectives

The major objectives are to detect fish from a picture accurately. To identify the share of freshness and number of days after catching by taking its eye and gill image with a mobile application [6]. Then to assess whether the sample contains any formaldehyde chemicals which is employed for preserving the food. It should be ready to detect the area of interest, the eyes of a fish, gills, and complexion [8]. It also have manual method of option submission to give the feedback on stiffness of body, sliminess of skin, meat color, overall appearance of fish, etc., can be feed to the model. The app predicts the percentage of freshness, whether fish is consumable or not and the fish and decide independently by consumer whether to buy or not. The formalin detector kit determines the detection of formalin in ppm. Formalin is carcinogen

AUTHOR	SYSTEM	METHOD	REMARK
Makoto Egashira, Yasuhiro Shimizu,Yuji Takao. (2007)	Meat freshness Inspection system.	Based on odour sensor	Using metal oxide semiconductor based e-nose.
Xingyi Huang,Junwei Xin, Jiewen Zhao. (2011)	Fish freshness assessment.	Multi sensor based on volumetric electrodes	Using array of volumetric sensors, chemically modified with phthalocyannies.
Fang ski Han,Xingi Huang,Ernest Teye,Feifei Gu & Haiyang Gu. (2013)	Fish freshness assessment.	Computer screen photo assisted gas sensor array.	An array of mettalophophynes id used to classify fillets of thawed fish
S. Ghosh, D Sarker b,T.N. Misra. (2014)	Fish freshness monitoring.	E-tongue based on polypyrole.	Using array of volumetric sensors, chemically modified based on screen printed electrodes
Amin Taheri Garavand, Amin Nasiri , Ashkan Banan, Yu-Dong Zhang. (2020)	Freshness monitoring.	Computer vision based technique	Non-destructive method using skin tissue.
Anandhu suresh,Arathi vinayachandran,Chinju Philip, Jithu George . (2020)	Fresko Pisces.	Deep learning technique.	Fish eye, gill & skin image are trained on pre-trained CNN.

Fig. 1 Comparison with other model

used as preservative to store the fish for long days not comprising their appearance. The pungent smell of formalin is not easily detectable by human but the sensor kit can detect the presence. So, the process is based on checking the fish sample for formalin by using the formalin detection kit, and then, the mobile application is used to identify its freshness. So, the confusion of freshness after adulteration can be avoided totally. To classify the attention condition accurately from reactive and bulged eyes to non-reactive and cloudy eyes, the gill color change accurately, either pink or red and yellowish red in color. Image pre-processing can fundamentally expand the unwavering quality of an optical examination. A few channel tasks, which heighten or diminish certain image entities empower a neat and quicker evaluation. Image resizing or image interjection happens when resizing or changing our picture starting with one pixel network then onto the next. Picture resizing is significant in the wake of expanding or diminishing the entire number of pixels finished [7].

4 Proposed Work

Fresh fish could be a product with an awfully short commercial life and a high variability. Those characteristics make it very difficult to use the identical systems utilized in other foods. The target of quality assessment of fish is to help the upkeep or improvement of portability by minimizing customer complaints about quality.

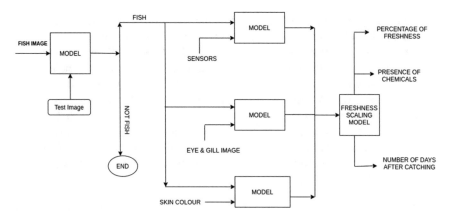

Fig. 2 Data flow diagram

Developing techniques for location of fish freshness incorporate sensory assessment strategy, microbiological technique, physical and chemical technique, and so on [8]. But these strategies are for the most part incapable to accomplish fast recognition of fish freshness. Because the freshness of fish depends upon numerous characteristics, only the distinguished and relevant features are extracted; i.e., bulginess of eyes, change in gill color, and change in skin color are the features mainly taken into consideration. An information set was prepared by incorporating the above-mentioned features. Sometimes the fish may be stored in chemicals like formaldehyde in many days so as to preserve its freshness. Such cases can be caught by testing with foramldehyde detection kit provided with the proposed work. The initial block involves the classification deep learning model VGG-16 which takes the input of fish image and if the image is not of a fish, output will be sent as not fish. If it is detected as fish, it gets carried away to the next section. This model takes in the sensory values for detecting the presence of gases, specifically the eye and gill part of fish, and also the skin appearance in a detailed manner. All these features are then combined and calculated in freshness scaling model, where a specific function scales the features as fresh or stale, whether it contain any harmful chemicals and determines number of days from catching. Finally, the trained model is converted to tensor flow lite and it is loaded onto an android mobile application in product phase [9] (Fig. 2).

5 Modules

The system design consists of mainly three modules: (1) User module (2) Admin module and (3) Mapper module. An android mobile application is used here. It is free and access to all the people with smartphone. The eye and gill photo of the fish can be taken at anywhere even at the supermarket and then it can be uploaded to the proposed model using the Fresko Pisces application. The user module is the user interactive page of the system and admin module does all the controlling functionality of model and database of the system. The mapper module maps the user with admin in terms of request and acknowledgement [10]. The model will give the output as the percentage of freshness, number of days after catching and a comment stating whether the fish is consumable or not. Apart from this, the system is divided into automatic deep learning detection system and a manual freshness finding system. In former case, the percentage is calculated using test accuracy of an image that has been tested with the trained model as shown below. In latter case, the opinion chosen by user for each parent feature is taken into account for finding the percentage. Finally, both these values are combined in an equal manner to provide a final output.

6 Methodology

The whole methodology starts with a collection of whole amount of data required for building the dataset. After the collection of data, the data is preprocessed according to the model requirement for building the CNN model was implemented. Python 3 language and libraries like Pandas, Scipy, Scikit is used to learn in each phase of the project. New fish may be an item with an awfully brief commercial life and a tall inconstancy. Those characteristics make it exceptionally troublesome to apply the same frameworks used in other nourishment. The objective of enabling a quality appraisal of fish is to help the upkeep or advancement of productivity by minimizing the client complaints on quality [11]. Strategies developed for the location of fish freshness incorporates a tactile assessment strategy, microbiological technique, chemical and physical techniques, and so on. Be that as it may, these procedures are generally not ready to acknowledge quick areas of freshness point for use as far as its simplicity of use and pertinence exist. After which it is finally made due with 2 key features our computation should have. The essential being that it should have the option to work regardless of the gadgets used to get the image of the test for example back various camera characteristics and the minute being that it should not require a broad data set for getting ready and should be quick to set up. The technique thinks about two features for the confirmation to get ready. The features being the fish gill deviation, the eye shadiness and staining of the fish. These features make a numerical outcome whether the fish is stale or not. The model was finally converted to a weight (h5) and then to tensor flow lite which was loaded to a mobile application hence became a product for public use. Below subsections depict various steps and features for implementation of the system.

6.1 Freshness Detection Using Images

In this section of implementation, the freshness is identified using images taken from sample. Initially, the user have to take the image of both eye and gill images from the sample and upload it to the system for testing purpose. The system will then match the image with already trained set of images and produce the resultant output. The system is designed such that the resultant will be fresh if and only if both eye and gill image is fresh, otherwise not fresh will be the output. The percentage calculated from the above method is test accuracy on testing the sample with trained data. Since there is two models, the average of both test accuracy is taken for calculation further.

For example, let the test accuracy for a sample image of eye model in order of [Fresh; Not-Fresh] is [0:96; 0:23], then system will compare the both the values and consider the larger value,i.e.,0.96. Likewise, the same is done for gills model. Finally, average of selected values is calculated as follows. Let the dominant value from eye model and gill model be (V1fresh) and (V2fresh), respectively, then.

$$Result = \frac{V\,1\,fresh + V\,2\,fresh}{2}$$

Result will have high percentage value of freshness and a comment stating that fish is worth for money.

6.2 Freshness Detection Using Quality Index Method

The QIM is traditional approach for determining the unknown by method of indexing. Here, this method is used to ascertain the value found out using the earlier detection. There are features which cannot be determined by images, it requires suggestion from user and it can be easily labeled just by examining the sample. For each quality, a determined index is substituted and finally the average of all index are taken to send the corresponding result. For the quality stiffness of fish body, the qualities vary from highly stiff to loose skin tissue. These are substituted with different indexes and finally based on the input provided by user. The final QIM value is calculated and converted to percentage. Now, the percentage identified from the automatic method and QIM is combined to get the average and it is the final percentage of freshness; it is the choice of developer to choose the quality index for each feature and then on wards it is set as a standard.

Let the system have five major features and each one have 4 sub features, assigning 4 as maximum and 1 as minimum index, total index would be $5*4 = 20$. If the user chooses the following options, result by QIM method would be,

$$Result = \frac{f1 + f2 + f3 + f4 + f5}{20}$$

Hence, the final percentage combining the image and QIM method can be calculated as.

$$Final\ Percentage = \frac{Result(image) + Result(QIM)}{2}$$

Fig. 3 Fresh eye

Fig. 4 Not-fresh eye

6.3 Fish Eye Cloudiness

Estimating the darkness of the point eye can too be used as a standard for fish quality. However, the entirety of darkness recorded by a camera sensor will rely upon a piece of parts, for example, camera quality, natural lighting in the midst of recording the observations, extra lighting (streak) in the occasion that given and the kind of fish. This fundamentally makes us inadequate to use inside and out qualities to qualify the fogginess of the eye. At the point when a fish ruins, the eyes ceaselessly force and go from being level to inward (indented), getting to be shady and smooth and the cornea as murky [12]. In next step, an example based on the methodology used where the pictures of known new tests are sometime taken as a late use and store their qualities as reference. The unused tests are assessed so as to the reference esteems (Figs. 3, 4, 5 and 6).

6.4 Gill Color Change

The change in gill color from bright pink color to dark red color or yellowish red color marks the transition from fresh to stale. So this change in color is traced by difference in RGB values extracted from the picture.

Fig. 5 Fresh gills

Fig. 6 Not-fresh gills

6.5 Skin Discoloration

The overall discoloration of skin from bright and shiny color to fade color will result in detection of freshness. Fresh fish will have bright and shiny skin, whereas not fresh fish will have less shine and fade color. The brightness and color combinations are extracted by the model by analyzing the RGB matrices generated from the image [13].

6.6 Slimness of Skin

This Feature is identified using manual methods by examining the fish skin and giving the corresponding option on mobile application whether mucus is present on large scale and it is a manual method detection. There are four options on various phases of mucus presence that can be seen on fish body and user can be easily identified by examining the body of fish.

6.7   Stiffness of the Fish

Stiffness of the body of fish is another feature to predict the freshness. As the fish starts spoiling, the stiffness reduces and body becomes loose. These is also a manual method of testing and the rate of stiffness in each phase of fish sample is provided and user can choose the best one which suits most.

6.8   Color of the Flesh

The color of flesh which determines freshness changes from cream color to slightly yellowish and to brown then blue color of the fish determines the freshness in terms of flesh color. RGB color identification by CNN technique is applicable in this case also.

7   Classification

The feature extraction task is made by a convolutional neural network that is additionally called CNN. CNN is a deep learning algorithm that accepts user input, allots significance (learnable loads and biases) to different viewpoints in the image and have the option to separate one from the other. Visual graphics group (VGG) is a network model, which won top 5 accuracy in VLIRC challenge and is still considered as best network. In convolutional models, the inception model, mobile net, VGG-16, image net models were utilized and transfer learning was done [14]. Out of the models chosen, VGG-16 show the most exactness in the wake of preparing thus the design of VGG-16 is as appeared underneath (Fig. 7).

8   Result

The final result of proposed paper is that an application is introduced to calculate the percentage of freshness, estimated remaining shelf life of fish, number of days from catching of fish, and the percentage of formaldehyde present in the fish sample. A mobile application will be acting as the user interface with the option to take photo from gallery or at real time and a submit button. The above-mentioned result will be listed in an order in the result section of application. The eye, gill, or skin images at each case produces the corresponding result. The model used under different scenarios are trained and tested with the same dataset of samples, which have produced the analysis that VGG-16 has produced the most acceptable result for our dataset. The peculiarity of the model is used instead of having a large number of

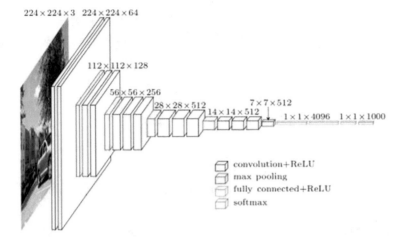

Fig. 7 CNN VGG-16 architecture

Fig. 8 Model comparison

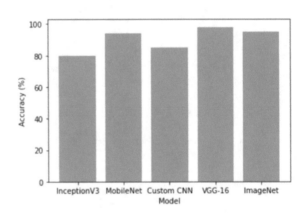

hyper parameters, it focuses on a convolutional layer of 3 × 3 filter with stride 1 and always use same padding and maxpool layer of 2 × 2 filter of stride 2. It follows this arrangement consistently with around 138 million parameters on untrained model and 119 million parameters on trained model. Hence this model is considered as one of the excellent vision model architecture till date and used for this work (Figs. 8, 9, 10, 11 and 12).

9 Conclusion and Future Work

It is concluded from our proposed work that, the fishes are devoured in expansive amounts each day, and it is necessary to guarantee some wellbeing guidelines [15].

Fig. 9 Model accuracy

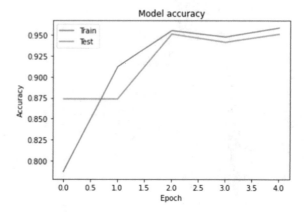

Fig. 10 Model loss

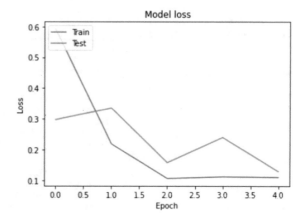

Fig. 11 Train versus test accuracy and loss

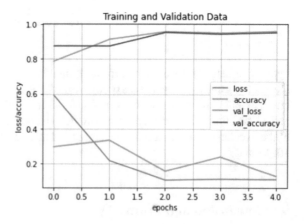

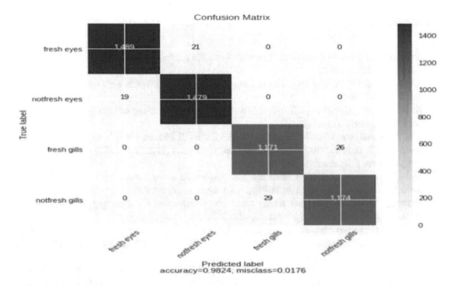

Fig. 12 Confusion matrix

Fish quality may be a complex concept by including a few variables such as safety, nutritional quality (due mainly to high-quality fats, edible proteins, and essential minerals), accessibility, comfort, and astuteness, eating quality, physical attributes of the species, measure and freshness. Almost the data dealing with capacity strategies, counting time/temperature histories that can influence the freshness and quality of the items is very important for all the parts of the nourishment chain. Also, regular condition that has the impacts of angling grounds, capture methods, and the event of different surrenders that impacts the general quality. One of the foremost special characteristics of fish is to consider it as a food, where it could be a highly perishable product. Subsequently, the time passed after the capture and the temperature of fish are monitored frequently with the key variables by deciding the final quality characteristics of a fish item. The foremost commonly utilized strategy to determine their freshness is by visual examination, which is performed by an individual. The result of such investigation relies heavily on the individual performance analysis and is indeed unreasonable for huge sets of tests. Numerous ponders have utilized the distinctive processing methodologies to attain the assignment of judging the freshness of a fish [16]. The proposed work has attempted to accompany the users with a specialized errand of recognizing the freshness of fish based on novelty approach. All the calculation, observations, and identification for this work is based on experimental study conducted on different environments and under strict conditions. For all the research conducted over this work, it is concluded that it should adhere the above stated as fixed standard. The future work of FRESKO PISCES is that the whole system can be incorporated into device so that it can be implemented over a conveyor belt to automatically distinguish the huge samples of fish and can be made on large basis in a more effective way.

References

1. Vijayakumar T (2019) Comparative study of capsule neural network in various applications. J Artif Intell 1(01):19–27
2. Bashar A (2019) Survey on evolving deep learning neural network architectures. J Arti Cial Intell 1(02):73–82
3. Sowmya B, Sheelarani B (2009) Colour image segmentation using soft computing techniques. Int J Soft Comput Appl 4:69e80
4. Das T, Pal AK, Chakraborty SK, Manush SM, Sahu NP, Mukherjee SC (2005) Thermal tolerance, growth and oxygen consumption of Labeo rohita fry (Hamilton, 1822) acclimated to four temperatures. J Therm Biol 30:378e383
5. Dubey SR, Jalal AS (2014) Fruit disease recognition using im-proved sum and di erence histogram from images. Int J Appl Pattern Recognit 1(2):199e220
6. Barbedo JGA (2013) Digital image processing techniques for detecting, quantifying and classifying plant diseases. Springer Plus 2013(2):660
7. Barbedo JG (2014) Computer-aided disease diagnosis In aquaculture: current state and perspectives for the future. Brazilian Agricultural Research Cor- poration (Embrapa)Revista Innover 1, Numero 1
8. Bashish DA, Braik M, Bani-Ahmed S (2011) Detection and classi cation of leaf diseases using K-means-based segmentation and neural-networks-based classification. Inf Technol J 10(2):267–275. ISSN 1812–5638. https://doi.org/10.3923/itj.2011.267.275 @ 2011 Asian Network for Scienti c Informa-tion
9. Navotas IC, Santos CNV, Balderrama EJM, Candido FEB, et al Fish identification and freshness classi cation through image processing using arti cial neural network. J Eng Appl Sci 13(18):4912–4922
10. Issac A, Dutta MK, Sarkar B, Burget R (2016) An encient image processing based method for gills segmentation from a digital fish image. In: 2016 3rd international conference on signal processing and integrated networks (SPIN), Noida, pp 645–649
11. Xue G, Yu W, Yutong L, Qiang Z et al (2019) Construction of novel xanthine biosen-sor by using Zinc Oxide (ZnO) by biotemplate method for detection of fish freshness. Analytical Methods 11(8). https://doi.org/10.1039/C8AY02554A
12. Karagoz A (2013) Fish freshness detection by digital image processing, Dokuz Eylul University, Izmir, Turkey, Available https://acikerisim.deu.edu.tr
13. Olafsdottir G, Martinsdottir E, Oehlenschlager J, Dalgaard P, Jensen B et al (1997) Method to evaluate fish freshness in research and industry. Trends Food Sci Technol 8(8):258–265
14. Chen Z, Lin Y, Ma X, Guo L, Qiu B, Chen G, Lin Z (2017) Multicolor biosensor for fish freshness assessment with the naked eye. Sens Actuators B Chem 252:201–208
15. Lougovois V (2005) Freshness quality and spoilage of chill-stored fish. ISBN 1–59454--5, pp 35–36
16. Raj JS, Vijitha Ananthi J (2019) Recurrent neural networks and nonlinear prediction in support vector machines. J Soft Comput Paradigm (JSCP) 1(01):33–40

Face-Based Alumni Tracking on Social Media Using Deep Learning

Deepti Prasad, Rasika Gaiwal, Shivani Bharambe, Supriya More, and Ashwini Jarali

Abstract Today's world is connected due to social media platforms. Our system can be used by universities to attract, interact and stay connected with the alumni by using various networks and engaging content. The system enables to retrieve an individual's information from social media sites based on the facial image. A proper study of various face recognition algorithms was done and found that a convolutional neural network was most promising. Also, various web scraping tools and techniques were studied to retrieve the individual's information. Our proposed system takes an input image of alumni and searches for alumni's name from the database (college database). The name and image are then used to search the active profile of the alumni from social media sites such as Facebook, Instagram and LinkedIn. Thus, the paper provides a model for uploading a photo and searching the person on a social media site by using face recognition and web scraping.

Keywords Feature extraction · CNN · Image encoding · Face recognition · Web scraping

D. Prasad (✉) · R. Gaiwal · S. Bharambe · S. More · A. Jarali
Department of Computer Engineering, International Institute of Information Technology, Pune, India
e-mail: deepti.prasad7@gmail.com

R. Gaiwal
e-mail: rasikagaiwal21@gmail.com

S. Bharambe
e-mail: shivani7798@gmail.com

S. More
e-mail: supriyamore0789@gmail.com

A. Jarali
e-mail: ashwinij@isquareit.edu.in

1 Introduction

For universities, the love and commitment of alumni are often a strong tool for fundraising opportunities, marketing and brand awareness. However, to get those benefits, a way to maintain a relationship with our students is needed, even after they receive their degree. Earliar, direct mail campaigns, phone calls and student reunions were most of the best practices for alumni engagement. These standard ways used to gather alumni information do not have a good collaboration rate, which leads to an interrupt while gathering information. Now, in the digital age, universities can engage and interact with their alumni in a more consistent and modern basis. Social platforms play a key role in connecting and gathering information about alumni. The details can be tracked by using the internet/social media. This can be done by using a face recognition technique, in which a person is searched based on his/her photo on social platforms like Facebook. Hence, can get data regarding the present activities of the alumni and directly establish a connection with them. Furthermore, using social media tools to have an interaction with students can create a long-lasting relationship between universities and students after their graduation. So, come up with a system that automatically searches an alumnus on social media sites using face recognition and web scraping. Face recognition is the process of extracting the facial features of an individual and comparing it with the existing features from the database to recognize an individual. Various methods can be used to deploy face recognition. Convolutional neural network method for face recognition is widely used as they ensure higher accuracy than other methods [1]. After face recognition module, its output is given as input to the web scraping module. Web scraping is the process of extracting data from webpages in a structured format. It is found to be the best technique to gather information from webpages [2].

In our system, firstly, alumni's image is uploaded which is passed to the face recognition algorithm. The face recognition algorithm labels all the faces present in the image with their corresponding name. These names are obtained from the college database by matching image encodings during the execution of the algorithm. The alumni image is now stored with the same name as obtained in the result. For web scraping, firstly the alumnus name is searched on the social media website. Then profile links and profile photos are scrapped from the accounts which appear in the search results. These scraped profile photos are given to the face recognition algorithm and the one which most accurately matches the alumnus's face is selected. This profile photo and its corresponding profile link are ultimately given as output. The output is displayed via an HTML page and also stored in a CSV file. Hence, this system can be used to automatically track an alumnus on social media. Thus, this system can be useful for universities to build their alumni network with less manual interference and time.

2 Literature Survey

2.1 Real-Time Face Recognition System

This work [3] presents a real-time application system using a face recognition algorithm. This proposed work implemented for face recognition and it is used for security applications.

2.2 Deep Learning-Based Image Recognition

CNN-based face recognition system is utilized and implemented the different algorithms. This work studies about the differentiation among the deep learning-based image recogniotion and CNN-based face recognition systems. The results are comapared based on the parameters include data agumentation, pixel value type and similarity indices [4].

2.3 Gathering Alumni Information from a Web Social Network

This work [5] presents a system which gathers alumni information from various webpages on a social network. It takes a list of names of alumni from an undergraduate program and a set of relevant web pages on which the alumni can be searched. The system searches the alumni on the set of pages using a searching mechanism and then discards the pages which are not much informative. It stores the informative web pages and then extracts information from them using an extractor. A website has a fixed page structure that is the features of pages are the same so a single manual expression is enough for retrieving information from various pages. After the extraction of information from the pages, it is stored in a database and can be used for further work. The drawback of this system is that it searches based on name and undergraduate program and there can be multiple persons with the same name in the same program. So accuracy will not be satisfactory in this scenario.

2.4 FindFace App

The FindFace app was studied briefly and found that it does not always match the image with the correct profile. Its creators claimed that it works 70% of the time. Also, ten similar people profiles are extracted by the app. Hence, this app is not efficient when it comes to an emergency scenario where getting the information of an individual is very important.

2.5 An Overview on Web Scraping Techniques and Tools

This work [2] presents an overview of different web scraping techniques and tools. It concludes that web scraping is the most effective way among all other techniques to automatically gather information from webpages. Using web scraping, users can collect structured data instead of the unstructured data from the webpage.

3 Proposed System

In our system, the user has to upload an image of the alumni. This image is then sent to the server. The server contains a face recognition module which detects the face in the image and recognizes that person by matching it with the images present in the database of college. The images will be stored in the form of encodings in the database to increase the security of the system. No unauthorized person will be able to access the images as it will be stored in the form of encoding. The encoding of images is generated using a CNN model. Then the corresponding name of the input photo is obtained as the output of the face recognition module. Once the identity is obtained, then the server performs the profile search on social sites using web scraping. In web scraping, firstly, social media accounts with the same name are searched and then the profile photo and profile links are scraped from them. Next, face recognition is used to match the input alumnus image with the fetched profile photos. The most accurately matched profile photo and its corresponding profile link is given to the output in the GUI. The obtained profile links are stored in an HTML page and CSV file which can be used by the college to connect with their alumni. This will automate the process of searching an individual on social sites and skips the process of manually searching for an individual. By doing so, a huge amount of time can be saved. Following is the architecture diagram showing how the system works (Fig. 1).

Various functionalities of our system (Fig. 2):

$T = \{F1, F2, F3, F4, F5\}$.

F1 - > Uploading the image of the person and sending it to the server.

F2 - > Detect the face from the image.

F3 - > Face recognition to get the identity of the individual.

F4 - > Web scraping to get the exact profile from social media.

F5 - > Provide profile links in GUI.

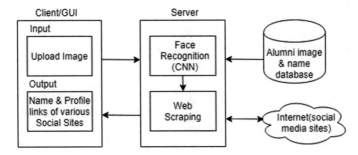

Fig. 1 Architecture diagram of our system

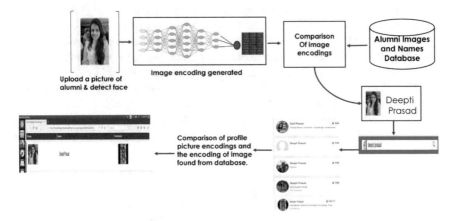

Fig. 2 Detailed flow diagram

The system consists of the following modules.

3.1 Client/GUI Module

This module lets the user (College Admin) upload the image of alumni and this image is sent to the server for further computations.

3.2 Server and Database Module

Once the image is collected from the client module, the server module comes into the picture. The database in this module is the college database which has records of students' name and their image. The image database is stored in an encoded form so that no unauthorized access will be possible.

Following are the actions taken by the server module:

- Face Recognition:

Face recognition is the first step of our system's server. In this, the faces of persons are identified and label it with their name. Face recognition has various steps, namely face detection, feature extraction and feature matching [4].

Detailed face recognition steps are as follows:

(1) Locate the uploaded image and find all the faces in it. Face detection is done by using a trained CNN model (Fig. 3).
(2) Focus on each face and cut-out only face pictures from the image (Fig. 4).
(3) Now, our algorithm will pick out the unique features of each of the faces detected; the model is trained to be able to find 68 specific points on any face (Fig. 5).
(4) Using these features, this complicated raw data (image) is reduced to a list of 128 computer-generated numbers which are known as encodings.
(5) Neural Networks are trained using these encodings so that the model formed can generate 128 encoded values for any new image (Fig. 6).
(6) Finally, compare the unique features of input face to all the faces in the database, and hence determine the person's name. A comparison of encodings is done by

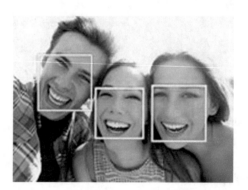

Fig. 3 Upload photo

Fig. 4 Locate faces

Fig. 5 Facial feature points located

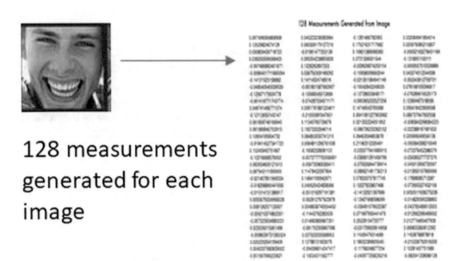

Fig. 6 Encoded values of an image

calculating the Euclidean distance between them. The image which has the least Euclidean distance value between the input image and itself is considered as the most similar face (Fig. 7).

Fig. 7 Labelled faces

Fig. 8 Architecture diagram
of the face recognition
process of our system

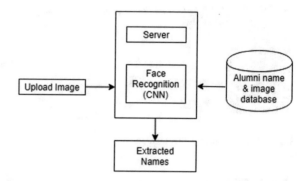

In our system, the alumnus image is given as an input to face recognition algorithm, which performs computations on the input image and searches the image similar to the input image in the college database. Once the person is found in the database, then his/her name is taken from the database. The extracted photo is stored with the same name as that obtained from the face recognition algorithm. After the name of the person is got, web scraping based on this name is performed (Fig. 8).

- Web Scraping

Web scraping is an automated process to gather information from a webpage. Web scraping is also called screen scraping as all the contents on the screen are gathered. There are different ways [5] in which web scraping can be done (like HTTP programming, HTML parsing, DOM parsing, etc.). Webpages have a specific page structure; thus, its HTML code can be analysed and parsed for scraping its contents. This process is known as HTML parsing [6] (Fig. 9).

In our system, firstly a web browser is connected with an automation tool using a web driver. This automation tool enters the social media site URL (the one on which wanted to search the alumnus) in the web browser. Once the webpage is successfully opened, the location of the search box on the webpage is found out using HTML parser. The alumnus name is entered in the search box using the automation tool (Fig. 10).

Fig. 9 Architecture diagram
of web scraping process of
our system

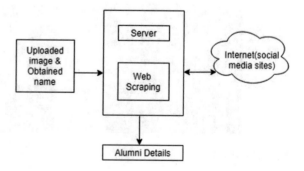

Fig. 10 Automated searching of alumni name in the search box on a social media website

Then all the profiles having the same name as the alumnus name will be displayed on the screen (Fig. 11).

From these profiles, their corresponding profile links and profile pictures are fetched [7] using HTML parser and a list is maintained. Next, all the fetched profile pictures are given to the face recognition module. Face recognition module then matches [8] those profile pictures with the alumnus photo and the profile link of the most accurately matched profile picture is displayed as output.

Fig. 11 List of accounts
with the same name as an
alumnus

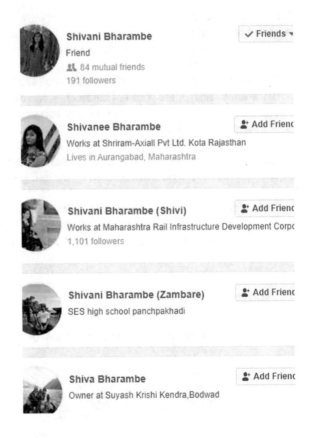

3.3 Output Module

The output of our system is maintained as an HTML page and CSV file. The CSV file will store the name and profile link of the alumni. The HTML page will show the output obtained from the web scraper in the form of a table containing alumni image, name and profile photo.

With the help of web scraping, the profile link of the person will be got and can be used to track the person. Hence, the information of the alumnus can be extracted.

4 Results

The system was tested by giving the same person's photo with different ages as input and was found that it accurately detects the person. Various tests with a person wearing a cap and earrings were performed to check whether the system outputs the exact profile link or not (Fig. 12).

Following are the results of three persons computed by the system (Fig. 13).

In the above image, the photo of alumni is given as input to the system (the first column in the table) and the system rightly extracted the name and social media profile picture of that alumni.

In the above image, the social media profile links are provided which will directly open the Facebook account link of that alumnus. In this way, structured data of the required profile links which can be used by the college to connect and interact with their alumni is got (Fig. 14).

For performance analysis, ten photos of five persons each were given as input to the system. The following graph illustrates the analysis of the number of times the correct social site profile link was fetched by the system (Fig. 15).

5 Applications

- This system can be very useful to track alumni on social sites.

Deepti Prasad.jpg

Shivani Bharambe.
jpg

Sushmita Prasad.
jpeg

Fig. 12 Alumni name and photo obtained as output from face recognition module and given as input to web scraper

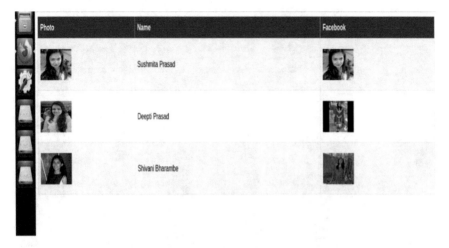

Fig. 13 Output containing alumni name, image and profile picture

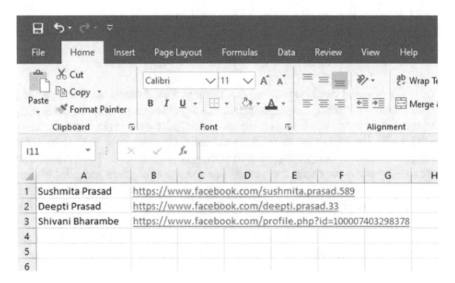

Fig. 14 Output CSV file having alumni name and social media profile links

- Also, it can be used for tracking various criminal activities on social sites if a criminal database is available.
- Can be used in hospitals to get the information of unknown patients if the name of the patient is obtained.
- Can also be used for getting the details of the employees for employee verification.

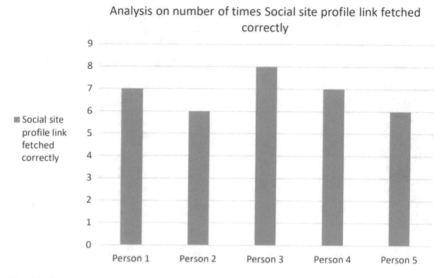

Fig. 15 Graphical representation of performance analysis of system

6   Conclusion and Future Work

The usage of social media is growing at a fast rate. Social media has also become a major source of collecting information. It has also given us the privilege of connecting and interacting with people around the globe due to its widespread network. Thus, social media has been used to track an alumnus. Our system is built to help the colleges/universities for getting the information of their alumni. The system is an automated process, so minimum manual interference is required and a lot of time is saved. The output of the system is presented in an HTML page and also stored in a CSV file, so even persons with a non-technical background can easily use this system.

Future work may include research on increasing the computation speed of the system using various methods like the principal component analysis.

References

1. Coúkun M, Uçar A, Demir Y (2017) Face recognition based on convolutional neural network. In: 2017 IEEE. 978–1–5386–1750–2/17.
2. Saurkar AV, Pathare KG, Gode SA An overview on web scraping techniques and tools. Int J Future Revol Comput Sci Commun Eng 4(4):363–367. ISSN: 24544248
3. Haji S, Varol A (2016) Real time face recognition system (RTFRS). In: 4th international symposium on digital forensics and security (ISDFS 16). Little Rock, AR
4. Hu G, Yang Y, Yi D, Kittler J, Christmas W, Li SZ, Hospedales T (2015) When face recognition meets with Deep Learning: an evaluation of convolutional neural networks for face recognition.

In: IEEE International Conference on Computer Vision Workshop (ICCVW)
5. Goncalves GR, Ferreira AA (2014) Gathering alumni information from a web social network. In: Latin American Web Congress (LaWeb), Brazil
6. Ashiwal P, Tandan SR, Tripathi P, Miri R (2016) Web information retrieval using Python and BeautifulSoup. Int J Res Appl Sci Eng Technol (IJRASET), 4(VI)
7. Narashima P, Akshata A, Meghana B, Karuna G (2015) Crawling through Web to Extract the Data from Social Networking Site—Twitter. In: 2015 national conference on parallel computing technologies (PARCOMPYECH)
8. Bruce V, Young A (1986) Understanding face recognition. Br J Psychol 77(3):305–327

Lane Detection Using Sliding Window for Intelligent Ground Vehicle Challenge

Sahil Samantaray, Rushikesh Deotale, and Chiranji Lal Chowdhary

Abstract In today's world the self driving car industry is booming and is expected to reach a global revenue of 175 billion dollars by the end of 2025. The self driving car employs various combinations of sensors to navigate the car autonomously. The cars in present times are equipped with driver assistance features like positioning the car within the particular lane. This requires a proper lane detection technique which is robust to errors and gives an accurate output. There are several algorithms being used currently in the industry for detecting the lanes so as to direct the driverless-car not to deviate from it. Researchers use traditional computer vision algorithms or also come up with deep learning models which are trained on huge annotated datasets to precisely predict the lanes. This article describes an approach to lane detection specifically used by automated bots participating in the Intelligent Ground Vehicle Challenge (IGVC). The algorithm uses openCV functions mostly and follows an image pipeline to properly segment and return the direction vector in terms of magnitude so that it can be sent to the bot as steering messages. We verified the results by performing navigation.

Keywords Lane lines · Thresholding · Sliding window · Autonomous vehicle · IGVC · Colour spaces

1 Introduction

The Intelligent Ground Vehicle Competition (IGVC) is an international robot- ics competition for teams of undergraduate students to design and build a fully auto-mated vehicle which can navigate the course and also avoid obstacles. In IGVC, the autonomous robot must navigate through a course which consists of lanes with white lines on a grass field. The competition has different types of penalties for not main-taining lanes and hitting obstacles etc. Therefore maintaining lane lines is important.

S. Samantaray (✉) · R. Deotale · C. L. Chowdhary
VIT University, Vellore 632014, Tamil Nadu, India
e-mail: sahil13399@gmail.com

© The Author(s), under exclusive license to Springer Nature Singapore Pte Ltd. 2021   871
J. S. Raj et al. (eds.), *Innovative Data Communication Technologies and Application*,
Lecture Notes on Data Engineering and Communications Technologies 59,
https://doi.org/10.1007/978-981-15-9651-3_70

Over the years different researchers have tried different methods to detect lane lines. The most commonly used is the one which generates a binary image of lane lines and then uses hough transform to fit the lane. This can be achieved through simple OpenCV functions. The image is passed through a pipeline with grayscale image followed by gaussian blur and then a canny edge detection. The canny edge detection gives us the edges from the images but it gives a lot of other edges besides lanes which aren't our interest. Thus to find lines we use hough transforms,in hough transforms the lines are represented in polar coordinates. After transforming the image into polar space we deduce lines present in an image. The advantage of using hough transforms after applying canny edge detection is that it can easily deduce straight lines from the image. Extrapolation of left and right lane lines is accomplished using gradients. The main advantage of using simple OpenCV methods is that they are not intricate and give computationally faster responses in real-time which is important in our case. But the problem of using these methods lies in their ineffectiveness in producing good results in edge cases and inability in taking decisions for new test cases. Other methods of evaluating lane lines are using deep learning approaches, these are segmentation models which segment out lanes from an image. Segnet model is used for lane segmentation, the model takes live image stream from the webcam and produces binary images which have lanes segmented out. Segnet is a convolutional neural network for pixel-wise classification of an image. The images are preprocessed before they are passed on to the neural network. The output from the network, a segmented image, undergoes some techniques to enhance the output lanes. There are many pixel-wise image segmentation models present currently and research has been going on in the area. The advantage of using deep-learning methods is the high accuracy and to generate results on newer cases which are cut above methods using simpler algorithms. Since deep learning methods require superlative computation power and even after using high performance computers deep learning models take time to produce results in real-time, this is a challenge which is to be overcomed. But due to the inability of using segmentation models in real-time we need to use simpler computer vision based methods and focus on improving accuracy of computer vision algorithms.

Our main motivation for this project is to build a simple lane detection algorithm which does not use any advanced algorithm for sensing the direction of the vehicle. This algorithm for getting direction, called the direction vector method is efficient and helps us in making small vehicles automated.

The paper is organised in the following way. In Sect. 2 we discuss about existing work which is present in the area. In Sect. 3.1 we give an overview of our image processing pipeline. Section 3.2 describes about the camera calibration. Sections 3.3 and 3.4 describe the image prepossessing steps we performed to get the desired results. In Sect. 3.5 we discuss about histogram techniques which we used to get lane lines on the image.Results have been discussed in Sect. 4. Finally we discuss about our work and give the future scopes the pipeline.

2   Literature Survey

In this paper [1] the author is trying to detect the lane lines present on the Intelligent Ground Vehicle Challenge course. The entire lane detection is done using a stereo vision camera which returns a feed of image along with its depth map. This 3D visualization of the surrounding can help us use the disparity map to segment the image with lane lines and obstacles based on distances of the reprojected image in a 3D space.

The authors of this paper [2] try to present an advanced lane detection system in real time. The entire methodology is divided into two parts, one which has the image preprocessing pipeline to segment the road lines and then it uses Extended Kalman Filter to track those lines in real space. This is a very efficient road line detection method using Hough Transform, but due to the lack of robustness of the pipeline it may not work in all scenarios.

The article [3] gives us another method of road lines using edge detection. The algorithm relies on distinct pixel differences between road and lane lines. The authors have compared several edge detection algorithms like Canny edge detector, Roberts, Prewitt and Sobel on the basis of time for the processing time and accuracy. However such a method of lane lines detection is bound by many constraints like speed, processing resources and accuracy. The algorithm may not be appropriate for rustic roads.

Here the authors [4] try to shift from traditional Hough Transform used for lane detection to a faster Hough Transform. The traditional algorithm uses all the points in the region of interest like in [2] which is time consuming but in this it considers the vanishing points and previous frame lane line measurements because they do not change drastically. This makes the process much faster. The problem in this algorithm is that there is no method to remove the interference of the other edges detected apart from lane lines.

In this paper [5] the authors have proposed a new method for robust lane detection which is applicable to complex scenarios. The algorithm uses a combination of image processing techniques, a convolutional neural network and Random Sample Consensus algorithm or RANSAC for the lane lines fitting. Initial preprocessing was achieved by first smoothing the image using Gaussian Blur and then followed by edge detection with the help of a hat-type kernel. For simpler roads the RANSAC algorithm was sufficient to fit the lane but in complicated scenarios they used a combination of CNN and RANSAC. The algorithm hence gives accurate results on various test data.

The paper [6] gives a way of using instance segmentation for lane detection which can be a very efficient method to detect lanes considering a dynamic scenario. The authors have developed an architecture called LaneNet which returns an instance map depicting the lane. It labels every pixel classified as lane with a respective lane-id. The lanes are then transformed for lane fitting using a transformation matrix given by a trained model with custom loss function called H-Net. The H-Net is used to shift from conventional approach of 'birds-eye view' to fit a polynomial on the lane.

The researchers [7] at NVIDIA have made an algorithm to get steering instructions by training a CNN which gives steering instructions. This end to end approach works very well with minimum training data and also gives accurate results. Similarly [8] in this paper the authors suggest an end to end technique for ego lane detection which uses two transfer learning approaches to segment lanes individually. Another paper [9] suggests a method to maintain the car within the lanes by obtaining steering instruction through which we can compute using a deep learning model trained on actual on roads dataset with ground truth values of steering.

3 Methodology

3.1 Overview of the Proposed Methodology

We use several image processing techniques for the entire lane detection which accurately detects the lane in different weather conditions. The image feed from the wide lens camera is made to go through a series of OpenCV libraries. This customized lane detection algorithm is very computationally efficient due to its simplicity. For more efficient lane detection and higher accuracy we place it a height which is appropriate based on the parameters defined in the practical implementation of the lane detection algorithm. This is a complete 8 step process Fig. 1, where we first have our camera calibration followed by perspective transformation which gives us a 'birds-eye view' of the lanes thus it will be easier to plot the lines using the sliding window approach.Then we choose the color space which is best for binarization for that particular environment then comes the thresholding which gives us a binary image of the lanes. Finally we proceed with plotting the lane lines by filtering out lane pixels. Then we determine the change in direction by calculating the curvature of the lane lines or the direction vector. This change in direction can be given to the vehicle as steering instruction.

3.2 Camera Calibration

Camera's don't create perfect images especially wide angle lens cameras can distort the image of objects around the edges. This will cause a huge problem in our case as we wont get our proper output we need. Due to irregularities in camera our images can get stretched or skewed in many ways thus giving incorrect representation of lane lines and hence we will be plotting an incorrect function on top of the lane lines. Thus to undistort our distorted camera image we use an OpenCV function which takes in input of calibration images of chessboard to find out object points and image points. It then uses these returned values to calibrate the camera and compute undistorted images. The camera calibration is not automated; we use the ROS camera calibration

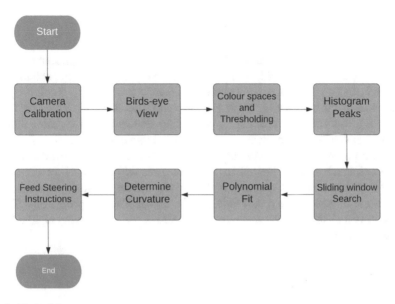

Fig. 1 Methodology

module. In this module we show the camera a picture of Chessboard and start the calibration module. The camera then starts detecting the chessboard corners from all angles that is shown to it. It then calculates the deviations and adjusts for that correction.

3.3 Birds-Eye View

For getting a birds eye view of the lane lines we use perspective transformation using OpenCV functions to help us plot the line much more easily and accurately. Perspective transformation means to see image from a different viewpoint. So when we have our lane lines image we choose certain desired coordinates and the result we get is a new image from a new perspective or in this case a top down view. The image we get will represent the lane lines such that they seem parallel to each other. This will be helpful for us later on to compute the curvature. In this we first use an OpenCV function getPerspectiveTransform which takes in the source and destination points as arguments and returns a Transformation matrix. We can interchange the source and destination points to get an inverse transformation matrix which will return the image from top-view to the original view. This transformation matrix along with the image is given as arguments to another OpenCV function warpPerspective which does the final computation.

3.4 Colour Spaces and Thresholding

An image comprises Red, Blue and Green components. We can represent these individual components as individual colour spaces along with a few other colour spaces like HSV, YCbCr and HLS. This raw image feed is converted to 6 colour spaces. The colour space which gives us a clear output of lane lines is selected for further processing. Selecting one channel for processing reduces the computation time and removes most of the noise. Next we apply thresholding to the image to highlight the pixels with high intensities. The thresholding we use is binary thresholding on the selected channel which is the red channel as it gives us the most distinction. Pixels below a certain threshold are set to black and rest are set to white. Thresholding is not enough to remove the minor noises in the image. Thus, a median blur filter of a certain kernel size is used to remove these noises. All this vividly highlights the white lane lines on grass.

3.5 Lane Detection Using Sliding Window

The histogram of the image is computed along the column of the image thus we get the histogram peaks where the lane lines are present. There will be prominent two peaks depicting the left and right lanes respectively. This gives us the X position of the lane lines on the image from where we can start searching for lanes to plot. Using the starting point we can implement sliding window search to determine the lanes as we iterate upwards. After we set up the parameters of our window we iterate through the image and search for pixels that are activated which essentially are lane lines. After iterating through all windows we need to fit a polynomial line on the lanes that have been detected. The polynomial fit will be of second order.

$$f(y)_l = A_l \, y^2 + B_l \, y + C_l \tag{1}$$

$$f(y)_r = A_r \, y^2 + B_r \, y + C_r \tag{2}$$

Equations 1 and 2 represent the polynomial line that fits the left lane and the right lane respectively.

3.6 Measuring Curvature

The general idea of measuring the curvature of lines is to calculate the radius of curvature at any point x on the function $x = f(y)$ using:

$$R_{\text{curve}} = \frac{[1 + (\frac{dx}{dy}^2)]^{\frac{3}{2}}}{|\frac{d^2x}{dy^2}|} \quad (3)$$

But this kind of curvature needs to be transformed to real world coordinates because this equation calculates the radius of curvature based on pixel values. This gives us steering instruction with respect to the image. To convert this into real world space we need to project this image using the physical distance of the lanes which can differ from place to place. Some authors in [10] have worked on lane keeping by calculating the virtual road curvature by which they get steering and angles and using the bicycle model they maintain lanes. Thus we use another efficient and a generalized method to determine the curvature which is to calculate the direction vector of the lanes. This is calculated by finding the direction vector which is the middle of the plotted lane lines. Then we compute the mean of the this direction vector which gives us the general heading or the direction of the lane. This value of direction vector is then given to the vehicle as steering instruction.

4 Results and Discussion

For our initial analysis we take a still image and run the algorithm on it. As we compute the birds eye view first that result can be seen in Fig. 2. From this output we select the red channel out of all the channels because it gives the best output after thresholding. We can then find the output after choosing the red channel followed by thresholding in Fig. 3.

But we find that there are minor noises in the image. We thus apply a median filter to the image to remove such noises. Such noises can affect the lane detection algorithm here because they rely on pixel densities thus it can cause problems while plotting lane lines. The output of median blur is in Fig. 4.

Fig. 2 Birds eye view

Fig. 3 After thresholding

Fig. 4 Noise removal

Further, we calculate the histogram along the image columns and perform a sliding window search for the lanes. We set the number of windows here to be 9 and the minimum pixel 20px. After running the search we get Fig. 5 this output.

These detected lanes are warped back to our original image using the inverse transformation matrix by interchanging source and destination points, to get a clearer view of the detected lane lines which can be seen in Fig. 6.

After verifying our algorithms on still images we ran it for a video file of lanes which was taken by a logitech C270 camera. The result can be found in Figs. 7 and 8. We faced a few incorrect predictions due to the influence of sunlight. These appear as bright patches on the ground thus there is difficulty in detecting the lanes. Further problems arise due to white noises which were still not cleared after applying the median filter. Another problem which led to incorrect lane predictions was when there is excessive steering with respect vehicles position thus one lane crosses the half of the image thus it affects lane detection because the histogram computed is done along the image therefore this interference will affect our results.

Fig. 5 Sliding window search

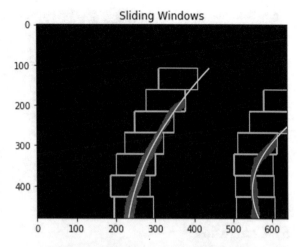

Fig. 6 Wrapping back to original image using inverse transformation

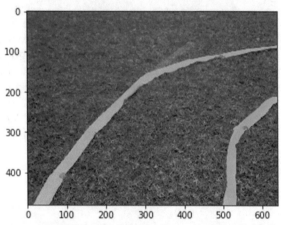

Fig. 7 Result for straight lanes

Fig. 8 Result for curved
lanes

5 Conclusion

In this paper, we had proposed a method of lane detection for Intelligent Ground
Vehicle Challenge course using sliding window technique. Several basic image processing algorithms were used in combination thus giving us a unique pipeline for this
purpose. After the preprocessing we use sliding window search to find the lanes. Our
results show that the lanes are being detected. After detection we calculate the shift
in direction vector which is given as steering instructions or twist messages to the
vehicle to autonomously navigate. This algorithm does not use any database instead
it is a real time based algorithm which can help navigate a small automobile device.

6 Future Work

The future work includes making the model robust to sunlight through some filtering
technique because the influence of sunlight has a considerable impact on the lane
detection. Also there can be an inclusion of techniques to detect lanes from any
camera angle. We can use machine learning to find the source and destination points
so we can automate the process of selecting points for perspective transformation to
give us better results during lane detection.

We can try to predict the lanes of the most likely region where it can possibly be
present and minimize our search area to improve the timing. Like in this paper [11]
the author has used beam search algorithm for improving the process of generating
a caption for an image and in [12] the authors try to use greedy randomized adaptive
search for the multi target and multi sensor tracking problem. We can try to incorporate the same strategy to find generalized regions of lanes based on possibility.

References

1. Kawatsu C, Li J, Chung CJ (2014) Obstacle & lane detection and local path planning for IGVC robotic vehicles using stereo vision. Robot intelligence technology and applications, vol 2. Springer, Berlin, pp 667–675
2. Li M, Li Y, Min J (2018) Lane detection based on connection of various feature extraction methods. Adv Multimedia
3. Phueakjeen W, Jindapetch N, Kuburat L, Suvanvorn N (2011) A study of the edge detection for road lane. In: The 8th electrical engineering/electronics, computer, telecommunications and information technology (ECTI) association of Thailand-Conference 2011. IEEE, New York, pp 995–998
4. Wei X, Zhang Z, Chai Z, Feng W (2018) Research on lane detection and tracking algorithm based on improved Hough transform. In: 2018 IEEE international conference of intelligent robotic and control engineering (IRCE). IEEE, New York, pp 275–279
5. Kim J, Lee M (2014) Robust lane detection based on convolutional neural network and random sample consensus. In: International conference on neural information processing. Springer, Berlin, pp 454–461
6. Neven D, De Brabandere B, Georgoulis S, Proesmans M, Van Gool L (2018) Towards end-to-end lane detection: an instance segmentation approach. In: 2018 IEEE intelligent vehicles symposium (IV). IEEE, New York, pp 286–291
7. Bojarski M, DDel Testa M, DDworakowski M, Firner B, Flepp B, PGoyal B, D Jackel L, Monfort M, Muller U, Zhang J et al (2016) End to end learning for self-driving cars. arXiv preprint arXiv:1604.07316
8. JKim J, Park C (2017) End-to-end ego lane estimation based on sequential transfer learning for self-driving cars. In: Proceedings of the IEEE conference on computer vision and pattern recognition workshops, pp 30–38
9. Chen Z, XHuang Z, (2017) End-to-end learning for lane keeping of self-driving cars. In: 2017 IEEE intelligent vehicles symposium (IV). IEEE, New York, pp 1856–1860
10. Ho ML, Chan PT, Rad AB (2009) Lane change algorithm for autonomous vehicles via virtual curvature method. J Adv Transp 43(1):47–70
11. Chowdhary CL, Goyal A, Vasnani BK (2019) Experimental assessment of beam search algorithm for improvement in image caption generation. J Appl Sci Eng 22(4):691–698
12. Murphey RA, Pardalos PM, Pitsoulis LS (1997) A greedy randomized adaptive search procedure for the multitarget multisensor tracking problem. Network Des: Connect Facilities Location 40:277–302

An Enhanced Packet Delivery Ratio for Mobile Ad Hoc Networks Using DCCHSM Algorithm

J. Vijayalakshmi and K. Prabu

Abstract Clustering is widely preferred in wireless ad hoc networks in order to provide scalability for the high quantity of nodes that are mobile. The intend of strategies of clustering is pretty complicated, appropriate to topology formation of such networks. The proposed system used a simulator clustering to create a neighbor table. And it uses hybrid clustering involving k-means and k-medoids for its information routing. The nodes with the maximum energy level are always opted to be the cluster head. This proposed DCCHSM algorithm provides better performance compared to the existing cluster algorithm least cost clustering (LCC), flexible weight-based cluster algorithm (FWCA) and double cluster head (DCH) and also improving increased packet delivery ratio for mobile ad hoc networks (MANET).

Keywords MANET · DCCHSM · LCC · FWCA · DCH · CH · Simulator clustering

1 Introduction

MANET is one of the specific types of ad hoc network that are connected wirelessly. They have dynamic topology characteristics because of the mobile nature of the nodes and their capability to move independently. The intercommunication between the nodes takes place either directly or through the nodes present in between the source and the destination nodes and acts as the forwarding nodes. Examples are portable devices mobile, laptops, etc. [1, 2]. Though the MANET are capable of self-organizing the network, the cluster framing and cluster head selection remain as

J. Vijayalakshmi · K. Prabu (✉)
Research Department of Computer Science, Sudharsan College of Arts & Science, Pudukkottai, Tamil Nadu 622104, India
e-mail: kprabu.phd@gmail.com

J. Vijayalakshmi
e-mail: kv.anandeesh@gmail.com

tedious job as it have to complete the transmission operation with the limited energy availability of the nodes, prolonging the life expectancy of the network framed.

2 Clustering

The nodes with the similar attributes are grouped together; this is termed as clusters; the clusters are usually formed with the cluster head and cluster members; the cluster heads are always interrelated paving way for a communication that is more reliable using limited energy resources [3, 4]. Cluster head selection has two variants,

1. Minimum distance: All nodes in a cluster should be at certain minimum distance from the CH which is closer to it.
2. Average distance selection: According to this, each cluster should be in an average space within clustering members.

2.1 Simulator Clustering

MANET data transmission takes place through a multi-hop mode for reducing mobile node energy consumption. For multi-hop mode data transmission, the topology and routing protocols play a key role in discovering a quality route and delivering packets efficiently.

This is a challenging task for routing protocols due to their limited resources like memory, battery power, dynamic topology, no centralized authority. Since, MANET mobile nodes are self-organizing networks. Since their energy gets depleted, they need energy management for a longer life in the network. Clustering techniques are used in network simulations as they are better than other techniques [5–7].

Firstly, they can reduce energy consumption in mobile nodes and secondly, they prevent data duplication. Figure 1 explains a simulated clustering.

The goal of any routing protocol is finding an optimized and efficient path for information transfer between nodes. Most routing protocols are based on topology using existing network link information for packet forwarding. Current routing protocols for their reliability use position information to locate destination and neighbor nodes. The topology of MANETs is dynamic and Gaps exist in this dynamic configuration.

3 DCCHSM Algorithm

The main feature of the proposed protocol is that it classifies neighboring cells into different priority levels according to their distances from the node. This classification facilitates the selection of the packet forwarders and in each hop, neighboring cells that are closer to the destination cell are preferred. As a next step in protocol,

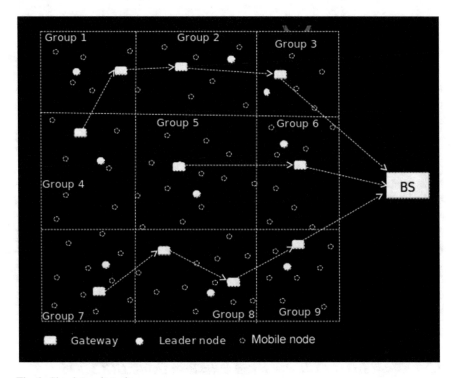

Fig. 1 Simulator clustering

formation of this hash tables are created. Hash tables help in effective retrieval of information from mobile nodes. By hashing node co-ordinates or meta-data, perfectly designed hash keys are used to map co-ordinates using the key value of same range. Keys are then stored at the mobile nodes based on a geographic proximity to the sink (Bose et al. 2001).

3.1 DCCHSM

The proposed dynamically clustering-based cluster head selection in MANET (DCCHSM) protocol is aimed at lesser energy consumption and faster information transfer between nodes. It is an efficient protocol which defines its routing paths by assessing network node positions, uses a distance measure to connect these nodes, and arrives at shortest routing paths between clusters.

The basic aim of a routing protocol is to route information from a source to a destination by identifying the correct route. A major hurdle in routing is the network characteristics that may have variations. MANET routing algorithms can be categorized as proactive, on-demand, and hybrid routing algorithms. The proactive or table-driven algorithm's nodes have a routing table with information on nodes routing information and are similar to wired routing algorithms.

The node within a group delivers their data to their group's cluster head (CH). These CH's in a hierarchical arrangement communicate to the sink nodes for transmitting information. The beacon inhibition mechanism is used by DCCHSM to assess the time taken to transmit a beacon and also estimate the threshold values of the nodes.

The estimated threshold value of a neighborhood (previously grouped nodes) is then segmented based on the threshold values. A neighbor table is created and priorities are updated by transmitting a HELLO packet to neighbor nodes. Hash table creation in DCCHSM is based on beacon values. In the neighbor table, if the nodes beacon value is less than average threshold value of node range, then the nodes are moved for construction of the hash table of each node [8, 9].

Clustering is then applied to the node attributes using a hybrid technique involving k-means and k-medoids algorithms. Cluster assignments are then created using a cluster group's average beacon value range of a node. The cluster assignment heads thus form the main points of action for DCCHSM protocol in MANET communications. The algorithm for cluster head selection using DCCHSM algorithm is as below.

3.2 Algorithm for Cluster Head Selection Using DCCHSM

Step 1   : Initialize with cluster center, that is selected randomly
Step 2   : Repeat for l = 1; 2; : : :
Step 3   : Compute the distances

$$D_{ik}^2 = (x_k - v_i)^T (x_k - v_i), \quad 1 \le i \le c, \quad 1 \le k \le N.$$

Step 4   : Select the points for a cluster with the minimal distances, they belong to that cluster.

Step 5   : Calculate cluster centers $v_i^{(l)} = \frac{\sum_{j=1}^{N_i} x_i}{N_i}$

Step 6   : Until $\prod_{k=1}^{n} max \left| v^{(l)} - v^{(l-1)} \right| \ne 0$

Step 7   : Calculate the partition matrix
Step 8   : Set the point energies.

$$f(i) = \sum_{i=1}^{N} point(1: dim\_num, i) - cluster\_center(1: dim\_num, j)^2$$

Step 9   : Set the cluster energies.

$$clust\_energy(j) = clust\_energy(j) + f(i)$$

Step 10 : Adjust the point energies by a weight factor.

$$f(i) = f(i) \frac{cluster\_population(j)}{cluster\_population(j) - 1}$$

Step 11 : Assign weighted node f(i) as Cluster Head (CH).

Where x= data point (node location) ; v = neighbor distance node ; N= total number of nodes

3.3   Protocol Formation and Neighbor Table Creation

The routing process is done to select a suitable trail for the information to travel. This process is actually tedious which depends on the network type, channel characteristics, and performance metrics. Protocol formation in this research work starts with the construction of neighbor tables before packets are sent and received. Information about adjacent neighbors is stored in a table holding the details of the neighbors. It has a list of adjacent neighbors, neighboring interface, address and additional details about each neighbor. The construction of neighbor table is shown in Fig. 2. The table for the neighbor holds the current state of the nodes and the packets transmitted holds the time span.

The protocol after creating a neighbor table creates a priority table. In this present work, to create a clustering nodes and subsequent selection of cluster heads were done by following the concept of k-means and k-medoids algorithms for clustering nodes and selecting cluster heads. The nodes consume lesser energy when cluster heads (CHs) are selected dynamically and keep changing based on a rotation mechanism. Though any mobile node qualifies to be a CH, this work selects a CH based on the average distance between nodes. Nodes which have many nodes nearer to them in terms of average distance between them become the CH. MANET's have extraordinary characteristics include dynamic network topology and moving nodes. MANET is very helpful in regions for networks that require ad hoc connectivity and have devices with mobility [10, 11].

This was done by hybrid technique using k-means and k-medoids for a secure routing and efficient delivery of information in MANETs called DCCHSM. Next level of investigation is concerned with identification of cluster head. This was done with the neighborhood distance measurement. The same 100 nodes were used. By adopting the above principle, five cluster heads were identified.

Address	Interface	Holdtime (secs)	Uptime (h:m:s)	Q Count	Seq Num	SRT (ms
172.16.100.1	Ethernet1	13	0:00:41 0	0	11	4
172.16.110.2	Ethernet0	14	0:02:01	0	10	12
172.16.110.2	Ethernet0	12	0:02:02	0	4	5

Fig. 2 Neighbor table

4 Simulation Results and Analysis

The results described Table 1 present with various parameters followed for the simulation. DCCHSM uses a Routed Topology Dataset for its implementation on NS2 simulating the proposed topology. The number of nodes taken in the requirement phase was 100 nodes to be grouped into five clusters.

Cluster-based routing protocol for double cluster head (DCH) is utilized in the existing system, flexible weight-based clustering algorithm (FWCA), least cluster change (LCC) [12–14]. This algorithm is not built in energy efficient-based concepts of cluster head selection. But the proposed work is efficiently selecting the CH using dynamic clustering-based cluster head selection in MANET (DCCHSM) algorithm [15]. This algorithm performance measurement the packet delivery ratio is increased than compared to existing algorithm.

Packet delivery ratio: is the total number of received packets for the number of packets sent from the source.

In Fig. 3 the proposed DCCHSM algorithm provides better performance compared to existing algorithm LCC FWCA and DCH also increased packet delivery ratio with mobility is increased.

In Fig. 4, the proposed DCCHSM algorithm provides better performance compared to existing algorithm LCC, FWCA, and DCH and also increased packet delivery ratio with transmission range is increased.

In Fig. 5, the proposed DCCHSM algorithm provides better performance compared to existing algorithm LCC, FWCA, and DCH and also increased PDR for number of nodes is increased.

Table 1 Simulation parameters

Parameters	Value
Simulation	NS2
MAC protocol	IEEE 802.11
Examine protocol	DCCHSM, DCH, LCC, FWCA
Mobility model	Random waypoint
Number of nodes	100
Transmission range	250 m
Size of network	500 m * 500 m
Pause time	25 s
Hello interval	5.0 s
Simulation duration	500 times

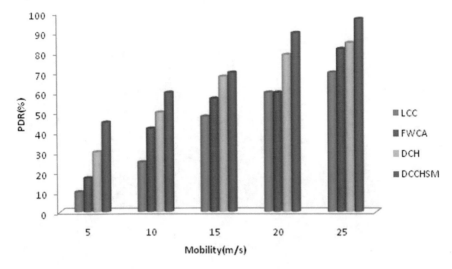

Fig. 3 PDR (%) versus mobility (m/s)

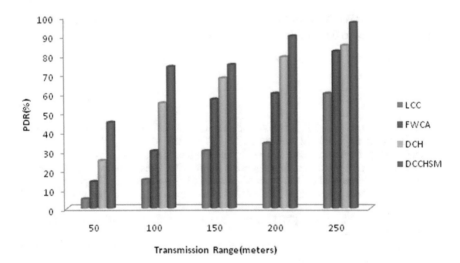

Fig. 4 PDR (%) versus transmission range (meters)

5 Conclusion

In this paper, energy efficient cluster head selection and is energy based on packet transmission using NS2 simulation. This article presents the simulator clustering, neighbor table creation, and uses hybrid clustering k-means and k-medoids for its information routing. This DCCHSM algorithm improves the performance compared to existing clustering algorithm least cost clustering (LCC), flexible weight-based

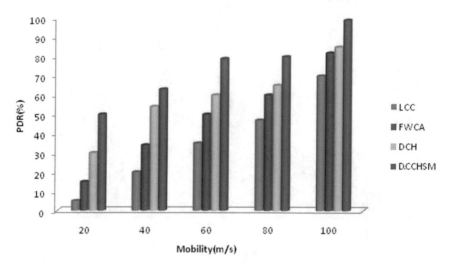

Fig. 5 PDR (%) versus no. of nodes

cluster algorithm (WCA) and double cluster head (DCH) and also improved increased packet delivery ratio for mobile ad hoc networks.

References

1. Cha HJ, Kim JM, Ryou HB (2014) A study on the clustering Schemes for node mobility in mobile ad hoc networks. Adv Comput Sci Appl 279(1):1365–1369
2. Rathika SKB, Bhavithra J (2012) An efficient fault tolerance quality of service in wireless networks using weighted clustering algorithm. Bonfring Int J Res Commun Eng 2(1):1–6
3. Abdulsaheb GM, Khalaf OI, Sulaiman N, Zmenzm HF, Zmezm H (2015) Improving Ad hoc network performance by using an efficient cluster based routing algorithm. Indian J Sci Technol 8(30):1–8
4. Balamurugan M, Kavi priya C (2015) Energy based cluster head selection algorithm in MANET. IJCSET, 5(8):312–315
5. Popli R, Garg K, Batra S (2015) A hierarchical classifications of various clustering schemes for MANET. IJIRAE 2(7):110–113
6. Vijayalakshmi J, Prabu K (2018) Performance analysis of clustering schemes in MANETs. In: International conference on intelligent data communication and technologies and internet of things (ICICI), pp 808–813
7. Ramesh S, Smys S (2017) Performance analysis of heuristic clustered (HC) architecture in wireless networks. In: 2017 international conference on inventive systems and control (ICISC). IEEE, pp 1–4
8. Perkins CE, Bhagwat P (1994) Highly dynamic destination-sequenced distance-vector routing (DSDV) for mobile computers. Comput Commun Rev 24(4):234–244
9. Mohd JH, Muntjir M, Abu Sorrah H (2015) A comparative survey of computations of cluster head in MANET. Int J Comput Appl 118(3):6–9
10. Jacquet P, Muhlethaler P, Clausen T, Laouiti A, Qayyum A, Viennot L (2001) Optimized link state routing protocol for ad hoc networks. In: Proceedings of international multitopic conference (IEEE INMIC), pp 62–68

11. Singh M, Singh G (2015) A secure and efficient cluster head selection algorithm for MANET. JNCET 2(2):49–52
12. Kaur S, Kumari V (2015) Efficient clustering with proposed load balancing technique for MANET. Int J Comput Appl 111(13):21–26
13. Pearlman MR, Haas ZJ (1999) Determining the optimal configuration for the zone routing protocol. IEEE J Sel Areas Commun 17(8):1395–1414
14. Sharma SK et al (2014) Routing protocols and security issues: a survey 04(4)918–924
15. Vijayalakshmi J, Prabu K (2017) Overview of multicast routing protocols for mobile ad-hoc networks. In: Proceedings of the international conference on intelligent sustainable systems (ICISS 2017) IEEE xplore compliant - part number:CFP17M19-ART. ISBN:978-1-5386-1959-9, pp 256–260
16. Chiang C-C (1997) "Routing in clustered multihop", Mobile wireless networks with fading channel. Proc IEEE SICON 97:197–211

Mrs. J.Vijayalakshmi received her MCA and M.Phil from Bharathidasan University, Tiruchirappalli in the year of 2007 and 2013. She is now Research Scholar in Bharathidasan University, Tiruchirappalli, Tamilnadu, India. Her Research interest is Ad hoc Networks. She has published more than 9 papers at various National/ International Conferences and Journals.

Dr. K. Prabu received his MCA and M.Phil from Annamalai University, Tamilnadu, India. He received his Ph.D. Degree in Computer Applications from Manonmaniam Sundaranar University, Tirunelveli, India. He is now working as an Associate Professor in PG & Research Department of Computer Science, Sudharsan College of Arts & Science, Pudukkottai, Tamilnadu, India. He is a Reviewer of 06 National/International Journals. His Research interest is Adhoc Networks, Wireless & Mobile Computing, and Wireless Sensor Networks. He has published more than 75 technical papers at various National/International Conferences and Journals. He is a life member of ISTE, IACSIT, and IAENG.

Retraction Note to: A Cluster-Based Distributed Cooperative Spectrum Sensing Techniques in Cognitive Radio

N. Shwetha, N. Gangadhar, L. Niranjan, and Shivaputra

Retraction Note to: Chapter "A Cluster-Based Distributed Cooperative Spectrum Sensing Techniques in Cognitive Radio" in: J. S. Raj et al. (eds.), *Innovative Data Communication Technologies and Application*, Lecture Notes on Data Engineering and Communications Technologies 59, https://doi.org/10.1007/978-981-15-9651-3_20

The Volume Editors and authors retract this conference paper [1] because it has substantial overlap with a PhD thesis by a different author [2].

N. Shwetha, N. Gangadhar, and Shivaputra agree to this retraction; L. Niranjan has not responded to any correspondence from the publisher about this retraction.

[1] Shwetha N., Gangadhar N., Niranjan L., Shivaputra (2021) A Cluster-Based Distributed Cooperative Spectrum Sensing Techniques in Cognitive Radio. In: Raj J.S., Iliyasu A.M., Bestak R., Baig Z.A. (eds) Innovative Data Communication Technologies and Application. Lecture Notes on Data Engineering and Communications Technologies, vol 59. Springer, Singapore. https://doi.org/10.1007/978-981-15-9651-3_20

[2] Babu G., (2019) Analysis and Evaluation of Distributed Cooperative Spectrum Sensing Techniques in Cognitive Radio Sensor Networks, St. Peter s Institute of Higher Education and Research; 2019 http://hdl.handle.net/10603/280064

The retracted version of this chapter can be found at
https://doi.org/10.1007/978-981-15-9651-3_20

Author Index

Printed in the United States
by Baker & Taylor Publisher Services